D1060738

Fundamentals of Formation Evaluation

Fundamentals of Formation Evaluation

Donald P. Helander

OGCI Publications
Oil & Gas Consultants International, Inc.
Tulsa

4554 South Harvard Avenue
Tulsa, Oklahoma 74135

Printed in the United States of America

Library of Congress Catalog Card Number: 82-062331
International Standard Book Number: 0-930972-02-3

Preface

Many years ago I wrote a manual for teaching formation evaluation in industry short courses and at the University of Tulsa. That first manual has undergone continuous revision and updating. This book is the culmination of another revision. Of course, as with any technical book, one realizes that the work is never finished and, hence, the author can never be completely satisfied. In bringing the work to this point, however, I would like to express my appreciation to all of my former students, both at the University of Tulsa and in my industry short courses, who have contributed toward the distillation of this work from its early beginnings. In teaching this material I have learned a great deal from my students about how to express the concepts in a clear and logical manner, which I hope is reflected here. I would also like to express my appreciation to the service companies and many authors whose contributions appear in this book. I would also like to say that I have certainly not referenced all the outstanding work published in the area of formation evaluation, and to those authors whose work is not referenced, this certainly is not due to a lack of value in their work.

Since this book was never intended to be a complete treatise on formation evaluation, it will be apparent to the reader that not all aspects of formation evaluation are included. Rather, it is written to provide a basis for an introductory course in formation evaluation and should be considered in that light. As a book designed for a first course in formation evaluation, suggestions and comments from those who might use this book for that purpose are certainly welcomed. As in any course, it is also designed to be supplemented by problems in order to point out the important concepts.

A second book is in preparation to be co-authored by my friend and colleague, Mr. Frank Millard. This book will be directed toward a more advanced level of formation evaluation. As now projected, it will include log calibration and quality control, complex carbonate and shaly sand analysis, various techniques used for fracture detection, and discussions of additional wireline measurements not covered here, such as natural gamma ray spectroscopy logs, neutron activation spectral logs such as the silicon log, carbon-oxygen log, and litho density log.

Finally, I would again like to thank all of my past students who have been instrumental in enabling me to present this material in what I hope is a lucid manner through their comments and suggestions for presenting ideas and concepts in a manner more readily understood by them. Last, but not least, I would like to thank my family for their patience with regard to the time I spent in writing and rewriting the material ultimately compiled in this book.

Donald P. Helander

Contents

1 **Introduction to Formation Evaluation** 1

Mud Resistivity 4
  Function of Mud Type 4
  Function of Temperature 4
Downhole Temperature 6
Filtrate Invasion 7
Measurement of R_{mf} and R_{mc} 7
Resistivity Profiles 8
Diameter of the Invaded Zone 8
Surface Logging Systems 11

2 **Coring and Core Analysis** 13

Types of Coring 13
  Conventional Coring 13
  Diamond Coring 14
  Wireline Coring 14
  Sidewall Coring 14
  Special Coring Tools 14
Factors Affecting Cores 14
  Flushing 14
  Pressure and Temperature Reduction 15
Coring Program 15
  Empirical Approach 16
  Statistical Approach 16
Core Analysis Technique 18
  Conventional or Plug Analysis 18
  Whole-Core Analysis 18
  Sidewall Core Analysis 18

Basic Measurements 18
Residual Saturation 21
Retort Method 21
Porosity 21
Bulk Volume Measurements 22
Pore Volume and Grain Volume Measurements 22
Absolute Permeability 25
Supplemental and Special Core Analysis 26

3 **Mud Logging** **29**
Mud Systems 29
Mud Properties 30
Drilling Time Log 31
Cuttings Analysis 32
Oil Staining 36
Hydrocarbon Odor 37
Fluorescence 37
Cuts 38
Cuttings Gas Analysis 40
Acid Test 41
Hydrocarbon Mud Log 41
Sampling Unit 44
Detectors 44
Gas Chromatography 46
Mud Analysis for Oil 47
Interpretation of Hydrocarbon Mud Log 47
Analysis of Liberated Gas 49
Analysis of Recycled Gas 50
Analysis of Partial Liberation 50
Influence of Penetration Rate Change 50
Analysis of Produced Gas Influence 51
Mud Log Format 52
Drilling Data Analysis 53
Shale Bulk Density 55
Shale Factor 57

4 **Basic Relations** **59**
Rock Resistivity 59
Influence of Formation Water Resistivity 59
Influence of Rock Structure on Rock Resistivity 63
Influence of Hydrocarbons on Rock Resistivity 69
Influence of Clay on Rock Resistivity 71
Flushed Zone Resistivity 71

Rock Permeability 72
Porosity-Permeability Relationship 72
Porosity-Residual Water Saturation-Permeability Relationship 73
Resistivity Gradient Approach 74
Shaliness-Permeability Relationship 74

5 **The Spontaneous Potential, SP** **77**
Electrochemical Potential 77
Membrane Potential 79
Liquid Junction Potential 80
Total Electrochemical Potential 80
Electrokinetic Potential 81
Measurement of SP 84
Factors Affecting Shape and Amplitude 84
Ratio of R_{mf}/R_w 84
Additional Factors 85
Uses of the SP 87

6 **Resistivity Measurement** **99**
Nonfocused Resistivity Measurements 99
Normal Resistivity 99
Lateral Resistivity 102
Electrical Survey 104
Micro-Resistivity Measurement 104
Induction Resistivity Measurements 109
Factors Affecting Induction Resistivity Measurements 111
Multiple Resistivity Measurements 116
Dual Induction-Laterolog 8 and Dual Induction-SFL 117
Focused Current Resistivity Measurements 121
Guard Electrode System 122
Point Electrode System 127
Factors Affecting Focused Current Resistivity Measurements 128
Micro-Focused Current Resistivity 130
Proximity Log 133
Dual Laterolog-R_{xo} Log 135

7 **Acoustic Logging** **143**
Elastic Wave Propagation 143
Transmitters and Receivers 148
Acoustic Logging Tools 148
Porosity Evaluation 152

Transit-Time-Resistivity Relation 155
Transit Time as an Abnormal Pressure Indicator 155
3-D Log Applications 157
  Fracture Detection 158
Elastic Moduli Evaluation 158
Borehole Televiewer 160

8 **Radioactivity Logs** **167**

Nature of Radioactivity 167
  Alpha Particles 168
  Beta Particles 168
  Gamma Rays 169
  Neutrons 169
Transformation Series 169
Gamma Absorption 171
Neutron Production 172
  Capsule Sources 172
  Accelerator Sources 173
Neutron Interactions 173
Nuclear Cross-Section 175
Radiation Detectors 177
  Ionization Chamber 177
  Geiger-Mueller Counter 177
  Proportional Counters 178
  Scintillation Counters 178
Gamma Ray Log 178
  Factors Affecting Gamma Ray Log Response 179
  Application of the Gamma Ray Log 181
Neutron Log 182
  Factors Affecting Standard Neutron Tool Response 183
  Depth of Investigation 184
  Porosity Evaluation 185
  Neutron Deflection-Resistivity Relation 185
  Sidewall Neutron Log (SNP) 188
  Dual-Spaced Neutron Log 189
Density Log 191
  Factors Affecting Density Log Response 194
  Bulk Density-Resistivity Relation 196
  Bulk Density as an Abnormal Pressure Indicator 197
Pulsed Neutron Decay Logs 197
  Factors Affecting Lifetime Measurements 198
  Interpretive Relationships 199
  Influence of Shale Component 202
  Pulsed Neutron Decay Logs as Abnormal Pressure Indicators 204
  Applications 205

9 **Interpretation Methods** 207

Standard Log Interpretation Methods 209
- Conventional Method 209
- R_{wa} Method 210
- F_R Comparison Method 211
- Movable Oil Method 212
- R_{xo}/R_t vs. E_{ssp} Method 213
- S_{or}/S_o Method 215

Crossplotting Methods 215
- Crossplots for Water Saturation Analysis 215
- Porosity Log Crossplots 230
- M vs N Type Crossplot 239
- MID Plot Method 241

Shaly Sand Interpretation 249

Appendices 253

A. Basic System Classification 253
B. Guide to Selected Drilling Fluids 255
C. Standard Abbreviations for Lithologic Descriptions 261
D. Geologic Digital Terms 267
E. Method for Computer Processing of Sample Data 279
F. Letter and Computer Symbols for Well Logging and Formation Evaluation 289
G. SI Metric System of Units 305

Index 325

1 Introduction to Formation Evaluation

FORMATION evaluation is the process of using borehole measurements to evaluate the characteristics of subsurface formations. It applies to many areas of engineering where various rock properties are needed. In this book, however, our efforts will be directed toward the identification and evaluation of commercial hydrocarbon-bearing formations. Formation evaluation represents the expenditure of a considerable sum of money each year. In each individual well the evaluation cost may range up to 20% of the total well cost. A wide variety of in-situ measurements are available for evaluating formations in an individual well. These measurements may be grouped into four categories:

Drilling Operation Logs (mud logs): cuttings analysis, mud analysis, and drilling data collection and analysis

Core Analysis: qualitative measurements (visual lithology, presence of shows, etc.) and quantitative measurements (porosity, permeability, formation factors, etc.)

Wireline Well Logs: electrical (spontaneous potential, SP, nonfocused current resistivity, focused current resistivity, induction), acoustic (transit time, full-wave train, borehole televiewer), and radioactive (gamma ray, neutron, density, neutron lifetime, spectral)

Productivity Tests: formation tester, drill stem tests, and production tests

Obviously, not all of these measurements will be made in any single well. Rather, a judicious selection of specific measurements is made in order to completely identify and evaluate the commercially productive hydrocarbon-bearing zones. The problem is to select the minimum cost combination of measurements providing a definitive evaluation. Also, in evaluating commercially productive hydrocarbon zones, a wealth of information of great value to petroleum geologists, geophysicists, and engineers might be obtained. A partial list of applications of borehole measurements is shown as follows. This list includes some auxiliary applications of the data directed toward solving the primary problem in formation evaluation.

1. Estimating recoverable hydrocarbons (primary application)
2. Estimating hydrocarbons in place (primary application)
3. Rock typing
4. Abnormal pressure detection
5. Evaluating rock stresses
6. Locating reservoir fluid contacts
7. Fracture detection
8. Identifying geologic environments

Determination of recoverable hydrocarbons, or at least of hydrocarbons in place, is the primary goal in the selection of measurements run in a specific well. Any additional information generated from the array of data obtained is usually considered a bonus. The evaluation program is, therefore, designed to provide reliable estimates of the following expressions for hydrocarbons in place (see Fig. 1-1):

$$N = \frac{7758\ Ah\phi\ S_{oi}}{B_{oi}}$$

$$G = \frac{43{,}560\ Ah\phi\ S_{gi}}{B_{gi}}$$

where N = initial oil in place, stb
A = drainage area, acres
h = productive interval thickness, ft
ϕ = effective porosity, fraction
S_{oi} = initial oil saturation, fraction
B_{oi} = initial oil formation volume factor, $\frac{\text{reservoir bbl}}{\text{stb}}$
G = initial gas in place, scf
S_{gi} = initial gas saturation, dimensionless
B_{gi} = initial gas formation volume factor, $\frac{\text{ft}^3}{\text{scf}}$

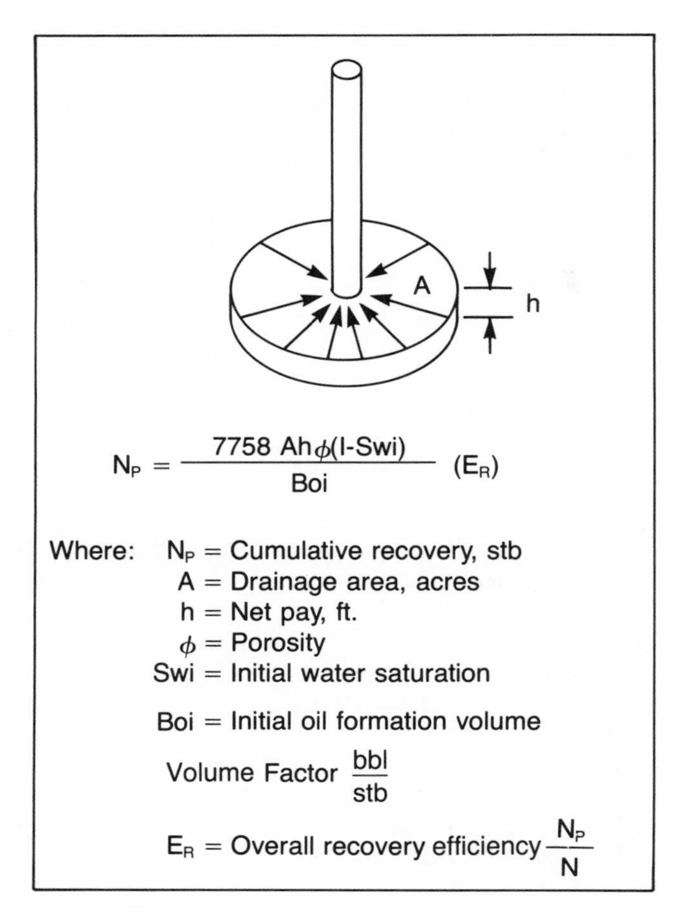

Fig. 1-1. Illustration of evaluation goal

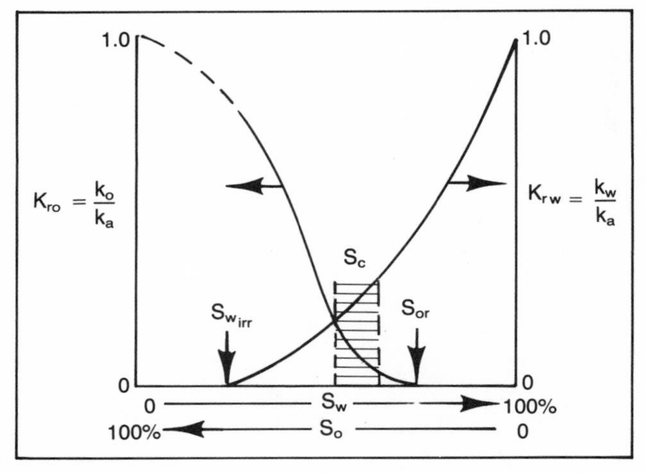

Fig. 1-2. Typical water-oil relative permeability curve

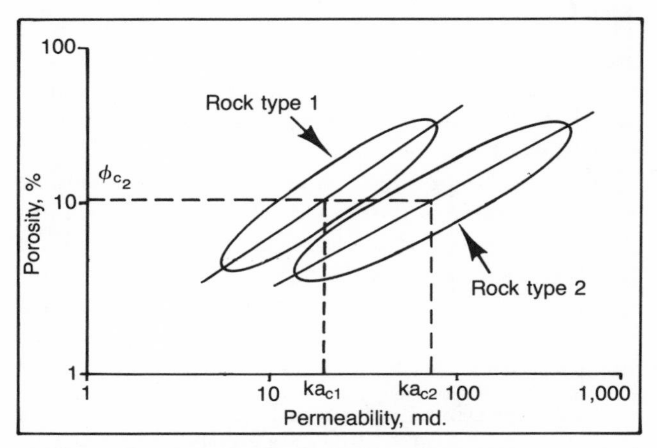

Fig. 1-3. Typical porosity vs. permeability trends for different rock types

Recoverable hydrocarbon volume might be projected if an approximation of recovery efficiency, E_R, can be made so that $N_P = NE_R$ where N_P = cumulative produced oil in stb, as shown in Figure 1-1.

With proper selection of borehole measurements, it is possible to obtain quantitative estimates of three of the parameters shown in equations 1 and 2: S_{hi} (S_{oi} or S_{gi}), ϕ, and h. Even if drainage volume, Ah, and recovery factor, E_R, were unknown, an evaluation of the three parameters S_{hi}, ϕ, and h can provide information of value since $N_P = f(S_{hi}, \phi, h)$.

The most important of these parameters is hydrocarbon saturation, S_h. It represents not only a volumetric quantity, but is also related to the ability of the rock to transmit fluids. This can be seen in the typical relative permeability versus saturation curve shown in Figure 1-2. As the hydrocarbon saturation, S_h, approaches some *critical* or *cut-off* saturation value, S_c, the ability of the hydrocarbons to flow decreases rapidly, and, conversely, the ability of water to flow increases rapidly. In a practical sense, therefore, a hydrocarbon saturation in excess of the critical saturation must exist to have recoverable hydrocarbons. Usually the critical saturation is about 50% and nearly always falls in the range from 30% to 70%.

Most sedimentary rocks also exhibit a relationship between single phase permeability, k_a, and porosity, ϕ, as shown in Figure 1-3. This leads to a second condition, *cut-off* porosity, which varies with rock type. *Cut-off* porosity is the minimum porosity above which an economically acceptable single phase permeability is probable. This means that only zones having porosity greater than cut-off porosity can be considered when estimating recoverable hydrocarbons. Figure 1-4 summarizes these basic concepts of formation evaluation.

The process of estimating hydrocarbon saturation, S_h, porosity, ϕ, and producing interval thickness, h, can be an arduous task.[9, 10] Complications result because (1) desired parameters must be inferred from measurements indirectly, (2) empirical relations of a statistical nature must be used, and (3) economics must be considered.

First consider the problem of indirectness in obtaining the required parameters. To obtain hydrocarbon

BASIC CONCEPT OF FORMATION EVALUATION

Determine N_P based on:

$$N_P = \frac{7758\ Ah\phi(1-S_{wi})}{B_{oi}} (E_R)$$

where $E_R = f(k_o)$

But require an economic k_o where:

$$k_o = k_{ro}k_a$$

Therefore:

(a) $S_{oi} > S_c$ for large enough K_{ro}
(b) And possibly $\phi > \phi_c$ for large enough k_a

Fig. 1-4. Illustration of basic formation evaluation concept

saturation, S_h, we estimate water saturation, S_w, assuming that $S_h = 1 - S_w$. However, S_w is in turn a function of five variables: bulk resistivity, R_t, water resistivity, R_w, porosity, ϕ, and two empirical constants, m (cementation exponent) and n (saturation exponent). These, in turn, may require other measurements. If the acoustic log is used for estimating porosity it is necessary to account for the effects of lithology, effective stress, and grain structure on the log-measured value of interval transit time, ℓ. This problem is illustrated in Figure 1-5.

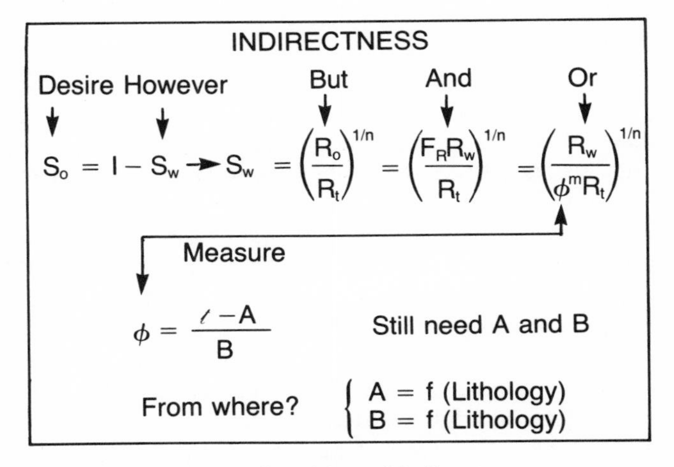

Fig. 1-5. Illustration of problem of indirectness

Next, to convert the indirect measurements to ones of interest, it is necessary to use empirical relationships, since theory cannot completely predict the relations between the indirect borehole measurements and the parameters needed in formation evaluation. The empirical relationships used are, for the most part, rather simple and include:

1. Fundamental Equations:

$$I_R = S_w^{-n} = \frac{R_t}{R_o} = \frac{R_t}{F_R R_w}$$

$$F_R = \phi^{-m} = \frac{R_o}{R_w}$$

2. Log Response Equations:

$$\ell = A + B\phi$$

$$\rho_b = \rho_f\phi + \rho_{ma}(1-\phi)$$

$$ND = C + D \log \phi,\ \phi_{SNP} \approx \phi \text{ and } \phi_{CNL} \approx \phi$$

$$\Sigma_{log} = \Sigma_{ma}(1-\phi) + \Sigma_w\ \phi\ S_w + \Sigma_{hc}\ \phi\ (1-S_w)$$

$$SSP = -K_c \log \frac{R_{mf}}{R_w}$$

$$R_a = f(R_m,\ d_h,\ R_i,\ d_i,\ R_s,\ h,\ R_t)$$

The nomenclature for the above equations and the nomenclature that will be used throughout this book is presented in Appendix F. Each of these equations will be discussed in more detail later.

The statistical nature of these empirical relations as illustrated in Figure 1-6 will affect our choice of evaluation tools. In this case, the plot of interval transit time, ℓ, data from an acoustic log and core porosities indicates

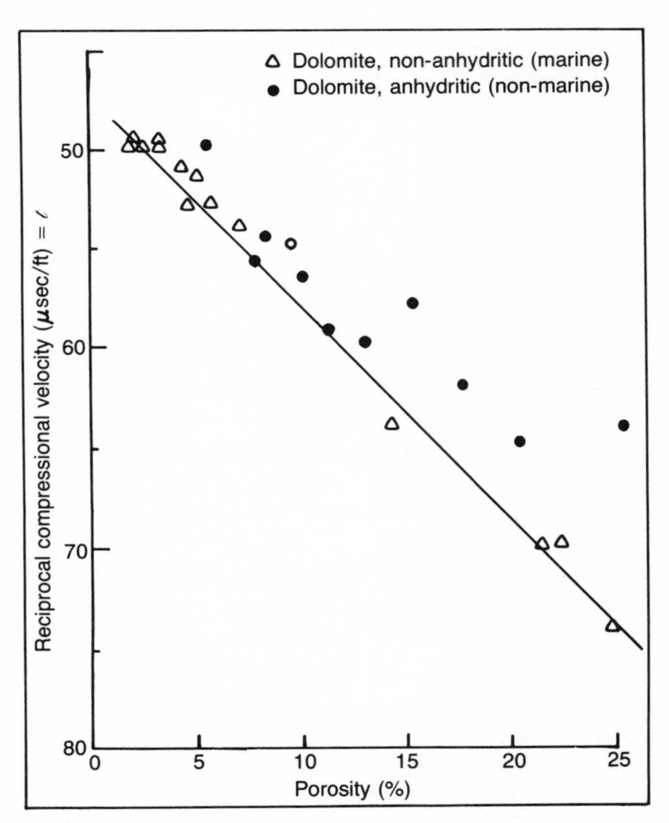

Fig. 1-6. Transit time vs. porosity for a typical carbonate tidal flat section. Permission to publish by the Society of Professional Well Log Analysts.[10]

two trends. The trend for the nonanhydritic dolomite can be fit reasonably well, but the anhydritic dolomites show appreciable scatter around any average correlation that could be used. Thus, any porosity evaluation in this section must include a method to distinguish anhydritic dolomite from nonanhydritic dolomite. This might be done by use of good cuttings samples, or through the use of another porosity tool not affected by anhydrite in the same manner as the acoustic log. This second porosity tool could be a neutron log where a simultaneous solution of the two response equations could differentiate the anhydritic from the nonanhydritic dolomites.

Finally, economics influences the choice of evaluation tools. Based on evaluation cost per-foot, cores may cost 500 times more, and "mud logs" may cost five times more, than wireline logs. Productivity tests are usually quite expensive, costing thousands of dollars per test. Economically, it is desirable to design an evaluation program using wireline logs as the principal source of information.

The ultimate objectives of well logging are the location of oil or gas formations and their quantitative evaluation. These require that the logging program give adequate information on: (1) formation lithology, (2) depth and thickness of productive zones, (3) formation porosity and fluid saturations, and (4) reservoir geometry and continuity through correlation with other wells, leading to a determination of recoverable hydrocarbon or hydrocarbons in place.

To design an evaluation program for estimating recoverable hydrocarbons, it is necessary to first consider past experience, theories, and supplementary geological and geophysical data to establish the specific set of empirical relations, which when solved will provide recoverable hydrocarbon volume, N_P. Second, enough measurements must be combined to solve these relations with satisfactory accuracy at minimum cost. To minimize costs the program ideally should:[9]

1. Use wireline logs as basic tools where possible
2. Supplement with "mud logs" (cuttings samples and possibly borehole fluid logs)
3. Use cores for calibration of logs and for other geologic data
4. Use productivity tests to obtain R_w and assist in evaluating important borderline cases that cannot be satisfactorily resolved from the above

Due to the great scope of information required from well logging operations, along with the restrictions imposed by various borehole conditions and formation characteristics, a wide variety of logging tools have been developed and are currently used. Changes in the composition and character of formations that occur both geographically and with depth require different logs in different areas and often in different sections of the same well. All the wireline logging tools commonly used will be discussed in subsequent chapters with specific emphasis on their response characteristics and applicable response equations, limitations, and area of application.

In order to understand these tool responses, however, it is first necessary to understand the hostile environment under which these tools are operated. All logs are affected in one way or another by the type of mud used. The factors with which we are generally concerned in well logging are the mud properties (such as mud resistivity) and the water loss into the formation, along with its associated build-up of mud cake (invasion process).

MUD RESISTIVITY

It is important to know mud resistivity, R_m, since it completes the circuit between the logging tool and the formation. We can classify borehole muds into two groups, those that are conductive and those that are non-conductive. The non-conductive muds include air, gas, and oil-base fluids having infinite resistivity. When logging in this fluid type, it is necessary to use a logging tool that does not depend upon borehole conductance, such as the induction, acoustic, or radioactive type logging devices. Since this mud group is infinitely resistive, we are generally concerned with those muds in the first category, namely the conductive water-base type muds. Current flow in conductive muds varies depending upon two factors—the type of mud and the temperature.

Function of Mud Type

Mud conductivity depends first on water content, which is approximately the same in all water-base muds and, second, upon the number of dissociated ions in solution. The number of dissociated ions varies greatly with mud type. Fresh-water muds have few dissociated ions. This is particularly true if the use of thinners and clay stabilizers is minimal. Conversely, saltwater muds may contain large amounts of calcium, magnesium, and sodium ions due to additives or contamination from formation fluids. It can be seen, therefore, that water-base muds may have widely divergent resistivities depending upon the type of makeup water, the types of additives used, and, in some cases, amount and type of contamination from formation fluids.

Function of Temperature

It can generally be stated that mud resistivity, R_m, varies inversely with temperature. As temperature increases, mud resistivity decreases. To measure mud resistivity, a uniform or homogeneous sample of the

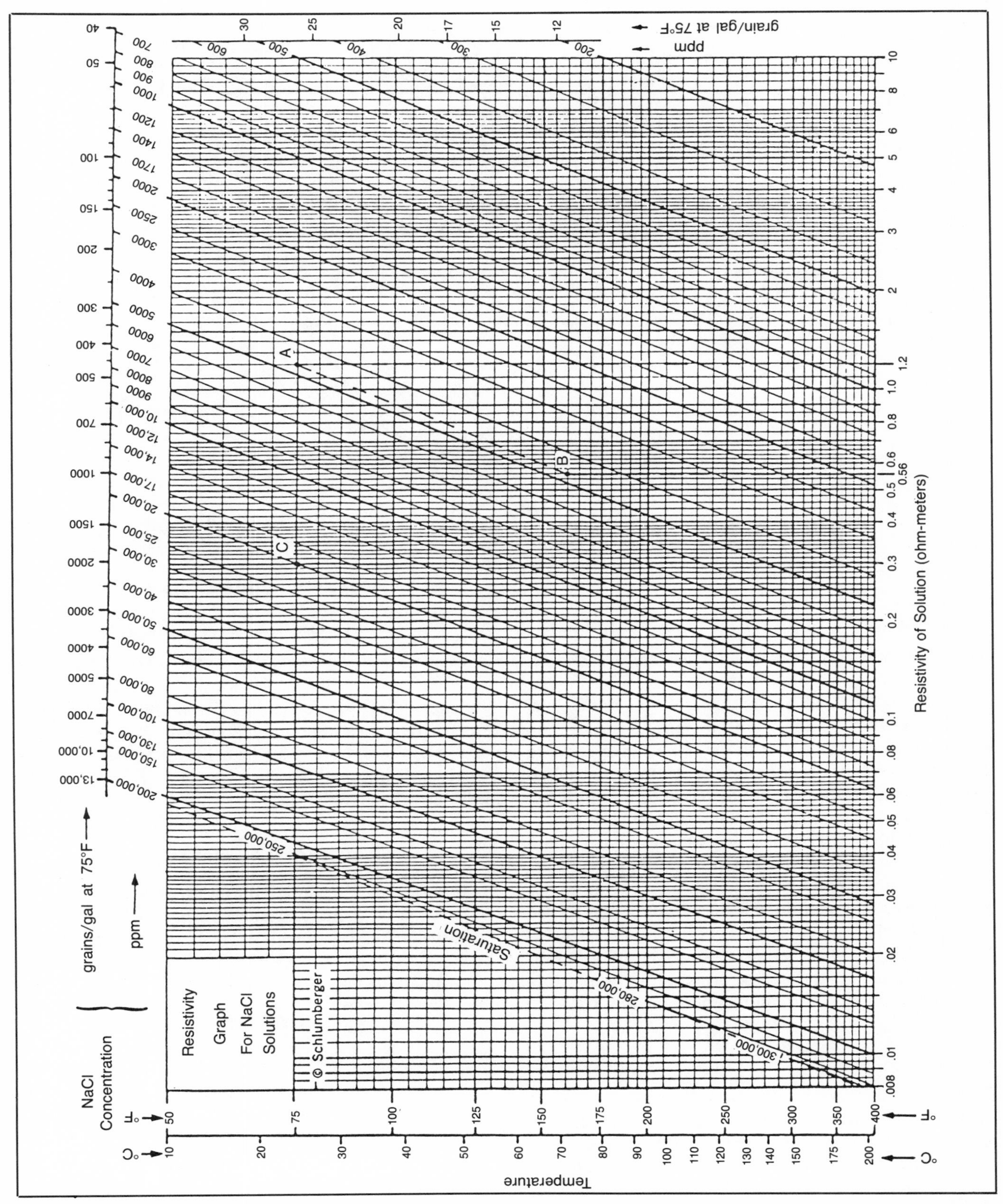

Fig. 1-7. Resistivity of Sodium Chloride (NaCl) solutions as a function of temperature and salinity. Permission to publish by Schlumberger Ltd.[8]

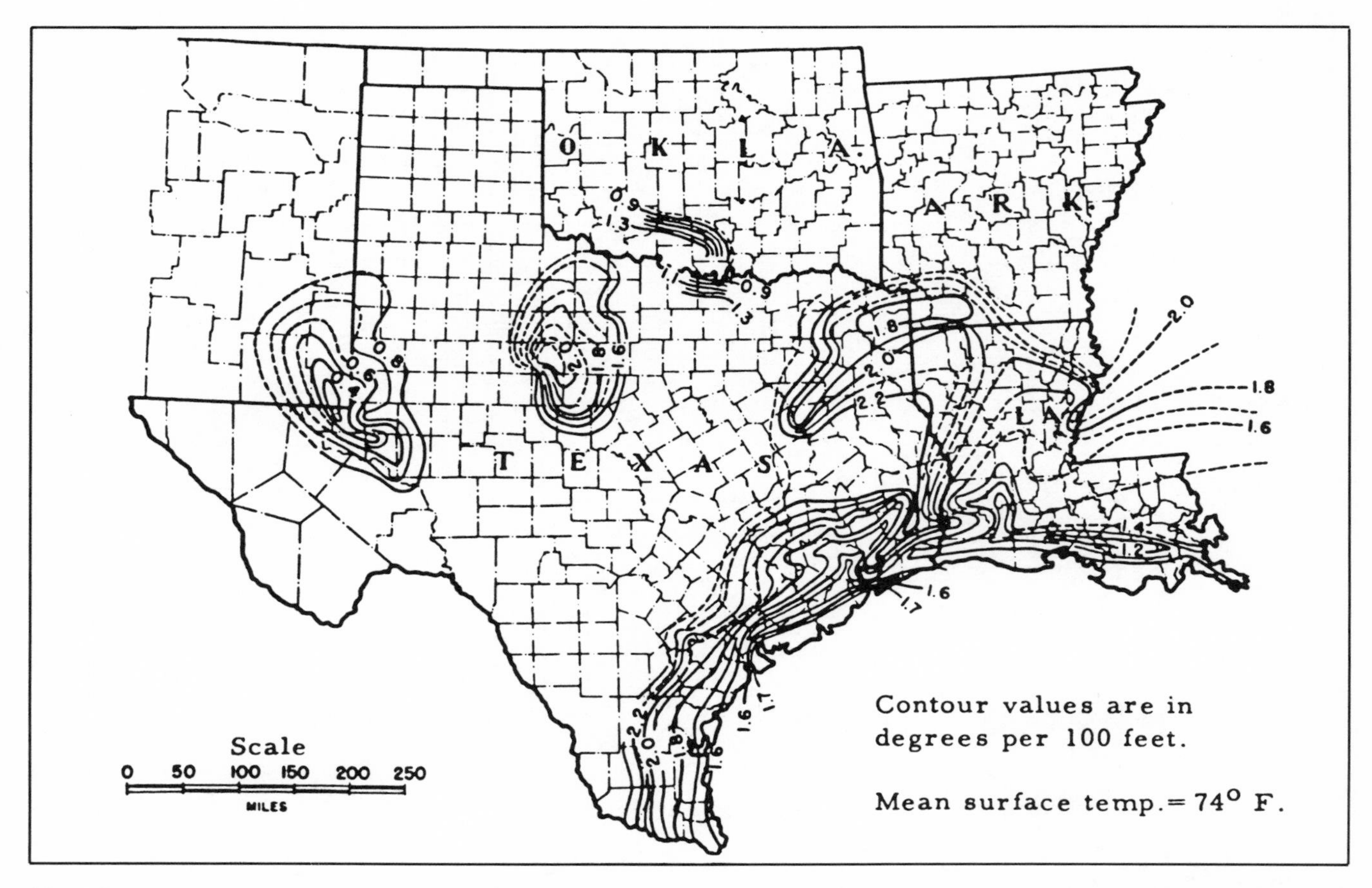

Fig. 1-8. Geothermal gradient map of the southwestern portion of the United States. Permission to publish by World Oil.[5]

mud must be obtained, preferably from the return line, or, if not available, then from the mud pit near a mud gun. A four-electrode resistivity cell is then used to measure resistivity at a particular temperature. This resistivity value must next be converted to mud resistivity at the temperature existing opposite the various formations of interest. The most desirable approach would be to measure mud resistivity at more than one temperature. The usual approach, however, is to assume that the mud acts as a sodium chloride solution and to use the sodium chloride resistivity chart (see Fig. 1-7) to determine R_m at other temperatures. The relationship between resistivity and temperature (°F), graphically presented by Figure 1-7, can be approximated by:

$$R_2 = R_1 \frac{T_1 + 6.77}{T_2 + 6.77}$$

Of course, the fallacy in this method is readily apparent since all muds are not sodium chloride solutions. As mentioned previously, water-base muds may contain calcium, magnesium, potassium, bicarbonate, carbonate, sulphate, etc. Work by Lynn[4] showed that temperature-resistivity relations varied for the muds studied in a linear manner as sodium chloride solutions. In many cases the differences were small. However, some differences were quite large and, therefore, application of the sodium chloride behavior for temperature corrections of mud resistivity may lead to relatively large errors.

DOWNHOLE TEMPERATURE

Temperatures recorded during well logging operations are usually the bottomhole temperature, T_{bh}, and the flowline temperature, T_{fl}. The borehole temperature at any depth can be determined using the temperature gradient, g, usually expressed in °F/100′ where:

$$g\ (°T/100') = \frac{(T_{bh} - T_{fl})100}{D_t}$$

where D_t = Total Depth

The borehole temperature at any desired depth, T_{fD}, can then be calculated from:

$$T_{fD} = T_{fl} + g\left(\frac{D}{100}\right)$$

If flowline temperatures are not available, it is possible to estimate T_{fD} by using empirical data such as that shown in Figure 1-8.

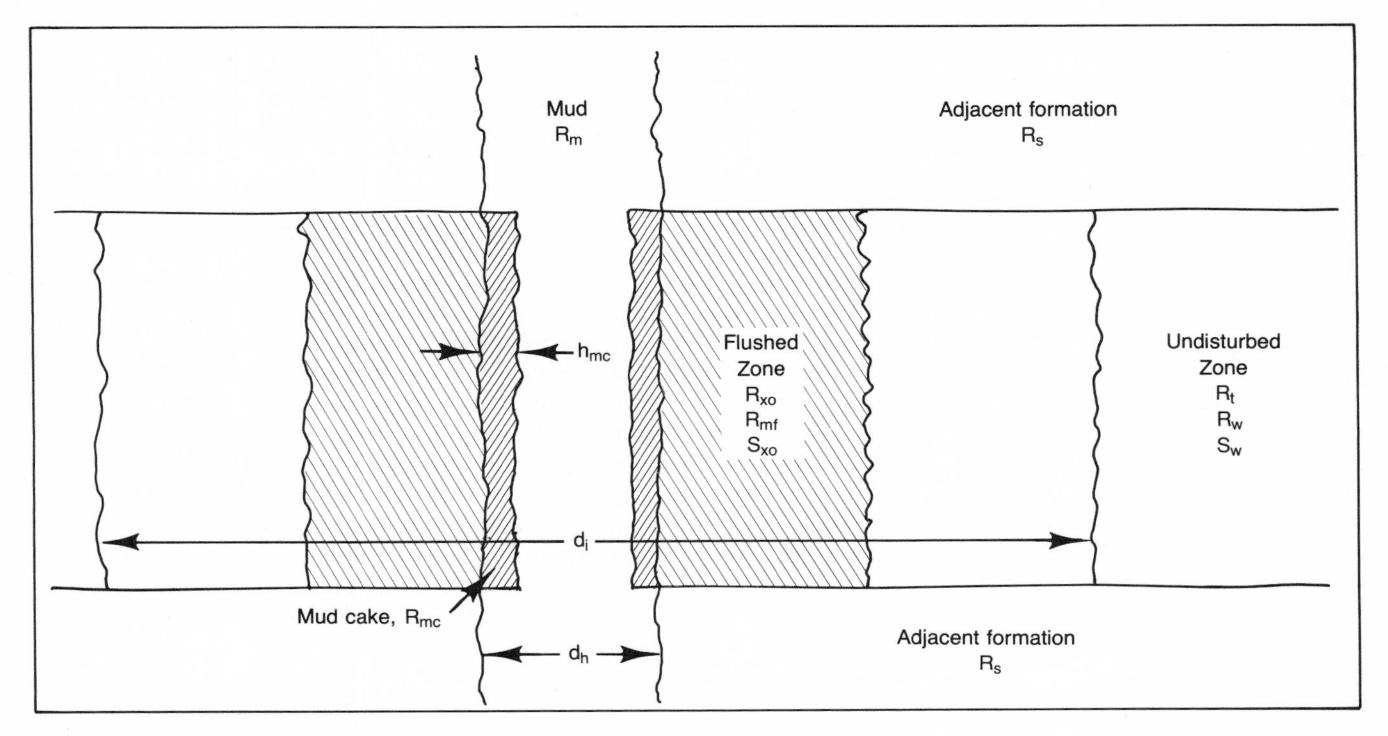

Fig. 1-9. Cross-section of porous and permeable formation

FILTRATE INVASION

The second consideration relative to the effects of mud influence on tool response is that of mud filtrate invasion or water loss into the porous and permeable formations. This process of filtrate invasion creates significant problems in log interpretation but, at the same time, provides a unique method of log interpretation.

As the filtrate invades the formation, it creates a zone of varying resistivity, as shown in Figure 1-9. There are four distinct zones of resistivity. The first zone nearest the borehole is that of the mudcake, which has moderate resistivity. This cake consists of highly compacted solids generally having very low permeability (10^{-5} millidarcies) and a thickness generally between ⅛ to ¾ in. The second zone, the flushed zone, contains mud filtrate and, if an oil sand, will also contain residual oil. The filtrate saturation in this zone is usually referred to as S_{xo}, where: $S_{xo} = 1 - S_{or}$ if oil-bearing or $S_{xo} = 1 - S_{gr}$ if gas-bearing. The third zone is a transition zone saturated with a mixture of mud filtrate and formation fluids. The final zone is the undisturbed zone, which has true formation resistivity, R_t.

The depth of invasion ranges from less than one foot in high porosity formations to perhaps 10 to 15 ft in low porosity formations. Since depth of invasion varies, it is difficult to assign specific values for the width of the flushed and invaded zones, but it can generally be stated that the flushed zone, even in very porous formations, will penetrate from three to four inches.

MEASUREMENT OF R_{mf} AND R_{mc}

For proper log interpretation, the value of mud filtrate resistivity, R_{mf}, and mud cake resistivity, R_{mc}, are necessary. These values should be measured on the well site by the logging engineer. Mud filtrate and mud cake samples are usually obtained using the API mud filter press. However, numerous errors are inherent in obtaining R_{mf} and R_{mc} values in this manner, with some of the apparent problems listed as follows:

1. Filter paper is used to represent porous rock
2. Differential pressure downhole is usually assumed to be 100 psi
3. Mud cake usually disturbed (a cell to minimize this problem was developed by Lynn)[4]

Usually R_{mf} and R_{mc} are measured at only one temperature. In order to determine R_{mf} and R_{mc} at other temperatures, it is usually necessary to assume that *both* the mud filtrate and the mud cake behave as sodium chloride solutions. This is a gross assumption when considering the mud cake. Variations of R_{mf} and R_{mc} due to temperature are then obtained from the resistivity graph shown in Figure 1-7, which relates resistivity and temperature for sodium chloride solutions.

Many times, actual measurements of R_{mf} and R_{mc} are not available, and empirical approaches must be used to

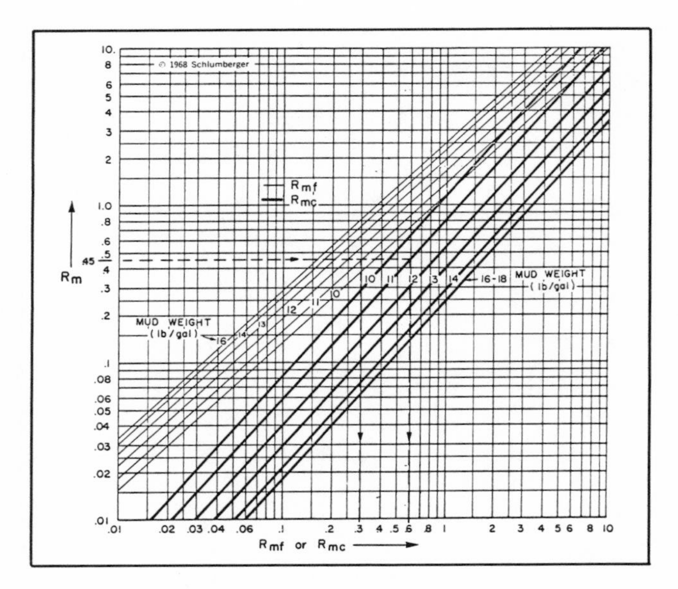

Fig. 1-10. R_{mf} and R_{mc} as a function of R_m and mud weight. Permission to publish by Schlumberger Ltd.[8]

provide approximate values. One approach uses correlations such as presented in Figure 1-10, based on data obtained by Overton and Lipson.[3] Figures 1-11 and 1-12 show correlations of R_{mf} and R_{mc} as functions of R_m for lime and gyp muds.[6]

A desperation approach when neither of the first two methods can be used is the "rule of thumb" method. Typical relationships were presented by Johnson[2] and are $R_{mf} = 0.88\ R_m$, and $R_{mc} = 1.11\ R_m$.

RESISTIVITY PROFILES

To illustrate the effect of invasion on formation resistivity, two examples of resistivity profiles are shown in Figures 1-13 and 1-14. These profiles show the variation of resistivity with distance from the borehole for both a water sand and an oil sand, respectively. As shown, there is a considerable difference between the two profiles. Specifically, the oil sand (see Fig. 1-14) contains a zone of low resistivity just inside the undisturbed zone. This zone occurs when the initial water saturation is low (less than about 50%) and is referred to as the resistivity annulus. It contains an abnormally high formation-water saturation. The creation of this annulus can be visualized as follows. The mud filtrate penetrates the formation radially, sweeping the movable oil and formation water ahead of it. For beds that have a rather large oil saturation, the relative permeability to oil is appreciably greater than that to water. Therefore, the oil moves faster, leaving a zone enriched in formation water behind it. This flow is exceedingly small compared with the reservoir volume, and the saturation in the uncontaminated zone remains undisturbed. It seems likely that due to the effects of diffusion, capillary pressure, gravity, etc., the existence of a well-defined annulus is a transitory phenomenon, although field log experience seems to show that the annulus may exist when the logs are run.

DIAMETER OF THE INVADED ZONE

Factors affecting the diameter of the invaded zone include: (1) type of mud, (2) differential pressure between mud and formation, (3) formation permeability, (4) formation porosity, (5) drilling process and exposure time, and (6) gravity segregation.

We have already mentioned that mud filtrate invasion depends upon the water loss characteristics of the particular mud.

The second factor, differential pressure between the hydrostatic pressure of the mud column and formation fluid pressure, is highly important since it is related to the amount of filtrate injected into the formation. A reasonable value for this differential pressure is 100 psi. This is the pressure used in determining fluid loss in the mud cell.

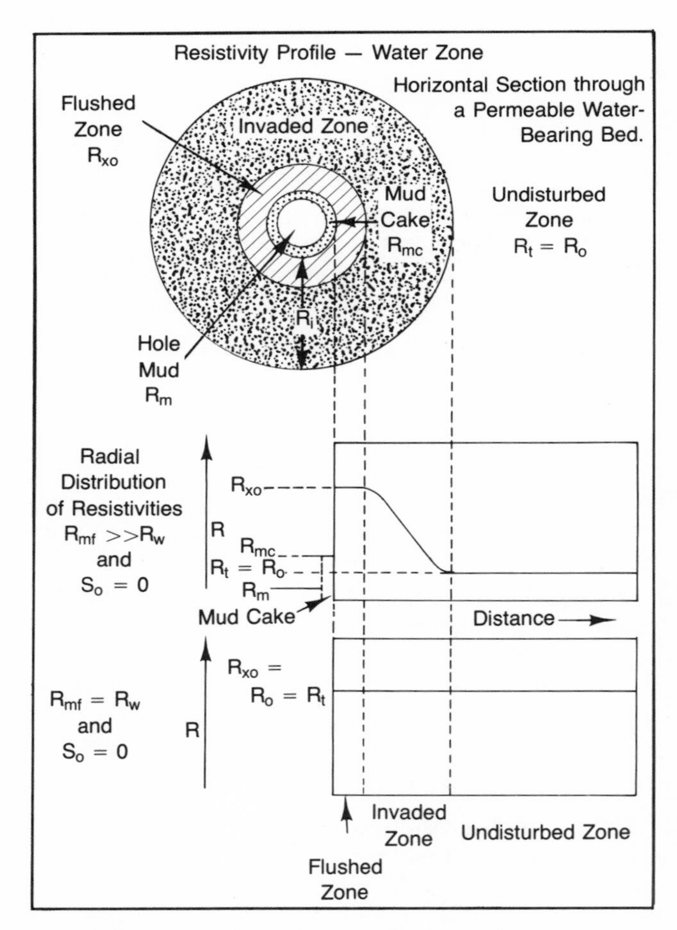

Fig. 1-13. Resistivity profile—water zone

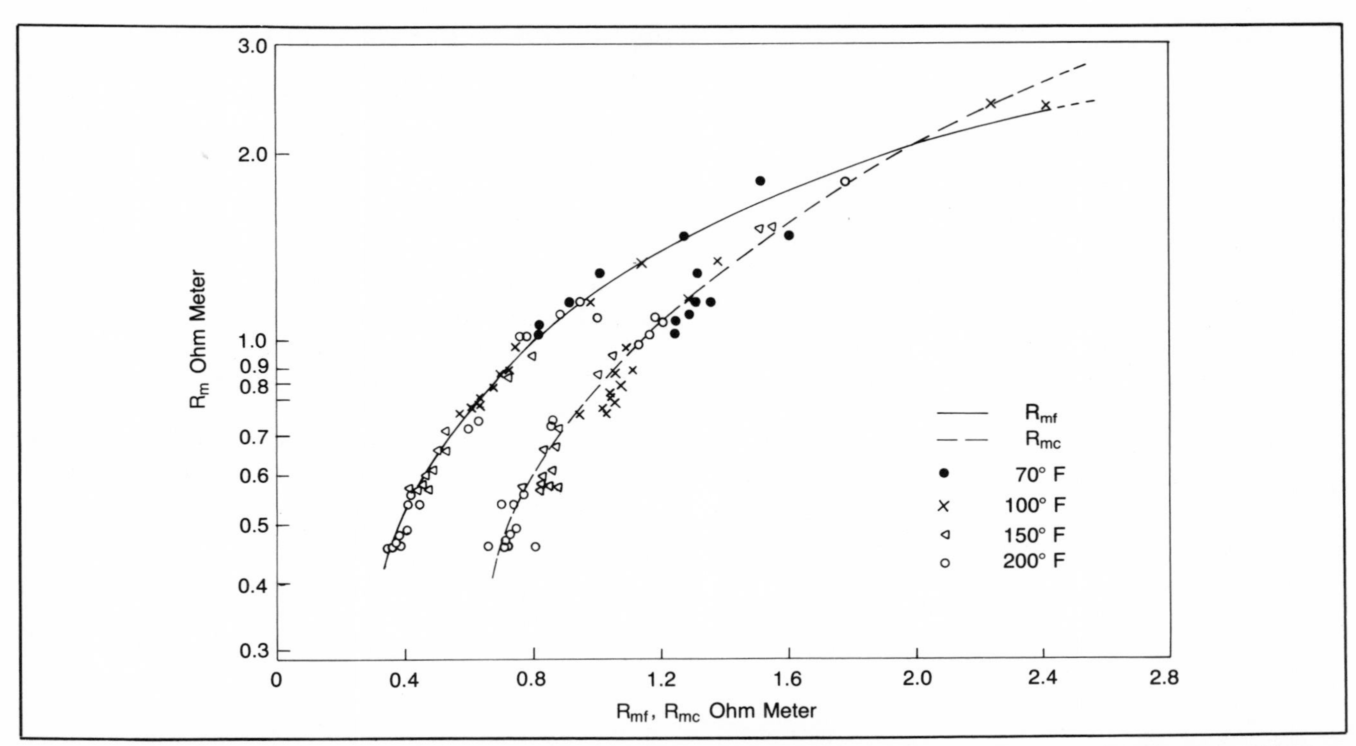

Fig. 1-11. R_{mf} and R_{mc} as a function of R_m for lime base muds. Permission to publish by the Society of Professional Well Log Analysts.[6]

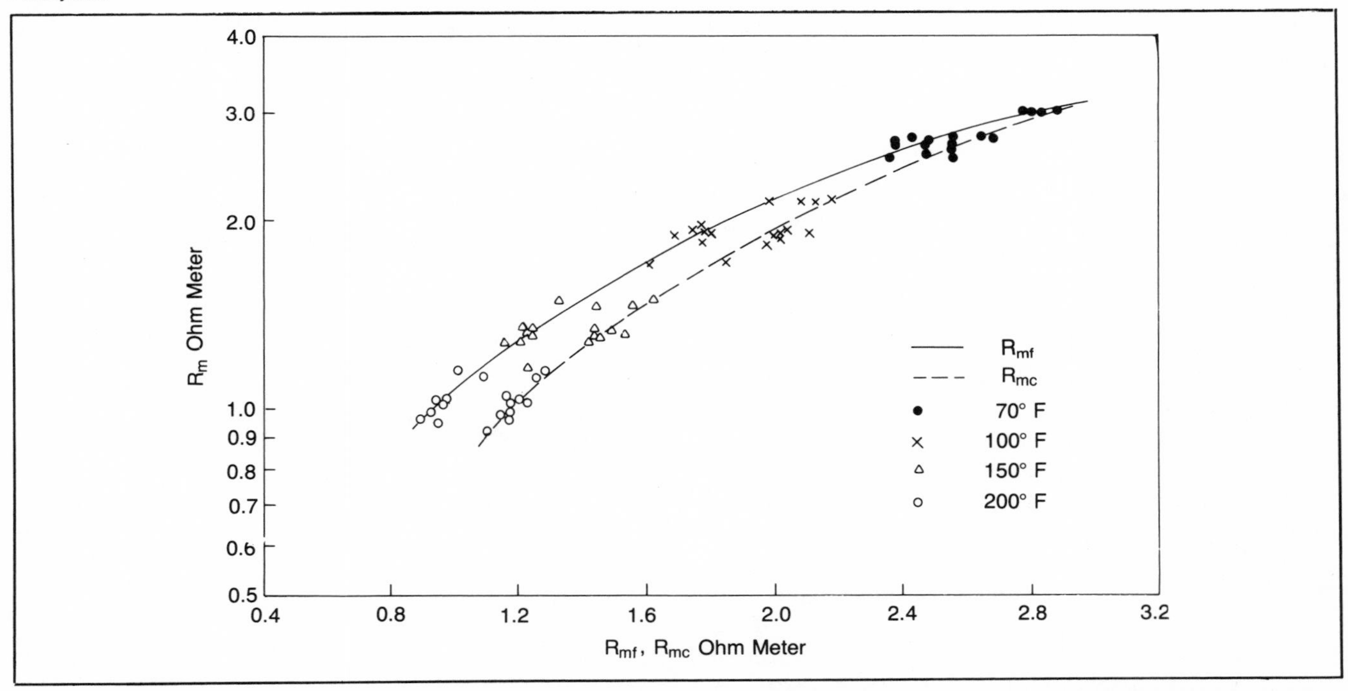

Fig. 1-12. R_{mf} and R_{mc} as a function of R_m for gyp base muds. Permission to publish by the Society of Professional Well Log Analysts.[6]

The third factor, permeability, has little bearing on the ultimate depth of invasion, but it is related to the time it takes the filtrate to move a certain distance into the formation.

The fourth factor, porosity, is the deciding factor in the depth of invasion. We can simply state that as the porosity increases, the depth of invasion decreases. Consider a single formation that is being drilled

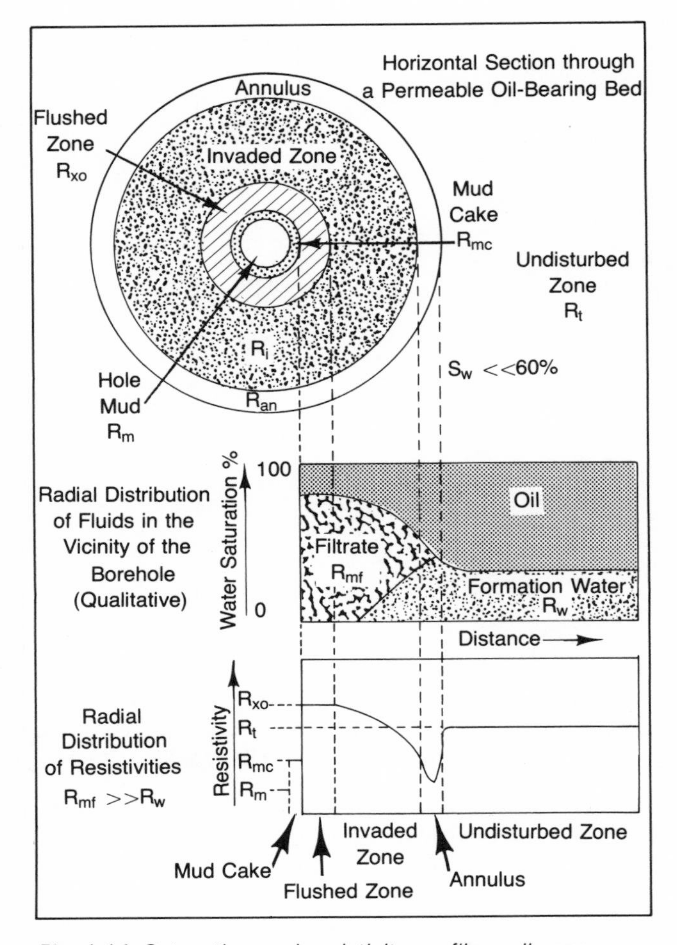

Fig. 1-14. Saturation and resistivity profile—oil zone

through. When the mud comes in contact with the permeable formation, the liquid phase moves into the rock, but the mud solids are filtered out on the formation face, producing a mud cake. The mud cake build-up and the volume of fluid lost to the formation depends on the type of mud system as well as some dynamic factors; however, the volume of fluid lost to the formation can be considered a relatively constant value. In other words, a certain volume of mud filtrate will enter the formation in the process of sealing off the formation. This volume is not dependent upon the formation itself but rather upon the mud. It can be seen that the greater the pore volume per foot of depth from the borehole, the greater the storage capacity of the rock. Thus, greater porosity provides greater storage capacity with distance from the borehole and, therefore, a shallower invaded zone.

The dynamics of the drilling process and fluid circulation alters the depth of the invasion since, with continued drilling, mud cake on a specific zone can be partially or totally eroded off. If this occurs, the invasion process is again initiated to form a new mud cake, and the invaded zone enlarges accordingly. With continued drilling and hence longer exposure time, greater filtrate invasion can be expected.

Gravity segregation generally takes place after the mud filtrate has been forced through the mud cake. This gravity segregation (normally the lighter filtrate rises) definitely alters the invaded zone shape with time (see Fig. 1-15). The mud filtrate is often less saline and, therefore, less dense than interstitial pore water. Consequently, in the lower part of the formation, the mud filtrate rises obliquely from the wall of the hole and the invasion is shallow, and in the upper part, the mud filtrate accumulates below the upper boundary and the invasion is deep. If filtration stops completely, all the filtrate eventually will gather along the upper boundary,

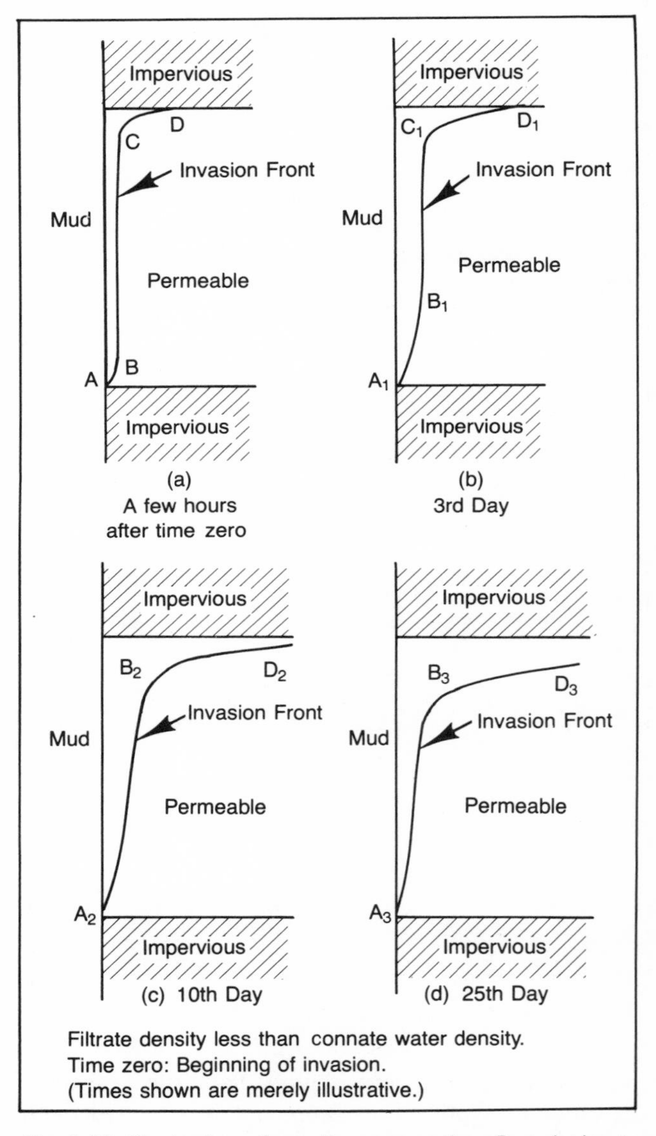

Fig. 1-15. Illustration of gravity segregation. Permission to publish by Petroleum Engineer International.[1]

and the invaded condition will disappear elsewhere in the bed. Of course, the rate at which this takes place depends upon the vertical permeability present within the bed and the difference in densities between the fluids. This rate can be approximated by the equation:

$$v_z = 2.74 \frac{k_z}{\mu \phi} (\rho_w - \rho_{mf}) g$$

where:
v_z = vertical flow rate, ft/day
k_z = vertical permeability, darcies
μ_w = water viscosity, cp
ϕ = porosity
ρ_w, ρ_{mf} = fluid densities, gm/cc
g = acceleration of gravity, cm/sec^2

Diameter of the invaded zone, d_i, is difficult to determine since no tool has been developed to measure it. It may be possible to approximate an electrically equivalent value for d_i using combinations of resistivity logs having different depths of investigation. In essence, this develops a resistivity profile that reflects d_i.

SURFACE LOGGING SYSTEM

The surface equipment required for well-logging operations can be used for lowering and operating a variety of downhole tools. Generally, any electrically-operated, wireline tool can be run into the well and operated from the same surface equipment.

Well Setup. There are three basic setups used, depending on the wellsite and type of downhole tool. If the drilling rig is still on location, a setup similar to that shown in Figure 1-16 is used. The cable is threaded through a lower sheave anchored at floor level and then through an upper sheave connected to a strain gauge weight indicator, which, in turn, is coupled to the traveling block. When the drilling rig has been moved off the well, the setup depends on the type of downhole tool being used. For large, heavy tools, it is necessary to have a mast so the tool can be raised to the vertical position over the well. A portable, hydraulic mast is usually used for this purpose. Finally, small, easily handled downhole tools can be run into the hole simply by setting up a single sheave at the wellhead.

Logging Unit. The logging unit is the control center for all operations. It can be either truck, barge, or platform-mounted (for offshore operations).

Hoisting equipment. The power source for operating the hoisting drum varies. The hoisting drum is driven by a variable-displacement hydraulic pump with a reversible hydraulic motor. The hoist drum, which carries the specially constructed cable, is constructed of nonmagnetic material and contains a set of slip-rings, which maintain electrical connections to all cable conductors while the drum is turning.

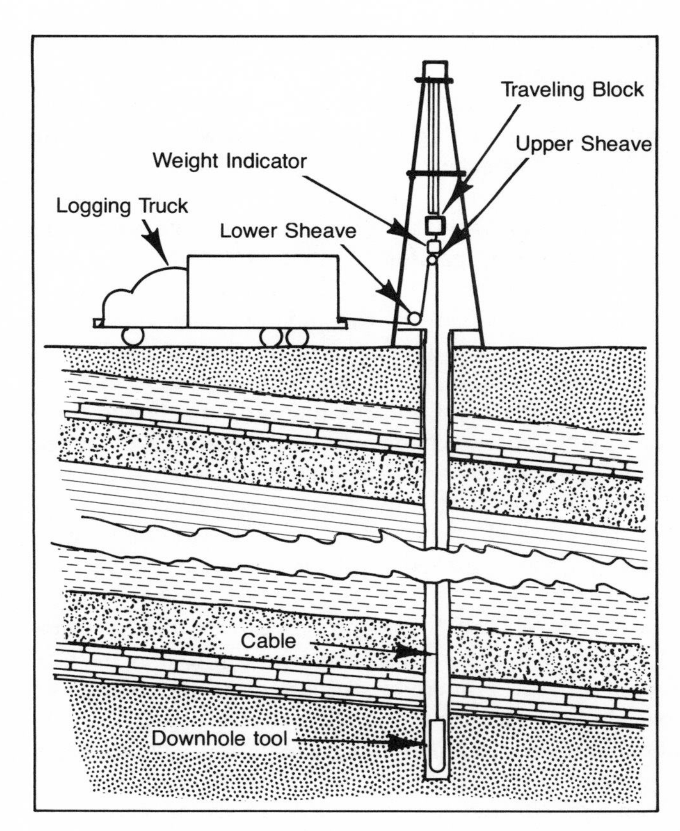

Fig. 1-16. Schematic of logging unit setup. Permission to publish by Oil, Gas and Petrochem Equipment.[12]

Cable Construction. The size and design of the cable spooled on the drum primarily depends on the number of electrical conductors desired. A seven-conductor cable is used in electrical-logging operations, while in perforating operations, a one- or three-conductor cable can be used. (The fewer the conductors, the more limited the application is.) A typical logging cable is shown in Figure 1-17. It consists of seven rubber-insulated, symmetrically-spaced, stranded copper wires. A cloth-braid wrapping separates the conductors from the outer steel jackets. High tensile strength and protec-

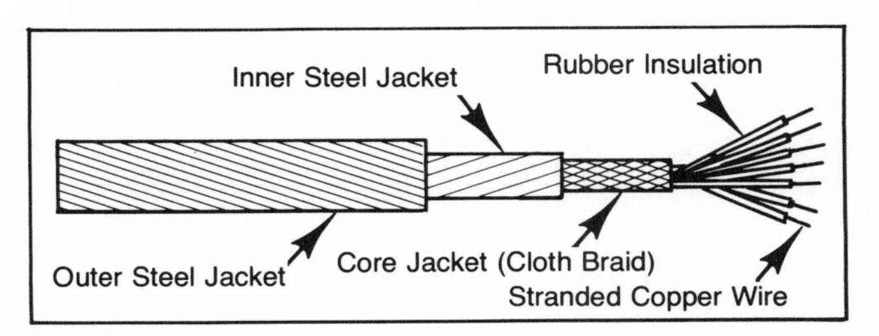

Fig. 1-17. Schematic of seven-conductor logging cable. Permission to publish by Oil, Gas and Petrochem Equipment.[12]

tion for the conductors are provided by the two oppositely wound, spiral-supporting jackets. Each layer contains 24 strands of steel wire. The downhole end of the cable is fitted with an adapter head to which the downhole tool is attached.

Depth measurement. Accurate depth measurements are obtained by passing the logging cable over a calibrated measuring wheel. The grooved periphery is designed to conform to the cable without appreciable slippage and with minimum distortion of the cross-section of the cable. Any cable operating at great depths under varying conditions of load and buoyancy, however, will undergo changes in length with respect to the length passed over the measuring wheel. To assure accuracy, the cable stretch must be considered. Stretch correction curves based on experimental data can be used to adjust the measured depths indicated by the measuring wheel.

The use of a self-correcting gear differential, mechanically coupled to the measuring wheel, provides an automatic correction method. All stretch of the cable, tool weights, running speeds, and temperature differences are compensated for by the differential-gear assembly. The gear differential is, in turn, mechanically coupled with a selsyn transmitter matched to selsyn receivers in the logging unit. These receivers are mechanically coupled to the recorder chart drive and an odometer.

Corrections are usually slight since the reverse spiralling of the two layers of steel wire reduces the amount of stretch to a very small fraction of the amount commonly encountered with wire rope. Under typical logging conditions, the measurement of depth in a 10,000-ft well will not differ more than ± 1 ft from the depth obtained by careful steel-tape measurement of drill pipe.

Operating and recording section. Panel units control the operation of a specific downhole tool. Each panel unit will have an AC power supply. A DC power supply may also be available for operating specific downhole tools.

The chart, or film, on which the log is recorded is driven in exact synchronism with the passage of the cable over the measuring wheel. The usual practice is to represent 100 ft of depth by one, two, five, or twenty-five inches of chart or film. Recorders are also designed such that resistivity responses can be recorded linearly or logarithmically. Logarithmic recording of multiple resistivity measurements as obtained from the Dual Induction SFL and the Dual Laterolog-R_{xo} tools is standard for the five-inch per 100-ft log. Linear resistivity responses can, however, be obtained for the one-inch or two-inch per 100-ft logs for correlation purposes.

REFERENCES

1. Doll, H. G.: "Filtrate Invasion in Highly Permeable Sands," *Petroleum Engineer* (Jan. 1955).
2. Johnson, H. M.: "The Importance of Accuracy in Basic Measurements for Electric Log Analysis," paper presented at the SPWLA Third Annual Conference on Well-Logging Interpretation, McMurray College, Abilene, TX, Oct. 15-17, 1958.
3. Overton, H. L., and Lipson, Leonard B.: "A Correlation of the Electrical Properties of Drilling Fluids with Solids Content," *Trans.*, AIME (1958) 213.
4. Lynn, R. D.: "Effect of Temperature on Drilling Mud Resistivities," paper SPE 1302-6 presented at the SPE 34th Annual Meeting, Dallas, Oct. 4-7, 1959.
5. Moses, Phillip L.: "Geothermal Gradients Now Known in Greater Detail," *World Oil* (May 1961) 79.
6. Wieland, Denton R., Wang, George C., and Helander, Donald P.: "Laboratory Resistivity Evaluation of Lime, Gyp, and Chromate Lignite-Chrome Lignosulfonate Muds," paper presented at the SPWLA Third Annual Logging Symposium, May 17-18, 1962.
7. Krueger, Roland F.: "Evaluation of Drilling-Fluid Filter-Loss Additives Under Dynamic Conditions," *J. Pet. Tech.* (Jan. 1963).
8. *Log Interpretation Chart Book*, Schlumberger Ltd., New York (1969).
9. Pickett, G. R.: "Principles for Application of Borehole Measurements in Petroleum Engineering," *The Log Analyst* (May-June 1969).
10. Pickett, G. R.: "Applications for Borehole Geophysics in Geophysical Exploration," *The Log Analyst* (Feb. 1970).
11. Campbell, John M., Helander, Donald P., and Van Poolen, H. K.: *Petroleum Reservoir Property Evaluation*, Campbell Petroleum Series, John M. Campbell Co., Norman, OK (1973).
12. Helander, Donald P.: "Guide to Surface Equipment for Well Logging," *Oil, Gas and Petrochem Equipment* (formerly *Oil and Gas Equipment*), PennWell Publishing Co. (May 1965) 11.

2 Coring and Core Analysis

CORING and core analysis are an integral part of formation evaluation and provide vital information, unavailable from either log measurements or productivity tests. Core information includes detailed lithology, microscopic, and macroscopic definition of the heterogeneities of the reservoir rock, capillary pressure data defining fluid distribution in the reservoir rock system, and the multiphase fluid flow properties of the reservoir rock, including directional flow properties of the system. Also, selected core data are used to calibrate log responses, such as the acoustic, density, or neutron logs used to determine porosity. As a result, core data becomes an indispensable source in the collection of basic reservoir data directed toward the ultimate evaluation of recoverable hydrocarbons in the reservoir.

The process of obtaining the basic reservoir data required for evaluation is followed by the problem of generating this information at minimum cost. In order to do this, a number of questions need to be resolved:

1. How many wells must be cored in any given reservoir?
2. What types of core data are required?
3. What types of coring fluids are necessary under the reservoir conditions with regard to the type of core data to be obtained?
4. How should these cores be handled in preparation for these analyses?
5. How many core samples should be analyzed?
6. How can the coring and logging programs be coordinated to minimize the coring requirements and make maximum effective use of the cheaper log information?

Because coring data can be obtained at minimum expense during the development of the reservoir, the formation evaluation program and the coordination of both coring and logging programs should be planned shortly after the discovery well. The required core information will vary from reservoir to reservoir; however, general guidelines can be established, with the basic approach remaining the same, although conditions vary. Additionally, specific coring techniques can be evaluated and their limitations discussed. This chapter will be devoted to a discussion of the coring program and the application of core analysis data as it relates to formation evaluation.

TYPES OF CORING

It has long been recognized by the oil industry that an examination of subsurface rock samples is essential in evaluating potential hydrocarbon reservoirs. Early-day evaluation consisted primarily of the examination and qualitative evaluation of cuttings; however, the unreliability of the information obtained from cuttings analysis, and subsequently the need for larger core samples, was recognized as early as 1875.[1] However, in the early part of this century the impetus for more quantitative information led to the development of coring techniques, which have since become more sophisticated and reliable. Today there are several types of coring used in conjunction with rotary drilling, which include conventional coring, diamond coring, wireline coring, and sidewall coring. Each of these methods has its advantages, disadvantages, and special uses; however, at present most coring is done using full-gauge diamond core bits and conventional barrels.

Conventional Coring

The conventional core barrel assembly consists of a cutter head, an outer barrel, a floating inner barrel, and a finger-type "catcher," which keeps the core in the barrel when the assembly is raised. Mud circulates from the drill pipe between the two barrels to the cutter head. Ordinarily, the conventional core barrel will hold a 20-ft core, but shorter and longer lengths can be accommodated. Advantages of conventional coring include (1) the large-diameter core obtained for a given hole size, (2) a high percentage recovery of the formation cored, (3) adaptability to most formations, and (4) no additional surface drilling equipment required. The disadvantage is the necessity of pulling the drill pipe to recover the core after each core has been cut.

Diamond Coring

The use of diamond core bits, in addition to improving coring, can often improve penetration rates over conventional drilling bits in hard formations. The diamond core barrel is very similar to the barrels used in "conventional" coring. Advantages of diamond coring include (1) longer bit life, (2) the possibility of cutting up to 90 ft of core at one run, (3) high percentage recovery, and (4) the economic penetration of hard, abrasive formations. Disadvantages include (1) high initial expense for the barrel and bits, (2) the requirement of precise operating conditions, and (3) supervision by a person knowledgeable in diamond coring.

Wireline Coring

In wireline coring, a hoisting assembly, including a wireline reel, sheave, and wireline lubricator, is needed in addition to the usual surface drilling equipment. Additional subsurface equipment includes a special core drill collar and bit, a core barrel and bit, and a wireline guide and overshot. The core drill collar and bit are run on the drill pipe with a bit plug inside. Prior to coring, the bit plug is pulled with the wireline overshot. The core barrel with cutter head and core catcher is dropped inside the drill pipe and automatically latches into place in the drill collar. After the core has been cut, the barrel with core inside is pulled with the wireline overshot. Advantages of wireline coring include (1) cutting and recovery of consecutive cores without pulling the drill pipe, (2) alternate coring and drilling without making a trip with the drill pipe, and (3) usually lower coring costs. Disadvantages include (1) an appreciable amount of additional surface equipment, (2) use of this method in relatively soft formations only, (3) recovery of cores smaller in diameter than in conventional coring, and (4) usually lower core recovery. Taking more cores improves the qualitative analysis in that lithology can be studied better over longer intervals at reasonable cost; however, the disadvantages of limited-size cores considerably reduce the quality of special core analysis test.

Sidewall Coring

In sidewall coring, a sample is obtained from the wall of a previously drilled open hole at chosen depths. The most widely used tool is a percussion type that is run on a logging truck wireline, and resembles a perforating gun having hollow open-nosed bullets. The bullets are attached to the gun body by short cable wires, shot into the borehole wall, and withdrawn by the cable wires. Usually this tool is used in conjunction with open-hole logging in which points for sidewall sampling are picked from the log. Sidewall coring is restricted to soft and medium hard formations, and the cores recovered usually suffer from compaction microfracturing or intergranular disarrangement. Advantages of sidewall coring include (1) sampling at any depth after the hole has been drilled and (2) the possible aid or confirmation supplied for log interpretation. Disadvantages include (1) samples too small for complete laboratory analysis, (2) badly altered samples, and (3) the formations sampled being flushed by drilling mud filtrate.

Special Coring Tools

Core orientation tools provide information on the direction and dip of the formation cored. One method grooves the core as it is cut, and the groove is related to the compass bearing by a downhole recording system.

A rubber sleeve core barrel is available to improve core recovery from unconsolidated sands when conventional barrels fail. In addition to retaining the core until it is brought to the surface, the barrel provides a packaged core, which is convenient for handling and transportation to the laboratory.

FACTORS AFFECTING CORES

The ideal in core recovery would be, of course, to obtain a sample of the rock as it exists in-situ in the undisturbed state. This, however, is impossible since, during the drilling process and the subsequent removal to the surface, the core and its contained fluids are irretrievably altered. Three factors are responsible for this core alteration: (1) flushing of the rock ahead of the bit by the drilling fluid, (2) pressure reduction, and (3) temperature reduction, which, along with pressure reduction, occurs while bringing the core to the surface. All three factors are involved in saturation changes that occur within the core during recovery from in-situ to surface conditions. An example of these saturation changes is shown in Figure 2-1, which illustrates the gross fluid changes which can be encountered.

Flushing

In essence, flushing by mud filtrate tends to reduce the hydrocarbon content and increase the water content within the rock sample. Both pressure and temperature reductions allow dissolved gas to come out of solution

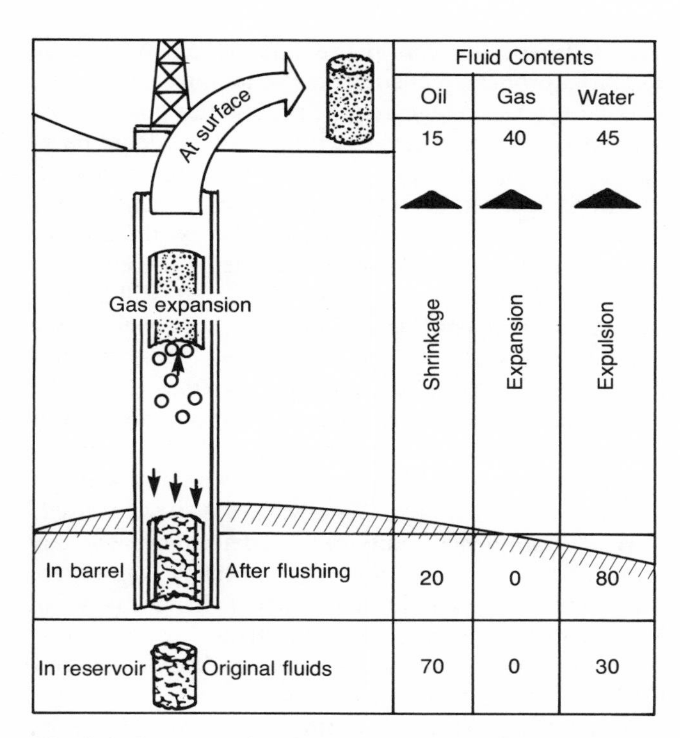

Fig. 2-1. Example of typical saturation changes occurring in core from in situ to surface conditions. Permission to publish by the Oil and Gas Journal.[7]

from the oil in place, which acts as a miniature dissolved gas drive, forcing liquids, both oil and water, to be flushed from the pore space. As a result, surface saturations will consist of some volume of residual oil, a volume of total water consisting of both filtrate water and reservoir water, and a volume of evolved gas. Of course, although the fluids contained in the surface core sample are not representative of in-situ saturation conditions, the saturations are still normally measured at the surface and recorded since, in a qualitative sense, these data can have important applications in the evaluation of any potential hydrocarbon reservoir.

Pressure and Temperature Reduction

The pressure and temperature reductions that occur are also responsible for rock stress changes that can affect the measurement of important rock properties such as permeability, porosity, and resistivity measurements used to define the formation resistivity factor, cementation factor, and saturation exponent. For some rock types, the stress relief is of such magnitude that important rock properties should be measured under simulated in-situ pressure and temperature conditions, particularly in-situ pressure conditions. A number of studies[3, 5, 8, 14, 15, 16] have investigated the importance of pressure reduction on rock properties, as illustrated in Figures 2-2 and 2-3.

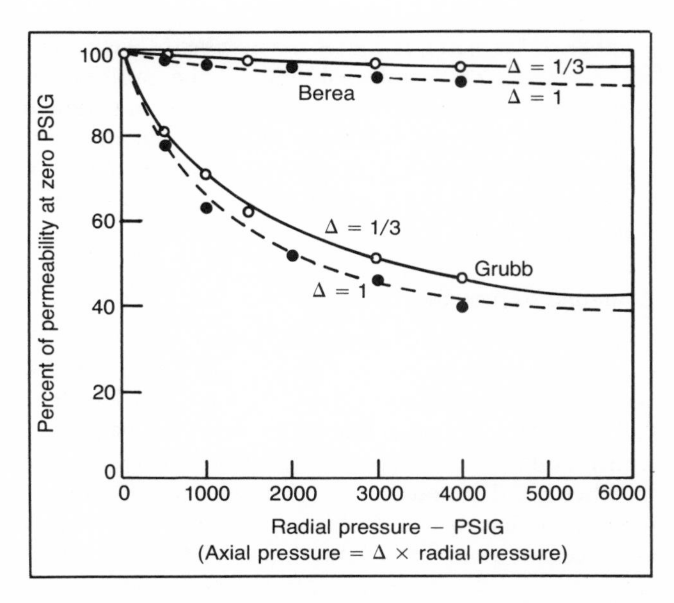

Fig. 2-2. Permeability reduction in Berea and Grubb sandstone as a function of radial and axial loading. Permission to publish by the Society of Petroleum Engineers of AIME.[15] *Copyright 1963 SPE-AIME.*

The influence of stress relief with respect to the measurement of formation factor and cementation exponent is discussed in more detail in Chapter 4 since these particular measurements are of prime importance in log calibration and subsequent formation evaluation work.

CORING PROGRAM

As stated earlier, one of the principal goals in designing a coring program is to minimize the number of wells that have to be cored and to rely on well logs to supply the required evaluation data in the uncored wells. The

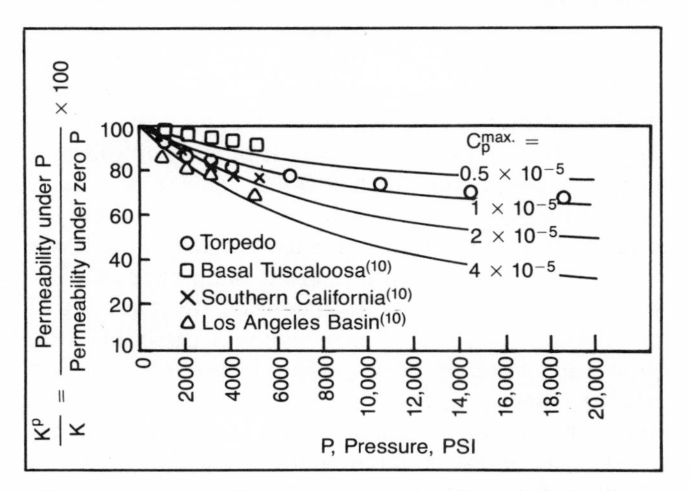

Fig. 2-3. Permeability reduction in sandstones as a function of net overburden pressure. Permission to publish by the Society of Petroleum Engineers of AIME.[14] *Copyright 1962 SPE-AIME.*

number of key wells, and thus the number of wells to be cored, offers no simple solution to the problem of designing a coring program. One might say that enough wells should be cored to adequately define the reservoir; however, the problem still remains. Two possible approaches to designing a coring program are outlined as follows.

Empirical Approach

Clark and Shearin[6] point out that successful programs have been based on one key well per 640 acres, while others require one well per 320 acres or less. The Canadian Conservation Board, for instance, has established minimum coring frequencies in several provinces: in Alberta and Saskatchewan, one well per section is required, while in British Columbia two wells per section are required. A review of published field case histories indicates a wide range of coring frequencies, but it seems that probably one well should be cored per section (every 640 acres). If, as a result of this coring frequency, cores and log analyses indicate statistically significant areal changes in rock properties, the cored-well spacing should probably be reduced. If the core and log interpretations show that areal variations in rock properties are insignificant, key well spacing might be increased to one every 1,000 or even 2,000 acres.

Statistical Approach

Obviously a problem exists in the attempt to represent a large volume with a very small volumetric sample. Intuition tells us that the reliability of average rock properties will increase as the amount of data used to obtain that average increases. It is interesting to visualize the percent bulk volume to be analyzed from a whole core analysis and what it represents in relation to the bulk volume of a single well spacing. For example, if one takes a three-inch diameter whole core one foot long and compares the volume of this core to a well drainage area of 20 acres one foot thick, the bulk volume sample can be seen to represent $5.6 \times 10^{-10}\%$. This sample size is further reduced as well spacing is increased and as plug analysis from the whole core is used in lieu of the whole core itself. Although sample size is diminutive, statistical procedures are available to describe the reliability of any average rock property. By applying these statistical procedures to reservoir properties, it can be determined whether these properties are changing significantly in areal or vertical directions. Obviously, if significant areal variation exists, additional cores will probably be required. If areal variation is not significant, coring frequency might be reduced.

The statistical approach utilizes the concept of confidence limits in designing an optimum coring program. The width of the confidence interval on an average reservoir property depends on the number of samples making up the average and the probability that the true average of the reservoir lies within the confidence interval. The confidence interval is narrow for low probability and is broader for high probability. If there is truly no significant areal variation in a property, the average for half of the wells will lie outside the 50% confidence limits, the average for a quarter of the wells will lie outside of the 75% confidence limits, and the average for only one out of 20 wells will lie outside of the 95% confidence limits. Because confidence limits are used to determine whether a significant variation exists from well to well, the confidence level should be high enough that the average property for a well normally will not fall outside of the limits. Usually the 95% level is used.

While confidence limits can be determined for any measureable reservoir property, porosity values are most used since, in addition to cored wells, porosity can be determined for uncored wells from log interpretation. This procedure can also be applied to net pay thickness and to water saturation.

An interesting study discussing the potential degree of areal variation in permeability, porosity, and gross sand thickness was presented by Greenkorn and Johnson.[11] One objective of their study was to determine the areal variation of these three parameters within a small segment of a natural sandstone reservoir. The data was obtained from a brine-saturated upper Pennsylvanian

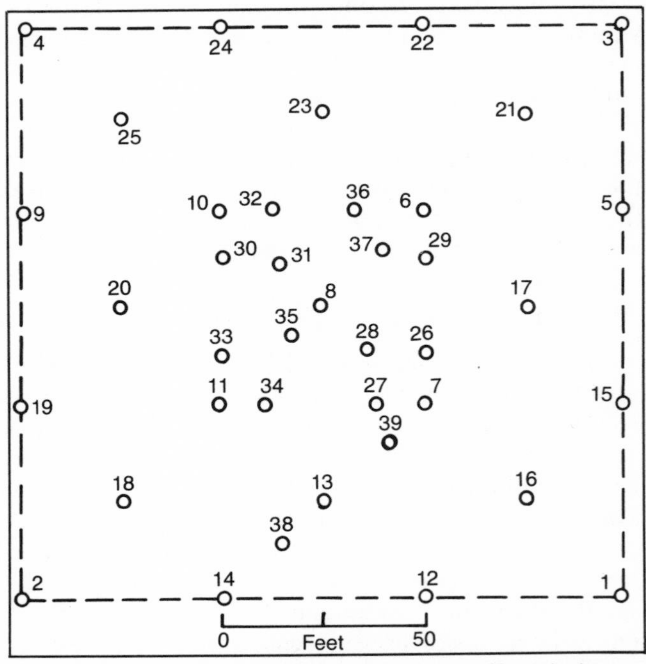

Fig. 2-4. Test area showing location of wells. Permission to publish by the Society of Petroleum Engineers of AIME.[11] Copyright 1960 SPE-AIME.

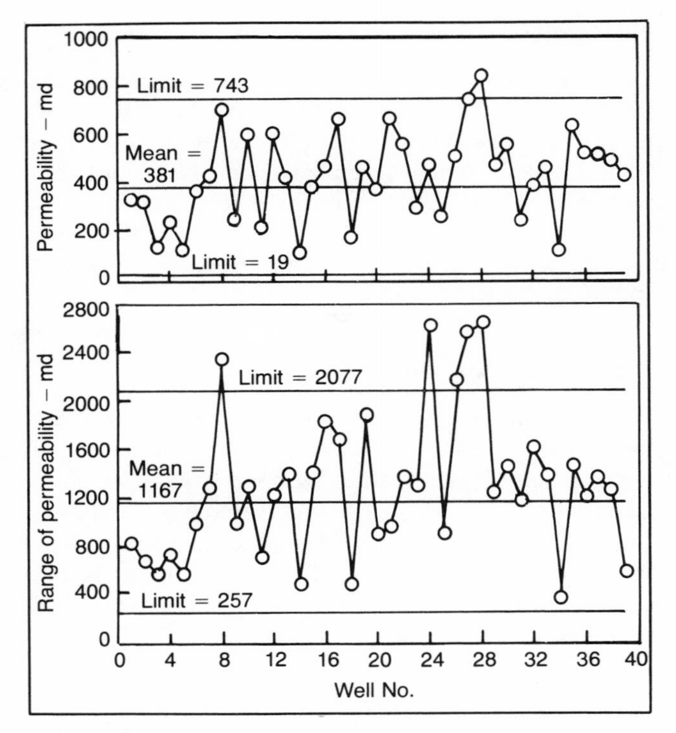

Fig. 2-5. Permeability and permeability variation in test segment. Permission to publish by the Society of Petroleum Engineers of AIME.[11] Copyright 1960 SPE-AIME.

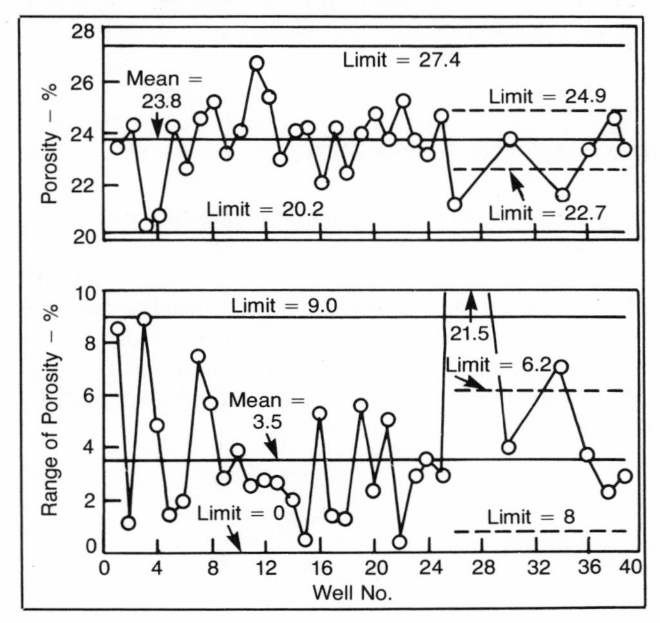

Fig. 2-6. Porosity and porosity variation in test segment. Permission to publish by the Society of Petroleum Engineers of AIME.[11] Copyright 1960 SPE-AIME.

sandstone at a depth of 250 ft. Thirty-nine complete cores were obtained and analyzed within a 150-ft square of this sandstone, as shown in Figure 2-4.

A second objective was to analyze the variation of these three parameters to establish the magnitude of the changes of variation. The rock segment may be quite heterogeneous, and yet the distribution of heterogeneities (change in variation of these parameters) may be quite uniform so that, over all, the segment acts in a homogeneous manner or conversely is so changeable from place to place that the homogeneities dominate the reservoir behavior. The results of this study are shown in Table 2-1 and Figures 2-5, 2-6, and 2-7. The probable limits are: avarage well permeability, 19–743 md, average well porosity, 20–27% and gross reservoir thickness, 2.3–14.5 ft. Additionally, all three parameters showed significant areal variation within this small reservoir segment. The areal variation exhibited by this reservoir is typical of many, if not most, petroleum reservoirs.

TABLE 2-1
Arithmetic Mean Values for Reservoir Properties of Individual Wells

Wells	*Gross Reservoir Thickness (ft)*	*Permeability (md)*	*Porosity (%)*
1	8.0	331	23.5
2	9.0	315	24.3
3	13.8+	132	20.3
4	10.8	233	20.9
5	10.4	122	24.3
6	7.4	368	22.7
7	7.2	424	24.6
8	7.6	700	25.2
9	8.5	249	23.2
10	7.4	592	24.1
11	7.7	205	26.7
12	7.2	598	25.4
13	7.6	408	23.0
14	8.4	100	24.1
15	6.7	373	24.2
16	6.1	460	22.1
17	6.4	714	24.2
18	9.7	168	22.5
19	10.0	456	23.9
20	9.7	364	24.7
21	7.8	657	23.8
22	9.0	553	25.2
23	7.2	292	23.7
25	9.0	252	24.7
26	7.2	502	21.3
27	7.6	738	–
28	7.5	867	–
29	7.0	460	–
30	6.0	546	23.8
31	7.5	245	–
32	7.6	386	–
33	7.1	453	–
34	6.9	106	21.7
35	7.4	650	–
36	8.4	517	23.4
37	8.2	527	–
38	7.5	488	24.7
39	7.4	427	23.5

Permission to publish by the Society of Petroleum Engineers of AIME.[11]

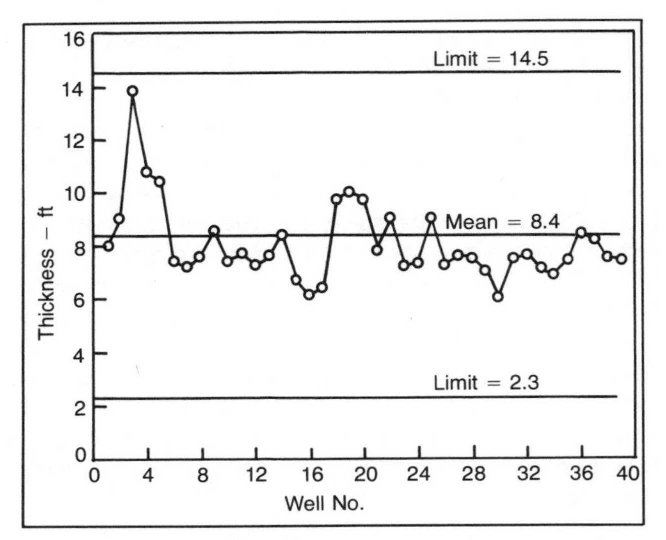

Fig. 2-7. Gross sand thickness in test segment. Permission to publish by the Society of Petroleum Engineers of AIME.[11] Copyright 1960 SPE-AIME.

CORE ANALYSIS TECHNIQUES

There are three types of core analysis techniques: (1) conventional or plug analysis, (2) whole-core analysis and (3) sidewall core analysis. The technique used depends on the coring method, the type of rock to be analyzed, and the type of data to be obtained.

Conventional or Plug Analysis

The plug analysis method is used most frequently. In this method, a small plug sample, which is easy to work with in the laboratory, is cut at selected intervals from the whole core. The data obtained from the small plug are then assumed to represent the reservoir rock properties of the sampled interval. The validity of this approach is increased as the rock type becomes more uniform.

It is also necessary to make a decision on the number of samples required for analysis. It is a generally accepted practice to determine the basic rock properties such as porosity and permeability on a frequency of one sample per foot with fluid content possibly being determined less frequently, for example, one for every two to five feet of core. In many instances, this may turn out to be more data than needed; the indications are that in many reservoirs the average permeability and porosity determined from one sample for every two feet did not differ significantly from those determined from one sample per foot.

An additional factor to consider is the sampling procedure since it is most important to avoid any bias in selecting the samples to be analyzed. A number of techniques can be used that will eliminate the possibility of selecting the best looking samples.

Whole-Core Analysis

The whole-core analysis method is used when the plug analysis method becomes invalid because of the presence of heterogeneities such as fractures or vugs. This method uses the whole core for rock property measurement in as long a length as possible. The technique requires larger equipment in the laboratory, and not all commercial laboratories are equipped to perform this type of analysis.

Sidewall Core Analysis

Considering the process under which these cores are obtained and the sample size of the core, the measured data will have limited value. Of course, in some areas and in some situations, this rock sample is all that is available. It is, therefore, desirable to look at the relative value of rock properties as determined from sidewall samples and those obtained from conventional cores. Several studies have been conducted[9, 10, 13, 17] with the results briefly summarized here. These studies indicate in general that (1) percussion sample porosities in softer, looser sands are only slightly higher than those of conventional cores, (2) sidewall sample permeabilities are decreased in higher permeability formations, and (3) water saturations from the sidewall cores are lower and oil saturations slightly higher than conventional core data. Table 2-2 provides an extensive comparison of sidewall and conventional core analysis data from the U.S. Gulf Coast region for three different formations. Based on studies to date, different limiting values and standards of interpretation have been shown to be necessary when using sidewall sample analysis rather than core analysis. In most areas where sidewall coring is widely used, however, it appears that suitable relations could be developed to permit reliable qualitative and possibly even quantitative reservoir evaluations to be obtained. Certainly the sidewall core data and well log data would be mutually complementary in the evaluation process.

BASIC MEASUREMENTS

The quality of basic data obtained from core analysis is dependent upon not only the proper sampling procedure, but also upon proper handling and preservation of the core as it is retrieved from the well. As soon as the core is removed from the core barrel, the core should be wiped clean of the drilling fluid without washing, and then immediately sealed from the atmosphere to prevent fluid losses. A description of core lithology and other pertinent characteristics of the core, such as the presence of fractures and apparent staining, should be

TABLE 2-2
Comparison of Sidewall and Conventional Core Analysis Data

Gulf Coast		*Number Samples*	*Average Depth*	*Perm.*	*φ*	*Res. Oil*	*Total Water*
Combined Data	Conventional	3178	8041	851	28.1	6.4	61.1
	Sidewall	2160	7255	239	28.5	4.7	69.6
Compared by Production							
Condensate	Conventional	967	9131	797	27.7	1.4	53.2
	Sidewall	526	7536	177	27.7	1.1	63.7
Oil	Conventional	1111	7035	936	29.2	15.0	53.4
	Sidewall	601	6815	384	29.8	14.0	57.8
Water	Conventional	1101	8100	822	27.5	2.3	75.8
	Sidewall	1033	7367	185	28.1	1.1	79.4
Compared by Formation							
Miocene	Conventional	1044	9149	853	27.3	7.9	58.3
	Sidewall	557	8160	292	28.4	5.3	69.5
Frio	Conventional	1097	7933	1197	27.7	4.3	61.5
	Sidewall	785	6763	233	28.3	5.4	68.3
Cockfield	Conventional	1037	7050	493	29.5	7.2	63.6
	Sidewall	818	7110	208	28.7	3.7	70.9
Compared by Formation and Production							
Miocene							
Condensate	Conventional	378	10830	726	27.5	1.9	53.0
	Sidewall	143	8835	274	28.3	1.9	64.5
Oil	Conventional	393	7605	1109	28.1	17.1	51.1
	Sidewall	181	7259	334	29.1	13.1	61.1
Water	Conventional	273	9044	659	26.0	2.9	76.0
	Sidewall	233	8445	271	28.0	1.3	79.0
Frio							
Condensate	Conventional	403	8083	1103	27.4	0.9	51.1
	Sidewall	220	7071	143	27.3	0.8	63.8
Oil	Conventional	224	7471	1246	28.1	15.5	49.0
	Sidewall	208	6389	483	30.2	16.9	54.3
Water	Conventional	470	8025	1255	27.6	2.0	76.2
	Sidewall	357	6791	142	27.7	1.6	79.2
Cockfield							
Condensate	Conventional	186	7947	279	28.6	1.3	58.1
	Sidewall	163	7024	137	27.8	0.9	62.9
Oil	Conventional	493	6397	658	30.7	13.1	57.2
	Sidewall	212	6854	329	29.9	11.9	58.5
Water	Conventional	358	7481	377	28.4	2.1	75.2
	Sidewall	443	7265	176	28.5	0.5	79.8

Permission to publish by the Society of Professional Well Log Analysts.[14]

recorded as a function of depth in as detailed a description as is deemed necessary.

Suitable methods for preserving cores are (1) by packaging in aluminum foil and sealing in low melting point paraffin or other wax, (2) by tightly wrapping and double sealing in plastic wrapping materials, or (3) by quick freezing. The quick freezing process has been found to be a desirable method for preserving the fluid content in the cores.

The basic measurements made on the retrieved core include porosity, absolute permeability, and residual fluid saturations. In conjunction with the detailed lithologic descriptions these data are then presented in both tabular and graphical formats versus depth, as shown in the typical core analysis log of Figure 2-8.

The determination of rock porosity and permeability cannot be accomplished, however, without first removing the residual fluid contained within the core sample. This is normally accomplished by using a solvent extraction method, for example, a modified ASTM (American Society for Testing and Material) extraction apparatus, as shown in Figure 2-9, or a Soxhlet extractor, as shown in Figure 2-10.

Use of the modified ASTM extraction apparatus allows a measurement of the water volume removed from the core during the cleaning process. Once all the

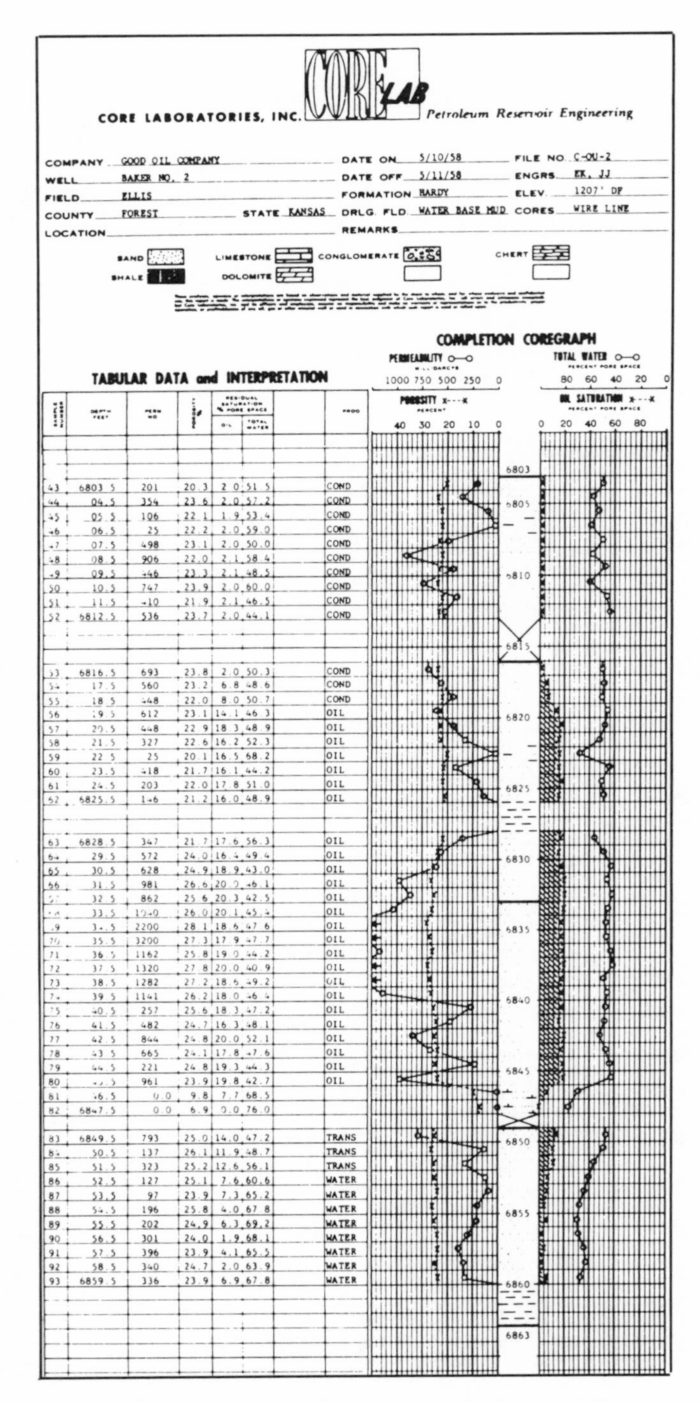

CORE LABORATORIES, INC. CORE LAB Petroleum Reservoir Engineering

COMPANY GOOD OIL COMPANY DATE ON 5/10/58 FILE NO. C-OU-2
WELL BAKER NO. 2 DATE OFF 5/11/58 ENGRS. EK, JJ
FIELD ELLIS FORMATION HARDY ELEV. 1207' DF
COUNTY FOREST STATE KANSAS DRLG. FLD. WATER BASE MUD CORES WIRE LINE
LOCATION REMARKS

SAND SHALE LIMESTONE DOLOMITE CONGLOMERATE CHERT

COMPLETION COREGRAPH
PERMEABILITY o—o MILLIDARCYS 1000 750 500 250 0
POROSITY x---x PERCENT 40 30 20 10 0
TOTAL WATER o—o PERCENT PORE SPACE 80 60 40 20 0
OIL SATURATION x---x PERCENT PORE SPACE 0 20 40 60 80

TABULAR DATA and INTERPRETATION

Sample Number	Depth Feet	Perm MD	Porosity %	Residual Saturation % Pore Space: Oil	Residual Saturation % Pore Space: Total Water	Prod
43	6803.5	201	20.3	2.0	51.5	COND
44	04.5	354	23.6	2.0	57.2	COND
45	05.5	106	22.1	1.9	53.4	COND
46	06.5	25	22.2	2.0	59.0	COND
47	07.5	498	23.1	2.0	50.0	COND
48	08.5	906	22.0	2.1	58.4	COND
49	09.5	446	23.3	2.1	48.5	COND
50	10.5	747	23.9	2.0	60.0	COND
51	11.5	410	21.9	2.1	46.5	COND
52	6812.5	536	23.7	2.0	44.1	COND
53	6816.5	693	23.8	2.0	50.3	COND
54	17.5	560	23.2	6.8	48.6	COND
55	18.5	448	22.0	8.0	50.7	COND
56	19.5	612	23.1	14.1	46.3	OIL
57	20.5	448	22.9	18.3	48.9	OIL
58	21.5	327	22.6	16.2	52.3	OIL
59	22.5	25	20.1	16.5	68.2	OIL
60	23.5	418	21.7	16.1	44.2	OIL
61	24.5	203	22.0	17.8	51.0	OIL
62	6825.5	146	21.2	16.0	48.9	OIL
63	6828.5	347	21.7	17.6	56.3	OIL
64	29.5	572	24.0	16.4	49.4	OIL
65	30.5	628	24.9	18.9	43.0	OIL
66	31.5	981	26.6	20.0	46.1	OIL
[illegible]	32.5	862	25.6	20.3	42.5	OIL
[illegible]	33.5	1040	26.0	20.1	45.4	OIL
[illegible]	34.5	2200	28.1	18.6	47.6	OIL
70	35.5	3200	27.3	17.9	47.7	OIL
71	36.5	1162	25.8	19.0	44.2	OIL
72	37.5	1320	27.8	20.0	40.9	OIL
73	38.5	1282	27.2	18.6	49.2	OIL
74	39.5	1141	26.2	18.0	46.4	OIL
75	40.5	257	25.6	18.3	47.2	OIL
76	41.5	482	24.7	16.3	48.1	OIL
77	42.5	844	24.8	20.0	52.1	OIL
78	43.5	665	24.1	17.8	47.6	OIL
79	44.5	221	24.8	19.3	44.3	OIL
80	[illegible]	961	23.9	19.8	42.7	OIL
81	46.5	0.0	9.8	7.7	68.5	
82	6847.5	0.0	6.9	0.0	76.0	
83	6849.5	793	25.0	14.0	47.2	TRANS
84	50.5	137	26.1	11.9	48.7	TRANS
85	51.5	323	25.2	12.6	56.1	TRANS
86	52.5	127	25.1	7.6	60.6	WATER
87	53.5	97	23.9	7.3	65.2	WATER
88	54.5	196	25.8	4.0	67.8	WATER
89	55.5	202	24.9	6.3	69.2	WATER
90	56.5	301	24.0	1.9	68.1	WATER
91	57.5	396	23.9	4.1	65.5	WATER
92	58.5	340	24.7	2.0	63.9	WATER
93	6859.5	336	23.9	6.9	67.8	WATER

Fig. 2-8. Typical core analysis presentation. Permission to publish by the John M. Campbell Company.[18]

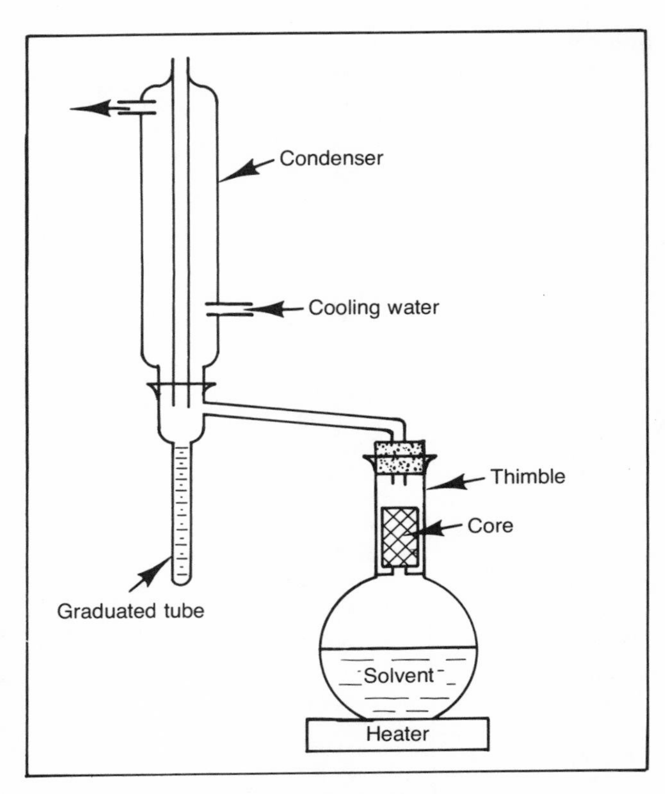

Fig. 2-9. Schematic of modified ASTM extractor

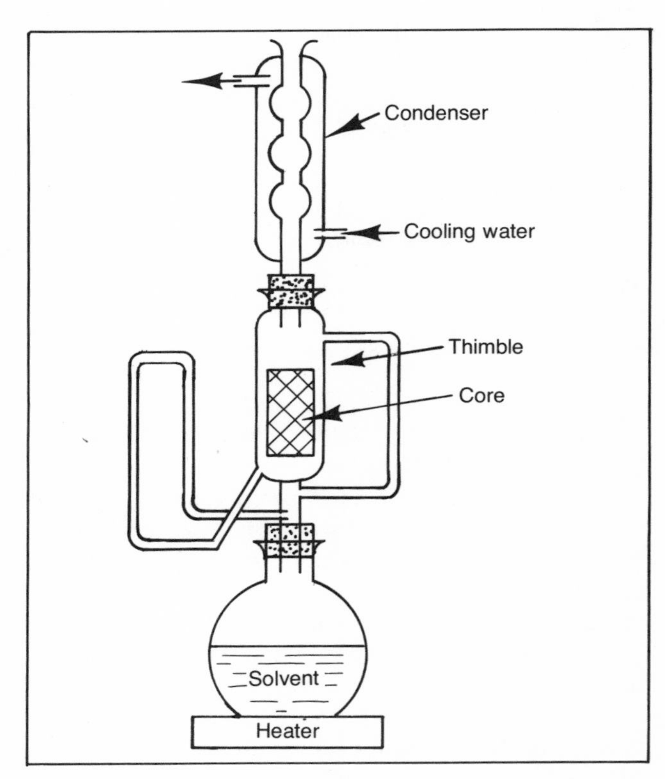

Fig. 2-10. Schematic of Soxhlet extractor

water is removed, the core can be placed in the Soxhlet extractor (see Fig. 2-10) in order to complete the cleaning process. The mechanics of the Soxhlet extractor are essentially the same as the ASTM apparatus except that no receiving vessel is supplied for trapping water.

Residual Saturation

As previously mentioned, the residual water volume can be determined during the cleaning process when using the modified ASTM apparatus. In this method the core is placed so that a vapor of tolulene, naphtha, or other suitable solvent rises through the core, is condensed, and subsequently refluxes back over the core. This process leaches both the oil and water from the core. The water (immiscible with the solvent) and extracting fluid (solvent and miscible oil) are condensed and collected in the graduated receiving tube. The water settles to the bottom of the graduated tube because of its greater density, with the extracting fluid refluxing back into the main heating vessel once the graduated tube fills. This process continues until no more water is collected in the receiving tube. Residual water saturation can then be determined, using the water volume measurement and the porosity.

The measurement of oil saturation is an indirect determination. It is necessary to note the weight of the core sample prior to extraction. Then, after the core has been cleaned and dried, the sample is again weighed. The oil saturation as a fraction of pore volume is given by:

$$S_{or} = \frac{(\text{wt of wet core}) - (\text{wt of dry core}) - (\text{wt of water})}{(\text{pore volume}) \times (\text{density of oil})}$$

Retort Method

One of the most popular means of measuring the initial saturations is the retort method. This method takes a small rock sample and heats the sample to vaporize the water and oil, which is condensed and collected in a small receiving vessel. A sketch of an electric retort is shown in Figure 2-11.

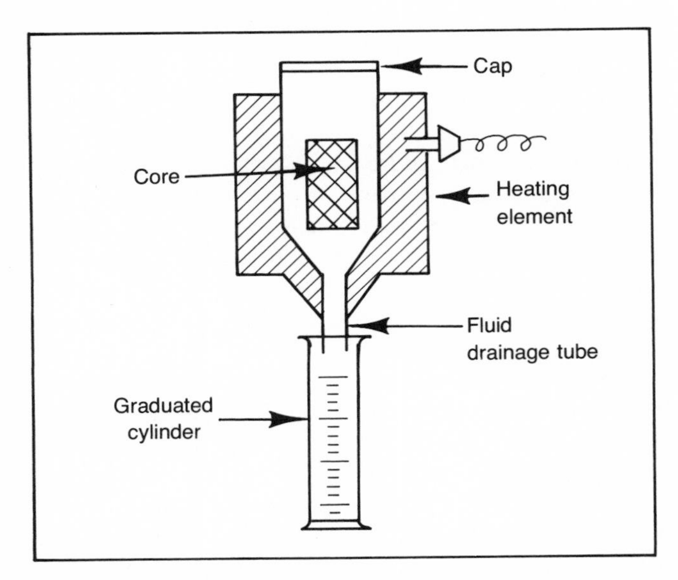

Fig. 2-11. Schematic of retort distillation apparatus

The retort method has several disadvantages as far as commercial work is concerned. First, in order to remove all the oil, it is necessary to approach temperatures on the order of 1000° to 1100°F. At these temperatures the water of hydration and crystallization is driven off, causing the total water-recovery values to be high. The recovered water volume being taken to the first plateau, as observed on a graph of water recovery versus time, compensates for this effect. The waters of hydration and crystallization are expelled at higher temperatures that occur at later times.

The second error that occurs from retorting samples is that when heated to high temperatures the oil itself has a tendency to crack and coke. This change of a hydrocarbon molecule tends to decrease the liquid volume and in some cases coats the internal walls of the rock sample itself. Thus a fluid correction must be made on all sample data obtained with a retort.

Before retorts can be used, calibration curves must be prepared on various gravity fluids to correct for the losses from cracking and coking with the various applied temperatures. Another correction curve can also be obtained that correlates recovered API oil gravity with initial API oil gravity. It is normal for the oil gravity of the recovered liquid to be less than the oil gravity of the original sample liquid. These curves can be obtained by running "blanks" (retorting known volumes of fluids of known properties). The retort method is rapid and when using the corrections, satisfactory results are obtained. It gives both the water and oil volumes, so that the oil and water saturations can be calculated from the following formulas:

$$S_{wr} = \frac{\text{water vol}}{\text{pore vol}}$$

$$S_{or} = \frac{\text{oil vol}}{\text{pore vol}}$$

$$S_{gr} = 1 - S_{wr} - S_{or}$$

Porosity

Porosity is measured as either total or effective porosity. Total porosity is the percentage of rock bulk volume that is void space, whether the individual pores are interconnected or not. It is an impractical term in consolidated rock since the "dead" pores will not contribute to production. Effective porosity, the void space available for production, is the porosity of interest to the log analyst. The analyst using a porosity value must always be aware of which value he is using. In nearly all cases, however, laboratory equipment and procedures are designed for measuring effective porosity.

In most sandstone samples, the total and effective porosity are approximately equal. This indicates that most pore spaces are interconnected. Limestone and dolomite formations, however, can have widely varying porosities. Porosity in limestones and dolomites has been classified as original or secondary porosity. Original porosity is the porosity occurring at or soon after deposition, while secondary porosity is largely due to the solvent action of ground waters. The majority of porous voids found in limestone and dolomites are of a secondary nature.

Because of the extreme heterogeneity of limestones and dolomites, porosity designations are useful in describing them: intercrystalline, intergranular, vuggy, cavernous, and fracture. Intercrystalline porosity occurs between extremely fine crystals of matrix rock and may be the only porosity present, or it may be present as porosity in the matrix with other types of porosity. Intergranular porosity is found in limestones composed of fairly regular grains. Oolitic limestone exemplifies this type. Vuggy porosity gets its name from the occurrence of irregular-shaped holes having the appearance of worm holes, while cavernous porosity is in the form of irregular and larger-sized pores up to several feet or greater in diameter. Both these types of porosity are caused by the dissolution effects of percolating ground waters. Fractured porosity caused by earth stresses is in the form of sedimentary or irregular fractures and joints or pattern fractures.

Porosity measurement requires the measurement of two out of three properties of a sample: bulk volume V_b, grain volume V_{gr}, and pore volume V_p. The bulk volume is the volume of solid matrix rock plus all pore space and is synonymous with gross or total volume. The grain volume is the total volume of solid material (grain volume) making up the matrix of the sample. The pore volume is the total of the pore space within the sample. Depending on which two volumes are determined in the laboratory, the porosity can be determined as follows:

$$\phi = \frac{V_p}{V_b} = \frac{V_p}{V_{gr} + V_p} = \frac{V_b - V_{gr}}{V_b}$$

A number of laboratory techniques have been developed for measuring each of these three properties of a core. For the porosity ranges usually encountered, the acceptable methods for determining porosity are accurate to within ± one-half of a porosity percent.[12] The most common techniques used in the laboratory, their advantages, and limitations are briefly summarized as follows.

Bulk Volume Measurements

The measurement of bulk volume as presented by Jenkins[12] can be determined by calipering the dimensions of a specifically shaped sample and applying the proper geometric formula; by measuring the displacement of a non-wetting fluid such as mercury; by the application of Archimedes' principle, e.g., the sample is weighed in air and weighed again suspended in a wetting liquid; or, by directly adding the independently measured grain volume and pore volume values. Where the bulk volume is measured by displacement of a wetting liquid, the sample must be saturated with the displaced liquid to the extent that no displaced liquid will be imbibed into the sample.

The technique used for the determination of bulk volume may vary in any one laboratory depending upon the size and type of the sample analyzed and other factors such as the surface texture. Calipering samples for calculation by a geometric formula is the least accurate method, resulting in routine errors of 1%, and not uncommonly they are 2% and 3% of the bulk volume value. The displacement techniques are normally the best, yielding routine errors of less than 0.5% of the bulk volume. The accuracy of the summation of grain and pore volumes method is dependent upon the accuracy of the grain and pore volume measurements.

Pore Volume and Grain Volume Measurements

The pore volumes and grain volumes of core samples are obtained by a variety of methods. These methods have come to be known as the actual porosity determination methods, even though separate bulk volume measurements are also necessary in most of them.

Boyle's Law Method. A Boyle's Law Porosimeter consists of a steel bomb, which is filled with a gas, such as helium or air, under closely controlled pressure, and a graduated burette that receives the gradually expanding gas and accurately measures its volume (see Fig. 2-12). A duplicate test is then made with a core sample being placed in the bomb. As the pore space of the sample is penetrated by the gas, the difference between the volumes of gas determined by the two tests is a measure of the sand grain volume. Where the "dead" gas volume between the sample and the bomb wall is eliminated, the porosimeter can be made to measure pore volume directly. The method is routinely capable of measuring grain volume with errors of ±0.1% to 0.5% of the grain volume. Where the pore volume is measured directly, the error routinely is less than ±2% of the pore volume.[12]

The Boyle's Law method yields effective porosity values. A truly isolated pore will have the same influence on the porosity measurement as if it were a solid. The method is ordinarily rapid once the core sample has been prepared, but it can be very slow if the sample has low permeability, since pressure equilibrium must be attained after compressing or expanding the gas before an accurate measurement can be made. Failure

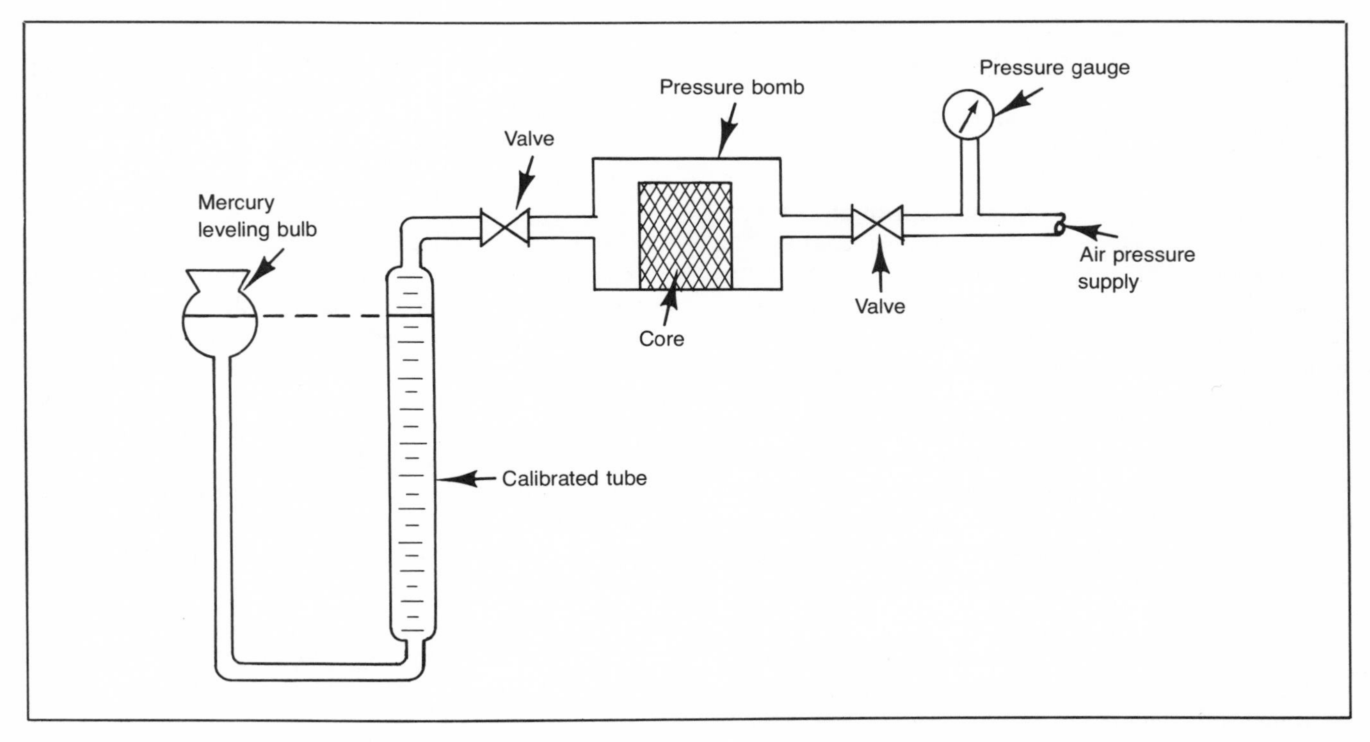

Fig. 2-12. Schematic of Boyle's Law porosimeter

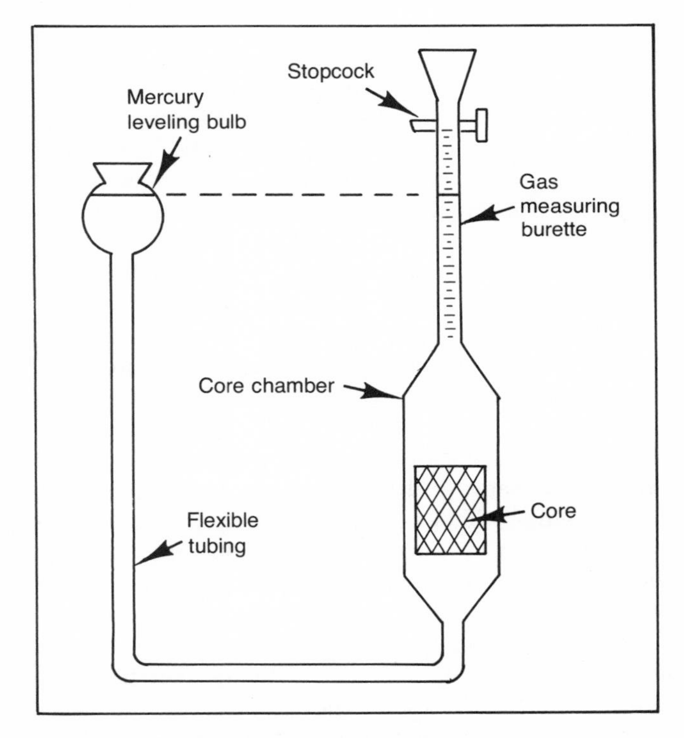

Fig. 2-13. Schematic of Washburn-Bunting porosimeter

to wait for pressure equilibrium causes low porosity values to be measured. On the other hand, adsorption of the gas on the pore surfaces will cause an erroneously high value to be measured. Inert gases such as helium are sometimes used to minimize the adsorption problem, but they do not eliminate it. Because of this, Boyle's Law porosity values are normally considered to be slightly high.

Washburn-Bunting Method. The Washburn-Bunting technique is a vacuum extraction method. The pore volume is determined by displacing air from the core sample with the mercury-containing apparatus shown in Figure 2-13. This equipment is first calibrated in order to determine the amount of air that clings to the glass walls inside the apparatus. This is done using a glass plug of approximately the same dimensions as the core. The plug is placed in the core chamber, the stopcock opened, and the mercury run to just above the stopcock. Then the stopcock is closed, and the mercury level lowered, thus pulling a partial vacuum and allowing air to expand and rise. The mercury level is then raised and the amount of trapped air read on the graduated tube. This is the "correction volume." The same procedure is repeated with the core in place until all the air has been removed from the core and replaced by mercury. The total volume of void is measured as the sum of the volumes of air displaced each time the mercury bulb is raised and lowered.

The porosity is then calculated by:

$$\phi = \frac{\text{Total Air Volume} - \text{Correction Volume}}{V_b}$$

This method measures effective porosity and is sometimes referred to as a multiple-expansion method of porosity determination. The primary objection to the method is that the core is of no further value because it contains mercury after the test. The apparatus was subsequently modified by Stephens to provide a sample chamber that allows air to be removed from the core but prevents contamination of the core by mercury. The length of time required for this method is a deterrent to its use. Potential errors include leaks in the system or incomplete evacuation due to tight core. A dirty instrument will cause high pore volumes to be measured, yet the instrument is difficult to keep clean on a routine basis. The technique is poor for measuring low permeability and porosity samples. Routine accuracy of pore volume measurement to within ±5% of the actual pore volume is claimed for this method. A scrupulously clean instrument and careful laboratory technique will allow more accurate measurements.

Resaturation or Gravimetric Method. As described by Jenkins,[12] an extracted and dried sample is weighed, saturated with a liquid of known density, either water or hydrocarbon, then reweighed. The weight increase divided by the density of the saturating liquid yields the pore volume of the sample.

This technique also yields effective porosity. The use of the technique must be restricted to samples whose saturated weights can be determined. Samples with solution cavities on the surface cannot be handled by the resaturation technique. Some analysts feel that the pore volume can be measured within ±2% of the actual pore volume, but most analysts feel that the extracted and dried samples can seldom be completely resaturated. Pore volumes determined by this technique commonly are 10% lower than pore volumes determined on the same samples by Boyle's Law or other techniques. The technique requires considerable time, but many samples can be handled simultaneously.

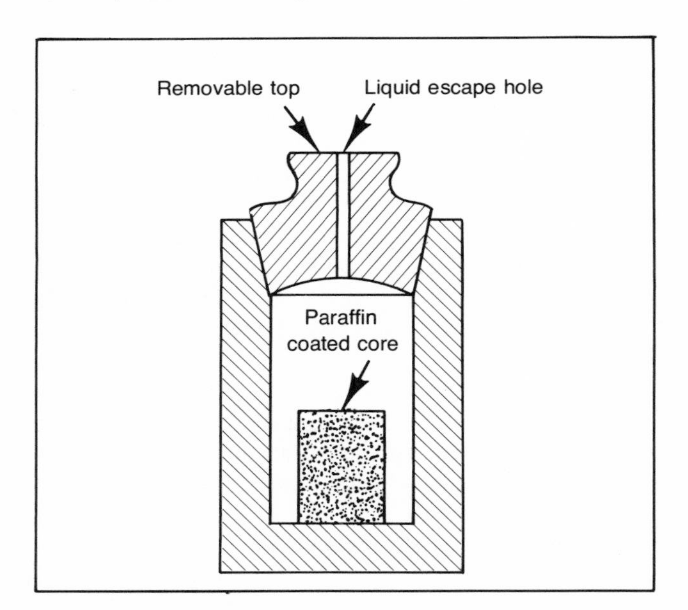

Fig. 2-14. Schematic of liquid pycnometer

Grain Density Method. In this method, the cleaned sample is weighed, given a coating of paraffin, and weighed again. Its volume is then determined by the weight of water displaced from a pycnometer and the density calculated. A liquid pycnometer is shown in Figure 2-14. The coating is then removed and the sample crushed to grain size, transferred to a liquid pycnometer, weighed, and the grain volume calculated from the weight of liquid displaced. The sand grain density, ρ_{gr}, is then calculated. The bulk density, ρ_b, and grain density are related to bulk volume and grain volume, and the porosity is calculated as follows:

$$\phi = \frac{\rho_{gr} - \rho_b}{\rho_b}$$

This method measures only total porosity. Other objections to its use are that the several necessary analytical weighings are time-consuming and that the core is destroyed during the test procedure. Potential errors involve the possible loss of sand grains in crushing. When the liquid displacement alternative is used, the grain density can routinely be determined within ±.005 grams/cc., which will enable the determination of grain volume of the uncrushed sample to within ±.5% of its value.[12]

Summation of Fluids Method. The procedures discussed thus far are run on extracted and dried core samples. However, the summation of fluids method uses the sample just as it is received in the laboratory. The sample is split into two portions for the porosity determination, and each sample is weighed. Bulk density and gas volume are measured on Sample No. 1, and the oil and water saturation are measured on Sample No. 2. Sample No. 1 is placed in a pycnometer and the bulk volume is measured; from this measurement and the weight of the sample, bulk density is calculated. The sample is then placed in a cell to which a calibrated mercury pump is attached and mercury forced into the core at about 750 psi. The volume of mercury, read from the pump, gives the gas volume in the sample.

Sample No. 2 is placed in a retort and heated to drive off the water and oil for saturation determinations. The volume of gas measured on Sample No. 1 is converted so that it is based on the same weight of original rock as represented by the Sample No. 2. The volume of oil, water, and gas are added together to get the pore volume of Sample No. 2. Assuming both samples have the same bulk density, the bulk volume of the second sample is calculated and the porosity evaluated by:

$$\phi = \frac{V_o + V_w + V_g}{V_b} \times 100$$

This procedure may be more representative of the porosity of the foot sampled because more of the sample may be used. The disadvantage, however, is that the core may not be homogeneous over the entire cored foot and, therefore, the bulk density may not be constant, resulting in inconsistency in porosity measurements.

In some modifications the summation of fluids method yields effective porosity. However, when a high temperature retort is used to distill the oil and water from the samples, the pore volume is closer to being total pore volume than effective pore volume. It is a simple and rapid method, which lends itself very well to routine laboratory determinations. Its disadvantage is that a distinction in the recovered water must be made between pore water and water of hydration, and a correction must be made to the observed oil volume.

The gas volume is usually determined to within ±2% of its value. Except where the pore water and water of hydration or absorbed water are very difficult to distinguish, the water content is determined within ±2%. This method, then, gives pore volume values essentially accurate to within ±2% of the correct values.[12]

Unconsolidated Sand Porosity Measurements. Unconsolidated sands or poorly cemented sands present a very difficult problem to the laboratory technician because the sand is so loose that a test plug from the core will crumble while being cut or during the test procedure. Some procedures have been used by laboratories, such as spraying or coating a core plug with plastic or cementing material or embedding the plug in sealing wax to prevent the plug from disintegrating. The technique used depends upon the degree to which the core is unconsolidated.

If the sand is too loose to handle in any manner as a plug, the total porosity of the sand may be approximated by a sand pack. The sand is dried and a sand pack is prepared by adding increments of kerosene or tetrachlorothane and sand until the pack is completed. The sand is tamped with a glass rod after each incremental amount of sand and liquid is added. The bulk volume is the volume of the sand pack, and the pore volume is the volume of the liquid added.

A second method involves determining the bulk density of a dry sand pack by weighing a known volume of dry sand pack in a graduate of known weight. With grain density of the sand determined in a liquid pycnometer, the porosity is calculated by the grain density method.

A comparison of these various methods was presented by Jenkins.[12] Table 2-3 is from the study by Jenkins using data taken from 80 formations from the Gulf Coast, the Arkansas-Louisiana-Texas area, the West Texas and Mid-Continent area, the Rocky Mountain area, Canada, and Iran. The number of samples per formation varied from 3 to 380, with the average number being 40 per formation. Based on the trends indicated from these data and other comparisons, he concluded that the Boyle's Law, summation of fluids, and grain density methods yield porosity values accurate to ±0.5 porosity percent. The Washburn-Bunting method normally yields values accurate to ±1.0 porosity percent. The resaturation technique routinely yields lower porosity values by 2-10% than do the other methods.

Absolute Permeability

About 1856, Darcy made studies of water flowing through sand filters and as a result formulated the law of fluid flow bearing his name. The absolute permeabil-

TABLE 2-3
Comparison of Porosity Measuring Methods

Number of Samples	Type	Porosity Range	Washburn-Bunting ϕ%	Σ Fluids ϕ%	Resat. (Water) ϕ%	Boyle's Law ϕ%	Grain Density ϕ%
232	Clean Sand	8–35			15.7	17.6	
85	Sl. Shaly	4–28			17.5	19.5	
29	Conglomerate	6–15			10.1	11.4	
70	Carbonate	6–25			18.9	20.0	
95	Clean Sand	8–35		17.8		18.6	
390	Sl. Shaly	4–28		21.1		20.4	
15	Conglomerate	6–15		10.2		10.9	
112	Carbonate	3–20		14.8		14.8	
137	Sand	1–22				5.55	5.42
210	Sand	1–22	14.0		11.3	12.8	
3220	All Types	2–35		14.6		14.6	

Permission to publish by the Society of Professional Well Log Analysts.[12]

ity of a porous medium to fluids is defined as "the rate of flow of a specified fluid through a unit cross section of the porous medium (100% saturated with that fluid) under a unit pressure gradient and conditions of viscous flow." This law is stated symbolically as follows for a steady state (rate does not vary with time) horizontal flow:

$$q = \left(\frac{k_a}{\mu}\right)\left(\frac{A}{L}\right)\Delta P$$

where: q = rate of fluid flow, cc/sec
A = cross-sectional area open to flow, cm^2
μ = viscosity of flowing fluid, cp
ΔP = pressure drop, atm
L = length between pressure taps, cm
k_a = constant of proportionality called absolute permeability, Darcies

The absolute permeability of porous rock is measured on a cylinder or cube plug cut from a core of the rock. The cross-sectional area available for flow is accurately measured. The plug is then placed in a core holder, and fluid flows through the plug. The upstream and downstream pressures and the volume of fluid per unit of time that flows through the plug are measured. With the viscosity of the fluid known or determinable, the permeability of the plug can then be calculated. If air is used, the following equations are generally applied to account for its compressibility:

$$k_a = \frac{q_m \mu L}{A(P_1 - P_2)}$$

where: q_m = cc/sec at $P_m = \dfrac{P_1 + P_2}{2}$

(from Boyle's Law: $P_1 V_1 = P_m V_m = P_2 V_2$) and

$$k_a = \frac{2 q_a \mu L}{A(p_1^2 - p_2^2)}$$

where: q_a = cc/sec at 1 atm pressure and flowing temperature.

The air-permeability apparatus is generally used by core analysis laboratories, although the types of measuring devices and core holders vary considerably. The practice has been to measure permeability both vertically and horizontally with the bedding planes. For this purpose, samples are prepared so that direct comparison of the two measurements can be made on the same core. Generally, vertical permeability is less than horizontal permeability because of restrictions offered to flow by the manner in which sand grains were laid down. Sometimes the vertical permeability is only a small fraction of the horizontal permeability; however, generally it runs about 50 to 90% of horizontal permeability.

Heterogeneous cores, and particularly those of limestones and dolomites, require the examination of large samples to provide representative data unobtainable on the small one-inch diameter plugs that are usually used for testing sandstone cores. Methods have been developed to determine permeability on these large samples by means of radial flow, and when the samples are large enough these permeability measurements may be representative of the formation. Darcy's law applied to radial flow is as follows:

$$k_a = \frac{q\mu \ln\left(\frac{r_e}{r_w}\right)}{2\pi h(P_e - P_w)}$$

where: k_a = absolute permeability, Darcies
q = liquid flow cc/sec or gas flow evaluated at $\dfrac{P_e + P_w}{2}$
r_e = outer radius, cm
r_w = inner radius, cm
h = vertical thickness, cm
P_e and P_w = pressures at r_e and r_w respectively, atm

The values for permeability determined by the use of fresh water may differ from the correct absolute permeability as a result of clay swelling in the core. Except for changes in the pore system brought about by the liquids involved, absolute permeability is independent of the fluid flowing.

An apparent discrepancy between liquid and air permeabilities was explained by Klinkenburg.[2] He showed that permeability to gas is a function of the mean free path of the gas molecules, and thus depends on the factors that influence the mean free path such as pressure, temperature, and the nature of the gas. When the mean free paths are small, as at high pressure, the permeability of the rock to gas approaches that of liquids. Absolute permeability is determined by measuring gas permeability at two or more mean test pressures and plotting measured permeability versus the reciprocal of mean pressure and extrapolating to the zero point on the scale of reciprocal pressure, as shown in Figure 2-15.

SUPPLEMENTAL AND SPECIAL CORE ANALYSIS

Many additional measurements can be made on the retrieved core. Some of these measurements are supplemental and can be obtained at relatively little cost or time in the process of initial analysis. These tests include chloride content determination, oil gravity evaluation, fluorescence, permeability to water, and vertical permeability. Other tests that might fall into the special area of prime importance are also somewhat costly and time-consuming and require special laboratory apparatus and techniques. Included in this group are such important

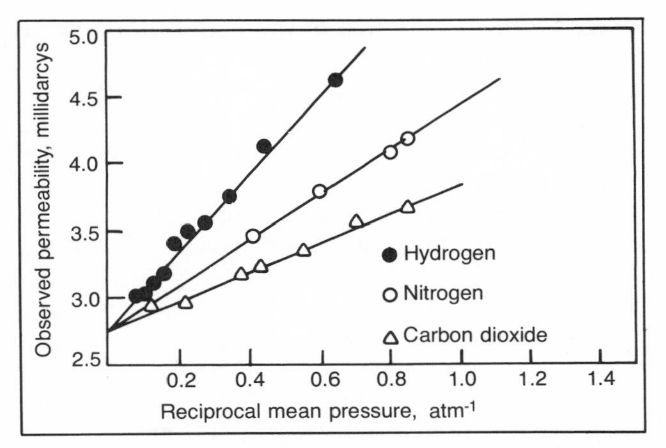

Fig. 2-15. Klinkenburg correction for gas slippage: permeability constant of core sample "L" to hydrogen, nitrogen, and carbon dioxide at different pressures (permeability constant to iso-octane, 2.55 Md). Permission to publish by the American Petroleum Institute.[2]

measurements as an evaluation of electrical rock properties for the determination of the cementation exponent, m, and the saturation exponent, n, evaluation of capillary pressure behavior, and the evaluation of relative permeability such as the oil-water, gas-water and gas-oil behavior. The supplemental tests will be briefly described here, as well as the measurement of relative permeability. A discussion of the electrical properties of rock for determining the m and n exponents will be presented in Chapter 4.

Supplemental Analysis

Chloride Content. The degree of salinity is determined quantitatively by measuring the amount of chloride ion that can be leached from the sample. The leached water is titrated with silver nitrate. The test is made on the assumption that there are no undissolved chlorides in the sand itself. Tests have shown that this assumption is correct.

The chloride ion concentration of the water obtained from the core may be calculated in the following manner:

$$C = \frac{KM_s}{V_w}$$

where: C is the concentration of chloride ion, expressed as grains of sodium chloride per gallon of core water (can also be expressed in PPM).

M_s is the amount of chloride ion expressed as milligrams of sodium chloride per 100 grams of dry sand.

V_w is the volume of core water expressed as milliliters per 100 grams of dry sand.

K is a factor for converting milligrams per milliliter to grains per gallon. For sodium chloride solutions, this factor equals 58.5.

This test is most useful as an indicator to the degree of core flushing by the mud filtrate.

Oil Gravity. The small quantities of oil extracted from the core can be used to estimate oil gravity. The various empirical means used provide an estimate of the oil viscosity, which can be valuable in early recovery estimates.

Fluorescence. Ultraviolet light examination of the core samples can provide useful qualitative information to the analyst. It can give an indication of the presence of hydrocarbons, the type of hydrocarbon (oil, dry gas, or condensate) and the degree of flushing.

Permeability to Water. The susceptibility of the rock to waters of different salinity can be evaluated by running water permeability tests on the core using various salinity waters. Formations containing swelling clays can show significant permeability reductions when contacted with fresher waters. A knowledge of the magnitude of this permeability reduction is important in assessing the potential formation damage due to filtrate invasion and in designing subsequent water injection operations.

Vertical Permeability. Most formations have a lower permeability normal to the bedding plane than horizontal to the bedding plane. Measurement of a vertical permeability on the core provides a measure of the influence of the bedding plane, shale laminations, or other natural obstructions to vertical flow and is most important in many rock types for assessing the vertical movement of fluids in the reservoir. Vertical flow considerations include potential water or gas coning and gravity segregation of gas and oil during production.

Special Analysis

In addition to the special measurements mentioned earlier, it might be desirable to measure acoustic velocities, compressibility, matrix capture cross-section, as well as the measurement of relative permeability, described as follows.

Multiphase flow measurements have been conducted in the laboratory for many years. Basically, two techniques have evolved: (1) steady state approach and (2) unsteady state approach. Simply stated, the steady state approach consists of injecting two fluids at a set ratio into a saturated core and measuring the pressure drop once steady state conditions (same flow rate of injected and produced fluids) is established. A series of steady state tests are conducted over the complete saturation range. This method is experimentally slow but the relative permeability relation is obtained directly. The unsteady state approach, on the other hand, consists of injecting one fluid into a saturated core and measuring the produced volumes continuously with time during the displacement process. Evaluation of the relative permeability behavior requires solution of a displacement equation proposed by Welge.[4] This method is experimentally quick, and with computers the relative permeability behavior can be quickly determined. The displacement equations, however, have introduced some assumptions that may not be entirely applicable, but comparative studies indicate that the two approaches usually give similar results.

Coring and core analysis is an integral part of formation evaluation. It provides the only direct analysis of the rock of both a qualitative and quantitative nature. To maximize the value of this data it is important that the analyst understand the coring procedure, the quality of the data, the limitations of the measurements obtained, and the application of this data with regard to the evaluation program.

In summary, one might make a partial listing of the applications of core data that tend to accentuate the importance of this information in the overall evaluation program. Some of these applications are: (1) permeability magnitude, (2) porosity magnitude, (3) position of permeable zones, (4) permeability profiles, (5) position of vertical flow barriers, (6) permeability distribution, (7) determination of net pay, (8) lithology, (9) presence or absence of oil, (10) characteristics of pore structure, m and n determinations, (11) indication of fluid contacts, (12) indication of transition zones, (13) indication of type of fluid production, (14) possibility of water or gas coning, (15) capacity of productive interval, (16) indication of formation damage, (17) calibration of well logs.

These and other applications present a formidable array of information that may be obtained from core analysis. The information presented in this chapter and the wealth of information existing in the literature with respect to core analysis can provide a basis for the judicious application of core data toward the design of an optimum evaluation program.

REFERENCES

1. Williams, W.C., and Phillipi, P.M.: "Diamond Coring in the Pennsylvania Oil Fields," *Oil and Gas J.* (Jan. 19, 1939).
2. Klinkenburg, L.J.: "The Permeability of Porous Media to Liquids and Gases," *Drilling and Production Practices*, API, Dallas (1941).
3. Fatt, I., and Davis, D.H.: "The Reduction in Permeability with Overburden Pressure," *Trans.*, AIME (1952) 195.
4. Welge, Henry J.: "A Simplified Method for Computing Oil Recovery by Gas or Water Drive," *Trans.*, AIME (1952) 195.
5. Fatt, I.: "The Effect of Overburden Pressure on Relative Permeability," *Trans.*, AIME (1953) 198.
6. Clark, N.J., and Shearin, H.M.: "Formation Evaluation of Oil and Gas Reservoirs," paper SPE 582-6 presented at AIME Formation Evaluation Symposium, Houston, 1955.
7. Elmdahl, Ben A.: "How to Use Core Analysis to Find Oil," *Oil and Gas J.* (Feb. 27, 1956) 54.
8. Fatt, I.: "Effect of Overburden and Reservoir Pressure on Electric Logging Formation Factor," *Bull.*, AAPG (1957) 41.
9. Reudelhuber, Frank O., and Furen, John E.: "Interpretation and Application of Sidewall Core Analysis Data," *Trans.*, Gulf Coast Association of Geological Societies (1957) 7.
10. Webster, Gerald M., and Dawson-Grove, G.E.: "The Alteration of Rock Properties by Percussion Sidewall Coring," *J. Pet. Tech.* (April 1959).
11. Greenkorn, R.A., and Johnson, C.R.: "Variation of a Natural Sandstone Reservoir Element: An Objective Analysis of Core Measurements," paper SPE 1577-G presented at the 35th Annual SPE Meeting, Denver, 1960.
12. Jenkins, Ralph E.: "Accuracy of Porosity Determinations," *Trans.*, First SPWLA Logging Symposium, Tulsa, 1960.
13. Koepf, E.H., and Granberry, Raymond J.: "The Use of Sidewall Core Analysis in Formation Evaluation," paper presented at the Formation Evaluation Symposium, Houston, Nov. 1960.
14. Dobrynin, V.M.: "Effect of Overburden Pressure on Some Properties of Sandstones," *Soc. Pet. Eng. J.* (Dec. 1962) 360–366.
15. Gray, D.H., and Fatt, I.: "The Effect of Stress on Permeability of Sandstone Cores," *Soc. Pet. Eng. J.* (June 1963) 95–100.
16. Helander, Donald P., and Campbell, John M.: "The Effect of Pore Configuration, Pressure, and Temperature on Rock Resistivity," *Trans.*, 7th Annual SPWLA Logging Symposium, Tulsa, 1966.
17. Fertl, Walter H., Cavanaugh, Robert J., and Hammack, Gregory W.: "Comparison of Conventional Core Data, Well Logging Analysis and Sidewall Samples," *J. Pet. Tech.* (Dec. 1971) 23.
18. Campbell, John M., Helander, Donald P., and Van Poolen, H.K.: *Petroleum Reservoir Property Evaluation*, Campbell Petroleum Series, John M. Campbell Co., Norman, OK (1973).

3 Mud Logging

THE term "mud logging" is actually a misnomer since this form of logging encompasses a larger spectrum of data than that derived only from the mud system. A mud log consists of the continuous monitoring of the drilling operation, including the drilling mud and cuttings returns, and includes a wide variety of data. The data collected, including computed parameters, are usually presented in analog form versus depth. The final log output is quite variable in content and format and could probably be more descriptively entitled a drilling operations log.

Drilling data has been recorded as a function of depth since the infancy of the oil industry, with published references to drilling time logs available from the 1880s. From these early beginnings, the sophisticated and indispensable drilling operations log (mud log) has developed. Today's logs are generally the continuous presentation of data from three primary sources: (1) drilling operations, (2) formation cuttings, and (3) mud.

The principal advantage of the mud log is that it is the first—and possibly the only—log available. It is, therefore, a necessity on an exploratory well, but it does have other applications. In fact, for all wells drilled today some form of drilling operations data will have been recorded. This information should be an integral consideration in the formation evaluation program. Therefore, the formation evaluator should be directly concerned (and involved) with the drilling program, since it will affect his analysis. This is particularly important when cuttings and/or mud analysis are planned as part of the drilling program. It is surprising how often cuttings and mud analysis logs are planned (at considerable expense) without considering the influence of the mud program on these analyses. Clearly, the formation evaluator should assist in the design of the mud program in order to optimize the quality of the cuttings and mud analysis data.

MUD SYSTEMS

Mud systems are usually designed by drilling fluid engineers whose primary goal is to optimize the drilling operation; unfortunately, this is all too often done with little regard given to optimizing the evaluation program. While it is easy to fault the drilling engineer for not considering the potential problems of the evaluator, it can also be said that the problems might have been eliminated or minimized with input from the evaluator, who should always communicate his needs to the drilling engineer and if necessary "educate" him in regard to his operation. It follows, then, that a good evaluator should also be knowledgeable of mud systems and drilling problems.

The drilling fluid program is designed to allow maximum drilling rate at minimum cost. In some cases, the mud costs may exceed 10% of the total well costs. In designing an effective mud system, there are at least ten important design criteria that must be met:

1. Removing the cuttings from the bottom of the hole and carrying them to the surface
2. Cooling and lubricating the bit and drill string
3. Sealing porous and permeable zones with an impermeable cake
4. Controlling subsurface pressures
5. Holding cuttings and weight material in suspension when circulation is interrupted
6. Releasing sand and cuttings at the surface
7. Supporting part of the weight of drill pipe and casing
8. Reducing to a minimum any adverse effects upon the formation adjacent to the hole
9. Ensuring maximum information about the formations penetrated
10. Transmitting hydraulic horsepower to the bit

A deficiency in performing some of these functions

properly could have a serious adverse effect on the quality of the data obtained at the surface and, therefore, on the ability of the evaluator to analyze this data. The general outline for developing and applying an economically engineered mud program follows.[19]

1. Review the well beforehand to determine problems that will be encountered
2. Isolate and identify specific problems and their depth of occurrence
3. Prepare a "problem profile"
4. Design the mud program: select the sequence of mud systems, conversion depths and procedures; schedule the pattern of mud properties and program the changes to be made as drilling proceeds
5. Delegate responsibility and authority for field application—well site supervision and monitoring procedures
6. Prepare a post-well analysis and make recommendations

Although development wells, where there is good data from offsetting wells, present relatively easy mud design problems (developing improved mud systems), wildcat wells pose a much more difficult problem. In each case, however, input from the evaluator is desirable. Wildcat programs are monitored closely so that any adjustments can be quickly made before the drilling program is affected. This should also apply to any problems arising with the mud logging program. The criteria normally considered to indicate the need for a mud system redesign, at least from the drilling engineer's viewpoint, are listed as follows.[19]

High Drilling Costs: time-related (mechanical, geological), problem-related, such as a stuck pipe with or without abnormal pressure, sloughing shale (quantity and type), excessive trip time (tight hole, balled drilling tools, staging in or out of hole), abnormal drill pipe torque, time lost circulating and conditioning mud, mud instability (with gain in solids—shale problems—and without gain in solids—contamination and high temperatures, etc.);

High Mud Costs: chemicals, barites, or both;

Erratic Drilling Time: between wells or intervals within the well;

Erratic Mud Costs: between wells or intervals within the well;

Reports of inconsistent cuttings recovery.

Most wells are drilled with water-base mud systems, although oil-base, air, gas, or mist systems are also used. Since water-base mud systems predominate, this discussion will focus on operating conditions and problems associated with water-base mud systems. The application of oil-base muds is economically limited, but may have special advantages and, therefore, could be used for the conditions shown below. (See Appendix A for the basic mud system classifications.)[20]

1. In drilling troublesome shales
2. In drilling deep, hot holes
3. In drilling and coring pay zones
4. In drilling salt, anhydrite, carnallite, and potash zones
5. As a directional drilling fluid
6. As a slim hole drilling fluid
7. In drilling hydrogen sulfide (H_2S) and carbon dioxide (CO_2) bearing formations
8. As a perforating and completion fluid
9. As a spotting fluid to free stuck pipe
10. As a packer fluid
11. As a workover fluid
12. For corrosion control

Mud Properties

Once a mud system is selected, a pattern of specific physical properties can be defined. In order to develop an efficient mud system, however, it is necessary to understand the function of various mud properties as related to the drilling process. The relation of mud properties to mud system design was presented by Weiss.[19]

Mud Weight: It is essential that the differential between mud hydrostatic pressure and formation pressure be carried as low as possible to maximize penetration rate. Rate of penetration is reduced rapidly as pressure differential increases, due to a mechanism referred to as chip hold-down. To maximize drilling rate, it is necessary to equalize the hydraulic pressures around the cutting chip at the instant of fracture, so that it will be released. Any solids which bridge the fracture will produce a pressure differential and hold the chip in place, thus reducing bottomhole cleaning and inhibiting penetration. Hydraulic pressure equalization efficiency explains why drilling rates decrease from air or gas to clear water to mud and get progressively worse as mud solids increase.

Rheology: To obtain best bottom-hole cleaning, rheology should be maintained to deliver maximum hydraulic energy at the formation face. Shear sensitive systems not only assist in this requirement, but permit design of an adequate cuttings transport fluid with a minimum of solids.

Gel structure: A preferred fluid should possess a gel structure that forms quickly, but to a fragile gel of limited total strength.

Fluid loss: This is undoubtedly one of the most controversial subjects in drilling fluid technology. Several facts, however, warrant consideration when designing any mud system.

API and high temperature-high pressure fluid loss values have no engineering value since no correlation can be established between them and dynamic filtration. It is the pattern of change in these values that can be significant.

Dynamic filtration is a balance between pressure differential, depositional forces, erosive forces across the cake and erodibility of the cake's matrix. For these reasons cake toughness rather than thickness or fluid loss should be a more significant criterion for evaluating dynamic filtration changes. At this time, however, there is no way to objectively measure cake toughness, only gross changes can be observed.

Materials that tend to granulate a cake such as diesel oil or crude lignites have been found to increase dynamic filtration while lowering API values. Mud filter cakes possess permeability of 10^{-3} md or less. Therefore, filtration pressure drop is a near wellbore phenomenon.

Material balance studies on actual drilling wells have shown that water requirement for a mud system is primarily related to the amount of new hole drilled, and relatively independent of open-hole exposed from previous drilling. It has been proposed that cake is also deposited within an annulus of the permeable matrix surrounding the wellbore. Such a stable, mechanically protected filtration controlling membrane would be compatible with the above observations and explain the lack of filtration problems associated with repeated trips. Based upon these controls, differential pipe sticking should be expected to be most prevalent in recently drilled formations before a stable, mature matrix membrane has formed.

Shales do not possess permeability; therefore shale problems occur as a result of wellbore stresses within shale or wetting by the liquid mud phase. Neither of these conditions will be affected by fluid loss.

Concern over filtration is often out of proportion to its importance in the overall mud program. Care in regard to hydrostatic pressure differentials and drilling practices is essential to reduce cases of differential pipe sticking. If recurrent shale instability is encountered, which is not an abnormal pressure phenomenon, a change in the mud system should be considered rather than filtration lowering. Large amounts of money can be spent maintaining needlessly low filtration values. Awareness of trends in filtration values, observations of cake texture, and close liaison with the driller contribute the most effective approach to filtration control.

To control these and other properties, a variety of additives are commercially available (see Appendix B). Because additives used to control mud properties might have a detrimental effect on the quality of the data being logged, one should test the additives for such factors as fluorescence, cuttings retention, gas retention, etc.

DRILLING TIME LOG

The drilling time log is one of the earliest logs recorded and is simply the time required to drill a linear unit of formation. This log also provides the first data available on the formations penetrated since cuttings and mud log information have a lag time before observation at the surface. Initially, drilling rate was simply obtained by marking the kelly in one-foot intervals and recording the time required for each mark to reach the kelly bushing—a technique ripe for errors. In 1943 a mechanical method was introduced, ensuring that drilling rate recorders would provide dependable data. Figure 3-1 shows a portion of a typical drilling time log on which the recorder marked at one-foot intervals on a time-actuated chart. Spacing of the marks indicates how fast the hole was drilled. The close spacing observed in interval A indicates fast drilling, which in this case was a porous sandstone. The second line, parallel to the drilling time, is a record of drilling operations. Raising the kelly causes the pen to move to the right, which is what occurred at 12:07 (6580 ft). When drilling is resumed the pen returns to its normal position.

Drilling time is a function of two major factors, rock type and drilling conditions. When drilling conditions do not change drastically, changes in drilling time tend to reflect rock type. For example, porous sands tend to drill faster than shales; in shale-sand sequences the drilling time log tends to correlate with the SP (spontaneous potential) or gamma ray curves. In limestone sections, on the other hand, porous limestones drill faster than nonporous limestones, and the drilling time log tends to correspond to porosity responses.

The data recorded in Figure 3-1 is usually presented in analog form, as shown in Figure 3-2 where the drilling time is recorded in min/ft on the same scale as the electrical survey (ES) log. Note the similarity between the drilling time log and the SP log through this shale-sand-shale sequence. The same data (see Fig. 3-1) could also be recorded as drilling rate, for example, in ft/hr. This presentation averages the data and may be preferred when drilling rapidly. The drilling time log, however, will give a more detailed insight into lithology.

Probably the most important use of the drilling time log is the immediate response at the surface to porous zones. If drilling time decreases (drilling break), drilling can be immediately halted and cuttings circulated out. If cuttings analysis (and hydrocarbon mud analysis) are

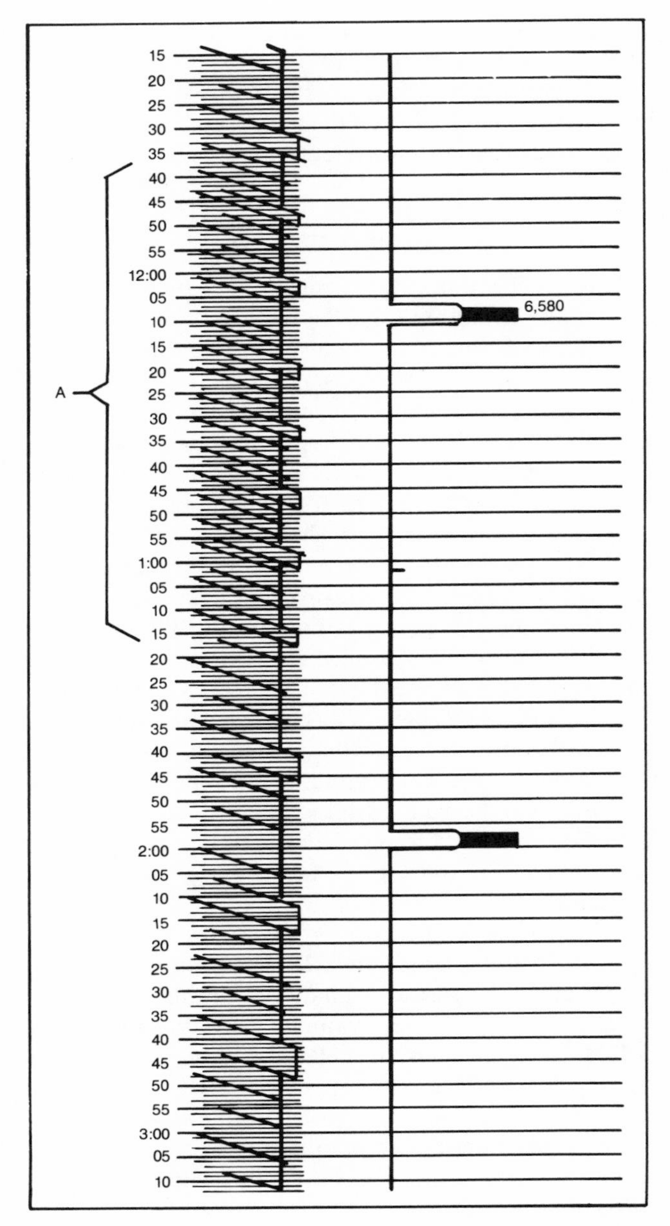

Fig. 3-1. Typical drilling time log. Permission to publish by the Society of Petroleum Engineers of AIME.[3] Copyright 1962 SPE-AIME.

favorable, a drill stem test of the drilled interval could be made immediately and/or the remainder of the zone can be cored. Applied in this manner, the drilling time log would be particularly useful in exploratory wells.

There are a number of other uses for the drilling time log, such as immediate correlation work, determining sample lag, etc. These uses and its value in conjunction with other recorded data will be expanded later.

CUTTINGS ANALYSIS

Cuttings samples for lithology, porosity indications, and hydrocarbon shows have been analyzed and recorded as a function of depth probably since the first wells were drilled. Presentation of the geologic sample log has become more refined over the years; today, efforts are being made to quantify this information for computer usage. Although to some degree the sample log has become standardized, variations still exist since it is individually hand-prepared.

The geologist's sample log finds its greatest application in "hard" rock areas (generally pre-Cretaceous) where sediments contain many limestones, dolomites, and anhydrites in addition to sands and shales. In "hard" rock areas the drilling rate is slower, and as a result the cuttings are larger (small chips), and the cuttings depth can be more readily identified. In "soft" rock areas—typically sand-shale sequences where the potential pay zones contain unconsolidated or poorly consolidated sands—the drilling rate is much faster, resulting in poor cuttings returns (often only sand grains), which are difficult to classify with respect to depth.

Sample collection in rotary drilling operations is obtained by diverting a small portion of the mud stream into a unit for separating cuttings from the mud. Separating units include rotary screens, vibrating screens, and small settling tanks. A cuttings sampler (Cuttings Sample Master) that automatically separates cuttings from a portion of the drilling mud, described by Kennedy,[11] is shown in Figure 3-3. The integral components of the Cuttings Sample Master are listed as follows.

1. Separating cylinder
2. AC motor and speed reducer, which rotates cylinder
3. AC motor, which powers eccentric for vibration of trough
4. Trough mount
5. Eccentric mounted weights, which vibrate trough
6. Trough frame
7. Inlet hose from flow line
8. Cans to operate water valve
9. Supporting rollers for separating cylinder
10. Sample container—standard Tyler sieve, 8″ diameter, 80 or 100 mesh
11. Adjustable gate to regulate amount of sample into sieve
12. Classifier, which is attached to and rotates with separating cylinder
13. Hole size, which regulates size of cuttings to sieve
14. Trough mount
15. Vaporproof light

When using the Cuttings Sample Master, a portion of the drilling fluid from the flowline flows under gravity to the separating cylinder through a 2-in hose or pipe. In the reduced velocity of the separating cylinder, the solid drill cuttings that have a higher density than the drilling

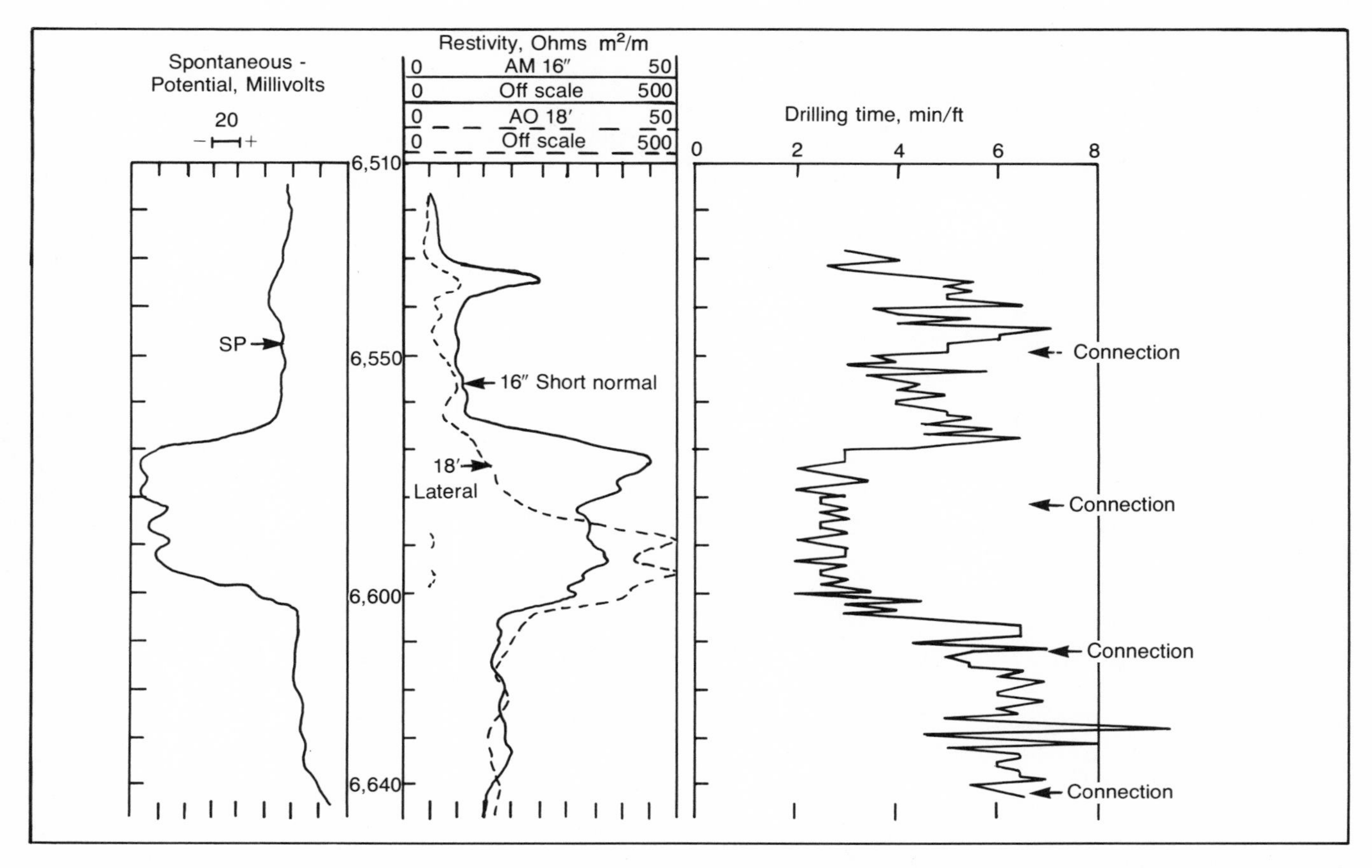

Fig. 3-2. Drilling time log correlated with electrical survey. Permission to publish by the Society of Petroleum Engineers of AIME.[3] *Copyright 1962 SPE-AIME.*

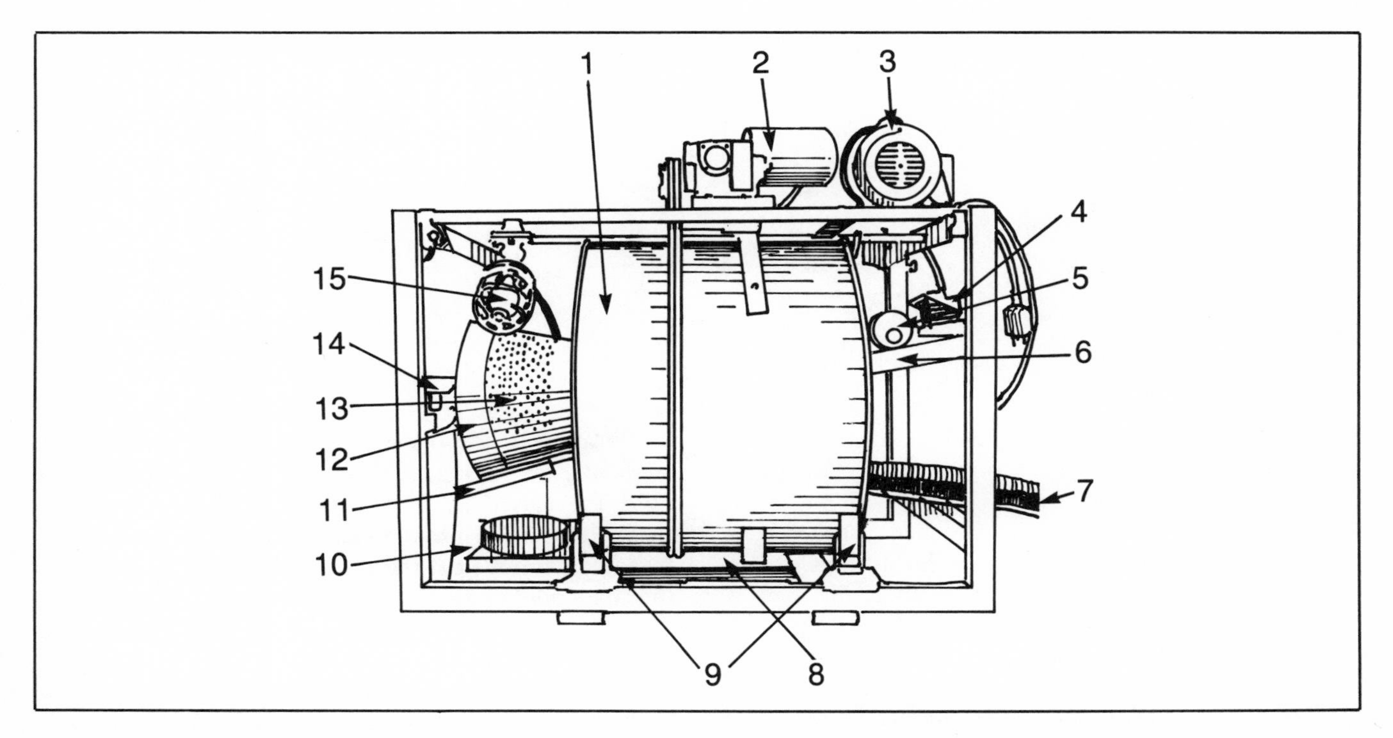

Fig. 3-3. Cuttings Sample Master. Permission to publish by the Oil and Gas Journal.[11]

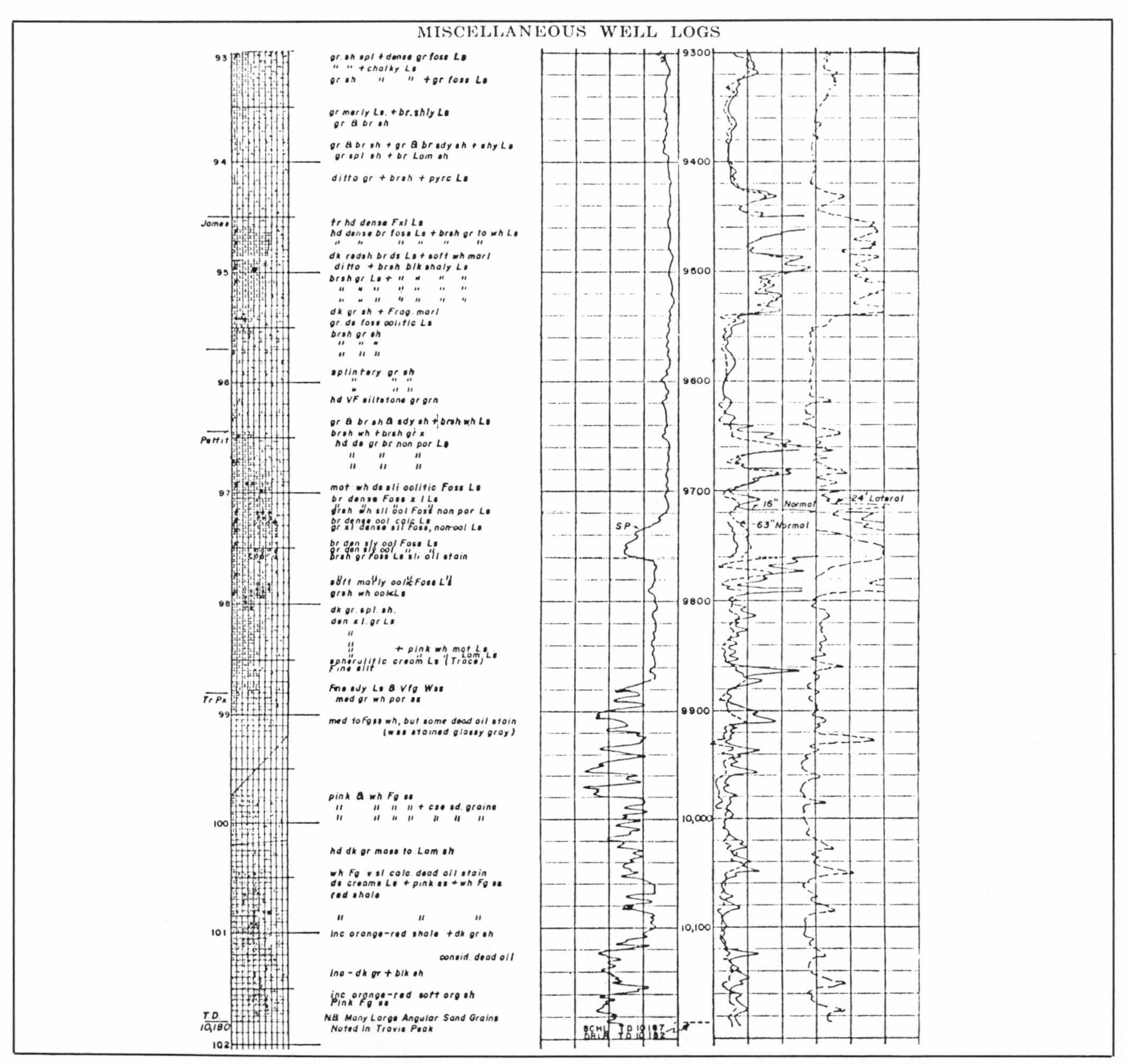

Fig. 3-4. Sample log correlated with electrical survey. Permission to publish by the Society of Petroleum Engineers of AIME.[3] Copyright 1962 SPE-AIME.

mud drop out. The cuttings settle to the lowest point in the cylinder where they are "scooped up" by longitudinal vanes in the cylinder.

As the cylinder rotates, the attached vanes decant most of the liquid scooped up with the cuttings. When each vane is near the top of the cylinder, the vane is tilted at an angle that allows the cuttings to drop from the vane and fall onto the trough that extends through the cylinder. Water jets mounted on the trough are actuated by a cam, and water sprays the vane clean. The water also washes the mud from the cuttings.

Vibration of the trough and the action of the water move the cuttings downward out of the cylinder and onto the classifier. There, the finer fragments pass through the holes into the sample container. The coarser fragments tumble off the end of the classifier and onto the shale pile. Normally, the separating cylinder makes about ½ rpm, but velocity can be changed by changing the sheave size.

The sampling device may be located over the mud tank so that all of the drilling mud and all of the wash water return to the system. Or, it may be positioned so that most of the wash water goes over the side of the mud tank to the reserve pit.

With this unit, human error in collecting samples is almost entirely eliminated, and the unit operates independently of the shale shaker. Cuttings are washed, classified, and the important fines saved for lithologic analysis. Coarse fragments may be discarded or saved for paleontological analysis. Sample fragments that pass through the classifier and into the sample container range in size from approximately 0.13 in to clay-sized particles. A drilling crew member can collect, bag, and store a good sample in less than two minutes. The advantages of this unit are:

1. It can recover from the drilling mud a continuous sample of all cuttings sizes down to 100 mesh.
2. It operates completely independently of a shale shaker, so friable rock fragments are not broken apart on the shaker screen.
3. Lost-circulation material in the mud does not interfere with the machine's operation.
4. The crewman does not have to climb below mud-tank level to collect a sample.
5. Time required to collect and bag a good sample is normally less than two minutes, so in fast drilling, sample collection intervals can be more frequent.
6. Samples are washed and ready to bag with low viscosity mud. A washing hose is also included for additional sample washing if necessary.

It should be pointed out that it will not operate efficiently without a continuous water pressure. To use the device, it is also necessary to have sufficient flow-line height above the top of the mud tank to allow gravity flow of mud into the machine. Normally, twelve inches is sufficient height. Also, the machine adds a small amount of water to the mud system. However, with oil-base mud, diesel oil could replace water as a washing agent.

Samples are collected each time a specified interval of hole is drilled; in "hard" rock country, samples are usually caught in 5 ft or 10 ft intervals, although intervals as small as one foot may be used in potential pay zones. In "soft" rock areas, samples are usually caught over intervals of from 20 to 50 ft.

Samples obtained at the surface, however, must be depth-correlated since there is a lag time from the time the sample is cut until it reaches the surface. One method is to use a tracer material (preferably of the same size and density of the cuttings) that is timed for the round trip after being added at the surface. By subtracting the time from surface to bit, the lag time can be determined. At times, the lag time can be accurately determined by using the drilling time log. When a drilling break occurs, drilling can be halted; the time required for the samples to be circulated out is a good measure of sample lag.

Some causes of poor samples include contamination with rock samples that have already been drilled (i.e., shale cavings), insufficient mud gel strength to lift the cuttings so that they are ultimately ground very fine by redrilling, and lost circulation problems where the samples are unrepresentative of the section drilled and also generally highly contaminated. The greatest problem in poor sample quality, however, may be caused by carelessness or human errors.

Evaluation of the recovered samples includes the geologic analysis of rock type, analyses for oil and gas shows, and analysis of rock properties such as shale density. The quality of this information depends on the geologist and his local experience.

The samples are microscopically examined in either the dry or wet state. Wet examination can assist in defining crystal formation and oolitic structure by differential refraction. Two techniques are practiced: the interpretive method in which the geologist selects representative cuttings and the percentage method in which the geologist uses all the cuttings in the sample (except cavings) and describes the zone in a composite manner. The sample log is a presentation of the geologist's description and is generally shown both graphically and with an accompanying description (see Fig. 3-4). The lithology symbols used on the sample log are shown in Figure 3-5. The geologic sample log can also be presented using a color code to identify the various lithologies. The problem with color coding is that it is expensive to reproduce but makes an effective visual display.

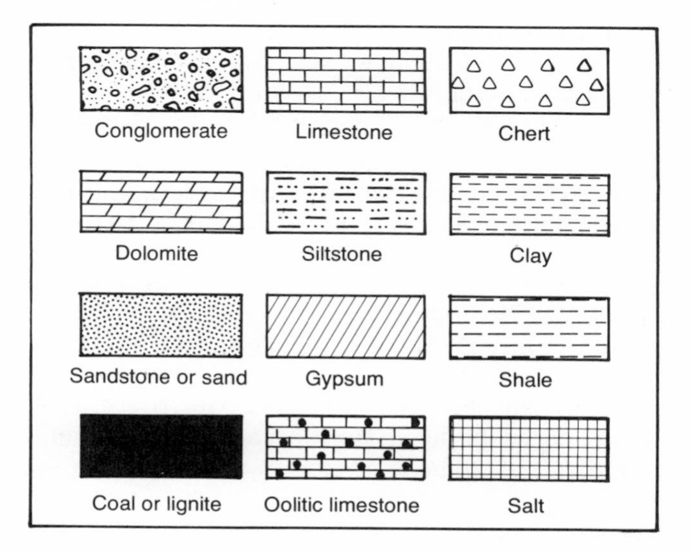

Fig. 3-5. Lithology symbols. Permission to publish by Oil, Gas and Petrochem Equipment.[8]

In addition to the graphic presentation, a detailed written description of the samples is also included, which describes the rock properties, briefly outlined by Pearson[3] as follows.

1. *Shales*
   Color: red, green, etc.
   Texture: waxy, velvety, papery
   Fabric: laminated, splintery, flaky, fissile, blocky
   Accessory minerals: micaceous, sandy, calcareous, bentonitic, fossiliferous, carbonaceous, glauconitic, bituminous
2. *Sands*
   Color: brown, gray, etc.
   Texture: very fine, fine, medium, coarse, very coarse-grained
   Shape of grain: rounded, subrounded, angular, subangular
   Sorting: well-sorted, poorly sorted
   Secondary material: clean, shaley, calcareous, dolomitic, silty, phosphatic, siliceous, tuffaceous, micaceous, glauconitic, pyritic, fossiliferous, carbonaceous, sideritic, ferruginous
   Degree of cementation: friable, light, dense
   Cementation material: calcareous, siliceous, shaley, ferrous
   Porosity and permeability: poor, fair, good
3. *Limestones and dolomites*
   Color: white, brown, etc.
   Texture: dense, chalky, sucrosic, oolitic, oolicastic, coquinoid
   Crystal size: mat, cryptocrystalline, microcrystalline, megacrystalline
   Secondary material: shaley, sandy, dolomitic or limey, silty, cherty, siliceous, fossiliferous, ferruginous, anhydritic
   Luster: dull, earthy, resinous
   Type of porosity: intergranular, intercrystalline, pinpoint, oolicastic, vugular, fracture
   Degree of porosity and permeability: poor, fair, good

These descriptions should be presented using the standardized abbreviations shown in Appendix C, which are compiled by the Society of Professional Well Log Analysts for lithologic description and engineering abbreviations. A glossary of geological terms is shown in Appendix D.

With the increased application of computer processing in formation evaluation, there is a need to present the lithologic data obtained from sample data (as well as all data obtained during drilling) by computer processing. One such method proposed recently is shown in Appendix E.

Sample analysis, with respect to hydrocarbon shows, is another important part of sample description. A number of tests are used to indicate hydrocarbon presence in samples including: (1) oil staining, (2) hydrocarbon odor, (3) oil fluorescence (percent, intensity, and color), (4) cut (visible cut and cut fluorescence), (5) cuttings gas analysis and (6) acid test in carbonates and calcareous sands.

Oil Staining

Oil stains can generally be determined in the common reservoir rock types (limestones, dolomites, or sandstones), but may be difficult to recognize on dried samples. The color of the sample is always stated in the description of lithology so the effect of oil would appear, for example, as "oil sand, dark brown...," or "sand, gray with slight tarnish coat...," etc. Once a certain color associated with hydrocarbons has been established then the percent of stain in the total representative sample can be estimated. Suspected caving should not be included in the sample. This percentage may be translated to a show number, as described in Table 3-1 by Wyman and Castano[14] or as described by Kennedy.[15]

A (+) or (−) after the show number would point to which end of the range the percentage fell; for instance, nearly 90% might be 3−, 95% a 3, and 100% a 3+. Or if accuracy were warranted, percentages could be translated as shown in Table 3-2.

Table 3-2 is based on a cubic equation that emphasizes the importance of very high or very low percentage staining (or fluorescence) (see Fig. 3-6). Although designating a percentage stain (or fluorescence) in the broad categories is certainly better than nothing, a trained

TABLE 3-1

% Stain	Show Number
None	0 (No stain)
0–40	1 (Poor stain)
40–85	2 (Fair stain)
85–100	3 (Good stain)

Permission to publish by the Society of Professional Well Log Analysts.[14]

TABLE 3-2

Percentage	Show No.	Percentage	Show No.
0	0.0	60	1.8
1 (trace)	0.1	70	2.0
3	0.2	75	2.2
5	0.3	80	2.4
10	0.5	90	2.8
20	0.9	95	3.0
25	1.1	97	3.1
30	1.3	99	3.2
40	1.5	100	3.3
50	1.65		

Permission to publish by the Society of Professional Well Log Analysts.[14]

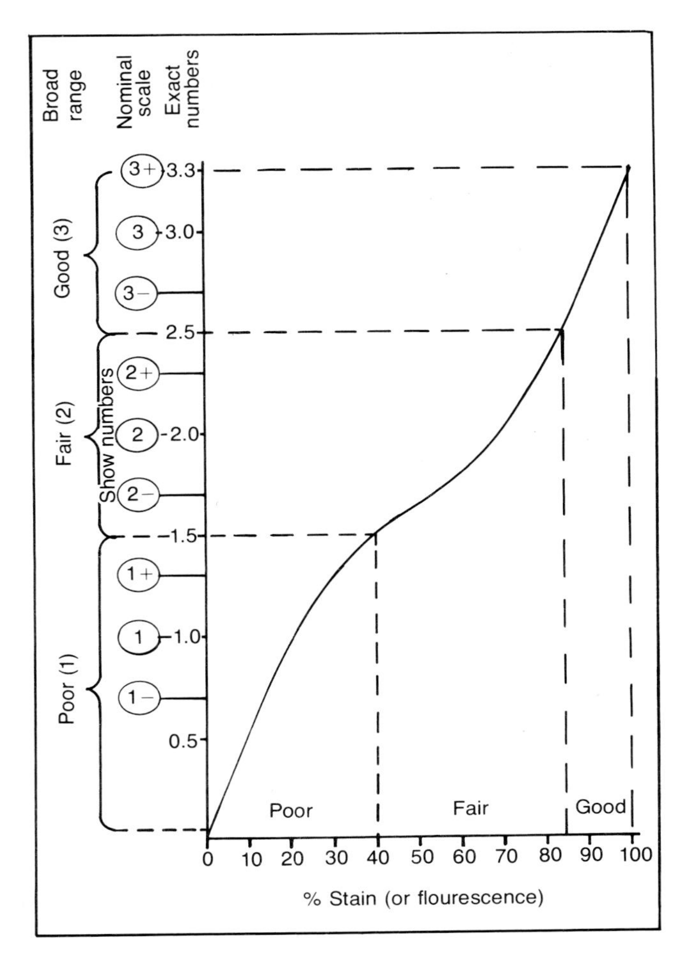

Fig. 3-6. Relation between show and stain or fluorescence. Permission to publish by the Society of Professional Well Log Analysts.[14]

observer should try to distinguish further; for instance, a small stain of 10% (show number 0.5) should be distinguished from a stain of 30% (show number 1.3) even though both might be considered poor staining (or roughly, show number 1).

The coloration of an oil sand is a function primarily of the color of the oil, with low gravity oils tending toward black and high gravity oils tending toward colorless. In a high gravity oil sand, determination of whether the sand is stained at all may be difficult; even if the sand has a uniform fluorescence, it may appear gray. Such information is valuable, even though the results may appear incongruous, such as percent oil stain = 0 and percent oil fluorescence = 34. In addition to recording the percent of oil staining (which applies to percent fluorescence also), it is helpful to note the distribution of this staining (if less than 100%). This notation can be abbreviated by placing an "a" or "b" after the percentage show number where:

"a" represents staining or fluorescence separated into distinct areas (typical where oil saturation is controlled by permeability). Staining seen only in vugs or confined to a different type of lithology within the sample being examined would fall in this category.

"b" represents staining or fluorescence distributed throughout sample—may be mottled or spotty.

Figure 3-7 illustrates some examples of different percentages and distributions of staining (or fluorescence). Where free oil or staining is observed on fractures, vugs, bleeding, cores, etc., its presence should be fully described.

Hydrocarbon Odor

One of the most simple tests for the presence of hydrocarbons is odor detection, which should be noted as soon as the ditch sample container is opened. Other odors, such as sulphurous, may also be noted. The presence of hydrocarbon odor could be classified as: none (0), slight (1), fair (2), and good (3).

Fluorescence

In many cases, the most valuable test for oil shows is to note the amount, intensity, and color of fluorescence due to hydrocarbons. To evaluate natural fluorescence, place a representative sample into the sample tray and observe it under the ultraviolet light. Note the percent fluorescence in the total representative sample and record the associated show number. Do not include any suspected caving-in representative sample. If only certain chips fluoresce, careful note should be made of the lithology giving the fluorescence. The percentage of oil fluorescence usually will be approximately equal to the percentage of oil staining on the same sample; however, this is not always the case. The show number may be assigned in the same manner as for the percent of oil staining. Again, if the fluorescence is less than 100%, it may be important to note the distribution of the staining, i.e., "a" for area type distribution or "b" for mottled or spotty distribution. Since fluorescence may be caused by certain minerals, such as secondary calcite, or a contaminant such as pipe dope, care must be taken to avoid confusing these with true formation hydrocarbons. A mineral fluorescence will not leach in a solvent; therefore, no cut fluorescence will be seen.

The intensity of fluorescence may yield important clues on the fluid content. For instance, even though a series of samples are uniformly fluorescent, a lessening of intensity may indicate a transition from oil to water-producing sands. As with odor, the rating of intensity may be fairly qualitative but can usually be classified as weak (1), fair (2), or strong (3). Finer subdivisions can

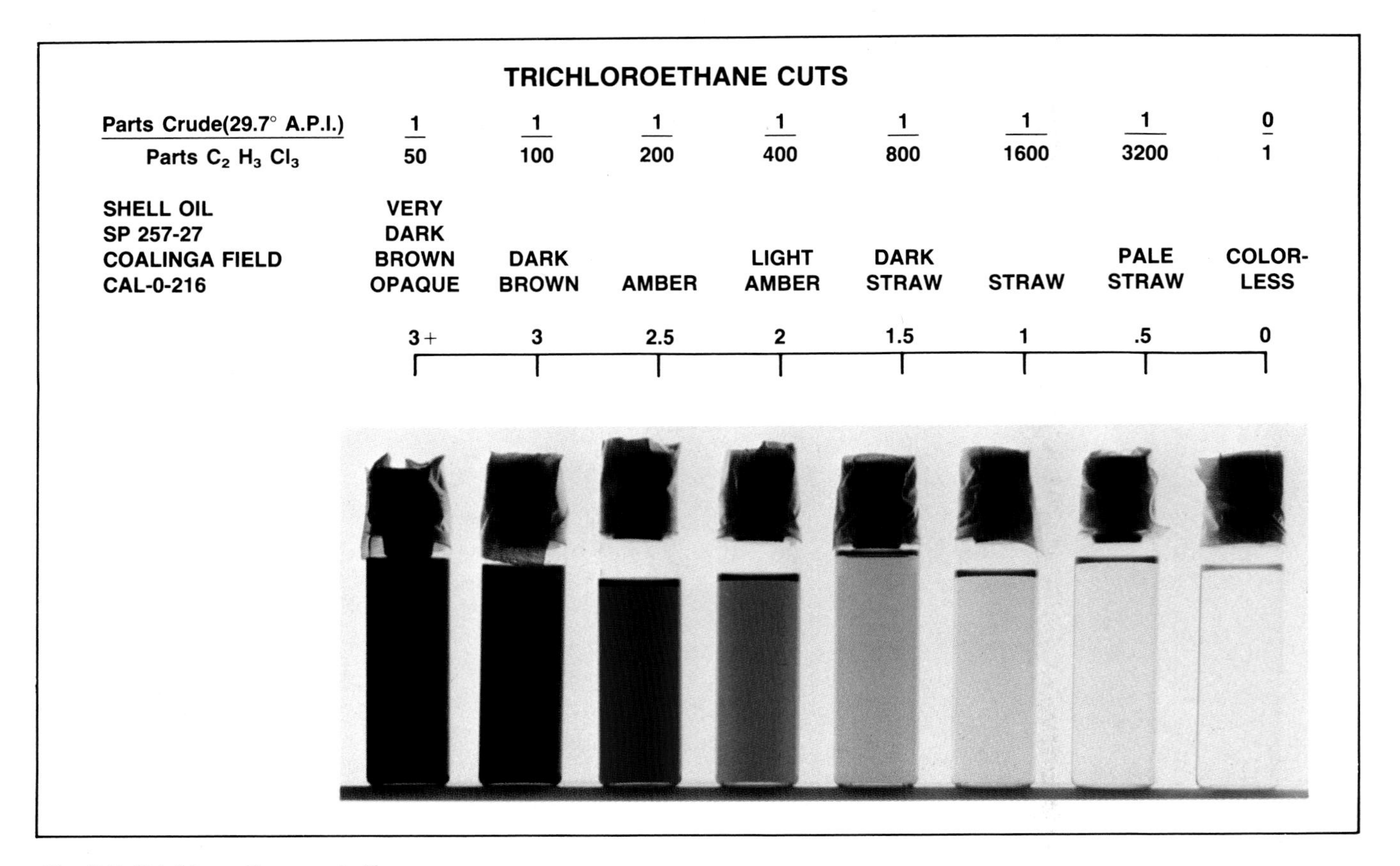

Fig. 3-8. Trichloroethane cuts.[14]

easily be made when comparing several samples at the same time.

It has long been recognized that the color of oil fluorescence can be used to make qualitative identification of the crude. Colors range from brown to gold to green-yellow to blue-white, with a variety of colors and shades in between. Although the particular color will not give a clue to whether or not the formation will produce, it does generally indicate the approximate gravity of the crude where the darker colors (browns and oranges) are associated with low gravity crudes and the lighter colors (yellow-white and blue-white) are indicative of high gravity oils. Refined oils such as dieseline (sometimes used as a mud additive) will give a bluish-white fluorescence.

Cuts

Although there are a great number of solvents available to test for oil (e.g., carbon tetrachloride, toluene, benzene, and ether), trichloroethane is generally preferable as a reasonably safe yet effective solvent (Chlorothene, which is commonly used, is a 1, 1, 1-trichloroethane).

Visible Cut. There are several techniques available for testing oil cuts with solvent. An effective but simple technique is to observe the color of the cut through a glass vial (4 oz, ¾ in diameter) using consistent ratios of material and solvent. This can be done as follows.

1. Grind up the selected sample until individual grains are separated. Care must be taken to avoid any contamination of the sample, particularly if sands are gray or if they can be expected to yield only a faint straw-colored cut.
2. Fill the test vial one-third full of the pulverized sample. If the sand is damp or "fluffy," pack the sample by tapping the bottom of the vial.
3. Pour in a sufficient amount of solvent (trichloroethane) to saturate the sand sample and fill an additional one-third of the vial (mixture of sand and solvent would then fill two-thirds of the vial; a different amount of sample may be used as long as the ratio of solvent to sample is kept constant).
4. Shake thoroughly to wash all staining from the rock surface.
5. Allow the sediment to settle and report the color of the fluid as soon as possible after cutting. Colors and their associated show number may be described as follows: colorless—0, pale straw—0.5, straw—1,

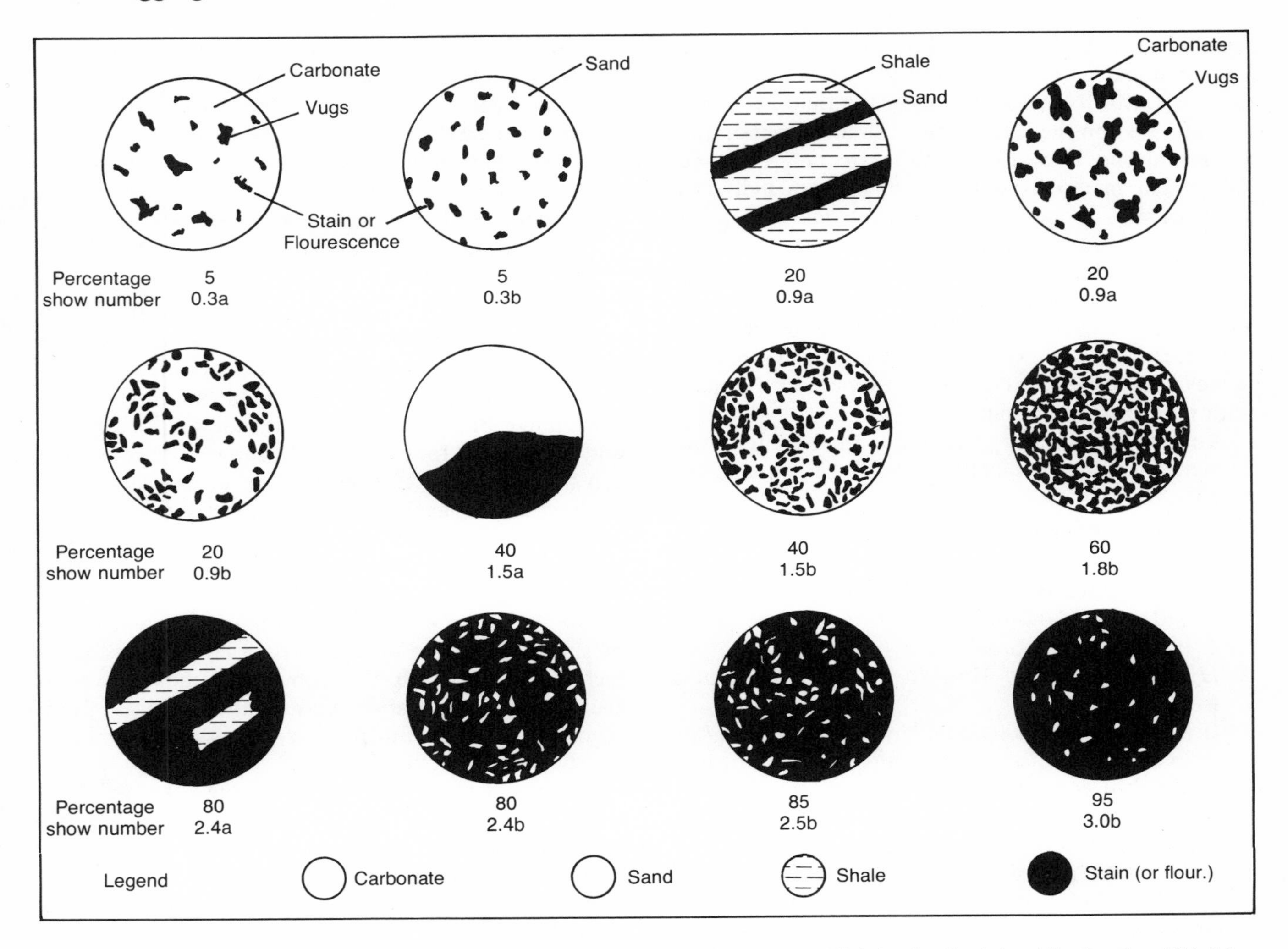

Fig. 3-7. Examples of percentage oil staining or fluorescence. Permission to publish by the Society of Professional Well Log Analysts.[14]

dark straw—1.5, light amber—2, amber—2.5, dark brown—3, very dark brown—3+.

Some crudes such as those found in the Gulf Coast of the United States have a green cast. A natural color photograph of cuts is helpful to distinguish the colors. A color reproduction of such a photograph is shown in Figure 3-8. A darker-colored cut (which may be from a low gravity oil reservoir) does not mean it is more likely to produce oil than a light cut (which may be from a condensate); however, in the same zone where the same type of oil is expected, a darker cut would usually be more encouraging.

Cut Fluorescence. The main value of cut fluorescence is to identify weak oil shows, i.e., shows that have a visual cut between colorless and pale straw (0–0.5). In addition, it serves to give added numerical weight to the worth of a visible cut (which is considered an important test for the show number average). The amount of cut fluorescence can be broken into four general categories: None (show number 0), slight—transparent when viewed through a cut bottle (show number 1), medium—translucent when viewed through a cut bottle (show number 2), strong—opaque when viewed through a cut bottle (show number 3).

The associated colors may vary from very pale yellow or blue to a dark green, gray-yellow, or blue-white, depending on the type of oil. Although the actual intensity of the cut fluorescence does not appear to increase for visible cuts above a show number of 0.5, the amount of substance in the cut solvent giving off a fluorescence does appear to increase with darker visible cuts; this is indicated under the ultraviolet lamp by less translucence of the solvent. Experience has shown that the color or amount of cut fluorescence is usually given about the same show number as the color of visible cut when the number is above 1.

An easy, qualitative, yet positive indication for very

small oil saturations of light hydrocarbons is to leach an uncontaminated formation chip placed in a porcelain spot tray with solvent and observe the fluorescent ring concentrated on the sides of the dish after evaporation of the solvent. It is also interesting to note if the cut fluorescent materializes by streamers from individual grains or vugs as opposed to a relatively fast "bloom" of fluorescence in the solvent.

When making tests for cut fluorescence, checks should be made to ensure the solvent is free from contamination. This is easily done by making sure the clean solvent has *no* fluorescence. Quantitative and more sensitive measurements of cut fluorescence can be made with the aid of a fluorometer.

A method to quantify the results of these tests was presented by Wyman and Castano[14] as summarized below.

1. *Oil Stain.* What % of a fresh surface is stained with oil?
   If *0%*, record *0* under "% oil stain" on the core record form; if *0–40%*, record *1*; if *40–85%*, record *2*; if *85–100%*, record *3*.
2. *Hydrocarbon Odor.* How strong is the odor of a fresh surface?
   If it is *none*, record *0* under "hydrocarbon odor" on the core record form; if *slight*, record *1*; if *fair*, record *2*; if *good*, record *3*.
3. *Sample Oil Fluorescence.* What % of a fresh surface fluoresces from oil under the ultraviolet light?
   If it is *0%*, record *0* under "% fluorescence"; if *0–40%*, record *1*; if *40–85%*, record *2*; if *85–100%*, record *3*.
   How bright is the fluorescence?
   If it is *weak*, record *1* under "Fluorescence intensity"; if *fair*, record *2*; if *strong*, record *3*.
   Describe the color of the fluorescence as accurately as possible.
   If it is some shade of *brown* (e.g., gold-brown), record *1 GB* under "Fluorescence color"; if it is closer to *orange, gold, or yellow* (e.g., yellow-orange), record *2 YO*; if it is a *light* color (e.g., blue-white), which is indicative of higher gravity oils, record *3 BlW*.
4. *Cut.* Pour trichloroethane solvent over the crushed sample in a cut bottle (the height of the solvent in the bottle should be twice that of the sample).
   What color is the cut?
   If it is *colorless*, record *0* under "color of cut"; if it is *straw-colored*, record *1*; if *light ambered*, record *2*; if *dark brown*, record *3*.
   What is the color of the cut fluorescence (observe the solvent under the ultraviolet light)?
   If it *does not fluoresce* at all, record *0* under "cut fluorescence"; if it has a *slight fluorescence* (the color may be pale yellow or blue-white) so that it appears transparent when viewed through the cut bottle, record *1*; if it has a *medium fluorescence* (the color may be a green to gray-yellow or blue-white) so that it appears translucent when viewed through the cut bottle, record *2*; if it has a *strong fluorescence* (the color may still be green to gray-yellow or blue-white) so that it appears opaque when viewed through the cut bottle, record *3*.

Show number averages were also proposed by Wyman and Castano[14] to assist in classifying the test results. A show number is obtained by simply adding the show numbers for any one sample and dividing by the number of observations made (add 0.3 for each + and subtract 0.3 for each −). The show number average can serve several useful purposes. It gives a quick idea of the general magnitude of the show and a number to compare the worth of shows in the same type formation. Also, it can serve as an arbitrary classification guide for summary word and symbol designations.

The use of the show number average for comparison of shows should be used only as a cursory means of identifying the worth of a show. One should be very cautious about comparing show number averages of one sand with those of a different zone without examining the supporting details.

The use of the show number average as a guide to classification is an arbitrary means of defining the difference between word and symbol summaries. Each geologist or engineer may have his own idea of the difference between an oil sand and an oil-stained sand, a "two dot" show and a "one dot" show (as it appears on a lithologic or summary log), but in order to standardize the meaning of these terms an arbitrary division can be made by means of show number averages. Show number averages can be conveniently subdivided (see Table 3-3). A summary of the show classification system is given in Table 3-4.

Cuttings Gas Analysis

The analysis of cuttings for gas is principally a matter of removing the gas from the sample. This is done by placing the sample in a blender, covering it with water, and pulverizing the sample for a short duration. The blender is sealed during the pulverizing operation, although there is a connection from the blender cap to a gas analyzer. The air-gas mixture produced during the pulverizing process is analyzed for hydrocarbons; any gas in the cuttings is referred to as microgas. If microgas determination is negligible, but there are other indications of the presence of oil, the washed cuttings can be put in a windowed vacuum cell, covered with water, and the cell put under vacuum. The evolution of gas can be observed with a microscope, and if any is noted, it is referred to as vacuum gas.[8]

Acid Test

Slight oil stains on small cuttings of carbonate rocks or calcareous sands often can be detected by immersion in dilute (15%) HCl. The reaction of the acid on the cuttings, stained however faintly, may form relatively large bubbles lasting long enough to float the specimen. The bubbles that form on the surface of a similar specimen, which is not oil-stained, normally break away before they become large enough to float it. The greater firmness of the bubbles on the oil-stained cuttings is due to the tough elastic skin formed around the bubbles by the oil.

The acid test is useful in detecting microscopic oil stains in limestone, dolomite, and calcareous sandstone. Although this test is used primarily on ditch cuttings in limestone areas, it can also be applied to small chips from calcareous cores.

TABLE 3-3
Convenient Subdivision of Show Averages

Range	*Show Symbol*	*Word Summary*	*Remarks*
0–0.5	X	sand, gray	No oil shows or no significant shows. May be a gas sand but would not be expected to produce oil.
0.5–1.5	•X	sand, slightly oil-stained	Slightly oil-stained sand; gas or condensate sand. May produce gas, high gravity oil, condensate, or water but rarely medium to low gravity oil.
	•X		Same as above except the percent stain and/or fluorescence is 85–100% (i.e., slightly better chance for hydrocarbon production).
1.5–2.5	•	oil-stained sand	Oil-stained sand may produce oil, water, or both. Very encouraging in higher oil areas.
2.5+	••	oil sand	Oil sands have a good chance of producing oil, usually in the medium to lower gravity range.

Permission to publish by the Society of Professional Well Log Analysts.[14]

HYDROCARBON MUD LOG

The hydrocarbon mud log was introduced into the oil industry in 1939; its principal measurement is the amount and type of gas contained in the mud. Additionally, the mud returns can be analyzed for the presence of oil as well as periodic or continuous evaluation of mud properties. This information is then presented in analog form (see Fig. 3-9). The mud log includes information obtained from sample analysis and drilling data analysis. It might be pointed out that the drilling time log and geologic sample log can be run even if the hydrocarbon mud log is not run. Hydrocarbon mud analysis, however, uses more mechanical aids than the geologist uses in detecting hydrocarbons during sample analysis; the hydrocarbon mud log, detecting minute hydrocarbon

TABLE 3-4
Summary of Show Classification System

			Sample Oil Fluorescence			*Trichloroethane Cut*			
Show Number	*% Oil Stain*	*Hydrocarbon Odor*	*%*	*Intensity*	*Color*	*Color of Cut*	*Cut Fluor.*	*Show Number Average*	*Show Symbol*
0	None	None	None	None	None	None	None	0 to 0.5	X
1	>0–40	Slight	>0–40	Weak	Brown (B), Orange-Brown (OB), Brown-Gold (BG)	Straw	Slight (Trans-parent)	0.5 to 1.5; 0.5 to 1.5 Stain and/or Fluor. over 85%	•X; •X (underlined)
2	40–85	Fair	40–85	Fair	Orange (O), Gold (G), Yellow-Orange (YO), Yellow (Y)	Light Amber	Medium (Slightly Translucent)	1.5 to 2.5	•
3	85–100	Good	85–100	Strong	White-Yellow (WY), Green-Yellow (GnY), Yel Wht (YW), Blu Wht (Bl W)	Dark Amber, Dark Brown	Strong (Opaque)	2.5 to 3	••

A blank space indicates the particular test was not made or else it is not applicable.
Permission to publish by the Society of Professional Well Log Analysts.[14]

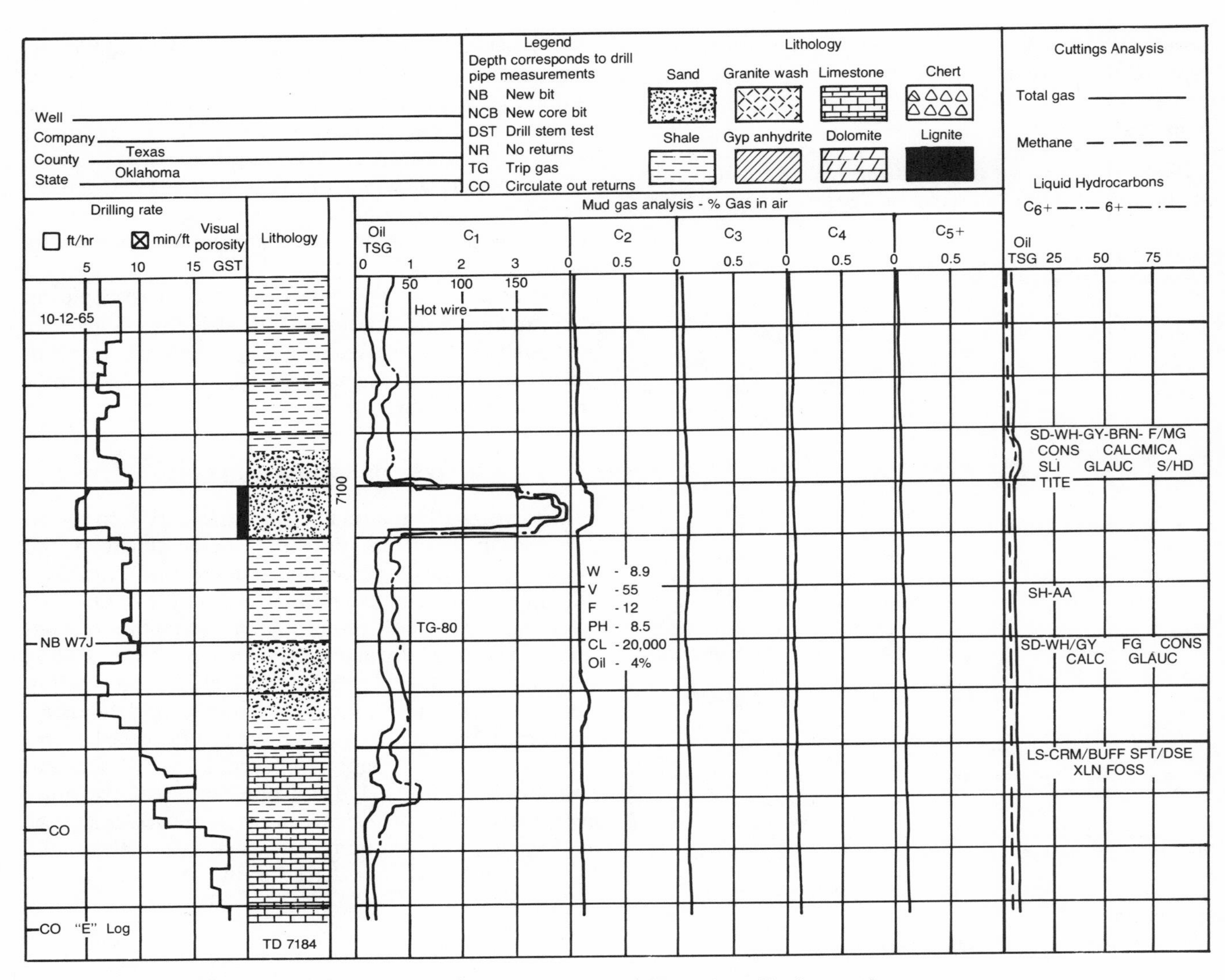

Fig. 3-9. Typical mud log. Permission to publish by Oil, Gas and Petrochem Equipment.[7]

presence, can sense hydrocarbon presence the geologist might miss.

A standard mud logging format has been proposed by API, but it seems that no one has accepted it, since at least fifty different forms are being used by the several hundred hydrocarbon well logging companies operating in North America.[15, 18] These mud logging formats do tend to have one thing in common, that the data presented includes information from the three major data sources available during the drilling process: (1) drilling data, (2) cuttings data, and (3) mud data. Another hydrocarbon mud log labeled to show the typical data is presented in Figure 3-10. Although it is an older log, it is not untypical of a hydrocarbon mud log that might be obtained today.

The two basic components of a gas-analysis system are the sampling unit and gas detector. While there are variations in the type of equipment used, Figure 3-11 shows a typical system.[8]

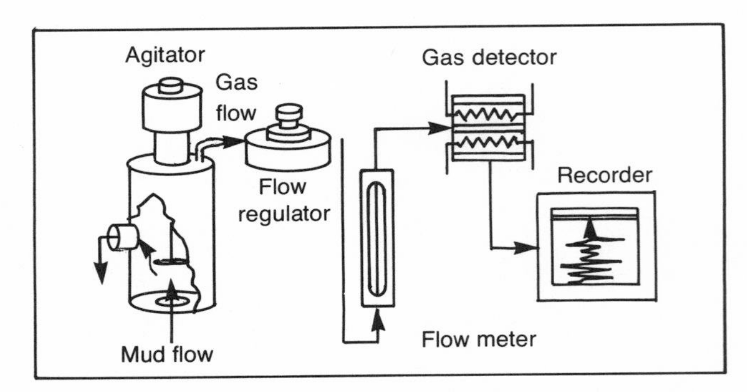

Fig. 3-11. Components of typical hydrogen mud analysis system. Permission to publish by Oil, Gas and Petrochem Equipment.[8]

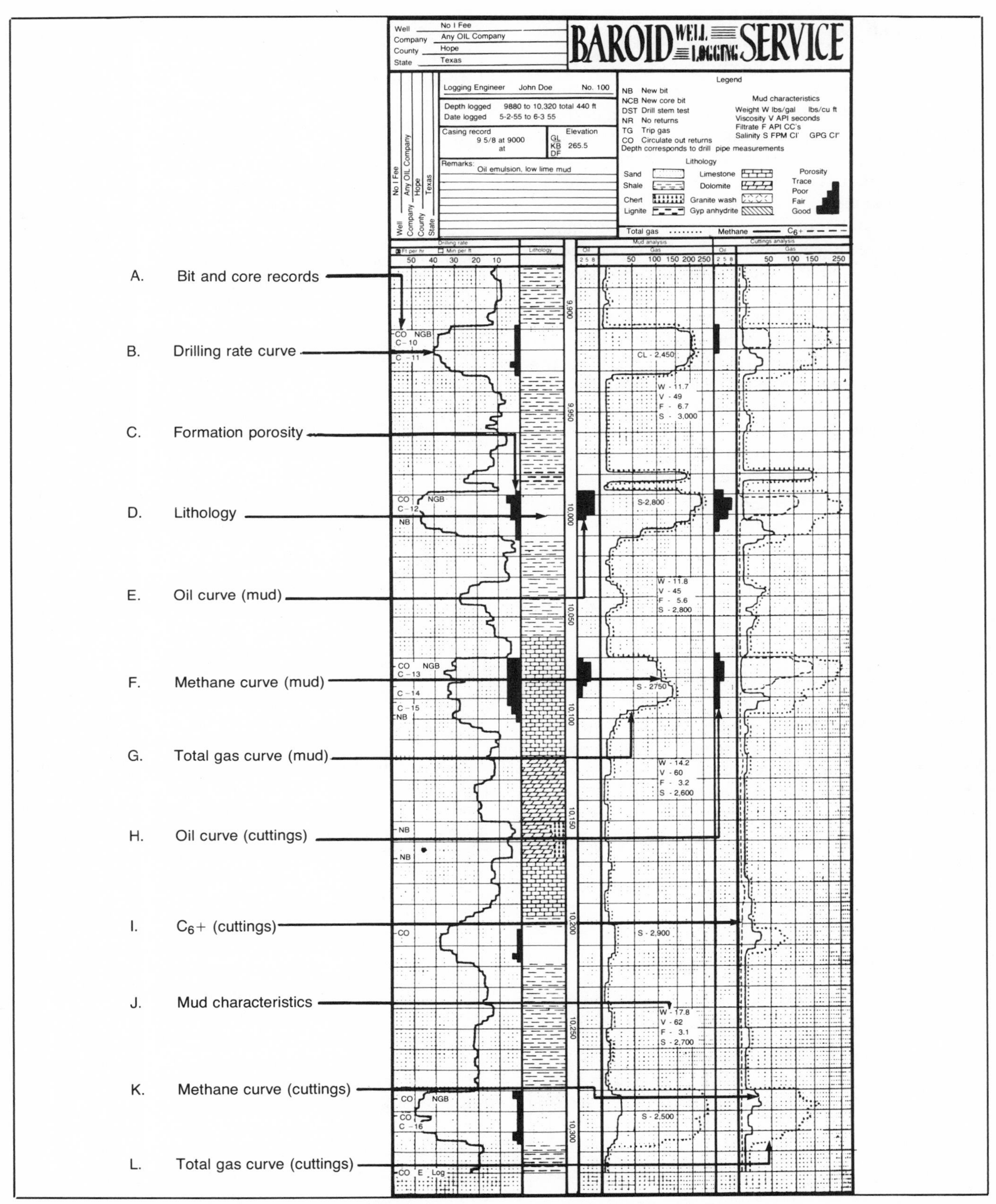

Fig. 3-10. Labeled mud log showing data source. Reprinted from Subsurface Geology in Petroleum Exploration *by J.D. Haun and L.W. Leroy by permission of Colorado School of Mines Press.*[2]

Sampling Unit

It is important in analyzing mud returns for gas that a representative mud sample be obtained for dependable results. The usual technique is to sample the mud-return line at the bell nipple, pumping this sample through an agitator. The agitator is efficient for releasing methane from the mud, but it is not dependable in releasing heavier components such as ethane and propane.

The steam still[7] provides a method for obtaining a much more complete separation of the low-boiling hydrocarbons from the mud. Virtually 100% of the methane and 95% of the butanes are collected by this technique, even from oil muds in which all hydrocarbon gases are highly soluble. Variables such as mud properties and trap efficiency, which affect ordinary separation techniques, are eliminated. The only limitation in using the steam still technique is that the separation process is not continuous. In this batch operation, a 5-cc mud sample is placed in the still and then swept with 1,000 cc of steam. The gases released from the mud are then condensed and analyzed.

Detectors

Several kinds of gas detectors have been used in hydrocarbon mud logging operations, including the catalytic filament detector, CFD, thermal conductivity filament detector, TCD, flame ionization detector, FID, infrared analyzer, IA, and to a limited extent the mass spectrometer.

The catalytic filament detector was for many years the standard used in mud logging for hydrocarbon detection. This type of detector cell is used in a wheatstone bridge circuit (see Fig. 3-12) and is called a hot wire analyzer. Two of the resistors of the wheatstone bridge are replaced by matched platinum filaments (the reference and detector cells). One filament is open to the air, and the other is in a cell through which the gas sample passes. The filaments are heated by passing current through the bridge. The bridge is in balance when air is in the sample cell. However, when the sample cell contains hydrocarbons, there is catalytic oxidation of the hydrocarbon gases at the filament. This increases filament temperature and its resistance, unbalancing the bridge. The bridge unbalance is indicative of the amount of hydrocarbon gas in the sample. This gas detector can be used for continuous sampling and is simple in construction and inexpensive.

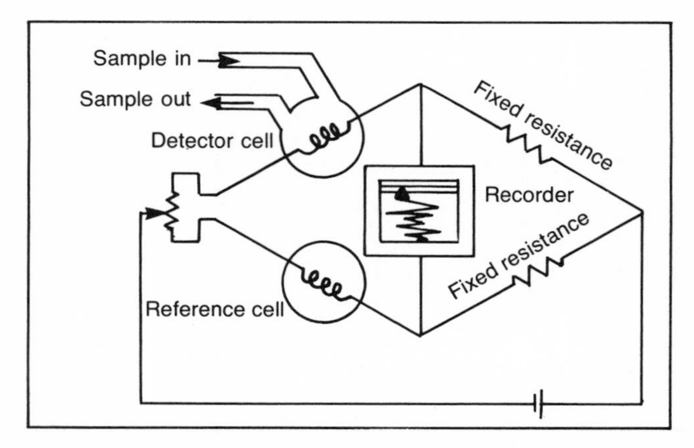

Fig. 3-12. Elements of a hot wire analyzer. Permission to publish by Oil, Gas and Petrochem Equipment.[8]

Usually two bridge circuits are used. In one circuit the detector filament is heated to a low temperature that is insufficient to ignite methane. In the second circuit the detector filament is heated to a temperature high enough to ignite all hydrocarbon gases including methane. Thus two measurements are made: (1) all hydrocarbons except methane and (2) all hydrocarbons including methane.[3]

The hot-wire detector is very sensitive. It will react to hydrogen, hydrogen sulfide, or any other combustible gas. Under certain conditions this may provide a high erratic background through which it is difficult to detect the presence of hydrocarbon gases. In addition, the detector filament may become "poisoned," rendered insensitive by the deposition of sulfur or carbon on the filament. If the gas volume is very high, some of the gas may not be burned by the high-voltage detector filament and actually may tend to cool the filament (in some cases record zero gas concentration). Thus the magnitude of the reading may not at all be indicative of the amount of hydrocarbon gas present. Likewise, if a large amount of heavy hydrocarbons are present, they may heat the low-voltage detector filament to a temperature high enough for methane to ignite. In this case, both the high-voltage and low-voltage detectors will read the same. The principal advantage of the hot-wire technique is its simplicity. The equipment is inexpensive and does not require extensive personnel training for satisfactory operation.

The thermal conductivity detector is also used in a wheatstone bridge circuit; its application is more recent than the CFD and usually only found in wellsite chromatography. As discussed by Mercer,[5] the CFD and TCD are quite similar in function, especially when air is used as the carrier gas in chromatography. The basic difference is that only combustible gases are detected by the CFD since its internal temperature, and consequently its resistance, depend on the results of combustion in proportion to the concentration of gas present. When the CFD is in an environment not conducive to combustion (i.e., lack of oxygen, etc.), it performs as a TCD. The TCD, on the other hand, varies its internal temperature and resistance as a function of the thermal conductivity of gases in comparison to some standard gas. In chromatography, the TCD has the advantage of detecting all gases with a different heat loss capacity than the carrier gas (usually helium) and is considerably more sensitive than the CFD.

Both the catalytic combustion filaments and the thermal conductivity filaments are subject to serious limitations in wellsite application. A partial list of these difficulties is given as follows:[5]

1. Particular dependence on power supply stability, something almost impossible to obtain at the wellsite
2. Immediate onset of filament deterioration and consequent limited filament life
3. Possible filament burn out with slight excesses in power supply
4. Variations in individual filament performance when compared with another
5. Sensitivity to environmental changes, particularly temperature fluctuations
6. Lower limit of sensitivity in the 500 PPM range for the CFD and 100 PPM range for the TCD (approximate in wellsite equipment)
7. Comparatively poor linearity
8. Very limited concentration range—cannot handle "rich" samples
9. Filament replacement relatively expensive

The use of flame ionization detection in wellsite gas analysis both for total gas detection and chromatographic gas analysis has not been used very long. Mercer described the FID and its advantages over the CFD and TCD methods, as discussed here.[5] Figure 3-13 schematically shows the characteristics and elemental circuitry of the FID as a component of a hydrogen flame chromatograph.

In this detector, a very small (one-sixteenth to one-eighth inch) flame is maintained at the flame tip by a constant flow of hydrogen fuel. Oxygen for combustion and air to evacuate the flame cell are both provided by a low volume purge air jet. The column effluent is mixed with hydrogen before reaching the flame tip. Wellsite hydrogen flame chromatographs usually use air as a carrier, which provides additional oxygen for combustion at the flame tip. As the effluent gases are burned in the flame, sufficient thermal energy is released to ionize the organic solutes passing through it. The actual mechanism of ionization is still unknown, but several theories have been advanced to explain the phenomenon. The chimney screens the flame from drafts and spurious electrical signals, and the ignitor coil provides an automatic electrical means for flame ignition.

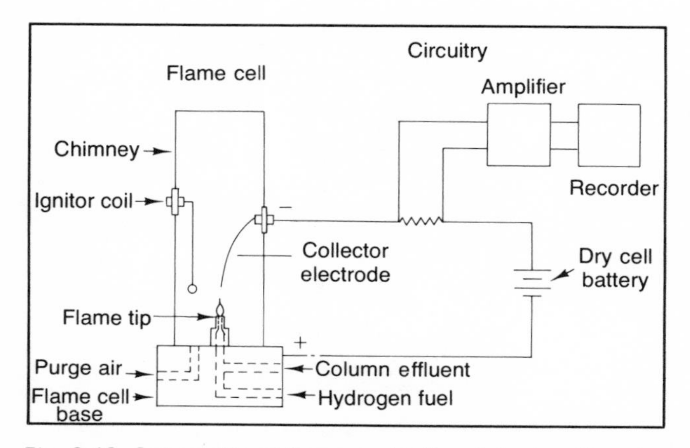

Fig. 3-13. Schematic of flame ionization detector. Permission to publish by the Canadian Well Logging Society.[5]

The action upon which ionization detectors depend is the conduction of electricity by gases. If electrically charged molecules (ions) are present in an applied electrical field, their motion in a fixed direction causes them to conduct. The number of ions produced in a flame of pure hydrogen burning in air is greatly affected by the presence of volatile carbon compounds. Gas molecules normally do not conduct, and the increased conductivity due to the presence of a very few ions can be observed and explains the great sensitivity of ionization detectors. To establish the required electrical field, a voltage drop of approximately 300 volts (this value is not critical) is maintained across the flame from the flame tip to the collector electrode. The ions generated in the flame are collected at the electrode, completing the circuit (see Fig. 3-13). The resulting ion current is monitored by measuring the voltage drop across a very high resistance in series. Advantages of the FID with respect to the CPD and TCD include:[5]

1. Ease of construction.
2. Increased stability—dry cell battery, not rig power.
3. Increased sensitivity—ion current measurement. Lower limit approximately .5 to 1 PPM in wellsite applications to date.
4. Insensitive to contamination such as water vapor.
5. Responds to all organic compounds except formic acid and the response is generally greatest with hydrocarbons.
6. Insensitive to all fixed gases, H_2S, CO_2, SO_2, CO, N_2, He, etc. (if any of these gases are anticipated, another detector must be present in addition).
7. Insensitive to temperature fluctuations and can be operated satisfactorily at room temperature.
8. Wide linear dynamic range in response to concentration (approximately 10^5).
9. No deterioration of flame cell during operation.

The infrared analyzer can provide a continuous analysis for only one component, usually methane. This instrument (see Fig. 3-14) has two Nichrome filament sources heated by an electric current that provides a continuous and uniform spectrum throughout the infra-

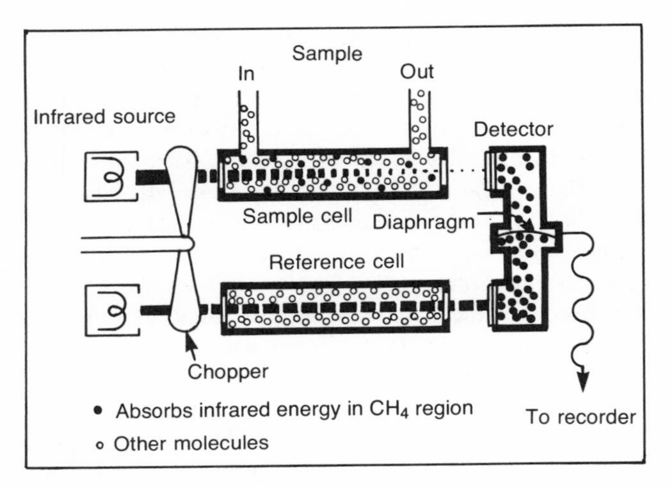

Fig. 3-14. Elements of infrared analyzer for methane. Permission to publish by Oil, Gas and Petrochem Equipment.[8]

red range. The infrared beams pass through an air-filled reference cell and a sample cell, and into the detector filled with the material being analyzed, such as methane.

In operation, the air in the reference cell absorbs some of the radiation passing through it but does not absorb energy in the same band as that of the methane gas in the detector. The radiation arriving at the detector undergoes further absorption and causes heating of the detector gas. Energy passing through the sample cell undergoes the same absorption when the sample cell contains air or any gas other than methane. If the gas volumes on both sides of the diaphragm in the detector are heated equally, there is no movement of the diaphragm. However, if some of the gas in the sample cell is methane, part of the critical energy band is absorbed in the sample cell, resulting in a greater heating of the reference side of the detector than of the sample side, and the diaphragm will be displaced.

The metal diaphragm acts as one plate of an electrical condenser. Any movement changes the capacity of the condenser. This produces an electrical signal proportional to the amount of gas in the sample. This analyzer is vibration sensitive so it must be mechanically damped, and it must have a power supply capable of supplying constant frequency and voltage.

Application of the infrared and gas-chromatography analyzers provides an efficient mud-analysis technique. The mud gas sample can be continuously monitored for methane by the infrared analyzer. When a good show is indicated, a sample can then be run through the gas chromatograph for a complete analysis of the components.

Gas Chromatography

The function of gas chromatography is to analyze each hydrocarbon accumulation for the identity and relative proportion of each component. This information can be of exceptional value in predetermining the character of the hydrocarbons in a reservoir and aid in determining the relative significance of hydrocarbon shows. Additional uses are the differentiation of formation hydrocarbons from others pre-existing in the drilling fluid, and the computation of open hole gas production during air drilling.[5]

By virtue of its design, the chromatograph is an intermittent stream analysis tool and should not, therefore, replace the basic total hydrocarbon detector. The chromatograph is normally positioned between the drilling fluid gas trap and the total gas detector. It extracts a constant volume of the unknown gas-air mixture for analysis at a preselected interval. In general, a new sample cannot be introduced until the chromatogram of the previous sample has been completed. Hydrocarbons from cuttings may be collected by the usual procedures and transferred by syringe to the chromatograph for analysis.

The wellsite chromatograph usually comprises the following components:[5]

1. Sample collection apparatus: sample valve, sample loop, injection port, programmer
2. Carrier gas: helium, hydrogen, nitrogen, air
3. Column: one-eighth to one-fourth inch metal tubing, diatomaceous solid support phase, nonvolatile liquid solvent phase
4. Detector: flame ionization cell (FID), thermal conductivity filament (TCD), catalytic combustion filament (CFD)
5. Amplifier
6. Recorder

Figure 3-15 shows a schematic diagram of a partition gas chromatograph. In this unit, a sweep gas (helium, air) flows continuously through a column packed with

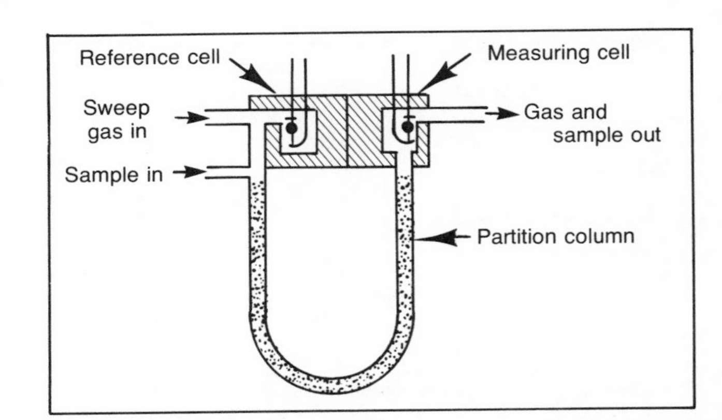

Fig. 3-15. Components of a partition gas chromatograph. Permission to publish by Oil, Gas and Petrochem Equipment.[8]

an inert solid coated with a nonvolatile organic liquid. At the inlet end of the ¼ in diameter tube column, a small measured volume of the unknown sample is injected into the sweep gas stream. The heavier sample components tend to be absorbed and pass slowly through it, but the lighter components (methane) are relatively insoluble and pass through rapidly. At the outlet, the various components appear separately, and each is detected with a gas analyzer (thermal conductivity, hot-wire). The component is identified by the length of time it takes to travel through the column. Each component is recorded on a strip-chart recorder as a peak. The area under each peak gives the volume of that component in the sample.

MUD ANALYSIS FOR OIL

The presence of crude oil in mud is detected by using a fluorescent-light viewing box. In this instrument, the mud sample is exposed to ultraviolet radiation, causing the fluorescence of materials containing hydrocarbons. Fluorescence is caused by the absorption of ultraviolet radiation, causing atoms in the material to reach an excited state. In order to return to the lower energy state, excited electrons emit radiation equal to the difference between the two energy states. This radiation, seen as fluorescence, is produced in most crude oils when viewed under a suitable ultraviolet light. The usual source of ultraviolet light has a wavelength of 3,600 anstrom units, which causes fluorescence of nearly all crude oils. The color of fluorescence given off by a crude oil is generally characteristic of its gravity (see Table 3-5). Fluorescence-light units enable the detection of concentrations of less than 10 ppm oil-in-mud.

The presence of refined rig oils creates a problem in distinguishing the presence of light crude oils since refined oils fluoresce white or blue-white, which, as shown in Table 3-5, is in the high-gravity crude-oil range. However, in this gravity range, the light ends of crude oil can be readily detected by gas analysis.

TABLE 3-5
Fluorescence Color of Crude Oils

Gravity, °API	*Color of Fluorescence*
Below 15	Brown
15–25	Orange
25–35	Yellow to Cream
35–45	White
Over 45	Blue-White to Violet

Permission to publish by PennWell Publishing Company.[8]

INTERPRETATION OF HYDROCARBON MUD LOGS

The basic premise of hydrocarbon evaluation of the drilling mud, and primarily gas detection, is related to the observation of gas accumulations, which are liberated from the hydrocarbon and detected when a hydrocarbon-bearing formation is drilled. As illustrated by Pearson,[3] a 9-in. hole drilled through a 20-ft oil sand "drills up" about 9 cu ft of rock. Assuming 20% porosity, 80% initial oil saturation, a formation volume factor of 1.6, and a solution gas-oil ratio of 1,000 cu ft/bbl, this 9 cu ft of rock contains about 7 gal of oil and 167 standard cu ft of gas. Much of the oil and gas present in the rock never enters the mud stream. Perhaps 80% is driven out of the drilled rock by invasion of mud filtrate ahead of the bit. If 50% of the remaining oil stays in the cuttings, we have only 0.7 gal of oil and 33 standard cu ft of gas mixed in the mud stream. If it takes 60 min to drill the 20-ft oil sand, this oil and gas may be mixed with 24,000 gal of mud.

These small quantities may not be evident to the human senses but can be detected using sensitive instruments. Limits of detection for the hydrocarbon-mud log are usually estimated to be 0.01 cu ft of gas per hour or 1 part oil in 100,000 parts mud.

The illustration is only an example of possible conditions. Actual conditions of oil in place, drilling rate, amount flushed, etc., vary greatly. This example does show that small but significant quantities of gas and oil are liberated in the mud stream (and remain in the cuttings).

If liberated gas were the only source of gas, then interpretation of the mud gas measurement would be rather simple. This is not the case, however, since numerous other sources of gas can exist during drilling. These sources might be classified as:

1. Liberated gas: gas mechanically liberated by the bit as the formation is drilled
2. Produced gas: gas produced from a zone due to underbalanced pressure conditions (effective hydrostatic pressure less than formation fluid pressure)
3. Recycled gas: gas contained in the mud which has been pumped down the hole and appears a second time at the surface
4. Contamination gas: gas artificially introduced into the mud from a source other than the rock formations

Some sources of gas introduction into the drilling mud are schematically shown in Figure 3-16. Some of these sources might be briefly described as follows.[13, 16, 17]

1. Connection gas: Quite frequently, swab effect of raising the kelly when making connections brings small (but often significant) amounts of gas into the borehole.
2. Trip gas: The general term applied to produced gas which characteristically occurs within one lag time after a trip is completed and circulation has been resumed. Three basic factors influence the presence,

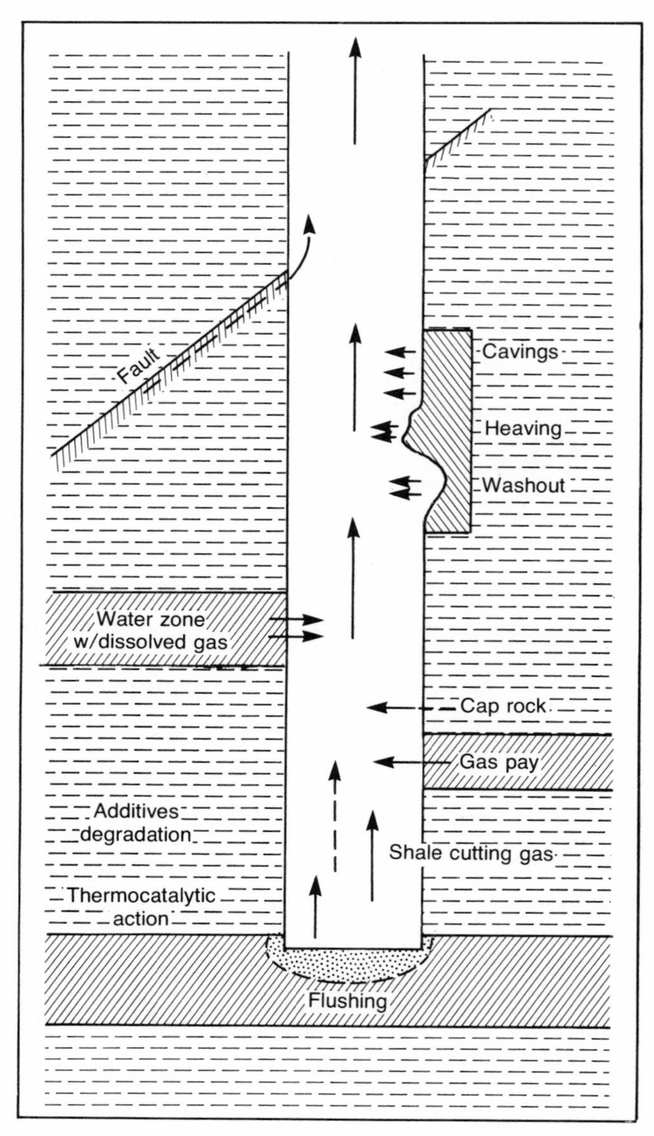

Fig. 3-16. Schematic of some gas sources during drilling. Permission to publish by World Oil.[16]

location, and magnitude of the trip gas kick: (1) the loss of the annular pressure drop; (2) the effect of bit swabbing the entire hole (this effect is influenced by such factors as the speed at which the pipe is tripped out of the hole, variations in hole size, the configuration of a packed hole assembly, and tripping out with a full hole core barrel); (3) the time over which these factors influence the static mud system. The most significant difference between trips and connections is the extreme accentuation of these influences during a round trip as compared to the relatively minor influence of a connection.

3. Kelly air: A simple mechanical operation such as unscrewing the kelly for pipe addition can introduce air into the drill pipe. While a continuous mud weight recorder or chromatograph allows its detection at the wellhead after a complete circulation cycle, the kelly-air effect is usually insignificant.
4. Downtime: When mud pumps are shut down for trips, rig repairs, or logging operations, increased gas shows may appear when circulation is reestablished.
5. Degradation of mud additives: Gases such as H_2S and CO_2 may also originate from mud additives degrading in hot wells. For example, degradation of modified lignosulfonates at temperatures exceeding 400°F produces significant amounts of H_2S and CO_2.
6. Faults: Boreholes often cut faults that may channel gases, resulting in localized gas flow into the wells and gas cutting.
7. Thermodynamic processes: Clays exhibiting catalytic activity are present in both formation and drilling mud. Field tests show that gaseous hydrocarbons also originate as a result of the grinding action by the drill bit and the subsequent temperature increase, in the presence of a catalyst (clay mineral) and organic matter in the rock. This effect may become significant in the interpretation of mud-gas reading.

In order to provide a clearer understanding of mud gas data in relation to the sources of gas as they occur in the mud, Mercer[13, 17] used a simple drilling model to illustrate the impact of bit penetration through hydrocarbon accumulations. The series of cases presented by Mercer show variations in the configuration of the mud gas data, indicating differences in the response of the hydrocarbon-bearing zone to bit penetration, and subsequent rig operations. It should be pointed out that references to gas kicks refer to high gas responses by the detector and not a high volume drilling gas kick.

The simple drilling model analyzed and discussed by Mercer shows that the geometry of the gas kick recorded by the instrumentation and plotted with respect to time is directly related to significant characteristics of the hydrocarbon zone, as well as the impact of concurrent drilling operations. It is apparent that the configuration of the gas kick as recorded directly from the drilling mud is of greater interpretive significance than the magnitude of the gas kick. When recorded instrument chart data versus time is digitized and plotted in graph format versus depth, the magnitude of the gas kick may be faithfully reproduced, but the configuration of the kick is usually lost. Thus, it is obvious that a basic and vital interpretation must derive from a plotted graph. The function of the plotted graph should be to collate, according to depth, pertinent data produced from various sources. This graph then provides a broader understanding of the hydrocarbon accumulation and a convenient means for future reference.

To illustrate these concepts, a diagramatic technique is used that graphically relates the gas detector response plotted versus time to the actual penetration of the rock by the drilling bit through the penetration rate curve plotted versus depth. This technique allows direct comparison of the geometry of the gas kick to actual rock penetration.

Analysis of Liberated Gas

Figure 3-17 illustrates a typical full hole drilled through a hydrocarbon-bearing zone in which the total bottom pressure (TBP) is greater than the formation fluid pressure (FP).

As described by Mercer, during penetration the bit continuously introduces into the mud system rock components of the cylinder defined by the hole size and interval thickness. As the bit penetrates, it mechanically creates pseudo-permeability and allows material contained in the absolute pore volume of the cylinder to enter the mud system and be transported to the surface. The term pseudo-permeability is used because the liberating action of the bit is purely mechanical and not directly related to inherent rock permeability.

If the pore volume contains hydrocarbons, those contained in the rock cylinder will be transported to the surface in various proportions of two possible forms: (1) liberated directly to the mud or produced into the mud from the cuttings as they are subjected to ever decreasing hydrostatic pressure and (2) retained by the cuttings chips themselves.

Thus, the primary source of hydrocarbons available to gas detection equipment under these conditions is from the cylinder of rock mechanically liberated to the mud system by bit action. This is a type-one gas kick.

In Figure 3-17, the penetration rate curve corresponding to the porous interval shows a characteristic drilling break from 20 minutes per foot to 5 minutes per foot. Such drilling breaks are often valuable in determining thickness of porous intervals. The hypothetical gas detector response shows hydrocarbon concentration in the mud versus time. Hydrocarbon concentration is primarily a function of penetration rate, absolute pore volume, and formation pressure. Other factors such as the BTU content and hole size also are involved. These additional factors are not included in Figure 3-17 but should be considered when evaluating the significance of a particular gas kick.

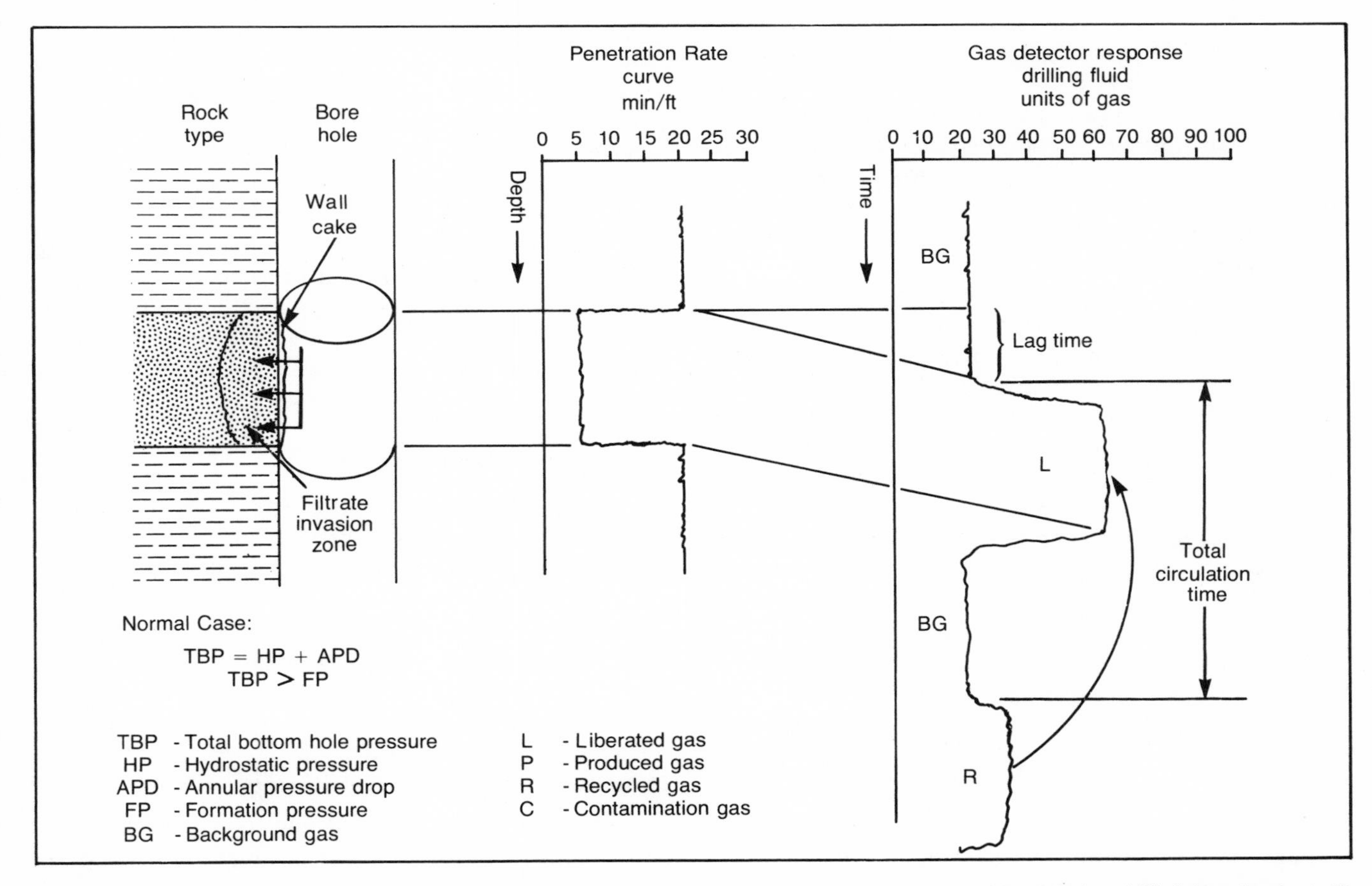

Fig. 3-17. Response to liberated and recycled gas. Permission to publish by the Society of Professional Well Log Analysts.[13]

Substantial increases in any of the three factors normally will have a visible effect on gas detector response. Normally, penetration rate is the most important single factor in determining gas kick magnitude. Its effect will be discussed in greater detail later.

If the bit penetrates the rock cylinder at a constant rate, and if porosity and formation pressure are constant throughout the interval, it can be assumed that an equilibrium is established between the volume of mud circulation through the bit and the volume of gas mechanically liberated to the mud system. This supposition is shown in Figure 3-17, but is rarely true. Lag time is the circulating time that commences when the bit contacts the porous interval and terminates when the gas increase commences. If the gas kick comprises only liberated gas, it can be concluded that the gas kick would begin to end one lag time after the bit ceases to penetrate hydrocarbon bearing porosity.

Since formation pressure normally is constant throughout a single porous interval, variations in liberated gas magnitude are clearly related to the remaining two parameters of penetration rate and porosity. Should penetration rate be relatively constant, kick magnitude variations may often be directly related to rock porosity with resolution better than one foot.

A type-one gas kick (see Fig. 3-17) is the normal case, because a margin of safety is always desired when penetrating possible blowout zones. Figure 3-17 shows the situation in which mud filtrate invades the porous formation while wall cake is deposited on the surface of the hole. Since the pressure differential across the hole/rock interface is positive, no additional hydrocarbons beyond those contained in the rock cylinder contribute to the liberated gas kick.

In unusual circumstances of high formation permeability, low formation pressure, and exceedingly high total bottom-hole pressure, mechanically liberated hydrocarbons may be pumped directly into the formation and not return to the surface. Another variation of this possibility may occur if filtrate invasion immediately preceded the bit; resident hydrocarbons may be flushed by the filtrate so that the bit mechanically liberates only mud filtrate and not hydrocarbons. While these two possibilities occur with extreme rarity, they should be considered in instances where gas kicks were normally expected but did not occur.

Three basic principles of interpretation emerge from this discussion of liberated gas:

1. In a type-one gas kick (FP is less than TBP), only mechanically liberated gas forms the kick.
2. The gas kick configuration is an early indication of the liberating interval's thickness, and possibly the quality of the porosity, as well as depth and thickness of the most porous interval.
3. Presence of a type-one gas kick gives no direct indication of the presence or absence of permeability. If there is no effective permeability when drilling a hydrocarbon-bearing zone, the liberated gas kick will still occur. If permeability is present and a sufficient hydrostatic overload is carried in the mud system, the liberated gas kick's configuration will remain relatively the same.

During coring, only a small portion of the rock cylinder normally drilled is exposed to mechanical liberation by the core bit. Thus, a much smaller quantity of liberated gas is introduced into the mud system per foot of penetration. Often, coring rates are slower than full hole penetration rates, further decreasing the liberated gas quantity per mud volume. These factors usually combine to result in the common phenomenon of considerably lower gas readings while coring.

Analysis of Recycled Gas

If mud gas is not completely volatilized in the settling pit, but is pumped back down the hole, the gas detector may record a second appearance of an earlier kick. In Figure 3-17, the liberated gas kick has recycled to the surface for the second time and is designated R.

Recycled gas may be identified by certain tests. The recycle should be no larger than the original kick, but should be similar in shape. Recycled kick composition may be misleading since more volatile hydrocarbons are often liberated to the atmosphere in the pits or degasser, resulting in the recycled kick showing a larger proportion of heavy ends.

A good indication of total mud system circulating time may be obtained by measuring time elapsed from the beginning of the primary gas kick to the beginning of the recycled gas kick. Such direct information often may be helpful in assuring the accuracy of an estimated lag time.

Analysis of Partial Liberation

Possible explanations for instances in which the duration of the gas kick does not seem to extend through the entire period of probable liberation are demonstrated in Figure 3-18. In a type-one gas kick, only liberated gas comprises the kick. If kick geometry is solved consistently with principles derived in Figure 3-17, a significant variation within the interval of the drilling break becomes apparent.

Influence of Penetration Rate Change

Figure 3-19 illustrates a type-one gas kick in which only liberated gas is present. If the porosity, formation pressure, and mud pump volume remain constant

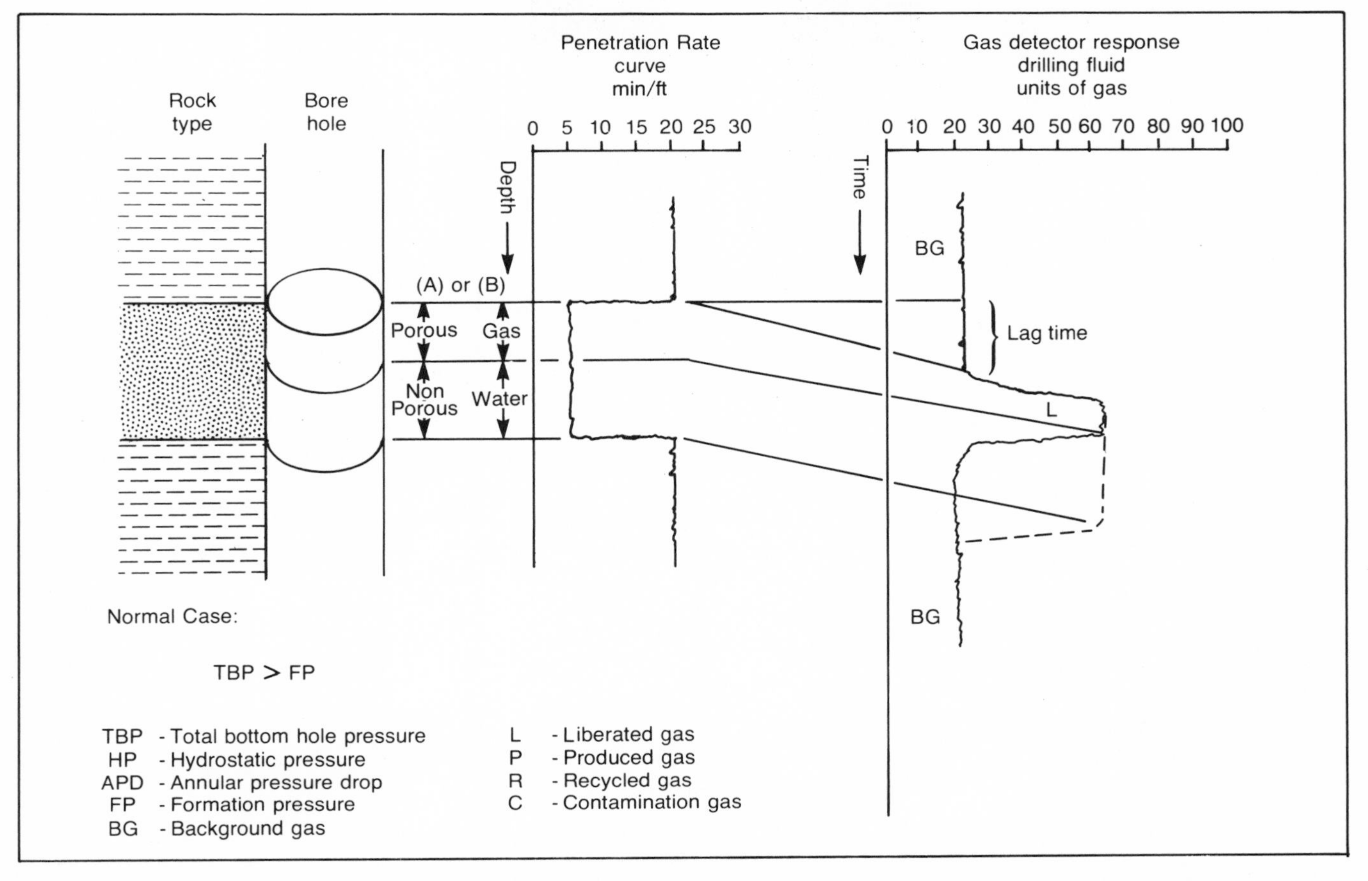

Fig. 3-18. Response and possible explanations to partial liberation. Permission to publish by the Society of Professional Well Log Analysts.[13]

throughout the zone, gas kick magnitude becomes a direct function of penetration rate. A substantial decrease in penetration rate (designated A in Fig. 3-19) should result in a corresponding decrease in liberated gas versus mud volume equilibrium (designated B in Fig. 3-19).

A decrease in penetration rate through a zone of given thickness requires a greater total period of penetration. This longer drilling interval results in a kick of decreased magnitude but of longer duration.

Consequently, during extremely slow penetration of hydrocarbon accumulations where no particular drilling break is evident and where rock porosity is especially low, it is possible that no recognizable gas kick will occur. The presence of liberated gas may be very difficult to detect over the carried background gas in the mud system at that time. Such a situation requires careful interpretation in conjunction with all other qualifying information before concluding that significant liberation did not occur.

Experience has shown that penetration rate is the most important single factor governing magnitude in a type-one gas kick, and this suggests that using kick magnitude as the single or primary criterion in judging significance is erroneous. Of particular concern is the dangerous combination of effects from various liberated gas zones in the same well. In many instances, permeable hydrocarbon zones may exist up the hole and be effectively contained by existing hydrostatic pressure. Extreme care must be taken not to liberate large quantities of gas from a thick downhole reservoir by drilling through it too rapidly. The result may be to decrease the effective hydrostatic head on the upper zone, due to gas cutting, and allow it to blowout. Penetration rate should be reduced to minimize liberated gas in the mud system, thereby maintaining sufficient effective hydrostatic pressure up the hole.

Analysis of Produced Gas Influence

An abnormal case in which TBP is less than FP is shown in Figure 3-20. The gas kick is characterized by significant differences from those previously discussed and is designated a type-two gas kick.

Figure 3-20 shows the usual situation in which the hole does not begin to make fluid immediately upon

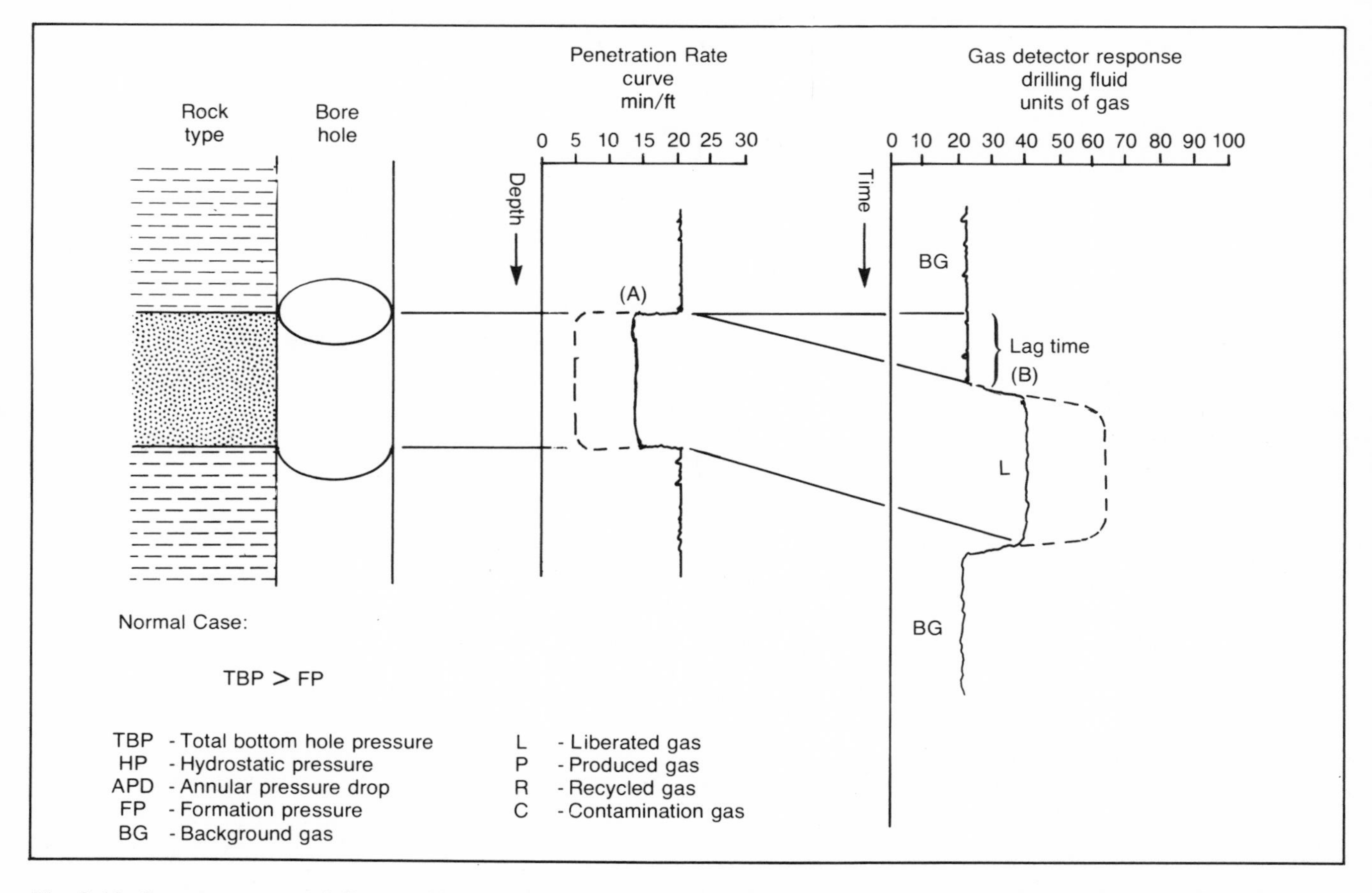

Fig. 3-19. Gas response as influenced by penetration rate change. Permission to publish by the Society of Professional Well Log Analysts.[13]

penetration, but the gas kick commences at one normal lag time. Such kicks are characterized by exceptional initial magnitude and continuation beyond the time normally anticipated for terminating a liberated kick. If the source zone is clearly defined by penetration rate and other available data, it is apparent that the formation is contributing additional hydrocarbons to the mud system beyond those mechanically liberated. Significant contrasts in interpretation result from a type-two gas kick.

There is now no direct relationship between mechanical liberation and mud circulation, thus definitive analysis of the source zone thickness and quality becomes extremely difficult. The gas kick magnitude no longer can be related to the general significance of the source zone in comparison with other type-one gas kicks.

The presence of produced gas demonstrates conclusively that at least some degree of effective permeability is present. This direct evidence of permeability is in contrast to the absence of any definitive evidence in a type-one kick where only mechanical liberation occurs.

Since produced gas generally is independent of mechanical liberation and its attendant controlling factors, it is reasonable to expect that the configuration and magnitude of type-two gas kicks encountered while coring would be generally independent of the mechanical characteristics of the coring operation.

Mud Log Format

It was mentioned earlier that mud logs have been and are presented in a number of formats. Kennedy[15] proposed a basis for encoding mud log data for computer processing compatibility and also suggested a mud logging format (see Appendix E).

McAdams and Mercer[18] suggest presenting the three most important data available: (1) rate of penetration (as an indicator of porosity), (2) lithology, and (3) total gas response lag corrected (for identification of hydrocarbon liberating zones) on a conventional API wireline format. This format is suggested to better integrate mud log data into a more readily usable formation evaluation tool. This presentation, referred to as a Detailed Hydrocarbon Log, generates a log that closely resembles any of the conventional wireline logs, in both format and significance. An example of this log is shown in Figure 3-21. McAdams and Mercer also pro-

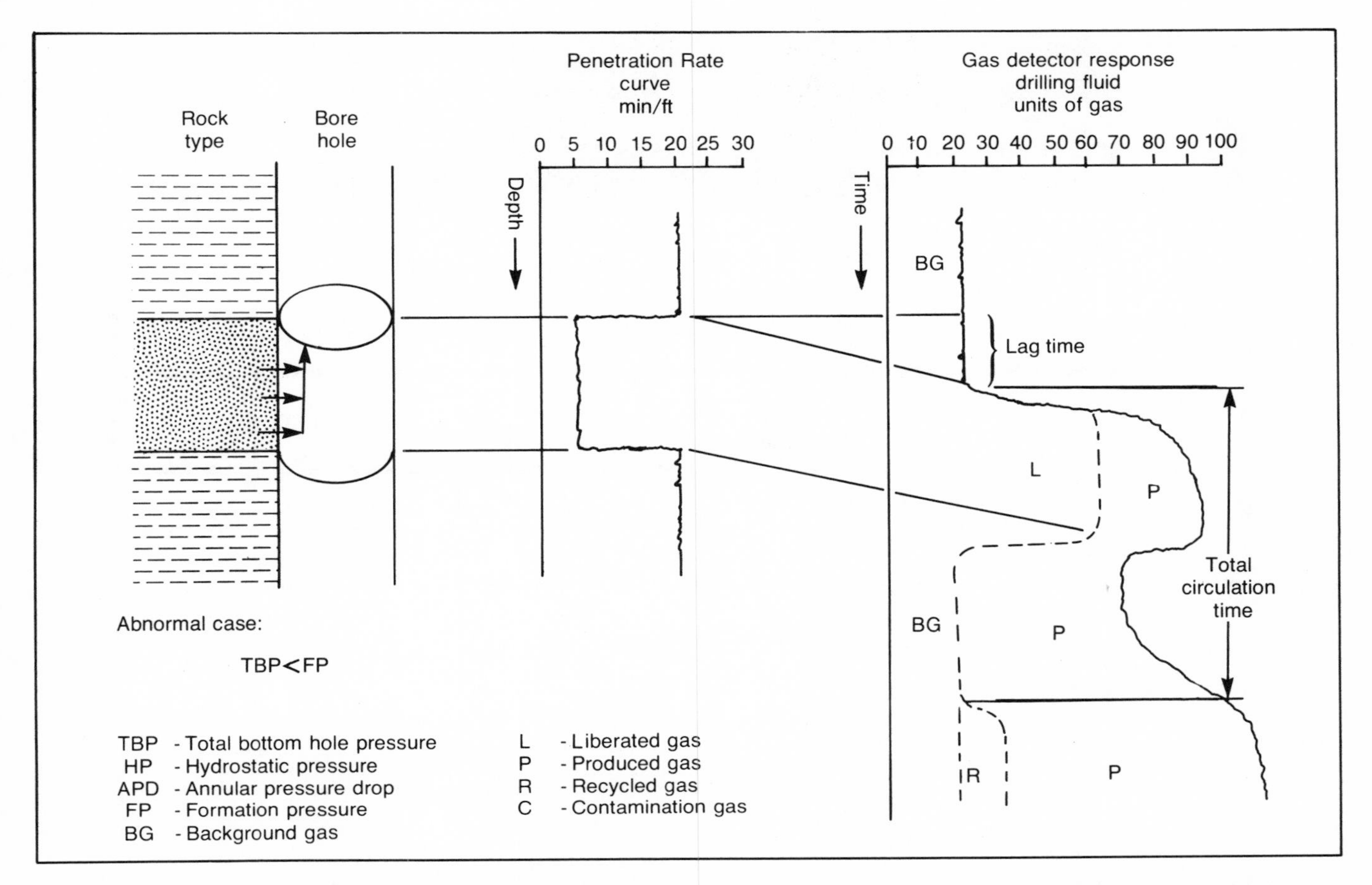

Fig. 3-20. Response as influenced by produced gas. Permission to publish by the Society of Professional Well Log Analysts.[13]

posed that the total gas units be calibrated at a standard of 100 units per 1% methane in air, which has been adopted in Canada.

DRILLING DATA ANALYSIS

Recent advances in computer hardware and rig instrumentation have spawned a rapid growth in the utilization of drilling data principally directed to drilling optimization and early warning of drilling problems. As early as 1973, over 80 drilling data collection and analysis units that utilized a computer were being used or were available.[12] Results of these efforts, however, provide the evaluator with valuable information applicable to formation evaluation.

One of the first major advances came in 1966 when Jorden and Shirley[4] proposed a mathematical method to normalize the rate of penetration, R. This dimensionless number, called the d-exponent, accounts for the variable influence of bit size, D in inches, weight on bit, W in pounds, and rotary speed, N in rpm. It is calculated from the following relation:

$$d = \frac{\log\left(\frac{R}{60N}\right)}{\log\left(\frac{12W}{10^6 D}\right)}$$

Of course, not all factors influencing the rate of penetration are accounted for, but it does provide a more understandable drilling response than simply that of the penetration rate. It is particularly useful in early detection of abnormal pressures. Mud weight variations, however, were still particularly troublesome, and a modified d-exponent has been introduced to normalize this effect:

$$d_c = d\left(\frac{MW_1}{MW_2}\right)$$

where: MW_1 = normal mud weight and MW_2 = actual mud weight used. A comparison of a d and d_c plot in the same well and their response to overpressure are shown in Figure 3-22.

Numerous drilling models have been developed that attempt to describe relationships between formation characteristics and drilling parameters. Applications of

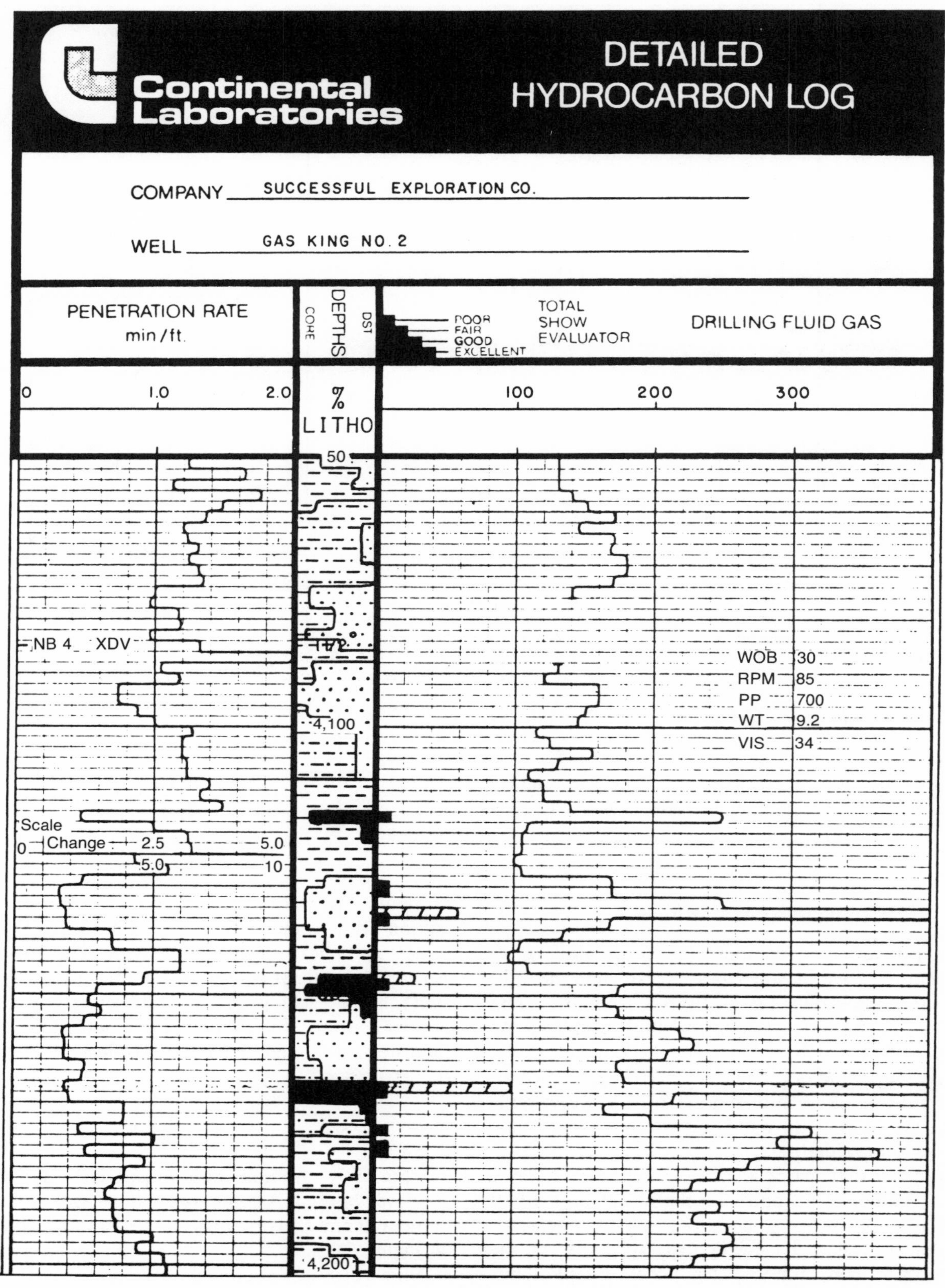

Fig. 3-21. Example of detailed hydrocarbon log. Permission to publish by the Society of Professional Well Log Analysts.[18]

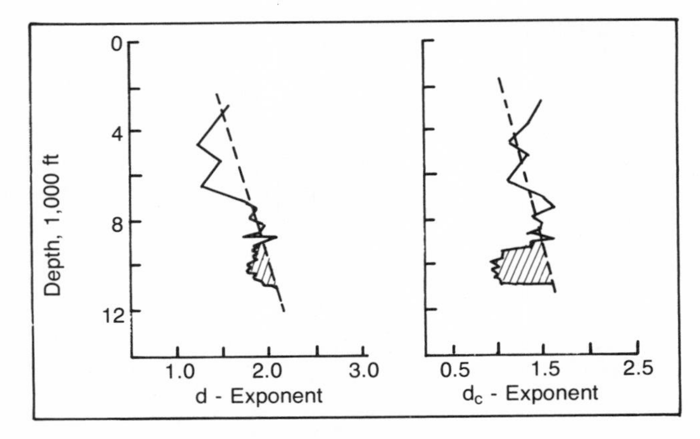

Fig. 3-22. Comparison of d and d_c exponents versus depth. Permission to publish by Elsevier Scientific Publishing Company.[16] Copyright 1967 SPE-AIME.

these models include predictions of formation porosity, drillability, pore pressure, etc. An example of porosity and pseudo-density as predicted by a model proposed by Bourgoyne[10] is shown in Figure 3-23.

Of course, many possible parameters can be monitored that can include not only drilling operations data but also a wide variety of sample mud data. Several of these have particular application as abnormal pressure indicators, such as shale bulk density and shale formation factor.

Shale Bulk Density

Bulk density increases in normally compacted shales with depth; observation of shale density with depth can be a good sensor of abnormal pressure. Bulk density of shales can be measured by a number of methods, such as by the high pressure mercury pump technique, fluid density gradient column, mud balance method, etc.

As discussed by Fertl,[16] the density gradient column method uses a mixture of two fluids of different densities in a graduated column such that the density of the mixture varies with column height. Shale cuttings density is determined every 5, 10, or 30 ft and even larger increments in soft, fast-drilling intervals by measuring the distance the selected drill cuttings settle in the density column. This value is then entered in proper calibration curves to obtain the corresponding shale density value.

Naturally, care must be taken in selecting and preparing the cuttings for analysis, which includes (1) washing, (2) screening to discard large cavings and/or smooth recirculated cuttings, and (3) air or spin drying until cutting surface has a dull appearance. Multiple cuttings have to be tested due to variance in sample data. The average density value, for a given time, is then plotted versus depth.

Mud balance measurements to determine shale bulk density are a faster and probably more representative method, because of the larger sample size involed. The procedure includes the following steps:

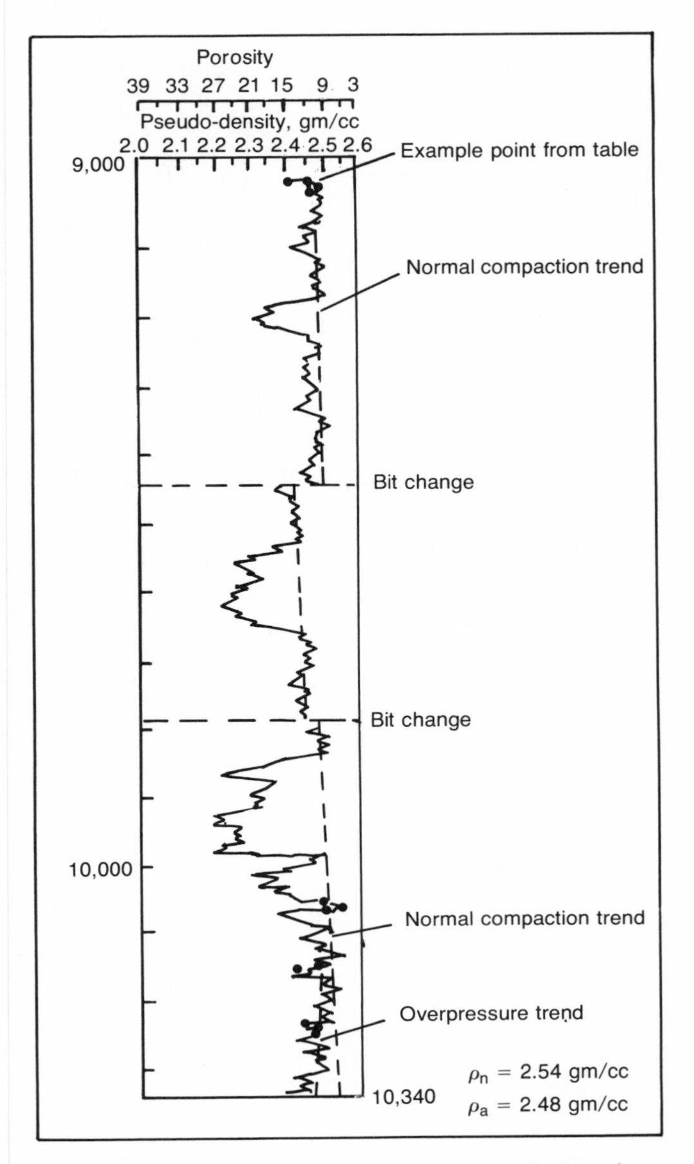

Fig. 3-23. Porosity and pseudo-density predictions from drilling data. Permission to publish by Petroleum Engineer International.[10]

1. Place shale cuttings into cup of the mud balance so that the balance indicates 8.3 lb/gal with the cap on;
2. Finish filling the cup with water and weigh again (W_2);
3. Shale bulk density then equals $[8.3/(16.7 - W_2)]$.

Shale bulk density values are then plotted versus depth, and normal compaction trend lines are established. These data can be used to forecast densities at greater depth. Since shale porosity increases in pressure transition and overpressure zones, any decrease in bulk

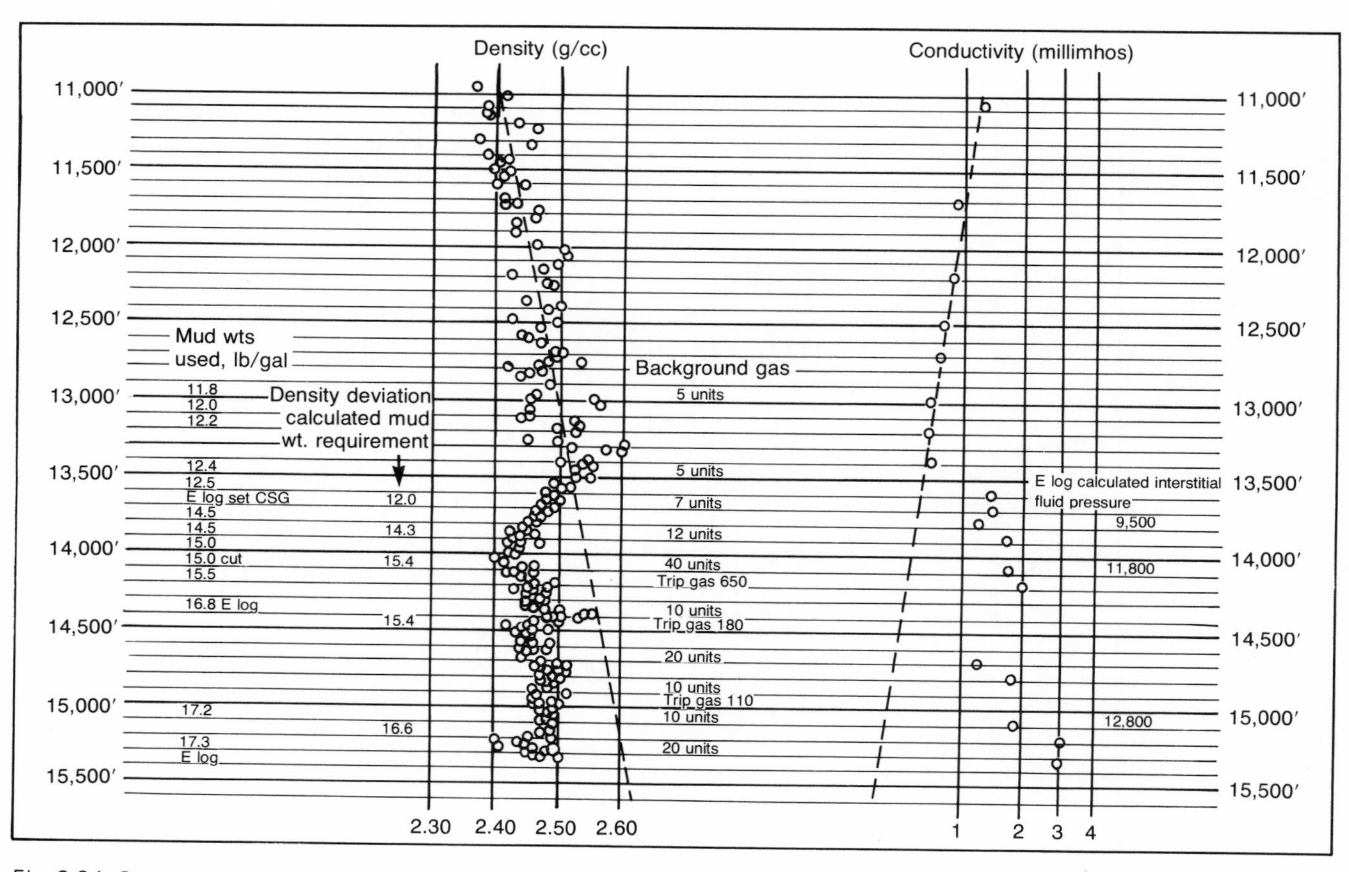

Fig. 3-24. Overpressure detection from shale density and log conductivity measurements. Permission to publish by World Oil *and Society of Petroleum Engineers.*[16]

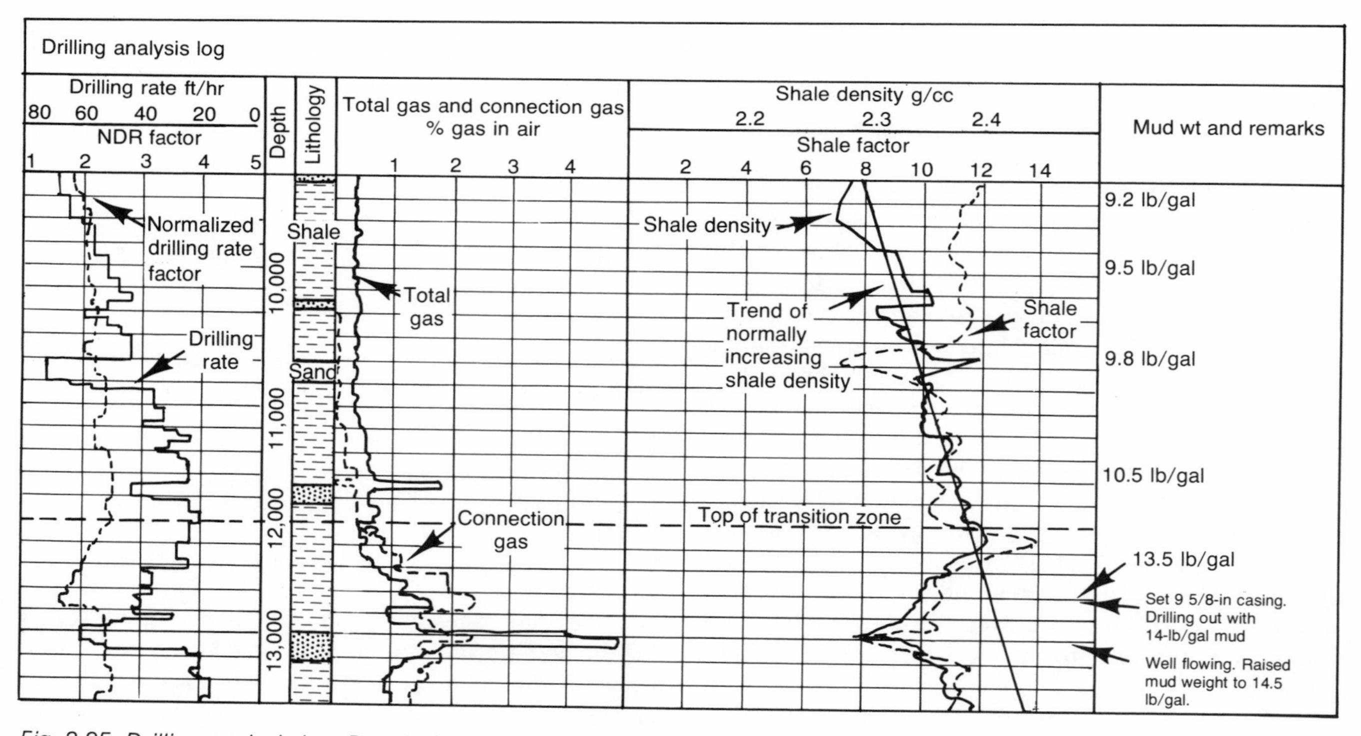

Fig. 3-25. Drilling analysis log. Permission to publish by the Oil and Gas Journal.[6]

density values may reflect the presence of such abnormal-pressure environments (see Fig. 3-24).

In principle, the cuttings density method is similar to bulk density measurements by well logs. However, the cuttings method has the advantage of a much shorter lag time, since the only delay is the bottom-up time during mud circulation.

Shale Factor

The methylene blue test, generally called the shale formation factor method, can provide useful information. Generally, this shale factor may be equated with the cation exchange capacity of solids carried out of the borehole, which in turn may be related to the montmorillonite content, and thus, the water-holding capacity of cuttings. The shale factor not only identifies clay composition, but it is also representative of the clay/sand abundance ratio in shale cuttings or, in general, the clay content of the section. Knowledge of the latter is of extreme importance when drilling thick carbonate sections with few shale streaks, where density data may easily give erroneous interpretations. The same is true in silty, sandy shales. More recently, shale factor logs have been used to locate formations that can cause severe borehole instability. The shale factor also appears to be a supplementary and useful method for the detection of the impermeable cap rock often found on top of high-pressure sections. An example of the shale density and shale factor applications in an abnormally pressured section is shown in Figure 3-25.

REFERENCES

1. *Recommended Practice Standard Hydrocarbon Mud Log Form*, API RP 34, American Petroleum Institute, Dallas (Nov. 1958).
2. Haun, John D., and LeRoy, L.W.: *Subsurface Geology in Petroleum Exploration*, Colorado School of Mines, Golden (1958).
3. Pearson, Alvin J.: "Miscellaneous Well Logs," *Petroleum Production Handbook, Volume 2, Reservoir Engineering*, Thomas C. Frick (ed.) Society of Petroleum Engineers, Dallas (1962).
4. Jorden, James R., and Shirley, Orval J.: "Application of Drilling Performance Data to Overpressure Detection," *J. Pet. Tech.* (1966).
5. Mercer, Richard F.: "The Use of Flame Ionization Detection in Oil Exploration," *Trans.*, CWLS (1968).
6. "New Drilling-Service Package Unveiled," Gene Kinney (ed.) *Oil and Gas J.* (Feb. 5, 1968).
7. Helander, Donald P.: "Drilling Mud-Logging Becoming More Important in Formation Evaluation," *Oil, Gas and Petrochem Equipment* (formerly *Oil and Gas Equipment*), PennWell Publishing Co. (Feb. 1969) 15.
8. Helander, Donald P.: "Drilling Mud-Logging Necessary for Periodic Well Evaluation," *Oil, Gas and Petrochem Equipment* (formerly *Oil and Gas Equipment*), PennWell Publishing Co. (March 1969) 15.
9. Helander, Donald P.: "Basic Concepts of Cuttings Analysis," *Oil, Gas and Petrochem Equipment* (formerly *Oil and Gas Equipment*), PennWell Publishing Co. (April 1969) 15.
10. Burgoyne, Adam T.: "A Graphic Approach to Overpressure Detection While Drilling," *Petroleum Engineer International* (Sept. 1971) 43.
11. Kennedy, John L.: "A New Device Aimed at Better Sampling of Drill Cuttings," *Oil and Gas J.* (Sept. 27, 1971).
12. Kennedy, John L.: "Drilling Issue," *Oil and Gas J.* (Feb. 1973) 71.
13. Mercer, Richard F.: "Liberated, Produced, Recycled or Contamination," *Trans.*, SPWLA (June 1974).
14. Wyman, Richard E., and Castano, J.R.: "Show Descriptions from Core, Sidewall and Ditch Samples," *Trans.*, SPWLA (June 1974).
15. Kennedy, Karl F., and Altman, T.D.: "Describing Lithology and Mud Log Data in Digital Form," paper presented at the CWLS 5th Formation Evaluation Symposium, Calgary, May 1975.
16. Fertl, Walter H.: *Abnormal Formation Pressures*, Elsevier Scientific Publishing Co., New York, 1976.
17. Mercer, Richard F.: "Gas-Cut Mud: Causes, Effects and Cures," *World Oil* (Sept. 1976).
18. McAdams, Joseph B., and Mercer, Richard F.: "Detailed Hydrocarbon Logs Enhance Formation Evaluation," *Trans.*, SPWLA (1977).
19. Weiss, Walter J.: "Drilling Fluid Economic Engineering," *Petroleum Engineer International* (Sept. 1977).
20. Wright, Thomas R.: "Guide to Drilling, Workover, and Completion Fluids," *World Oil* (Jan. 1977).

4 Basic Relations

RESISTIVITY is a physical property that can be defined as the electrical resistance of a cube of material, and can be represented by

$$R = \frac{rA}{l}, \ (\Omega m),$$

where: R = resistivity, Ωm
r = resistance, Ω
A = area, m^2
l = length, m

Cube dimensions, given in meters, provide rock and fluid resistivities having convenient magnitudes. Resistivity can also be stated in terms of conductivity where:

$$\text{Resistivity} = \frac{1}{\text{conductivity}}$$

and

$$\text{Conductivity, } C = \left(\frac{\text{siemens}}{\text{m}}\right).$$

Convenient conductivity units used in well logging are:

$$C\left(\frac{mS}{m}\right) = \frac{1{,}000}{R(\Omega m)}$$

ROCK RESISTIVITY

The conductivity of sedimentary rocks is produced by the movement of ions in the formation water. This conduction is electrolytic in nature and is different from metallic conduction, in which current flow is caused by the movement of electrons. Metallic conduction is approximately a million times better than electrolytic conduction. The basic laws of electricity apply to both cases, however.

Since rock conduction is essentially electrolytic in nature (metallic ore deposits that are electronic conductors are rarely encountered in oil and gas exploration), the well logs measuring resistivity can be considered to be salt water indicators. The resistivity of the undisturbed formation in situ is defined as the true formation resistivity R_t.

To understand the meaning of formation resistivity, however, it is imperative to understand the factors that influence rock resistivity, R_t: (1) formation water resistivity, R_w, (2) rock structure, and (3) the presence of hydrocarbons.

Influence of Formation Water Resistivity

In order for a water, and therefore a sedimentary rock, to conduct, ions must be present. The presence of ions is caused by the dissociation of salts, for example, NaCl dissociating into Na^+ and Cl^-. Current flow is created by subjecting the ions to an electrical potential. Since each ion carries a specific electric charge, the greater the number of ions in solution, the greater the current-carrying capacity; therefore, the greater the salinity of the water contained in a rock, the greater its conduction and conversely the lower its resistivity.

Temperature influences the electrolytic conductivity of a solution, since it greatly affects the movement of ions in solution. Also, ions are subject to a frictional resistance (viscous drag) as they move through the solution. The higher the water viscosity, the greater the frictional resistance or viscous drag on the ion, and the slower it moves. Water viscosity is a function of temperature, with increasing temperature lowering the viscosity. Therefore, when the temperature of salt water is increased, the ions can move faster. The conductivity of a rock containing salt water will, as a consequence, be greater as the temperature of the rock is increased. The relationship of resistivity, salinity, and temperature for a salt solution can be seen in Figure 4-1.

From the preceding discussion, the formation water

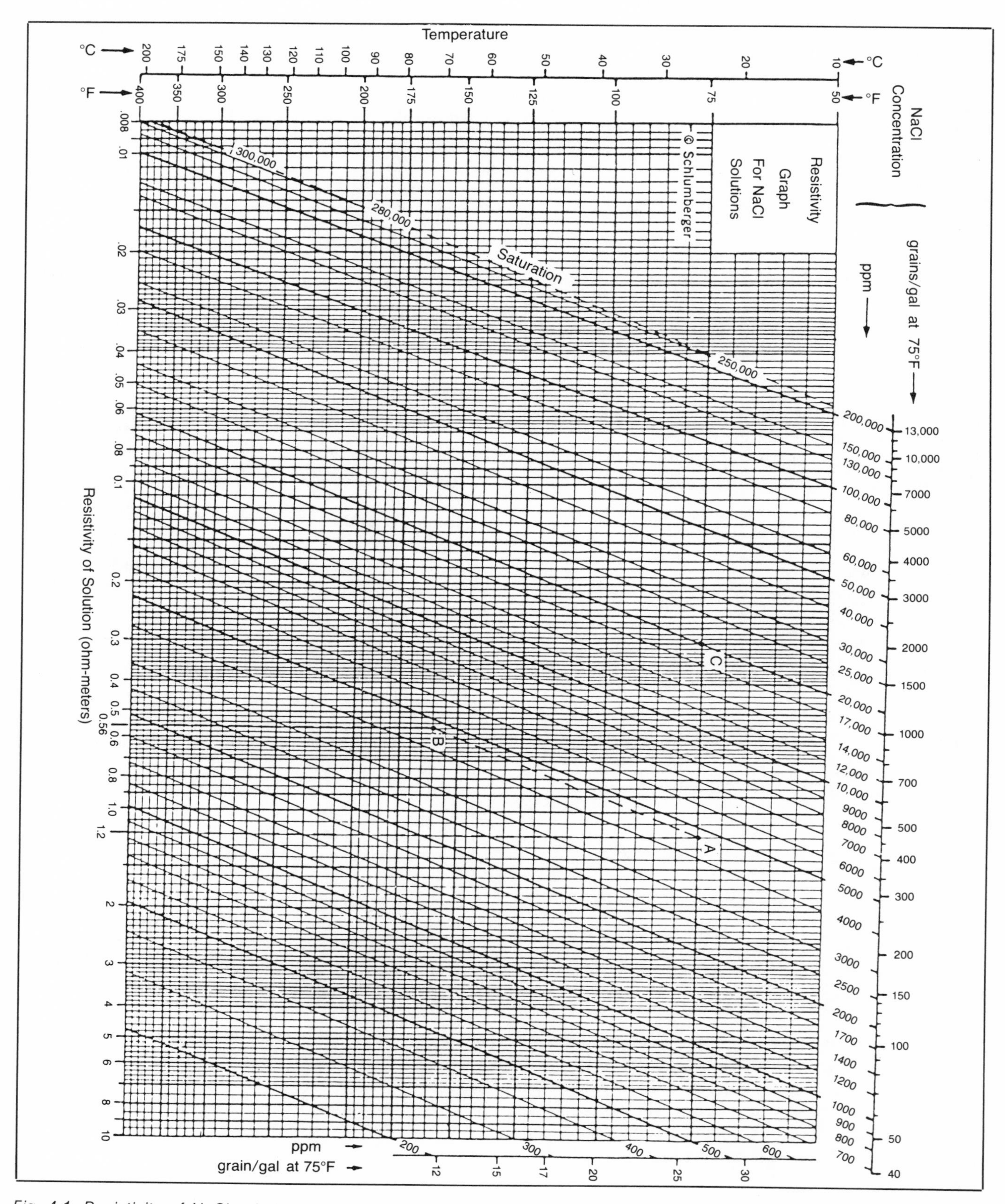

Fig. 4-1. Resistivity of NaCl solutions as a function of salinity and temperature. Permission to publish by the Oil and Gas Journal.[23]

resistivity, R_w, will directly affect the rock resistivity, R_t, and therefore:

$$R_t = f(R_w)$$

Since R_w directly influences R_t, it is desirable to know R_w. The formation water resistivity, R_w, may be determined in a number of ways: (1) by obtaining a good water sample and measuring R_w directly or converting a water analysis to R_w, (2) by calculating R_w from the spontaneous potential (see Chapter 5), (3) by determining R_w from water zone data (see Chapters 7, 8, and 9), or (4) by predicting R_w from catalogues of R_w data.

Direct Resistivity Measurement of Water Sample. Probably the most accurate method of determining R_w is by direct measurement of a good water sample. However, this is often not possible. In general, reliable water samples may be obtained from a separator or gun barrel of a high WOR well producing from the formation of interest or from the bottom joint of pipe on a drill stem test that produces a large amount of water. Gas wells, wells with casing leaks, or wireline formation testers would provide unreliable samples.

There are many instances in which a chemical analysis of the water sample is available but R_w was not measured. These analyses are usually reported in ionic concentrations, but occasionally molecular concentrations are presented. It is possible to predict R_w from

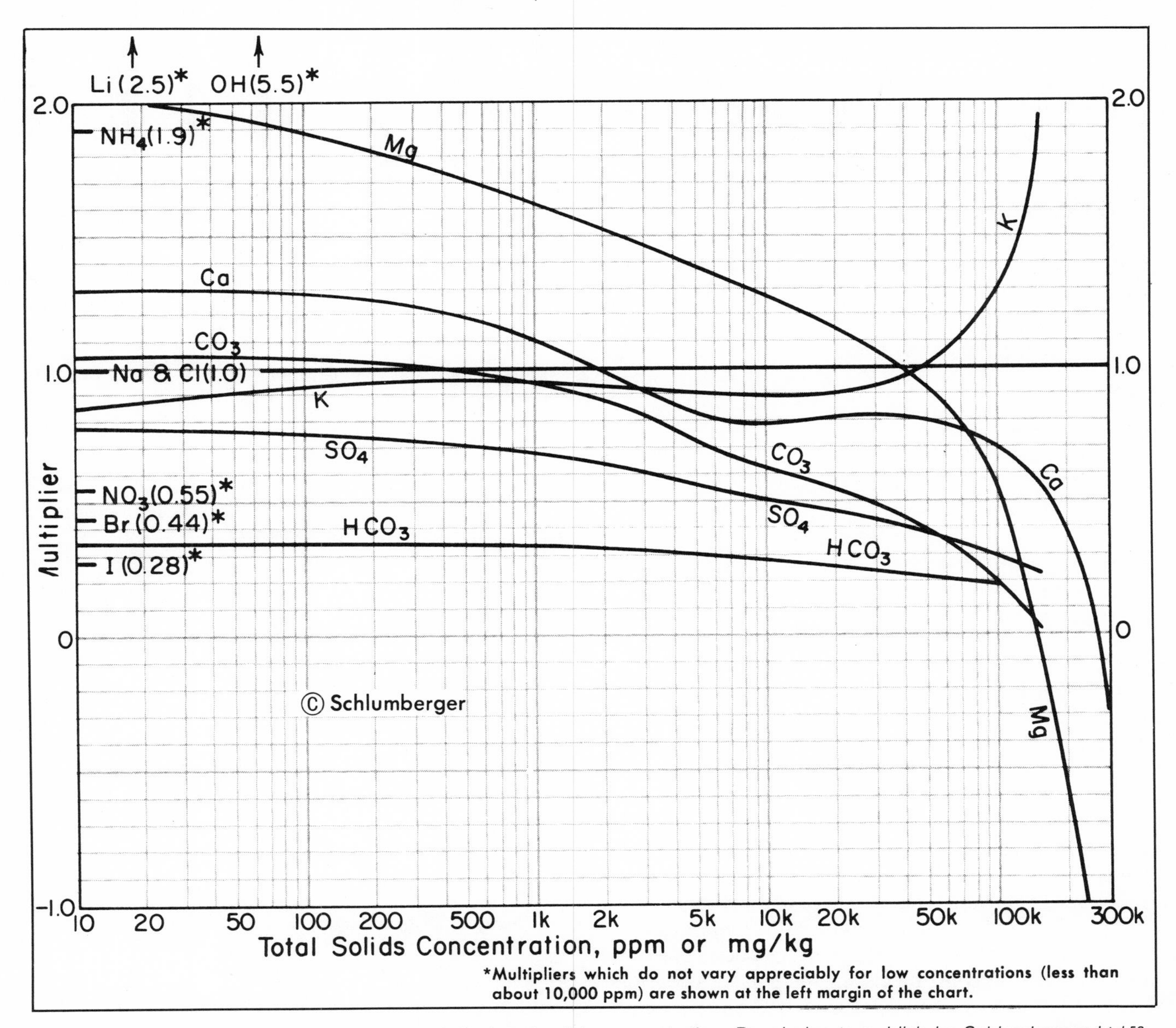

Fig. 4-2. Variable Dunlap multiplier as a function of total solids concentration. Permission to publish by Schlumberger Ltd.[53]

these data, using the variable Dunlap multiplier method or empirical equations.

Variable Dunlap Multiplier Method. Dunlap and Hawthorne[12] originally proposed a method in which the various ions reported in the chemical analysis were converted to an NaCl equivalent. This was accomplished by multiplying each ionic concentration by a conversion factor, resulting in an equivalent NaCl concentration in ppm from which R_w could then be determined where: ppm (parts per million) equals the ratio of the weight of the dissolved material to the weight of the solvent (water) multiplied by 10^6. Therefore, 10 grams of NaCl in 100 grams of water would be expressed as 100,000 ppm.

The conversion factors were determined by measuring resistivities of a number of simple solutions containing only two types of ions, one of which was always either sodium or chloride. The Dunlap and Hawthorne approach is limited, however, since, in solutions containing more than one salt, the contribution of the one salt to the total conductivity depends not only on the fractional concentration but also on the concentration of all other solutes. Therefore, the original Dunlap and Hawthorne technique has been modified by extensive research[47, 48, 52, 53] in order to incorporate the effect that varying ionic concentrations have on resistivity (see Fig. 4-2). This method can be illustrated in Table 4-1, in which R_w is determined at 170°F.

TABLE 4-1
Variable Dunlap Multiplier Method

Ionic Analysis	*ppm*	*Multiplier*	*Equivalent NaCl, ppm*
$Na^+ + Cl^-$	36,000	1.0	36,000
Mg^{++}	4,000	.92	3,680
Ca^{++}	3,500	.80	2,800
CO_3^{--}	2,500	.39	975
SO_4^{--}	4,000	.38	1,520
Total Solids =	50,000 ppm		44,975 ppm

$R_w = 0.068\ \Omega m$ @ 170°F (from Fig. 4-1)

The amount of dissolved minerals in formation waters varies from a few hundred ppm in basically fresh water to several hundred thousand ppm in the saltiest brines. Sea water has an average salinity of 35,000 ppm.

Empirical Equations. Calculating R_w from chemical analysis using a single equation is based on data obtained over a wide range of chemical concentrations.[43] The work of Hawkins et al.[41, 42] was selected as the basis for correlating the water resistivity with chemical analysis. Figure 4-3 is a plot of the concentration range for various chemical constituents for these data. The equations used for determining R_w are based on this range; extrapolation outside this range is not recommended.

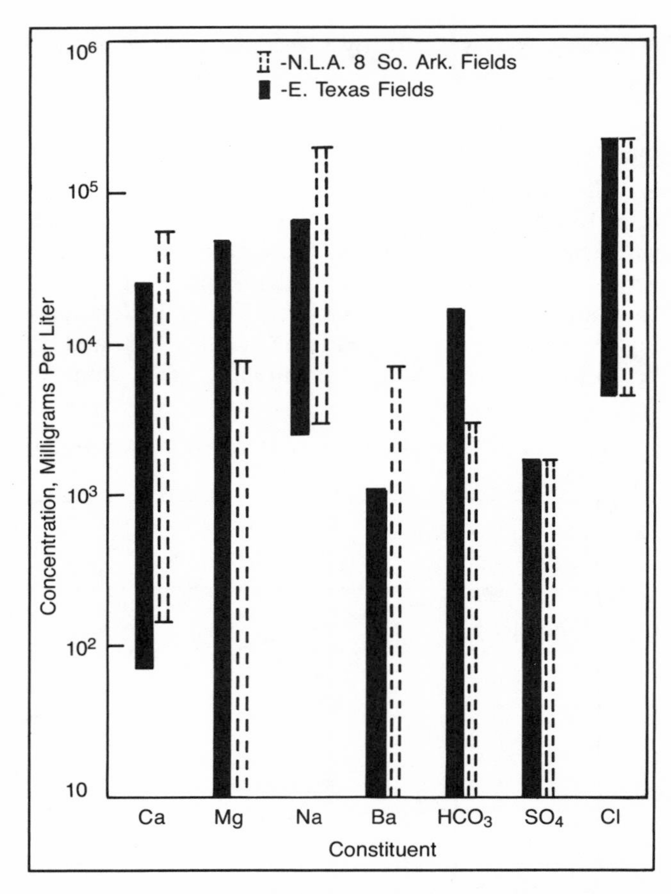

Fig. 4-3. Correlation range for empirical equation of Talash and Crawford. Permission to publish by the Society of Petroleum Engineers of AIME.[45] Copyright 1965 SPE-AIME.

Curve fitting the water resistivity, R_w, data shown in Figure 4-1 as a function of salinity and temperature, a relationship can be obtained from:

$$\text{Log } R_w = D + M_1 \log C + M_2 (\log C)^2 + \ldots$$

where R_w = water resistivity
C = NaCl concentration
M = slope of the straight line portion of the curve below 40,000 ppm
D = constant

This equation can be generalized to include all the mineral constituents found in a chemical analysis of brine, which include Ca, Mg, Na, Ba, HCO_3, SO_4 and Cl; their concentrations are denoted by C_1, C_2, C_3, C_4, C_5, C_6, and C_7, respectively. The equation can, therefore, be written as:

$$\text{Log } R_w = A_0 + \frac{A_1 \log C_1}{B_1(\log C_1)^2} + \frac{A_2 \log C_2}{B_2(\log C_2)^2} + \ldots + \frac{A_7 \log C_7}{B_7(\log C_7)^2}$$

Where the constants D, M_1, and M_2 are replaced by

TABLE 4-2
Generalized Resistivity Equation

R_w Antilogarithm $A_0 + A_1 \log eC_1 + A_2 \log eC_2 + A_1 \log eC_3$
$+ A_4 \log eC_4 + A_5 \log eC_5 + A_6 \log eC_6$
$+ A_7 \log eC_7 + B_1(\log eC_1)^2 + B_2(\log eC_2)^2$
$+ B_3(\log eC_3)^2 + B_4(\log eC_4)^2$
$+ B_5(\log eC_5)^2 + B_6(\log eC_6)^2 + B_7(\log eC_7)^2$

Constituent	*Symbol for Concentration*	*Coefficients**	
		$A_0 =$ 16.683128	
Ca	C_1	$A_1 =$ 0.11351271	$B_1 =$ −0.011957698
Mg	C_2	$A_2 =$ −0.027588759	$B_2 =$ 0.0
Na	C_3	$A_3 =$ 1.91185	$B_3 =$ 0.060854577
Ba.Sr	C_4	$A_4 =$ 0.010841613	$B_4 =$ −0.0016959557
HCO_3	C_5	$A_5 =$ 0.0	$B_5 =$ 0.0004970884
SO_4	C_6	$A_6 =$ 0.0	$B_6 =$ 0.0
Cl	C_7	$A_7 =$ 1.10986695	$B_7 =$ 0.051224888

* For 75°F.

Permission to publish by the Society of Petroleum Engineers of AIME.[45]

TABLE 4-3
Coefficients for the Generalized Equation with No Variables Eliminated

Constituent	*Symbol for Concentration*	*Variables*	*Coefficients**
—	—	—	$A_0 =$ 16.361336
Ca	C_1	$\ln C_1$	$A_1 =$ 0.13002054
Mg	C_2	$\ln C_2$	$A_2 =$ 0.066758057
Na	C_3	$\ln C_3$	$A_3 =$ 1.8343853
Ba.Sr	C_4	$\ln C_4$	$A_4 =$ 0.01145093
HCO_3	C_5	$\ln C_5$	$A_5 =$ −0.0032474232
SO_4	C_6	$\ln C_6$	$A_6 =$ 0.002358852
Cl	C_7	$\ln C_7$	$A_7 =$ −1.1003902
Ca	C_1	$(\ln C_1)^2$	$B_1 =$ −0.013120794
Mg	C_2	$(\ln C_2)^2$	$B_2 =$ 0.0032216627
Na	C_3	$(\ln C_3)^2$	$B_3 =$ 0.056792218
Ba.Sr	C_4	$(\ln C_4)^2$	$B_4 =$ 0.0017810766
HCO_2	C_5	$(\ln C_5)^2$	$B_5 =$ 0.0010285636
SO_4	C_6	$(\ln C_6)^2$	$B_6 =$ 0.000035688437
Cl	C_7	$(\ln C_7)^2$	$B_7 =$ 0.051451065

* For 75°F.

Permission to publish by the Society of Petroleum Engineers of AIME.[45]

the constants A_1 through A_7 and B_1 through B_7, it can be written as:

$$\text{Log } R_w = A_0 + \sum_{i=1}^{7} A_i \log C_i + \sum_{i=1}^{7} \overline{B_i(\log C_i)^2}$$

In order to find R_w directly, therefore, the equation becomes:

$$R_w = \text{Antilog}\left[A_0 + \sum_{i=1}^{7} A_i \log C_i + \sum_{i=1}^{7} \overline{B_i(\log C_i)^2}\right]$$

The coefficients A_0, A_i, and B_i can be obtained by using multiple regression analysis. Table 4-2 indicates the generalized resistivity equation and the coefficients obtained in the final regression analysis for the North Louisiana and South Arkansas data. The average deviation, using the coefficient in Table 4-2 for the generalized resistivity equation was ±3.77 percent. Some of the variables were eliminated as their coefficients turned out to be zero.

Table 4-3 indicates the coefficient for the generalized equation with no variables eliminated and is preferred over Table 4-2 if the SO_4 and HCO_3 concentrations are unusually high. The data used in the correlation analysis were obtained at 75°F. Conversion of R_w to other temperatures can be obtained by using the information in Figure 4-1.

Catalogues of Water Resistivity Data. Water resistivity data have been compiled and published by numerous sources; most companies compile their own water resistivity data. The application of these data to project R_w should be used cautiously since both salinity and temperature generally increase with depth.

Influence of Rock Structure on Rock Resistivity

The matrix of a rock is, for all practical purposes, nonconducting, the only conductor being the salt water solution contained in the pore space. Therefore, the conductivity of a rock depends upon its interconnected solution-filled porosity. As a result, the resistivity of a rock, R_t, would be expected to vary not only as the formation water resistivity, R_w, in the pore space varies but also as the pore space varies from rock type to rock type. We might state, therefore, that:

$$R_t = f(R_w, \text{Rock Structure})$$

The influence of rock structure (pore geometry) can be analyzed in the following manner. Consider a water-

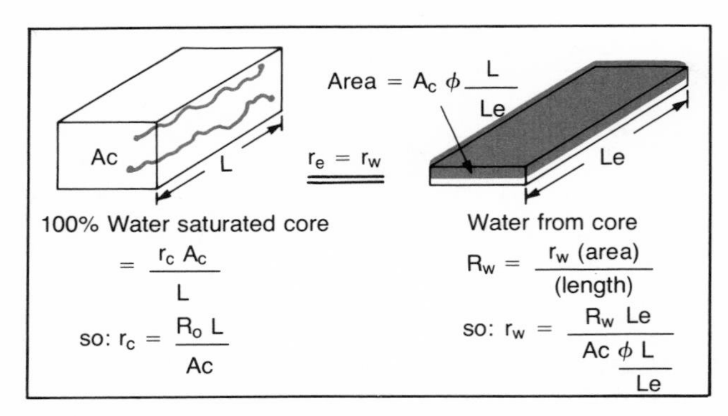

Fig. 4-4. Water-saturated core and equivalent water volume

filled core, as shown in Figure 4-4, and at the same time consider a "pseudo core" made up of water from the core that has the same resistance as the core, i.e., $r_c = r_w$. Now the true resistivity of this core saturated 100% with water would be:

$$R_t = R_o = \frac{r_c A_c}{L}$$

in which the true resistivity of the rock (in this case the core) saturated 100% with water is a special case of true rock resistivity and designated as R_o. The resistance of the core can be expressed as:

$$r_c = \frac{R_o L}{A_c}$$

The volume of the core is $V_c = A_c L$ and the volume of water from the core can be expressed as:

$$V_{cw} = V_c \phi = A_c L \phi = \underbrace{A_c \phi \left(\frac{L}{L_e}\right)}_{\text{Area}} \ \underbrace{L_e}_{\text{Length}}$$

Having an expression for the equivalent length and area of the water volume from the core we can express its resistance as:

$$r_w = \frac{R_w L_e}{A_c \phi \left(\frac{L}{L_e}\right)}$$

Now since $r_c = r_w$ the two expressions for resistance may be equated:

$$\frac{R_o L}{A_c} = \frac{R_w L_e}{A_c \phi \left(\frac{L}{L_e}\right)}$$

which results in:

$$\frac{R_o}{R_w} = \left(\frac{L_e}{L}\right)^2 \frac{1}{\phi}$$

The resistivity ratio R_o/R_w is defined as the formation resistivity factor, F_R, and the ratio L_e/L is defined as the tortuosity, τ. We can then write:

$$\frac{R_o}{R_w} = F_R = \left(\frac{L_e}{L}\right)^2 \frac{1}{\phi} = \frac{\tau^2}{\phi}$$

Two important things can be observed from this relationship: (1) The formation resistivity factor, F_R, is not only a function of the porosity, ϕ, but also the tortuosity, τ, of the rock. In other words, F_R is dependent upon the manner in which the pores are interconnected in addition to the total voidage; (2) The formation resistivity factor, F_R, should not be dependent upon the resistivity of the saturating fluid as long as it completely fills the void space. We should be able to fill the pore space with salt water, fresh water, or conductive coffee and get the same value of F_R.

A more general definition of F_R, therefore, would be:

$$F_R = \frac{\text{Resistivity of a Rock 100\% Saturated with a Conductive Fluid}}{\text{Resistivity of the Conductive Fluid}}$$

Therefore:

$$F_R = \frac{R_o}{R_w} = \frac{R_{xo}}{R_{mf}} = \frac{R_{(matrix+coffee)}}{R_{coffee}}$$

The range of F_R in rocks encountered when logging is from a minimum of about five to a maximum of several hundred in sandstone and from a minimum of about ten to values of many thousands in limestones and dolomites.

Natural rock systems, however, essentially preclude a mathematical treatment for the development of a definitive relationship between F_R and ϕ; therefore, the only avenue of approach has been the empirical analysis of experimental data. The first definitive work was done by Archie[2] in which he observed that the formation factor, F_R, was related to porosity, ϕ (see Fig. 4-5), in the form:

$$\log F_R = \log a - m \log \phi$$

$$F_R = a\phi^{-m}$$

or where m represents the slope of the linear trend of F_R vs. ϕ as observed for specific rock types when plotted on logarithmic graph paper and the intercept $a = 1$. Archie stated that the slope, m, appeared to vary for different rock types as a function of the degree of cementing of the rock. Hence, this slope, m, is generally referred to as the "cementation factor," which is a misnomer since it varies as a function of many factors.

The cementation factor, m, varies over a wide range of values, from 1.3 to approximately 2.8 (where $a = 1$). Table 4-4 shows the generalized range of cementation factors, m, for various rock types, and Figure 4-6 graphically illustrates the Formation Resistivity Factor, F_R, and porosity, ϕ, relationship for a wide range of rock types (wide range of m since the cementation factor is a rock typing parameter).

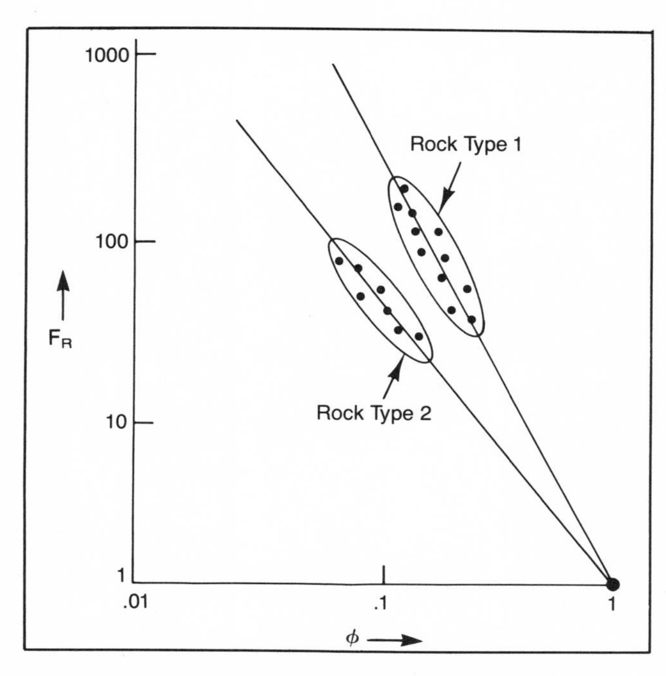

Fig. 4-5. Linear F_R vs. ϕ trends for different rock types

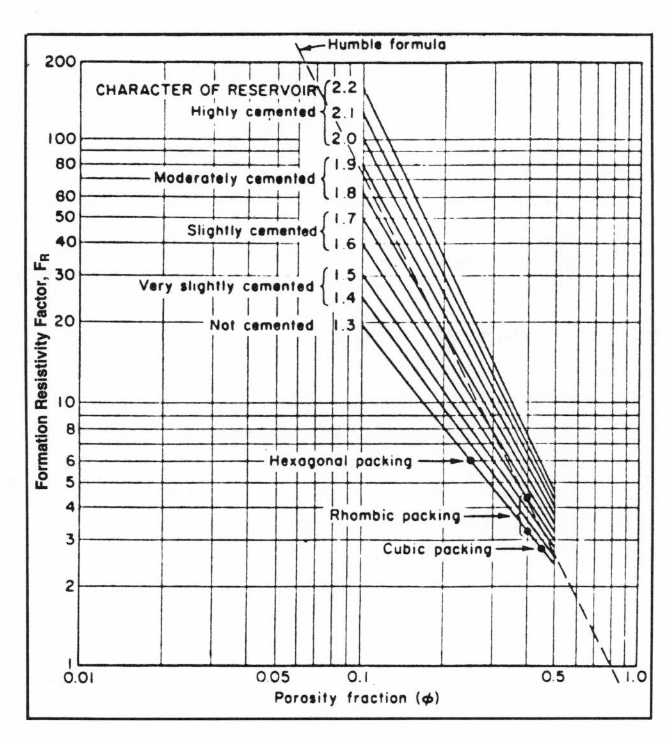

Fig. 4-6. Formation resistivity factor vs. porosity as a function of m. From Oil Reservoir Engineering *by Sylvain J. Pirson. Copyright 1958, McGraw-Hill Book Company. Used with the permission of McGraw-Hill Book Company.*[30]

For limestones and dolomites, the expression generally used is:

$$F_R = \phi^{-2} = \frac{1}{\phi^2}$$

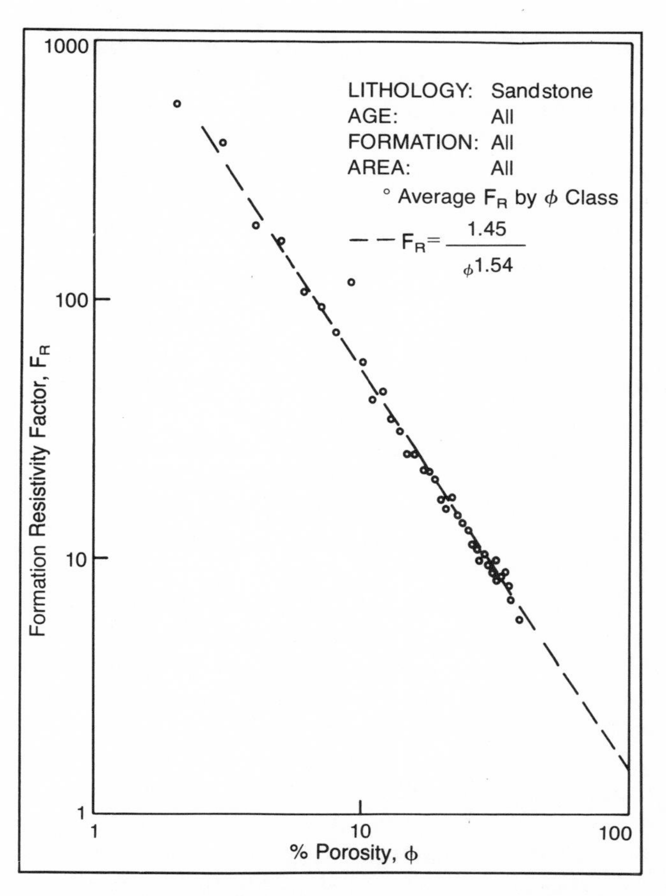

Fig. 4-7. Formation resistivity factor vs. porosity for sandstone cores. Permission to publish by the Society of Professional Well Log Analysts.[50]

TABLE 4-4
Cementation Factors for Various Rock Types

Lithology	*m*
Sandstones	
loose uncemented sand	1.3
slightly cemented sand	1.3–1.7
moderately cemented sand	1.7–1.9
well-cemented sand	1.9–2.2
Limestones	
moderately porous limestones	2
some oolitic limestones	2.8

The practical application of the $F_R = f(\phi)$ relation for a particular rock type is best accomplished by evaluating the cementation factor using laboratory-measured values of the formation resistivity factor and porosity. This data is plotted on log-log graph paper, and the best-fitting linear trend defines the slope, *m*, and intercept, *a*, as shown in Figure 4-7. An empirical relation that has also received widespread use is one developed by Winsauer, et al.[15] where *a* and *m* were defined such that:

$$F_R = 0.62\phi^{-2.15}$$

This relationship has been found to be particularly representative of average sandstones and has been extensively used for that purpose.

The reason for the observed variation in cementation factor has been attributed to a number of different factors: (1) degree of cementation,[3, 21] (2) shape, sorting, and packing of the particulate system,[21, 35] (3) type of pore system—intergranular, intercrystalline, vuggy, etc.,[13, 15, 33] (4) tortuosity of the pore system,[11, 14, 15, 28, 40] (5) constrictions existing in porous system[14, 40] (6) presence of "conductive solids,"[9, 10, 20, 55] (7) compaction due to overburden pressure,[25, 31, 32, 36, 37, 38, 39, 43, 46] (8) thermal expansion.[43, 46, 54]

The complexity of natural rocks is aptly indicated by this general listing. As implied by these factors, no single concept or factor adequately describes the relative behavior of the formation resistivity factor with porosity.

Tortuosity is one of the most popular concepts for explaining this variation in cementation factor. This term has some merit, however, since it attempts to define the pore passages in the medium. The tortuosity coefficient as defined by Pirson is "a conceptual dimensionless number representing the departure of a porous system from being made up by a bundle of straight-bore capillaries."[30] The tortuosity coefficient is, therefore, a measure of the tortuous path available for current flow, with respect to the direct path available in a conductive solution. Using this concept alone to explain the relationship of pore geometry to the cementation factor implies that the increased resistance in some rocks having the same porosity is due to one having more tortuous passages than the other. Hence, all increased formation factors can be accounted for merely by increasing the value of the tortuosity.

There are limitations to ascribing the higher formation resistivity factors solely to tortuosity. Two important contributions, which are primarily theoretical in approach, provide a greater insight into the influence of pore geometry. These studies by Owen[14] and Towle[40] utilized synthetic pore models that could be considered on a mathematical basis. Although the systems are greatly oversimplified, the significant concept of pore constriction, which is the effect on resistivity of variations in cross-sectional area of the conducting path, was pointed out. A porous system, employing both the concepts of tortuosity and pore constriction, is stated by Owen as being "more nearly analogous to conditions within a natural porous body than is a uniform diameter tube system which requires high tortuosity values to explain large formation factors."[14]

The particular utility of these interrelated concepts of tortuosity and pore constriction are not immediately apparent except for the more logical explanation of pore

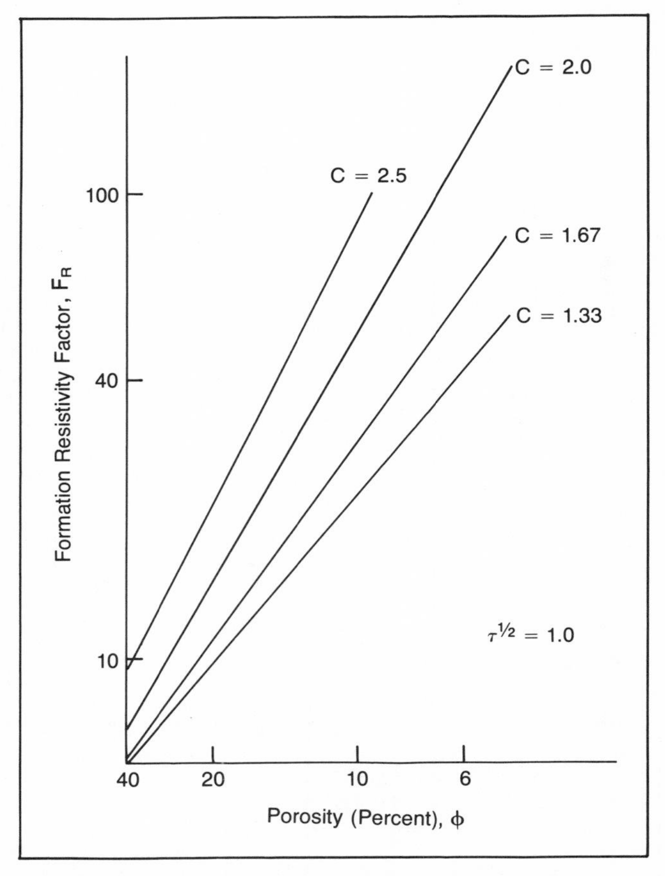

Fig. 4-8. Formation resistivity factor vs. porosity for constant tortuosity and various constriction factors. Permission to publish by the Society of Petroleum Engineers of AIME.[14] Copyright 1952 SPE-AIME.

geometry effects on formation resistivity. As pointed out by Owen, an infinite number of combinations of tortuosity and constriction factors can be chosen to give a particular porosity. Figure 4-8 illustrates how the formation resistivity factor increases with an increase in constriction factor at a constant value of tortuosity. The usefulness of this concept cannot be fully appreciated, however, until a logical explanation of the variation of formation resistivity under pressure is attempted.

The effect of "conductive solids" on the cementation factor has been noted by Patnode and Wyllie,[10] de Witte,[9] Winsauer and McCardell,[20] and Evers and Iyer.[55] The term "conductive solids" is a misnomer since, as pointed out by Winsauer and McCardell, the effect is actually caused by the double-layer conductivity associated with the highly charged clay surfaces. This double-layer conductivity reduces the cementation factor as the conductivity of the saturating solution decreases. This effect can be essentially negated, when measuring the formation resistivity factor at a static condition, by increasing the conductivity of the saturating solution. If the conductivity of the saturating solu-

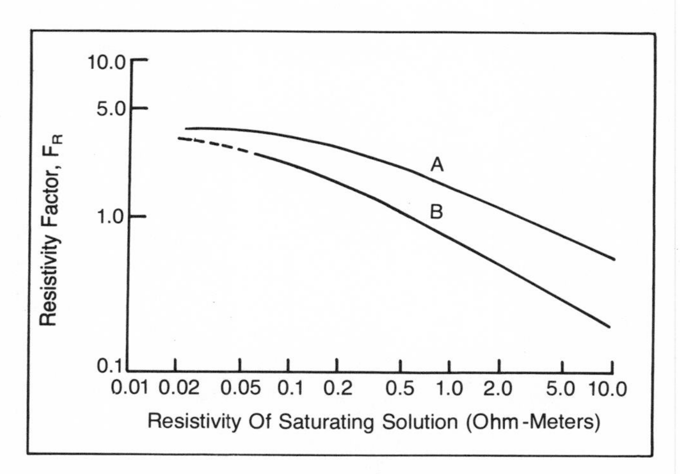

Fig. 4-9. Formation resistivity factor variations in shaly sands due to varying saturating solution resistivity. Permission to publish by the Academic Press.[28]

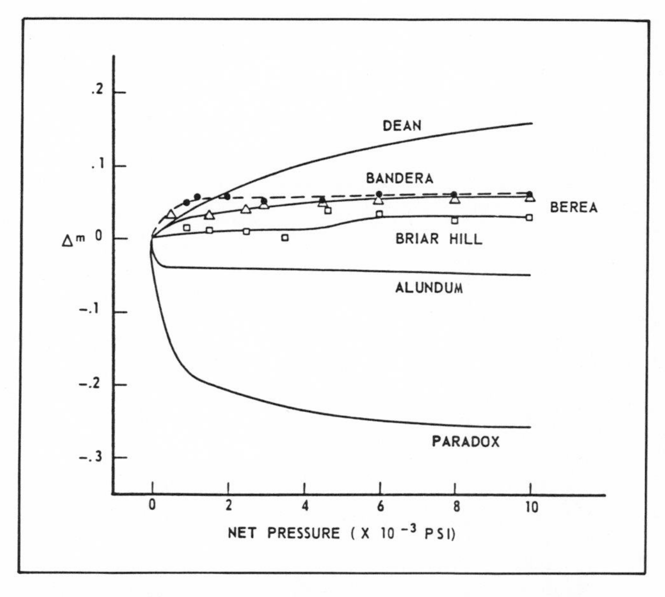

Fig. 4-10. The change in m ($F_R = \phi^{-m}$) with net pressure. Permission to publish by D.W. Hilchie.[43]

tion becomes high enough, a constant value of formation resistivity factor will be attained, and the resultant value of the cementation factor is assumed to reflect the true tortuosity and constriction factor of the formation. Figure 4-9 graphically illustrates the influence of the "conductive solids" on static formation resistivity factor measurements.

Up to this point, consideration has been given to the factors affecting the formation resistivity factor-porosity relationship at a static condition of pressure and temperature (for example, atmospheric conditions or in situ conditions). In other words, we have been referring to the effect of the geologic factors as applied to a particular state at which the core exists. Two other parameters, pressure and temperature, affect this formation resistivity factor-porosity relationship when comparing the results obtained on a sample at two different states. This is particularly important since experimental observation indicates that the change in formation resistivity factor with pressure and temperature is more rapid than can be accounted for solely on the basis of porosity change (see Fig. 4-10).

There are two primary reasons why the formation resistivity of subsurface samples is measured at atmospheric pressure and ambient temperature in the laboratory: (1) resistivity measurements at increased pressure and temperature are rather difficult and time consuming; (2) until recently, wells were relatively shallow, and the effects of pressure and temperature did not appear to be so critical.

Knowledge of formation resistivity behavior with pressure change is important because of the stress distortion occurring in the formation surrounding the wellbore. This wellbore stress distortion has been theoretically discussed by Hubbert[26] and its influence on quantitative log analysis briefly considered by Glanville.[31] The analysis by Hubbert indicates that the horizontal stress around the wellbore may be substantially relieved with respect to the stress actually existing in the formation. This distortion, however, may only exist a few hole diameters into the formation and could be misleading if a tool having a short depth of investigation is used in obtaining the basic resistivity data for a porosity determination. A tool having a deeper depth of investigation would probably see the formation in more or less its original state.

These problems could be relieved if all the formation samples taken for formation resistivity factor-porosity data were studied at elevated pressure and temperature. The most practical approach, however, would be to continue taking formation resistivity measurements at atmospheric pressure and ambient temperature, and then, in conjunction with additional data, correct the surface-measured resistivity values to existing formation conditions. However, at the present time no dependable, general correlations are available.

A second approach would be to study under elevated pressure and temperature only those samples that would probably exhibit a fairly significant change in formation resistivity factor at these increased conditions. Even to do this, it is necessary to have a reliable indicator to distinguish those samples in which there is variation in formation resistivity at elevated pressure and temperature conditions. For the most part, studies of the effects of pressure have been primarily qualitative. The information in Table 4-5 indicates the limited data in this area.

The "cementation" factor proposed by Archie represents the effect of all the numerous, interrelated param-

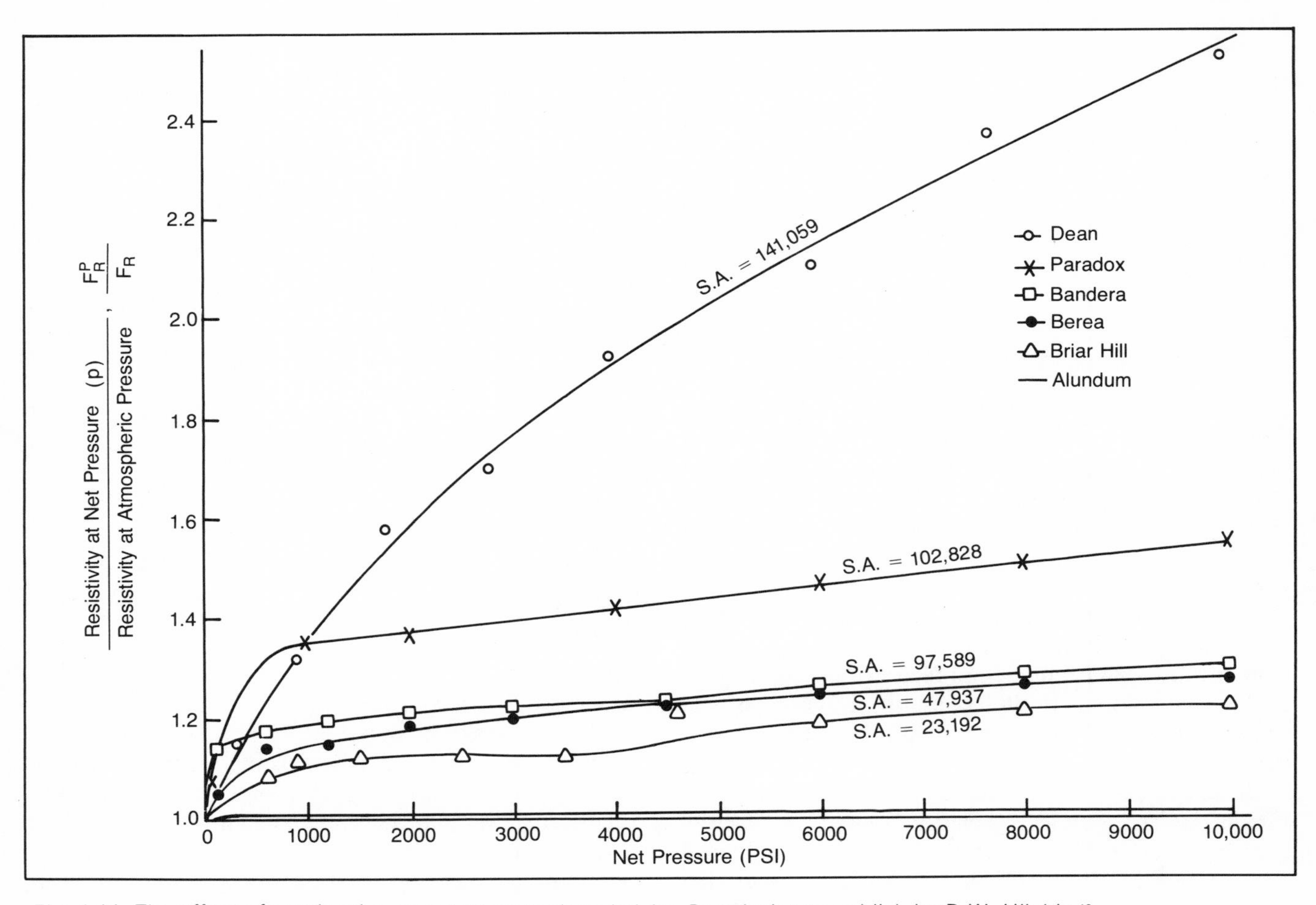

Fig. 4-11. The effect of overburden pressure on rock resistivity. Permission to publish by D.W. Hilchie.[43]

TABLE 4-5
Some Previous Studies of Rock Resistivity Under the Influence of Pressure

Investigator	*No. of Samples and Type*	*Net External Pressure Range Studied*
Fatt[25]	20 Sandstones	0–5000 psi
Wyble[32]	3 Sandstones	0–5000 psi
Glanville[31]	2 Sandstones	0–5000 psi
	3 Carbonates	0–5000 psi
Redmond[39]	4 Sandstones	0–20,000 psi
Glumov and Dobrynin[37]	1 Sandstone	0–350 Atmospheres
	1 Limestone	0–350 Atmospheres
Orlov and Gimaev[38]	2 Carbonates	0–400 Atmospheres
Dobrynin[36]	2 Sandstones	0–5000 psi
*Hilchie[43]	3 Sandstones	0–10,000 psi
	1 Shale	0–10,000 psi
	1 Limestone	0–10,000 psi
	1 Artificial (Alundum)	1–10,000 psi

* See Figure 11.

eters that are characteristic of natural rocks. The variation in the cementation factor, which is necessary to correlate the formation resistivity factor with porosity for consolidated medium, can be effectively ascribed to tortuosity (τ), pore constriction factor (PC), and electrical constriction factor (EC), where $m=f(\tau$, PC, EC) or $F_R=f(\phi, \tau$, PC, EC). A true formation resistivity factor that is representative of the pore volume may be obtained from a highly saline saturating solution, which negates the effects of the "conductive solids" (boundary layer). The saline solution will essentially nullify the influence of the electrical constriction and reduce the relationship to: $F_R=f(\phi, \tau$, PC).

Subjecting a rock to pressure will reduce the porosity, but the formation resistivity factor changes at a more rapid rate than can be accounted for solely by the change in porosity (see Fig. 4-11). This indicates that the other functional parameters (τ, PC, EC) are also changing. A prediction of porosity at another pressure condition, therefore, requires a method of predicting the influence of these three parameters.

It was found that a definite relationship existed between the theoretically calculated surface area and the change in relative formation resistivity factor.[46] The surface area was qualitatively shown to be closely associated with each of the three influencing parameters such that (τ, PC, EC) $= f$(SA). The separate effects of these three parameters, as associated with the microscopic mechanisms occurring during compression, cannot be dissociated from each other. Nevertheless, their combined effect can be satisfactorily ascertained by the surface area concept. This indicates that an equation of the form $F_R^P/F_R = f$(SA) will provide a suitable approach for predicting the effect of pressure on the formation resistivity factor.

The effects of temperature on the formation resistivity have been evaluated by Hilchie[43] and Sanyal, et al.[54] Hilchie found that as the temperature increased while the net pressure remained constant, the formation resistivity factor went through a minimum value after which it increased. He presented a method for predicting the effects of temperature if the percent pore volume, consisting of pores less than 0.5 microns, and the minimum temperature were known. This correlation should be used with caution, however, since the behavior of natural rocks under the influence of temperature is even less well defined than pressure. Further work in this area is required in order to properly evaluate this effect.

The combined effects of temperature and pressure were also studied by Hilchie.[43] One important observation was that, in general, the additive results of the separate pressure and temperature data were equal to the combined pressure and temperature data at low and moderate temperatures.

Influence of Hydrocarbons on Rock Resistivity

We have seen the influence of formation water resistivity, R_w, and rock structure on the true resistivity of a rock:

$$R_t = f(R_w, \text{Rock Structure})$$

and since the influence of rock structure is represented by the formation resistivity factor, F_R, this might be stated as:

$$R_t = f(R_w, F_R)$$

This certainly is the case for a 100% water saturated rock where $R_t = R_o$ since:

$$F_R = \frac{R_o}{R_w} \quad \text{or} \quad R_o = F_R R_w$$

When oil or gas replaces water in rock pores, the resistivity of the rock increases, since the conducting medium (water) is replaced by a nonconducting medium (oil or gas). As the percentage of oil or gas increases, the resistivity of the rock increases, although since most rocks are water-wet, the continuous coating of water on the rock surface will provide a conductive path even when large amounts of oil and gas are present. The presence of oil or gas makes the rock more resistive than

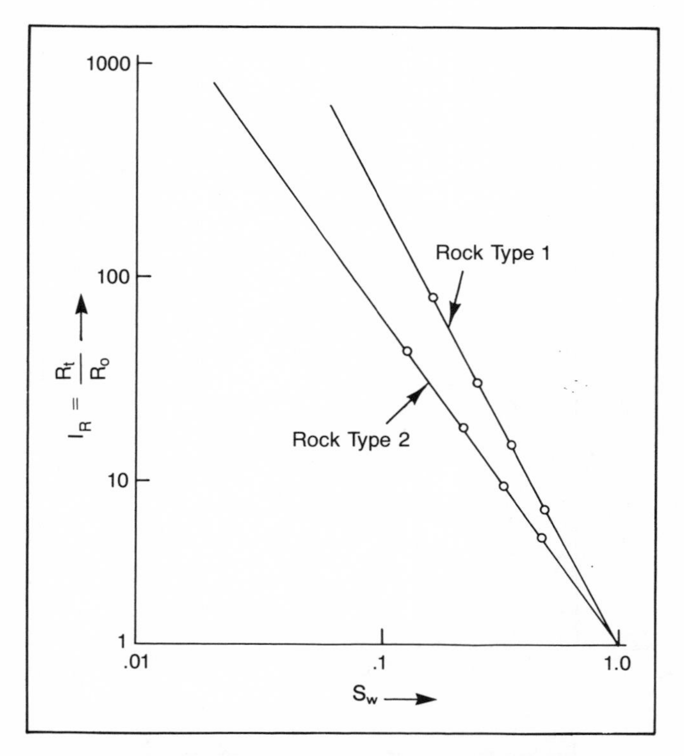

Fig. 4-12. Illustration of linear I_R vs. S_w trends for different rock types

when the same rock is water-saturated. Therefore, the functional relationship of the true rock resistivity should be expressed as:

$$R_t = f(R_w, F_R, \text{hydrocarbon influence})$$

Since the resistivity of a rock is of no significance with respect to oil or gas saturation unless the resistivity of the same rock when 100% water-saturated can be determined, a comparison of these two resistivities provides the key to the presence of oil or gas. Therefore, the Resistivity Index may be defined as:

Resistivity Index $= I_R$

$$I_R = \frac{\text{Resistivity of a Rock Containing Oil or Gas}}{\text{Resistivity of the Same Rock Containing Only Water}}$$

$$I_R = \frac{R_t}{R_o}$$

This relationship assumes that the salinity and temperature of the water are unchanged.

Experimental work by Archie[2] indicated that the Resistivity Index, I_R, was related to water saturation, S_w, in clean rocks (see Fig. 4-12). These linear trends can be expressed as:

$$\log I_R = \log \frac{R_t}{R_o} = -n \log S_w$$

since obviously the intercept is equal to one. For average water-wet rocks, the slope, n, is approximately 2 and is

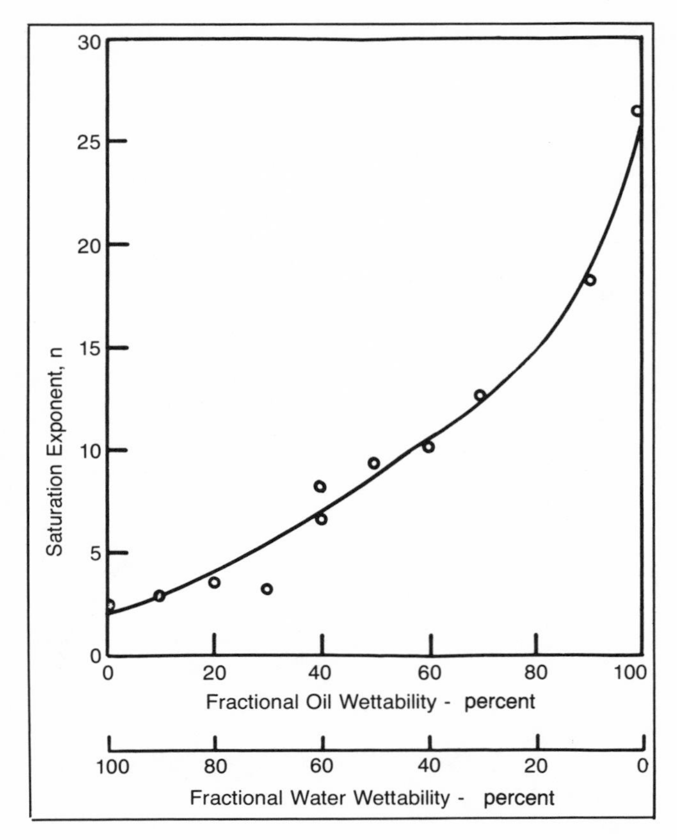

Fig. 4-13. Saturation exponent, n, vs. fractional wettability[34]

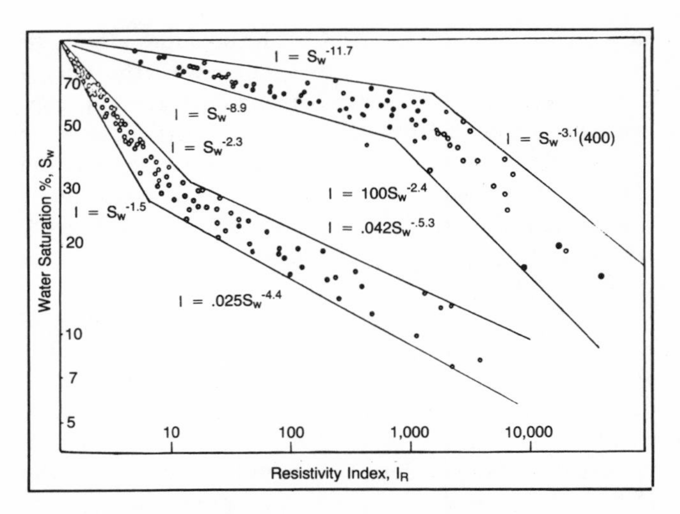

Fig. 4-14. Dependence of resistivity index on water saturation for water-wet and nonwater-wet cores[18]

normally referred to as the saturation exponent. Rewriting the above equation to:

$$I_R = \frac{R_t}{R_o} = S_w^{-n}$$

we can now define the true rock resistivity as:

$$R_t = f(R_w, \underbrace{\text{Rock Structure}}_{F_R = a\phi^{-m}}, \underbrace{\text{hydrocarbon influence}}_{I_R = S_w^{-n}})$$

An understanding of true rock resistivity gives the basis for formation evaluation, as stated below:

$$R_t = R_w F_R I_R$$

$$F_R = a\phi^{-m} = \frac{R_o}{R_w}$$

$$I_R = S_w^{-n} = \frac{R_t}{R_o}$$

These relations can be written many different ways, and can then be readily manipulated into the most convenient form for a particular application.

The relationship between water saturation and resistivity has been studied extensively by a number of authors; many factors have been shown to affect the magnitude of the saturation exponent, n:

1. Wettability of the rock surface[17, 18, 29, 34, 44]
2. Rock texture[11, 16]
3. Presence of clay[24, 29]
4. Measurement technique[6, 16, 19, 44]
5. Nature of displacing fluid[5, 6]

Saturation exponents have been reported in the literature from less than 2.0 to above 25. It appears that the factor having the greatest influence on the value of the saturation exponent, n, is wettability. Figure 4-13 indicates that as water wettability decreases the saturation exponent can be expected to increase. In fact, as indicated by the data of Keller,[17] (see Fig. 4-14) a value of $n = 2$ is only valid for water-wet cores in which water saturations, S_w, are in excess of 30%.

Measurement of the saturation exponent can be accomplished by a variety of techniques that can be generally defined as either static or dynamic. The static approach is illustrated in Figure 4-15. In this case, the core is progressively desaturated using an increasing gas pressure. At each pressure step, resistivity and water saturation are determined once equilibrium is established. With the dynamic method (see Fig. 4-16) the water saturated core is desaturated by injecting air at successively higher pressures while the outlet pressure remains at atmospheric pressure. When equilibrium is established at each injection, pressure, resistivity, and saturation measurements are obtained. Saturation is determined gravimetrically. It is also possible to simultaneously inject air and water mixtures through the core. With this approach, a different water saturation can be obtained by changing the air-water injection mixture.

The methods for measuring saturation exponents are rather tedious and time-consuming. It is, however, desirable to have these data measured in the laboratory on representative core samples.

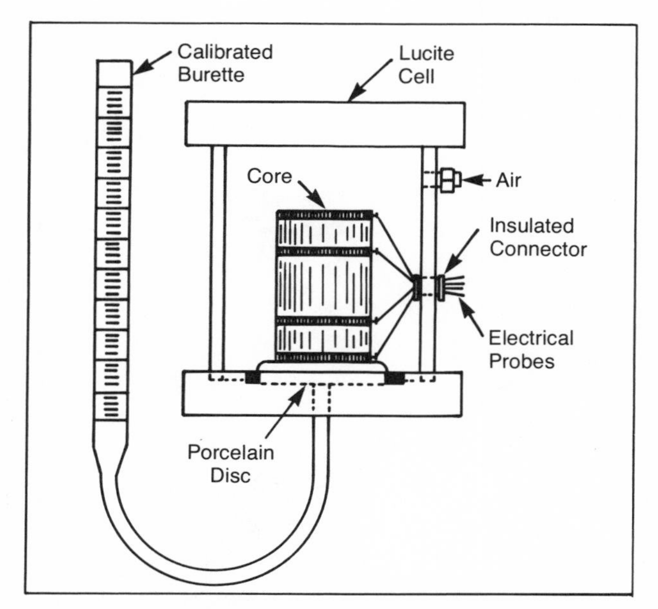

Fig. 4-15. Apparatus for measuring saturation exponents by static method. Permission to publish by the Oil and Gas Journal.[19]

Influence of Clay on Rock Resistivity

The presence of clay in a rock will influence the resistivity of the rock, depending on the quantity of clay and how it is dispersed. The clay acts as a separate path for current flow independent of the current flow path through the saline water. The presence of clay places an upper limit on the resistivity of the rock, which cannot be exceeded even though all of the saline water filling the pores may be replaced by nonconducting fluids (oil, gas, or distilled water). This creates very difficult problems in interpretation that have not yet been resolved, which will be discussed in Chapter 9 with respect to shaly sand interpretation.

Flushed Zone Resistivity

We have observed that the true resistivity of a rock can be expressed as:

$$R_t = R_w F_R I_R$$

In the same way we can express the true resistivity of the flushed zone as:

$$R_{xo} = R_{mf} F_R I_{xo}$$

where:

$$I_{xo} = S_{xo}^{-n}$$

and:

$$S_{xo} = 1 - S_{hcr} \qquad S_{hcr} = S_{or} \text{ or } S_{gr}$$

Expressing the above relation in terms of F_R we obtain:

$$F_R = \frac{R_{xo}}{R_{mf}}(1 - S_{hcr})^n = a\phi^{-m}$$

Determination of porosity based on flushed zone resistivity measurements will be discussed more fully in Chapter 6, "Resistivity Measurement," with respect to resistivity measurements with microdevices.

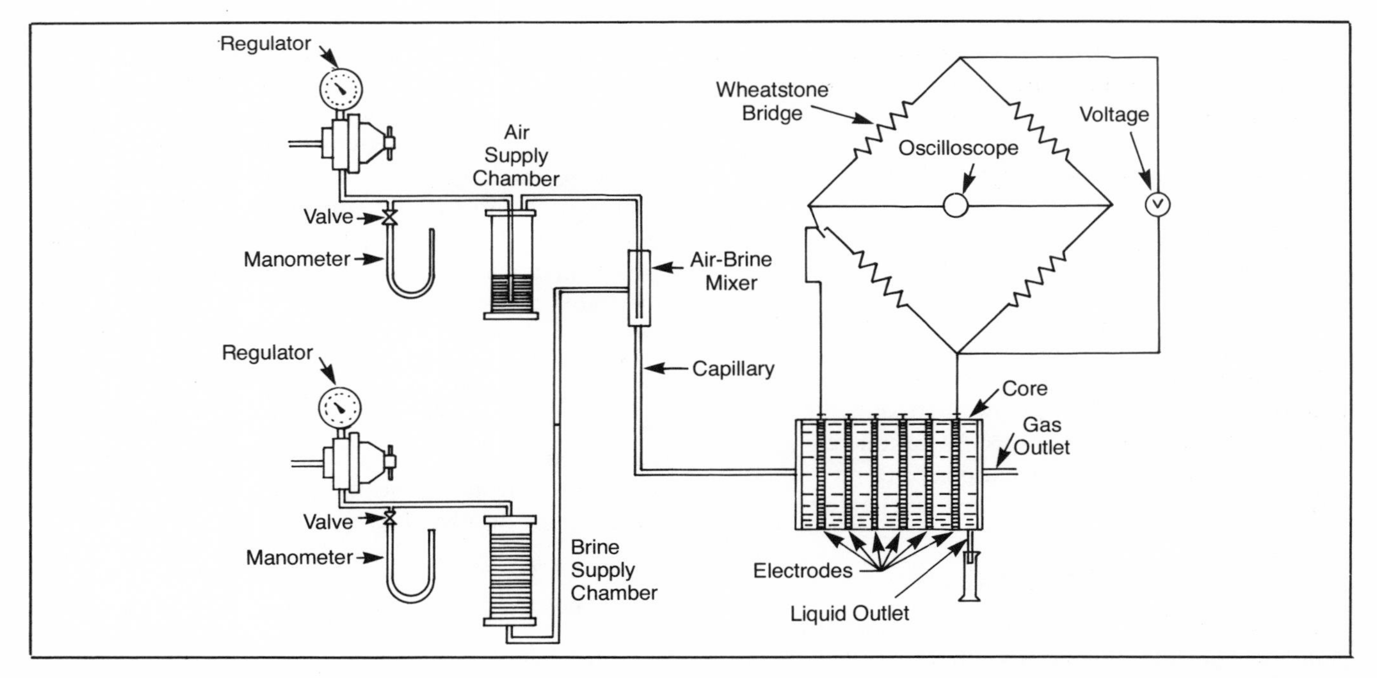

Fig. 4-16. Apparatus for measuring saturation exponents by dynamic method. Permission to publish by the Oil and Gas Journal.[19]

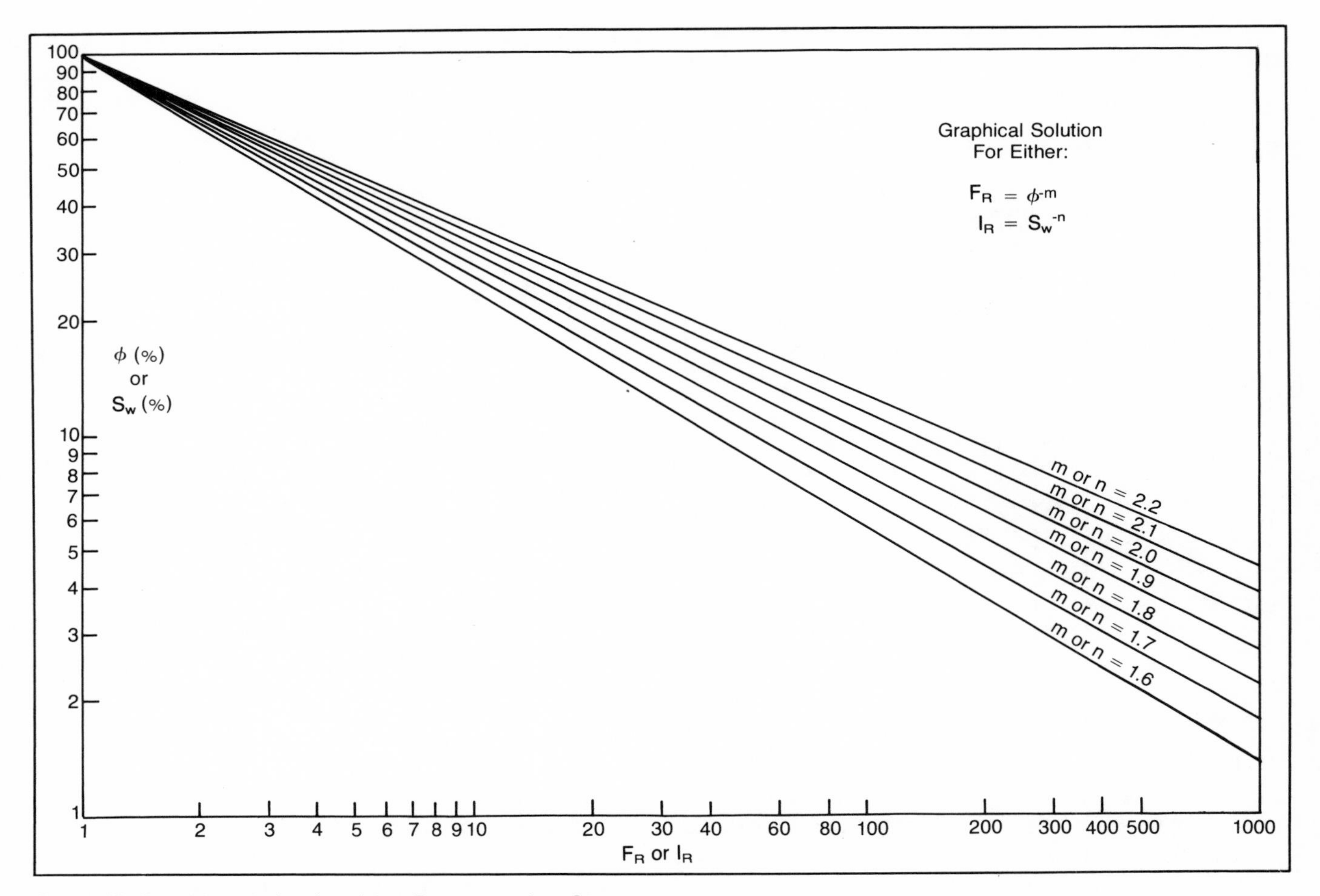

Fig. 4-17. Graphic solution for either $F_R = \phi^{-m}$ or $I_R = S_w^{-n}$

The two basic equations that form the cornerstone of well log analysis, sometimes referred to as the Archie relations, are:

$$F_R = a\phi^{-m} = \frac{R_o}{R_w}$$

$$I_R = S_w^{-n} = \frac{R_t}{R_o} = \frac{R_t}{F_R R_w}$$

The similarity between the expressions can readily be solved graphically using Figure 4-17.

The importance of the basic saturation equations can readily be illustrated in the flow chart shown in Figure 4-18. Determination of the specific parameters appearing in this schematic of conventional interpretation will be discussed in detail in the following chapters.

ROCK PERMEABILITY

In the preceding sections the fundamental relationships used in formation evaluation have been discussed. Two important reservoir rock parameters required in defining the potential productivity of a reservoir rock system, porosity, ϕ, and water saturation, S_w, are related to measurable rock properties; however, another rock property—permeability—is also necessary to evaluate the potential productivity of a formation. It would, therefore, be desirable to evaluate permeability using well logs. A number of indirect approaches have been devised for this purpose, some of which will be discussed here with other permeability indicators such as moveable oil techniques being discussed in more detail later.

A few of the techniques that have been successfully used to estimate permeability are the porosity-permeability relationship, the porosity-residual water saturation-permeability relationship, the resistivity gradient-permeability relationship, and the shaliness-permeability relationship.

Porosity-Permeability Relationship

It has been observed that most sedimentary rocks indicate a linear relationship when log k_a versus log ϕ or in some cases the log k_a versus ϕ is plotted.[8] The linear

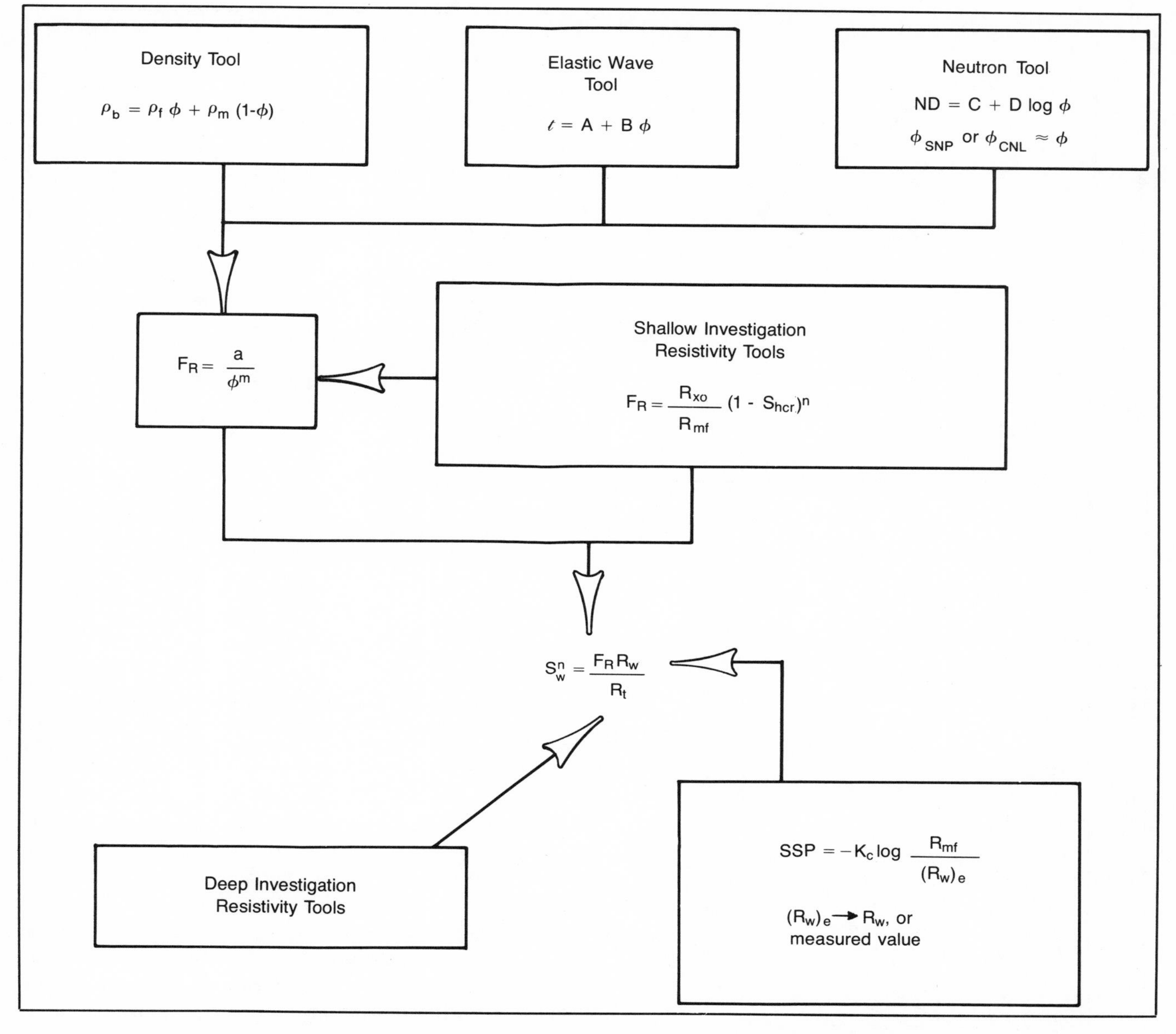

Fig. 4-18. Schematic of conventional interpretation. Permission to publish by John M. Campbell Co.

trends can vary from rock type to rock type, as shown in Figure 4-19. If such a trend can be established for a particular rock type, an estimate of absolute permeability can be obtained as a result of using measured log parameters directed toward the evaluation of porosity.

Porosity-Residual Water Saturation-Permeability Relationship

A number of empirical and semi-empirical relationships have been developed that relate permeability to other rock properties.[11, 22] One such relationship was presented by Wyllie and Rose[11] and later modified by Schlumberger[53] to:

$$k = 6.25 \times 10^{-4} \frac{\phi^6}{S_{wr}^2}$$

Timur[51] presented the statistical results of a study on 155 sandstone samples and found

$$k = 0.136 \frac{\phi^{4.4}}{S_{wr}^2}$$

which is graphically presented in Figure 4-20. Relationships such as this have been used extensively for estimating permeability. The Timur relationship is probably the best of this type in use at this time.

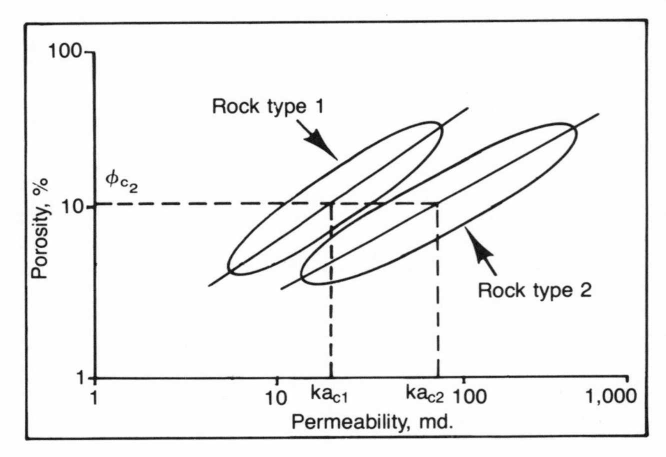

Fig. 4-19. Typical porosity vs. permeability trends for different rock types

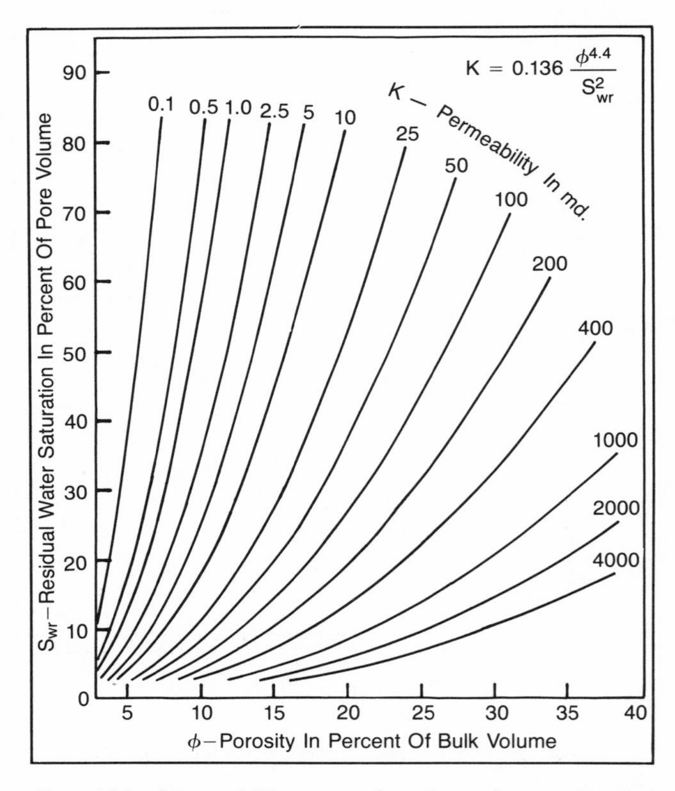

Fig. 4-20. Permeability as a function of porosity and residual water saturation. Permission to publish by the Society of Professional Well Log Analysts.[51]

Resistivity Gradient Approach

This method was first introduced by Tixier[8] in 1949 and is applicable to those formations that are hydrocarbon-bearing at the top, water-bearing in the bottom and exhibit a large transition zone between the two (see Fig. 4-21). This technique makes use of the empirical

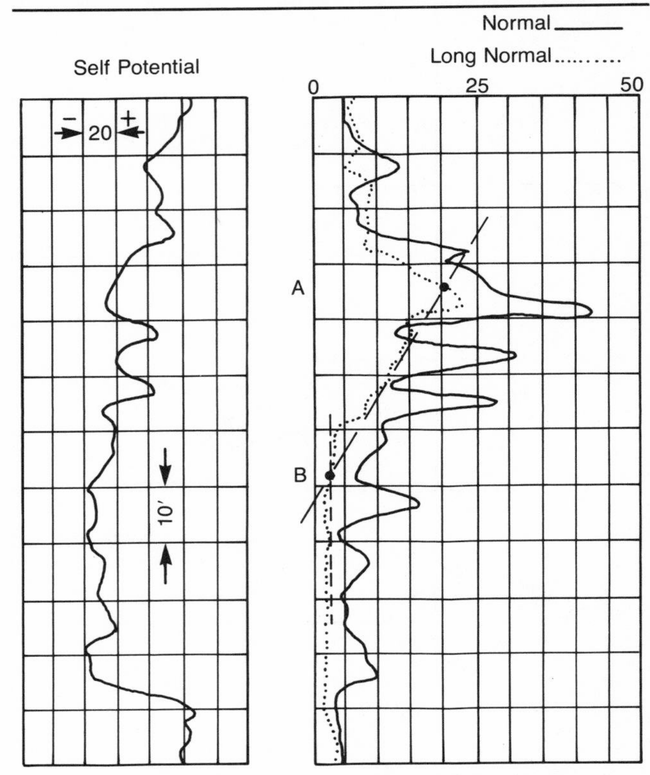

At "A," $R_{I,N} = 20$; at "B," $R_{I,N} = 2.5$, $R_t = 2$; then $\Delta R/\Delta L = (20 - 2)/34 = 0.531$, and $a = 0.531/2 = 0.265$. Average calculated permeability = 310 md., average laboratory permeability range = 200-700 md Mud resistivity = 0.6 at bottom-hole temperature, hole diameter = 9 in., AM = 16 in., AM_1 = 64 in. Salinity of water = sea-water, oil gravity 30° A.P.I.

Fig. 4-21. Example resistivity gradient calculation using electric log, Cisco Sand, North Texas. Permission to publish by the Oil and Gas Journal.[8]

relationships between (1) resistivity and water saturation, (2) water saturation and capillary pressure, and (3) capillary pressure and permeability. Figure 4-22 provides a quick solution for the permeability-resistivity relationship.

Limitations of the resistivity gradient approach are numerous, but probably the most important is its limited applicability. It is limited to those formations that show a large transition zone. Probably the weakest link is the equation relating capillary pressure to permeability at any saturation, which is a simplification of the general case presented by Leverett[1] that incorporates the *J*-function correlation.

Shaliness-Permeability Relationship

The technique presented by Rabe[27] for determining permeability illustrates a specific example, limited in scope, that enables permeability to be reasonably determined from well logs. It is included here to point out the

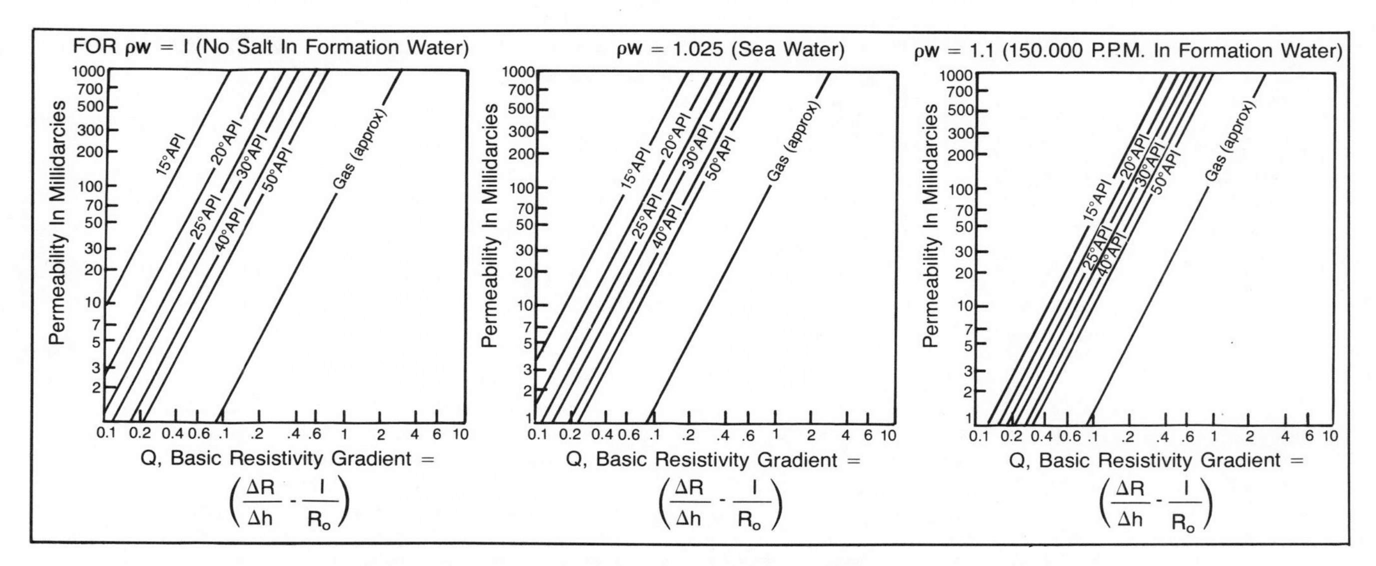

Fig. 4-22. Permeability vs. resistivity gradient. Permission to publish by the Oil and Gas Journal.[8]

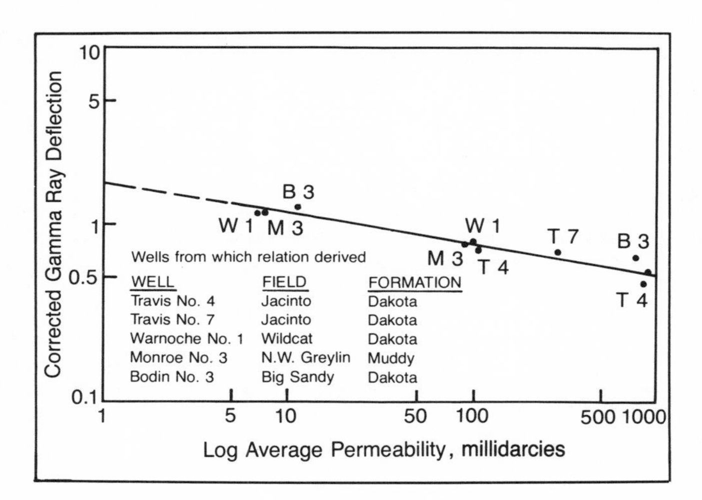

Fig. 4-23. Relationship between gamma radiation and permeability of Muddy and Dakota Sands in Denver-Julesburg Basin. Permission to publish by the Society of Petroleum Engineers of AIME.[27] *Copyright 1957 SPE-AIME.*

many potential methods that may be developed. This particular method developed for the Muddy and Dakota sands in the Denver-Julesburg Basin is based on the relationship that permeability is commonly a function of cementing material. Since in this basin the cementing material is generally clay and the amount of clay should be reflected in the response of the gamma radiation, a correlation was developed that related gamma ray radiation to permeability. The relationship between corrected gamma ray deflection and permeability is shown in Figure 4-23.

REFERENCES

1. Leverett, M.C.: "Capillary Behavior in Porous Solids," *Trans.*, AIME (1941).
2. Archie, G.E.: "The Electrical Resistivity Log as an Aid in Determining Some Reservoir Characteristics," *Trans.*, AIME (1942).
3. Guyod, Hubert: "Electrical Well Logging," *Oil Weekly* (Aug.-Dec. 1944).
4. Archie, G.E.: "Electrical Resistivity an Aid in Core Analysis Interpretation," *Bull.*, AAPG (Feb. 1947) 31.
5. Pirson, Sylvain J.: "Factors Which Affect True Formation Resistivity," *Oil and Gas J.* (Nov. 1, 1947).
6. Dunlap, Henry F., Bilhartz, Harrel L., Shuler, Ellis, and Bailey, C.R.: "The Relation Between Electrical Resistivity and Brine Saturation in Reservoir Rocks," *Trans.*, AIME (1949).
7. Rose, W., and Bruce, W.A.: "Evaluation of Capillary Character in Petroleum Reservoir Rock," *Trans.*, AIME (1949).
8. Tixier, Maurice P.: "Evaluation of Permeability from Electric Log Resistivity Gradients," *Oil and Gas J.* (June 16, 1949).
9. de Witte, L.: "Relations Between Resistivities and Fluid Contents of Porous Rock," *Oil and Gas J.* (Aug. 24, 1950) 49.
10. Patnode, H.W., and Wyllie, M.R.J.: "The Presence of Conductive Solids in Reservoir Rock as a Factor in Electric Log Interpretation," *Trans.*, AIME (1950).
11. Wyllie, M.R.J., and Rose, W.D.: "Some Theoretical Considerations Related to the Quantitative Evaluation of the Physical Characteristics of Reservoir Rock from Electric Log Data," *Trans.*, AIME (1950).
12. Dunlap, H.F., and Hawthorne, R.R.: "The Calculations of Water Resistivities from Chemical Analysis," *Trans.*, AIME (1951).
13. Archie, G.E.: "Classification of Carbonate Rocks and Petro-Physical Considerations," *Bull.*, AAPG (1952) 36.

14. Owen, John E.: "The Resistivity of a Fluid Filled Porous Body," *Trans.*, AIME (1952).
15. Winsauer, W.O., Shearin, H.M., Jr., Masson, P.H., and Williams, M.: "Resistivity of Brine Saturated Sands in Relation to Pore Geometry," *Bull.*, AAPG (Feb. 1952) 36.
16. Wyllie, M.R.J., and Spangler, M.B.: "Application of Electrical Resistivity Measurements to Problems of Fluid Flow in Porous Media," *Bull.*, AAPG (1952) 36.
17. Keller, G.V.: "Effect of Wettability on the Electrical Resistivity of Sands," *Oil and Gas J.* (Jan. 5, 1953).
18. Licastro, P.H. and Keller, G.V.: "Resistivity Measurements as a Criteria for Determining Fluid Distribution in the Bradford Sand," *Producers Monthly* (May 1953).
19. Whiting, R.L., Guerrero, E.T., and Young, R.M.: "Electrical Properties of Limestone Cores," *Oil and Gas J.* (July 27, 1953).
20. Winsauer, W.O., and McCardell, W.M.: "Ionic Double-Layer Conductivity in Reservoir Rock," *Trans.*, AIME (1953).
21. Wyllie, M.R.J., and Gregory, A.R.: "Formation Factors of Unconsolidated Porous Media: Influence of Particle Shape and Effect of Cementation," *Trans.*, AIME (1953).
22. Carmen, P.C.: *Flow of Gases Through Porous Media*, Academic Press Inc., New York (1956).
23. Hamilton, R.G., and Charrin, Paul: "Resistivity of Fluids," *Oil and Gas J.* (Dec. 3, 1956).
24. Moore, E. James: "Laboratory Analysis of the Electric Logging Parameters of the Weir Sand," *Producers Monthly* (July 1956) 20.
25. Fatt, I.: "Effect of Overburden and Reservoir Pressure on Electrical Logging Formation Factor," *Bull.*, AAPG (1957) 41.
26. Hubbert, M. King, and Willis, D.G.: "Mechanics of Hydraulic Fracturing," *Trans.*, AIME (1957).
27. Rabe, C.L.: "A Relation Between Gamma Radiation and Permeability, Denver-Julesberg Basin," *Trans.*, AIME (1957).
28. Wyllie, M.R.J.: *The Fundamentals of Electric Log Interpretation*, Academic Press, Inc. (1957).
29. Moore, E. James: "Laboratory Determined Electric Logging Parameters of the Bradford Third Sand," *Producers Monthly* (March 1958).
30. *Oil Reservoir Engineering*, Sylvain J. Pirson (ed.), 2nd ed. McGraw-Hill Book Co. Inc., New York, 1958.
31. Glanville, C.R.: "Laboratory Study Indicates Significant Effects of Pressure on Resistivity of Reservoir Rocks," *J. Pet. Tech.* (1959) 10.
32. Wyble, D.O.: "Effects of Applied Pressure on the Conductivity, Porosity and Permeability of Sandstones," *J. Pet. Tech.* (1959) 10.
33. Chambart, Louis G.: "Well Log Interpretation in Carbonate Reservoirs," *Geophysics* (1960) 25.
34. Sweeney, S.A., and Jennings, H.Y.: "The Electrical Resistivity of Preferentially Water Wet and Preferentially Oil Wet Carbonate Rocks," *Producers Monthly* (May 1960).
35. Atkins, E.R., Jr., and Smith, G.H.: "The Significance of Particle Shape in Formation Resistivity Factor-Porosity Relationships," *Trans.*, AIME (1961).
36. Dobrynin, V.M.: "Effect of Overburden Pressure on Some Properties of Sandstones," *Soc. Pet. Eng. J.* (1962) 2.
37. Glumov, I.F., and Dobrynin, V.M.: "Changes of Specific Resistivity of Water Saturated Rocks Under the Action of Rock and Reservoir Pressures," *Priklad Geofiz* (1962) 33.
38. Orlov, L.I., and Gimaev, R.S.: "The Influence of Rock Pressure on the Electrical Resistivity of Carbonate Rocks," *Priklad Geofiz* (1962) 33.
39. Redmond, J.C.: "Effect of Simulated Overburden Pressure on the Resistivity, Porosity, and Permeability of Selected Sandstones," PhD dissertation, Pennsylvania State U., University Park, PA (1962).
40. Towle, Guy: "An Analysis of the Formation Resistivity Factor-Porosity Relationship of Some Assumed Pore Geometries," *Trans.*, SPWLA (1962).
41. Hawkins, M.E., Dietzman, W.P., and Seaward, J.M.: "Analysis of Brines from Oil Production Formations in South Arkansas and North Louisiana," USBM R.I. 6292 (1963).
42. Hawkins, M.E., Dietzman, W.D., and Pearson, C.A.: "Chemical Analysis and Electrical Resistivities of Oilfield Brines from Fields in East Texas," USBM R.I. 6422 (1964).
43. Hilchie, Douglas, W.: "The Effect of Pressure and Temperature on the Resistivity of Rocks," PhD dissertation, University of Oklahoma, Norman, OK (1964).
44. Morgan, W.B., and Pirson, Sylvain J.: "The Effect of Fractional Wettability on the Archie Saturation Exponent," *Trans.*, SPWLA (1964).
45. Talash, A.W., and Crawford, Paul B.: "An Improved Method for Calculating Water Resistivities from Chemical Analyses," *J. Pet. Tech.* (Dec. 1965) 17.
46. Helander, Donald P., and Campbell, John M.: "The Effect of Pore Configuration, Pressure and Temperature on Rock Resistivity," *Trans.*, SPWLA (1966).
47. Moore, E. James: "A Graphic Description of New Methods for Determining Equivalent NaCl Concentrations from Chemical Analysis," paper presented at the SPWLA Seventh Annual Logging Symposium, May 8-11, 1966.
48. Moore, E. James, Szasz, S.E., and Whitney, B.C.: "Determining Formation Water Resistivity from Chemical Analysis," *Trans.*, AIME (1966).
49. Tunn, W.: "Determination of Some Important Petrophysical Data of Oil and Gas Reserves and Comparison of Values Obtained by Different Methods," *The Log Analyst* (May-June 1967) 8.
50. Carothers, J.E.: "A Statistical Study of the Formation Factor Relation," *The Log Analyst* (Sept.-Oct. 1968) 9.
51. Timur, A.: "An Investigation of Permeability, Porosity and Residual Water Saturation Relationships for Sandstone Reservoirs," *The Log Analyst* (July-Aug. 1968) 9.
52. Desai, Kantilal P., and Moore, E. James: "Equivalent NaCl Determination from Ionic Concentrations," *The Log Analyst* (May-June 1969) 10.
53. *Log Interpretation Charts*, Schlumberger Ltd. 1972.
54. Sanyal, S.K., Marsden, S.S., and Ramey, H.J., Jr.: "The Effect of Temperature on Electrical Resistivity of Porous Media," *The Log Analyst* (March-April 1973) 14.
55. Evers, J.F., and Iyer, B.G.: "Quantification of Surface Conductivity in Clean Sandstones," *Trans.*, SPWLA (1975).

THIS IS AN INTERLIBRARY LOAN

for Shafiq Zolieri

The library lending us this volume has stipulated that it may/may not be taken out of the University Library.

Please return it to the Interlibrary Loan Office no later than 10/18/85.

If you need the book beyond this date, please request an extension at least one week before the due date.

FA 188

5 The Spontaneous Potential

THE Spontaneous Potential curve was one of the first measurements made in a wellbore. Its primary purpose is to differentiate shale formations from nonshale formations. As a lithology curve, it is also used to determine bed boundaries and bed thickness and to qualitatively estimate the degree of shaliness of porous and permeable beds. As a quantitative tool, the spontaneous potential is used to determine water resistivity, R_w.

The spontaneous potential measurement records the change in naturally-occurring potentials as a function of depth in the borehole. There are two types of potential that may contribute to the SP, the electrochemical, E_c, and electrokinetic, E_k (electrofiltration or streaming) potentials. The electrochemical potential is always present when an E_{sp} is generated, whereas the electrokinetic potential may or may not be contributing to the measured SP. In fact, as will be discussed later, the contribution of the electrokinetic potential to the observed SP will probably be negligible. It is assumed in well log analysis that the observed SP response is due solely to the electrochemical component, with the presence of an electrokinetic potential producing an abnormal SP. Hence, $E_{sp}=E_c+E_k$ but usually $E_k \approx 0$ and $E_{sp}=E_c$.

ELECTROCHEMICAL POTENTIAL

The electrochemical potential is composed of two independent potentials referred to as the membrane potential, E_m, and the liquid junction potential, E_j. The membrane potential exists when two fluids of different activity (mud and formation water) are separated by a permeable, charged membrane (shale). The liquid junction potential arises between the contact of two solutions of differing activity (mud filtrate and formation water). These two potential cells in series produce the electrochemical potential.

Mounce and Rust[1] studied such a chain (formation water/shale/mud/mud filtrate/formation water) and found that current would flow in this system dependent upon the concentration of the two electrolytes. Since mud and mud filtrate are similar in activity, they can be considered as one electrolyte. Mounce and Rust performed their experiments using a circular trough divided into three sections by permeable walls of unglazed porcelain (see Fig. 5-1). When these three sections were filled with different electrolytes or with solutions of the same electrolyte having different concentrations, no current was observed in the trough, and potential differences were only detected between different sections. On the other hand, when one of the partitions was made of shale, large potential differences were observed in any one of the sections, thus indicating that current was flowing in the trough (potential differences noted in a single section). In addition, if one section was filled with shale, another with a low concentration NaCl solution, and the third with brine (high concentration NaCl solution), it was observed that a current was flowing from the shale to the low concentration NaCl solution, to the brine, and then back to the shale. This current flow could be reversed by reversing the two electrolytic solu-

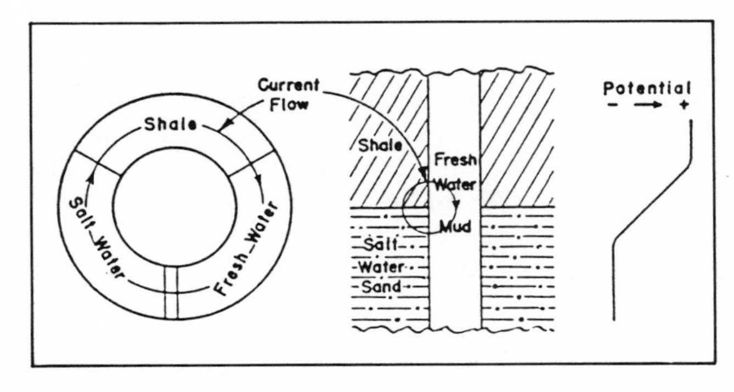

Fig. 5-1. Experimental system of Mounce and Rust. Permission to publish by the Society of Petroleum Engineers of AIME.[1] Copyright 1944 SPE-AIME.

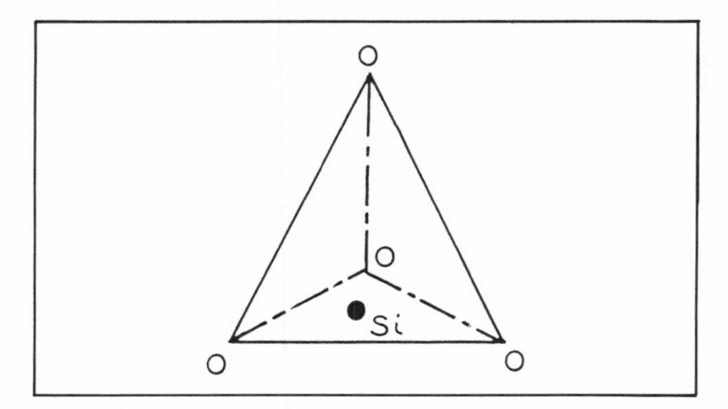

Fig. 5-2. SiO_4 tetrahedonal structure

tions. If the shale was replaced by a clean sand or a clean porous limestone, no current flow was observed. It was also noted that when the two electrolytes were identical in nature and concentration, no matter how large or small the concentration, there would be no current flow even if the shale were present. Mounce and Rust's study indicated that for a current to flow in the wellbore, it is necessary to have (1) a charged membrane and (2) two electrolytes of different concentrations (mud-mud filtrate, and formation water) forming a complete circuit with the charged membrane.

In order to understand the electrochemical potential and in particular the membrane potential, it is necessary to understand the nature of shales. The following discussion presents some of the theoretical concepts involved in the spontaneous potential.

Shales are mixtures of various clay minerals, which are generally composed of silicon, aluminum, and oxygen. The most common are kaolinite, montmorillonite, and illite, but they all have one feature in common: their crystalline structure is dominated by a tetrahedral disposition of the silicon and oxygen atoms and an octahedral arrangement of aluminum and oxygen atoms.

Silicon is a small atom with four electropositive charges (valence). Oxygen is a large atom with two electronegative charges. In all silicates, the two atoms combine in units of SiO_4, with the four oxygen atoms arranged symmetrically around the silicon atom. The oxygen atoms are the corners of a tetrahedron, with the silicon atom inside (see Fig. 5-2). The tetrahedron has a resultant negative charge (-4), and this is in turn neutralized by four more surrounding silicon atoms, leaving a net of 12 $(16-4)$ positive charges, which are again more than neutralized by other surrounding oxygen atoms.

Thus the characteristic SiO_4 group is repeated over and over, resulting in a layered framework or sheet of SiO_4 structures. Because oxygen has a weaker tendency than silicon to be surrounded by other atoms, the edges of these SiO_4 sheets are generally made of oxygen atoms, with the result that the edges are predominantly negative. They will attract positive ions (cations) such as K^+, Na^+, Ca^{++}, etc.

Furthermore, some of the silicon atoms in the tetrahedral network can be replaced by Al^{+++} without any change in structure (isomorphous substitution). Since this is a replacement of an atom with four positive charges by one of only three positive charges, an even greater excess negative charge will exist at the edges of the sheets, and cations will be even more strongly attracted.

In most clay minerals, aluminum, like silicon, prefers to surround itself with oxygen atoms and hydroxyl ions, since they are negatively charged. In this case, four oxygen atoms and two OH groups form the six corners of an octahedron, which encloses the Al^{+++} atom (see Fig. 5-3). This serves to link two sheets of SiO_4-tetrahedra, and the resulting structures as determined by X-ray diffraction may become very complicated. Divalent ions or magnesium may replace the trivalent aluminum isomorphously and add to the overall negative charge of the clay's framework.

The resulting negative charge loosely binds the cations to the clay and repels the anions. In effect, a shale bed acts as an ion sieve, allowing only cations to permeate or pass through its structure. This property gives rise to the shale potential.

When in a medium having a large dielectric constant such as water, electrostatic forces produced by charges on a clay are greatly reduced. The cations, loosely bound to the clay, become free to move around under thermal agitation, and to jump from one position to another on the lattice. They can also exchange places with other cations in the surrounding solution. By this process, cations such as K^+, Ca^{++}, or Mg^{++} that are present on the clay can be exchanged with cations such as Na^+ when the clay is left in contact with an NaCl solution for a long time. This ion exchange phenomenon is responsible for the conductivity of shales, the clay framework being otherwise nonconducting.

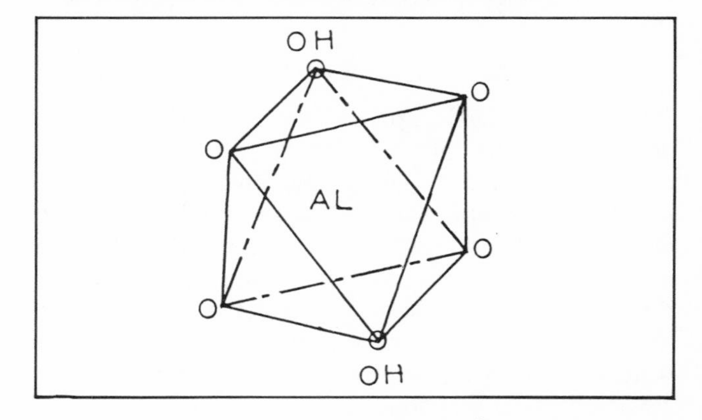

Fig. 5-3. Aluminum atom in octahedonal structure

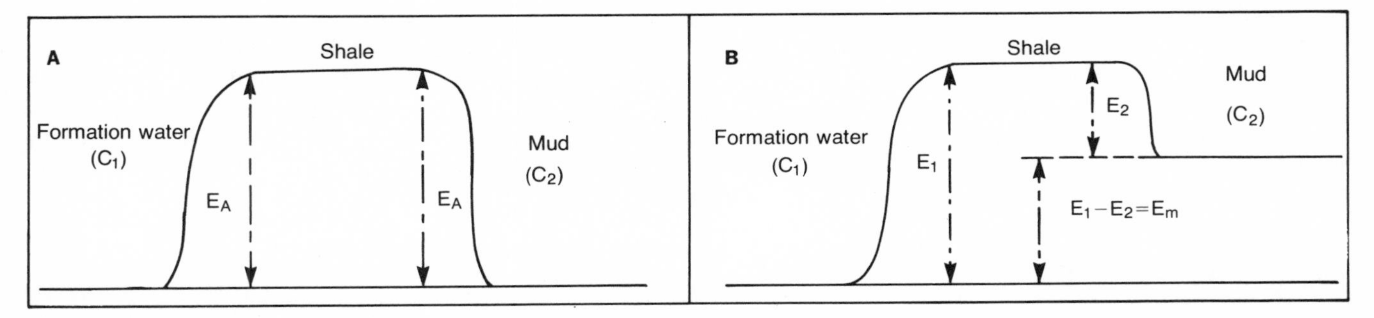

Fig. 5-4. (A) Schematic of initial potential barrier. (B) Schematic of equilibrium condition.

Membrane Potential

To visualize the generation of the membrane potential, E_m, consider a piece of shale placed between a concentrated salt solution (formation water) and dilute salt solution (fresh mud filtrate). The cations in the shale must have a certain energy in order to jump from one position to another in the lattice. That is, in order to leave its position and jump to the next exchange site, an ion must overcome a certain energy barrier, W_A. Since the electric charge on an ion is constant, and equal to zQ, where z is the valence and Q the charge on an electron, we may consider this energy barrier as a potential barrier, E_A, so $W_A = zQE_A$.

For ideal solutions, the number of cations in each solution that are able to go through this potential barrier is proportional to the total number of cations n_1 and n_2 present in each solution (that is, to the ionic concentrations c_1 and c_2) and to a factor $P(z, Q, E_A)$, which expresses the probability that an ion subject to the random thermal motion possesses the required energy: z, Q, E_A.

This probability factor is expressed by the common Boltzman relation of the form:

$$P(z, Q, E_A) = Ce^{[-zQE_A/KT]}$$

where: C is a constant (collision factor)
K is Boltzman's constant (equal to R/N, the gas constant divided by Avogadro's number), and
T is the absolute temperature.

For two solutions of the same salt at the same temperature, this factor is the same, and so the number of cations capable of overcoming the potential barrier in the formation water and in the mud are simply proportional to the ionic concentrations. If the formation water is much saltier than the mud, a larger number of ions from the formation water than from the mud will move into the shale. The net effect will be a flow of cations from the formation water through the shale to the mud.

Because of the predominantly negative charge of the clay framework, anions cannot migrate in the same way, so each cation transferred from the formation water to the mud leaves a negative charge behind in the formation water and adds a positive charge to the mud. This accumulation of charge makes the electrical potential of the formation water more negative and that of the mud more positive. This increases the potential barrier for the cations coming from the formation water to a value E_1 and lowers it for cations coming from the mud to a value E_2. A state of equilibrium is rapidly achieved in which the net transfer of cations is zero. The potential difference $E_m = E_1 - E_2$ between the two solutions is then equal to the electromotive force (emf) of the cell (see Fig. 5-4).

This emf is independent of the particular nature of the clay minerals so long as the only mobile ions in the clay are cations. For different kinds of clays, the magnitude of initial potential barrier (E_A) will differ, and a different number of cations in the solution will be able to cross it. However, the final emf, corresponding to zero net flux, will be the same in all cases because it corresponds to the two final potential barriers, E_1 and E_2, which in turn depend only upon the concentrations of the two solutions. Thus, for the first solution (formation water), we have (substituting for K its value R/N):

$$n(E_1) = n_1P(zQE_1) = n_1Ce^{[-zNQE_1/RT]}$$

and for the second solution (mud) we have:

$$n(E_2) = n_2P(zQE_2) = n_2Ce^{[-zNQE_2/RT]}$$

and for zero net flux, $n(E_1) = n(E_2)$, or

$$n_1e^{[-zNQE_1/RT]} = n_2e^{[-zNQE_2/RT]}, \text{ or}$$

$$\ln n_1 - \ln n_2 = \frac{zNQ}{RT}(E_1 - E_2)$$

But $NQ = F$, the Faraday, and,

$$E_m = E_1 - E_2 = \frac{RT}{zF} \ln \frac{n_1}{n_2}$$

where n_1 and n_2 are the number of cations in solutions 1 (formation water) and 2 (mud) and are proportional to their concentrations, c_1 and c_2.

where: c_1 = concentration of formation water
c_2 = concentration of mud filtrate

For monovalent cations, $z = +1$ and we obtain:

$$E_m = \frac{RT}{F} \ln \frac{c_1}{c_2}$$

For non-ideal solutions, the concentrations c_1 and c_2 must be replaced by the activities a_1 and a_2, but the overall picture is still the same.

Liquid Junction Potential

The liquid junction potential, E_j, can be explained in a similar manner, but here there is no shale to separate the two solutions, and anions as well as cations can transfer from one solution to the other. This ionic diffusion can also be visualized as a rate process, where each ion, making its way through the water molecules, overcomes a potential barrier.

Considering the formation water to be of a higher concentration, both cations and anions move from the formation water to the mud filtrate. The Na^+ ion is comparatively large, with a strong affinity for water (4.5 water molecules associated per ion), whereas the Cl^- ion is smaller, with less affinity for water (only 2.9 water molecules per ion) so that the Cl^- ions travel faster than the Na^+ ions; the result is a net positive charge left in the formation water. This increases the potential barrier, which restricts Cl^- migration relative to Na^+ migration.

Note that this liquid junction process differs from the shale membrane process in that the net positive charge left behind in the formation water and hence eventually the emf generated does not correspond to the total number of Cl^- ions passing over the potential barrier to the mud filtrate, but rather to the fraction of the Cl^- ions whose charges are not neutralized by the Na^+ ions leaving the formation water at the same time. It is apparent at once that the liquid junction potential is going to be less than the membrane potential.

For each gram-equivalent of NaCl transferred, the fraction of potential-producing ions is

$$\frac{U_{Cl} - U_{Na}}{U_{Cl} + U_{Na}}$$

where U is the mobility of the designated ion. This mobility ratio can be also expressed as $t_{Cl} - t_{Na}$ where t is the "transference number" of the ion. For any monovalent salt solution, $t^- + t^+ = 1$, by definition of transference numbers; hence $t_{Cl} + t_{Na} = 1$.

The liquid junction potential can be expressed in a form similar to that for the membrane potential; recall, however, that only a fraction $(t_{Cl} - t_{Na})$ of the total number of ions transferred are effective in producing it. So we may write:

$$E_j = (t_{Cl} - t_{Na}) \frac{RT}{F} \ln \frac{c_1}{c_2}$$

Notice that the expression for E_j is exactly the same as the expression for E_m except for the multiplier $(t_{Cl} - t_{Na})$. The value of t_{Cl} is very close to .6 and the value of t_{Na} is close to .4, so $t_{Cl} - t_{Na} = .2$. Therefore, the liquid junction potential, E_j, is only about one-fifth that of the shale membrane potential, E_m.

Total Electrochemical Potential

The total electrochemical potential, E_c, is the sum of the potential created by the two cells forming the chain for current flow, which is composed of the formation water/shale/mud/mud filtrate/formation water. Further subdivision gives the two cells discussed previously:

1. Membrane potential, which is composed of formation water/shale/mud.
2. Liquid junction potential, which is composed of mud filtrate/formation water.

Notice that in the original chain, another link, mud/mud filtrate, was measured for the sake of completeness. This link is omitted in the subdivision into two cells because the mud and its filtrate have practically identical electrochemical activities, although their resistivities may differ. As a result there is no significant difference at their contact. Since the total electrochemical potential is the sum of the two potentials, we can write:

$$E_c = E_m + E_j = \frac{RT}{F} \ln \frac{c_1}{c_2} + (t_{Cl} - t_{Na}) \frac{RT}{F} \ln \frac{c_1}{c_2}$$

and for non-ideal solutions we can write:

$$E_c = \frac{RT}{F} \ln \frac{a_1}{a_2} + (t_{Cl} - t_{Na}) \frac{RT}{F} \ln \frac{a_1}{a_2}.$$

Determination of the mean activities is possible; however, it is more convenient to replace the ratio of the activities of the electrolytes in the solution by the ratio of the conductivities of the two solutions. Although the relationship between conductivity and activity ratios becomes less accurate at higher concentrations, this approximation is permissible for dilute solutions. Therefore:

$$\frac{a_1}{a_2} \approx \frac{\text{(conductivity) 1}}{\text{(conductivity) 2}} = \frac{\frac{1}{R_1}}{\frac{1}{R_2}} = \frac{R_2}{R_1}$$

Therefore:

$$E_c = \frac{RT}{F} \ln \frac{R_2}{R_1} + (t_{Cl} - t_{Na}) \frac{RT}{F} \ln \frac{R_2}{R_1}$$

Since $(t_{Cl} - t_{Na}) = .2$, the equation for the total electrochemical potential may be written as

$$E_c = 1.2 \frac{RT}{F} \ln \frac{R_2}{R_1}$$

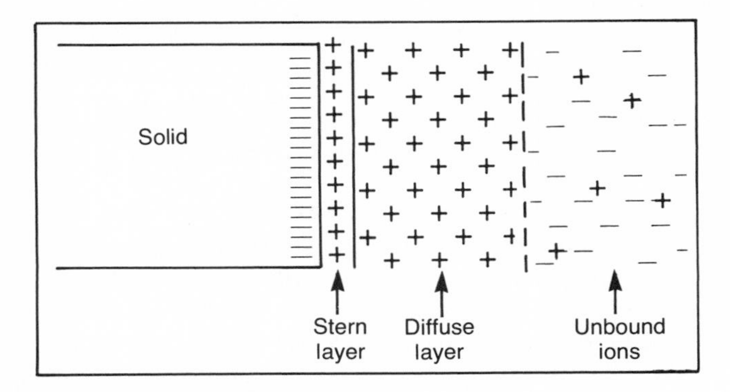

Fig. 5-5. Double layer concept[13]

Converting natural log to log to the base 10 and substituting R_{mf} for R_2 and R_w for R_1, then:

$$E_c = 1.2\frac{RT}{F}(2.303)\log\frac{R_{mf}}{R_w}$$

The term $1.2(RT/F)(2.203)$ is referred to as K_c and is a function of temperature and can be expressed as:

$K_c = 61 + 0.133T$ (°F) where K_c is in millivolts

$K_c = 65 + 0.24T$ (°C)

The electrochemical potential can thus be written as follows:

$$E_c = K_c \log\frac{R_{mf}}{R_w} = E_{ssp}$$

As stated earlier, it is usually initially assumed that the electrokinetic or streaming potential is negligible, such that: $E_{sp} = E_{ssp} = {}^{-}K_c \log(R_{mf}/R_w)$. The negative sign has been included since by convention a negative E_{sp} will be recorded when $R_{mf} > R_w$.

ELECTROKINETIC POTENTIAL

When a solid and a solution are in contact, a potential is developed at the boundary between the two substances. This potential may be the result of the adsorption of ions from the solution by the solid or it may be the result of ionization of the solid. For example, when a solid surface such as glass is wet with water, the surface becomes negatively charged, due to ionization of cations of the glass surface. The solution next to the glass will have a net positive charge as a result of the negative charge of the glass. This positive charge is referred to as the double layer, which consists of a first layer that is rigidly bound or fixed (Stern Layer) and a second diffuse layer of less rigidly bound ions, which is free to move, as illustrated in Figure 5-5. The potential difference between the bulk liquid, which is neutral, and the boundary between the fixed and diffuse layers, which has a net positive charge, is known as the zeta potential, E_z (see Fig. 5-6). If pressure is applied to the solution, the movable part of the solution will flow past the solid surface, carrying the zeta potential with it. Movement of the zeta potential by the liquid flow creates a potential difference between the two ends of the solid particle, which is the electrokinetic (streaming) potential.

The electrokinetic potential can be readily simplified for mathematical analysis by considering the double-layer as a simple condenser, as discussed by Lynch[9] and Althaus.[14] Based on electrostatics, the various factors can be related by:

$$\frac{E_T = 4\pi e x}{D}$$

where: E_T = potential between two plates
e = electrostatic units of charge per square centimeter
x = distance between the plates of the condenser
D = the dielectric constant of the water

Consider the situation in which the solid is a capillary tube of length L and diameter d. A pressure P applied at one end of the tube causes viscous flow of the liquid through the capillary. The velocity at any point in the liquid under such conditions can be expressed by Poisseuille's law as a function of the distance y from the center of the tube, the diameter of the tube d, the length of the tube L, the pressure drop P, and the viscosity μ.

$$V = \frac{P\left(\frac{d^2}{4} - y^2\right)}{4L\mu}$$

If the hypothetical condenser is at a distance of $(d/2 - x)$ from the center of the tube where x is very small compared to d, then the velocity at this point is

$$V_x \cong \frac{Pdx}{4L\mu}$$

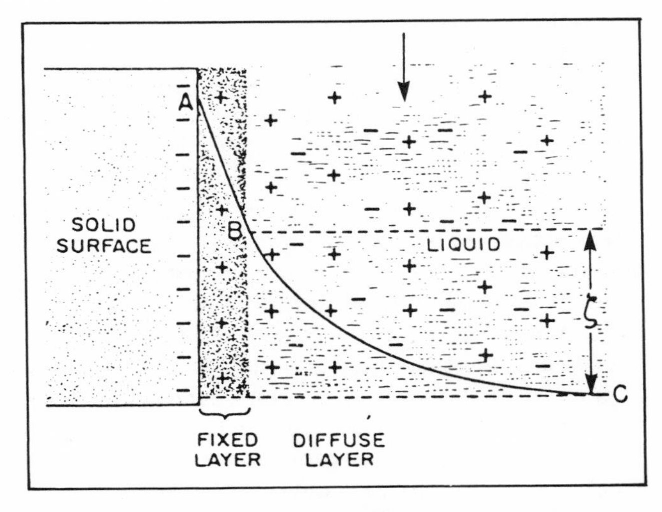

Fig. 5-6. Schematic of potential developed at a solid-liquid interface. Permission to publish by Harper and Row Publishers, Inc.[9]

The rate at which positive charge is being carried past a charged particle, then, is

$$i' = \pi d e V_x = \frac{\pi d^2 exP}{4\mu L}$$

The electrokinetic potential E_k, which is created by this flow of charge, causes electrical flow of current in the opposite direction, the quantity determined by Ohm's law ($i=E/r$). If the resistivity of the solution in the capillary is R_w, then the electrical flow of current is

$$i = \frac{E_k}{r} = \frac{E_k \pi d^2}{4R_w L}$$

At steady state, $i=i'$ and the electrokinetic potential is

$$E_k = \frac{exPR_w}{\mu} = \frac{E_T DPR_w}{4\pi\mu}$$

where: D = dielectric constant of water
P = pressure differential
R_w = resistivity of water
μ = viscosity of water

This equation indicates that the electrokinetic potential is independent of the length and diameter of the capillary. However, the potential should be directly proportional to the resistivity of the solution in the capillary. When mud filtrate invades a porous and permeable formation, it must flow through a negatively charged mud cake. This causes an electrokinetic potential with the low-pressure formation side of the mud cake assuming a positive charge relative to the high-pressure borehole side. Adapted to this system, the previous equation becomes

$$E_k = \frac{E_z DPR_{mc}}{4\pi\mu}$$

The resistivity of the mud cake has been substituted for the resistivity of the water.

This equation indicates that the electrokinetic potential should be dependent on the mud composition with zeta potential, the dielectric constant, and the viscosity being properties of the mud system. Experimentation has borne this out. Different types of mud give highly different potentials. The equation also indicates that the potential should increase in direct proportion to the differential pressure and the resistivity of the mud cake. Experimentation has shown these relationships to be exponential rather than direct, possibly because the character of the mud cake changes somewhat as the differential pressure changes, or possibly, as Truman[12] suggests, turbulence is created in the capillaries and/or the fluid is incapable of ionizing fast enough to replenish the ions in the system.

Wyllie[3] did extensive pressure-chamber studies on electrokinetic potentials using drilling muds and cores. He found the relationship between potential and pressure to be of the form $E_k = KP^y$

where: E_k = electrokinetic potential
P = differential pressure

K and y are constants for any particular mud system at constant temperatures. The constant K was found to be a function of the mud resistivity at 77°F, but the exponent y was not found to be related to any commonly measured mud property. Wyllie's work indicated that electrokinetic potentials could be significantly high in relation to electrochemical potentials under conditions in which the mud resistivity and/or differential pressure were high. This could cause large errors in log analysis.

Gonduin and Scala[6] also did pressure-chamber experiments and were able to produce electrokinetic potentials across shale cores as well as across sand cores. They pointed out that since the sand potential was always measured from the shale base line on logs, the shale electrokinetic potential would have the effect of shifting the shale base line in a negative direction and would, therefore, subtract from the sand electrokinetic potential. The total potential of a sand would be the electrochemical potential of the sand, plus the electrokinetic potential of the sand, minus the electrokinetic potential of the shale. They indicated that in most instances where the mud resistivities and the differential pressures are not severely high, the two electrokinetic effects very nearly cancel each other out and are not a problem in log analysis. Their observation in log analysis was that, in most instances where electrokinetic potentials could be developed, they usually were not, and the electrochemical formula would, therefore, usually yield the correct R_w, even under adverse circumstances.

Hill and Anderson[7] ran extensive pressure-chamber tests to establish the electrokinetic potentials of a large number of sands and shale samples in all of the common types of muds. They subdivided muds into seven types and constructed charts from their data showing the electrokinetic potential-pressure resistivity relationship for each of these types of mud. They also derived a chart showing that shale electrokinetic potentials average 12 mv per 1000 psi differential pressure for most of the resistivity range. This work is quite useful since it gives the most comprehensive data on the amount of electrokinetic potential that might be expected under a wide variation of mud and pressure conditions.

Truman[12] extended pressure chamber work up to 1600 psi and obtained electrokinetic potentials as high as 5500 mv with highly resistive fluids. In most of his experiments, the potential tended to approach a maximum and level out. To explain this, Truman expanded the idea of turbulence in capillaries and proposed the concept of relaxation, i.e., the inability of the fluid to furnish enough ions to keep the system functioning at a high rate of flow.

Althaus[14] studied electrokinetic potentials in south Louisiana tertiary sediments and made the following observations:

1. With the fairly low mud resistivities found in south Louisiana, there does not appear to be an electrokinetic potential problem with 500 psi or less of differential pressure.
2. With the conditions favorable for the generation of electrokinetic potentials, they still are not generated in some wells (or they may be cancelled entirely by the shale electrokinetic potential).
3. When electrokinetic potentials are developed in a well, they are developed in some sands and not in others. The magnitude of the potential varies from sand to sand.
4. The maximum potential generated increases almost linearly with an increase in mud resistivity. They also increase exponentially with an increase in differential pressure up to 3000 psi. Additional pressure has no apparent influence on the potentials. (Data are limited in this range.)

He concluded that electrokinetic potentials can be generated in south Louisiana but are highly unpredictable. In fact, electrokinetic potentials were not developed where theory predicted they should occur. Based on this work, it appears that the solution to the problem of electrokinetic potentials is not one of log analysis correction but one of prevention. This is best accomplished by drilling with the lowest possible mud weight, consistent with well pressure control, thus decreasing the differential pressure to the point where electrokinetic potentials are negligible. As the well is drilled deeper, the upper formations may be exposed to higher differential pressures, thereby increasing the possibility of generating electrokinetic potentials. If upper zones are of interest, these should be logged as early as possible.

If corrections for the streaming potential are to be made during log analysis, empirical corrections such as those presented by Hill and Anderson,[7] Althaus[14] or Halbert[11] should be used with caution. Probably the best approach is that suggested by Wyllie.[5] Defining an abnormal Spontaneous Potential (E_{spA}) as one in which a streaming potential, E_k, exists in addition to the electrochemical potential then for $R_{mf} > R_w$ and E_{spA} is negative:

$$E_{spA} = -E_c + E_k \quad \text{and} \quad E_c = K_c \log \frac{R_{mf}}{R_w}$$

now:

$$R_{mf} = \frac{R_{xo}}{F_R I_{xo}} \quad \text{and} \quad R_w = \frac{R_t}{F_R I_R}$$

but considering 100% water-saturated zones only where $I_R = I_{xo} = 1$ then E_c which is:

$$E_c = K_c \log \frac{\dfrac{R_{xo}}{F_R I_{xo}}}{\dfrac{R_t}{F_R I_R}} = K_c \log \frac{R_{xo}}{R_t}$$

substituting for E_c as follows:

$$E_{spA} = -K_c \log \frac{R_{xo}}{R_t} + E_k$$

Now if the E_{sp} values for a number of sands are plotted on semilog paper versus the ratio of R_{xo}/R_t, the plotted points would tend to pass through the origin if no streaming potential exists. It is assumed that most of the formations are water zones and that all interstitial waters and all shales are similar in chemical composition. If a streaming potential exists it would be indicated as the amount of E_{sp} displacement the trend of water data indicates when it intersects the ratio of $R_{xo}/R_t = 1.0$. Figure 5-7 shows an example of such a plot. This provides a basic approach for log interpretation since oil-bearing zones will fall below the normal trend. It was

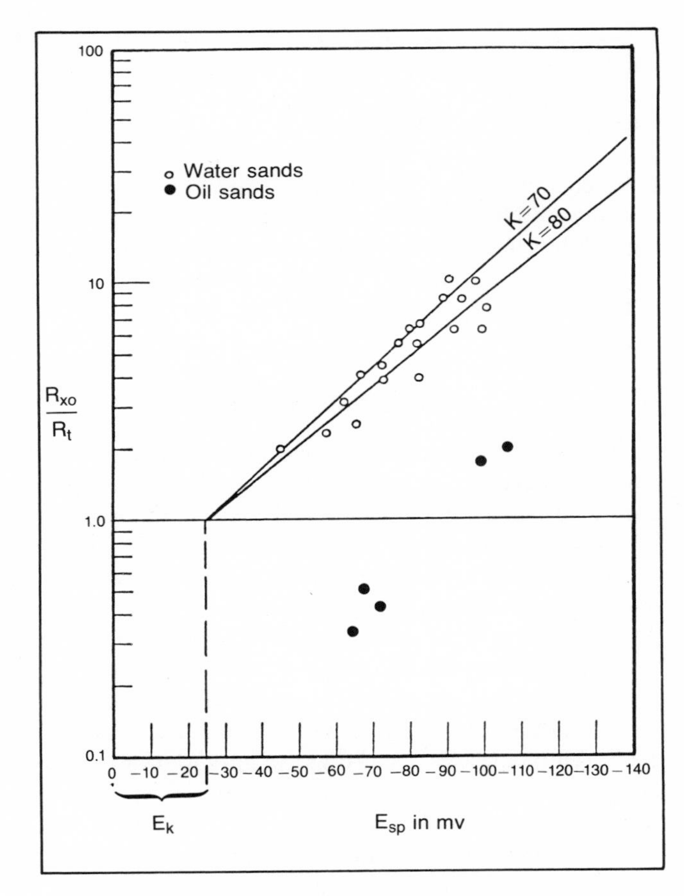

Fig. 5-7. Plot of R_{xo}/R_t versus SP. Permission to publish by the Academic Press, Inc.[5]

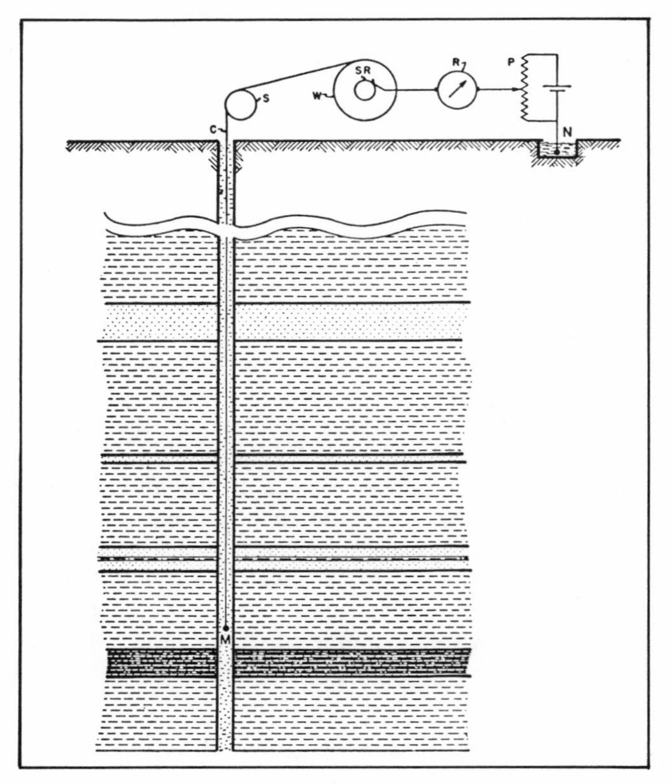

Fig. 5-8. Schematic circuit for SP. Permission to publish by the Society of Petroleum Engineers of AIME.[2] Copyright 1949 SPE-AIME.

assumed the sands were water-bearing and $R_o = R_t$ but for oil sands $R_t \gg R_o$ and, therefore, the value of R_{xo}/R_t is much lower than that which would be obtained if the sand were water-bearing. This interpretation technique will be discussed more fully in Chapter 9.

MEASUREMENT OF SP

The SP is recorded by measuring the difference in potential between an electrode in the borehole and a grounded electrode at the surface. A schematic of the SP measuring circuit is shown in Figure 5-8. The SP doesn't measure an absolute value but rather measures a change in borehole potential as a function of depth. The slope of the curve indicates the rate of change in potential. The SP opposite a thick shale is recorded as a straight line (no change in potential) until it approaches the bed boundary, at which time the current density in the borehole increases, causing the potential to change. The rate at which this potential change occurs depends on the rate at which the current density increases. This is shown schematically in Figure 5-9. The greatest rate of change in current intensity is at the bed boundary, past which all the current developed by the electrochemical cell must pass. It is at this point (bed boundary) that the maximum slope of the SP occurs; this is referred to as the inflection point. Immediately above the boundary in the shale, current is entering the borehole, causing the SP curve to deflect (usually toward a more negative value) from the normal shale potential. Immediately below the bed boundary in the sand, current is leaving the borehole, decreasing the current intensity and causing the slope of the SP to lessen. The shape of the recorded SP when passing a bed boundary (see Fig. 5-9) is a function of the relative resistivities of the shale-sand beds. A higher resistivity in one of the beds causes the current flow of that electrochemical cell to spread out more in the more resistive bed, which obviously affects the current distribution in the borehole and subsequently the shape of the SP. Bed boundaries must be selected at the inflection point.

As mentioned previously, the shale potential is recorded as a vertical straight line until it approaches a shale-sand boundary. This straight line is referred to as a shale baseline (see Fig. 5-10), and all SP changes are measured from this baseline. We are interested in the magnitude of the SP deflection from the shale baseline. The shale baseline is usually recorded on the first or second division to the left of the depth track with the maximum SP deflection recorded opposite clean formations usually reaching the second division of the track from the left. The maximum SP deflections registered usually define an imaginary line referred to as the sand baseline. The SP is measured in millivolts (mv), generally on a 10 or 20 mv per division scale, although in conductive muds 4 or 5 mv per division, scales may be used in order to obtain some definition in the SP.

FACTORS AFFECTING SHAPE AND AMPLITUDE OF SP

A number of factors affect the shape and amplitude of the SP, including R_{mf}/R_w ratio, bed thickness, bed resistivity, borehole diameter, invasion, and shaliness of porous and permeable bed.

Ratio of R_{mf}/R_w

This effect is predicted by the SP equation previously developed where:

$$E_{sp} = -K_c \log \frac{R_{mf}}{R_w}$$

As can be seen, this expression ideally predicts the SP to vary only as a function of the relative resistivities of the mud filtrate, formation water, and temperature (included in K_c). Of course, this is the basis on which we calculate R_w, but several important points are brought out by Figure 5-11.

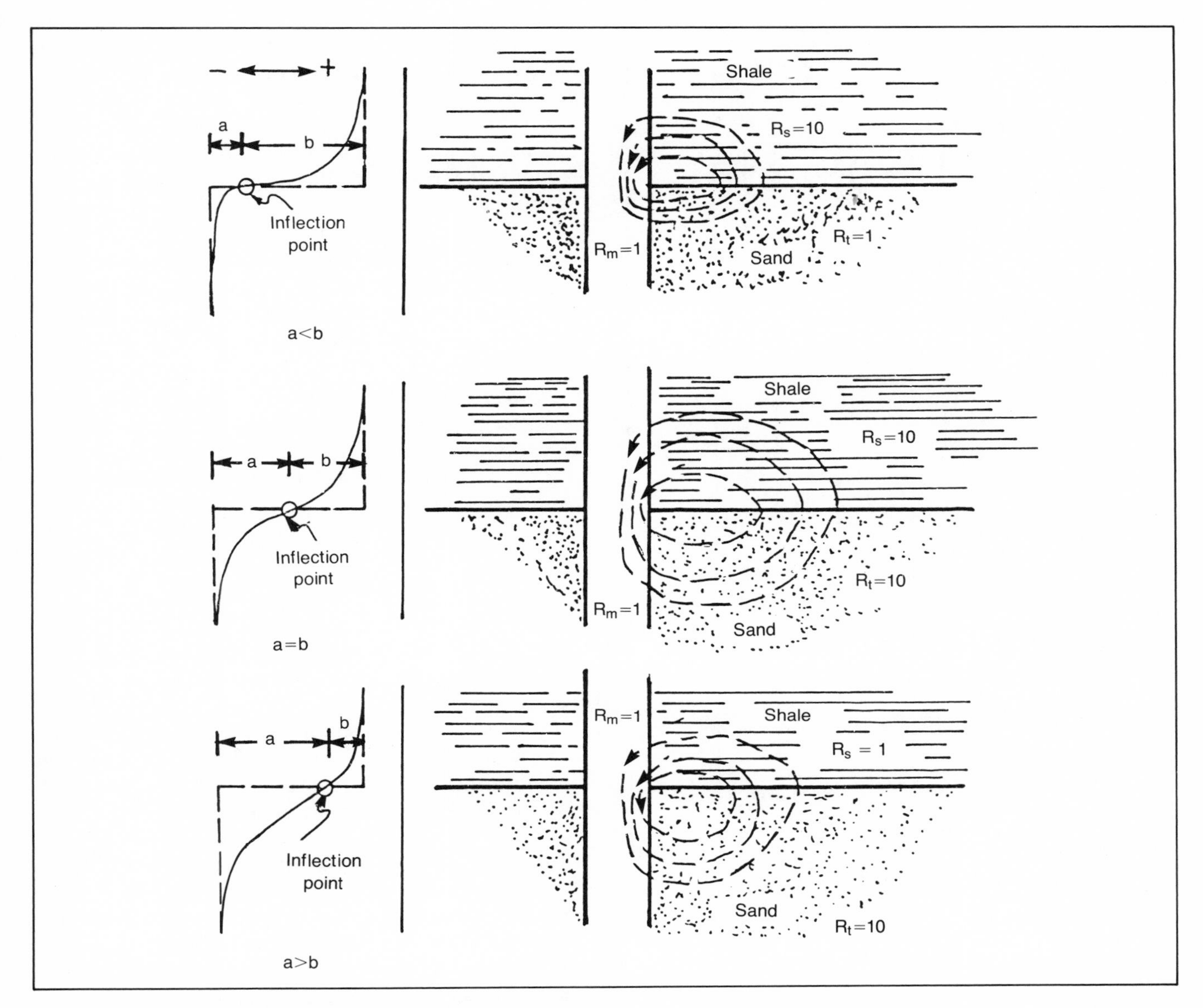

Fig. 5-9. SP shape as a function of current distribution

First, it is possible to get a reversed SP (SP deflects to right from shale baseline) that is a perfectly valid SP and from which R_w can be determined. Reversed SP's would usually be encountered in the shallower, fresh water zones or when highly conductive muds are used. The second point is that the magnitude of the SP can be controlled by controlling R_m and, therefore, R_{mf}. Use of salt muds may diminish or eliminate the SP response.

Additional Factors

The effects of other factors that affect the SP can be visualized by considering the SP quantitatively. The current flowing in the formation-borehole system follows Ohm's law as:

$$E_{ssp} = i(r_{sh} + r_{sd} + r_m)$$

where: r_{sh} = effective resistance of the shale bed
r_{sd} = effective resistance of the sand bed
r_m = effective resistance of the mud

This may be schematically represented as shown in Figure 5-12.

Since the actual SP is recorded only in the borehole, it can be expressed as:

$$E_{sp} = ir_m \quad \text{or} \quad i = \frac{E_{sp}}{r_m}$$

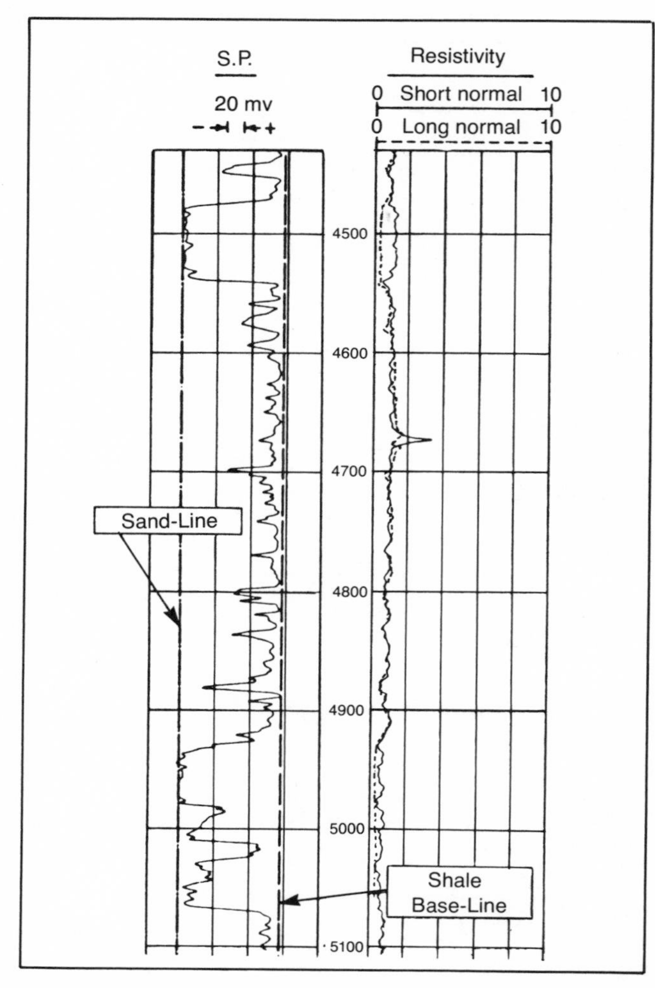

Fig. 5-10. Example of SP sand and shale baselines. Permission to publish by Schlumberger Ltd.[15]

By substitution, the following is obtained:

$$E_{ssp} = \frac{E_{sp}[r_{sh} + r_{sd} + r_m]}{r_m}$$

Now in the situation in which the sand and shale beds are thick, the relative resistance offered to current flow by the sand and shale is essentially negligible when compared to that offered by the borehole. In this case, r_{sh} and r_{sd} approach zero and:

$$E_{ssp} = \frac{E_{sp} + r_m}{r_m} \quad \text{or} \quad E_{ssp} \approx E_{sp}$$

Influence of Bed Thickness. The influence of bed thickness on the SP amplitude can now be readily visualized. For example, as the porous and permeable bed becomes thinner, the relative resistance of the sand, r_{sd}, increases with respect to r_m and:

$$E_{ssp} = \frac{E_{sp}[r_{sd} + r_m]}{r_m}$$

Since the static SP, E_{ssp}, remains the same, but now the resistance ratio is greater than 1.0, then the measured SP (E_{sp}) must decrease. Therefore, E_{sp} is less than E_{ssp}, as schematically presented in Figure 5-13. Of course, if the shale strata were becoming thinner within thick sand sections, the SP would not reduce to the shale baseline, as schematically shown in Figure 5-14.

Influence of Bed Resistivity. Bed resistivity effects on the SP can also be visualized. In this case, an increase in r_{sd}, for example, would again cause the measured E_{sp} to be less than E_{ssp} (see Fig. 5-15).

Influence of Borehole Diameter. Increasing borehole diameter reduces the amplitude of the SP. In this case, r_m decreases as hole size increases. For constant values of r_{sh} and r_{sd}, a decreasing value of r_m would tend to lead toward ratios of

$$\frac{r_{sh} + r_{sd} + r_m}{r_m}$$

greater than 1.0 and, therefore, measured values of E_{sp} would be less than E_{ssp}. Corrections for the influence of bed thickness, bed resistivity, and borehole effects could be made using the data in Figure 5-16.

Influence of Invasion. Invasion can be qualitatively considered to increase the effective diameter of the borehole, thereby decreasing the amplitude of the SP. It also increases the resistance of the porous and permeable bed, which would also lead to a decreased SP. The influence of invasion on the SP log can be seen in Figure 5-17. Invasion is also a time-varying factor such that a changing invasion profile can affect the shape and amplitude of the SP. As discussed in Chapter 1, gravity segregation can significantly alter the invaded zone. Impermeable streaks within the porous and permeable zone would present an invasion profile having more than one segregated zone, which would result in a sawtooth SP (see Figs. 5-18 and 5-19).

Influence of Shaliness in the Porous and Permeable Bed. Increasing shale content in a porous and permeable formation tends to decrease the SP. This can be visualized in the following manner. Consider the shale to be laminated within the porous and permeable bed so that we are considering a series of sand-shale layers. As these layers become thinner and thinner, the effect would be similar to that observed with single thin beds. The relative resistance of r_{sh} and r_{sd} would increase relative to r_m, and the measured E_{sp} would be less than E_{ssp}. This is illustrated in Figure 5-20.

The SP measured in a shaly sand is referred to as the pseudostatic SP, E_{psp}. The reduction in the SP as indi-

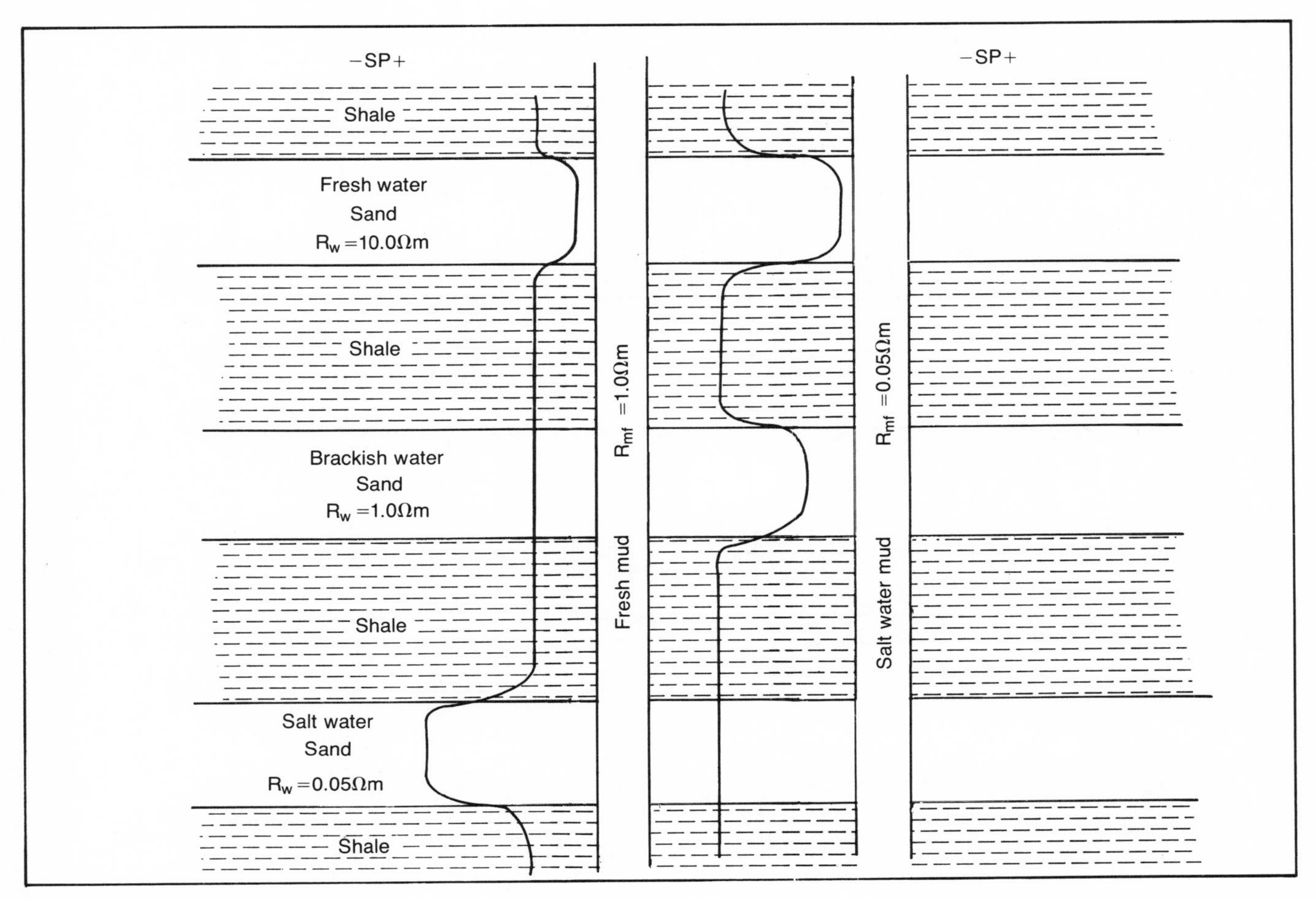

Fig. 5-11. Effect of R_{mf}/R_w on the SP

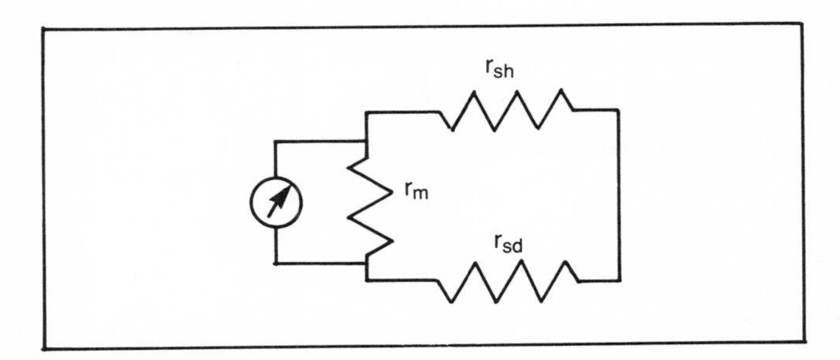

Fig. 5-12. Schematic of qualitative SP circuit

cated by the E_{psp} reasonably reflects the amount of shale, either laminated or disseminated, in the porous and permeable formation. This degree of shaliness can be defined by the ratio $\propto = E_{psp}/E_{ssp}$, which is referred to as the SP reduction factor.

USES OF THE SP

Bed Boundaries. Bed boundaries are determined from the inflection points of the SP curve, as previously discussed. In resistive formations, however, the SP becomes more difficult to interpret, especially when the conductive zones are thin and interbedded within thick resistive beds. The resistive beds tend to prevent the SP current from entering or leaving the borehole; because of this high resistivity, the bulk of the SP current leaves and enters the borehole via the more conductive zones. The resulting SP is difficult to evaluate in terms of the boundaries or thickness of the more porous and permeable zones. An example of the current distribution and resulting SP for a complex resistive system is schematically shown in Figure 5-21. Figure 5-22 gives an example electric log in a complex resistive system and outlines some guidelines to remember when analyzing the SP.

Net Sand. It is possible to determine net sand in some cases from the SP; however, it is generally more desirable to use a tool that responds to greater detail, such as one of the microresistivity devices. In complex, resistive systems, the SP would not be definitive for defining porous and permeable zones.

Shale Content. The SP can be used to evaluate the

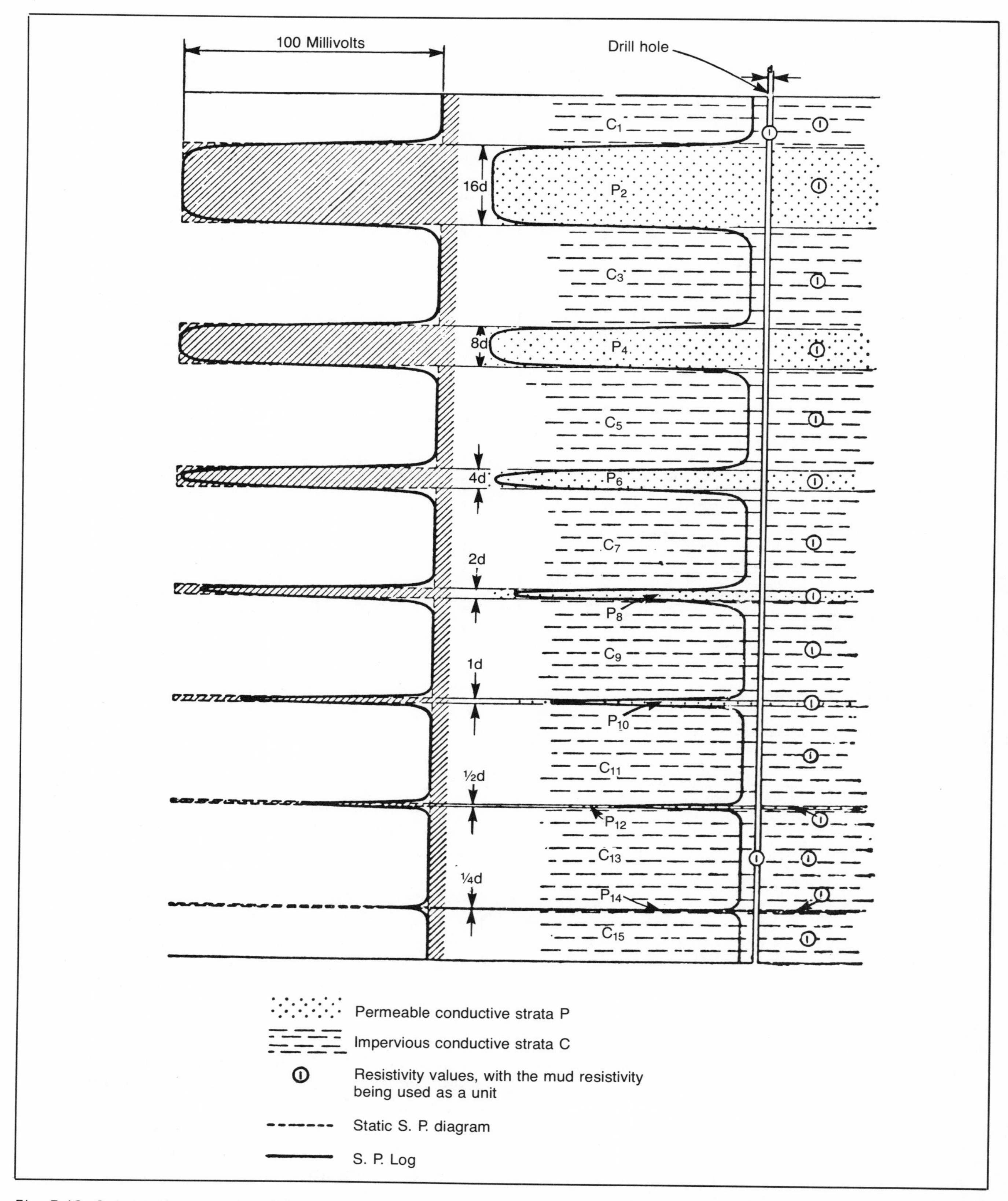

Fig. 5-13. Schematic example of bed thickness effects for varying thicknesses of permeable beds, $R_t = R_m$. Permission to publish by the Society of Petroleum Engineers of AIME.[2] Copyright 1949 SPE-AIME.

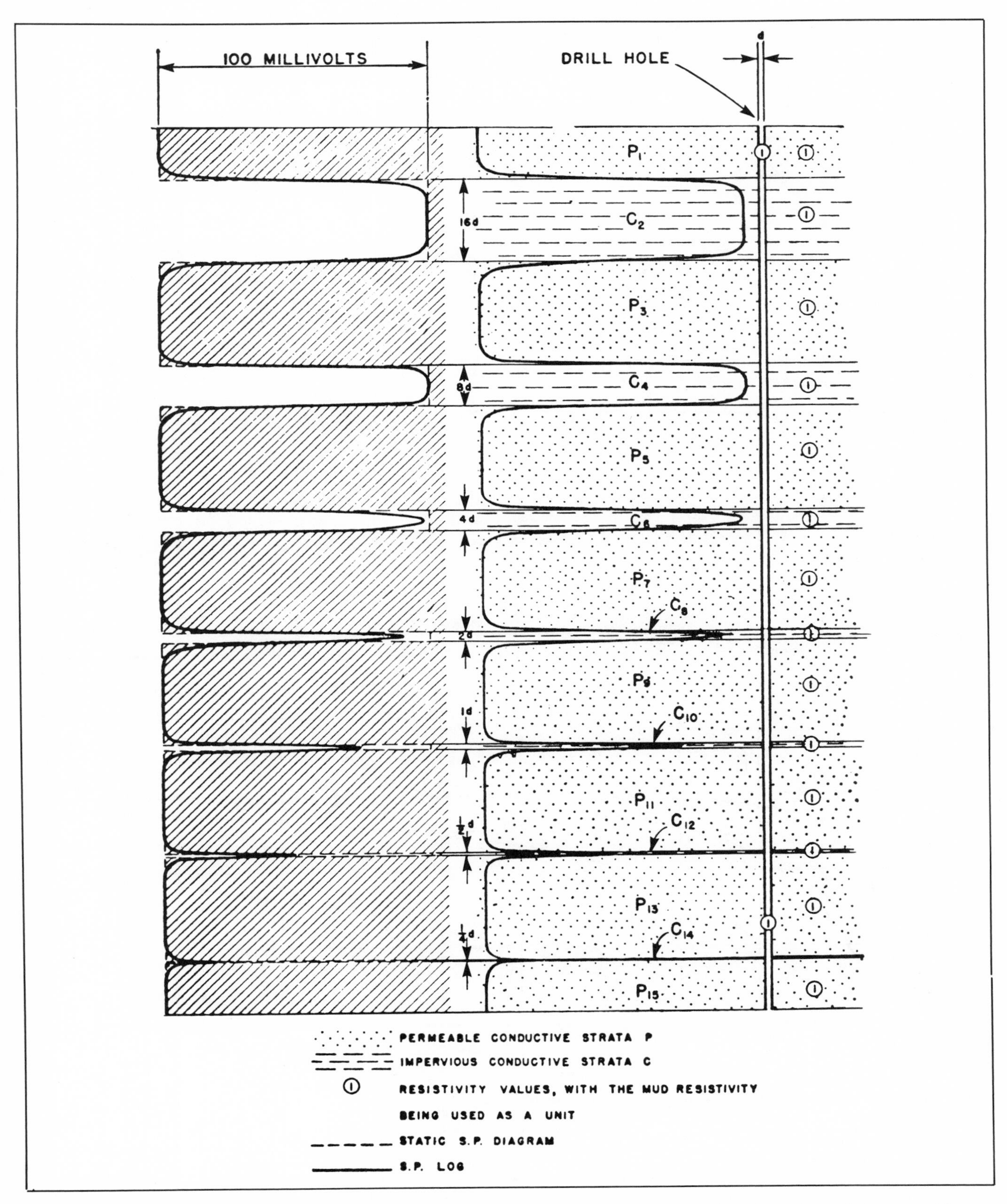

Fig. 5-14. Schematic example of bed thickness effects for varying thicknesses of impervious beds in a thick permeable section, $R_t = R_m$. Permission to publish by the Society of Petroleum Engineers of AIME.[2] Copyright 1949 SPE-AIME.

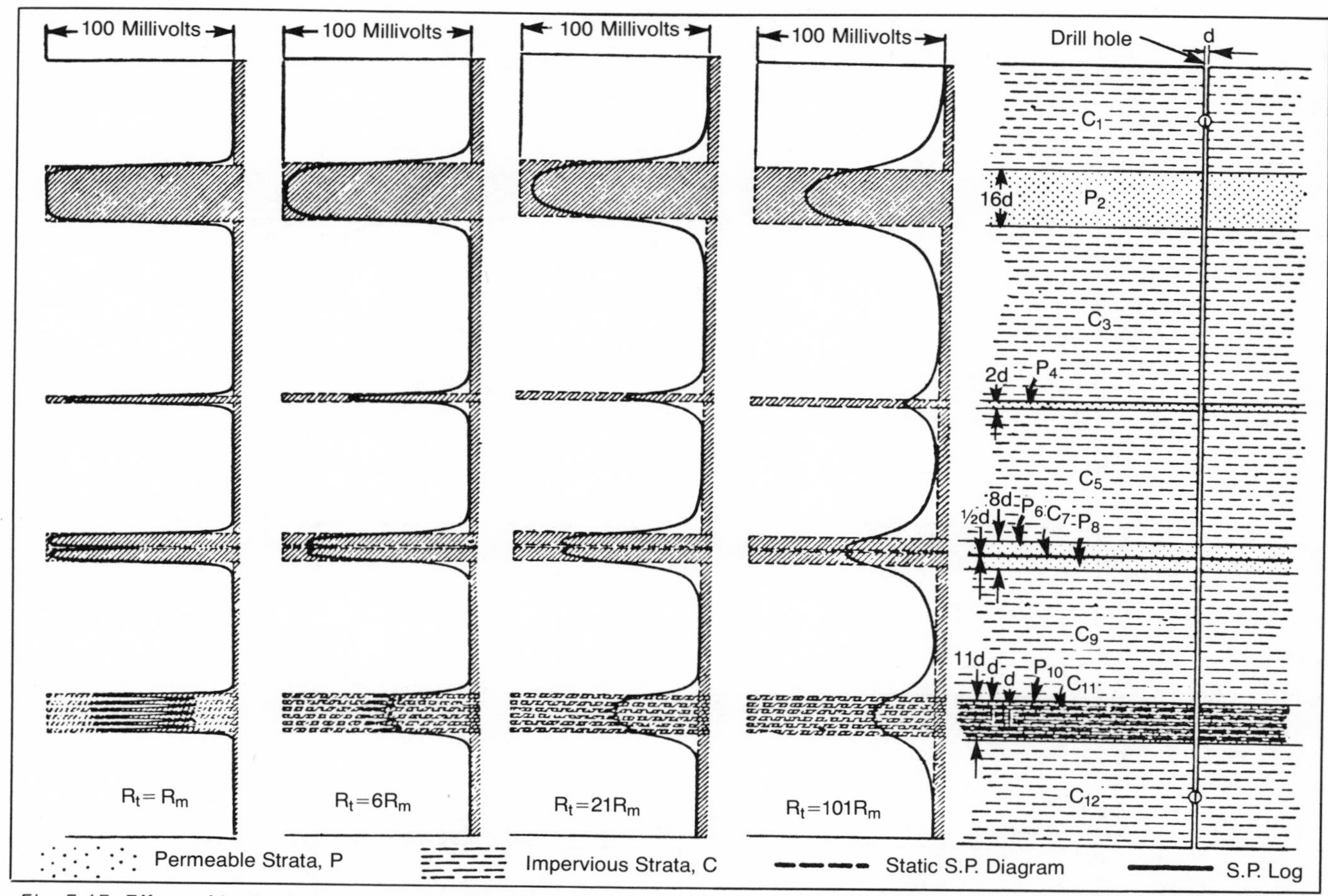

Fig. 5-15. Effect of bed resistivity on the SP log. Permission to publish by the Society of Petroleum Engineers of AIME.[2]

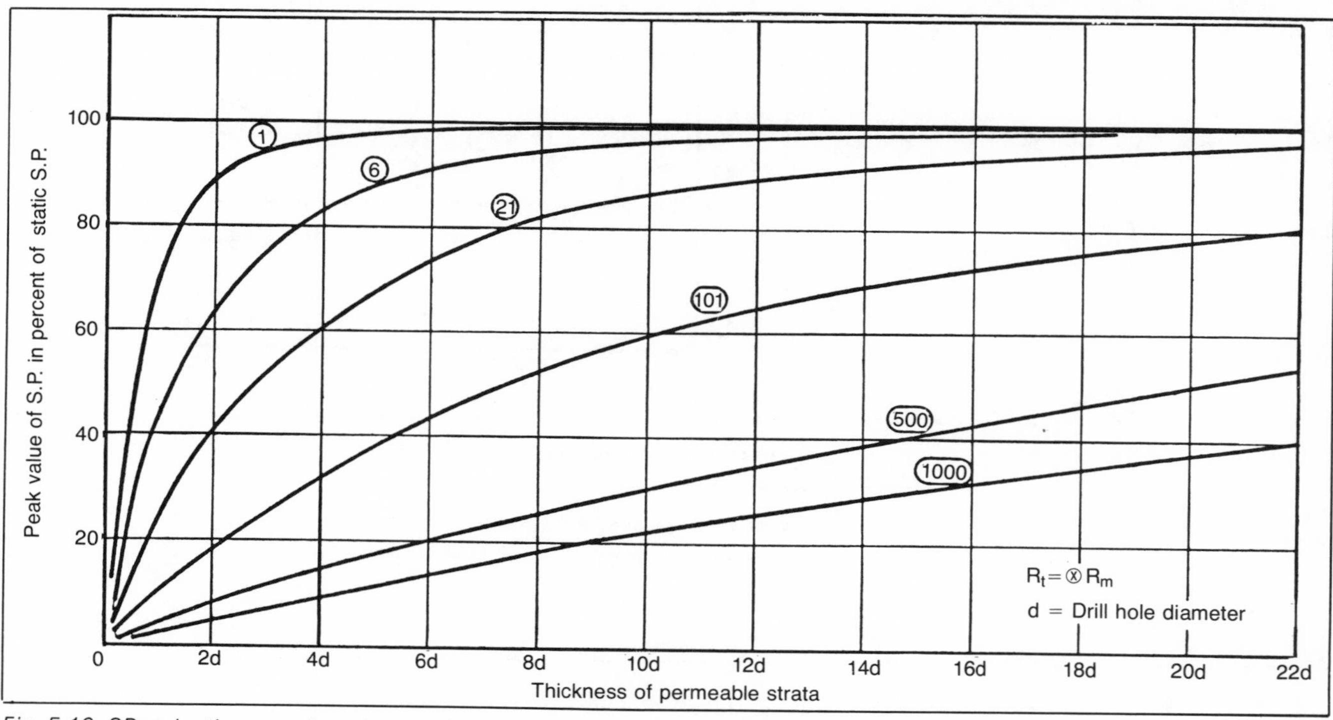

Fig. 5-16. SP reduction as a function of bed thickness and bed resistivity. Permission to publish by the Society of Petroleum Engineers of AIME.[2] *Copyright 1949 SPE-AIME.*

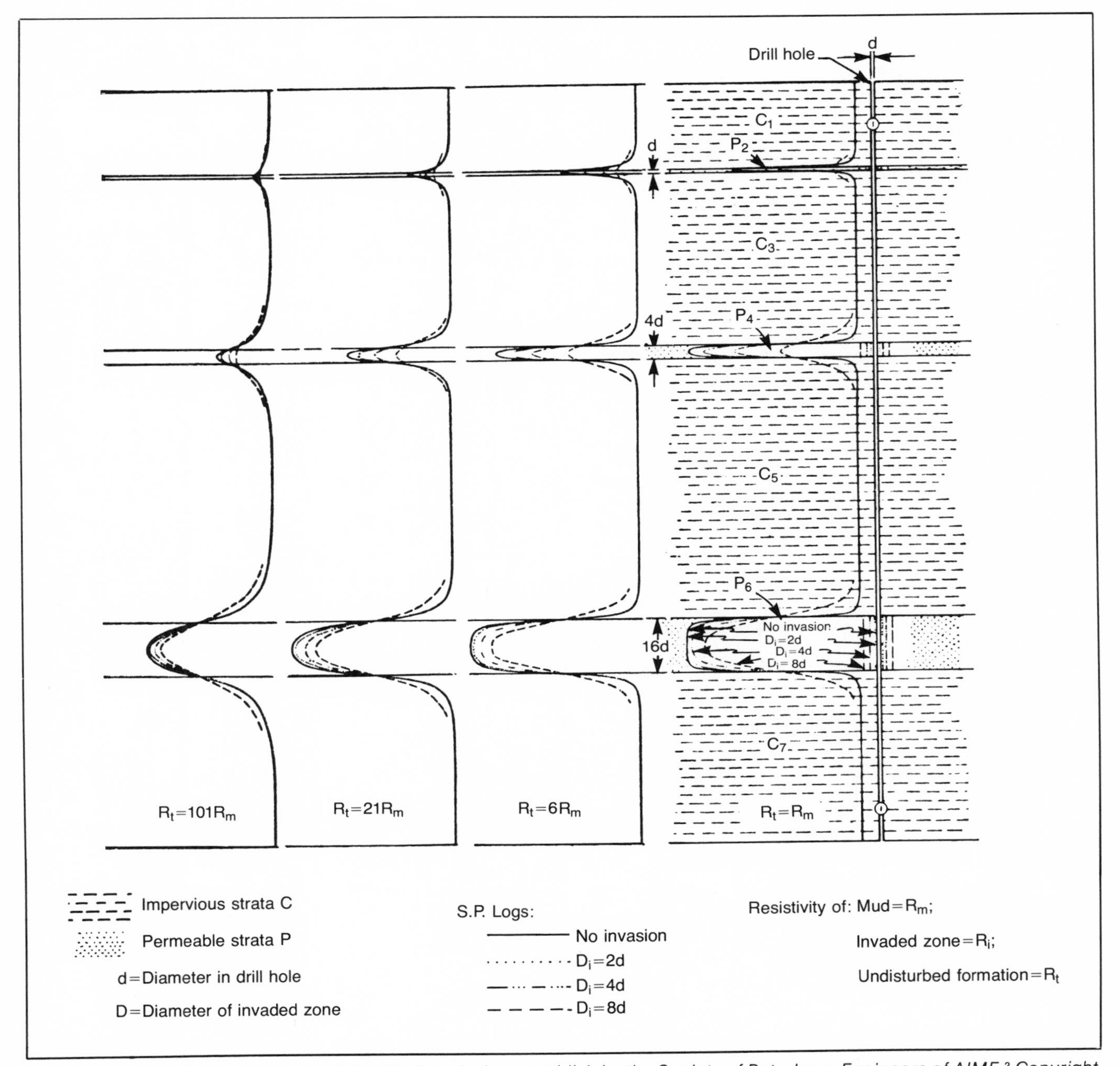

Fig. 5-17. Effect of invasion on the SP $R_i = R_t$. Permission to publish by the Society of Petroleum Engineers of AIME.[2] Copyright 1949 SPE-AIME.

amount of shaliness using the E_{psp} as determined from the shaly sand and comparing it to the E_{ssp} the sand should have if it were clean. The E_{ssp} may sometimes be assumed to be equal to that developed in a nearby clean sand, assuming the formation waters are the same. If this assumption seems valid, then $\propto = E_{psp}/E_{ssp}$ can be evaluated (if R_{mf} and R_w are known, then E_{ssp} can be calculated and $\propto$ determined).

Determination of R_w. A value for R_w can be determined from the SP if the formation is nonshaly, using:

$$E_{ssp} = {}^{-}K \log \frac{R_{mf}}{R_w}$$

In using this relationship, however, it is necessary to correct the measured SP for the influence of formation factors, which tend to reduce it from the maximum it would read in the clean sand. Figure 5-16 could be used to approximate the influence of bed thickness, bed resistivity, and borehole, but the correction charts of Figure

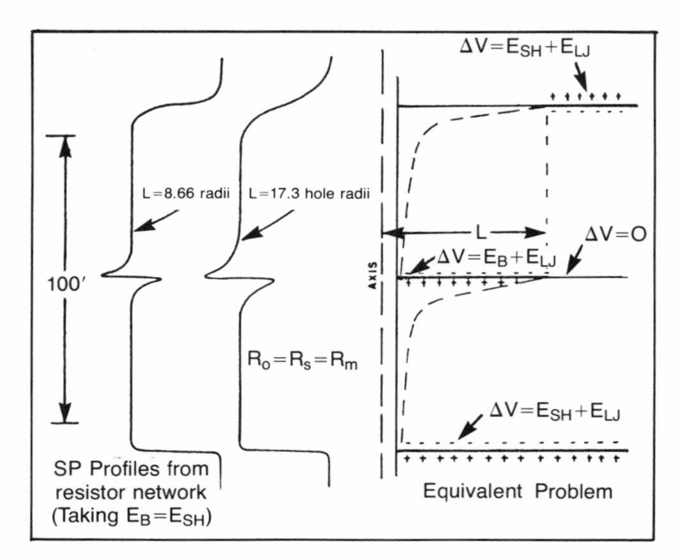

Fig. 5-18. Schematic of invaded zone producing sawtooth SP. Permission to publish by the Society of Petroleum Engineers of AIME.[8] Copyright 1959 SPE-AIME.

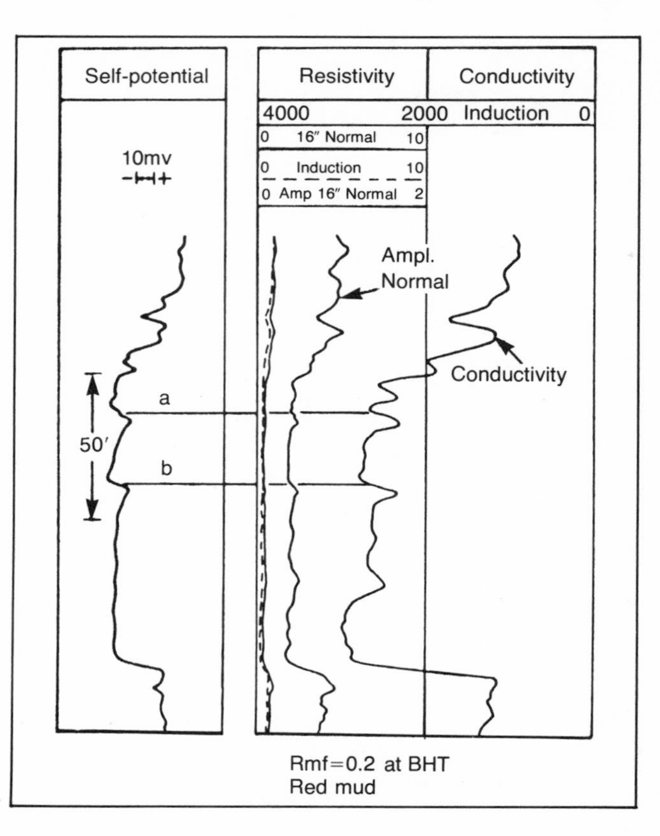

Fig. 5-19. Examples of sawtooth SP in a Gulf Coast water sand. Permission to publish by the Society of Petroleum Engineers of AIME.[8] Copyright 1959 SPE-AIME.

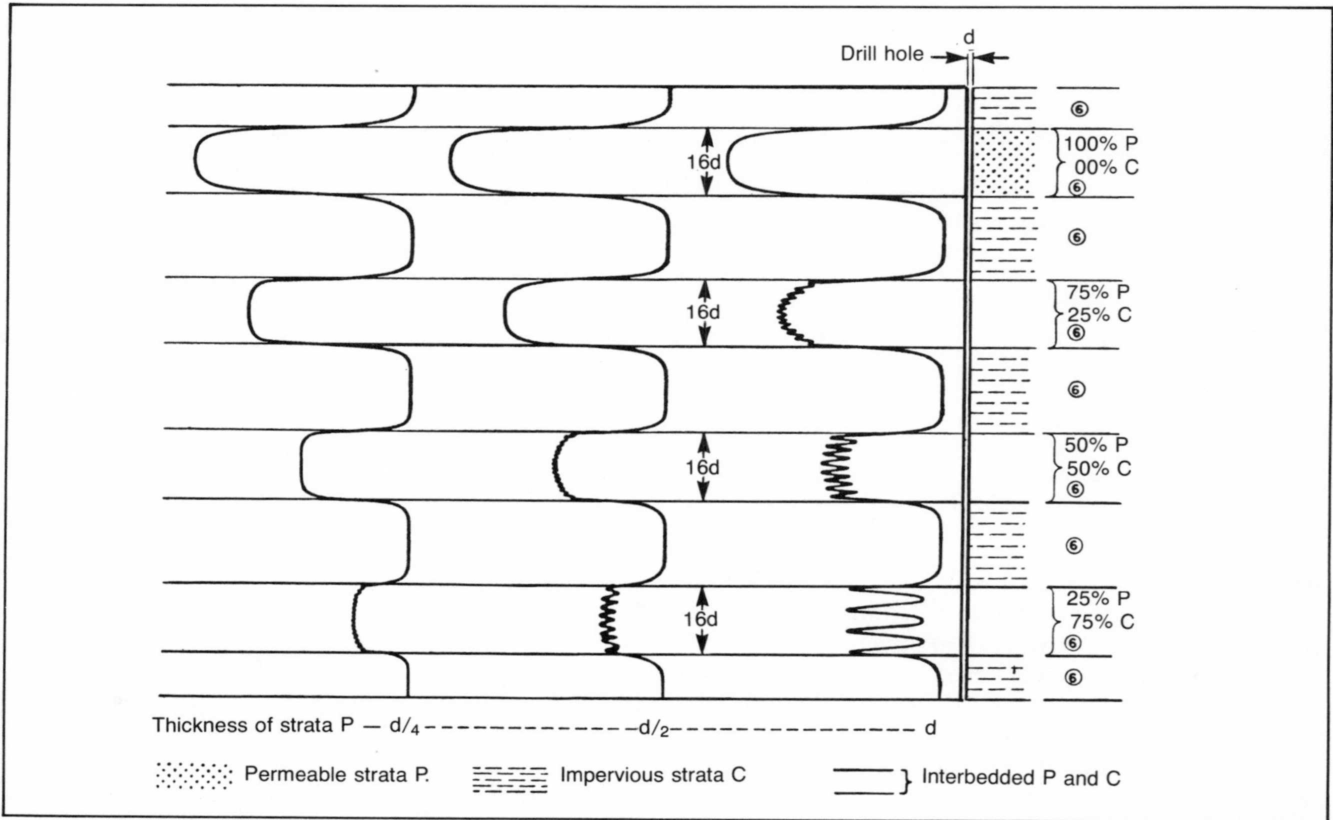

Fig. 5-20. Effect of shaliness on the SP. Permission to publish by the Society of Petroleum Engineers of AIME.[2] Copyright 1949 SPE-AIME.

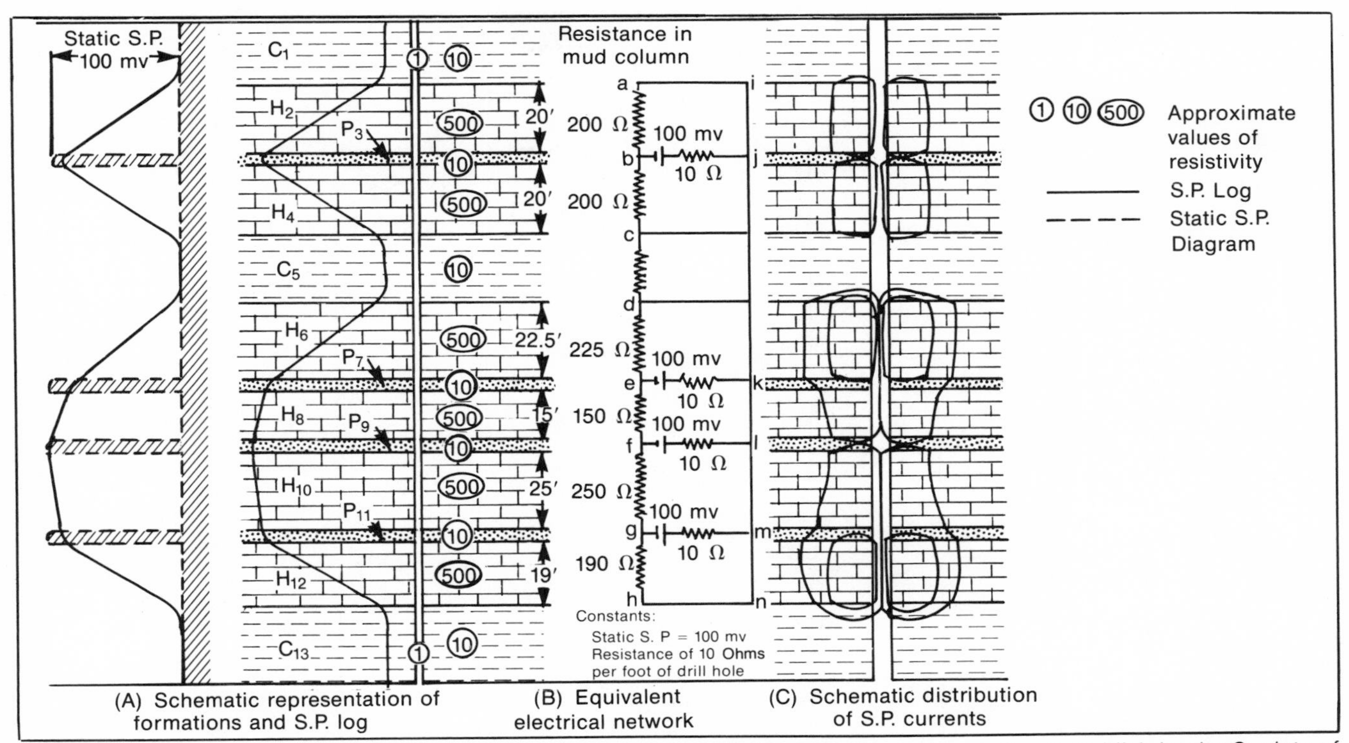

Fig. 5-21. Schematic of current distribution and resulting SP in resistive formations. Permission to publish by the Society of Petroleum Engineers of AIME.[2] Copyright 1949 SPE-AIME.

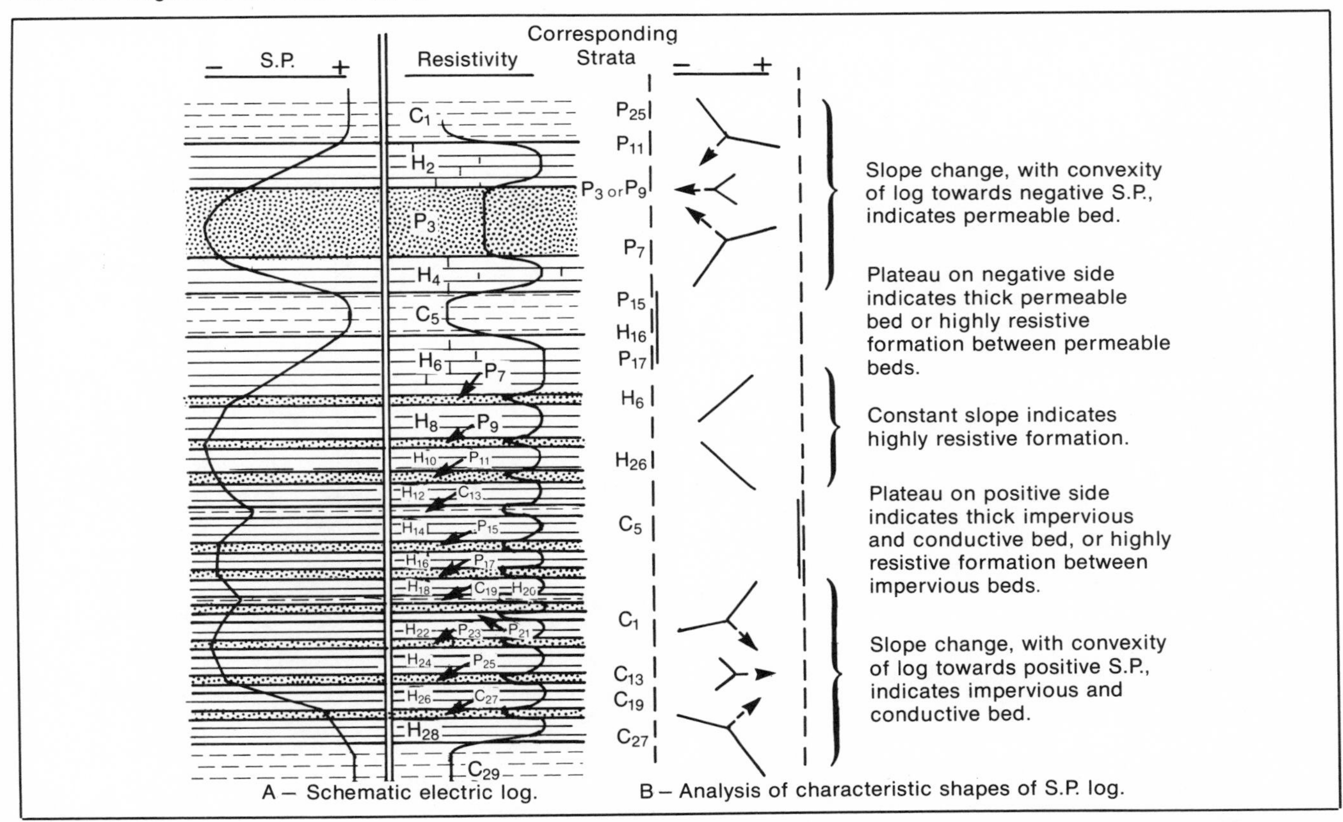

Fig. 5-22. Schematic of electric log in complex resistive system and guidelines for analysis. Permission to publish by the Society of Petroleum Engineers of AIME.[2] Copyright 1949 SPE-AIME.

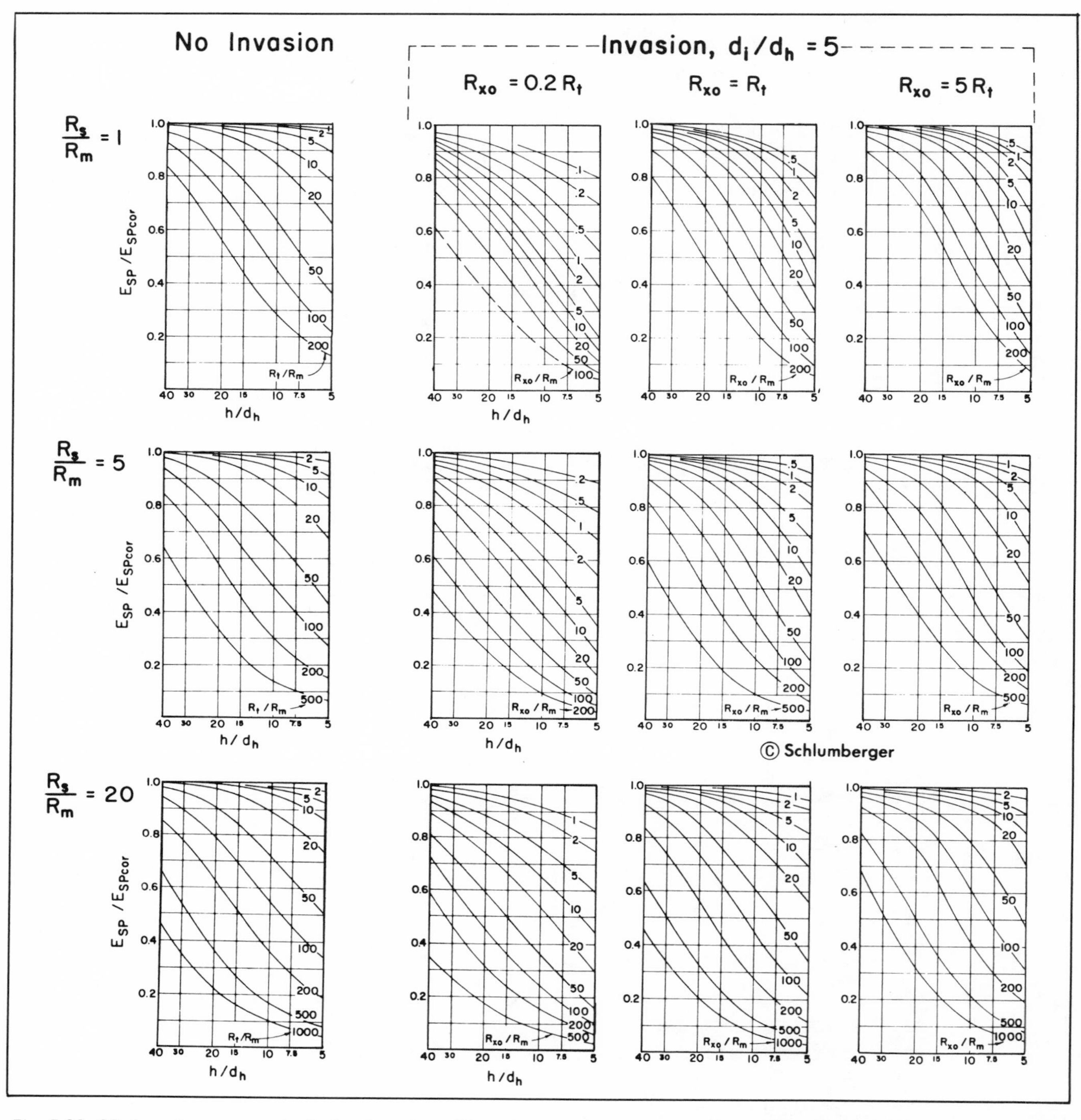

Fig. 5-23. SP departure curves including invasion effects for average cases. Permission to publish by Schlumberger Ltd.[16]

5-23 include the influence of invasion and are preferred. Therefore:

$$E_{sp} \xrightarrow[\text{departure curve}]{\text{Correct using}} E_{ssp} = -K_c \log \frac{R_{mf}}{R_w}$$

Although the SP departure curves shown in Figure 5-23 are the most comprehensive available, a simpler departure curve suitable for most analysis is shown in Figure 5-24.

Once a corrected value of the E_{sp} is obtained, then the SP expression can be solved for an apparent R_w. However, in the development of the SP equation, several major assumptions were made that are not always applicable in actual cases. It was assumed that the formation water and mud filtrate were NaCl solutions. It was also assumed that the activity ratio could be replaced by a resistivity ratio. As a result, the SP equation is used to calculate an equivalent water resistivity, R_{weq} as shown below:

$$R_{weq} = \frac{R_{mf}}{10^{(E_{ssp}/-K)}}$$

A comprehensive study by Gonduin, et al.[4] indicated that a better value of R_w could be obtained by using the empirical correspondence curves of Figure 5-25 when the above assumptions aren't valid. If R_{mf} is less than 0.1 Ωm at 75°F or in gyp muds, it is also desirable to correct R_{mf} at formation temperature to an R_{mfeq} for use in the above equation when solving for R_{weq}. Therefore:

$$R_{weq} \xrightarrow[\text{Chart}]{\text{Using Equivalency}} R_w$$

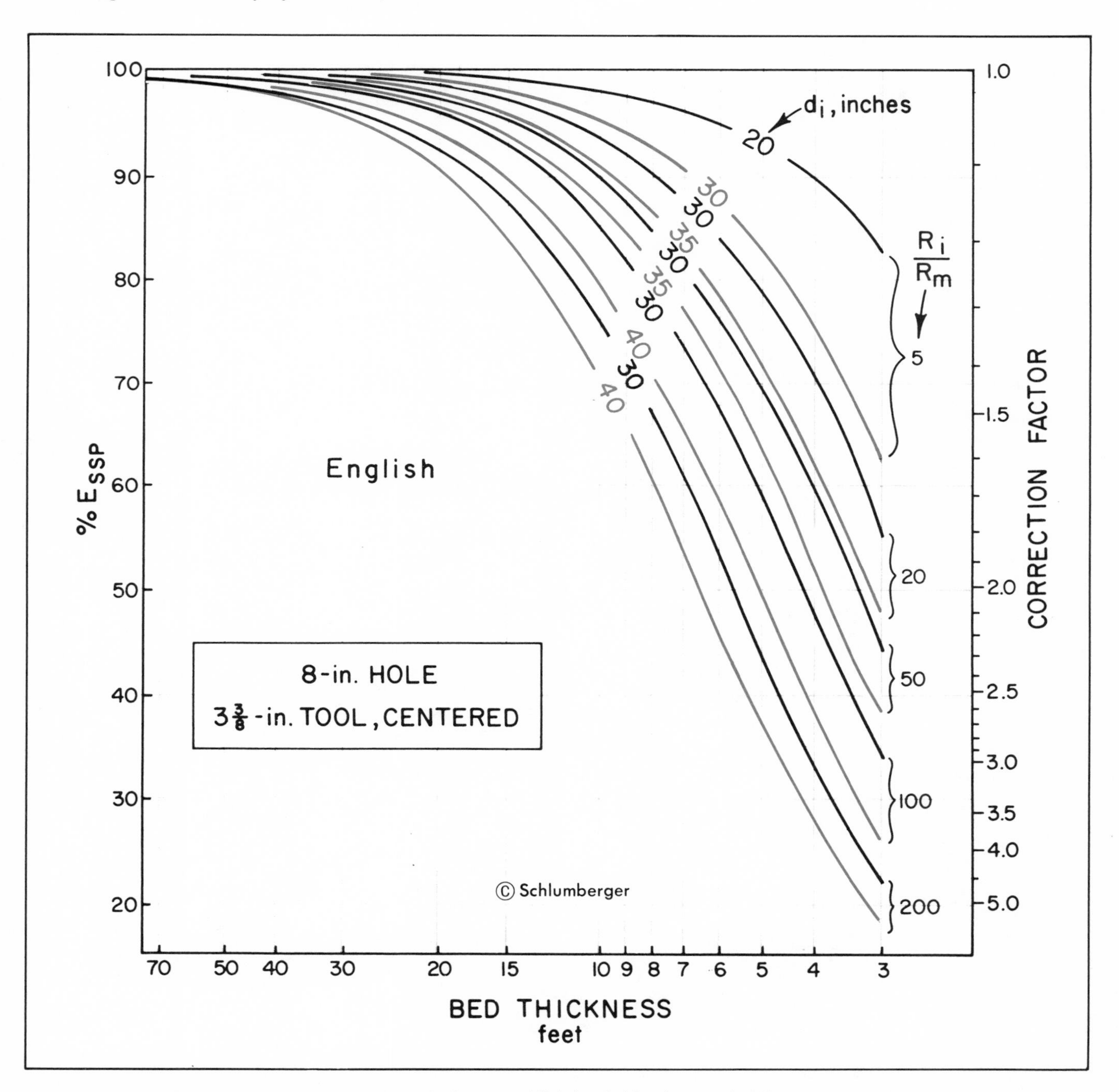

Fig. 5-24. Empirical SP departure curve. Permission to publish by Schlumberger Ltd.[16]

The method of processing an SP response to obtain R_w is outlined in Figure 5-26, and an example of this evaluation is shown in Figure 5-27. Finally, some obvious limitations should be pointed out: that the SP can't be run in nonconductive borehole fluids or in cased holes, and that the SP might not be diagnostic in salt mud logging systems where $R_{mf} \approx R_w$.

REFERENCES

1. Mounce, William D., and Rust, W.M., Jr.: "Natural Potentials in Well Logging," *Trans.*, AIME (1944) 155.
2. Doll, H.G.: "The S.P. Log: Theoretical Analysis and Principles of Interpretation," *Trans.*, AIME (1949).
3. Wyllie, M.R.J.: "An Investigation of the Electrokinetic Component of the Self Potential Curve," *Trans.*, AIME (1951).

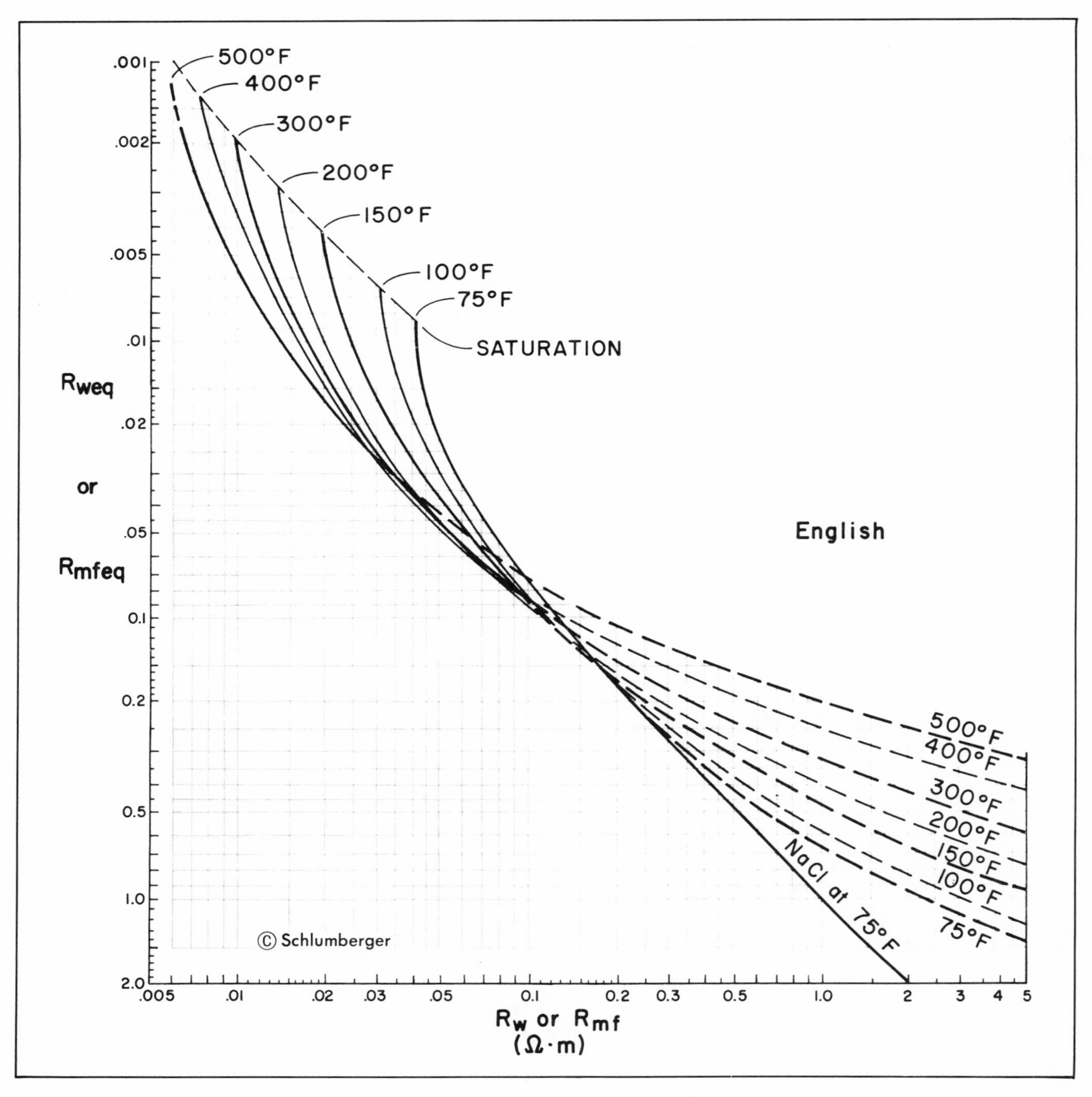

Fig. 5-25. Correspondence chart relating R_{weq} to R_w. Permission to publish by Schlumberger Ltd.[16]

$$E_{sp} \xrightarrow[\text{Fig. 16, 23, 24}]{\text{Correct using}} E_{ssp} = K_c \log \frac{R_{mf}}{R_{weq}}$$

$$K_c = 61 + 0.133\, T_{(°F)}$$

$$R_{weq} = \frac{R_{mf}}{10^{\left(\frac{E_{ssp}}{-K_c}\right)}}$$

$$R_{weq} \xrightarrow[\text{Fig. 25}]{\text{Correct using}} R_w$$

Fig. 5-26. Stepwise method of processing SP response for R_w

4. Gonduin, Michael, Tixier, Maurice P., and Simard, G.L.: "An Experimental Study on the Influence of the Chemical Composition of Electrolytes on the S.P. Curve," *Trans.*, AIME (1957).
5. Wyllie, M.R.J.: *The Fundamentals of Electric Log Interpretation*, Second edition, Academic Press Inc., New York (1957).
6. Gonduin, Michael, and Scala, Carl: "Streaming Potentials and the S.P. Log," *Trans.*, AIME (1958).
7. Hill, Harold J., and Anderson, A.E.: "Streaming Potential Phenomena in S.P. Log Interpretation," *Trans.*, AIME (1959).
8. Segesman, Francis, and Tixier, Maurice P.: "Some Effects of Invasion on the S.P. Curve," *Trans.*, AIME (1959).
9. Lynch, Edward J.: *Formation Evaluation*, Harper and Row, New York (1962).
10. Segesman, Francis: "New S.P. Correction Charts," *Geophysics* (Dec. 1962) 27.
11. Halbert, W.G., Jr.: "An Investigation of Streaming Potential in Pennsylvanian Age Limestones of North and West Texas," *The Log Analyst* (Feb. 1963) 3.
12. Truman, V.S.: "Streaming Potentials at Very High Differential Pressures," *J. Appl. Physics* (July 1963) 34.
13. Helander, Donald P.: "The Effect of Pore Configuration, Pressure and Temperature on Rock Resistivity," PhD dissertation, U. of Oklahoma, Norman, OK (1965).
14. Althaus, V.E.: " Electrokinetic Potentials in South Louisiana Tertiary Sediments," *The Log Analyst* (July 1967) 8.
15. *Log Interpretation Principles*, Schlumberger Ltd., New York (1969).
16. *Log Interpretation Charts*, Schlumberger Ltd., New York (1979).

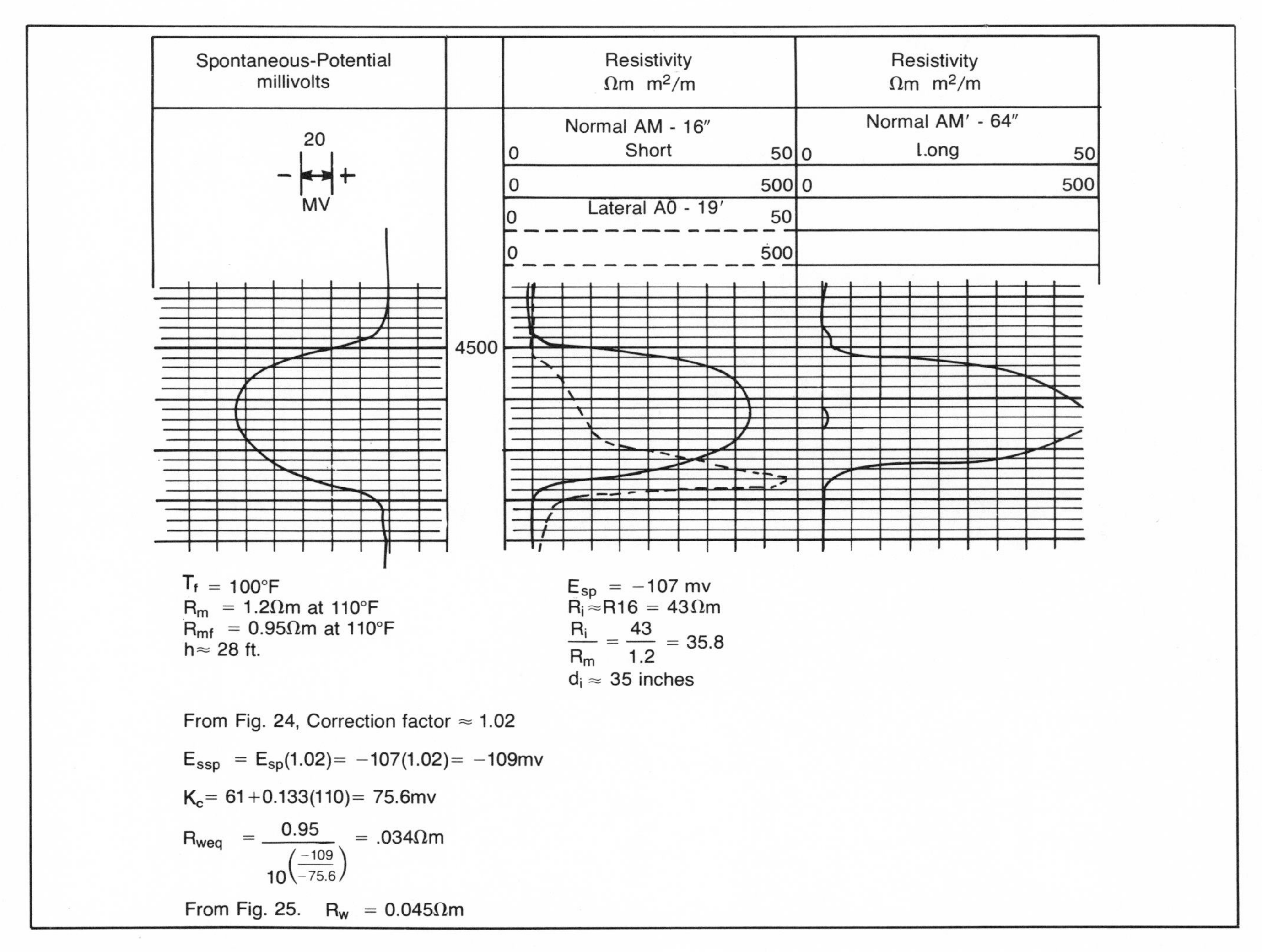

Fig. 5-27. Example of an R_w calculation from log data

6 Resistivity Measurement

DETERMINATION of the true formation resistivity, R_t (resistivity of the undisturbed formation) is required in order to quantitatively evaluate formations. A number of resistivity measuring tools have been developed in an attempt to determine R_t for the wide variety of conditions encountered in logging. The restrictions imposed by various borehole conditions and formation characteristics, even today, require different resistivity tools, since there is still no single resistivity logging tool applicable to all situations.

In addition to the deep-responding resistivity devices for determining R_t, a number of shallow-responding resistivity devices have been developed that are directed toward the measurement of R_{xo} and R_i.

This chapter will briefly summarize the three basic types of resistivity measurements. The chronological sequence in the development of these resistivity measuring methods is shown below:

1. Nonfocused Current Resistivity Measurements: normal resistivity, lateral resistivity, and micro-resistivity
2. Focused Current Resistivity Measurements: guard electrode resistivity, point electrode resistivity, micro-focused resistivity
3. Induction Resistivity Measurements: five-coil system and six-coil system
4. Multiple Resistivity Measurements: Dual Induction-Laterolog 8 or Dual Induction-SFL, and Dual Laterolog-R_{xo}

A combination of resistivity measurements with a lithology curve (SP or gamma ray) comprises the basic well log. These primary logs (surveys), listed in order of development, are:

1. Electrical survey, ES, which consists of an SP, normal resistivity (short and long), and the lateral resistivity curves.
2. Laterolog or Guard log, which consists of a gamma ray curve and a focused resistivity curve.
3. Induction log, which consists of gamma ray, conductivity, and reciprocated conductivity curves.
4. Induction Electrical Survey (IES or Induction-SFL), which consists of SP, conductivity, reciprocated conductivity, and short normal or SFL curves.
5. Dual Induction Laterolog-8 or Dual Induction-SFL; consists of an SP curve, two induction curves, and Laterolog-8 or SFL curve.
6. Dual Laterolog-R_{xo}, which consists of a gamma ray curve, two-point electrode focused resistivity curves, and a spherically focused micro-resistivity curve (R_{xo}).

These surveys may be considered the primary logs, since they generally provide the basic formation-depth record to which all other logs are referred. Usually all wells will have one of the above primary logs run in it.

NONFOCUSED RESISTIVITY MEASUREMENTS

Normal Resistivity

The normal resistivity measurement was one of the first to be run in a well and may still be recorded today. A schematic of the normal measuring device is shown in Figure 6-1. This device is considered to be a two-electrode system, although in practice three of the electrodes are usually in the hole.

If we consider electrode A to be a point source for current, and if it is in a homogeneous conducting medium of infinite extent, and if electrode B is at an infinite distance from electrode A, the current will flow spherically from A. A point electrode placed any distance from A will, therefore, define an equipotential sphere. A schematic of this ideal, spherical, electrical flow system is shown (two-dimensionally) in Figure 6-2.

Electrodes M and N, therefore, define two such equipotential spheres of radii lM and lN from electrode A, respectively. The potentials at these equipotential

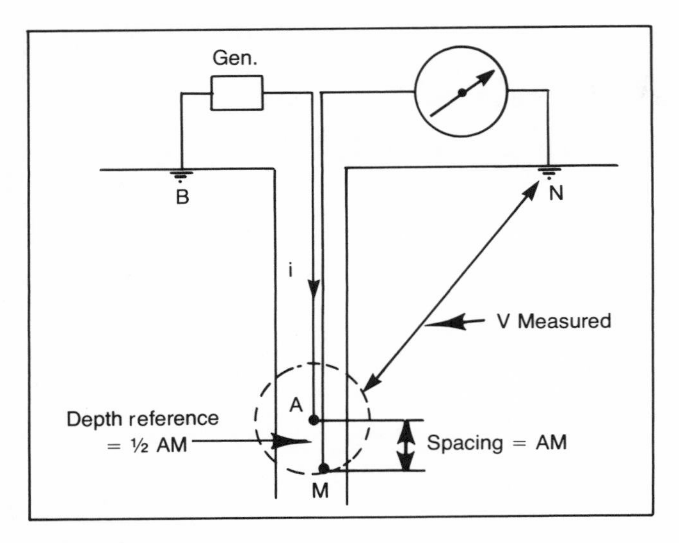

Fig. 6-1. Schematic of diagram of normal system

spheres defined by M and N are E_M and E_N respectively. Now for a volume element lying between M and N having a radius l and a thickness dl, the potentials will be E and $E-dE$ respectively. Writing Ohm's law [$E=ir=iR[l/A]$) for this volume element we have:

$$E-(E-dE)=\frac{iR[l-(l+dl)]}{4\pi l^2}$$

Since current flow is constant, we can integrate the following expression from M to N:

$$\int_{E_M}^{E_N} dE = -\int_{l_M}^{l_N} \frac{iRdl}{4\pi l^2}$$

and obtain:

$$E_M - E_N = \frac{iR}{4\pi}\left[\frac{1}{l_M}-\frac{1}{l_N}\right]$$

For the normal devices l_N is at electrical infinity and $l_M=AM$, which is referred to as the spacing. Considering the measured potential difference $E_M-E_N=V$ we can write:

$$V=\frac{iR}{4\pi}\left[\frac{1}{AM}\right]$$

or:

$$R=\frac{V}{i}(4\pi AM)$$

where $4\pi AM=K'$. K' is generally called the sonde constant dependent only on the spacing AM. Since current, i, is constant, $R \propto V$. In actual logging practice, galvanometers responding to potential changes are used to record resistivity.

Typical Curve Shapes. Extensive work on the response of electrical logging systems such as the normal resistivity tool has been done by Guyod.[1, 10] Typical resistivity response behavior for the normal electrode arrangement is shown in Figure 6-3.

These curves are based on the assumptions that there is no invasion, adjacent beds are very thick and $R_s=R_m$, borehole perpendicular to the formation, and the sonde is centered in the hole. It should be noted that the normal curve is symmetrical and that for beds less than the critical spacing (bed thickness, $h=AM$) the normal resistivity indicates a conductive bed (resistivity reversal).

Factors Affecting Normal Measurements. Obviously, we don't have an ideal system as shown in Figure 6-2, but one in which a number of factors influence the log response such that the recorded resistivity is an apparent resistivity, R_a, rather than the desired resistivity (either R_t or R_i). Factors that must be considered as influencing the measured potential in addition to the undisturbed zone are the borehole, the invaded zone, formation thickness, and the surrounding beds.

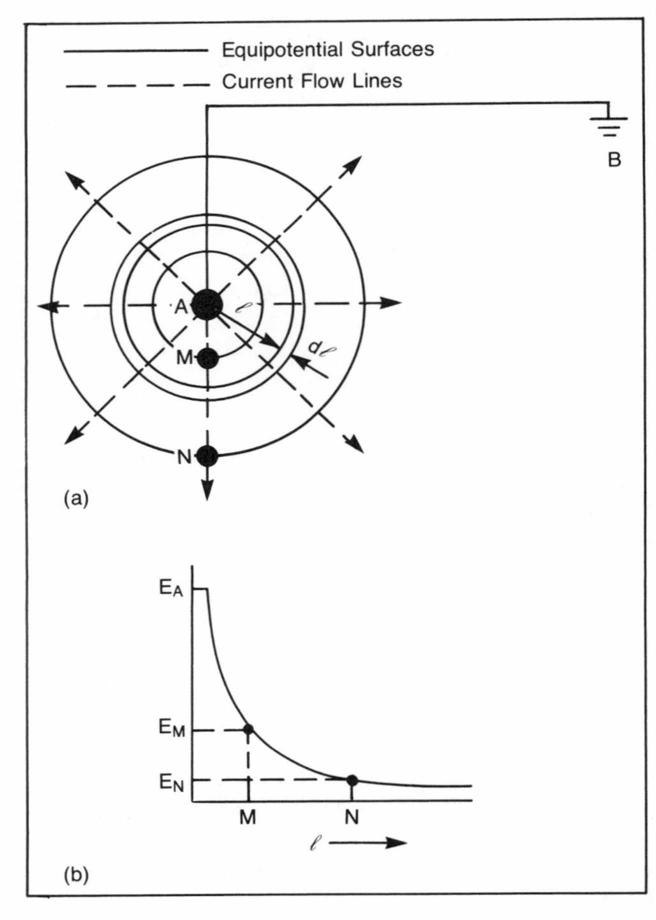

Fig. 6-2. Schematic of potential distribution for current flow in indefinite, homogeneous, conducting medium. Permission to publish by Harper and Row Publishing Co.[16]

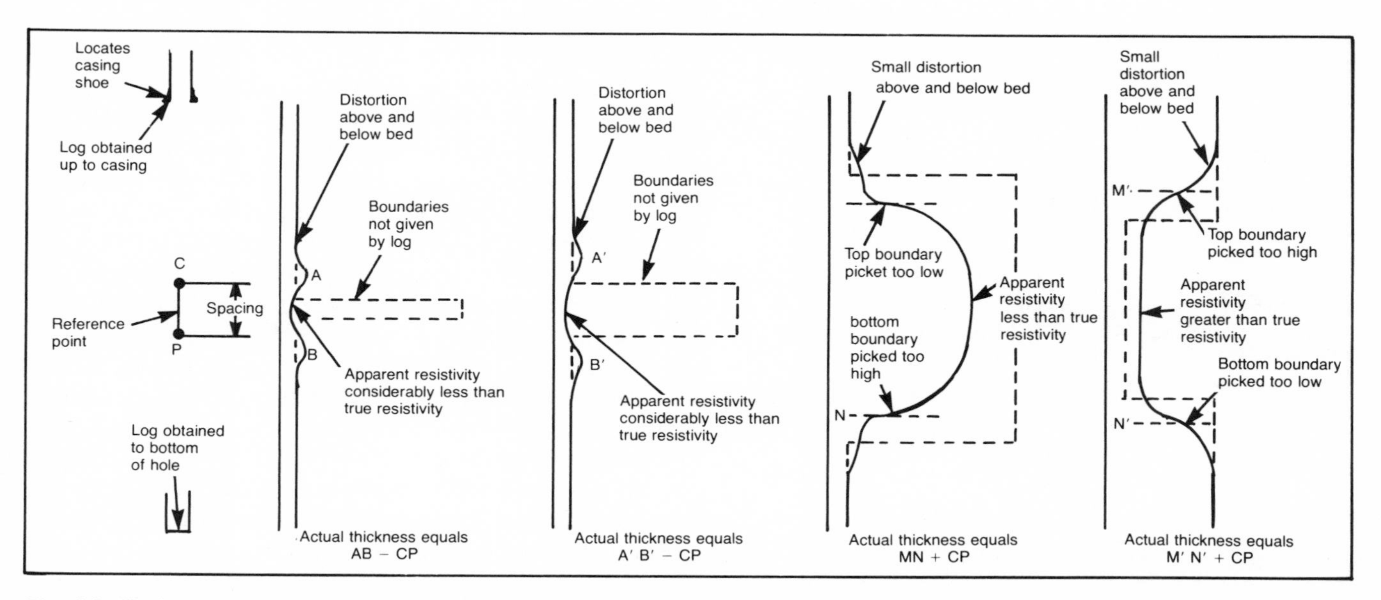

Fig. 6-3. Typical responses for normal measuring system. Permission to publish by Welex, a Halliburton Company.[1]

In the theoretical consideration of the normal system, the electrode system was assumed to be in a homogeneous, conducting medium of infinite extent. No consideration was given to the presence of the borehole, and since its resistivity is usually less than that of the formation of interest, considerable current distortion can occur. The magnitude of the borehole effect is a function of the size of the borehole and the relative resistivity between the mud and formation. Departure curves to evaluate the borehole effects are available and will be discussed later in this chapter.

Influence of the invaded zone is a function of the depth of invasion and its resistivity. If invasion is deep, most of the response is from the invaded zone, and in this case the response of the short normal, $R_{16''}$, can be corrected for other factors (borehole and surrounding beds) to read invaded zone resistivity, R_i. However, unless d_i is known, the response from a normal resistivity tool cannot be interpreted in terms of R_t, R_i, or R_{xo} using the available departure curves.

Spacing of the tool controls the vertical resolution of the normal device (see Fig. 6-3). It is apparent that as the spacing, AM, is increased to obtain deeper investigation, the bed thickness must be greater in order to obtain a response representative of the formation of interest. This is one of the reasons the 64 in. normal is seldom run today for R_t determinations.

The influence of surrounding beds depends upon their resistivity relative to the resistivity of the formation of interest. When resistivity contrast is high, the current flow is highly distorted, and the apparent resistivity recorded by the normal tool must be corrected.

Applications of Normal Measurements. The normal devices are used to determine bed boundaries and formation resistivities such as R_t and R_i.

Bed boundaries can be determined from the normal curves but are a function of the spacing, AM. The apparent thickness of the formation as selected from the curve where the inflection point occurs will be less than the true formation thickness by an amount equal to $2AM$ (see Fig. 6-3).

In order to determine the actual formation resistivity, it is necessary to correct for the effect of the borehole, invasion, bed thickness and surrounding beds on the normal responses. Two normal measurements have historically been run in electrical surveys, a short normal having a 16 in.-spacing ($AM = 16$ in.) and a long normal having a spacing of 64 in. ($AM = 64$ in.). The short normal is directed toward an estimate of R_i. However, the apparent response is subject to the influence of many factors as mentioned earlier such that:

$$R_a = R_{16''} = f(R_m,\ d_h,\ h,\ R_s,\ d_i,\ \underline{\underline{R_i}},\ R_t)$$

In order to obtain R_i the departure of the measured resistivity, R_a (in this case $R_{16''}$), should be evaluated. Under ideal circumstances this departure is small, and for all practical considerations $R_{16''} = R_i$ is usually assumed.

The long normal, on the other hand, is directed toward a measure of R_t. In this case, as will be the case for all resistivity responses, the apparent resistivity that is measured departs from the desired value due to a number of factors such that:

$$R_a = R_{64''} = f(R_m,\ d_h,\ h,\ R_s,\ d_i,\ R_i,\ \underline{\underline{R_t}})$$

As with the short normal, in order to obtain a specific resistivity, in this case, the true resistivity, R_t, the in-

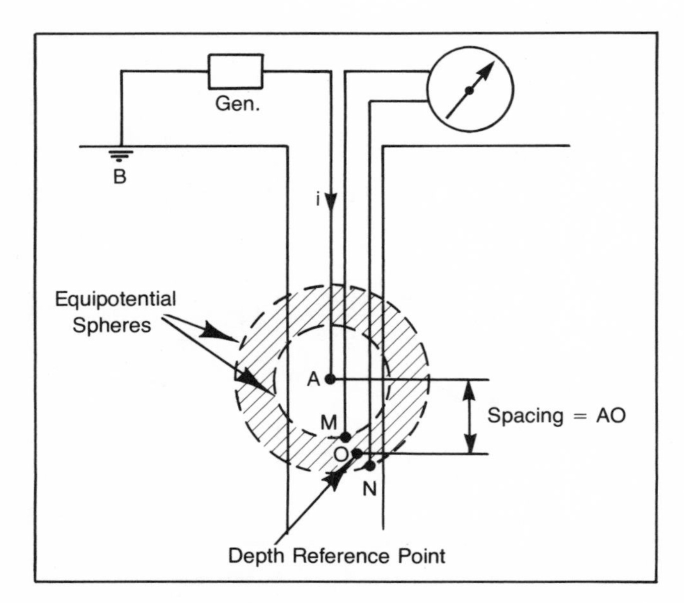

Fig. 6-4. Schematic diagram of lateral system

fluence of all these factors on the recorded response, $R_{64''}$, must be evaluated using departure curves developed for this purpose. Comprehensive departure curves have been developed by Guyod[10] and Schlumberger[3] based on mathematical analysis of simple geometries and electrical analogues. These departure curves are time-consuming to apply and, for most uses, simplified departure curves will give values within the range of accuracy desired. An approximate value of R_t from the apparent resistivity measured with the 64 in. normal can be obtained by using Table 6-1. For more precise evaluations of R_t from $R_{64''}$, detailed departure curves should be used.

Limitations of Normal Measurements. The normal curves cannot be used to determine formation resistivity under the following conditions:

1. Borehole fluid is nonconductive.
2. Borehole is cased.
3. Bed thickness is equal to or less than the spacing.
4. When R_t/R_m is high since the current distortion becomes too large to adequately correct R_a.

Lateral Resistivity

The lateral type of resistivity measurement is seldom run today since the ES log has been for the most part replaced by Induction-Electrical logs. Spacing of the lateral curve has varied from 6 to 24 ft as a function of the varying borehole and formation conditions encountered, although one common spacing is 18′8″ (19′). The long spacing is used for deep investigation such that the lateral curve is generally regarded as an R_t device. A schematic diagram of the lateral measuring device is shown in Figure 6-4. This type of device is considered a three-electrode system. The radius of investigation of this system is assumed to be equal to the AO spacing or as in the common case, 18′8″.

If we consider this electrode system in a homogeneous conducting medium of infinite extent as we did for the normal system, we can write the expression

$$E_M - E_N = V = \frac{iR}{4\pi}\left[\frac{1}{AM} - \frac{1}{AN}\right] = \frac{iR}{4\pi}\left[\frac{MN}{AM \times AN}\right]$$

or in terms of resistivity,

$$R = \frac{V}{i}\left[4\pi\left(\frac{AM \times AN}{MN}\right)\right]$$

where the sonde constant $K'' = 4\pi[(AM \times AN)/MN]$ is also only dependent on the electrode arrangement. Again, since current, i, is constant $R \propto V$ such that the potential drop occurring between M and N is recorded in terms of resistivity.

Typical Curve Shapes. Typical resistivity response behavior for the lateral electrode arrangement is shown in Figures 6-5 and 6-6. These curves are based on the same assumptions as presented for the normal curve shapes. It should be noted that the lateral curve response is unsymmetrical, such that only one resistivity value (R_a) will be considered to represent the formation resistivity (R_a value to be corrected to R_t). For the thick beds, it can also be seen that the lateral curve will define one of the bed boundaries depending on whether the standard or inverted electrode arrangement is used.

Factors Affecting Lateral Measurements. As with the

TABLE 6-1
Simplified Methods for Obtaining R_t from $R_{64''}$

Bed Thickness (h)	*Qualifications* *In Low Resistivity, when $R_{64''}/R_m < 10$ (invasion up to $2d_h$)*		*Device*	*Response*
$h > 20'$ ($> 4AM'$)			Long Normal	$R_{64''} = R_t$
$h \cong 15'(3AM')$	$R_m \cong R_s$	$R_{64''}/R_s \geq 2.5$	Long Normal	$R_{64''} = 2/3 R_t$
$h \cong 15'(3AM')$	$R_m \cong R_s$	$R_{64''}/R_s \leq 1.5$	Long Normal	$R_{64''} = R_t$
$h \cong 10'$ ($2AM'$)	$R_m \cong R_s$	$R_{64''}/R_s \leq 2.5$	Long Noral	$R_{64''} = 1/2 R_t$
$h \cong 10'$ ($2AM'$)	$R_m \cong R_s$	$R_{64''}/R_s = 1.5$	Long Normal	$R_{64''} = 2/3 R_t$

Permission to publish by Schlumberger Ltd.[29]

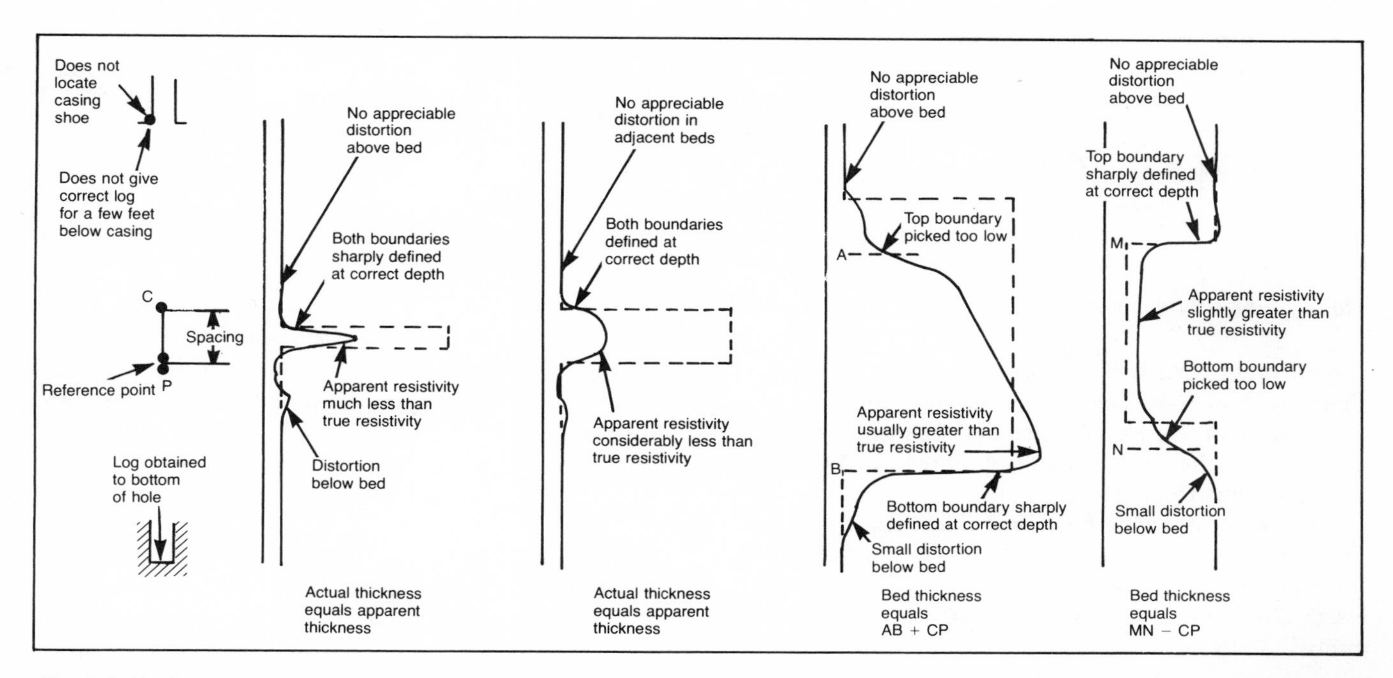

Fig. 6-5. Typical response curves for the lateral measuring system. Permission to publish by Welex, a Halliburton Company.[1]

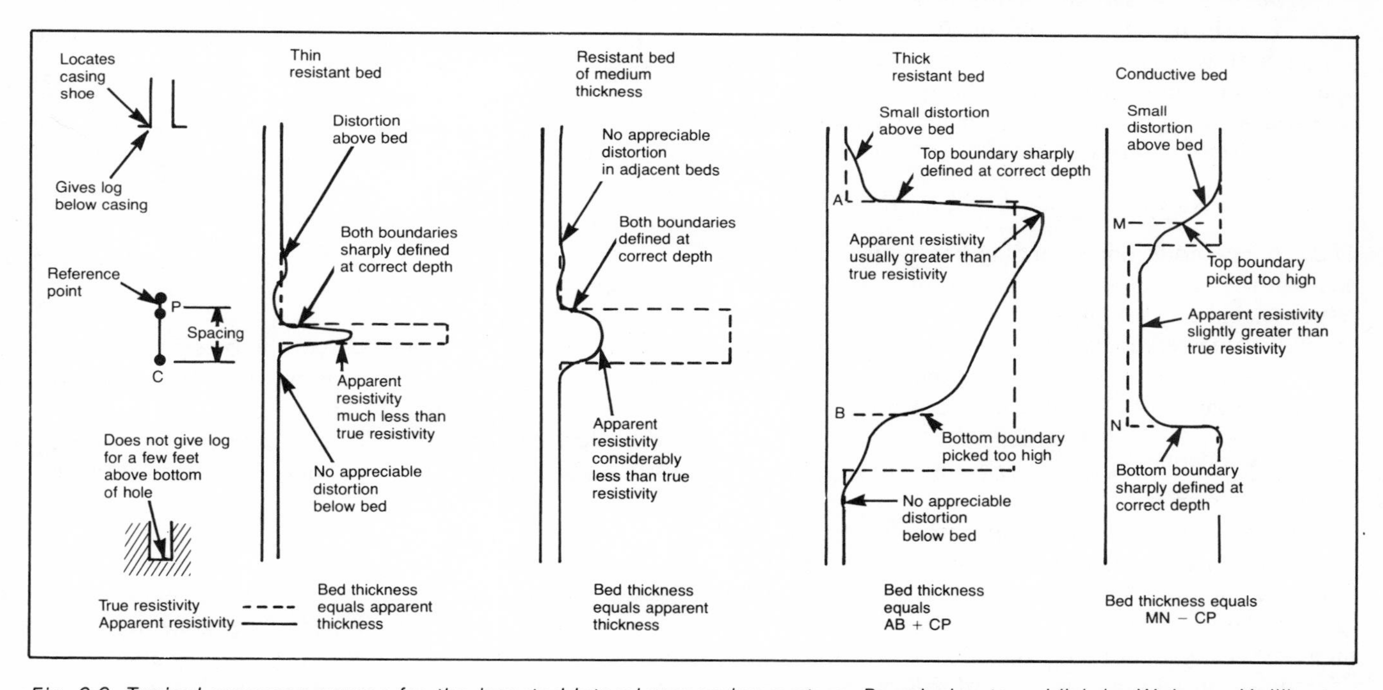

Fig. 6-6. Typical response curves for the inverted lateral measuring system. Permission to publish by Welex, a Halliburton Company.[1]

normal system there are a number of factors that influence the recorded log response: borehole, invasion, formation thickness, and surrounding beds.

Since the borehole was not considered in the theoretical development, current distortion will probably occur. However, because of the deep investigating characteristics of the lateral curve (18′8″ spacing), the effect is relatively insignificant until $R_a/R_m > 50$.

Due to the large radius of investigation of the 18′8″ lateral device and also since it measures only a spherical shell of material, the influence of the invaded zone is usually small. For cases where a high resistivity contrast does exist between R_i and R_t, and d_i is large, it may be necessary to correct for the effects of the invaded zone.

The response of the lateral curve to formation thickness is a function of the spacing and varies considerably

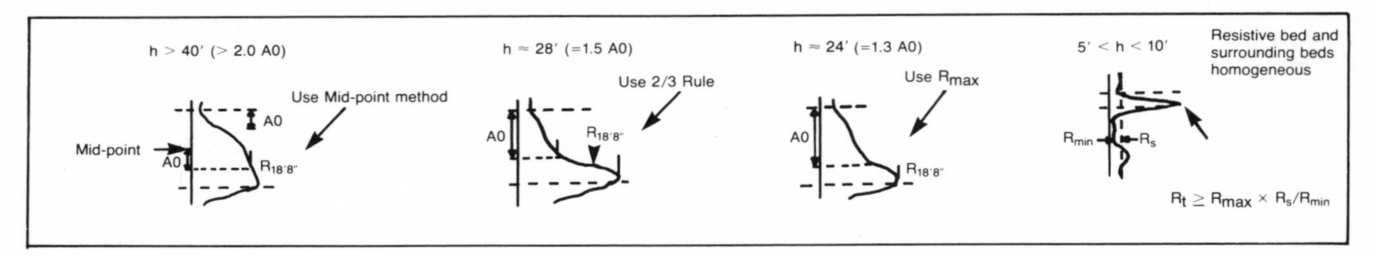

Fig. 6-7. Rules for estimating R_a from the 18′18″ lateral curve. Permission to publish by Schlumberger Ltd.[24]

for different formation thicknesses, as can be seen in Figures 6-5 and 6-6. This behavior of the lateral curve as a function of formation thickness makes the lateral curve one of the most difficult of all recorded borehole measurements to interpret.

For formations where the bed thickness is less than the *AO* spacing, the surrounding beds will have a large influence on the lateral response if a resistivity contrast exists. The curve becomes quite complex and essentially impossible to interpret.

Applications of Lateral Measurements. The lateral curve is principally used as an R_t measuring tool, although it is possible to determine one of the bed boundaries in thick formations. The apparent resistivity recorded with this electrode arrangement is also affected by many factors and as a result:

$$R_a = R_{18'8''} = f(R_m,\ d_h,\ h,\ R_s,\ d_i,\ R_i,\ \underline{\underline{R_t}})$$

There is one additional complicating factor, however, due to the unsymmetrical shape of the resistivity response. It becomes apparent that a problem exists as to what value represents R_a. A series of simple guidelines have been developed for estimating R_a and are shown in Figure 6-7. This might be expressed as:

$$R_a \xrightarrow[\text{Fig. 6-7}]{\text{Guidelines}} R_{18'8''} = f(R_m,\ d_h,\ d_i,\ R_i,\ \underline{\underline{R_t}})$$

Since $R_{18'8''}$ is not necessarily equal to R_t, it then becomes unnecessary to account for other factors influencing this response. This apparent resistivity value can then be corrected for borehole effects using a chart such as that shown in Figure 6-8. Where invasion is considered an important factor influencing the lateral response or more detailed log analysis is desired, more complete departure curves should be used.

Limitations of Lateral Measurements. The lateral curves cannot be used to determine formation resistivity under the following conditions:

1. Borehole fluid is nonconductive.
2. Borehole is cased.
3. Bed thicknesses between approximately 10 and 24 ft.
4. Series of thin beds (5–10 ft thick) such that the response is a complex combination of thin bed responses.

Electrical Survey

The electrical survey is most efficient when used in fresh mud, formations of low to moderate resistivity, and fairly thick beds. The hole must be uncased and filled with a conducting fluid. Hole size can range from 4¾ to 10 in. Maximum temperature and pressure conditions are generally 300°F and 10,000 psi, respectively. Logging speed is approximately 100 fpm. The electrical survey has been replaced by the more recently developed R_t measuring devices, but may still be run in some instances today since it is the cheapest type of primary log available. Familiarity with the curves run on this log is still essential, however, since many old ES logs are the only logs available for R_t determination.

Micro-Resistivity Measurement

Development of a nonfocused micro-resistivity measurement was dictated by the need to obtain a value of F_R leading to an indirect evaluation of R_o. At this stage in the development of the logging industry, a quantitative estimate of S_w required not only R_t but also R_o since:

$$I_R = S_w^{-n} = \frac{R_t}{R_o}$$

However, if a value of R_o isn't directly available from the recorded resistivity responses (water zone in formation of interest not penetrated), then the only alternative for estimating S_w is based on empirical relations such as the "Rocky Mountain" method developed by Tixier.[4]

In order to obtain F_R by some independent means, the fact that porous and permeable formations are invaded by mud filtrate was used. For this case, the usual problem of the presence of an invaded zone, and more specifically the flushed zone, provides the solution. A measure of R_{xo} could lead to F_R since:

$$F_R = \frac{R_{xo}}{R_{mf}}(1 - S_{or})^n$$

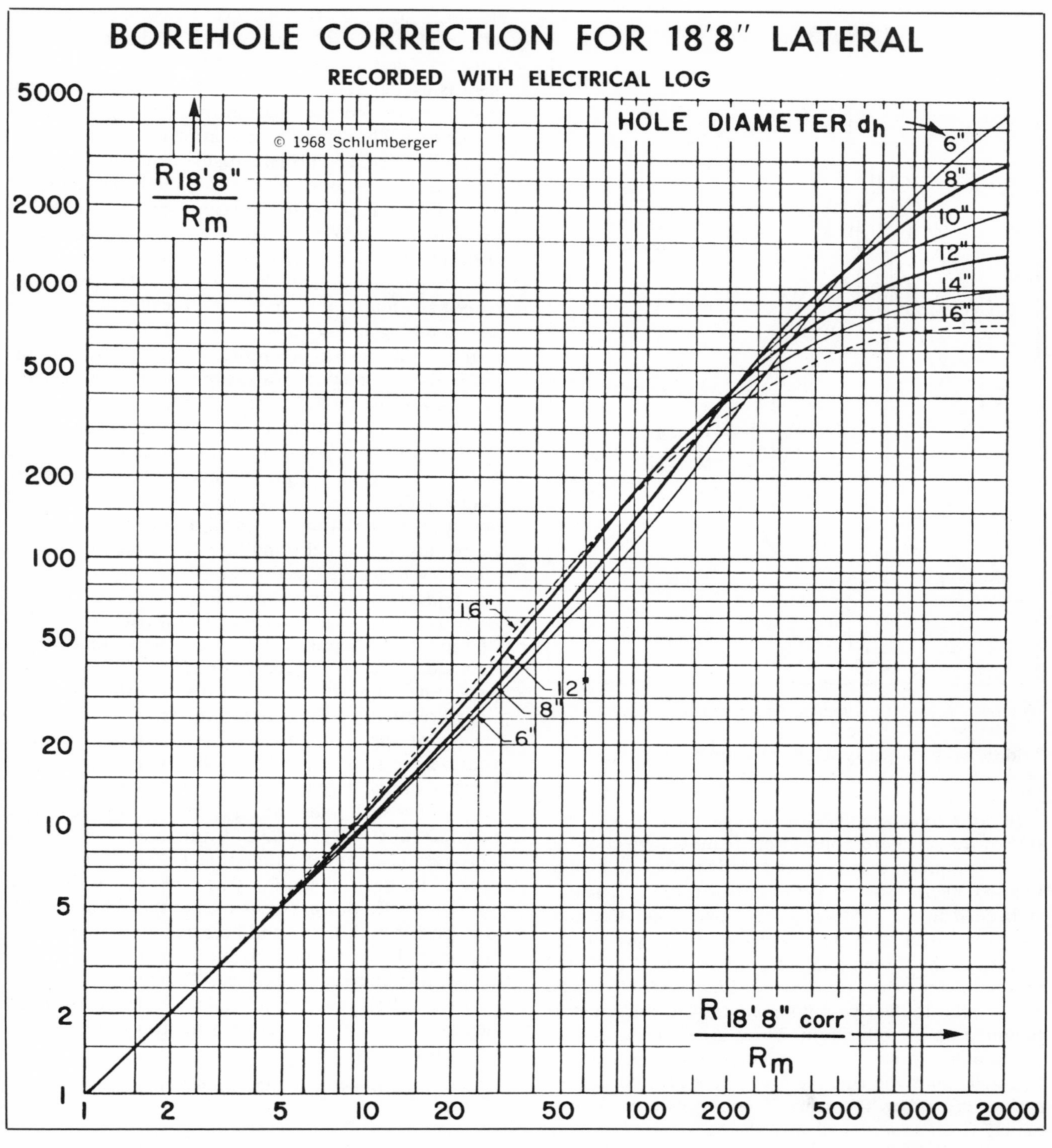

Fig. 6-8. Borehole departure curve for the 18'18" lateral recorded with the electric log. Permission to publish by Schlumberger Ltd.[24]

Using this value of F_R an estimate of S_w could be made since:

$$F_R = \frac{R_o}{R_w} \quad \text{or:} \quad R_o = F_R R_w$$

and then:

$$I_R = S_w^{-n} = \frac{R_t}{F_R R_w}$$

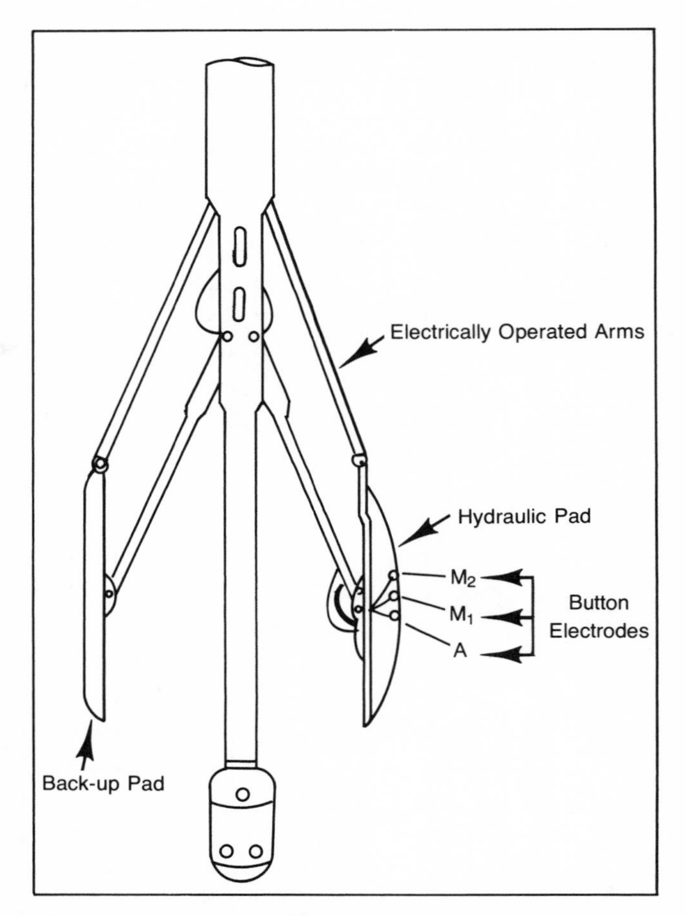

Fig. 6-9. Sketch of nonfocused microresistivity tool

Additionally, porosity might be estimated from:

$$F_R = \phi^{-m}$$

In order to accomplish this measurement of R_{xo}, very shallow resistivity responses were required, which led to the development of a contact resistivity tool. The nonfocused, micro-tool consists of button (point) electrodes embedded in a fluid-filled rubber pad. The electrodes are either flush or slightly recessed with respect to the surface of the rubber pad. This rubber pad is mounted on one of the arms, as shown in Figure 6-9. A second pad, similar to the first but not containing electrodes, is mounted on the other arm. These arms can be opened and closed by an electric motor. When opened, the hydraulic pads will ride the wall of the hole because of spring tension on the arms. These tools are approximately five feet long and with the pads closed have an external diameter of 4¾ in.

Three point electrodes, set one inch apart, are used for making the resistivity measurements. The combination of these three electrodes, A, M_1, and M_2, allows two different resistivity measurements to be made. The electrode combination, AM_1M_2, provides a very small, lateral type of resistivity measurement made between electrodes M_1 and M_2. This gives an electrode spacing, AO, of 1½ in. with the depth of investigation being assumed equal to the spacing. This resistivity measurement is usually referred to as a micro-lateral or micro-inverse curve and the apparent resistivity indicated as $R_{1.5}$ or $R_{1\times1}$. The second combination uses the A and M_2 electrodes, giving a normal measurement with a spacing of two inches and, therefore, a depth of investigation of four inches. This second measurement is usually referred to as a micronormal ($R_{2''}$) curve. Since operation of the electrical circuit requires a conductive fluid in the hole, the tool can only be used in holes containing water-base muds and, more specifically, is used for fresh mud logging systems where $R_{mf} \geq 3R_w$.

In addition to the two resistivity measurements, the tool is also designed to measure the distance between the faces of the two pads. This caliper recording provides a detailed log of the borehole diameter. The maximum expansion of the arms is 16 in., and the minimum hole size that can be logged is 6 in. The microcaliper, therefore, records borehole variations ranging from 5 to 16 in. and can measure variations as small as ⅛ in. The microcaliper log will usually indicate the presence of mud cake as well as caved sections.

A typical microlog is shown in Figure 6-10. As seen, the two resistivity curves, $R_{2''}$ *and* $R_{1''\times1''}$, are recorded to the right of the depth track with the caliper and bit size recorded to the left of the depth track. The deeper investigating curve, $R_{2''}$, is recorded as a dashed curve.

Two resistivity curves are recorded to evaluate the influence of the mud cake on the apparent resistivity measurements. The flushed zone is assumed to be deeper than the depth of investigation of the two resistivity curves such that both $R_{1''\times1''}$ and $R_{2''}$ are measuring a combination of R_{mc} and R_{xo}. By utilizing the two resistivity responses and R_{mc}, empirical charts can be constructed for determining R_{xo}. One such chart is shown in Figure 6-11. Empirical charts such as this are applicable to a specific pad design and hole condition and, therefore, care should be taken to ensure that the correct interpretation chart is being used. The chart is designed for an 8 in. hole size. If other hole sizes are encountered the chart can be used by multiplying the ratio $R_{1''\times1''}/R_{mc}$ by the following factors:

Hole Size	*Multiplying Factor*
4¾″	1.15
6.0	1.05
10.0	0.93

The influence of the mud cake is most important on the tool response. Determination of R_{xo} depends on the "normalized" response of the two apparent resistivities in terms of R_{mc}, and it is, therefore, important that an accurate value of R_{mc} be available. An inaccurate value

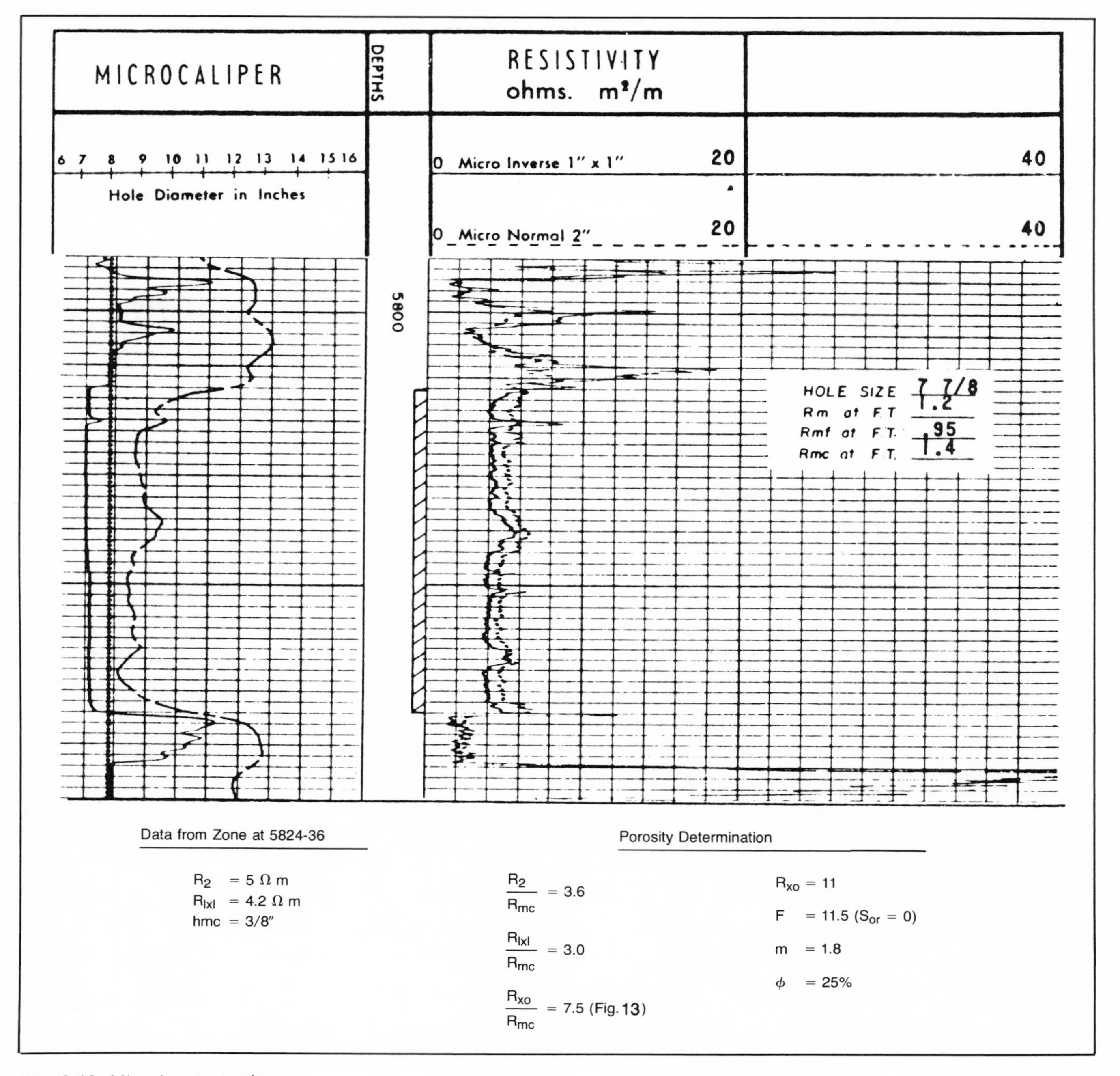

Fig. 6-10. Microlog example

of R_{mc} is probably the most frequent source of error in determining porosity from the microlog. In addition, if R_{mc} is low compared to R_{xo} there will be a great amount of current distortion. Current will tend to flow through the mud cake and not into the formation such that the tool has poor resolution for high R_{xo}. Also, when mud cake thicknesses are greater than ⅜ in., tool resolution is poor and R_{xo} values are unreliable.

Applications of Micro-Resistivity Measurements. Today this type of log is primarily considered a permeability indicator although it can be used to: (1) determine porosity, (2) determine bed boundaries, (3) recognize permeable zones, (4) recognize oil-water contacts, (5) recognize fracture and vug porosity, and (6) measure R_m from a mud log.

Using the value of R_{xo} determined empirically from normalized response charts (see Fig. 6-11) and assuming that the flushed zone contains a conducting fluid of only mud filtrate, R_{mf}, the formation factor and porosity can

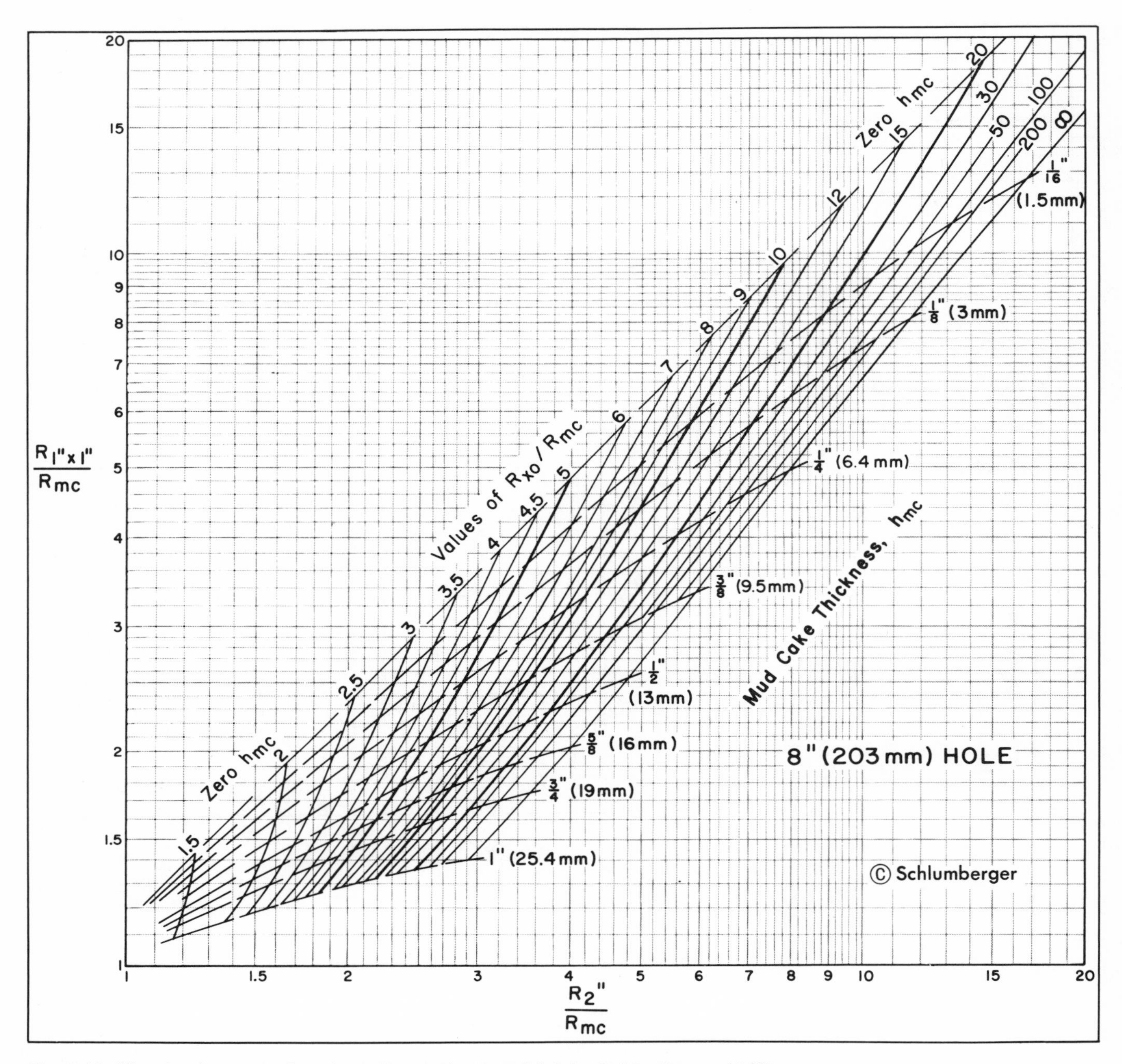

Fig. 6-11. Microlog interpretation chart. Permission to publish by Schlumberger Ltd.[24]

be determined by the relationship previously discussed:

$$F_R = \frac{R_{xo}}{R_{mf}}(1 - S_{or})^n = \phi^{-m}$$

It will be necessary to estimate S_{or}, which is another limitation to this technique for quantitative interpretation. This tool is most applicable when ϕ is 20% or greater and shouldn't be used when ϕ is less than 15%.

Since the micro-device has a very small electrode spacing, the tool has good vertical resolution. Accurate bed boundaries can, therefore, be picked from the resistivity curves.

Differences in the two resistivity measurements are referred to as separation. Positive separation occurs when $R_{2''}$ is greater than the $R_{1''\times1''}$. Negative separation occurs when $R_{1''\times1''}$ is greater than $R_{2''}$. In porous and permeable beds, separation will usually occur, and normally this is a positive separation if invasion is deep enough. Separation can be expected in porous and permeable zones since mud cake is present causing the two curves to measure different resistivities. By noting this separation, porous and permeable beds can be readily identified. In some cases shales indicate separation, but any confusion with porous and permeable beds can generally be resolved using the SP curve as well as the caliper curve where shales usually record greater than hole size.

It may be possible in some cases to see the oil-water contact. The residual oil saturation will approach zero, causing both curves to decrease.

The recognition of fracture and vug porosity is possible since significant differences in the apparent porosity would be indicated by successive runs through the same section.

The mud log is a measure of $R_{1''\times1''}$ and $R_{2''}$ while going in the hole with the microlog tool. The arms are collapsed and the lower 500–1000 ft are logged. In washed out zones, the two resistivity curves should be reading the same R_m at near-bottomhole conditions. This is a good check of the surface measured value of R_m after it has been temperature corrected.

Limitations of Micro-resistivity Measurements. Errors in the interpretation of this device can result from the following conditions:

1. Shallow invasion, which usually occurs in very porous and permeable beds, can appreciably affect $R_{2''}$ such that use of R_{mf} to determine F_R is not justified.
2. Very thick mud cake decreases the effect of the flushed zone on $R_{2''}$ such that the tool has very poor resolution. Mud cakes of ⅜ in. or more are detrimental to microlog interpretation.
3. High values of R_{xo} as compared to R_{mc} cause current to leak around the pad through the mud cake resulting in poor log resolution. Usually occurs in low porosity ($\phi < 15\%$) formations.
4. Pad not in close contact with formation face allows current leakage, poor log resolution, and spurious separation.
5. An incorrect value of R_{mc} will lead to erroneous values of R_{xo} as determined empirically. This is probably one of the most common errors encountered in microlog interpretation.
6. Shaly sands will lead to lower measured resistivities and, therefore, a lower value of R_{xo}. Porosity will, therefore, be calculated higher than that present.
7. Incorrect estimations of S_{or} will lead to incorrect porosity determination.

As a result of these limitations, particularly the inadequacy of our knowledge of R_{mc} and S_{or} and their influence on the resulting value of F_R, and the development of newer porosity responding devices, the micro-resistivity is little used for quantitative purposes today. However, it still provides an excellent measure of net pay and is an indicator of formation permeability.

INDUCTION RESISTIVITY MEASUREMENTS

The induction logging tool designed to measure formation resistivity was first introduced in the mid-1950's. It is a focused device specifically designed to provide a means of measuring formation resistivity when the borehole is filled with a nonconducting fluid such as oil base mud, water-in-oil emulsion mud, air, or gas. Both the nonfocused and, as we shall see, the focused electrical resistivity measurements are inoperative under these borehole conditions since they both require a conducting path between the tool and the formation. For nonconductive borehole conditions, the induction log is a superb resistivity tool; however, it was determined early in the application of the induction log that it also provided a superior resistivity measurement in fresh mud logging systems where the nonfocused type resistivity measurements were normally used. As a result, today the induction log is the primary resistivity tool in fresh mud logging systems (in addition to being the only resistivity device in nonconductive borehole systems), having essentially replaced the nonfocused type of resistivity measurement, i.e., the Electrical Survey (ES) log.

The induction-logging device is based on the principle of electromagnetic coupling between the logging tool and the formations. Thus no current flow between the two is required, and no electrodes are necessary. Figure 6-12 illustrates the principle of operation. For this over-

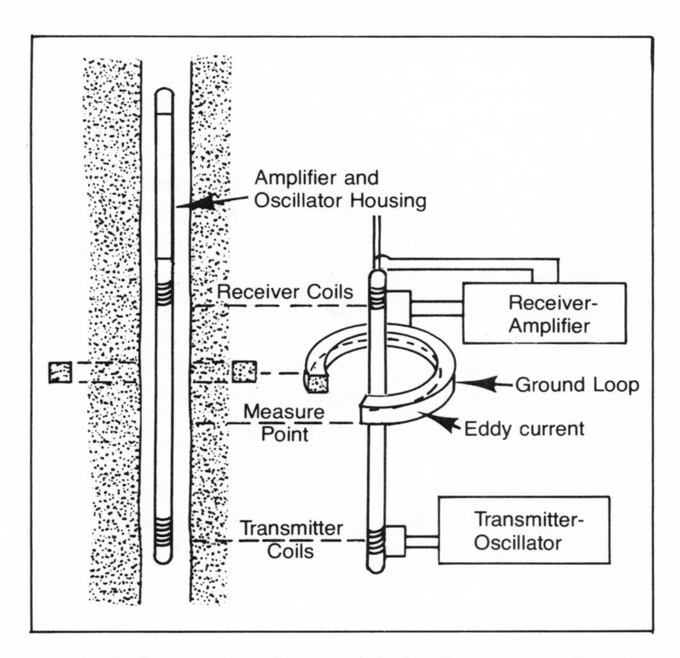

Fig. 6-12. Schematic of two-coil induction system. Permission to publish by Oil, Gas and Petrochem Equipment.[22]

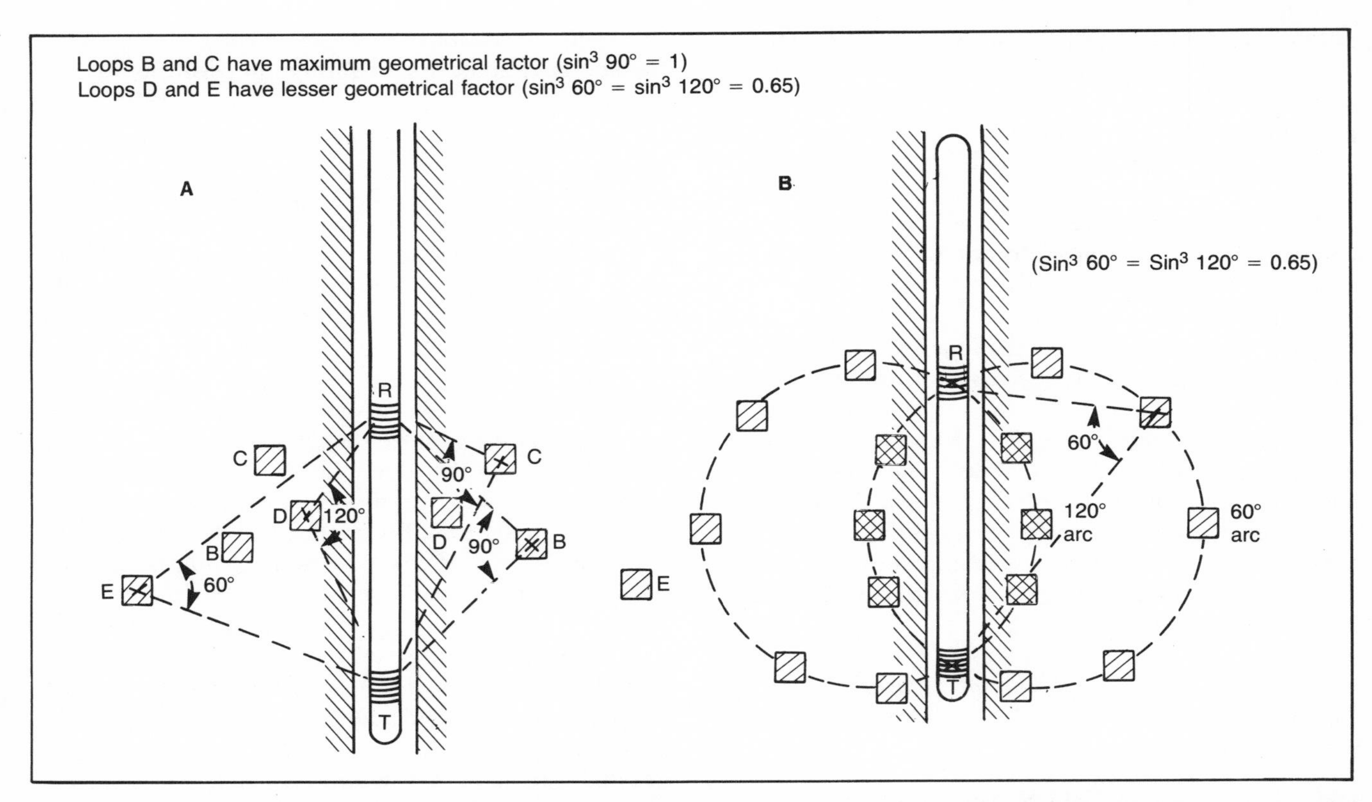

Fig. 6-13. (A) Unit ground loops and their geometrical factors. (B) Focus of unit cross-section ground loops having equal geometrical factors. Permission to publish by the Society of Petroleum Engineers of AIME.[2] Copyright 1949 SPE-AIME.

simplified system, a transmitter coil and a receiver coil are wound coaxially on a supporting mandrel. The distance between coils is called the spacing, usually 40 in., with the measure point located halfway between the two coils.

A 20-Khz alternating current of constant magnitude is fed to the transmitter coil from an oscillator. The energized transmitter coil generates an alternating electromagnetic field of this frequency in the formation surrounding the tool. This alternating field induces eddy currents that flow along coaxial ground loops around the borehole. The eddy currents, in turn, set up secondary magnetic fields that induce an emf in the receiver coil. The intensity of the current induced into the formation by the transmitter is proportional to its conductivity. The signal induced into the receiver is thus proportional to formation conductivity.

Focusing is accomplished by incorporating two to four additional coils in the tool. These coils are designed to minimize the borehole and adjacent bed effects and to reduce invasion effects. Their focusing action is due to the development of a cancelling signal, using auxiliary coils wound in opposite directions to the main coils. In this way, any signal developed in the receiver circuit due to coupling of a main coil with an auxiliary coil subtracts from the signal produced between the main coils.

Mutual inductance, and therefore the direct coupling between transmitters and receivers, is eliminated through the use of selected coil spacings and the direction and number of turns in the coil system. Elimination of mutual inductance is assured by suspending the tool in a nonconducting medium such as air and recording a zero signal.

The resulting voltage response for the induction tool was first defined by Doll[2] in 1949. His theoretical analysis of the total resultant signal is based on the geometrical factor concept, which summarizes the influence of each ground loop in the surrounding medium. This analysis is predicated on the basis that each ground loop (see Fig. 6-13) generates a signal, V_i, contributing to the total signal V. The ground loop signal is defined as:

$$V_i = K g_i C_i$$

where: K = apparatus constant
g_i = geometrical factor of individual ground loop
C_i = conductivity of individual ground loop

The total signal response corresponding to the whole space is:

$$V = K \int\int g_i C_i \, dr dz$$

For different homogeneous regions *A*, *B*, *D*, etc. the voltage response can be written as:

$$V=K\left[C_A\iint_A gdrdz+C_B\iint_B gdrdz+\ldots\right]$$

Defining then the integrated geometrical factors as:

$$G_A=\iint_A gdrdz$$

$$G_B=\iint_B gdrdz \quad \text{etc.}$$

Then the voltage response inducted in the receiver coil can be written as:

$$V=K[C_AG_A+C_BG_B+\ldots]$$

or:

$$V=KC_a[G_A+G_B+\ldots]$$

where: C_a = apparent conductivity of the whole space.

Now the integrated geometrical factors (these can be thought of as the influence of each zone) must add up to one, or:

$$G_A+G_B+\ldots=G=1$$

As a result, the apparent conductivity response of the induction tool is expressed as:

$$C_a=\frac{V}{K}$$

The geometrical factor concept introduced by Doll[2] provided a useful method for designing induction log coil systems and today provides a practical means for determining the true formation conductivity, C_t, from the apparent conductivity signal, C_a. This method of analysis, however, assumes that neither the shape nor the intensity of the induction field is in any way affected by the electrical characteristics of the formation. It was recognized early,[13, 14, 15, 17, 18] however, that under some formation conditions the performance of the induction tool could be substantially different from that predicted by the simplified geometrical factor concept. This anomalous performance was called the "Skin Effect" since it results from the same phenomenon that causes high frequency alternating currents to flow only near the surface of metallic conductors. For conductivities encountered in sedimentary rocks, however, a formation would not exhibit the sharply defined conducting skin indicated by metals. The gradual change of the current distribution within a formation as produced by the induction field of the logging device is actually the result of phenomena associated with propagation of the electromagnetic field through the formation and hence is referred to as a "Propagation Effect."

It can be stated that the transmitter coil develops an electromagnetic wave in the formation that is propagated outward at the velocity of a radio wave. In turn, the "eddy currents" generated by the wave in passing through the formation produce secondary electromagnetic waves that are similarly transmitted through the formation and in turn induce within each ground loop their own induced current flow. This results in a secondary signal contribution within the receiver circuit. If the velocity of the electromagnetic wave in the formation is approximately the same as the velocity of the wave in free space, i.e., 186,000 mi/sec and if the medium were infinitely resistive, the results derived from the geometric concept analysis would be valid.

However, the velocity of an electromagnetic wave in sedimentary rock systems is substantially lower than that in free space, which results in a reduction in the propagation velocity and, therefore, the absorption of energy from the electromagnetic wave by the formation. This absorption has two direct effects which influence the performance of the induction log: (1) it causes the amplitude of the wave to fall off more rapidly with the distance traveled and (2) it produces an appreciable phase shift within the field of investigation of the induction tool.[13] As a result of the attenuation and phase changes that occur within the electromagnetic field opposite a sedimentary rock system, the signal produced by the induction log can be expected to be different from those predicted using an analysis based upon the free space performance of the electromagnetic field, as is done using the geometric concept analysis. The propagation effect will, therefore, always reduce the recorded apparent conductivity to a value lower than that obtained from the geometric factor calculation. The propagation effect increases with increasing formation conductivity and increasing transmitter-receiver coil spacing and is nonlinear in its influence.[13] Since logging devices are designed based upon the geometric factor concept, some means for compensating for the propagation effect must be considered when using the apparent conductivity response. This can be done electronically or by using some form of departure curve. In addition, other electronic operations may be performed on the signal such as thin bed corrections and signal multiplication (to provide a true conductivity response as related to a specific calibration conductivity).

Factors Affecting Induction Resistivity Measurement

The apparent conductivity response, C_a, might then be expressed as a function of the conductivity contribution of the separate zones that influence the total signal. In a practical sense, then, these zones of influence might

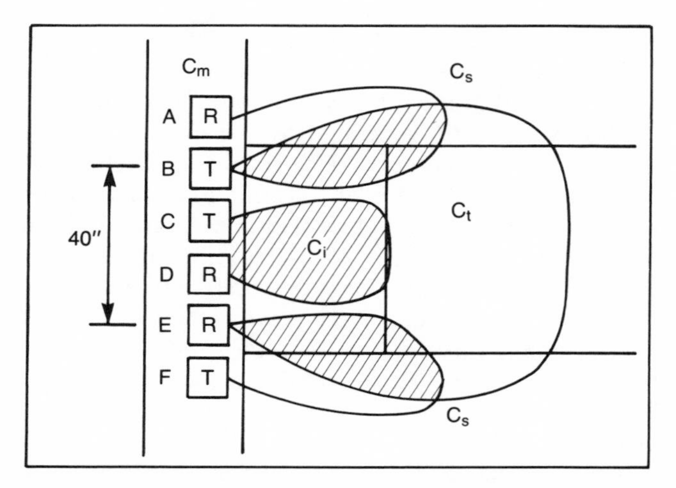

Fig. 6-14. Schematic of induction tool response showing concept of focusing and reduction of near-borehole contribution

be considered, as schematically shown in Figure 6-14, with the apparent conductivity expressed as:

$$C_a = C_I = f(C_m,\ C_s,\ C_i,\ \underline{\underline{C_t}})$$

Obviously, in nonconducting borehole systems where a nonconducting fluid is injected and formations are thick (5 ft or more for the induction log), the apparent conductivity approaches the true conductivity of the undisturbed zone. Applications for which it was originally designed (a nonconducting borehole system), therefore, provide ideal conditions for optimal response of this tool.

Subsequent applications in fresh water mud systems require consideration of the borehole and invaded zone influence as well as thin bed effects. In order to handle this problem in a practical manner, a stepwise approach to the problem is used. This involves first evaluating the influence of the borehole and removing the borehole signal contribution. In order to do this, the geometrical factor of the borehole, G_m, is required, and as can be expected, this will be a function of the hole size, d_h, and the annular space between the tool and the formation characterized by its "stand-off" (0 to 2.5 in. and given on the log heading). Figure 6-15 provides a means to evaluate either G_m or G_mC_m. Correcting for this borehole signal, a borehole corrected apparent conductivity, C_I', is obtained as follows:

$$C_I' = C_a - [C_mG_m] = f(C_s,\ C_i,\ \underline{\underline{C_t}})$$

It is normally found that the borehole signal is negligible unless the mud is highly conductive, the formation is quite resistive, or the borehole diameter is large.

The second step is to assess the influence of the surrounding beds, which is usually accomplished using departure curves, such as those shown in Figure 6-16a and 6-16b. To accomplish this, the borehole corrected conductivity, C_I', is converted to resistivity where:

$$R_a'(\Omega m) = \frac{1000}{C_I'\left(\frac{mS}{m}\right)}$$

Knowing bed thickness and the average surrounding bed resistivity, R_s, the appropriate chart can be used (interpolation may be necessary). As will be noted, different departure curves need to be used for different induction logging devices. Charts are included in this book for basically two induction tools, the 6FF40 (ILD) and the 5FF40 (ILM). The 6FF40 nomenclature means a 6 coil, focused tool having a 40-in. spacing between the main coil (transmitter-receiver) pair. Hence:

$$R_I' \xrightarrow[\text{Fig. 16a or 16b}]{h,\ R_s} R_I''$$

The bed thickness correction charts provide a twice corrected apparent resistivity, R_I'', or inversely a twice corrected conductivity, C_a'', and as a result:

$$C_I'' = C_iG_i + \underline{\underline{C_t}}G_t$$

It should be noted that for most cases where bed thickness is greater than about 4½ ft, the bed thickness correction is small.

As expressed in the preceding relation, the apparent conductivity is now only a function of an invaded zone of some electrically equivalent diameter, d_i, surrounded by an undisturbed zone of infinite diameter; both zones are considered to be infinitely thick. The relation expressed in terms of resistivity would be:

$$\frac{1}{R_I''} = \frac{G_i}{R_i} + \frac{G_t}{R_t} = \frac{G_i}{R_i} + \frac{(1-G_i)}{R_t}$$

In order to solve for R_t, G_i, $G_t = (1-G_i)$, and R_i must be determined. A shallow investigating resistivity measurement such as the short normal or SFL, where $R_{16''}$ or $R_{SFL} \approx R_i$, is run with the induction log in order to obtain R_i (see Fig. 6-17). With an apparent value of R_i one "only" has to define d_i (electrically equivalent)to obtain G_i from an induction response chart (see Fig. 6-18). If d_i isn't known (and it usually isn't), a range for d_i might be assumed, leading to a range in R_t and, therefore, a range in the predicted water saturation, S_w.

The induction log response chart as shown in Figure 6-18 also provides a generalized view of the depth of investigation of the induction logs. In fact, this concept is most useful in comparing the various resistivity logging responses, as shown in Figure 6-19, where pseudo-geometric factors for a number of different logging tools are presented.

The geometrical factor data shown in Figures 6-18 and 6-19 do not include the propagation effect (skin effect) and as such tend to be an oversimplification of the

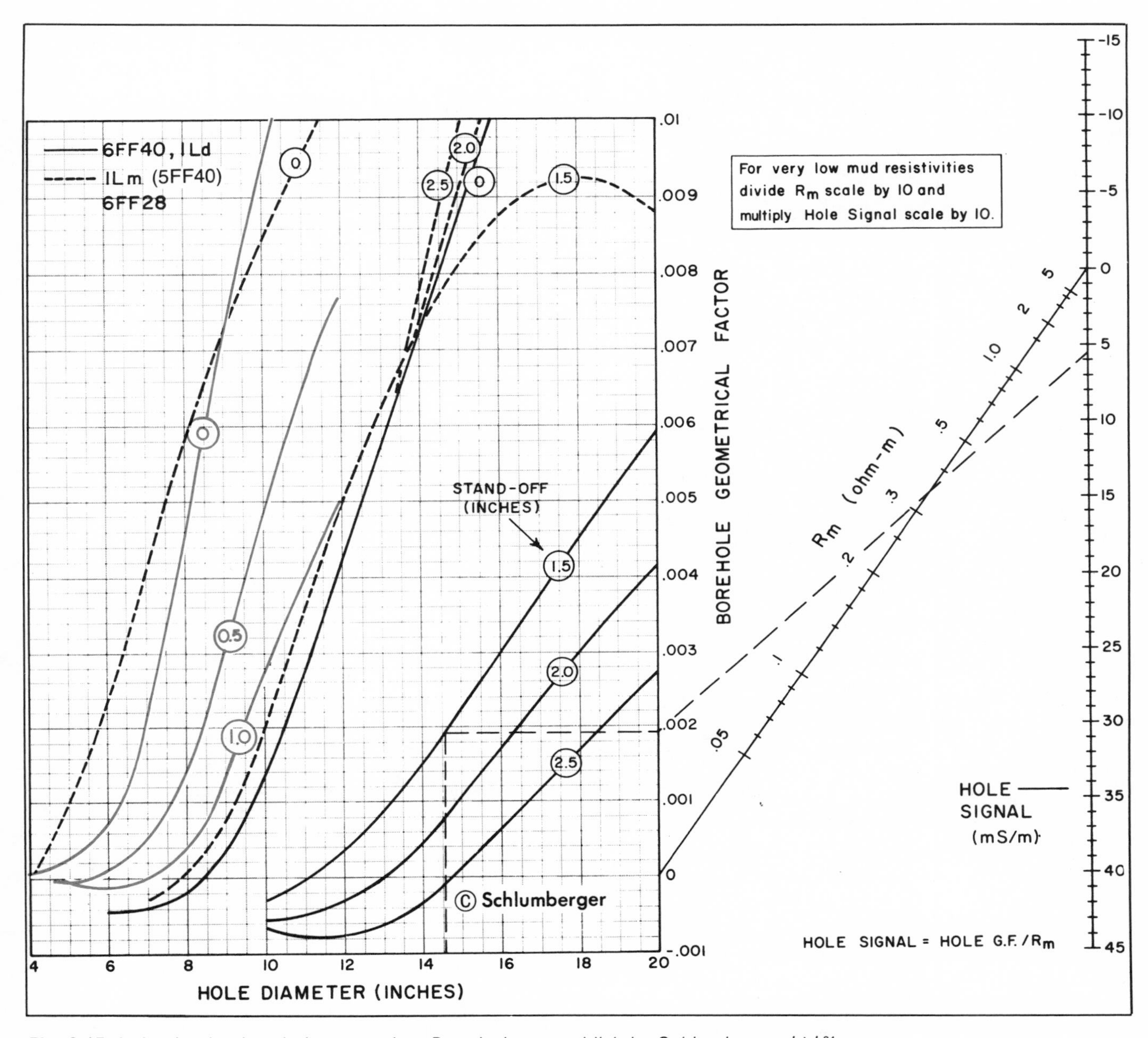

Fig. 6-15. Induction log borehole correction. Permission to publish by Schlumberger Ltd.[24]

radial response of the induction tools. Recent work[27] clearly indicates that the commonly published radial geometric factor curves describing the radial responses of the induction tools are not constant for all formation resistivity conditions. Propagation effects reduce the "depth of investigation" as formation conductivity increases with the radial response varying as the absolute values of R_{xo}, R_t, and d_i vary, assuming a step profile. As a result, commonly published induction log geometrical factors are only valid for high formation resistivities, specifically infinite resistivity. Pseudo-geometrical factors for the deep induction log (6FF40 and ILD) for noninfinite resistivity conditions are shown in Figure 6-20. The dashed curve represents the geometrical factor for the deep induction tool at infinite resistivity (no propagation effect considered). It can be seen that as formation resistivities increase (both R_{xo} and R_t), the radial response is less influenced by the propagation effect. Conversely, for the extreme case of $R_{xo} = 0.2$ Ωm, no useful signal is obtained beyond a diameter of 120 in. for any value of R_t. As a result, a hydrocarbon-bearing zone under these conditions would go undetected, obviously, a strong argument for using the laterolog under these conditions. This same data presented in a readily usable form is shown in Figure 6-21, which provides a solution to the equation:

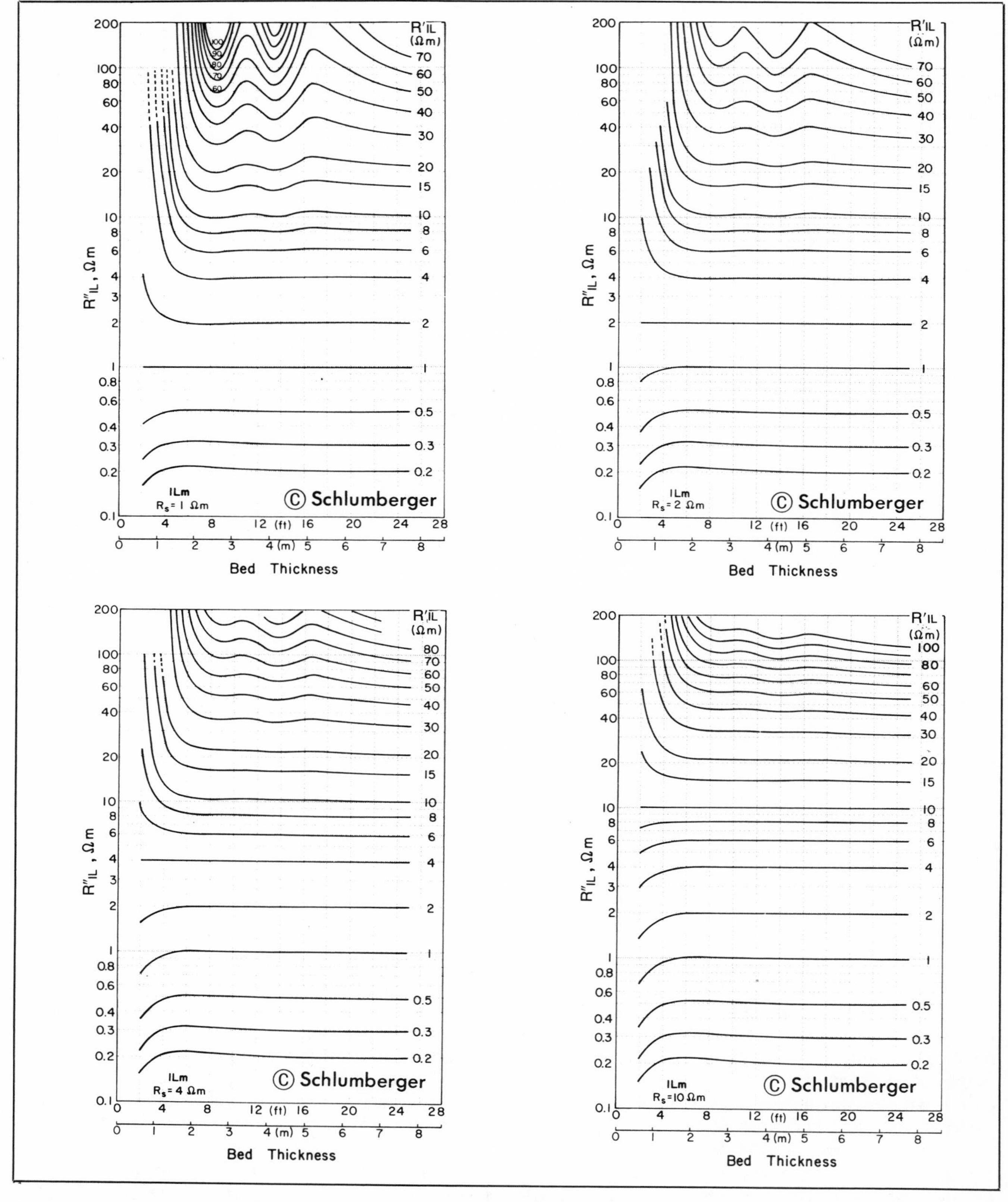

Fig. 6-16a. Induction Log (ILM OK if needed for approximately 5FF40 behavior) bed thickness correction. Permission to publish by Schlumberger Ltd.[29]

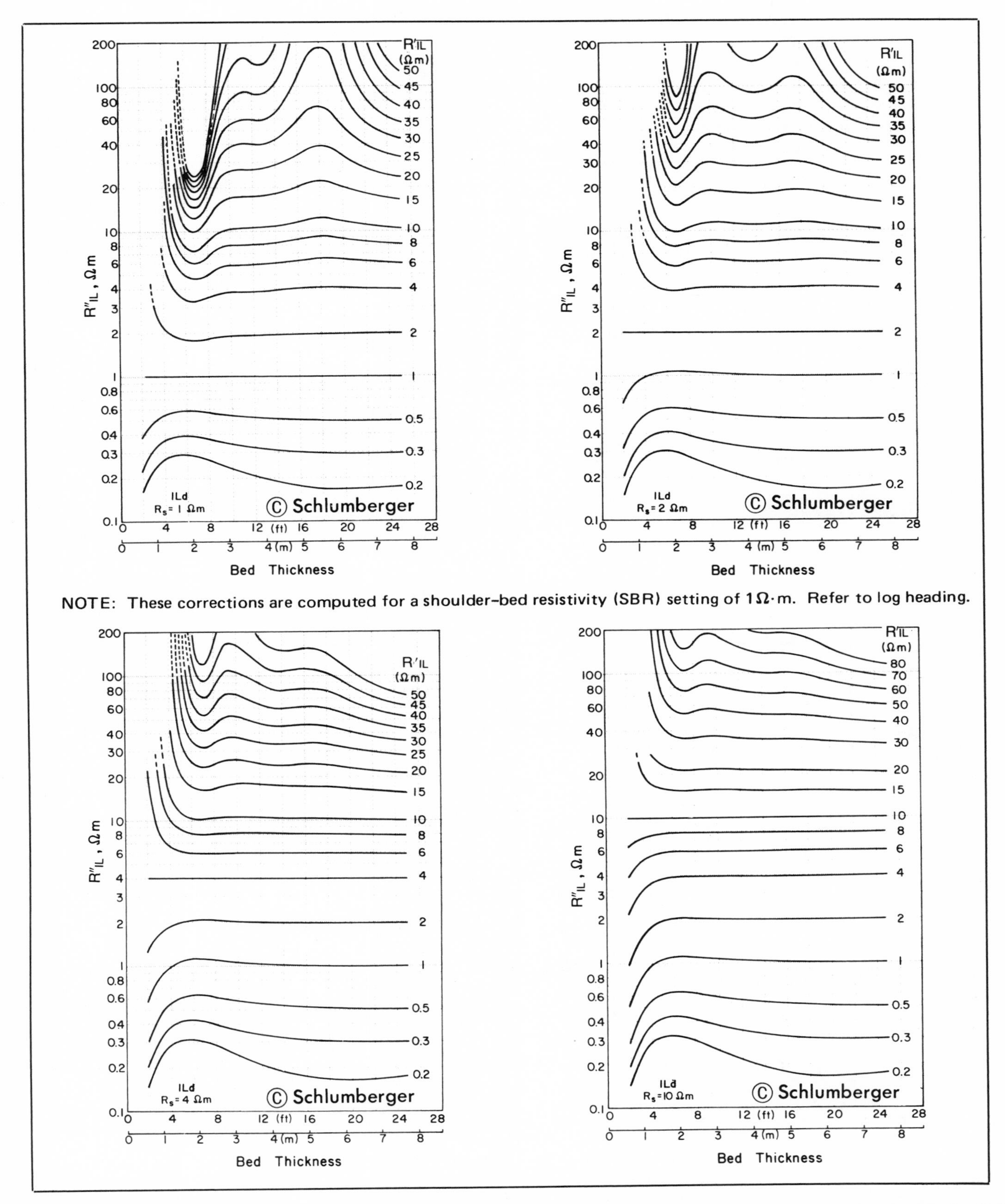

Fig. 6-16b. Induction log (6FF40 or ILD and 6FF28) bed thickness correction. Permission to publish by Schlumberger Ltd.[29]

$$\frac{1}{R_I''} = \frac{G_{xo}}{R_{xo}} + \frac{(1-G_{xo})}{R_t}$$

A typical induction log is presented in Figure 6-22, which has obviously been run in a fresh water mud system since both the SP and short normal, $R_{16''}$, have also been recorded. This is the common Induction-Electrical survey, IES. The conductivity response, C_a, is recorded in ms/m ranging from 0–500, which corresponds to a resistivity scale from ∞ Ωm at 0 ms to 2 Ωm at 500 ms/m. The conductivity response greatly expands the lower resistivities and greatly compresses the higher resistivities. Since this is the primary response of the induction tool, it can also be seen that the induction log response loses resolution in high resistivities. For high resistivities both the calibration problem and magnitude of borehole signal are such that absolute values of resistivity greater than 100 Ωm lead to large uncertainties. The uncertainty in zero readings for present induction logs is about ±2 ms/m, which in itself could lead to errors of 20% for formation resistivities of 100 Ωm. Additionally, a borehole signal of several ms/m would compound the problem. In an absolute sense, a formation resistivity of 100 Ωm is approaching the limit of absolute accuracy for induction type responses. Finally, as can be seen on Figure 6-22, the apparent conductivity as measured by this tool is also recorded in reciprocal form in Ωm. For resistivities exceeding about 15–20 Ωm, it is best to read $R_a = R_I$ rather than the conductivity curve where $C_a = C_I$.

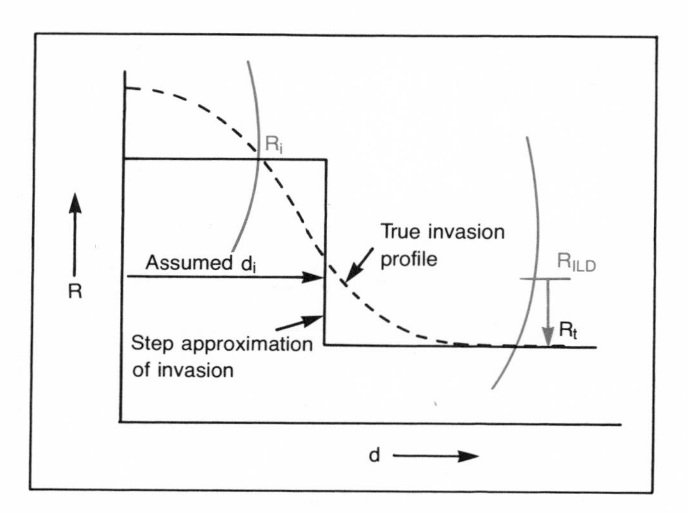

Fig. 6-17. Conceptual insight into invasion correction

MULTIPLE RESISTIVITY MEASUREMENTS

It was apparent in the 1950's that invasion of an oil zone by mud filtrate could produce a considerably dif-

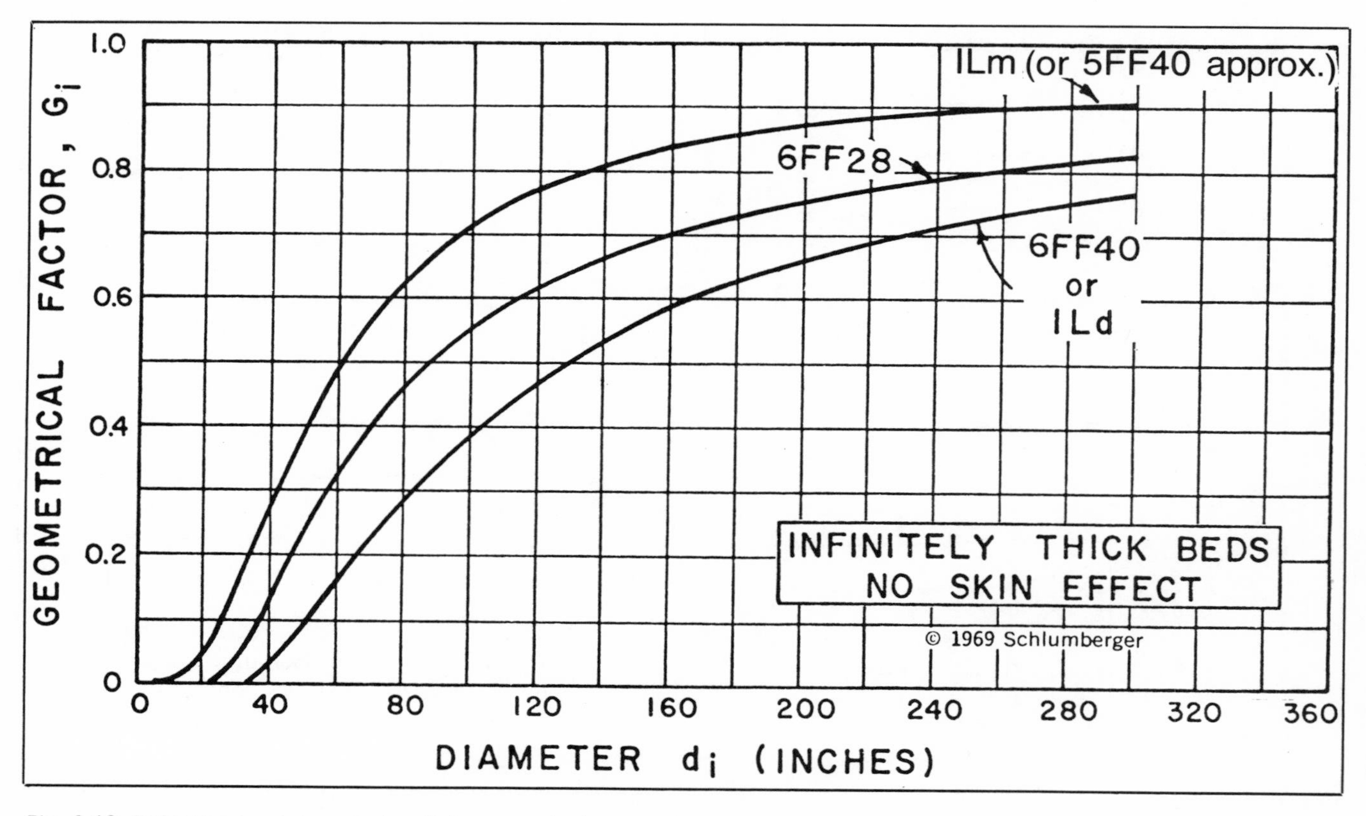

Fig. 6-18. Induction log integrated radial geometric factors. Permission to publish by Schlumberger Ltd.[24]

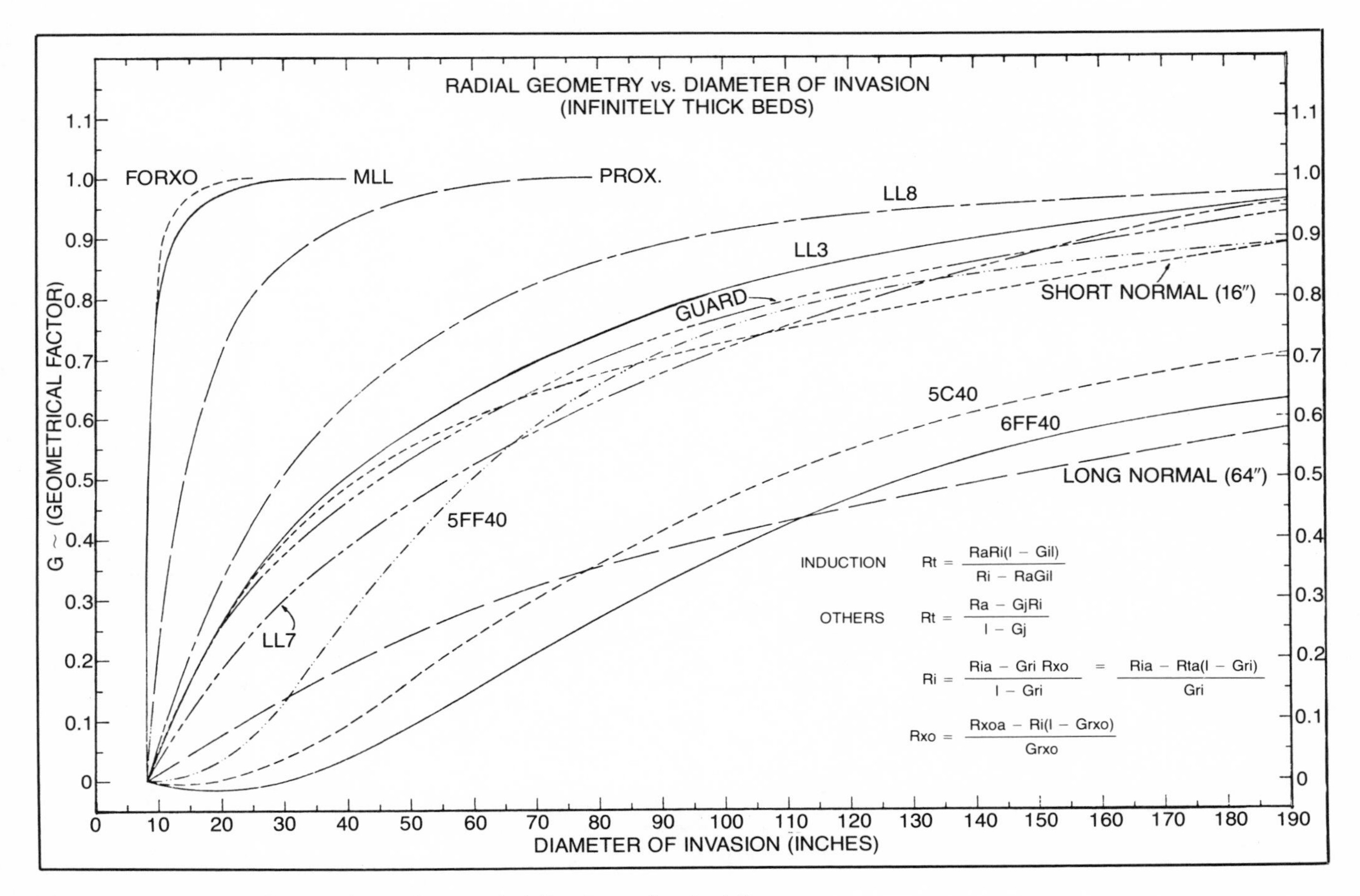

Fig. 6-19. Geometric factors for various resistivity measuring tools[28]

ferent resistivity profile than that produced by mud filtrate invasion of a water sand. As discussed in Chapter 1, resistivity profiles as shown in Figure 6-23 might be expected. Based on this difference, it was projected that if multiple resistivity measurements were made as a function of increasing "depth of investigation," a low resistivity zone should be observed in an oil-bearing zone. As a result, a multiple resistivity logging technique was pursued and referred to as "Displacement Logging."[8, 9] In fact, some successful displacement logs were run using a nonfocused electrical tool from which sequential resistivity measurements could be made with the sonde stationary in the borehole. Figure 6-24 is an electrical log over an interval containing both water and oil-bearing zones. Figure 6-25 shows the resistivity profiles obtained in the water zone, and Figure 6-26 shows the resistivity profiles obtained in the oil zone.

This positive approach as a "movable oil finder" had some limitations, however, which are most important. First, for practical application, a continuous multiple resistivity tool was needed. Second, and probably most important, was the fact that the absence of a low resistivity zone (i.e., the absence of an annulus) was not a positive indication of a zone having no movable oil for two possible reasons. One reason could be deep invasion where the annulus exists beyond the range of the deeper investigating resistivity responses. A more likely, and also more restrictive, reason is that the annulus is time dependent. In effect, it could have been formed, but at the time of logging it could have dissipated as a result of gravity segregation and ionic diffusion.

Dual Induction-Laterolog 8 and Dual Induction-SFL

The need for multiple resistivity measurements (and therefore a resistivity profile) was still apparent, not only to "see" hydrocarbon-bearing zones where an annulus still exists but for an even more important reason—if one does exist and only two resistivity responses (as with the IES and Induction-SFL) are obtained, the deep resistivity response could be adversely affected (R_t corrected $\ll R_t$) by the annulus, which would subsequently lead to a high estimation of water saturation, S_w. In the early 1960's the first commercial multiple resistivity tool, the Dual Induction-Laterolog-8 (DIL-LL8), was developed.[19] This tool provided three

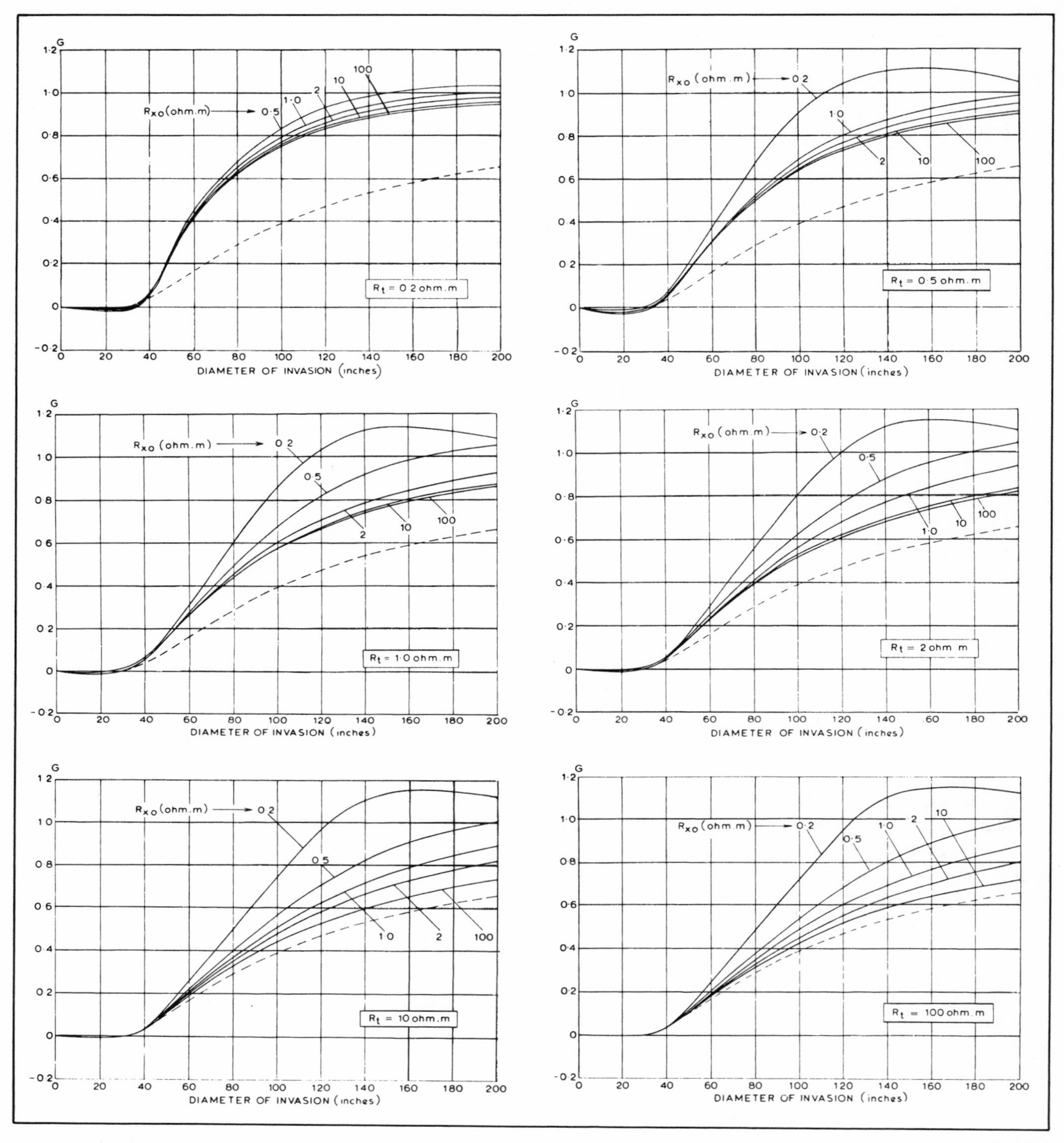

Fig. 6-20. Pseudo-geometrical factors for the deep induction log (6FF40 and ILD) for varying values of R_{xo} and R_t. Permission to publish by the Society of Professional Well Log Analysts.[27]

resistivity responses and the SP as a function of depth, all of which could be recorded continuously. A typical DIL-LL8 presentation is shown in Figure 6-27 with the lower portion having the resistivity curves presented logarithmically (5 in. = 100 ft) and the upper portion presenting all measurements linearly (1 or 2 in. = 100 ft). The relative radial depth of investigation for these three resistivity curves can be seen in Figure 6-18. The 6FF40 and 5FF40 responses correspond respectively to the ID and IM responses, as indicated on the log. The difference in notation is that when combining the two induction devices in one tool, a small difference in radial response exists as compared to these tools built separately. For all practical purposes, however, these differences in radial response are negligible.

Although this tool only provides one additional resistivity curve, it does enable the analyst to compensate for invasion effects on the deep resistivity response. The three resistivity curves, R_{LL8}, R_{IM} and R_{ID} can be used to automatically correct for invasion without having to make an estimate of d_i. This is accomplished by simulating the influence of the invaded zone on R_{ID} using a step type profile as shown in Figure 6-28. Since the radial response behavior of the three resistivity measures is known (see Fig. 6-18), the expected responses for R_{LL8}, R_{IM}, and R_{ID} can be generated as a function of varying

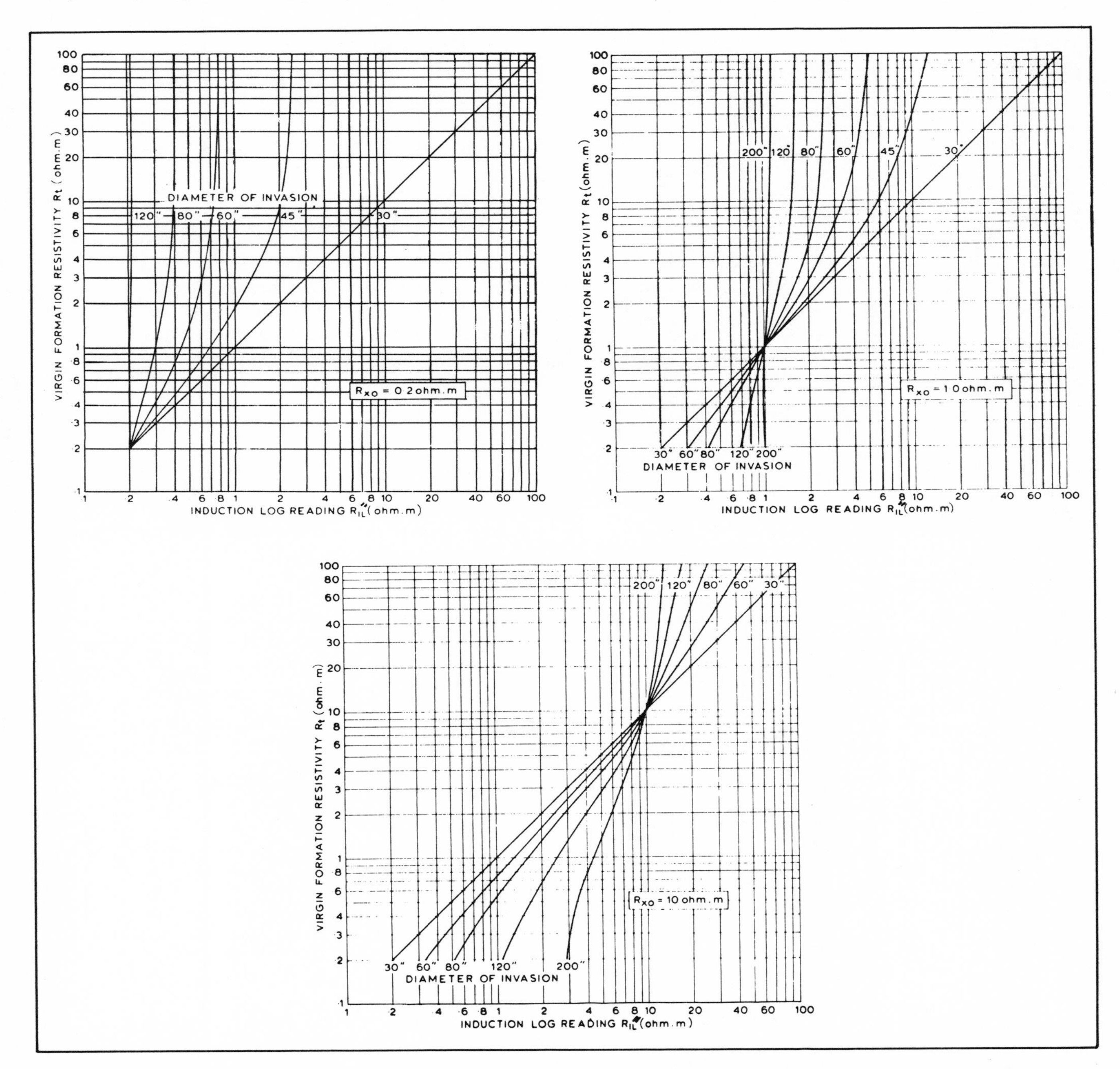

Fig. 6-21. True formation resistivity, R_t, from borehole and bed thickness corrected deep (6FF40 and ILD) induction log responses. Permission to publish by the Society of Professional Well Log Analysts.[27]

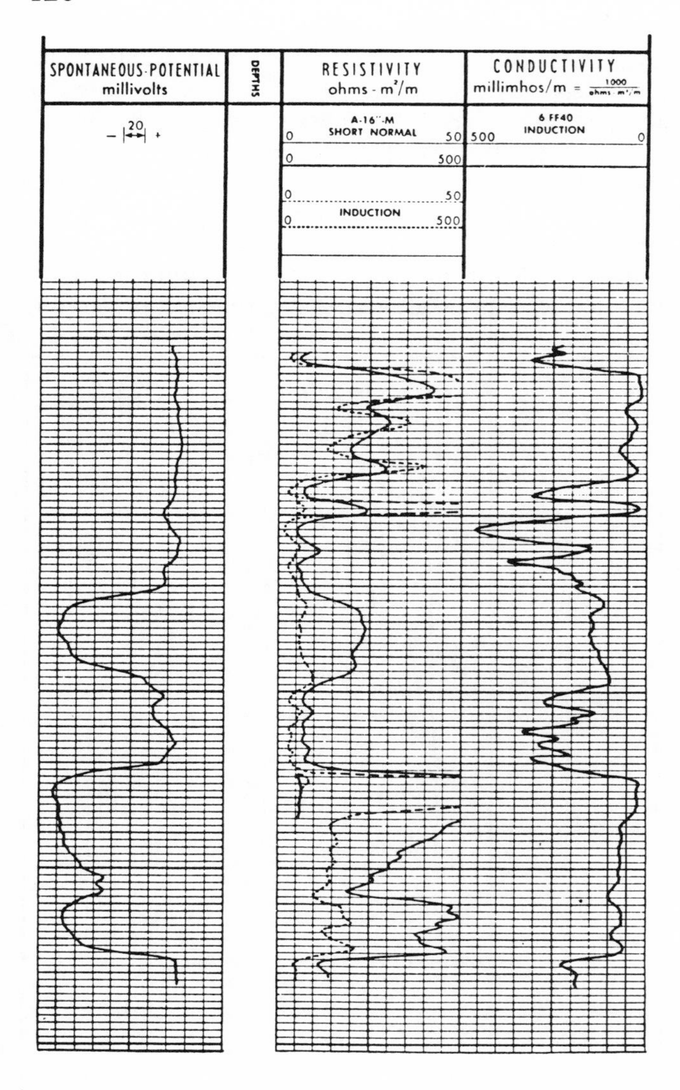

Fig. 6-22. Typical induction-electrical survey, IES. Permission to publish by Schlumberger Ltd.[25]

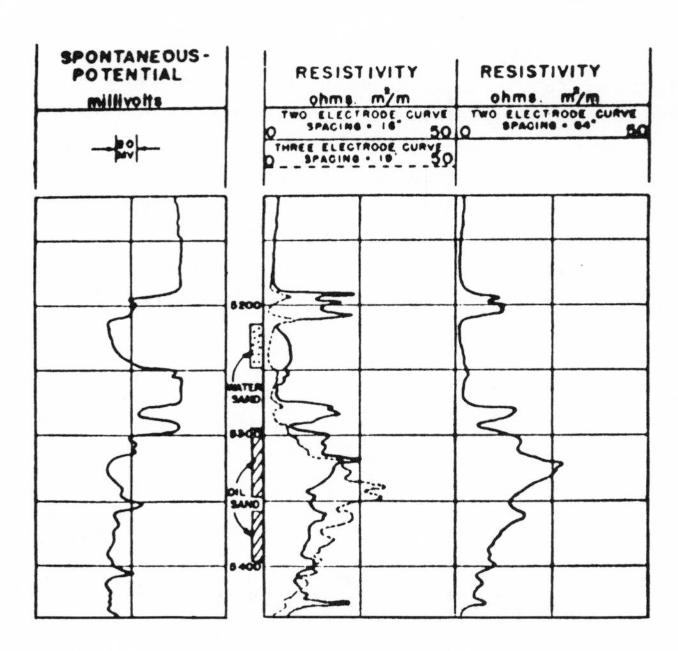

Fig. 6-24. Electrical survey, Atlantic Refining Co. well, Sholem Alechem Field, Stephens County, Oklahoma. Permission to publish by the Society of Petroleum Engineers of AIME.[8] *Copyright 1955 SPE-AIME.*

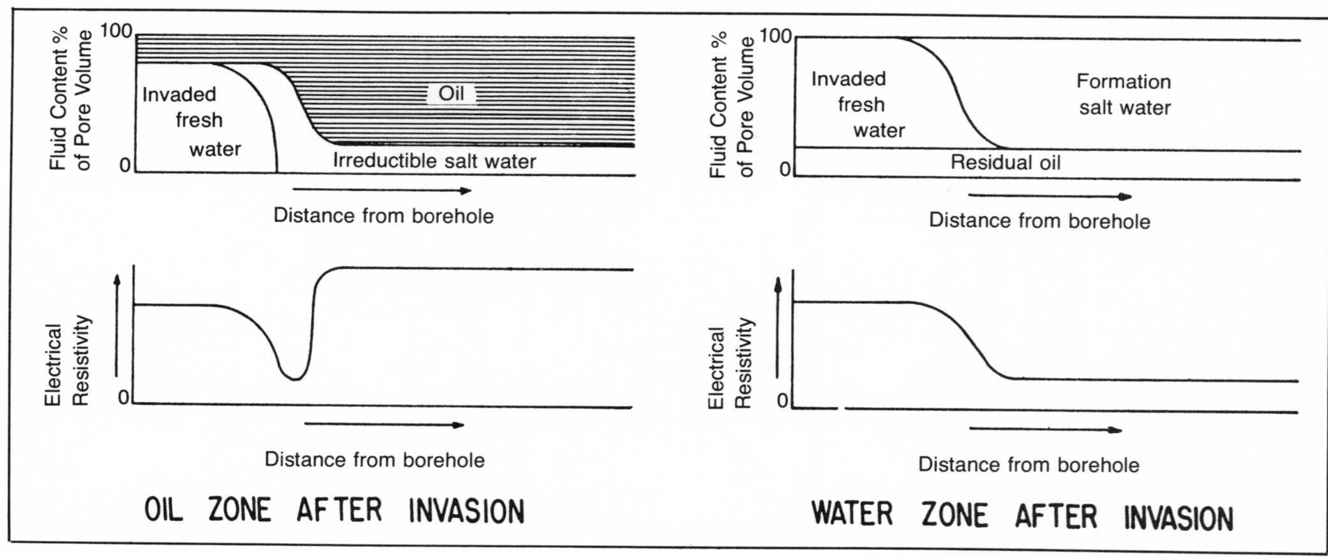

Fig. 6-23. Typical saturation and resistivity profiles for oil and water zones after invasion. Permission to publish by the Society of Petroleum Engineers of AIME.[9] *Copyright 1958 SPE-AIME.*

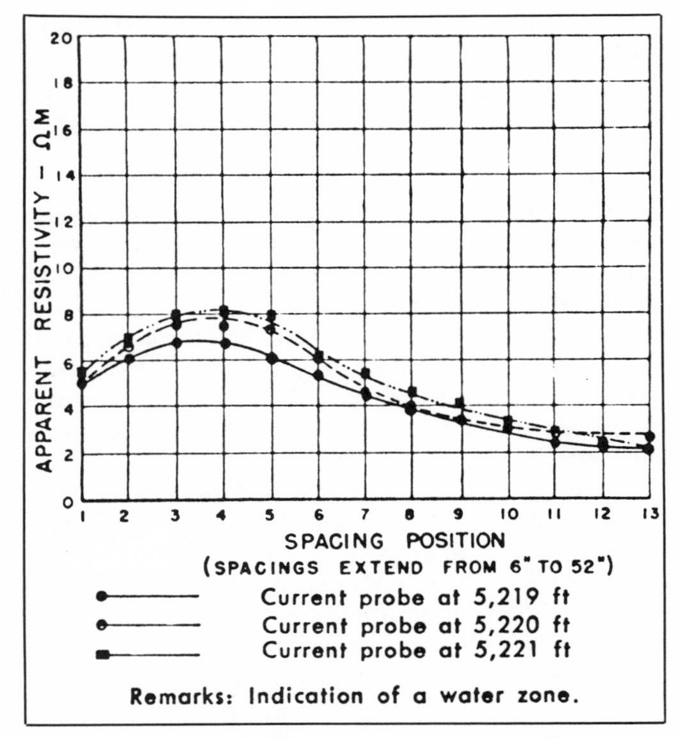

Fig. 6-25. Displacement logging results. Well located in Sholem Alechem Field, Stephens County, Oklahoma. Permission to publish by the Society of Petroleum Engineers of AIME.[8] Copyright 1955 SPE-AIME.

values of R_{xo}, R_t, and d_i for the step profile. Results of the simulation are shown in Figures 6-29, 6-30 and 6-31 for Schlumberger's Dual Induction tools, appropriately referred to as tornado charts. Selecting the appropriate tornado chart, a value for R_t/R_{ID} can be obtained using the two dimensionless ratios R_{LL8}/R_{ID} or R_{SFL}/R_{ID} and R_{IM}/R_{ID}, which reflect the three resistivity measures. If an invasion problem doesn't exist, i.e., very deep invasion or an annulus, the value then determined from the ratio R_t/R_{ID} should lead to an R_t from which a representative water saturation can be determined as follows:

$$\frac{R_t}{R_{ID}} \times R_{ID} = R_t \xrightarrow{\text{and}} S_{wA} = \left(\frac{R_o}{R_t}\right)^{1/n} = \left(\frac{FR_w}{R_t}\right)^{1/n}$$

For those cases where an invasion problem does exist (very deep invasion or an annulus), the projected value of R_t will be too low and hence S_{wA} too high. This will invariably be reflected in the step profile value of R_{xo}/R_t being abnormally high compared to the true value of R_{xo}/R_t. By comparing the step profile generated R_{xo}/R_t with the measured E_{SSP}, a second water saturation can be estimated. This can be accomplished using Figure 6-32 and projecting an apparent water saturation, S_{wR}, using the R_{xo}/R_t value generated by the step profile analysis. If $S_{wA} \geq S_{wR}$ then R_{ID} has been automatically corrected for invasion, and a good value for R_t has been obtained. Hence, $S_{wA} = S_w$. If very deep invasion or an annulus exists, it will be reflected in an abnormally high S_{wR} as predicted by the abnormally high R_{xo}/R_t value when compared to the E_{ssp}. When this occurs and $S_{wR} > S_{wA}$ (R_t too low), then S_{wR} can be used to make a reasonable estimate of the existing water saturation as follows:

$$S_w = S_{wA}\left(\frac{S_{wA}}{S_{wR}}\right)^{1/4}$$

FOCUSED CURRENT RESISTIVITY MEASUREMENTS

Focused current resistivity devices were developed to overcome the inherent problems associated with the

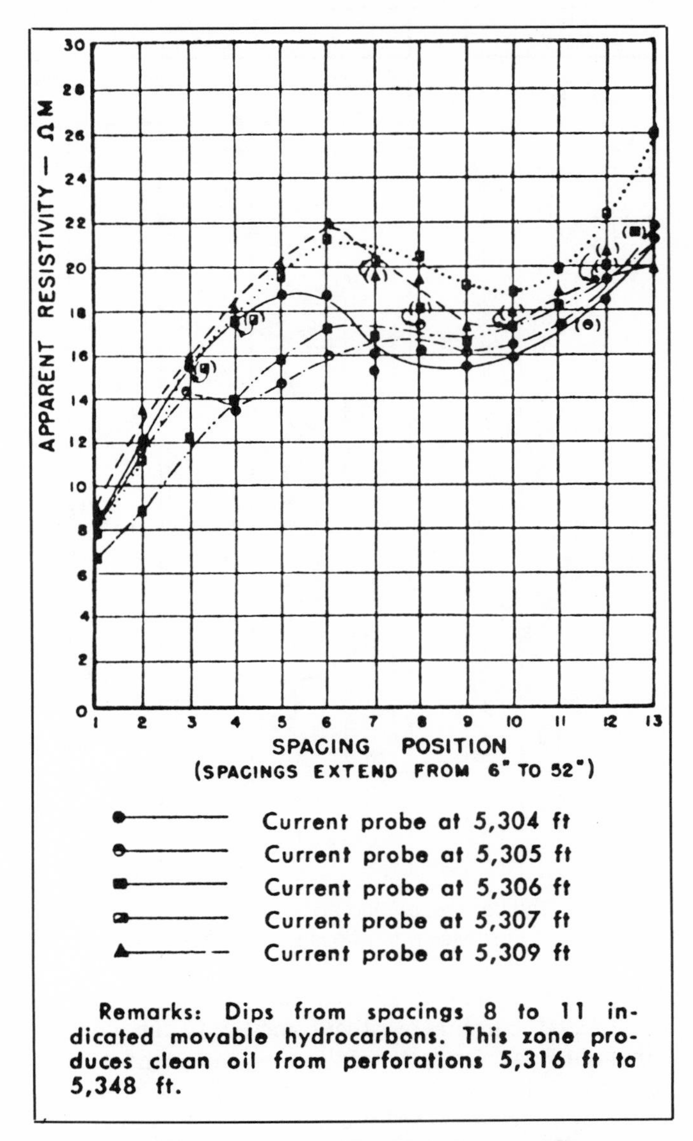

Fig. 6-26. Displacement logging results. Well located in Sholem Alechem Field, Stephens County, Oklahoma. Permission to publish by the Society of Petroleum Engineers of AIME.[8] Copyright 1955 SPE-AIME.

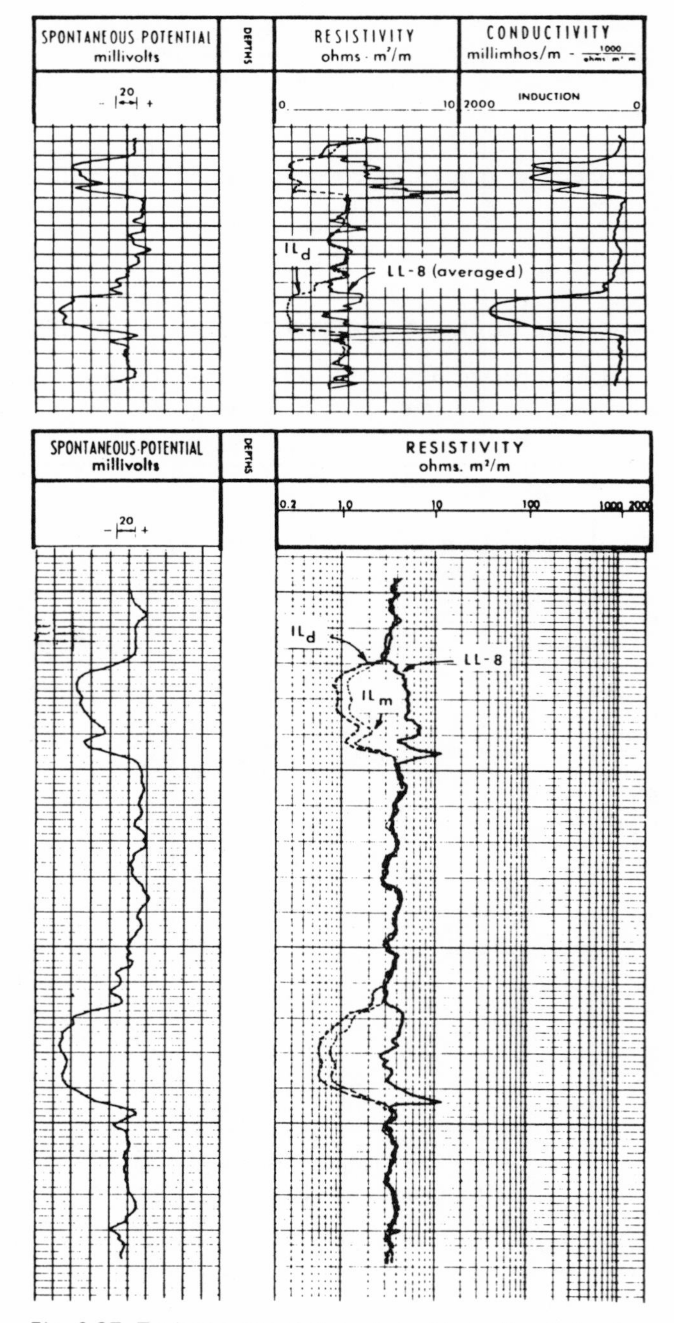

Fig. 6-27. Typical presentation of dual induction-Lateralog 8 (DIL-LL8). Permission to publish by Schlumberger Ltd.[25]

nonfocused current resistivity measurements.[6] With the nonfocused current resistivity methods, the current flows into the formation following the path of least resistance. The nonfocused system can, therefore, be seriously affected by the borehole and by the formations above and below the bed being measured. Salt mud systems (where $R_{mf} \leq 3R_w$) tend to short circuit the system, while surrounding beds of low resistivity distort the current distribution.

The focused current resistivity method uses electrode arrangements and an automatic control system to force surveying current through formations as a sheet of predetermined thickness. The resulting resistivity measurement thus involves only a portion of the formation, of limited vertical extent, and is practically unaffected by borehole mud. Its primary advantages over a conventional electrical log are sharper discrimination between different beds, more accurate definition of bed boundaries, and closer approximation of the true resistivity for thin beds, especially where salt muds are used. Principal applications include hardrock areas, all areas where salt muds are used for drilling, and those where highly resistive formations exist, even though fresh muds were used.

In any water-base mud, fresh or salt, the focused electrical method provides dependable logs for correlation purposes. It will, in every case, equal or excel the short normal curve in bed definition. Figure 6-33 compares the current distributions for normal and focused electrical systems and illustrates the greater adaptability of the focused electrical method in investigating thin, resistive beds.

Two types of focused resistivity devices have been developed, the guard-electrode system and the point-electrode system. The resistivity response of each of these systems is directed toward a measure of R_t.

Guard Electrode System

The basic guard system (see Fig. 6-34a) consists of two guard electrodes, 5–6 ft long, on each side of a short (3–6 in.) central electrode. The two guard electrodes control current flow from the center measuring electrode in a horizontal sheet. The current flowing to the

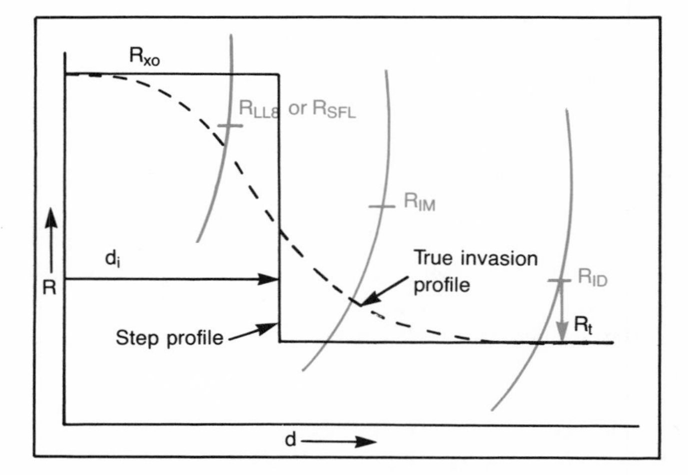

Fig. 6-28. Conceptual insight into invasion correction simulated by step profile using the three resistivity responses of the dual induction-LL8 or SFL

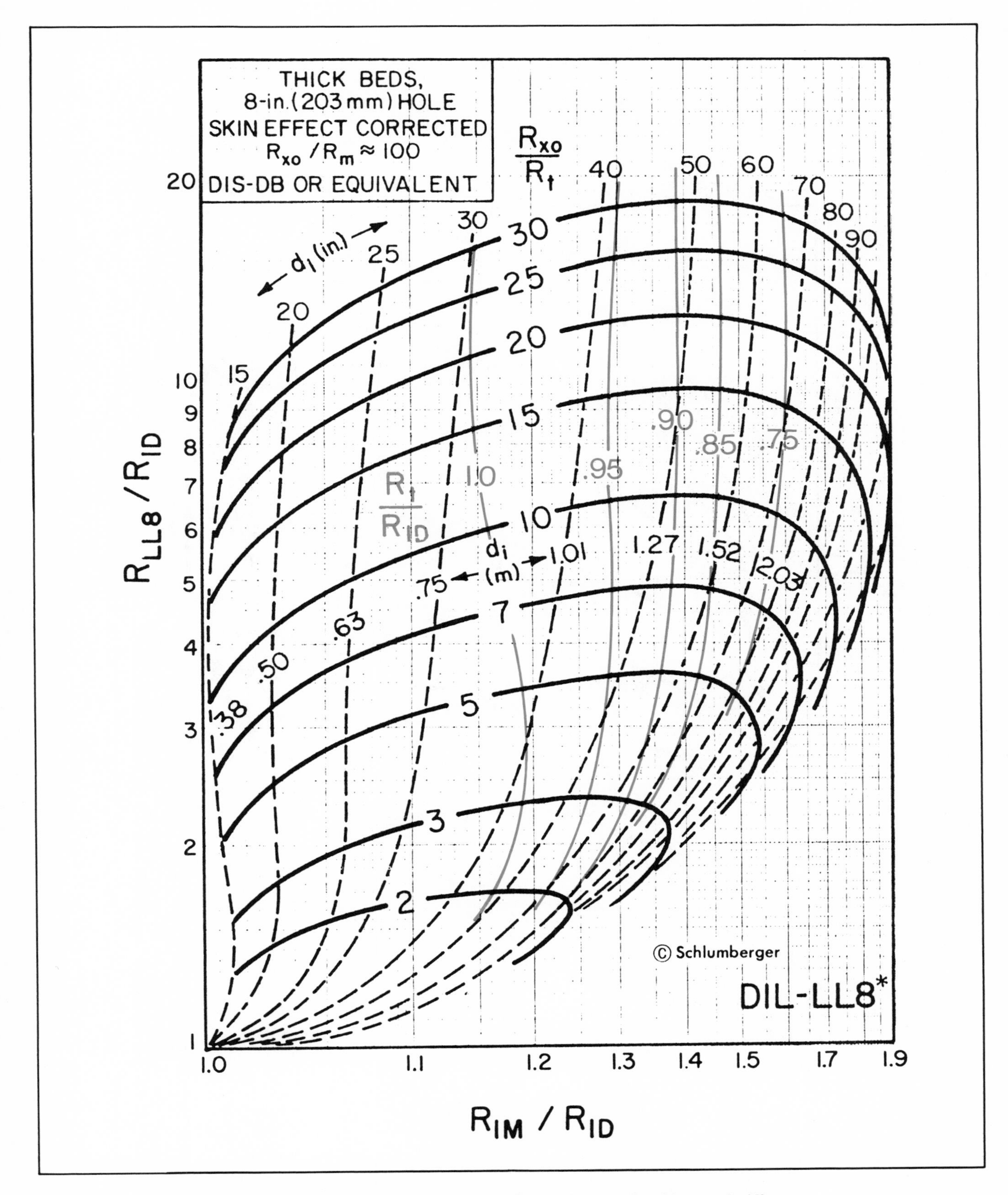

Fig. 6-29. Dual induction-Lateralog ILD-ILM-LL8. Permission to publish by Schlumberger Ltd.[29]

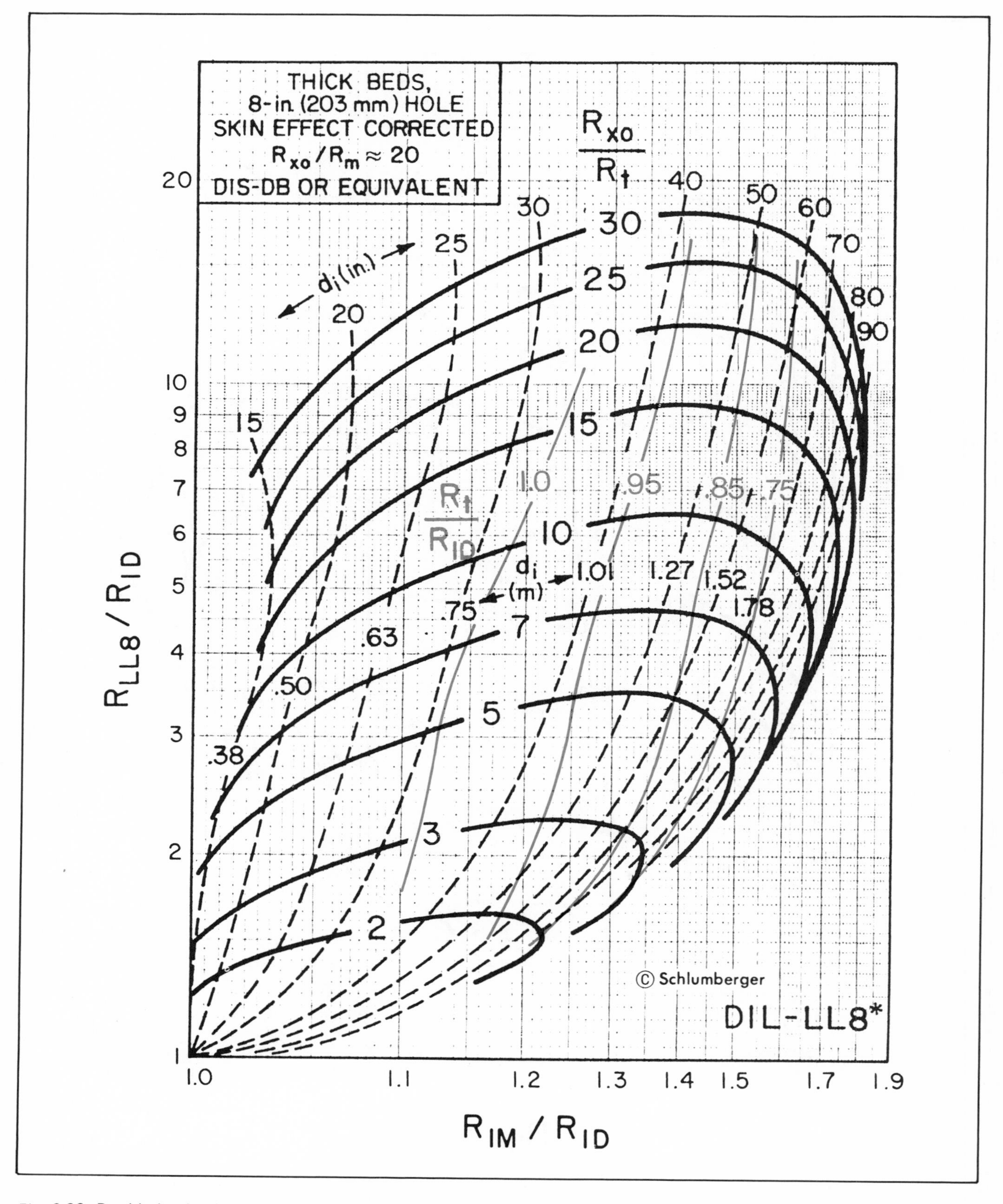

Fig. 6-30. Dual induction-Lateralog ILD-ILM-LL8. Permission to publish by Schlumberger Ltd.[29]

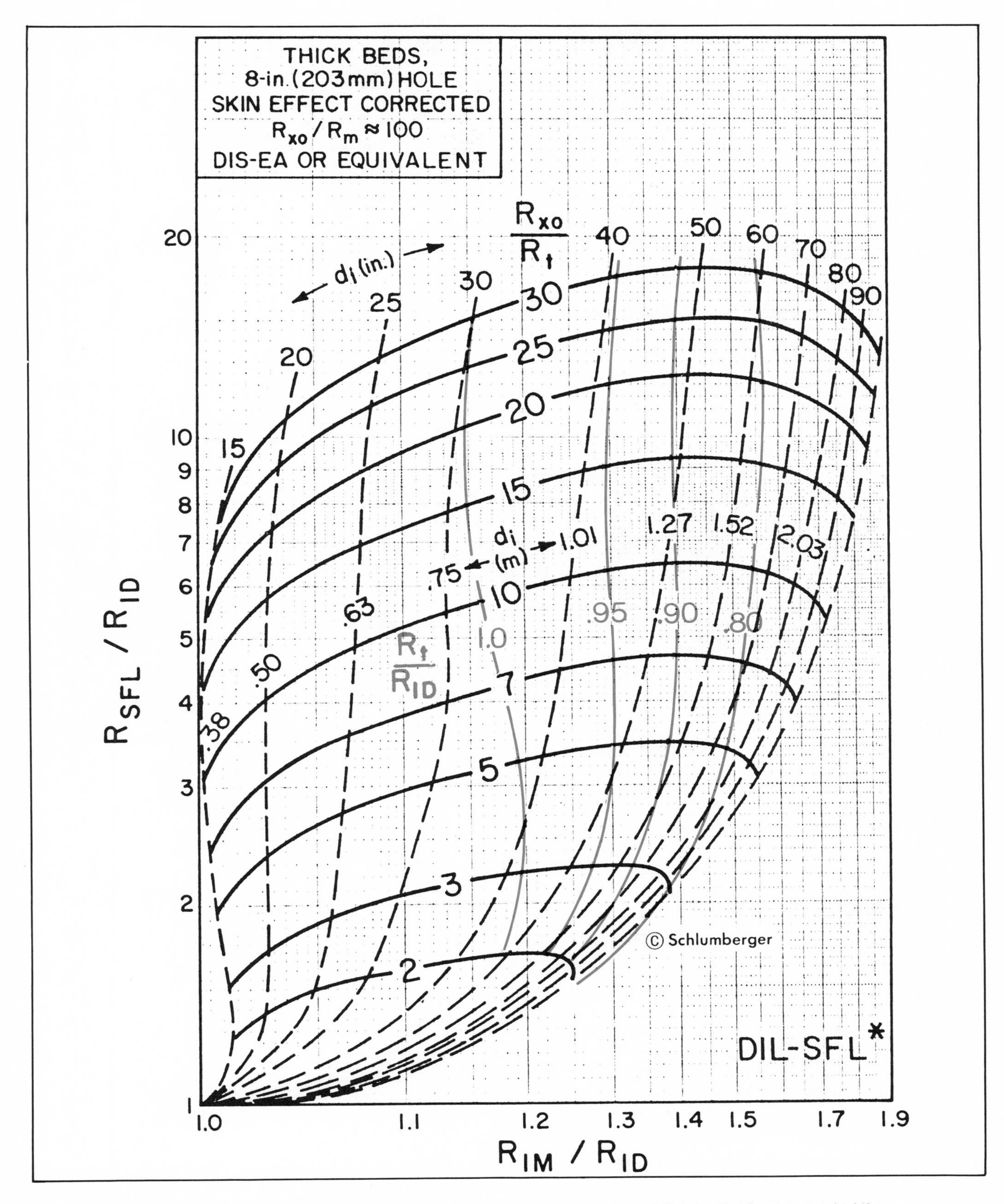

Fig. 6-31. Dual induction-spherically focused log ILD-ILM-SFL. Permission to publish by Schlumberger Ltd.[29]

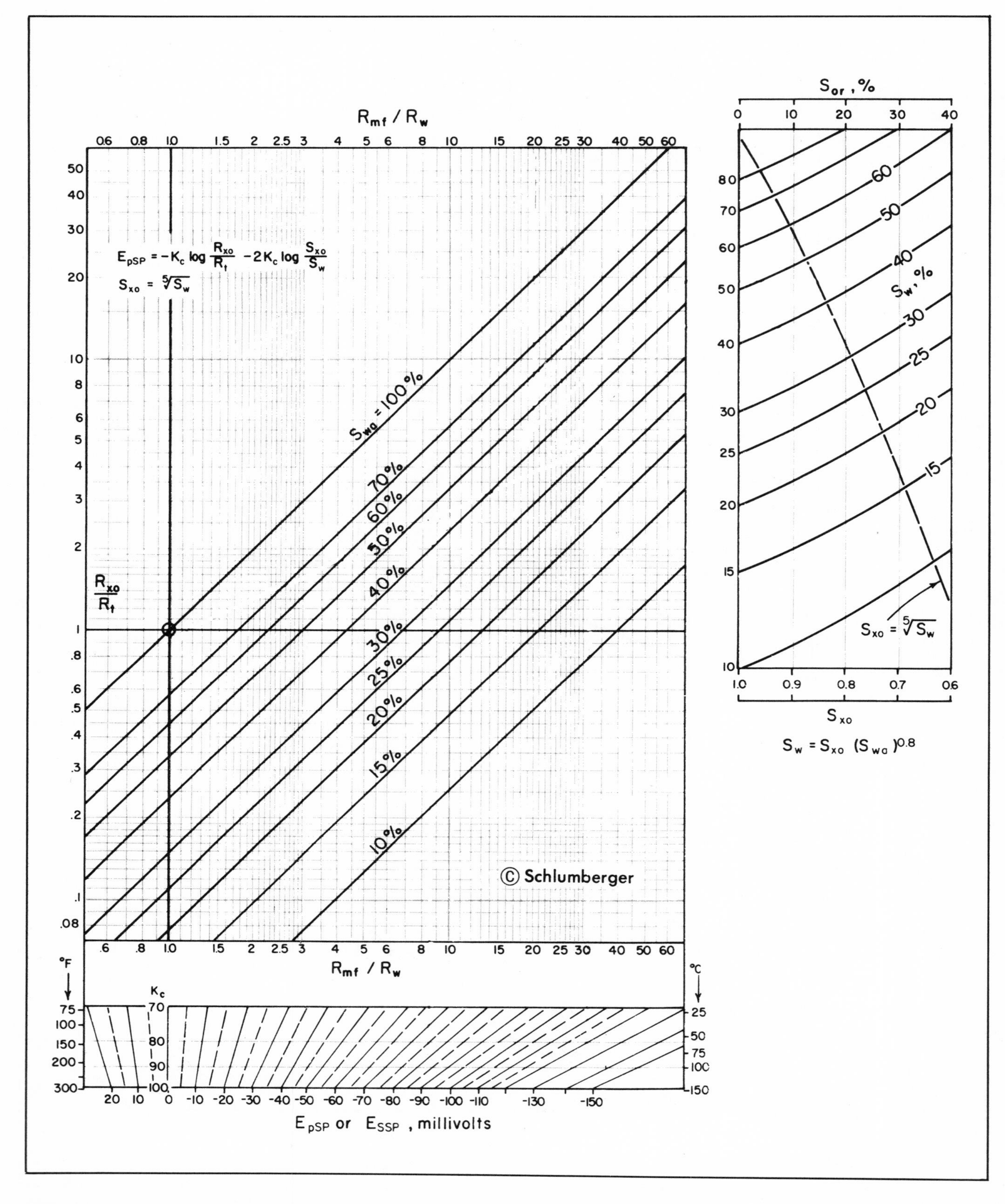

Fig. 6-32. Graphical section of E_{SSP} equation. Permission to publish by Schlumberger Ltd.[29]

central electrode and the potential of the electrode are measured, enabling formation resistivity to be determined.

Current focusing is achieved by applying a control current at guard electrodes A'_1 and A'_2 so the potential difference between A'_1 and the measuring electrode A'_0 is always zero. The generator supplies a constant current to electrode A'_0. The controller supplies an auxiliary current to the two guard electrodes, which are connected together and thus at an equal potential. By continuously comparing the potentials of A'_1 and A'_0, the controller can automatically adjust the auxiliary current to keep the potential difference between the two at a nil value. Since the two potentials are equal, current from A_0 cannot flow toward A'_1 and A'_2, and so must flow outwardly into the formation. The potential of the A'_0 electrode is recorded as a measure of formation resistivity.

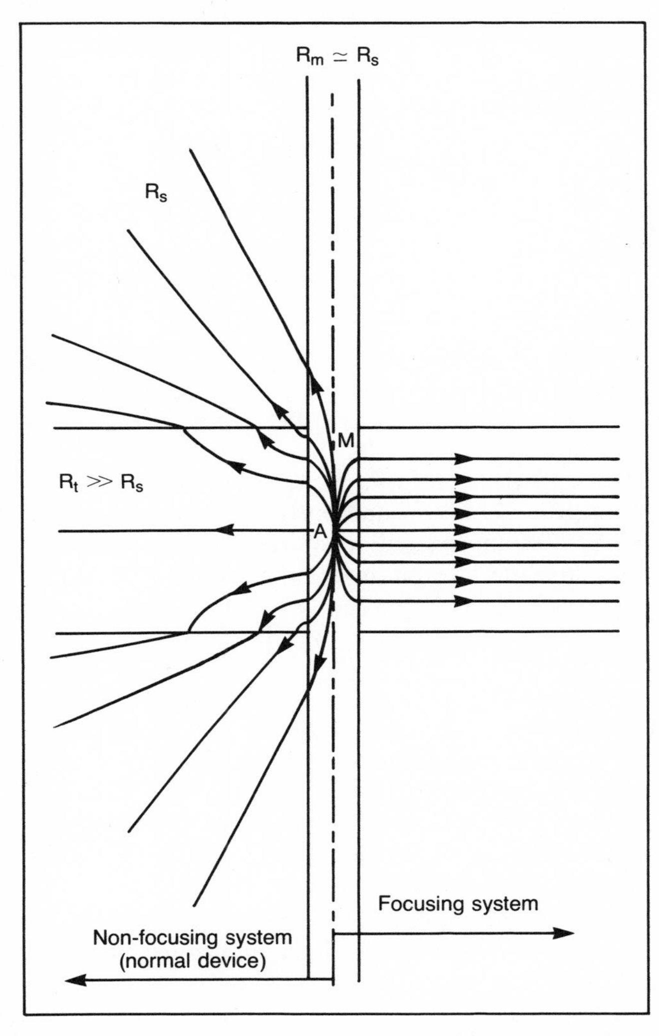

Fig. 6-33. Comparative current distribution of focused and nonfocused devices. Permission to publish by the Society of Petroleum Engineers of AIME.[6] Copyright 1951 SPE-AIME.

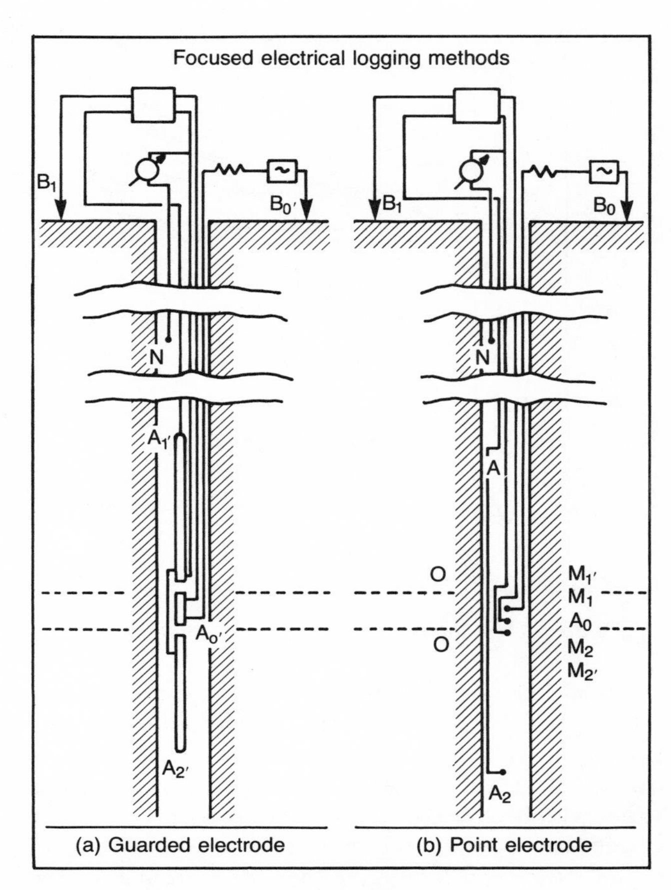

Fig. 6-34. Summary of focused logging devices. Permission to publish by Oil, Gas and Petrochem Equipment.[22]

The depth of investigation of this focused device is the distance from the wellbore at which the current sheet starts to deviate from a flat disk. An approximation of this distance can be obtained by multiplying the length of one guard electrode by three.

Point Electrode System

In this system, the focusing of the current beam is accomplished by a number of point electrodes. Figure 6-34b shows a seven-point electrode system. For this particular electrode arrangement, the potential sensing elements are the electrodes M_1, M'_1 and M_2, and M'_2. The current to electrodes A_1 and A_2 is adjusted by an automatic controller so that a zero potential difference will be maintained between the measuring electrodes M and M'. Since the current to electrode A_0 is constant, the potential of the M electrodes provides a measure of the resistivity of the formation opposite the center of the tool. The A_1A_2 spacing is 80 in. The O_1O_2 spacing is 32 in., thus the thickness of the current sheet is about 32 in.

The distance that the current sheet maintains at a con-

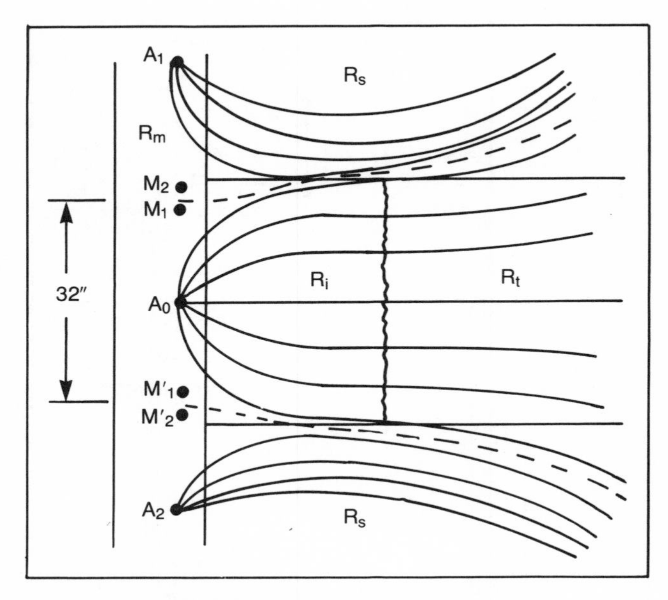

Fig. 6-35. Schematic of Lateralog 7 responses

stant thickness is determined by the electrode spread ratio, A_1A_2/O_1O_2. For the seven-point electrode system, the ratio is usually 2.5, and the current beam is controlled to a distance of about 10 ft, although this varies with both bed resistivity and thickness. Since the current-beam deviation occurs at about 10 ft, the depth of investigation will be somewhat less than that for the guard-electrode system.

The focused current resistivity measurements are usually called, unfortunately, Laterolog resistivity measurements and are not to be confused with the lateral resistivity measurement of the nonfocused current type.

Factors Affecting Focused Current Resistivity Measurements

As is the case with all resistivity measurements, a number of factors influence the resistivity response of these tools. The apparent resistivity will depart from the desired response, R_t, as a result of the influence of the borehole, invaded zone, and surrounding bed effects, schematically shown in Figure 6-35. These influences might be expressed as:

$$R_a = R_{LL} = f(R_m,\ d_h,\ h,\ R_s,\ d_i,\ R_i,\ \underline{\underline{R_t}})$$

In order to estimate R_t, the influence of the borehole (d_h, R_m), surrounding beds (h, R_s) and invaded zone (d_i, R_i) must be assessed. For logging situations best suited for the Laterologs, therefore, where the borehole is a salt water mud ($R_{mf} \leq 3R_w$) and the zone of interest is greater than the measure electrode spacing, then R_a approaches R_t. The stepwise process for using simplified departure curves for correcting the apparent resistivity measured by the Laterologs is outlined below.

The apparent resistivity measured by the focused current type resistivity methods can be expressed using the Laterolog-7 as an example:

$$R_{LL7} = f(R_m,\ R_s,\ R_i,\ \underline{\underline{R_t}})$$

The borehole effect is considered first and can be accounted for by using departure curves, as shown in Figure 6-36, where:

$$\frac{R_{LL7}}{R_M} \xrightarrow[\text{Using } R_m,\ d_h]{\text{Departure Curve Correction}} \frac{R'_{LL7}}{R_m}$$

and

$$\frac{R'_{LL7}}{R_m} \times R_m = R'_{LL7} = f(R_s,\ R_i,\ \underline{\underline{R_t}})$$

This borehole corrected resistivity R'_{LL7}, is now a reflection of only three influencing zones. It is next corrected for the surrounding bed effect using the departure curves shown in Figures 6-37 and 6-38 where:

$$\frac{R'_{LL7}}{R_s} \text{ and } h \xrightarrow[\text{Curve}]{\text{Departure}} \frac{R''_{LL7}}{R'_{LL7}}$$

and

$$\frac{R''_{LL7}}{R'_{LL7}} \times R'_{LL7} = R''_{LL7} = G_iR_i + G_t\underline{\underline{R_t}}$$

This borehole and surrounding bed effect corrected resistivity, R''_{LL7}, now reflects the composite influence of only the invaded and undisturbed zones, as schematically shown in Figure 6-39.

It is apparent that if an invaded zone resistivity, R_i, was measured and an appropriate d_i representing the step profile approximation of the invasion influence could be made, then an invasion corrected R_t could be estimated from:

$$R''_{LL7} = G_iR_i + (1 - G_i)\underline{\underline{R_t}}$$

The value of G_i is obtained from the estimated value of d_i using the radial response behavior (see Fig. 6-19) for the deep Laterolog resistivity. The major problem with this invasion correction is that when running a Laterolog 7 or 3, only one resistivity measurement is made and unless another resistivity tool is run in the borehole, it is necessary to assume $R''_{LL7} \approx R_t$.

A typical laterolog is shown recorded in two forms (see Fig. 6-40), which today have supplanted the linear type recording, which is not suitable for the resistivity ranges usually encountered when running these tools. One log presents the laterolog response on a hybrid scale (and conductivity scale) whereas the second is a recording through the same interval on a logarithmic scale. The hybrid scale is such that one half is linear and the

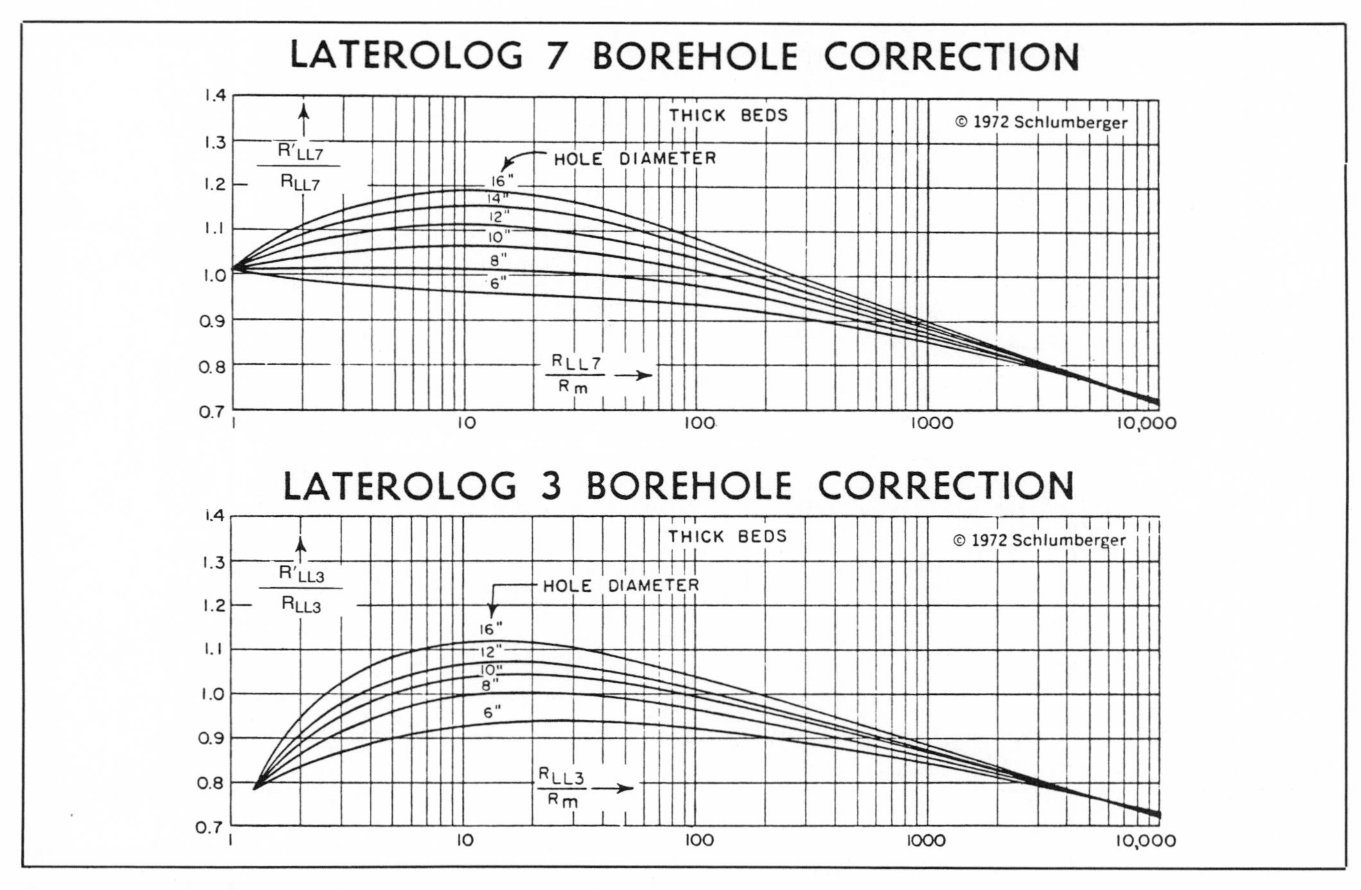

Fig. 6-36. Borehole departure curves for Lateralog 7 and 3. Permission to publish by Schlumberger Ltd.[24]

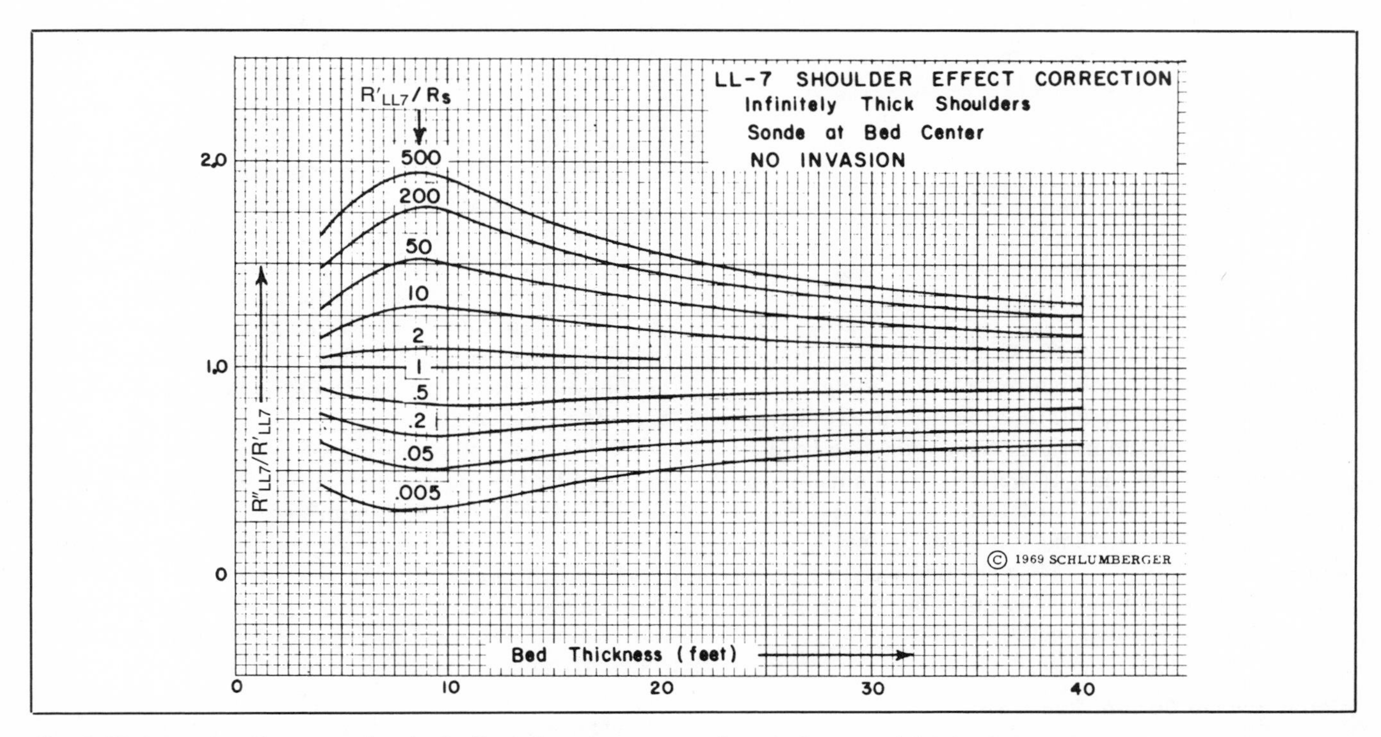

Fig. 6-37. Lateralog 7 surrounding bed effect departure curve. Permission to publish by Schlumberger Ltd.[30]

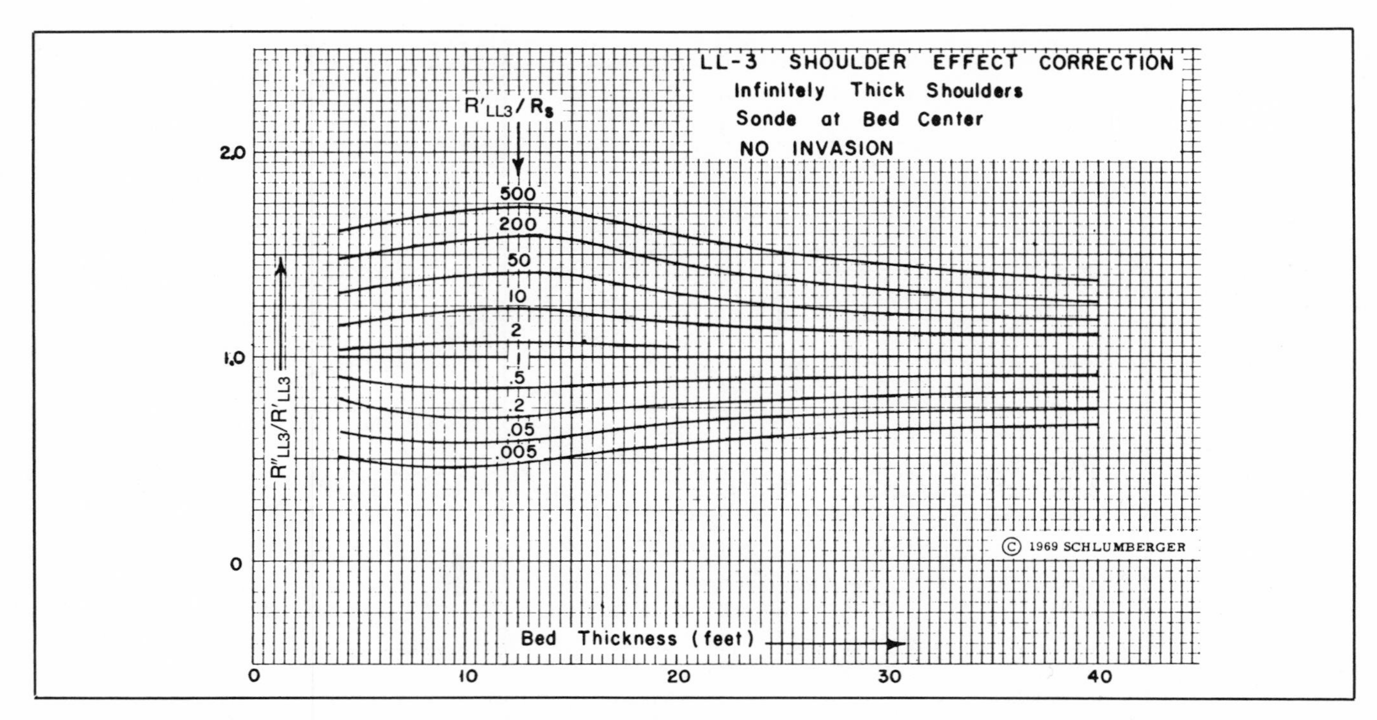

Fig. 6-38. Lateralog 3 surrounding bed effect departure curve. Permission to publish by Schlumberger Ltd.[30]

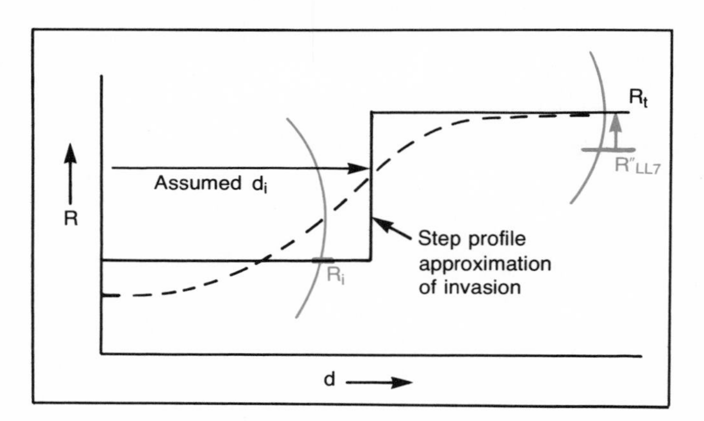

Fig. 6-39. Conceptual insight into invasion profile

other half is highly nonlinear (a reciprocated conductivity that compresses the high resistivity readings).

It should be pointed out that these focused resistivity devices cannot be run in either nonconductive borehole fluids (air, gas, oil base or water-in-oil emulsion mud systems) or cased holes. The laterologs, both point electrode and guarded systems, function optimally in salt mud systems ($R_{mf} \leq 3R_w$). They respond well to thin formations and in situations where the formations present high resistivity contrasts to the borehole.

Micro-Focused Current Resistivity

Although the deep-responding focused current resistivity tools provided a means to estimate R_t, the necessity to compare this value with the resistivity of the formation 100% water-saturated, R_o, still remains. In salt mud systems or high resistivity formations, the nonfocused current micro-resistivity tool suffers the same problem as the nonfocused current resistivity tools in that the current distortion becomes so great that its response to R_{xo} is very poor. In order to obtain a measure of R_{xo} and hence a value of F_R, the contact, focused current micro-resistivity tool was developed.[7]

The micro-focused tool is physically similar to the nonfocused tool shown in Figure 6-9. But its electrode arrangement and electrical circuitry are quite different. For this tool, small button electrodes are embedded in the hydraulic pad in a concentric circular arrangement. All the button electrodes within a particular ring are electrically connected, forming an essentially continuous ring electrode. Figure 6-41 shows the distribution of these ring electrodes.

The principle of operation of this tool is similar to that used by the focused current resistivity tool. In this system (see Fig. 6-41) A_0 represents the current electrode. An essentially zero potential is automatically maintained between the ring electrodes M_1 and M_2 by applying a bucking current at ring electrode A_1. In this way, the measuring current cannot flow past ring electrodes M_1 and M_2 and will enter the formation in the form of a cylindrical beam approximately 1.7 in. in diameter. The beam begins to flare out at a depth of about 3 in., which is considered the depth of investiga-

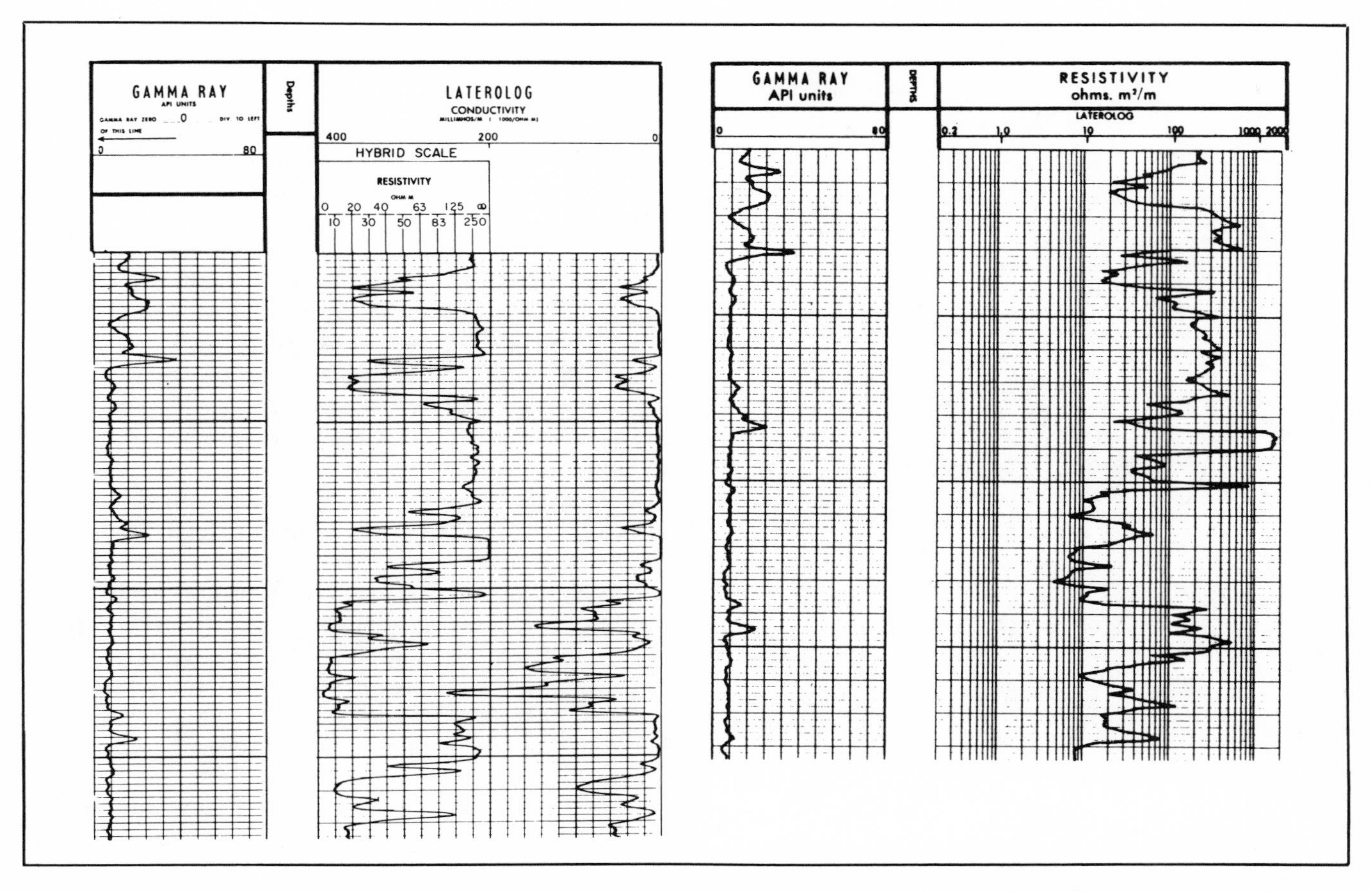

Fig. 6-40. Lateralog recorded on both hybrid and logarithmic scales through same zone. Permission to publish by Schlumberger Ltd.[25]

tion. Only one resistivity curve is recorded by this tool, although the caliper curve is also obtained. Figure 6-42a shows the focused current pattern for this tool. A typical focused current micro-resistivity log (in this case, the Schlumberger Microlaterolog) is shown in Figure 6-43 where the resistivity of the Microlaterolog, R_{MLL}, is shown on a compressed scale and on a linear scale.

This tool measures only one resistivity curve, which is influenced by the mud cake and R_{xo}. As with the focused current electrical logs (Laterolog and Guard logs), the apparent resistivity, R_{MLL}, is a series measurement of R_{mc} and R_{xo} (see Fig. 6-44).

In order to obtain R_{xo} from the measured R_{MLL}, the mud cake influence must be removed. Since the radial response behavior of the Microlaterolog is known, a departure curve as shown in Figure 6-45 has been generated, based upon a simple two-zone system as shown in Figure 6-44. The Microlaterolog resistivity, R_{MLL}, has been simulated for various values of R_{mc}, h_{mc} and R_{xo} and presented in the departure curve shown in Figure 6-45. The Microlaterolog rapidly starts to lose resolution in defining R_{xo} opposite thicker mud cakes ($h_{mc} > 1/2''$).

Applications. The focused current micro-resistivity tools can be used for the following: (1) define R_{xo} and hence F_R and ϕ, (2) define bed boundaries, (3) define net "pay," (4) provide R_m from a mud log, (5) indicate fracture and vug porosity, and (6) enable invasion correction to be made on Laterolog responses.

It is assumed that adequate invasion exists such that the tool is primarily measuring R_{xo} saturated with the conducting fluid, R_{mf}. Accurate values of R_{xo} are also dependent on a knowledge of R_{mc} and h_{mc}, and if these conditions are satisfied then, as with the Micr整log, the formation factor and porosity can be determined from:

$$F_R = \frac{R_{xo}}{R_{mf}}(1 - S_{or})^n = \phi^{-m}$$

An example of porosity determination is shown with Figure 6-43.

Again, it is necessary to estimate S_{or} for quantitative evaluation of F_R and ϕ. Figure 6-42 shows a comparison of the current distribution for the focused and nonfocused micro-tools. As can be seen from Figure 6-42b, the nonfocused tool tends to become short circuited by the mud cake if the mud cake is considerably less resis-

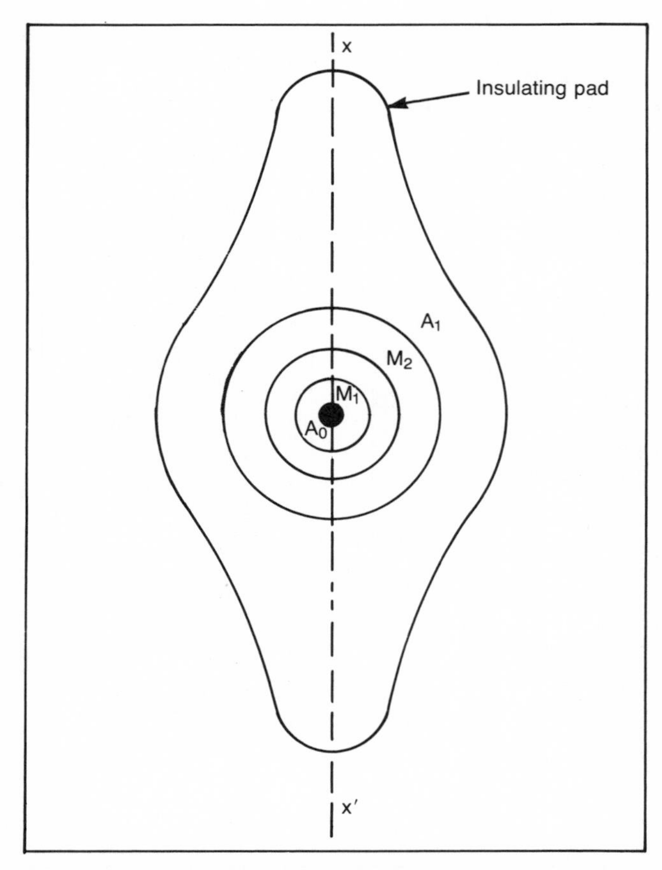

Fig. 6-41. Electrode arrangement of pad micro-focused logging tool. Permission to publish by Oil, Gas and Petrochem Equipment.[23]

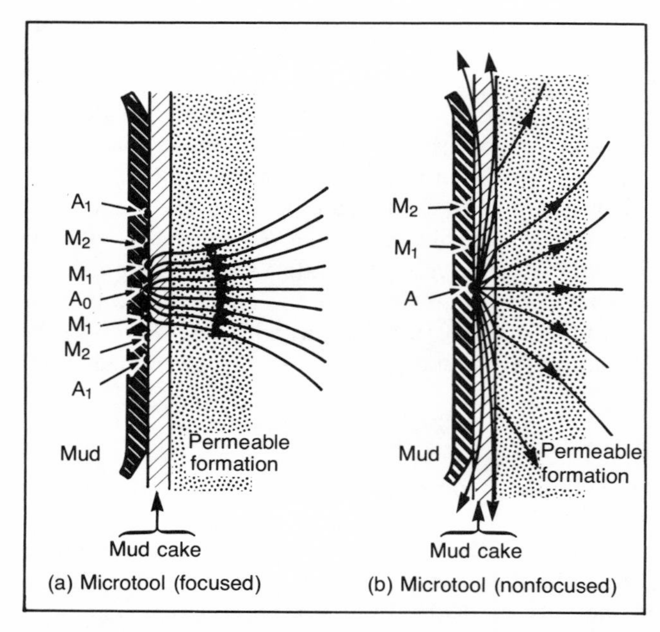

Fig. 6-42. Comparison of current distribution for focused and nonfocused micro-tools. Permission to publish by Oil, Gas and Petrochem Equipment.[23]

tive than the formation. This situation arises when a salt mud is used. The micro-focused tool, however, operates best in a salt mud since these muds have thin, low-resistivity mud cakes, and $R_{xo} \gg R_{mc}$. This is usually encountered in low porosity formations such that the micro-focused tool generally operates best where $\phi < 15\%$.

Since this tool also has a very small vertical resolution, bed boundaries can be accurately determined. However, since only one resistivity curve is recorded, it is more difficult to determine porous and permeable zones than with the Microlog since the characteristic separation isn't present. The caliper curve (and SP if also recorded) can be used to locate these zones of interest.

The resistivity curve in conjunction with the caliper and SP can be used to determine net pay, but care should be taken since the highly focused resistivity curve will respond to small irregularities such as shells, pebbles, etc. This could cause an underestimation of net pay.

It is also possible to measure R_m when going into the hole with the arms closed. With this tool, however, there is only one resistivity curve and, therefore, it becomes more difficult to evaluate the response as being only a function of the mud.

Repeat runs over the same interval may indicate the presence of fractures and vugs due to differing responses of the resistivity curves.

The value of R_{xo} generated from the Microlaterolog can also be used to approximate an invasion correction based on a step profile simulation of the invaded zone if d_i is assumed. This concept is shown in Figure 6-46.

In order to make the invasion correction, G_i must first be evaluated from Figure 6-19, using an assumed d_i (or a range of d_i values). R_t can then be estimated from the following expression:

$$\frac{1}{R''_{LL7}} = \frac{G_i}{R_{xo}} + \frac{(1 - G_i)}{R_t}$$

Limitations. Errors in evaluating R_{xo} can result from the following conditions:

1. Shallow invasion would lead to erroneous porosity determinations since it is assumed that the only conducting fluid affecting the measured rock resistivity is mud filtrate, R_{mf}.
2. Thick mud cake ($h_{mc} > ½''$) will cause the tool to have poor resolution.
3. Incorrect R_{mc} value will affect the value as determined empirically from the departure curve.
4. Incorrect estimation of S_{or} will lead to incorrect porosity determination.

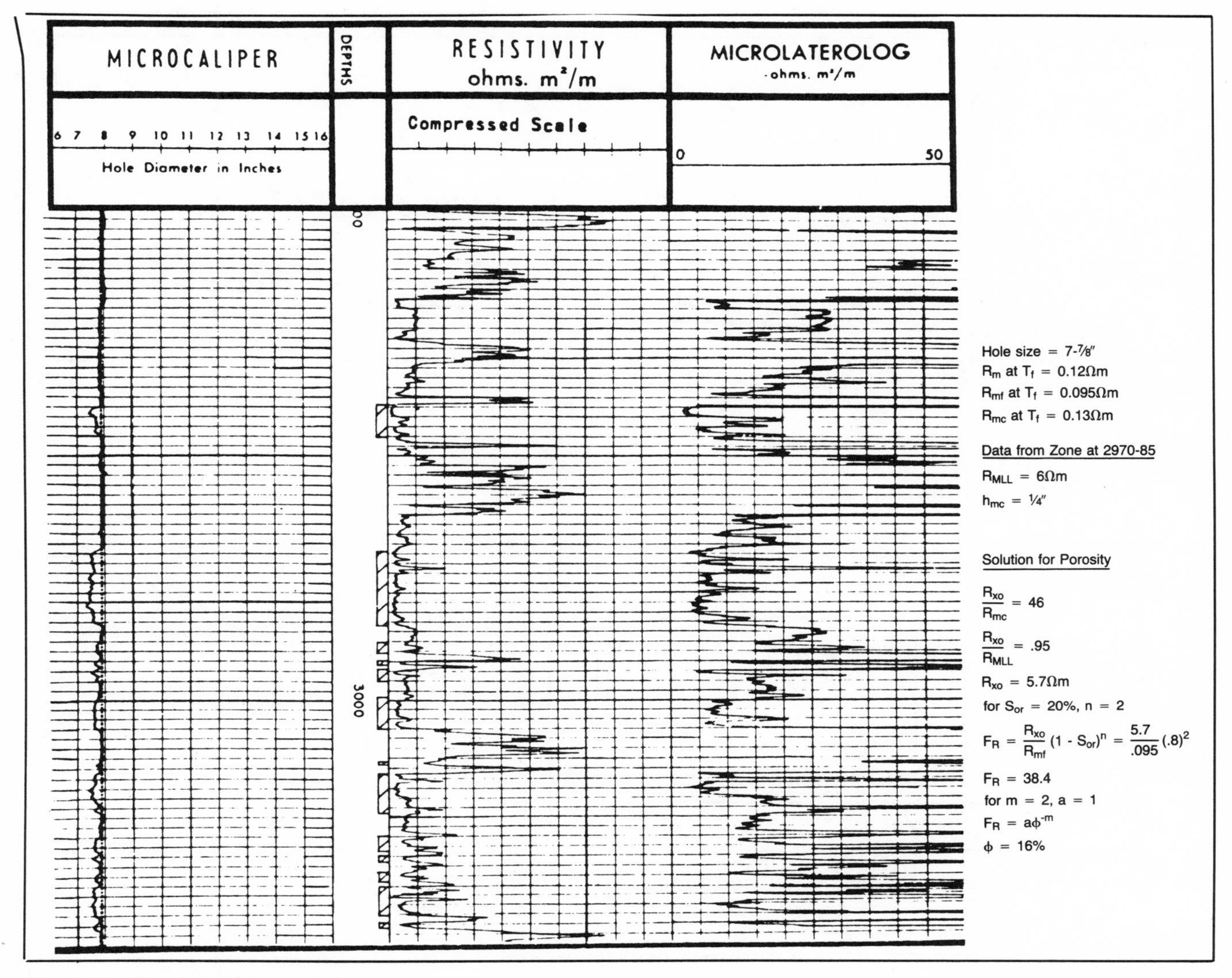

Fig. 6-43. Microlateralog example

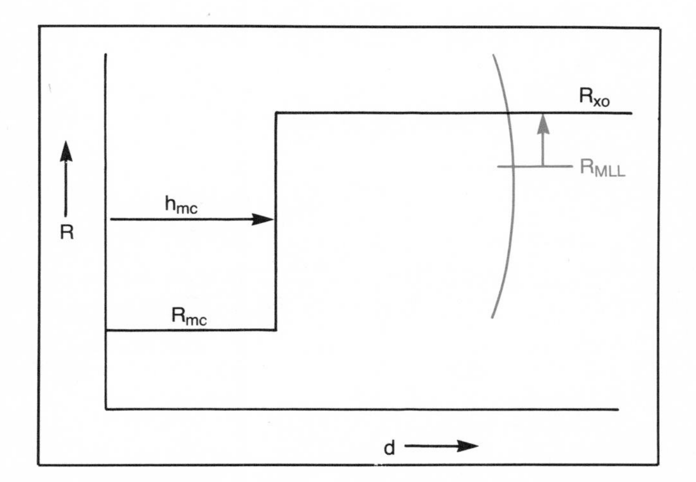

Fig. 6-44. Conceptual insight into mudcake-flushed zone resistivity profile

Proximity Log

The proximity log was developed to overcome the deficiencies encountered with the shallower sensing Microlaterolog when run opposite thick mud cakes. It is a contact tool having a focused current similar to the Microlaterolog but is a deeper sensor (see Fig. 6-19). Since it is a deeper sensor, it is influenced by the mud cake to a much less degree than the Microlaterolog (see the Proximity Log departure curve shown in Figure 6-47). As a deeper sensor, however, the rock resistivity reflected by the mud cake corrected response may not reflect R_{xo} since it can be influenced by the rock beyond the invaded zone (see Fig. 6-48). Deeper invasion enhances the probability that R_P will more nearly reflect R_{xo} once it is corrected for the mud cake effect, although the mud cake corrected Proximity log value, R_P, is always assumed to represent R_{xo}.

Approximately 90% of the signal for the Proximity

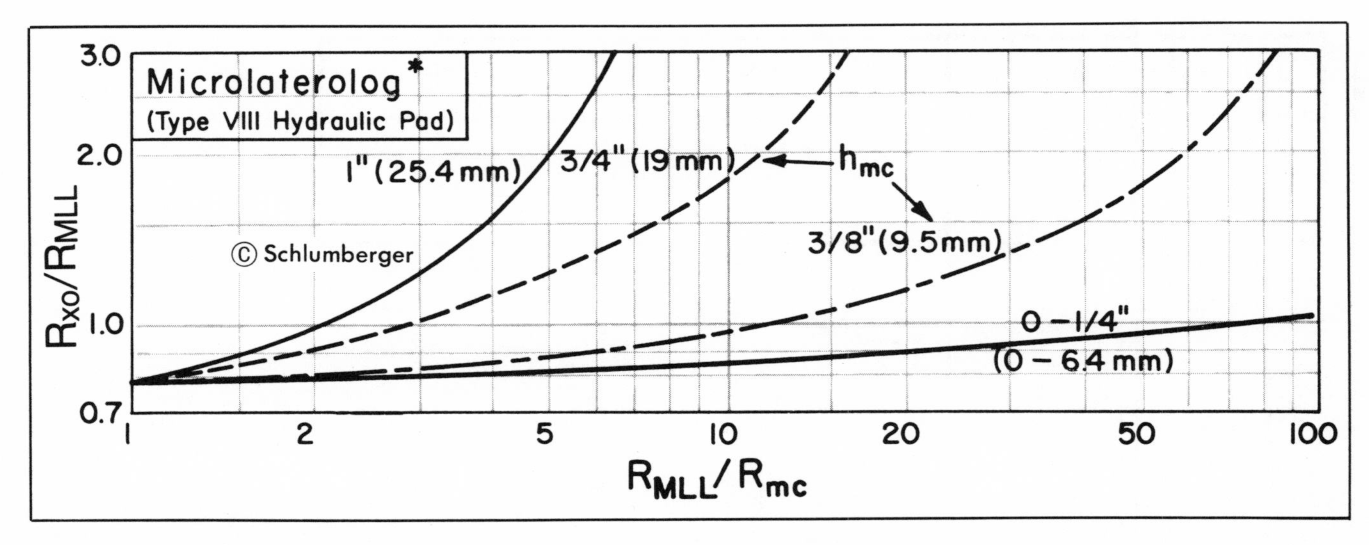

Fig. 6-45. Microlateralog departure curve for 8" hole. Permission to publish by Schlumberger Ltd.[29]

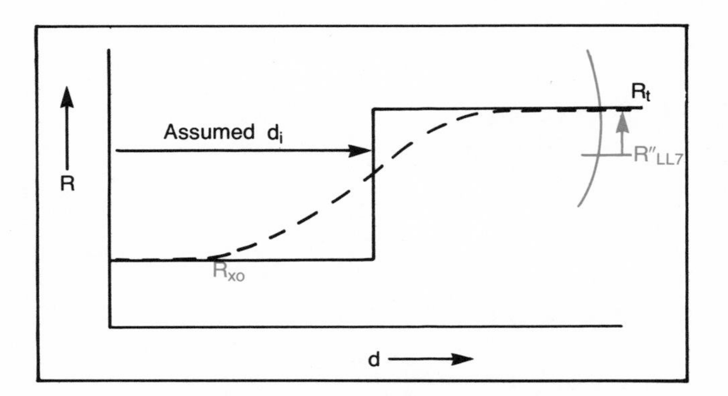

Fig. 6-46. Conceptual insight into invasion correction on R''_{LL7}

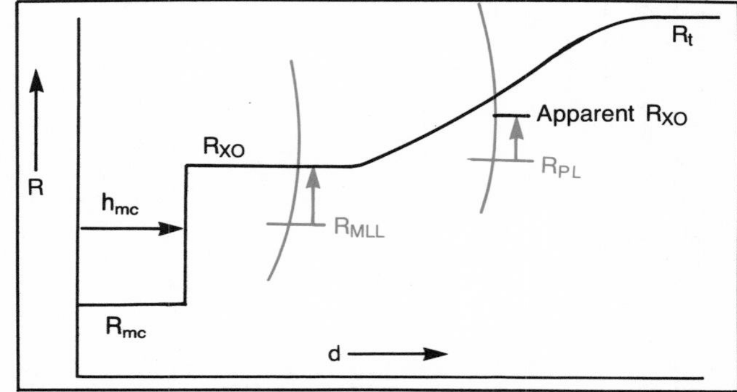

Fig. 6-48. Conceptual insight of Microlateralog and proximity log responses to R_{xo}

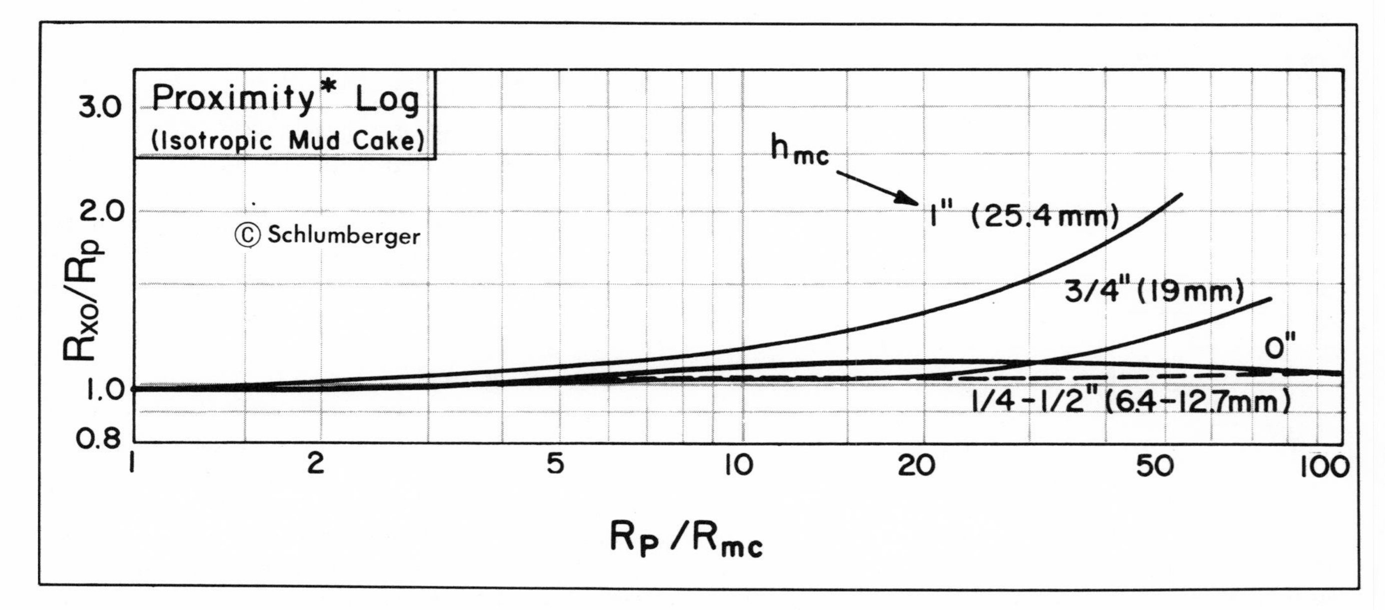

Fig. 6-47. Proximity log departure curve for 8" hole. Permission to publish by Schlumberger Ltd.[29]

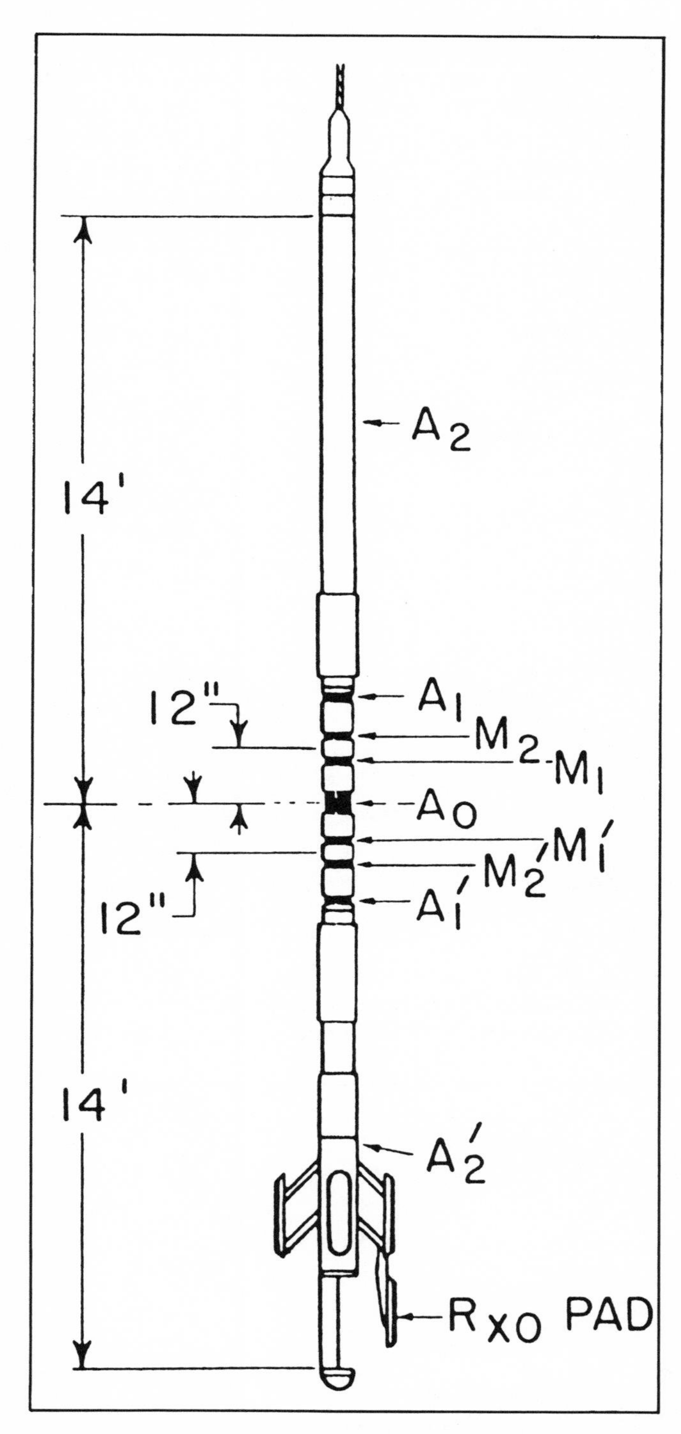

Fig. 6-49. Schematic of dual Lateralog-R_{xo} tool. Permission to publish by the Society of Petroleum Engineers of AIME.[26] Copyright 1972 SPE-AIME.

log is within 20 in. of the borehole. Vertical resolution of this tool is about 6 in., which, although greater than the Microlaterolog, is still quite small. Beds greater than one-foot thick give responses essentially unaffected by the surrounding beds.

Dual Laterolog-R_{xo} Log

Development of the Dual Laterolog-R_{xo}[26] now provides a means of measuring a resistivity profile as a function of depth in borehole systems not suited for induction type measurements, i.e., salt mud systems and where formation resistivity is high. This tool measures a deep Laterolog, LLD, and a shallow Laterolog, LLS, with the same electrode system and R_{xo}, MICROSFL, with a pad device attached to the lower portion of the tool (see Fig. 6-49). The current distribution for the deep and shallow Laterologs is shown in Figure 6-50. The two foot current sheet emanating from this tool provides the same vertical resolution for both Laterologs.

The pad type R_{xo} device is based on the principle of spherical focusing and is referred to as a MICROSFL. It is designed for minimum mud cake effect without requiring deep invasion since its depth of investigation is quite shallow. Figure 6-51 schematically shows the electrode arrangement on the pad and the current distribution in front of the pad.

This tool, therefore, provides three resistivity measurements, an R_t response, R_{LLD}, an R_i response, R_{LLS}, and an R_{xo} response, $R_{\mu SFL}$. These three resistivity responses are recorded simultaneously. In addition, an SP (or gamma ray) and a caliper response are also recorded. A typical Dual Laterolog-R_{xo} presentation is given in Figure 6-52. Additionally, a dual spaced neutron (CNL) or a compensated formation density (FDC) tool can be combined with the Dual Laterolog-R_{xo} tool, although the eccentricity required by these two tools tends to detract from the reliability of the shallow Laterolog, R_{LLS}, response.

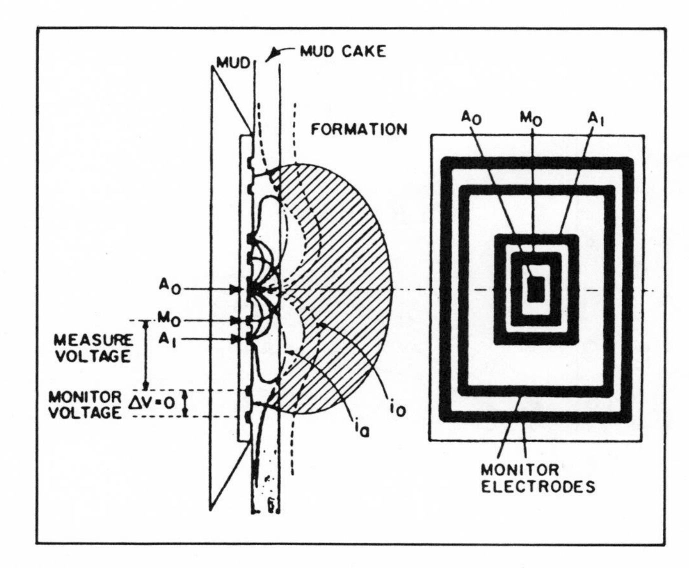

Fig. 6-50. Schematic of current patterns for both deep and shallow Lateralogs. Permission to publish by the Society of Petroleum Engineers of AIME.[26] Copyright 1972 SPE-AIME.

As with other resistivity devices, borehole and surrounding bed effects can influence the Laterolog responses and under conditions where these effects might be large, departure curves are available for making the desired corrections. Borehole departure curves are shown in Figures 6-53 and 6-54 respectively. The departure curves for making the mud cake correction on the MICROSFL are shown in Figure 6-55. Departure curves for correcting for the surrounding bed effects on the Laterologs deep and shallow are shown in Figures 6-56 and 6-57. As with all resistivity responses, the borehole correction is made first; then the borehole corrected resistivity, R'_{LL}, is corrected for surrounding bed effects to obtain R''_{LL}. These corrected resistivities are

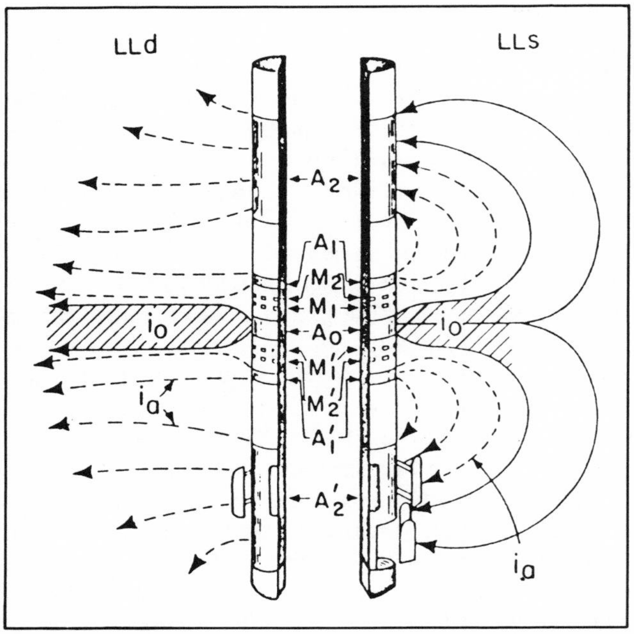

Fig. 6-51. Schematic of MICROSFL pad and current distribution. Permission to publish by the Society of Petroleum Engineers of AIME.[26] Copyright 1972 SPE-AIME.

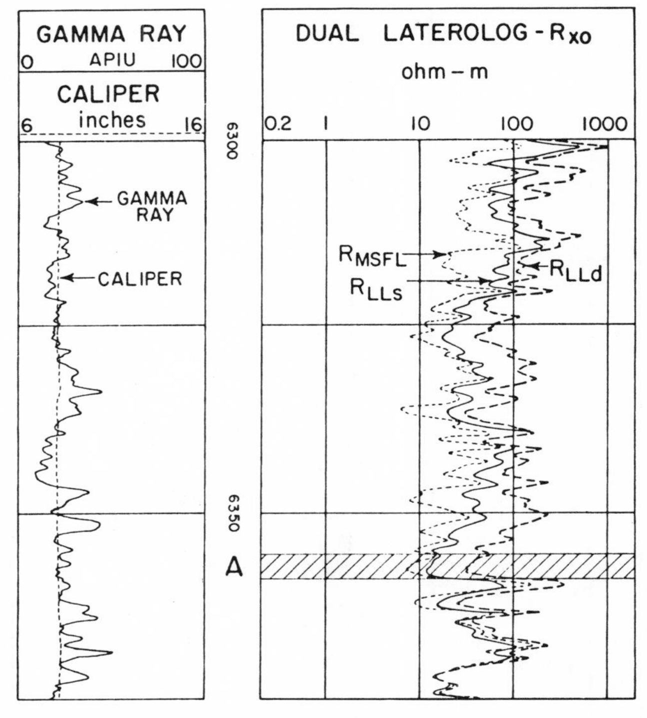

Fig. 6-52. Typical dual Lateralog-R_{xo} presentation. Permission to publish by the Society of Petroleum Engineers of AIME.[26] Copyright 1972 SPE-AIME.

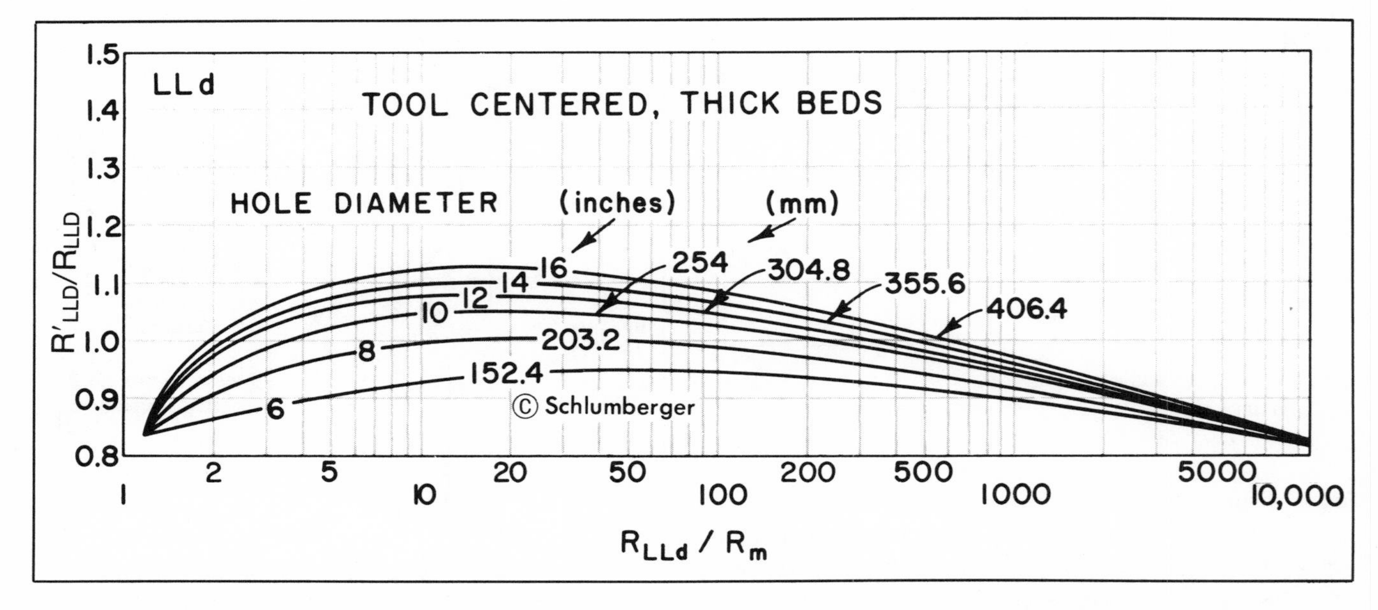

Fig. 6-53. Borehole departure curves for Lateralog deep, sonde centered. Permission to publish by Schlumberger Ltd.[29]

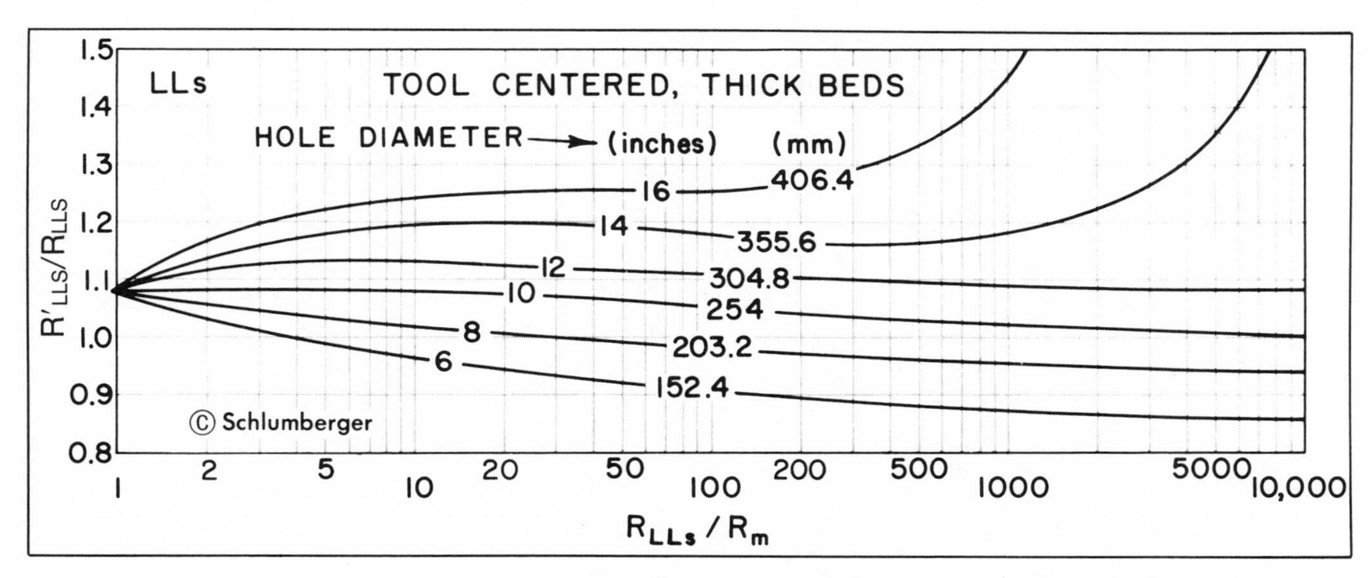

Fig. 6-54. Borehole departure curves for Lateralog shallow, sonde centered. Permission to publish by Schlumberger Ltd.[29]

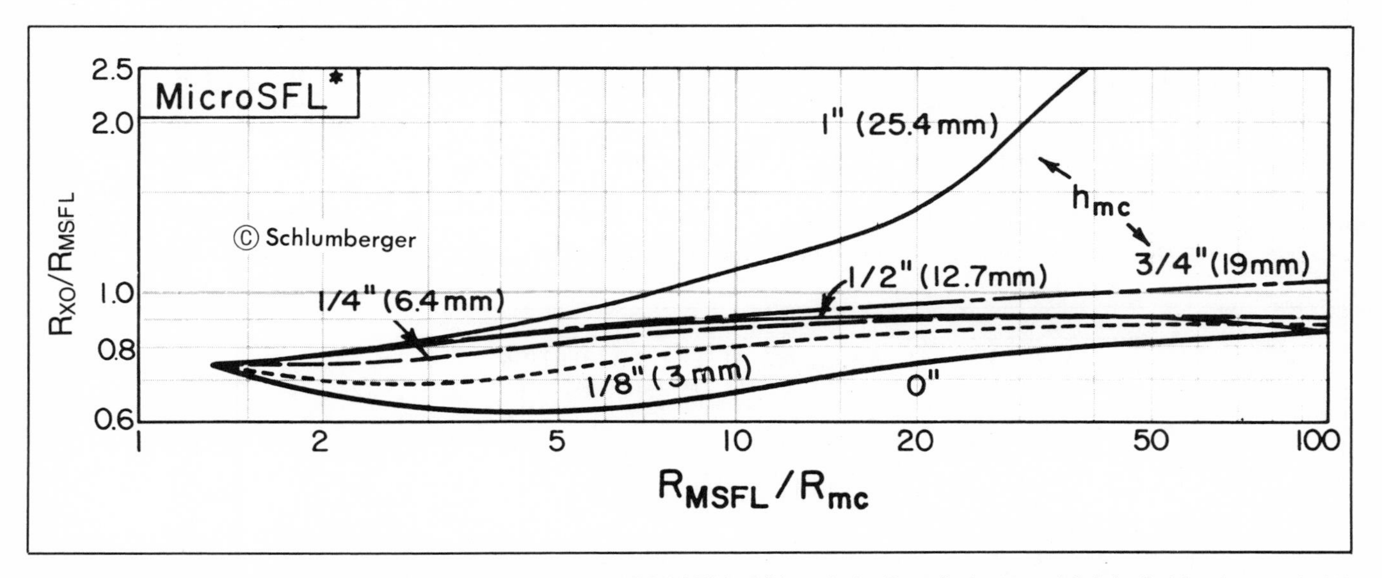

Fig. 6-55. Departure curves for mudcake correction for MICROSFL in 8" borehole. Permission to publish by Schlumberger Ltd.[29]

then used to account for invasion. The depth of investigation of these three resistivity responses is shown in Figure 6-58. The deep Laterolog, LLD, responds more deeply in the formation than either the LL_3 or LL_7. The shallow Laterolog, LLS, responds predominantly to the invaded zone, and the response of the MICROSFL is basically from the flushed zone, R_{xo}, and is little affected by mud cakes up to ¾ in.

Similar to the evaluation of the multiple resistivities recorded with the Dual Induction-LL_8 (SFL), the three resistivity responses measured with the Dual Laterolog-R_{xo} tool can be related to R_t, R_{xo} and d_i for a step invasion profile as shown in Figure 6-59.

As with the three Dual Induction responses, the three Dual Laterolog responses have been evaluated relative to a step profile. Since the radial response behavior of the three resistivity devices is known (see Fig. 6-58), the expected responses for $R_{\mu SFL}$, R_{LLS} and R_{LLD} can be simulated as a function of varying values of R_{xo}, R_t and d_i for the step profile. Results of this simulation are shown in Figure 6-60, which is also referred to as a Tornado chart. By entering this chart with corrected values of R_{LLD}, R_{LLS} and R_{xo} in the form of the two dimensionless ratios R_{LLD}/R_{xo} and R_{LLD}/R_{LLS}, a value for R_t/R_{LLD} can usually be defined, unless invasion is extremely deep. This automatic invasion correction doesn't require an estimate of d_i, considerably upgrading the determination of R_t in salt mud logging operations.

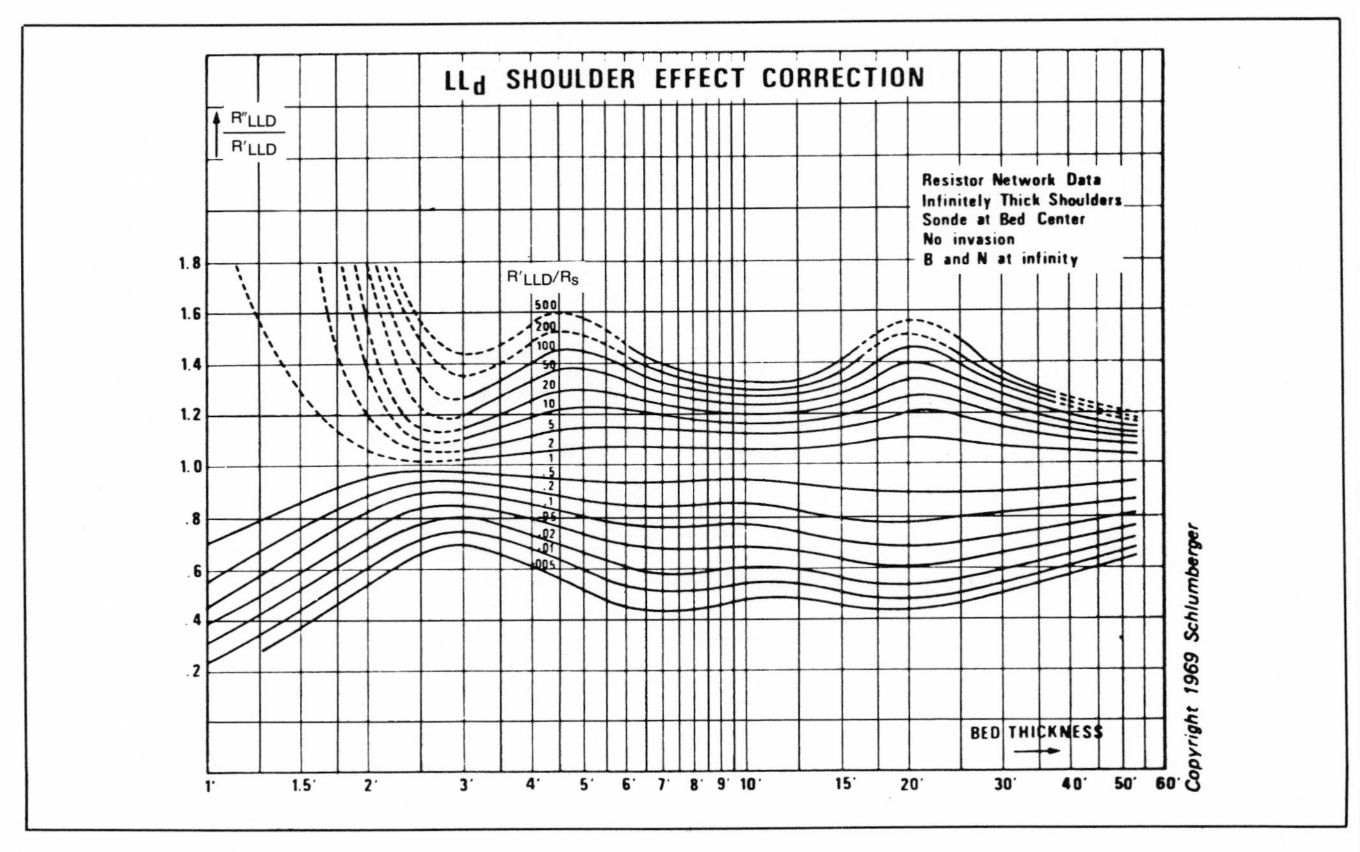

Fig. 6-56. Lateralog deep surrounding bed effect departure curve. Permission to publish by Schlumberger Ltd.[30]

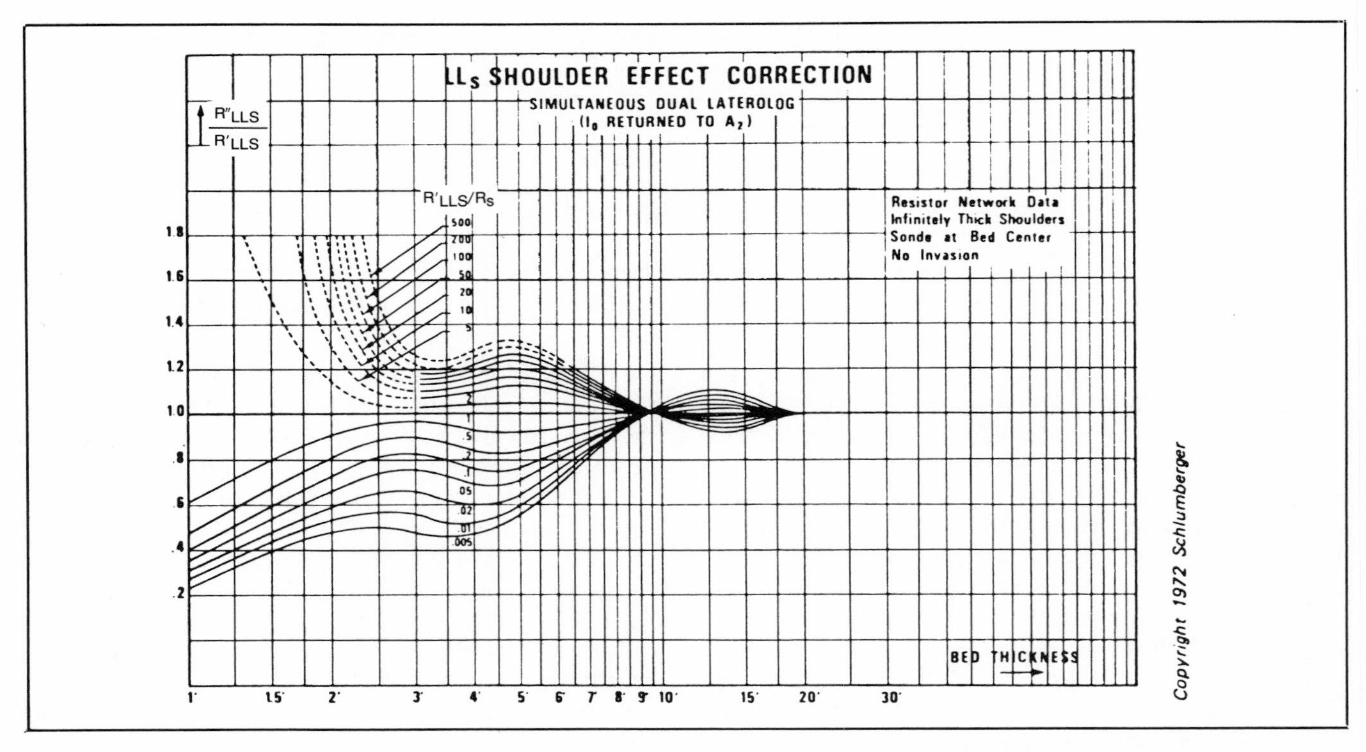

Fig. 6-57. Lateralog shallow surrounding bed effect departure curve. Permission to publish by Schlumberger Ltd.[30]

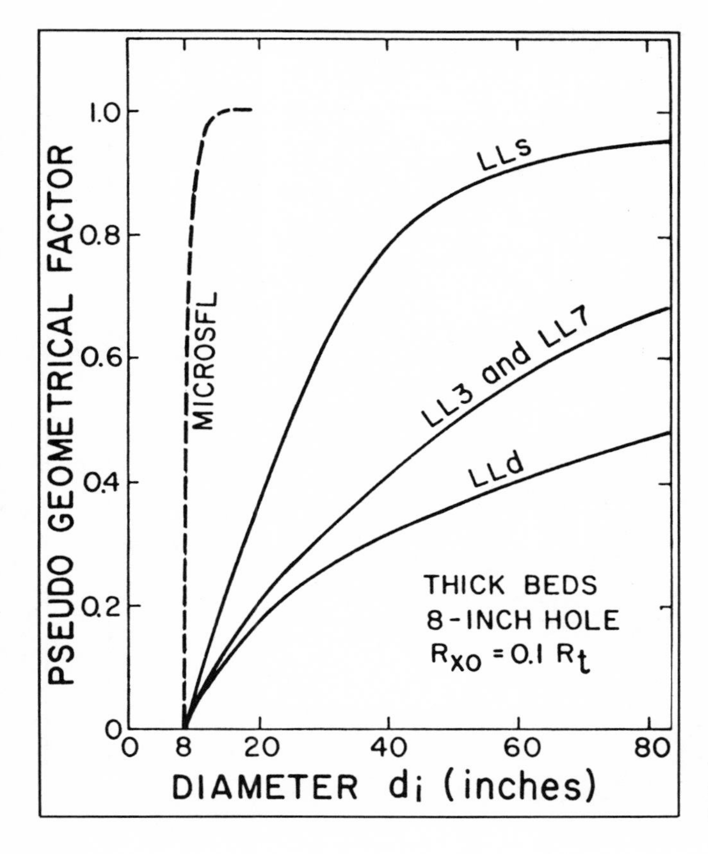

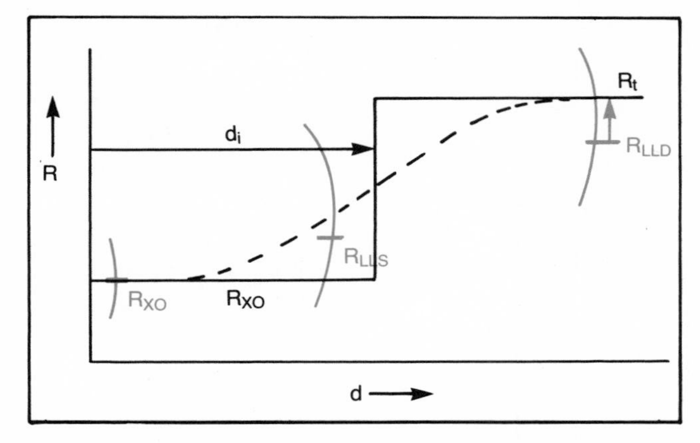

Fig. 6-59. Conceptual insight into invasion correction simulated by step profile using the three resistivity responses of the dual Lateralog-R_{xo} device

Fig. 6-58. Pseudo-geometrical factors for various Lateralogs. Permission to publish by the Society of Petroleum Engineers of AIME.[26] Copyright 1972 SPE-AIME.

REFERENCES

1. Guyod, Hubert: "Fundamental Properties of Resistance Curves," *Oil Weekly* (Aug.-Dec. 1944).
2. Doll, H.G.: "Introduction to Induction Logging and Application to Logging of Wells Drilled with Oil-Base Muds," *Trans.*, AIME (1949).
3. *Resistivity Departure Curves*, Schlumberger Well Surveying Corp., Houston (Dec. 3, 1949).
4. Tixier, Maurice P.: "Electric Log Analysis in the Rocky Mountains," *Oil and Gas J.* (June 23, 1949).
5. Doll, H.G.: "The Microlaterolog—A New Electrical Logging Method for Detailed Determination of Permeable Beds," *Trans.*, AIME (1950).
6. Doll, H.G.: "The Laterolog: A New Resistivity Logging Method with Electrodes Using an Automatic Focusing System," *Trans.*, AIME (1951).
7. Doll, H.G.: "The Microlaterolog," *Trans.*, AIME (1953).
8. Campbell, William M., and Martin, J.L.: "Displacement Logging —A New Exploratory Tool," *J. Pet. Tech.* (Dec. 1955) 233–239.
9. Winn, R.H.: "A Report on the Displacement Log," *J. Pet. Tech.* (Feb. 1958).
10. Guyod, Hubert, and Pranglin, J.A.: *Analysis Charts for Better Electric Log Combinations and Improved Interpretations*, Hubert Guyod, Publ., Houston (1959).
11. Doll, H.G., Dumanoir, Jean, and Martin, M.: "Suggestions for Better Electric Log Combinations and Improved Interpretations," *Geophysics* (Aug. 1960).
12. DeWitte, A.J., and Lowite, D.A.: "Theory of Induction Log," paper presented at the SPWLA Symposium, Dallas, May 1961.
13. Duesterhoft, W.C., Jr., Hartline, R.E., and Thomsen, H.S.: "The Effect of Coil Design on the Performance of the Induction Log," *J. Pet. Tech.* (Nov. 1961) 13.
14. Duesterhoft, W.C., Jr.: "Propagation Effects in Induction Logging," *Geophysics* (April 1961) 26.
15. Duesterhoft, W.C., Jr., and Smith, H.W.: "Propagation Effects on Radial Response in Induction Logging," *Geophysics* (Aug. 1962) 27.
16. Lynch, Edward J.: *Formation Evaluation*, Harper and Row Publ., New York (1962).
17. Moran, James H., and Kunz, Karl S.: "Basic Theory of Induction Logging and Application to Study of Two-Coil Sondes," *Geophysics* (Dec. 1962).
18. Zenor, H.M., and Oshry, H.I.: "Modification of the Induction Log Geometric Factor Due to Propagation," paper presented at the SPWLA Symposium, Houston, May 1962.
19. Tixier, Maurice P., Alger, Robert P., Biggs, W. Pat, and Carpenter, Bruce N.: "Dual Induction-Laterolog: A New Tool for Resistivity Analysis," paper SPE 713 presented at the SPE 38th Annual Meeting, New Orleans, October 1963.
20. Helander, Donald P.: "Guide to Surface Equipment for Well Logging," *Oil, Gas and Petrochem Equipment* (formerly *Oil and Gas Equipment*) PennWell Publishing Co. (April 1965) 11.

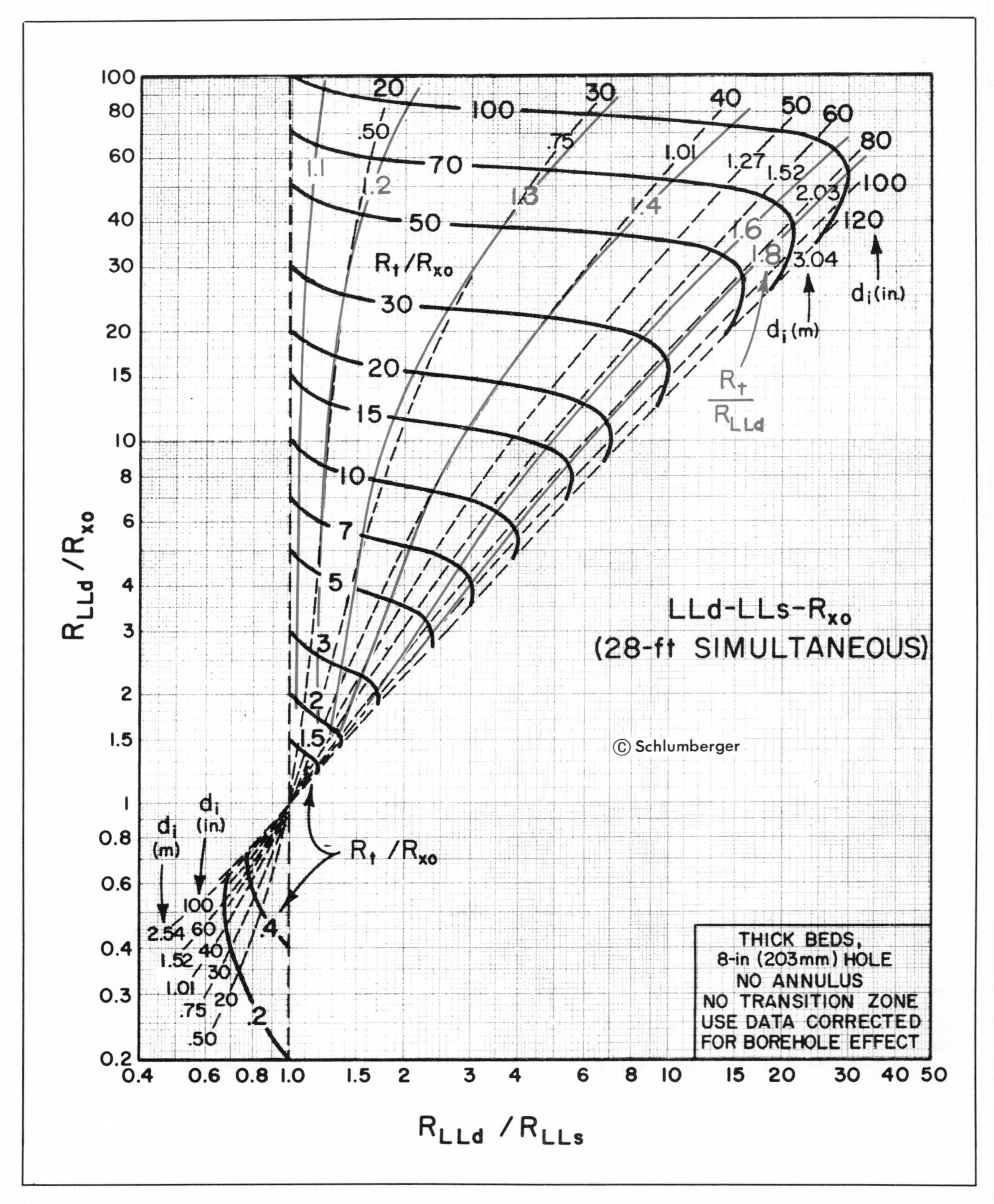

Fig. 6-60. Tornado chart for dual Laterolog-R_{xo} tool. Permission to publish by Schlumberger Ltd.[24]

21. Helander, Donald P.: "Logging Program and Primary Resistivity Tools," *Oil, Gas and Petrochem Equipment* (formerly *Oil and Gas Equipment*), PennWell Publishing Co. (May 1965) 11.
22. Helander, Donald P.: "Here's a Guide to Focused Electrical Logging Systems," *Oil, Gas and Petrochem Equipment* (formerly *Oil and Gas Equipment*), PennWell Publishing Co. (April 1965) 11.
23. Helander, Donald P.: "Here's a Guide to Contact Type Resistivity Tools," *Oil, Gas and Petrochem Equipment* (formerly *Oil and Gas Equipment*), PennWell Publishing Co. (June 1965) 11.
24. *Log Interpretation Charts*, Schlumberger Ltd., New York (1972).
25. *Log Interpretation. Volume 1—Principles*, Schlumberger Ltd., New York (1972).
26. Suau, Jean A., Grimaldi, P., Poupon, Andre, and Souhaite, Phillipe: "The Dual Laterolog-R_{xo} Tool," paper SPE 4018 presented at the SPE 47th Annual Meeting, San Antonio, Oct. 8–11, 1972.
27. Woodhouse, Richard, Threadgold, Phillip, and Taylor, P.A.: "The Radial Response of the Induction Tool," *The Log Analyst* (Jan.-Feb. 1975) 16.
28. Kelseaux, Ray (Cities Service Company): Private Communication.
29. *Log Interpretation Charts*, Schlumberger Ltd., Houston (1966).
30. *Log Interpretation Charts*, Schlumberger Ltd., Houston (1979).
31. *The Essentials of Log Interpretation Practice*, Schlumberger Ltd. (1972).

7 Acoustic Logging

ACOUSTIC velocity logging was introduced commercially as a continuously recorded wellbore measurement in the early 1950's by Seismograph Service Corporation.[12] The continuous velocity logging device, developed by Mobil Oil Company, consisted of one acoustic wave transmitter and one receiver and was used as a means of obtaining subsurface velocitites needed to interpret seismic records. The application of this logging device was greatly expanded, however, when it was determined that the acoustic velocity in a formation was related to formation porosity. Subsequent developments have led to a variety of downhole acoustic logs and the measurement of other formation characteristics in addition to porosity in both open and cased hole systems. This chapter will be devoted to acoustic logging applications related to formation evaluation in open holes.

ELASTIC WAVE PROPAGATION

In order to understand acoustic logging, it is first necessary to understand elastic wave propagation. When an elastic medium is deformed by a disturbing force, the particles within the medium surrounding the source are displaced from their mean equilibrium position. When the disturbing force is removed, the force of interaction between the particles of the medium tends to restore the displaced particles to their original positions. The inertia of the particles causes them to overshoot their rest positions, and as a result the restoring force again tries to return them to their rest positions. The interaction of inertia and restoring forces causes each particle to oscillate about its rest position, and this vibratory motion is transferred to nearby particles. The acoustic disturbance is propagated through the medium as an acoustic or elastic wave. Only the vibratory motion is transferred since the particles remain at their original positions except for small displacements that they undergo while oscillating.

A number of different elastic wave types can be produced when deforming an elastic medium by a disturbing force. The compressional wave (also referred to as a pressure or longitudinal wave) results from the oscillatory motion of the particles as they are displaced away from the source. This displacement creates zones of compression and rarefactions; the propagation of these zones constitutes the compressional wave. These waves can be propagated through all types of media—solid, liquid, or gas—since all these media oppose compression. Compressional waves, v_c, have the highest velocity of any of the elastic wave types.

The particle displacements of shear waves (also referred to as a transverse wave) are transverse to the direction of propagation. The vibratory motion of the particles is imparted to adjacent particles due to rigidity of the medium. As a result, only solids will propagate these waves since liquids and gases do not have enough rigidity, that is, resistance to shear stress, to support the propagation of this wave. The velocity of the shear wave, v_s, is less than for compressional waves but higher than the velocity of other elastic waves encountered in acoustical logging.

Boundary waves are propagated along the boundary between two media having different elastic properties. The boundary waves most commonly encountered in acoustical logging are the tube or Stonely waves propagated along the boundary between the formation and borehole fluid. These waves have lower velocities than either the compressional or shear waves.

The energy of the original disturbance is transferred through the medium by the advancing wavefront. In a homogeneous medium this wavefront can be represented by a spherical surface surrounding the acoustic source and expanding as the wavefront propagates. In a

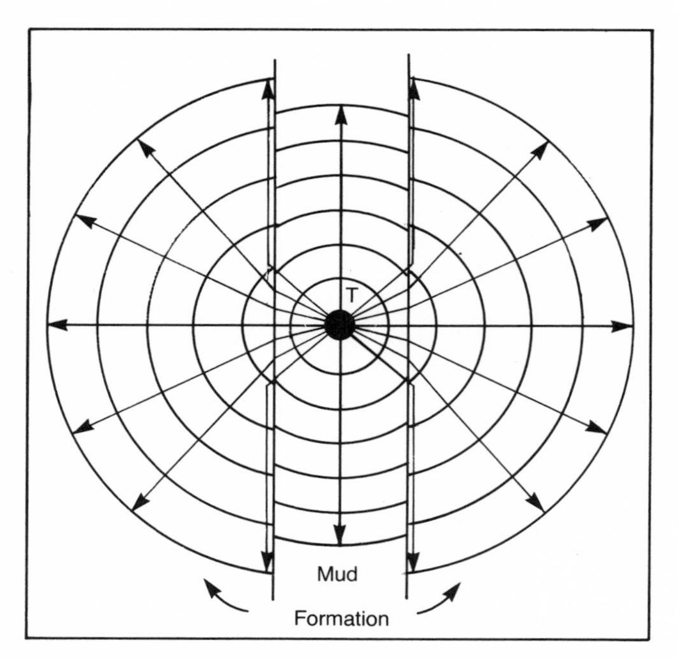

Fig. 7-1. Wavefronts and acoustic rays expanding from transmitter, T, through the borehole mud and formation. Permission to publish by Hubert Guyod.[21]

nonhomogeneous medium, as shown in Figure 7-1, the expanding wavefront is distorted as it propagates through the different media present. The area of the wavefront increases with the square of the distance from the source, and as such in a homogeneous medium the energy density will obey the "inverse square law" for radiant energy as:[32]

$$E_i = \frac{kE_o}{r_i^2}$$

where: E_o = energy of original disturbance
E_i = energy at point, i, a distance r_i from the source
k = proportionality constant

Since E_i is less than E_o, the peak displacement of the particles at r_i will be less than the peak displacement of the particles near the source. The amplitude of the elastic wave is a function of particle displacement, and the reduction of amplitude as the wave propagates from the source to r_i is the attenuation. Since no material is perfectly elastic, some of this energy is also dissipated as heat, which is another factor in acoustic wave attenuation. Additionally, when the medium is nonhomogeneous (fractures present), transmission losses occur at the velocity interfaces due to reflection, an important aspect in fracture detection.

The wavefronts that define elastic wave propagation are complex and difficult to visualize when the wave reaches an interface between two media of different velocities, as shown in Figure 7-1. As a result, it is convenient to describe wave propagation in terms of the acoustic ray. An acoustic ray is a vector directed from the acoustic source to a point at the wavefront normal to the surface of the wavefront. Obviously, there are a number of rays associated with a wavefront but usually only the "rays of interest" are considered. In acoustic logging, the "rays of interest" are those that represent the shortest acoustic path, i.e., shortest time, between a transmitter and receiver.

Behavior of an expanding wavefront in a complex medium can be described using the ray concept. When an incident ray, I, in a medium, M_1, having a velocity, v_1, impinges on a boundary with another medium, M_2, having a different velocity, v_2, both reflection and refraction of the ray occur, as illustrated in Figure 7-2a.

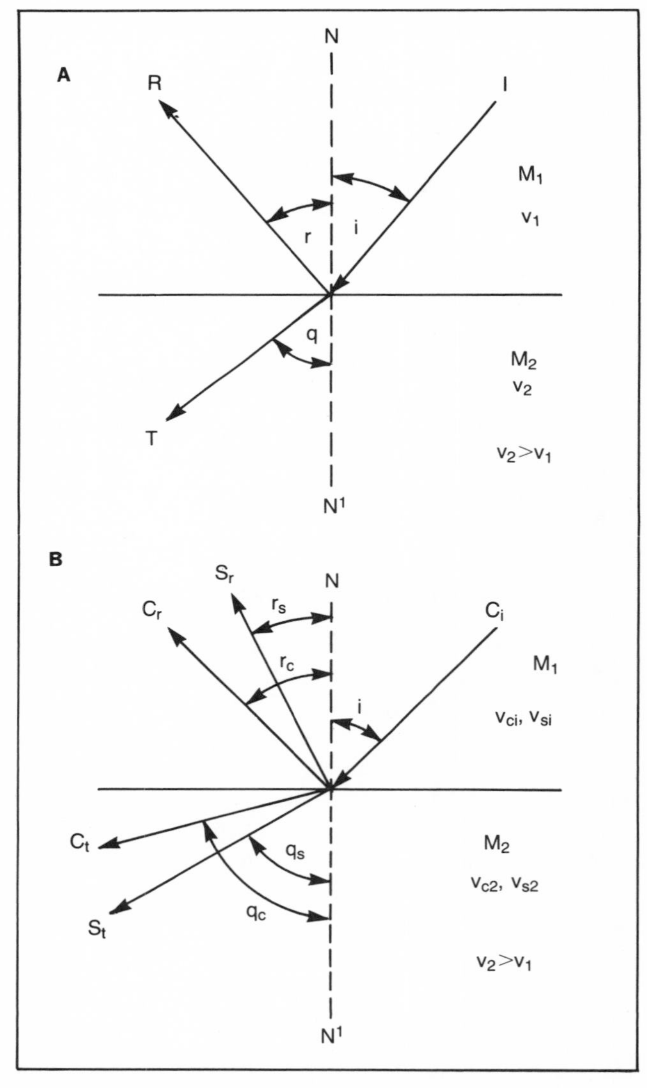

Fig. 7-2. Acoustic refraction of incident ray. Permission to publish by Birdwell Division, Seismograph Service Corp.[32]

The angle of reflection between the normal to the surface and the reflected ray is angle *r* and angle *r* equals angle *i*. The angle of refraction is related to the angle of incidence and the velocities of the two media by Snell's Law:

$$\frac{\sin i}{v_1} = \frac{\sin q}{v_2}$$

Figure 7-2a is an oversimplification, however, since when an incident ray impinges upon a boundary, the incident wave energy will be partially transposed into waves of different types; this transposition is known as mode conversion. A generalized form of Snell's Law relates the angle of incidence to the various angles of reflection and refraction that are obtained as:

$$\frac{\sin i}{v_i} = \frac{\sin a}{v_a}$$

where: i = the incident angle
v_i = velocity of incident ray
a = angle of reflection or refraction
v_a = velocity of reflected or refracted ray

Figure 7-2b illustrates this more complicated case, with four rays being produced: (1) reflected compressional (pressure or longitudinal) ray, c_r, (2) reflected shear ray, s_r, (3) refracted compressional ray, c_t, and (4) refracted shear ray, s_t.

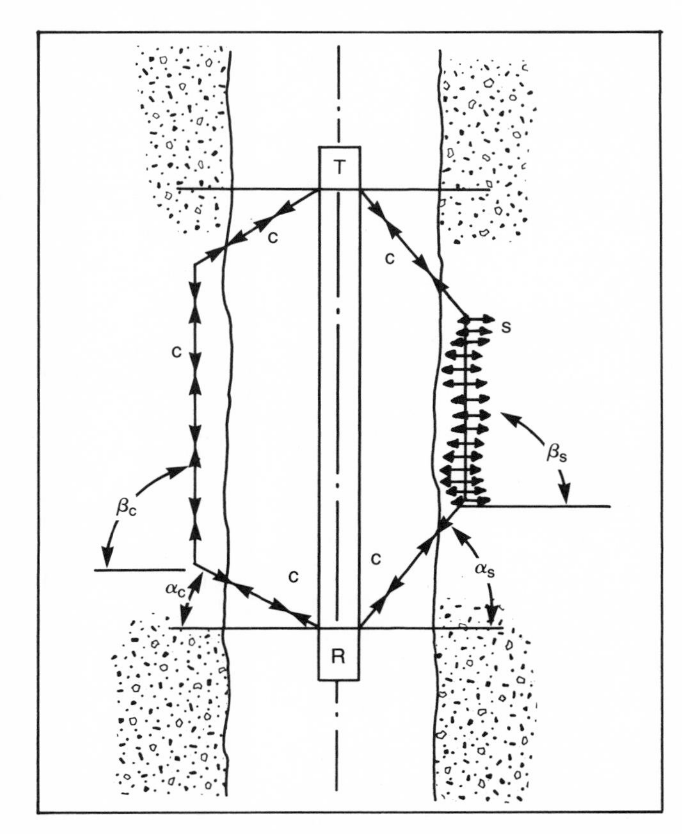

Fig. 7-3. Single receiver acoustic tool showing compressional and shear rays of interest. Permission to publish by the American Society of Mechanical Engineers.[24]

Now if a simple acoustical logging tool consisting of a transmitter and receiver operating in a borehole system (see Fig. 7-3) is considered, two rays of interest can be described that represent the formation compressional and shear wave propagation. Since the borehole contains a fluid, the incident ray is always a compressional wave. The rays of interest are those parallel to the borehole representing the quickest path in the formation to the receiver and, therefore, the ray whose angle of refraction, *B*, is 90°. These refraction angles are produced by the critical angles of incidence, as defined by:

$$\sin \alpha c = \frac{v_t}{v_p} \quad \text{and} \quad \sin \alpha s = \frac{v_t}{v_s}$$

where: αc = critical angle for refracted compressional wave
v_t = compressional wave velocity in the borehole fluid
v_c = compressional wave velocity in the formation
αs = critical angle for refracted shear wave
v_s = shear wave velocity in the formation

The rays of interest of the compressional and shear waves are different since v_c is always greater than v_s. Figure 7-3 does not include either the boundary waves (Stoneley wave propagation shown in Figure 7-4) or the compressional wave traveling directly through the fluid v_t. These waves will usually arrive at the receiver after the compressional and shear waves generated in the formation.

Now imagine that one ray could be displayed on an oscilloscope at the output of the receiver. This display could have a shape similar to the sine wave (solid curve) in Figure 7-5. In the physically real case, however, the radiated energy not only moves along the defined ray paths, but also along an infinite number of ray paths within the bounds of *BB'*, *CC'* and *SS'* (see Fig. 7-6), thereby forming a beam of acoustic energy. The wavefronts of all rays in each beam arrive at the boundary in phase, whereas, on the other side of the boundary, in the solid they will be out of phase. As a result of a phase wave assumption, all the energy within one beam add so the amplitude is greatly increased, but with the addition of phase distortion, a complex waveform will result, as shown by the dashed curve in Figure 7-5. If each pulse to the transmitter causes the oscilloscope to sweep once, the superimposed energy from the three beams producing the formation compressional, shear, and boundary waves will produce a complex waveform, which is referred to as a full-wave train (see Fig. 7-7).

The full-wave train recorded at one depth in the borehole contains a great amount of information since it

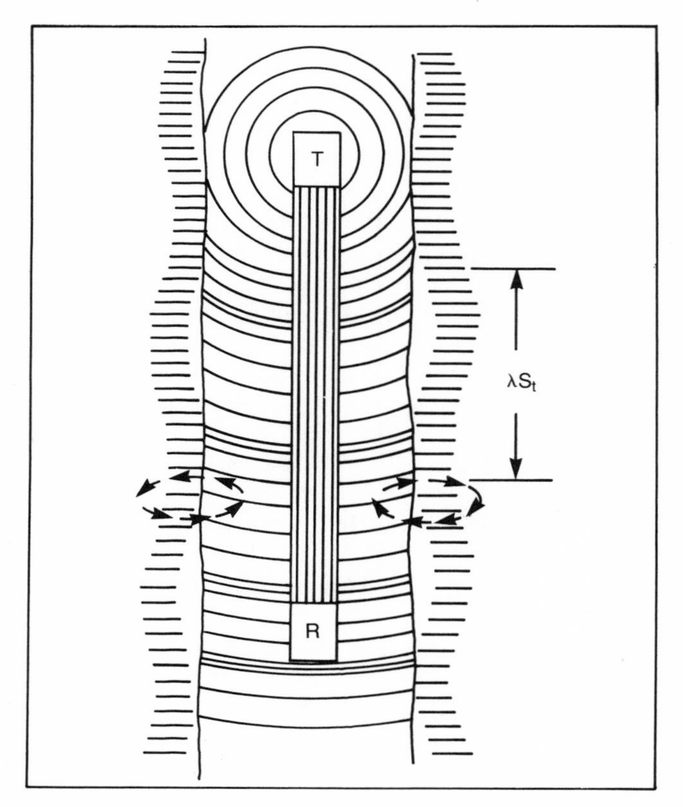

Fig. 7-4. Stoneley Wave Propagation. Permission to publish by the American Society of Mechanical Engineers.[24]

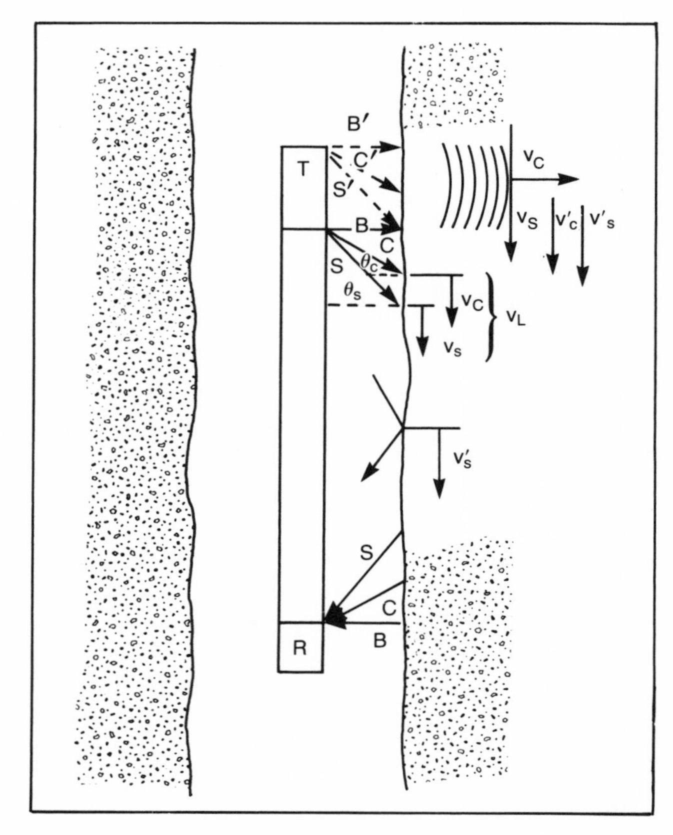

Fig. 7-6. Schematic of single receiver acoustic tool showing critical angle ray paths. Permission to publish by the Society of Professional Well Log Analysts.[11]

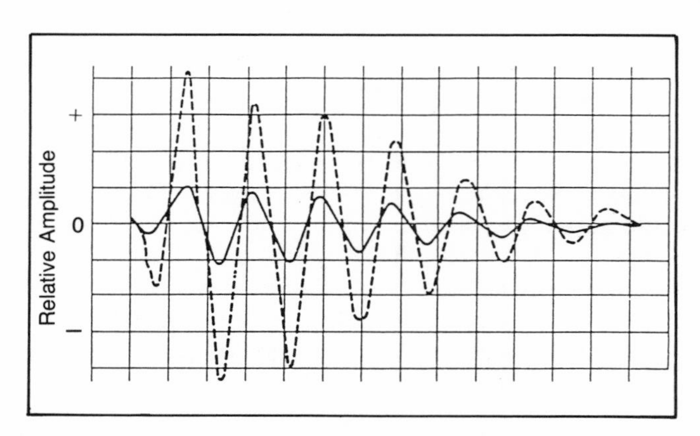

Fig. 7-5. Illustrative oscilloscope output for single ray and single energy beam. Permission to publish by the Northern Ohio Geological Society.[13]

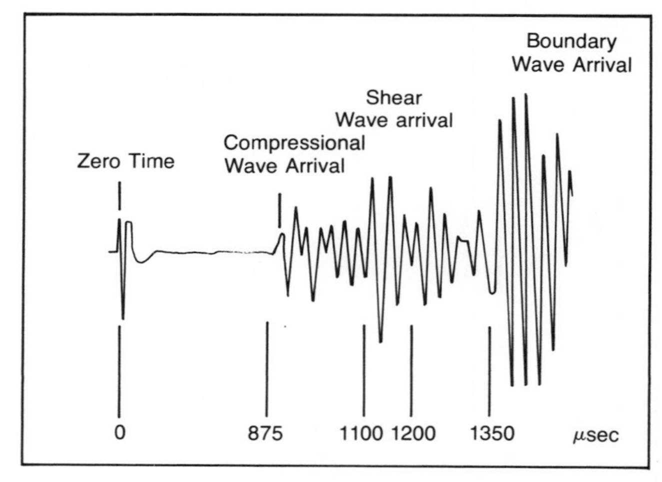

Fig. 7-7. Full-wave train. Permission to publish by the Northern Ohio Geological Society.[30]

represents the recording of basically three different wave types, each having four distinguishable parameters. These measurable parameters are (1) arrival time, (2) amplitude, (3) amplitude attenuation, and (4) frequency or its inverse, the period. The principal acoustic logging tool measures only one piece of this data, the transit time of the compressional wave, such that most of the time a large amount of latent information is discarded.

It is obvious that a recording of full wave trains as a function of depth to preserve all of the information possible would become a confusing array of overlapping wave trains (see Fig. 7-8c). In order to produce an easy-to-use, visual display of the full-wave trains, the variable intensity display was developed by Seismograph

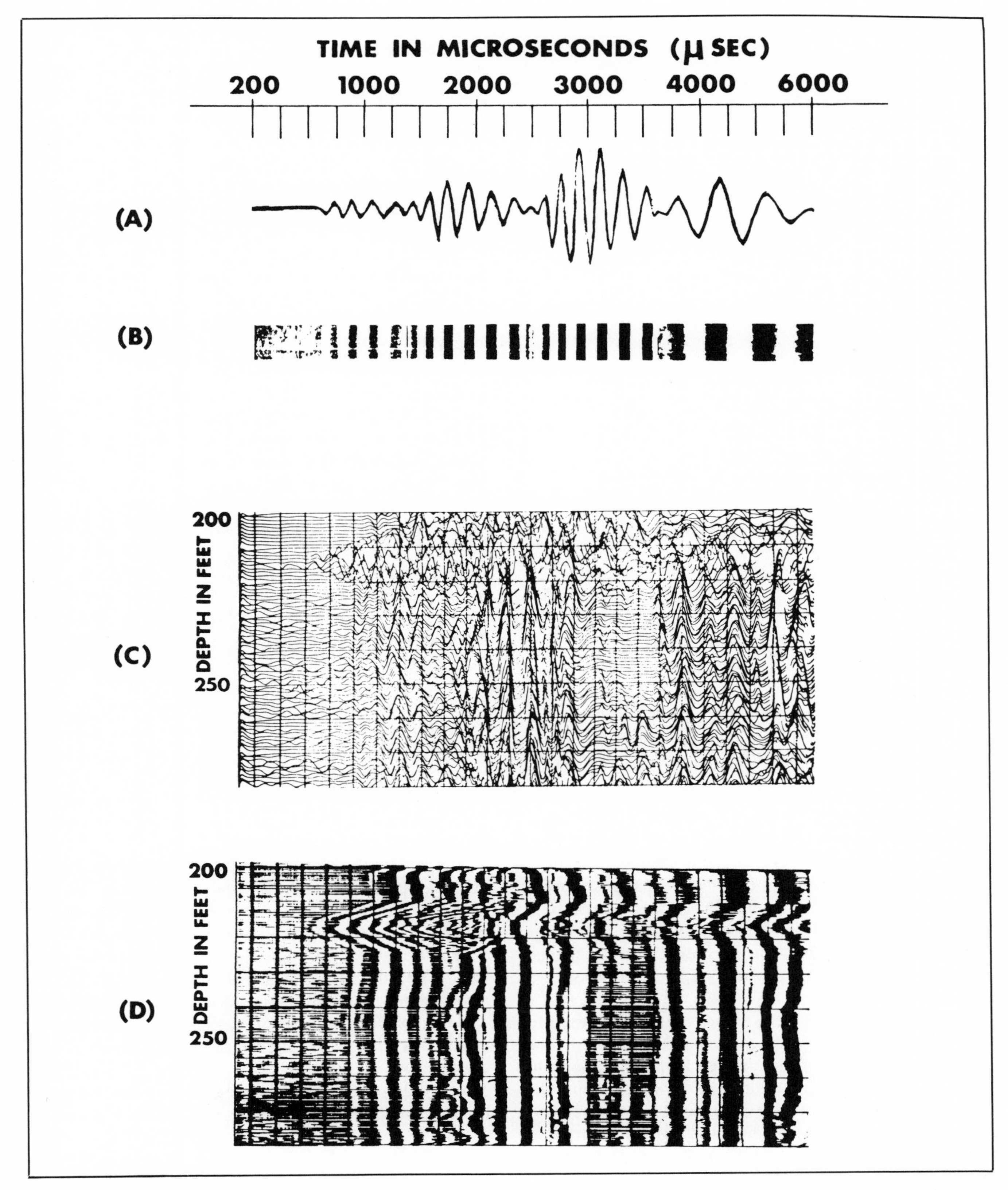

Fig. 7-8. Wave train processing and variable density display. Permission to publish by the American Society of Mechanical Engineers.[24]

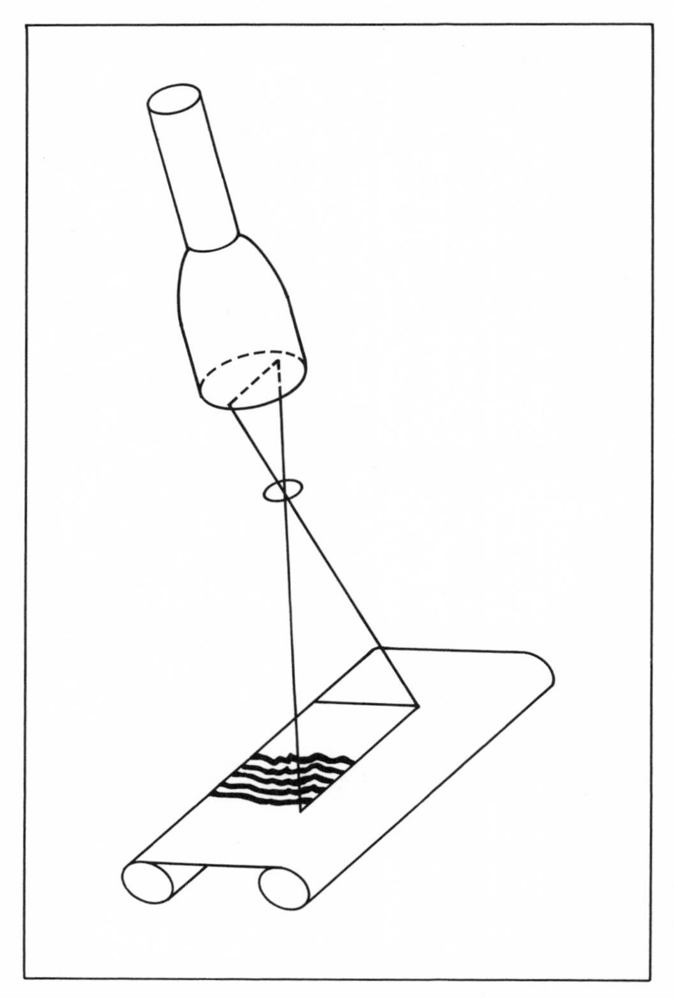

Fig. 7-9. Schematic of variable intensity recording. Permission to publish by the Northern Ohio Geological Society.[13]

Service Corporation. With this method of recording, the transmitter produces a pulse that initiates a timing circuit, and an oscilloscope sweep is started. The angle of sweep can be adjusted to display selected positions of the wave train ranging from 250 μsec to a maximum 25 m-sec. The variable intensity display is made by transmitting the image of the oscilloscope beam through a fiber optics system onto photographic film, as shown in Figure 7-9.

The signal from the receiver (see Fig. 7-8a) is used to modulate the intensity of the sweep (see Fig. 7-8b). The troughs of the signal produce high light intensities and result in dark zones on the film; the darker the zone, the higher the amplitude of the trough will be. Conversely, the peaks of the signal produce low light intensity and result in light zones on the film; i.e., the lighter the zone, the higher the amplitude of the peak. Figure 7-10 descriptively shows the method of depicting the wave train as an intensity modulated signal.

The variable intensity recording (see Fig. 7-8d) is a three-dimensional record of the full-wave trains, also recorded as "wiggle" traces in Figure 7-8c through the same formation interval. The three-dimensional aspect results from the display showing wave transit time and amplitude as a function of depth. This type of recording is commonly called the 3-D log.

Figure 7-11 shows a typical 3-D log presentation, indicating the first arrivals of the compressional, shear, and boundary waves. A common method of recording the time scale in the oil and gas industry is with time increasing to the left instead of the right to correspond to the standard of having porosity increase to the left.

TRANSMITTERS AND RECEIVERS

Acoustic energy sources, or transmitters, used in acoustical logging tools were originally made of magnetostrictive material, which deforms in the presence of a magnetic field. If this material is placed in an induction coil and an alternating current is passed through the coil, the material will alternately deform and regain its shape, producing an acoustic wave in the surrounding medium. The frequency of the acoustic wave is determined by the resonant frequency of the transmitter, which is usually about 20 KHz. These sources are called transducers since they convert electrical energy to acoustic energy. Most modern acoustical logging tools use piezoelectric ceramic crystals as transmitting transducers.

The acoustic energy that can be supplied by the small transmitting transducers used in acoustical logging tools is limited; therefore, the acoustic coupling between the transducers and formation must be good in order for any usable energy to reach the receiver. As a result, the borehole must be liquid-filled, since gas or air will cause too large a signal attenuation.

The receiver is also a transducer, which in this case converts acoustic energy into electrical energy. Since all of the waves reaching the receiver travel through the borehole fluid, the formation waves are converted and transmitted back through the fluid as compressional waves. Because of this, the receiving transducer must be sensitive to compressional waves. One compressional wave transducer, the piezoelectric crystal, produces an electrical voltage when deformed. A number of natural crystals such as quartz have piezoelectric properties, but most modern acoustical logging devices use manufactured ceramic crystals.

ACOUSTIC LOGGING TOOLS

The first acoustical logging tool used for formation evaluation purposes was a single receiver device de-

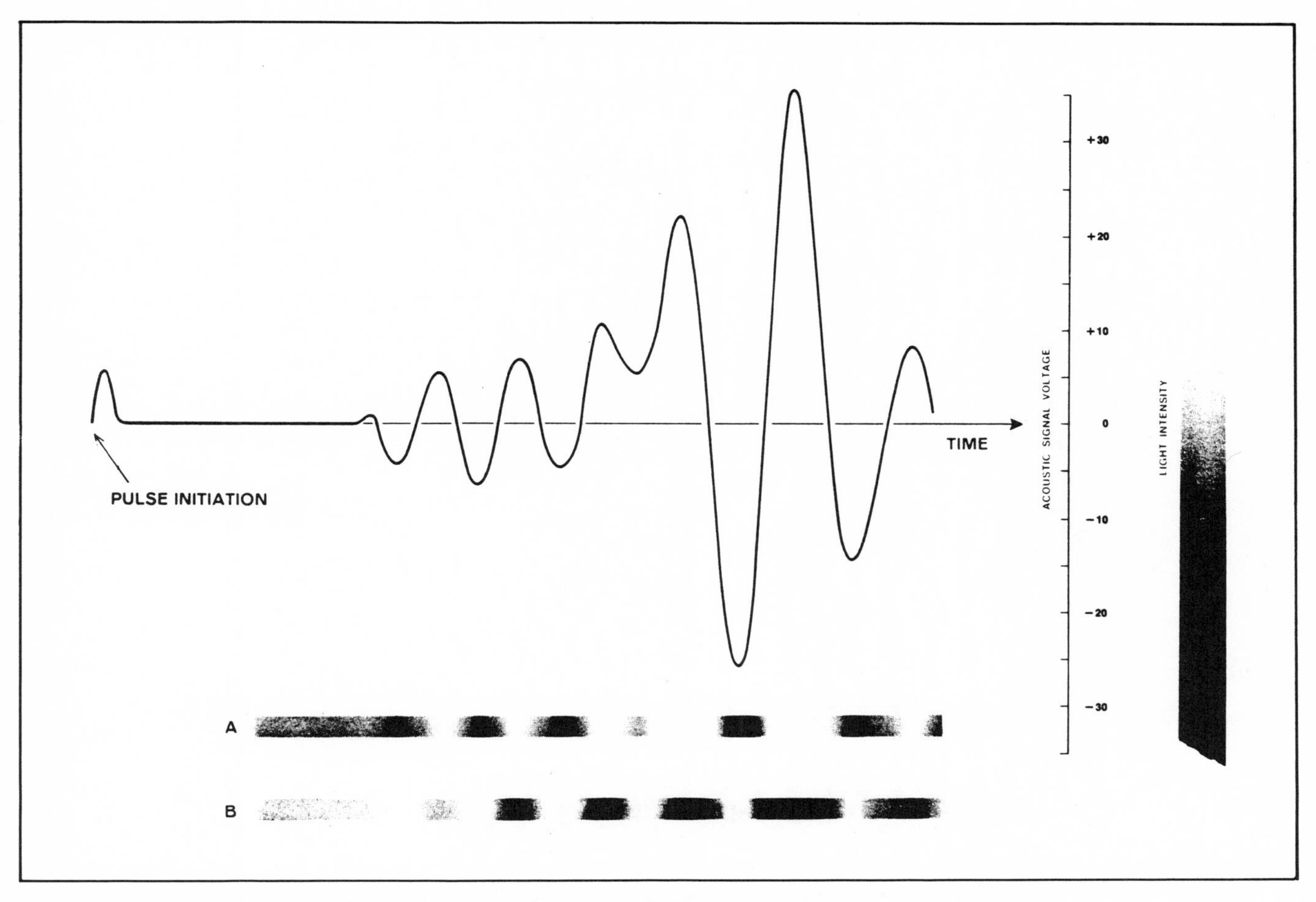

Fig. 7-10. Concept of variable intensity display. Permission to publish by Hubert Guyod.[21]

signed to measure the compressional wave transit time through the formation. This particular measurement uses only one of the many parameters that can be measured with acoustical logging devices. It was, however, a most useful measurement since compressional wave velocity is related to formation porosity. As a result, the important consequences of being able to measure compressional wave velocity, or the inverse, transit time of the formation enables an independent porosity value to be determined for evaluation purposes. This is still the single most important application of acoustic measurements.

Application of the single-receiver tool (see Fig. 7-12) is limited in the direct measurement of compressional wave transit time in the formation, since corrections must be made for the mud travel path from the transmitter to the formation and from the formation back to the receiver.

This single-receiver acoustic tool was soon replaced by the two-receiver acoustic tool for measuring compressional wave transit time in the formation. A schematic of the two-receiver tool and the compressional wave ray paths of interest is shown in Figure 7-13. The single-receiver configuration, however, is still used for the 3-D velocity log, as will be discussed later.

The two-receiver logging tool has a distinct advantage over the single receiver logging device since it provides a direct measure of compressional wave transit time in the formation. This direct measure can be obtained since all travel times will cancel except the transit time of the compressional wave over the distance *L* (spacing between receivers), as shown in Figure 7-13. This, of course, is based on the assumption that the tool is centralized and borehole size is constant.

In the two-receiver logging tool, the transmitter generates a short sinusoidal pulse, generally about 20 kHz/sec (about 20 times per second).[32] When the pulse is generated by the transmitter (at time zero), a timing circuit is started. The timing circuit is stopped by the arrival of the peak or trough of interest if the amplitude exceeds a sufficient level called the "threshold." The travel times for the compressional waves to traverse from the transmitter to each receiver is measured with respect to the same event, as shown in Figure 7-14. In

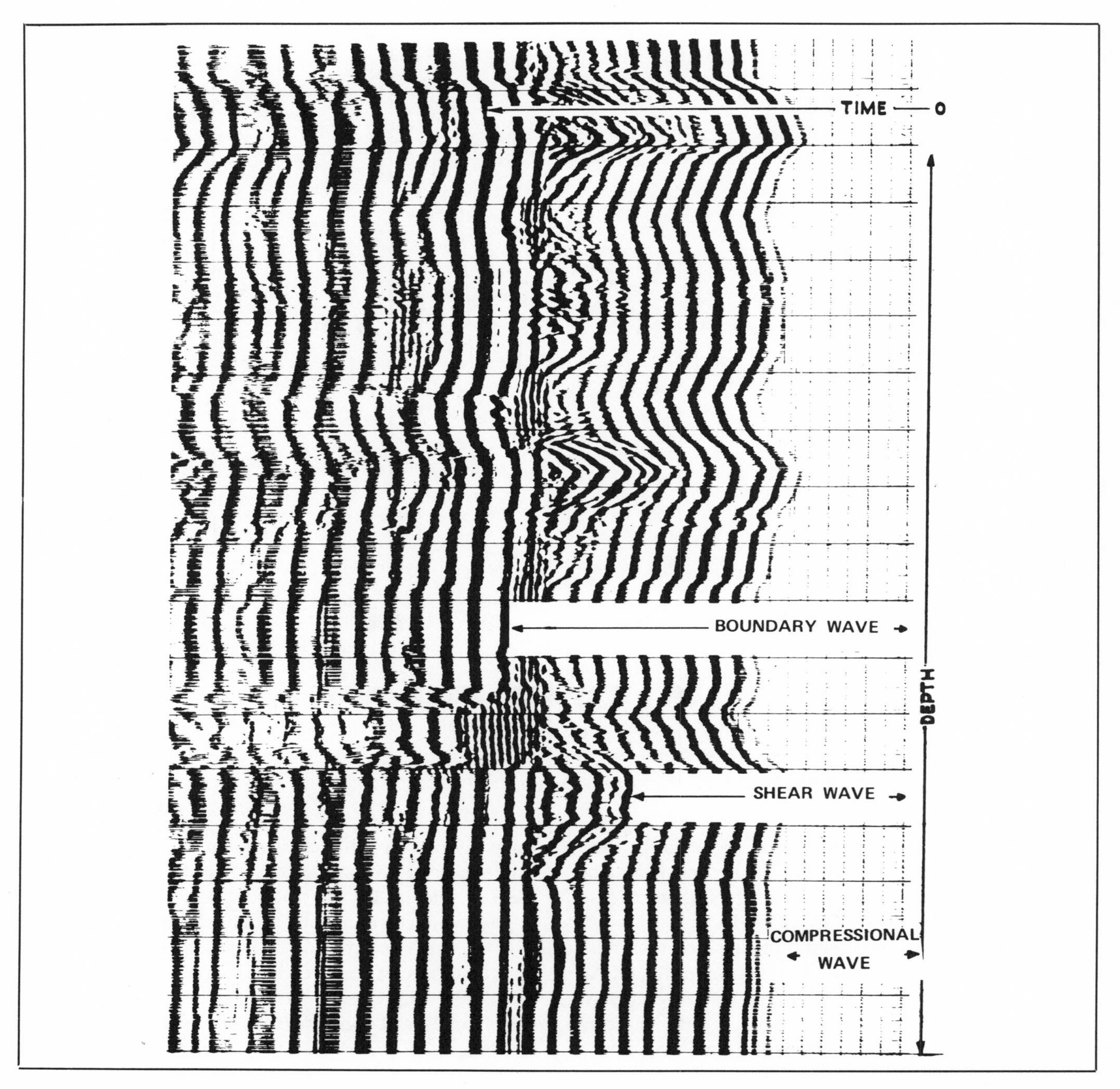

Fig. 7-11. Typical 3-D log. Permission to publish by Birdwell Division, Seismograph Service Corp.[25]

essence, the time measurement made with the two-receiver tool (Fig. 7-13) is the difference between the transit times of the same event of the acoustic signal as sensed by the two receivers. This can be shown as follows:

$$t_1 = t_{ro} + t_{Li} + t_{ro}$$

$$t_2 = t_{ro} + t_{Li} + t_L + t_{ro}$$

Therefore:

$$t_2 - t_1 = t_L$$

The time, t_L, is measured in μsec, and in order to put this time measure, t_L, on a time per unit distance, it is divided by L (the receiver spacing) such that the recorded log property is:

$$t = \frac{t_L}{L} \ (\mu\text{sec/ft})$$

This transit time measure relates to formation velocity, v, in ft/sec as:

$$t = \frac{10^6}{v}$$

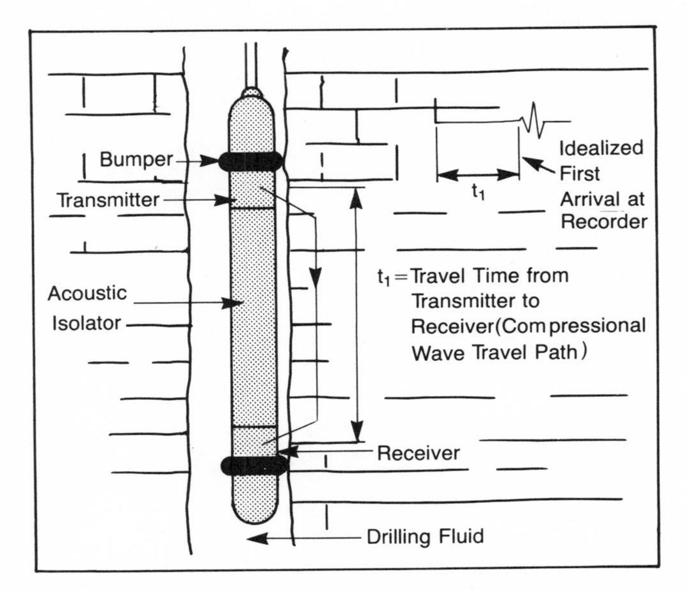

Fig. 7-12. Schematic of single receiver acoustic tool showing compressional wave, ray path of interest. Permission to publish by H.W. Lawrence and Robert W. Baltosser.[20]

Transit time is recorded at a reference point midway between the receivers, which are spaced from one to three feet apart. The transmitter is spaced from three to five feet from the first receiver with all transducers separated by rubber isolators to eliminate direct coupling.

In the measurement of transit time, amplitude attenuation might be of such a magnitude as to cause "cycle skipping." This can occur when the amplitude of the peak or trough is less than the threshold and the timing circuit is not stopped. As a result, the timing circuit continues to measure time until the amplitude of another peak or trough arrives that exceeds the threshold. Therefore, the timing circuit stops one or more cycles too late, which results in an abnormally large t. The error is almost invariably equal to one or more periods and usually can be easily recognized on the log.

The two-receiver acoustic logging tool, although an improvement over the single-receiver tool for measuring compressional wave transit time, was also affected by hole size changes and tool eccentricity. This led to the development of the borehole-compensated acoustic log for measuring compressional wave transit time in the formation.[16] This multiple transducer tool consists of two, two-receiver acoustic tools mounted in an inverted position to each other (see Fig. 7-15).

The span between receiver pairs is two feet. In operation a transit time is measured by each two-receiver system, sequentially. The two transit times are then averaged. This average transit time measurement tends to minimize spurious anomalies associated with hole

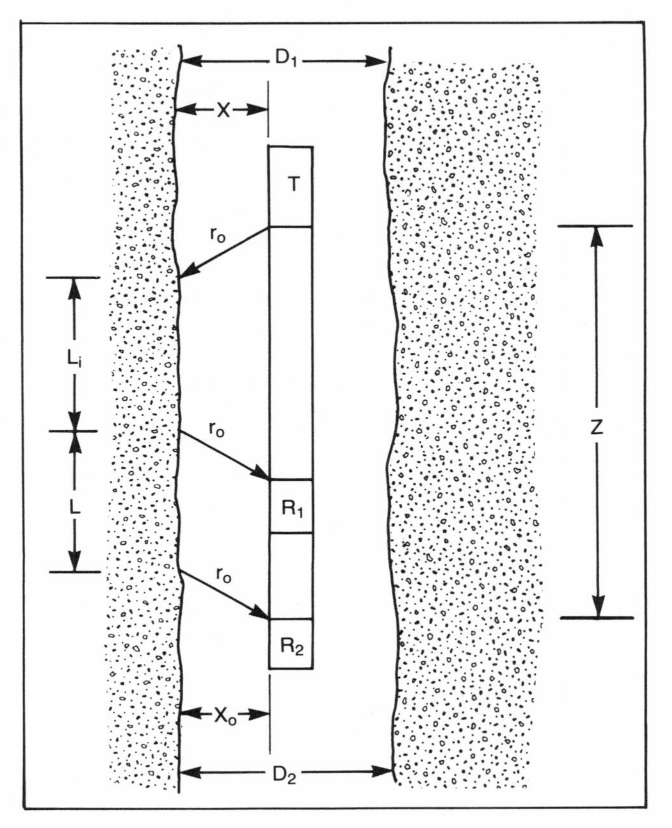

Fig. 7-13. Schematic diagram of two-receiver acoustic tool showing compressional wave ray path of interest. Permission to publish by the Society of Professional Well Log Analysts.[11]

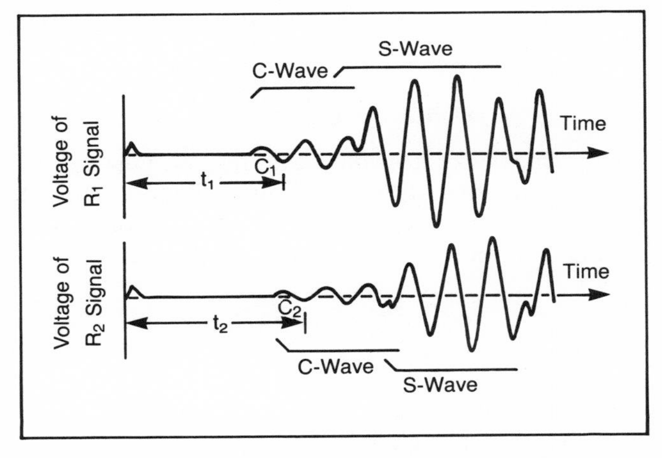

Fig. 7-14. Schematic of timing periods measured in two-receiver acoustic tool. Permission to publish by Hubert Guyod.[21]

size change, as shown in Figure 7-16, and the effects of tool eccentricity, as shown in Figure 7-17. This tool can be run with a gamma ray and a caliper log, and a computed porosity can be recorded using selected constants.

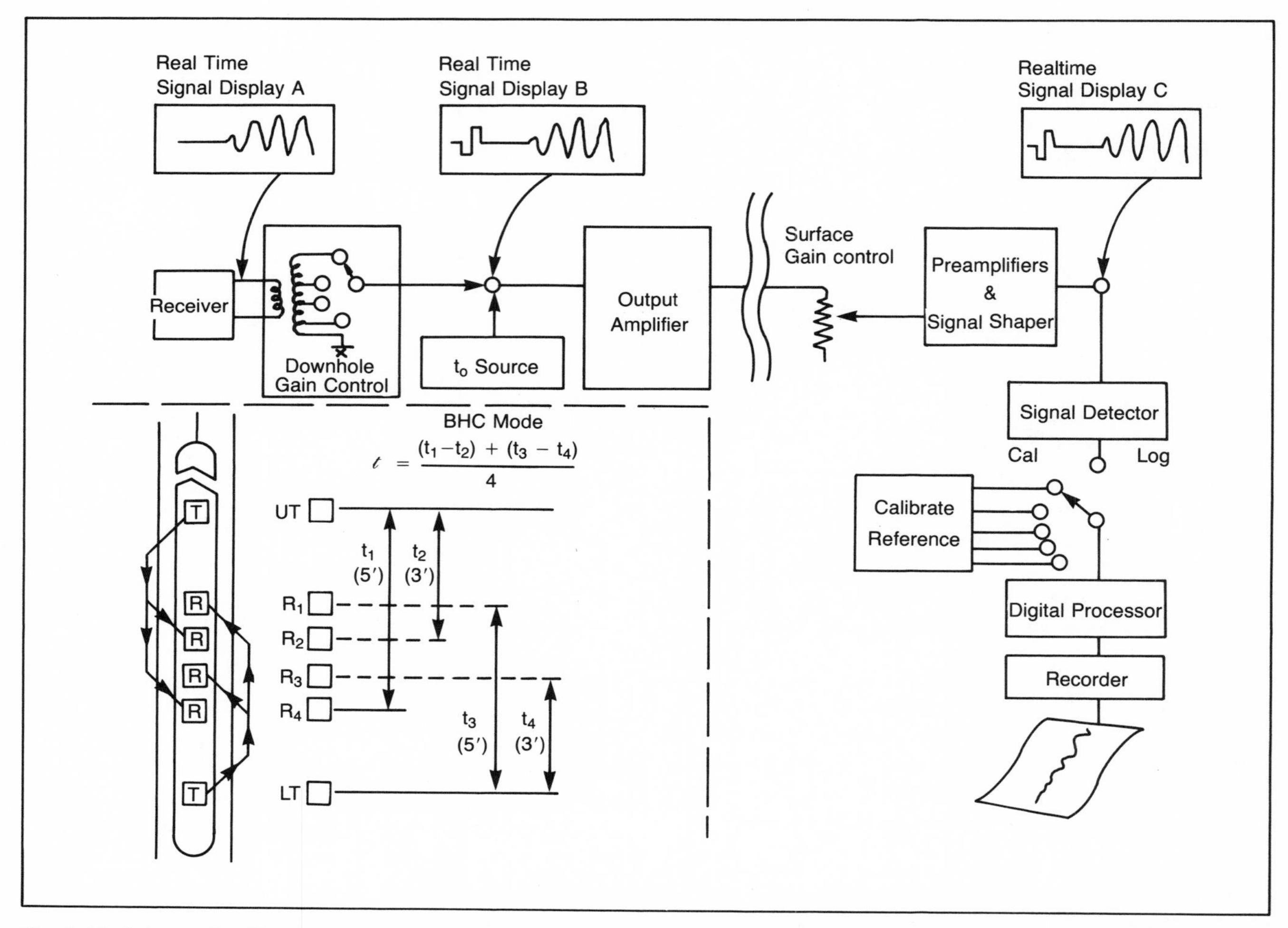

Fig. 7-15. Schematic of borehole compensated acoustic logging system. Permission to publish by the Society of Professional Well Log Analysts.[34]

POROSITY EVALUATION

The velocities of acoustic waves in a medium can be expressed in terms of the elastic moduli and density of the medium.[21, 28, 33] The velocity for the compressional wave in an extended medium can be expressed as:

$$v_c = \left(\frac{K + 1.33G}{\rho_b}\right)^{1/2}$$

where: K = bulk modulus
G = shear modulus
ρ_b = bulk density

The velocity for the shear wave, which is independent of the shape and dimensions of the medium, can be expressed as:

$$v_s = \left(\frac{G}{\rho_b}\right)^{1/2}$$

These relations are applicable to most substances encountered in logging, whether uniform or not, and are useful in the sense that they indicate the dependence of velocity on both compressibility and rigidity and, therefore, also porosity.

A number of studies have been done in an attempt to relate acoustic wave velocity in fluid-saturated porous media to more readily accessible rock properties.[3, 4, 5, 6, 9] These studies, while being of value in relating porosity and other properties to acoustic velocity, have produced relations too complex for practical application in formation evaluation. The work of Geertsma,[9] for example, based upon Biot's continuum theory, gives an expression for wave velocity in terms of porosity and the individual densities and elastic moduli of the rock matrix, pore fluid, and bulk rock. These studies do provide, however, the key to understanding the experimentally observed behavior of formation transit time versus porosity behavior.

It has been found that a linear relation exists between transit time, t, and porosity, ϕ, for distinct rock types

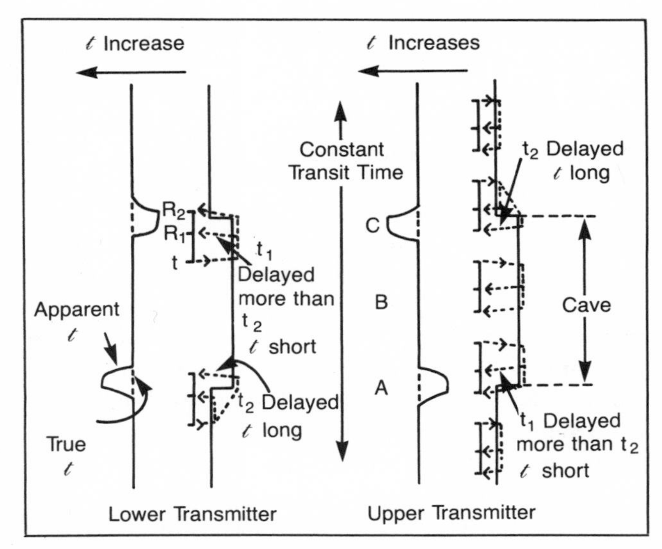

Fig. 7-16. Schematic of the cancelling effect of anomalies, caused by changing borehole size, when using the borehole compensated acoustic tool. Permission to publish by the Society of Petroleum Engineers of AIME.[16] Copyright 1965 SPE-AIME.

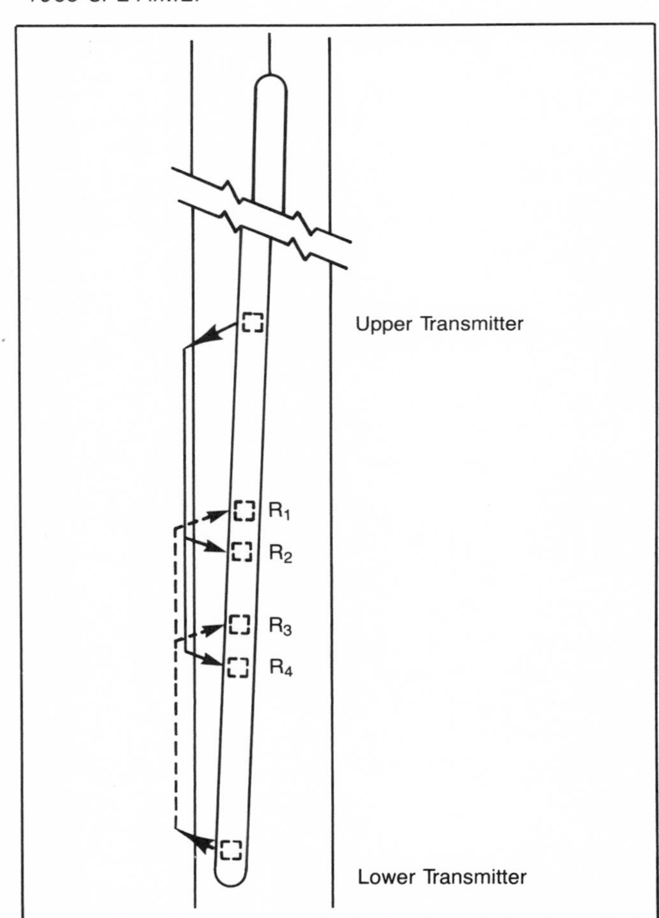

Fig. 7-17. Compensation for tool eccentricity. Permission to publish by the Society of Petroleum Engineers of AIME.[16]

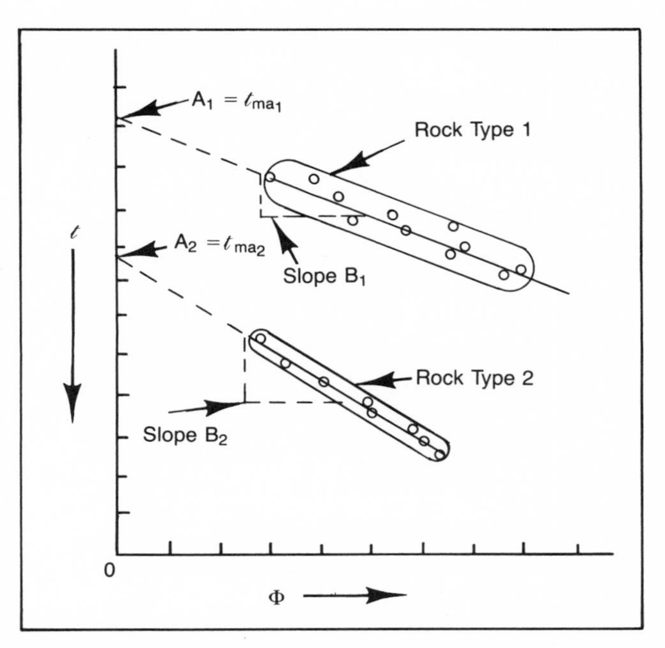

Fig. 7-18. Typical transit time-porosity crossplot

(Fig. 7-18). The general response equation relating transit time, ℓ, to porosity, ϕ, would be:

$$\ell = A + B\phi = \ell_{ma} + B\phi$$

where: A = intercept
B = slope

In relation to the more complex equation relating acoustic transit time to porosity, the intercept, A, is lithology dependent and B is dependent on such factors as lithology, effective stress, rock structure, etc. The observed ranges of A and B for common reservoir rock types are:

		$A\left(\frac{\mu sec}{ft}\right)$	$B\left(\frac{\mu sec/ft}{fractional\ \phi}\right)$
Sandstone		50→56	130→300
Limestone	Carbonate	45→50	60→120
Dolomite		42→48	

The intercept, A, and slope, B, values for sandstone and the carbonates are different, and as a result a ℓ vs. ϕ crossplot should distinguish these different rock types. Increasing slopes tend to relate to decreasing grain contact area, i.e., low value for the slope, B, tends to represent vuggy type rocks with larger slopes relating to the finer-grained rocks and shaly or silty rocks. When using the acoustic response for porosity evaluation, one should calibrate the acoustic response in terms of porosity using, for example, core porosity (one could use an independent porosity response such as the density or neutron response) to define the linear relation for the rock type of interest.

One relation for predicting porosity from compressional wave transit time was proposed by Wyllie et al.[6] and has been widely used in formation evaluation. This relation is the time-average equation, which can be stated as:

$$t = t_f \phi + t_{ma}(1-\phi)$$

where: t = compressional wave transit time measured by acoustic logging tool, (μsec/ft)
t_f = transit time of the fluid saturating the porous media near the wellbore, (μsec/ft)
t_{ma} = matrix transit time (transit time of solid rock system), (μsec/ft)
ϕ = fractional porosity

This relation can be rearranged such that:

$$t = \underbrace{t_{ma}}_{A} + \underbrace{(t_f - t_{ma})}_{B}\phi$$

As can be seen, this relation defines the intercept $A = t_{ma}$ in the same manner as defined in the general transit time response equation. The major problem with this expression is that the slope is defined as $B = t_t - t_{ma}$, which is not applicable to carbonates or for that matter, poorly consolidated or shaly sandstones. In order to use this relation in formations other than consolidated, clean sandstones, some means for correcting for an inadequately defined slope, B, must be used. In fact, its use for other than clean, compacted sandstones can be quite confusing, since for obtaining representative slopes in carbonates "pseudo" water velocities are many times used that do not exist in nature. For compacted, clean sandstones a prediction of porosity is possible, using the time average equation and assuming reasonable values of $t_f(v_f)$ and $t_{ma}(v_{ma})$. Representative velocities for various substances are shown in Table 7-1. A typical acoustic transit time log obtained with a borehole compensated tool is shown in Figure 7-19.

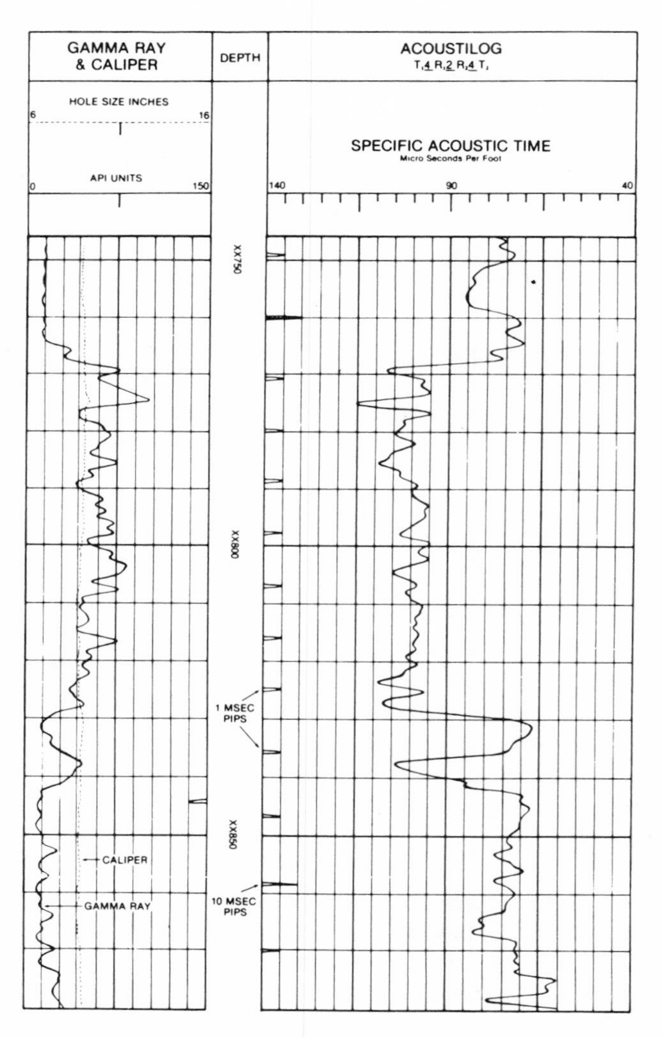

Fig. 7-19. Typical BHC acoustic log. Permission to publish by Dresser Atlas, Dresser Industries, Inc.[35]

TABLE 7-1
Velocities of Various Substances[21]

Material	*Porosity*	V_C *(ft/sec)*	V_S *(ft/sec)*
Non-porous solids			
Anhydrite		20,000	11,400
Calcite		20,100*	
Cement (cured)		12,000	
Dolomite		23,000	12,700
Granite		19,700	11,200
Gypsum		19,000	
Limestone		21,000	11,100
Quartz		18,900*	12,000
Salt		15,000	8,000
Steel		20,000	9,500
Water-saturated porous rocks in situ			
Dolomites	5–20%	20,000–15,000	11,000–7,500
Limestones	5–20%	18,500–13,000	9,500–7,000
Sandstones	5–20%	16,000–11,500	9,500–6,000
Sands (unconsolidated)	20–35%	11,500– 9,000	
Shales		7,000–17,000	
*Liquids***			
Water (pure)		4,800	
Water (100,000 mg/l of NaCl)		5,200	
Water (200,000 mg/l of NaCl)		5,500	
Drilling mud		6,000	
Petroleum		4,200	
*Gases***			
Air (dry or moist)		1,100	
Hydrogen		4,250	
Methane		1,500	

* Arithmetic average of values along axes (Wyllie et al., 1956).
** At normal temperature and pressure.
Permission to publish by Hubert Guyod.

TRANSIT TIME-RESISTIVITY RELATION

This independent approach for obtaining porosity from transit time measurements can readily be related to resistivity, providing a useful crossplotting approach for estimating water saturation with a minimum of data. The basic Archie relations:

$$F_R = \frac{R_o}{R_w} = \phi^{-m} \quad \text{and} \quad I_R = \frac{R_t}{R_o} = \frac{R_t}{F_R R_w} = S_w^{-n}$$

can be combined and written in the following form:

$$\log R_t = \log F_R + \log R_w + \log I_R$$

or:

$$\log R_t = -m \log \phi + \log R_w + \log I_R$$

This expression states that R_t can be plotted versus ϕ on log-log graph paper, and a linear trend will be observed for all zones within a constant lithology having a constant R_w and I_R. An example of such a plot is shown in Figure 7-20. As can be seen, the lowest observed resistivities are assumed to be 100% water-bearing ($I_R = 1$). This linear trend defined by the lowest observed resistivities is assumed to represent a locus of R_o values and when extrapolated to 100% ϕ would represent the apparent R_w. Additionally, this trend should define a slope of m (cementation factor) for the rock type. All points falling to the right of the 100% S_w trend have a resistivity $R_t > R_o$. These anomalies are assumed to have a higher resistivity as a result of the presence of hydrocarbons, which can be readily estimated as shown. This type of analysis has been used for many years in formation evaluation, including a variation of this approach using a graphical solution, which is linear in ϕ and nonlinear in resistivity, R_t.[7,19] The approach (see Fig. 7-20) is a powerful method for estimating S_w since it is not necessary to know R_w or m, and in fact they can be determined as shown.

This concept can be extended using the transit time-porosity relation discussed earlier where:

$$t = t_{ma} + B\phi \quad \text{or:} \quad \phi = \frac{t - t_{ma}}{B}$$

Substituting into the logarithmic expression for R_t, the following expression is obtained:

$$\log R_t = -m \log (t - t_{ma}) + m \log B + \log R_w + \log I_R$$

This expression states that a log-log plot of R_t vs. $(t - t_{ma})$ should exhibit a linear trend for all zones of constant lithology having constant R_w, B and I_R values. Such a graphical approach is shown in Figure 7-21. As can be seen from the illustrative crossplot in Figure 7-21, water saturation can be estimated for the anomalous data points and, additionally, m is defined by the slope of the linear trend representing constant values of I_R. Water saturation, S_w, and m can be obtained without a previous knowledge of B or R_w, and R_w can be determined if B can be estimated or is known. Using the graphical approach of linear $\phi = f(t)$ vs. nonlinear R_t, water saturation can also be estimated.[7] The application and implications of this crossplotting technique will be treated more fully later.

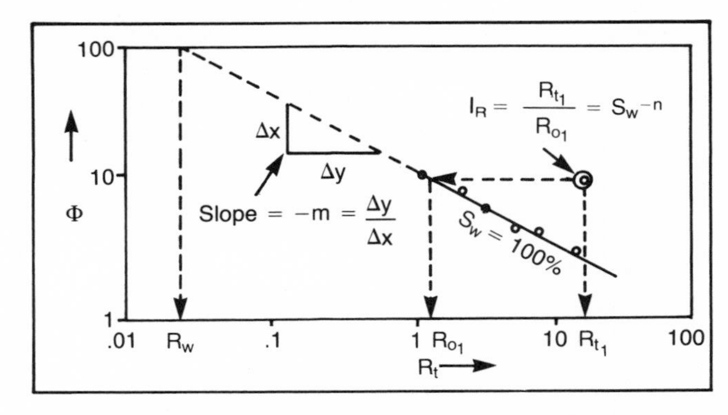

Fig. 7-20. Apparent resistivity vs. apparent ϕ crossplot

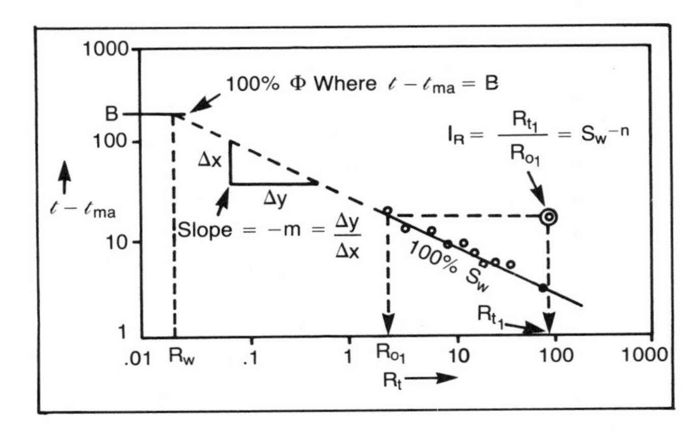

Fig. 7-21. Apparent resistivity vs. $(t - t_{ma})$ crossplot

TRANSIT TIME AS AN ABNORMAL PRESSURE INDICATOR

Since transit time is a function of porosity, it is an excellent indicator of abnormal pressure, since abnormally high pressure is closely related to abnormally high porosity existing in the rock system. The determination of abnormally high formation pressure is important because:

1. A knowledge of formation pressures will aid in design of drilling programs of offset wells, which in turn will usually permit lighter mud weights, reducing mud costs and increasing penetration rates.
2. Better pressure data will make selection of casing points better and casing string designs more economical.
3. Trouble caused by lost circulation and well kicks can be minimized by reducing the unnecessary usage of

heavy muds, thereby leading to a reduction in stimulation costs because of less filtration damage.

4. A more accurate selection of drill stem test equipment and well control equipment can be made.
5. Use of formation data can aid in long-range geological correlation over an area, as well as short-range correlation between different structures, [2,14] and reduce trouble, time, and cost.

Before we discuss the origin of abnormal pressure, perhaps we should define abnormal pressure as opposed to normal pressure. Normal pressures refer to formation pressures that are approximately equal to the hydrostatic head, a column of water extending from the reservoir depth to the surface that would balance the formation pressure. Cannon and Craze reported in 1938 that the normal pressure gradient in the Gulf Coast region is 0.465 psi per foot, pressure gradient defined as P/D, reservoir pressure P at depth D.[18] Abnormally pressured formations are those that contain pressures higher than hydrostatic and are encountered at varying depth. These formations are referred to as having abnormal pressures, geopressures, or overpressures, and some have been observed to be twice as much as hydrostatic pressure. Oilfield convention is to classify those formations that can be drilled with 10 lb/gal mud as normally pressured, those requiring more than 12 lb/gal as abnormally pressured, and those in between referred to as the transition zone.

The origin of abnormally pressured formations is considered to be primarily the result of the compaction phenomena. Shale compaction is controlled by its ability to lose water, the weight of the overburden pressure, and the time the shale is exposed to overburden pressures. As deposition progresses with more and more sediments being laid down, the underlying sediments become more deeply buried, and compaction results if contained waters can escape. Since the water is continuous with the water in the sediments and sea above, the pressure within the water is hydrostatic. Any reduction in volume must be accomplished by removal of shale waters. The process of squeezing out water becomes progressively more difficult because of continual reduction in permeability caused by compaction. Fine-textured shales, such as montmorillonite, within a water-rich environment, resist internal fluid migration to a much greater extent than the shales of the illite and kaolinite type.[18] Hubbert and Rubey[8] have also pointed out that rapid deposition can lead to abnormal pressures. Under lower rates of sedimentation, it is possible for fluid to be expelled quickly enough for hydrostatic equilibrium to be maintained. However, at rapid rates of sedimentation, with relatively impermeable shales, expulsion cannot take place at rates adequate to keep pace with the rate of sedimentation. Therefore, equilibrium is not maintained and pressures are abnormally high. Many of the zones now overpressured may later in geologic time reach equilibrium and become normally pressured.

Reservoirs with high pressures are usually isolated from those with normal hydrostatic pressure; otherwise the high pressures would be dissipated. This requires that porous reservoirs be sealed in all directions either by lensing or faulting. However, sand bodies in essentially shaly series are lenticular and erratic so that faulting is not necessary to maintain abnormal pressure. Zones of abnormal pressure commonly occur below the base of the main sand development in or below a major shaly series. High pressure may also be found in the main sand series where conditions are favorable for isolation of sand bodies by faulting or lensing-out of the sand.

The total weight of the overburden acts in a downward direction and is supported by the formation fluid pressure and the strength of the pore frame. The difference between the overburden pressure, P_o, and fluid pressure, P_f, is the net overburden pressure or the vertical rock frame stress, V_{rk}. This can be expressed by the equation:

$$V_{rk} = P_o - P_f$$

When formation pressures become higher than normal, the fluid is supporting a higher portion of the weight of the overburden; thus the weight supported by the rock frame stress is reduced. Usually a value of 1.0 psi/ft is used for the overburden pressure gradient, P/D.[14] This compaction process is shown in Figure 7-22.

A useful measure of the degree of compaction of a clay is its porosity. It is obvious that an estimate of clay or shale porosity as a function of depth will indicate the

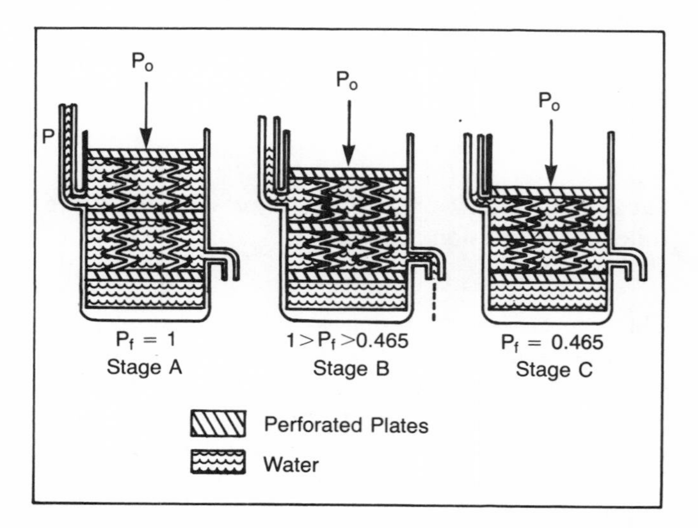

Fig. 7-22. Shale compaction model. Permission to publish by the Society of Petroleum Engineers of AIME.[15]

degree of compaction, since the porosity is reduced as water is expelled. Based on the work of Athy[1] and Hubbert and Rubey,[8] Foster and Whalen[14] have shown that shale porosity is an exponential function of depth such that a normal compaction trend would be linear when log ϕ is plotted versus depth, as shown in Figure 7-23. A section that is under-compacted, with respect to a given depth, will have abnormally high pressure.

Any change in a trend of increasing compaction with depth is an indication of abnormal pressure. Such changes may be recognized from all porosity-related logs by plotting the parameter related to compaction versus depth. To identify high pressure intervals and take preventive action, a log should be run slightly below the projected top of the high pressure shales. This depth can be estimated by projectively contouring the top of the high pressure zone of adjacent wells. If there are no existing wells nearby, frequent logging, possibly controlled through drilling time and density of shale cuttings, appears to be necessary. Abrupt changes can occur with increasing depth between normal and abnormal pressures.

Frequent logging can also signal the return to normal pressures, thus avoiding the danger of lost circulation caused by high mud weights, which are used to control overlying high pressure zones. If a return to normal pressure is noted, protective casing must be set because of the possibility of lost circulation.

Use of the acoustic transit time is one of the most reliable approaches for determining abnormal pressure because it is influenced by fewer parameters and will have a straight line normal trend on semilog paper (see Fig. 7-23). The actual example shown in Figure 7-24 clearly indicates the top of the overpressure. The normal and observed transit times might then be related empirically to pressure, using a correlation such as shown in Figure 7-25, for overpressured Miocene and Oligocene formations in Louisiana.

3-D LOG APPLICATIONS

The fact that a large quantity of data is recorded with the three-dimensional display of the full-wave train opens the door to a largely untapped area of potential applications in formation evaluation. Of course, this source of information has not been ignored since varied applications are being used, but the potential is only beginning to be realized with sophisticated current uses and new applications to aid the formation analyst. Some current uses are fracture identification and elastic

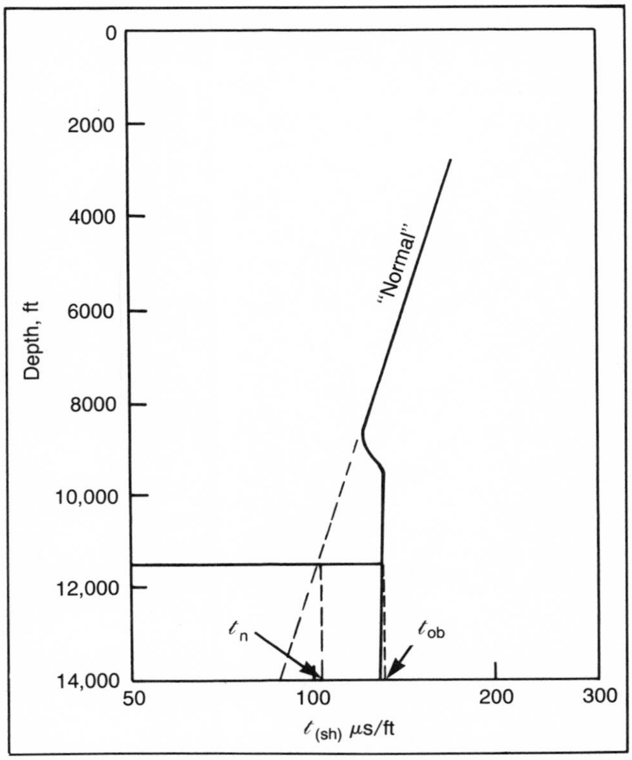

Fig. 7-23. Typical shale transit time vs. depth plot. Permission to publish by the Society of Petroleum Engineers of AIME.[15] Copyright 1965 SPE-AIME.

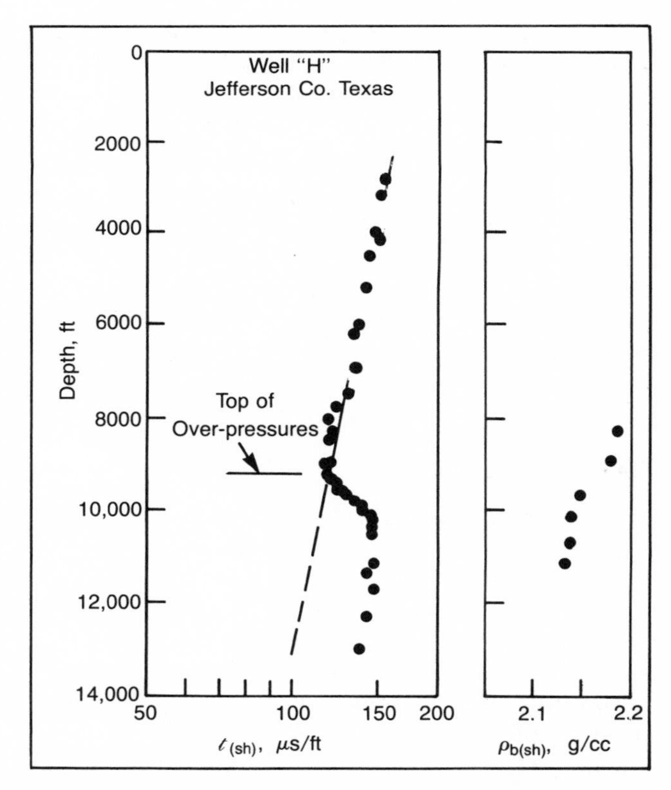

Fig. 7-24. Example of over-pressured zone as indicated by transit time data. Permission to publish by the Society of Petroleum Engineers of AIME.[15] Copyright 1965 SPE-AIME.

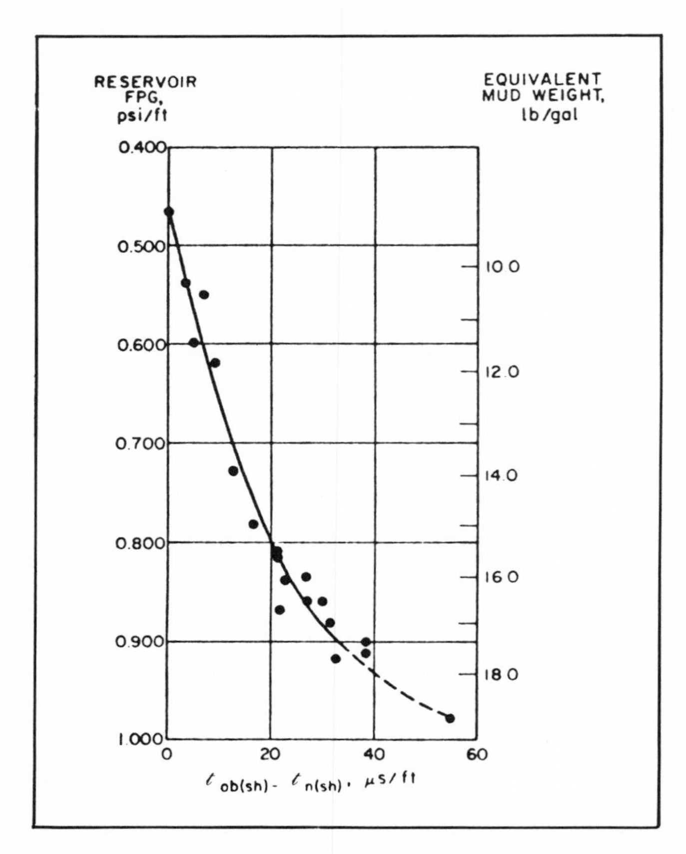

Fig. 7-25. Empirical transit time vs. pressure correlation. Permission to publish by the Society of Petroleum Engineers of AIME.[15] Copyright 1965 SPE-AIME.

moduli determination, as well as cement bond evaluation in cased holes. One study has indicated the possibility of using both the compressional and shear wave transit times to estimate porosity and lithology (in a two-mineral rock type).[31] This method is predicated on the linear relations noted by Pickett[10] when crossplotting compressional and shear wave velocities for the various rock types, as shown in Figure 7-26.

Fracture Detection

The 3-D acoustic log has been used in a qualitative way for fracture location. Two indications of the presence of fractures are the influence on shear and compressional wave amplitude and the presence of diagonal energy events resulting from the presence of a reflector plane. This is illustrated in Figure 7-27, which shows a 3-D acoustic log run through a fractured granite section. In zone C of this log, the compressional wave is not attenuated while the shear wave amplitude is severely reduced. This is probably due to a low angle or horizontal fracture, which would agree with the results of Knopoff's[26] study, which are shown in Figure 7-28. The strong energy arrivals for both compressional and shear waves in zone B indicate that this zone is not fractured. In zone A, however, both the compressional and shear waves are attenuated, probably due to an oblique fracture. A diagonal energy transmission can also be observed in zone C and on the lower portion of the log indicating a reflector plane (fracture). Fractures act as thin beds, or reflectors, filled with a fluid having a strong acoustic impedance mismatch to the rock. Figure 7-29 illustrates how these diagonal energy events occur. With the tool well below the reflecting horizon, and considering only those components traveling up the hole, one can visualize the energy waves being reflected back toward the receiver. As the tool moves up the hole toward the reflector, all the reflected arrivals will come in earlier, forming the diagonal pattern as a function of depth. When the transmitter reaches the reflector, only the normally transmitted energy arrives at the receiver, and this condition continues until the receiver has passed the reflector. When the receiver has passed the reflector, reflected energy again is observed, increasing in time as the tool moves away from the interface. The "dead" area on the log that shows no reflected events is, therefore, the tool spacing.

ELASTIC MODULI EVALUATION

Elastic rock properties can be determined from compressional and shear wave velocities (or transit times) and bulk density measurements.[28, 33] The elastic properties defined in terms of these logging parameters are:

$$\text{Poissons Ratio} = \mu = \frac{0.5\left(\frac{V_c}{V_s}\right)^2 - 1}{\left(\frac{V_c}{V_s}\right)^2 - 1}$$

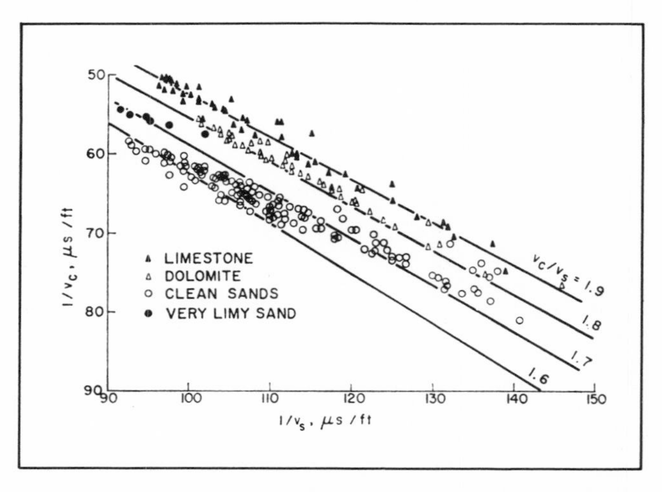

Fig. 7-26. Compressional versus shear wave transit times. Permission to publish by the Society of Petroleum Engineers of AIME.[10] Copyright 1963 SPE-AIME.

$$\text{Shear Modulus} = G = \rho_b v_s^2$$

$$\text{Bulk Modulus} = K = \rho_b v_c^2 - \tfrac{4}{3}\rho_b v_s^2$$

$$\text{Young's Modulus} = E = 2\rho_b v_s^2(1 + \mu)$$

(Modulus of Elasticity)

The ability to evaluate these moduli in-situ with logging tools has many ramifications, not only in the oil and gas industry but in numerous other areas such as the mining industry and civil engineering. Some applications within the petroleum industry would include the possiblity of predicting: (1) fracturing characteristics of various formations, (2) depth penetration for perforating charges, (3) sand production, and (4) subsidence.

Figure 7-30 shows the 3-D acoustic log and density log run through the Salinan Salt section of Silurian age in Michigan. The salt section was being solution-mined. Relatively low Young's Modulus values calculated for this section are indicated, and in fact actual subsidence took place in this brine field.

Figure 7-31 shows consecutively determined Young's Moduli determined from the open hole density log and cased-hole three-dimensional logs run at the times indicated. This logged section is through the roof-rock section above a solution salt mine. The four Young's Moduli have been superimposed on the same scale to observe the rate of variation on the Young's Modulus. It is interesting to note that the Young's Modulus decreased between run #1 in 1968 and run #4 in 1970, especially for the interval of zone A. The Six Arm Caliper survey indicated that a casing split in the interval could be due to a horizontal fracture at this interval. It is obvious, however, that the roof rock is weakening and will in time collapse, taking the casing with it and causing severe well damage. A prediction of the approximate time of roof-rock failure and where it will occur would be desirable because the casing above this point could be cut before collapse.

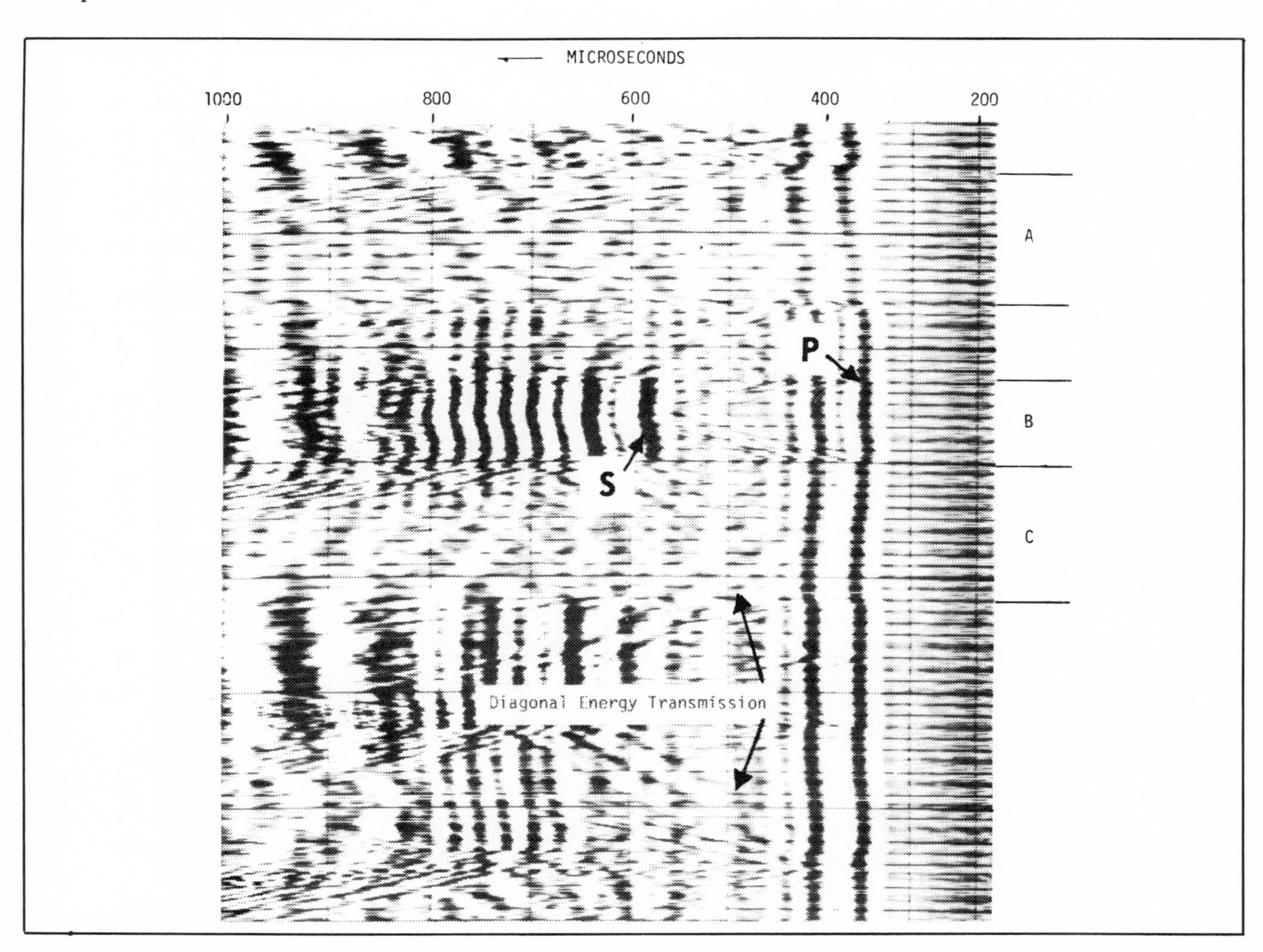

Fig. 7-27. 3-D acoustic log through fractured granite. Permission to publish by Birdwell Division, Seismograph Service Corp.[25]

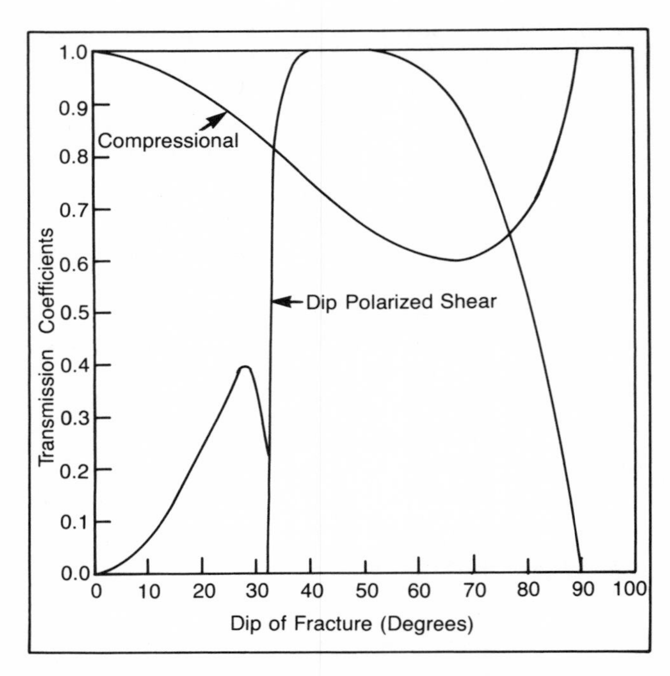

Fig. 7-28. Compressional and shear wave attenuation resulting from fracture dip. Permission to publish by the American Society of Mechanical Engineers.[26]

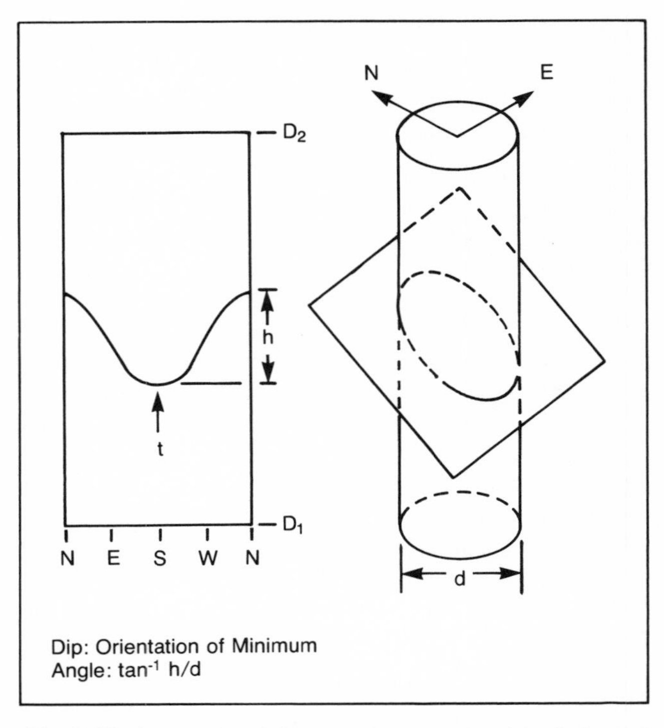

Fig. 7-32. Isometric of fracture intersecting borehole and corresponding borehole televiewer log. Permission to publish by the Society of Petroleum Engineers of AIME.[22] *Copyright 1970 SPE-AIME.*

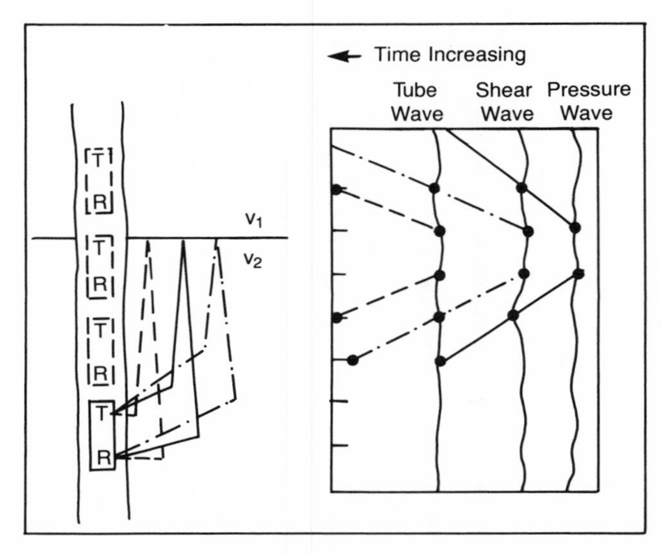

Fig. 7-29. Illustration of influence of reflector plane in producing diagonal energy events. Permission to publish by H. W. Lawrence.[17]

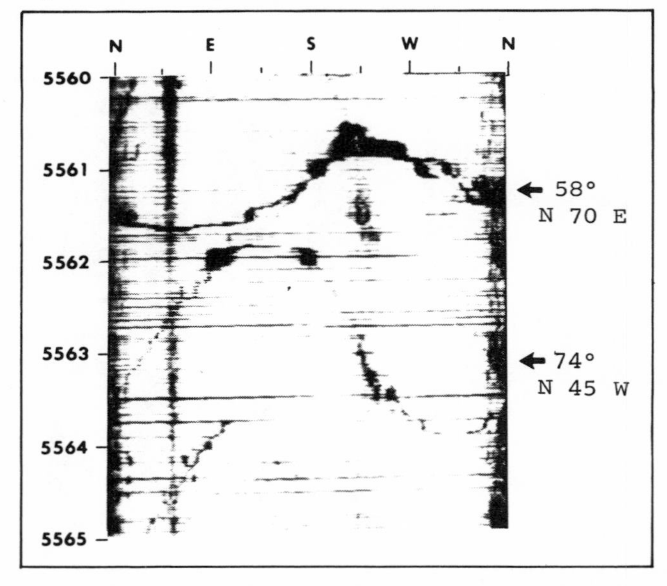

Fig. 7-34. Borehole televiewer indicating two high angle fractures intersecting the borehole. Permission to publish by the American Society of Mechanical Engineers.[26]

BOREHOLE TELEVIEWER

The borehole televiewer takes an oriented acoustic picture of the inside of the wellbore in the form of a continuous log. The log that is obtained is a representation of the borehole wall as if it were split vertically along magnetic north and laid out flat, as shown in Figure 7-32.

This tool consists of a downhole scanning instrument comprising a motor-driven acoustical transducer and a fluxgate magnetometer rotating at three revolutions per second. The log is oriented to a north marker signal from the fluxgate magnetometer. The transducer is pulsed at a rate of 2,000 times per second, with the beam focused on the borehole wall. The amplitude of the reflected signal depends on the acoustical impedance of the wall rock and associated physical properties of the wall. Fractured or highly disturbed zones are recognized by no reflected signal or a poor signal. A block diagram of the borehole televiewer system is shown in Figure 7-33. Advantages of this tool are that it can be operated in any type of borehole fluid and in holes from 2 to 12 in. in diameter.

An example of fracture location as well as the determination of dip angle and bearing is shown in Figure 7-34. Two high-angle fractures are seen extending over 40 ft of hole and nearly intersect within the borehole. The orientation and dip of these two fractures indicate that the two planes cross each other a short distance from the borehole.

Figure 7-35 compares a Borehole Televiewer log and a 3-D acoustic log run in the Clinton sand in Ohio. In the interval between 4000 and 4030 ft, the Borehole Televiewer shows a vertical fracture that strikes almost due east-west. This fracture is not observed on the 3-D acoustic log, which is not surprising since a vertical fracture would not be expected to affect either the compressional or shear wave amplitudes.

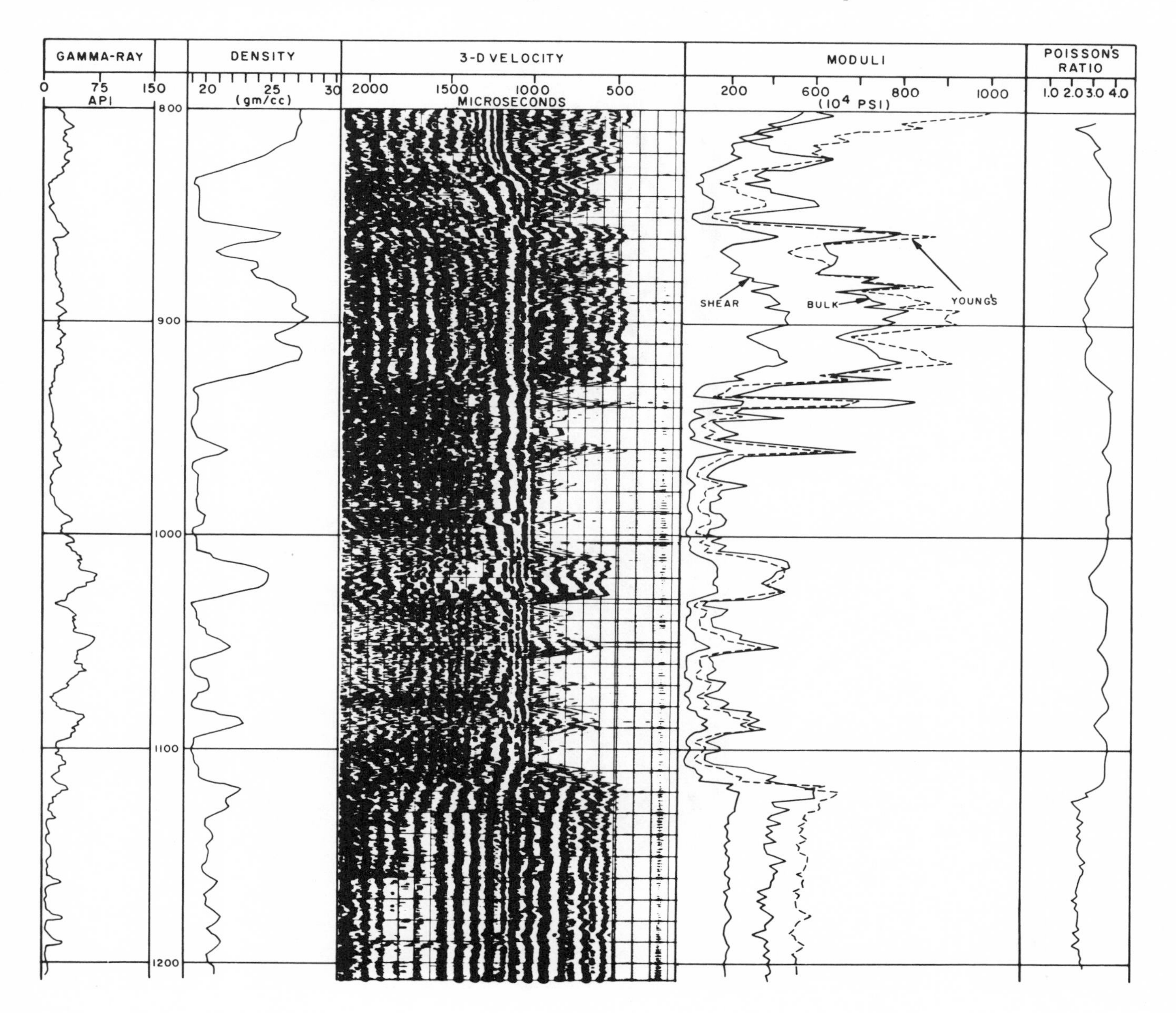

Fig. 7-30. Field logs and computed elastic moduli. Permission to publish by the Solution Mining Research Institute.[27]

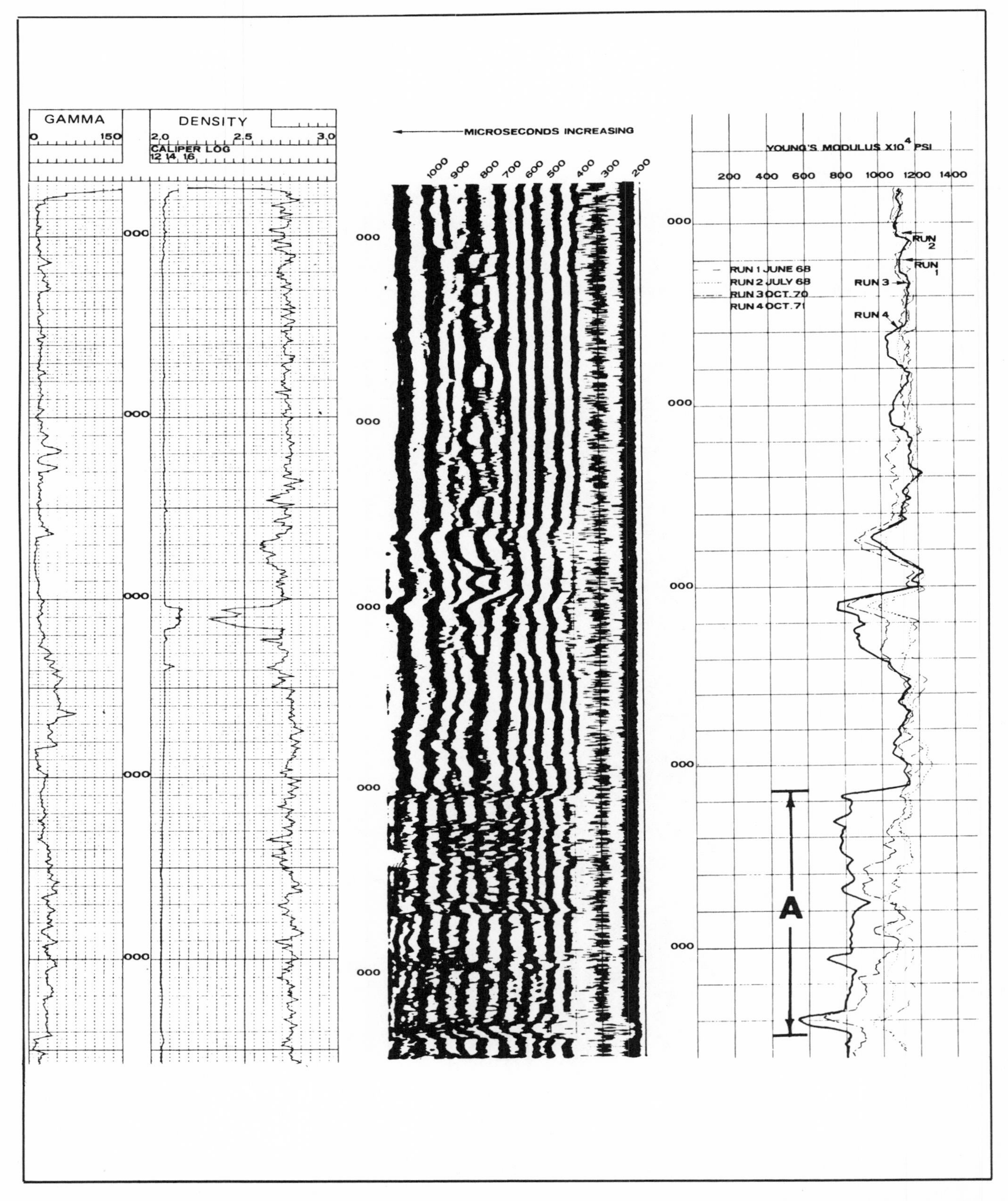

Fig. 7-31. Density and 3-D log and computed Young's Modulus in roof-rock section above solution salt mine. Permission to publish by the Northern Ohio Geological Society.[30]

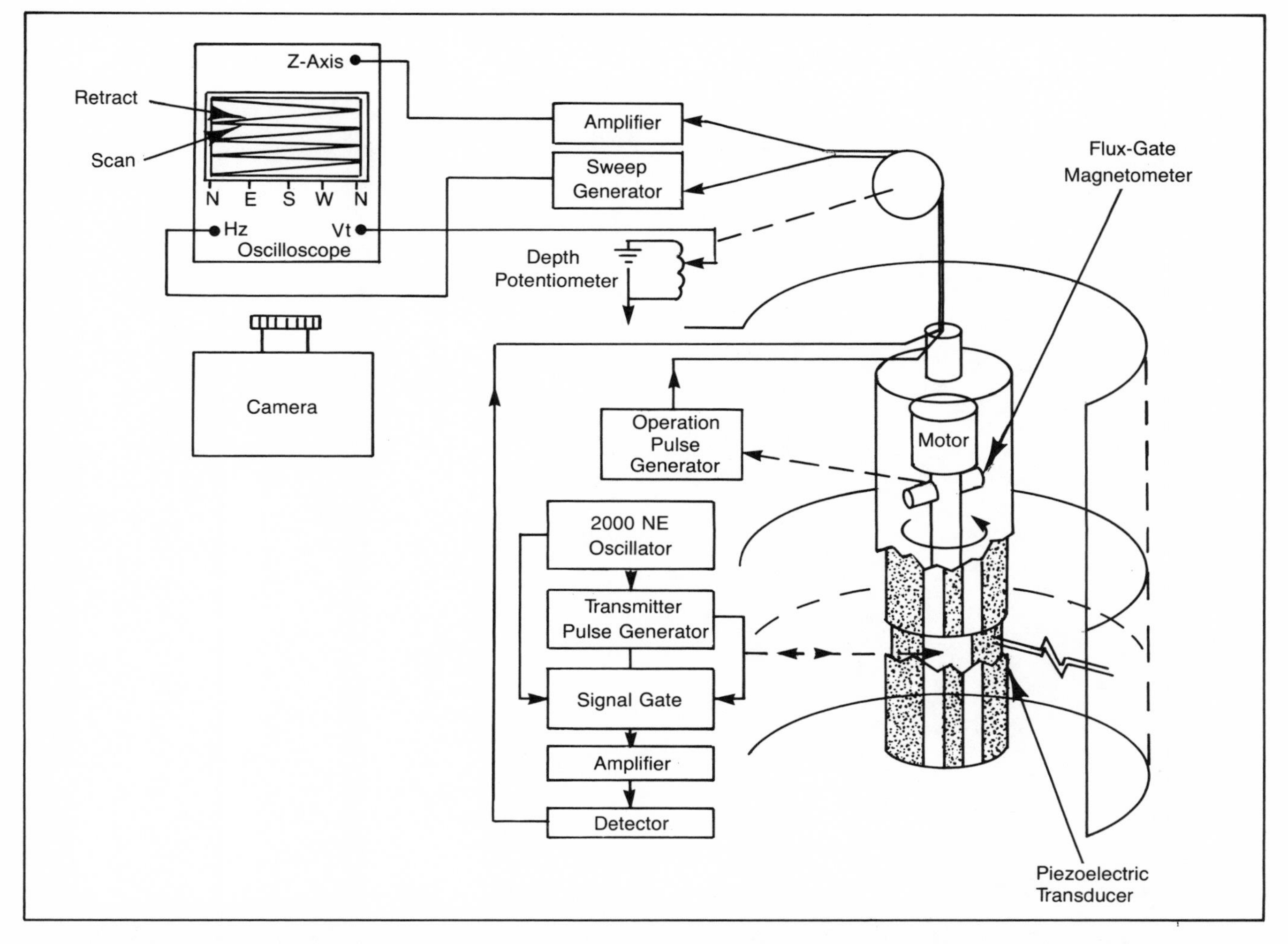

Fig. 7-33. Block diagram of borehole televiewer system. Permission to publish by the Society of Petroleum Engineers of AIME.[22] Copyright 1970 SPE-AIME.

REFERENCES

1. Athy, L.F.: "The Density, Porosity, and Compaction of Sedimentary Rocks," *Bull.*, AAPG (1930) 14.
2. Claudet, A.P.: "New Method of Correlation by Resistivity Values of Electric Logs," *Bull.*, AAPG (1950) 34.
3. Gassman, F.: "Elastic Waves Through a Packing of Spheres," *Geophysics* (1951) 16.
4. Brandt, H.: "A Study of the Speed of Sound in Porous Granular Media," *Trans.*, ASME (1955).
5. Biot, M.A.: "Theory of Propagation of Elastic Waves in a Fluid Saturated Porous Solid," *J. Acoustic Society of America* (March 1956) 28.
6. Wyllie, M.R.J., Gregory, Alvin R., and Gardner, L.W.: "Elastic Wave Velocities in Heterogeneous and Porous Media," *Geophysics* (Jan. 1956) 21.
7. Hingle, A.T.: "The Use of Logs in Exploration Problems," paper presented at the SEG 29th Annual Meeting, Los Angeles, Nov. 1959.
8. Hubbert, M. King, and Rubey, W.W.: "Role of Fluid Pressures in Mechanics of Overthrust Faulting," *Bull.*, Geological Society of America (1959) 70, 115-206.
9. Geertsma, J.: "Velocity Log Interpretation: The Effect of Rock Compressibility," *Soc. Pet. Eng. J.* (Aug. 1961) 1.
10. Pickett, George R.: "Acoustic Character Logs and Their Applications in Formation Evaluation," *J. Pet. Tech.* (June 1963) 15, 650-667.
11. Christensen, Dean M.: "A Theoretical Analysis of Wave Propagation in Fluid Filled Drill Holes for the Interpretation of the 3-Dimensional Velocity Log," *Trans.*, SPWLA Fifteenth Annual Logging Symposium, Midland, TX, May 1964.
12. Bird, James M.: "An Explanation of the '3-D' Acoustical Logging System for the Exploration Geophysicist," paper presented at a meeting of the Mexican Geophysical Society, Monterrey, Oct. 1965.
13. Christensen, Dean M.: "The Determination of the In-Situ Elastic Properties of Rock Salt with a 3-Dimensional Velocity Log," *Trans.*, Northern Ohio Geological Society, Second Symposium on Salt (May 1965).
14. Foster, Joe, and Whalen, Herbert E.: "Estimation of Formation Pressures from Electrical Surveys—Offshore Louisiana," paper SPE 1200 presented at the SPE Annual meeting, 1965.

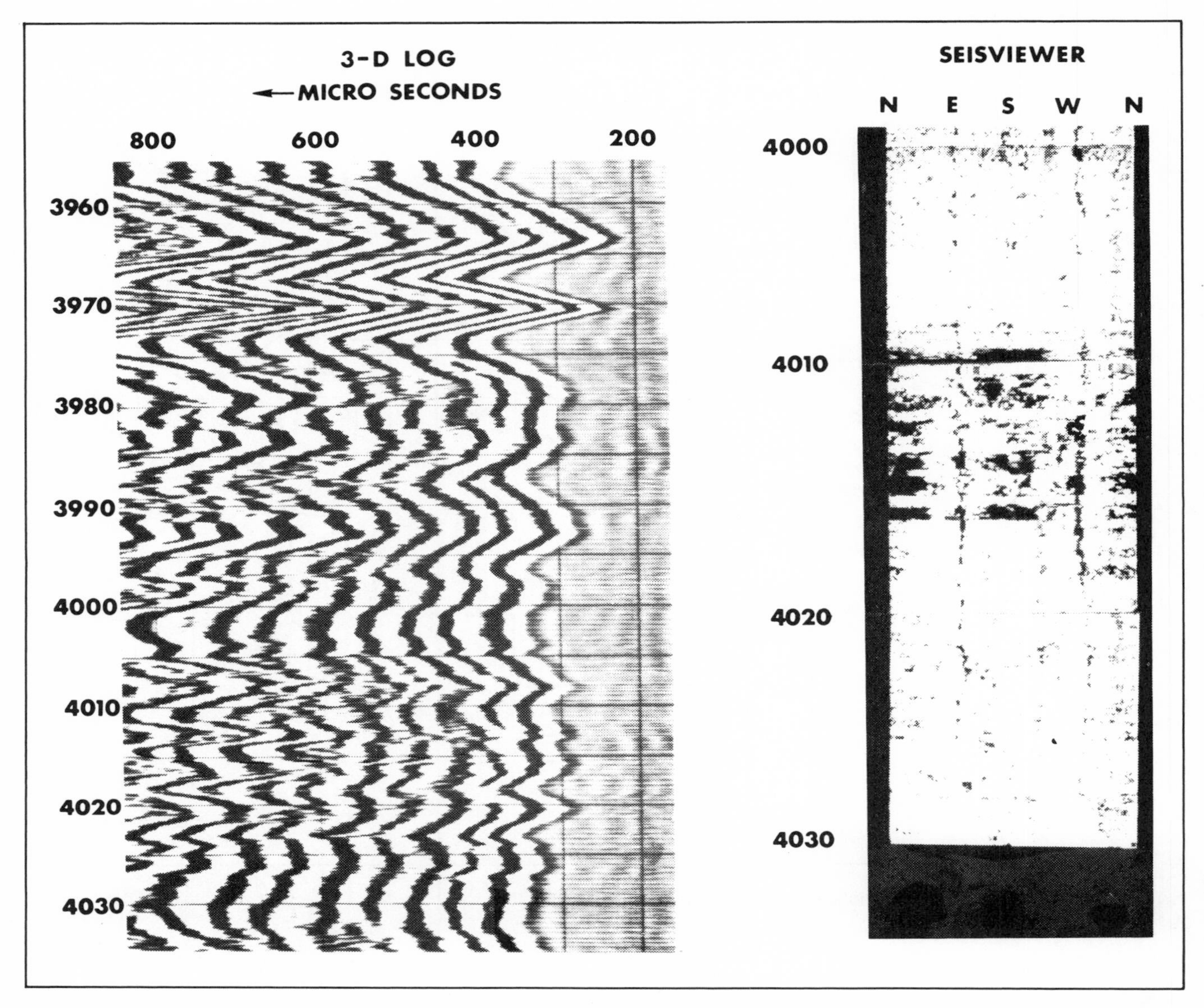

Fig. 7-35. Comparison of 3-D acoustic log and borehole televiewer (Seisviewer) opposite vertically fractured zone. Permission to publish by the American Society of Mechanical Engineers.[26]

15. Hottman, Clarence E., and Johnson, Robert K.: "Estimation of Formation Pressures from Log-Derived Shale Properties," *J. Pet. Tech.* (June 1965) 717–722.
16. Kokesh, F.P., Schwartz, R.J., Wall, W.B., and Morris, R.L.: "A New Approach to Sonic Logging and Other Acoustic Measurements," *J. Pet. Tech.* (March 1965) 282–286.
17. Lawrence, Homer W.: "Reflection, Refraction and Energy Mode Conversion as Seen on 3-D Velocity Logs," paper presented at the SEG 35th Annual Meeting, Dallas, 1965.
18. Fons, Lloyd, and Holt, Olin: "Formation Log Pressure Data Can Improve Drilling," *World Oil* (Sept. 1966), 70–74.
19. Pickett, George R.: "A Review of Current Techniques for Determination of Water Saturation from Logs," *J. Pet. Tech.* (Nov. 1966).
20. Lawrence, Homer W., and Baltosser, Robert W.: "Engineering Problems and Downhole Geophysical Solutions," paper presented at the Fourth Annual Idea Conference, New Mexico Institute of Mining and Technology, Socorro, May 1968.
21. Guyod, Hubert, and Shane, Lemay E.: *Geophysical Well Logging*, Hubert Guyod Publ., Houston (1969).
22. Zemanek, Joseph, Caldwell, Richard L., Glenn, Edwin E., Jr. Holcomb, S.V., Norton, L.J., and Straus, A.J.D.: "The Borehole Televiewer – A New Logging Concept for Fracture Location and Other Types of Borehole Inspection," paper SPE 2402, *J. Pet. Tech.* (Dec. 1970) 22.
23. Desai, Kantilal P., Helander, Donald P., and Moore, E. James: "Sequential Measurement of Compressional and Shear Velocities of Rock Samples Under Triaxial Pressures," *J. Pet. Tech.* (Dec. 1970) 22.
24. Geyer, Robert L., and Myung, John I.: "The 3-D Velocity Log: A Tool for In-Situ Determination of the Elastic Moduli of Rocks," *Proc.*, ASME Twelfth Symposium on Rock Mechanics, U. of Missouri, Rolla (Nov. 1970).
25. Myung, John I., and Sturdevant, William M.: *Introduction to the Three-Dimensional Velocity Log*, Birdwell Division, Seismograph Service Corp., Tulsa (1970).

26. Myung, John I., and Baltosser, Robert W.: "Fracture Evaluation by the Borehole Logging Method," paper presented at the ASME Thirteenth Annual Symposium on Rock Mechanics, U. of Illinois, Urbana, 1971.
27. Myung, John I., and Henthorne, Jay: "Elastic Property Evaluation of the Roof-Rocks with 3-D Velocity Log," paper presented at the Solution Mining Research Institute (a Technical Session), Atlanta, Dec. 13-15, 1971.
28. Myung, John I., and Helander, Donald P.: "Correlation of Elastic Moduli Dynamically Measured by In-Situ and Laboratory Techniques," *The Log Analyst* (Nov.-Dec. 1972) 13.
29. *Log Interpretation, Volume I—Principles*, Schlumberger Ltd., New York (1972).
30. Myung, John I., and Helander, Donald P.: "Borehole Investigation of Rock Quality and Deformation Using the 3-D Velocity Log," paper presented at the Northern Ohio Geological Society 4th International Symposium on Salt, Houston, April 1973.
31. Nations, Joe: "Lithology and Porosity from Acoustic Shear and Compressional Wave Transit Time Relationship," *The Log Analyst* (Nov.-Dec. 1974) 15.
32. England, Robert E.: *Well Log Interpretation, Volume I*, Birdwell Division, Seismograph Service Corp., Tulsa (1975).
33. Kowalski, John: "Formation Strength Parameters from Well Logs," paper presented at the SPWLA Sixteenth Annual Logging Symposium, June 1975.
34. Waller, William C., Cram, Milton E., and Hall, James E.: "Mechanics of Log Calibration," paper presented at the SPWLA Sixteenth Annual Logging Symposium, June 1975.
35. *Acoustic Logs*, Dresser Atlas, Dresser Industries Inc., Houston (1981).

8 Radioactivity Logs

RADIOACTIVITY logs have played an important role in formation evaluation since the first commercial gamma ray log was run in 1940. By 1941 the neutron log had developed to the point where it was also offered commercially. Before discussing the various types of radioactivity logs used in formation evaluation, however, a brief discussion of basic nuclear phenomena is necessary.

NATURE OF RADIOACTIVITY

Most definitions of radioactivity explain this phenomenon as the spontaneous disintegration of atoms accompanied by the emission of radiation. There are, however, some atomic nuclei that emit radiation according to the laws of radioactive decay but do not disintegrate. For all practical purposes, if any substance emits alpha particles, beta particles, or gamma rays without an apparent external cause, then that substance is radioactive.

The process of spontaneous disintegration is exponential in nature. This disintegration process can be expressed for N radioactive nuclei at time, t, as:

$$\lambda = -\frac{1}{N}\left(\frac{dN}{dt}\right)$$

where λ is called the disintegration constant. The probability per unit time for the decay of an atom is expressed by the right-hand side of the preceding equation. The basic assumption in this statistical theory of radioactive decay is that the probability is constant regardless of the age of the atom. Integration of this equation gives:

$$N = N_o e^{-\lambda t}$$

where N_o is the number of radioactive nuclei at zero time. The decay of a radioactive isotope that decreases in activity by a factor of 50% every four hours is shown in Figure 8-1.

The radioactive isotope loses 50% of its activity every four hours, so that eight hours from time zero only 25% of the initial (100%) activity remains. The time required for a radioactive isotope to lose 50% of its activity is called the half-life, $T_{1/2}$. The half-life is related to the disintegration constant as:

$$T_{1/2} = \frac{0.693}{\lambda}$$

The half-life of radioactive isotopes varies from millions of years down to the limits of measurement, presently in the millimicrosecond range (10^{-9} seconds). All of the naturally occurring radioactive elements have very long half-lives. The elements that are artificially made radioactive in observable quantities by nuclear transformation in reactors, accelerators, etc. usually have relatively short half-lives.

Three of the four types of radiation of interest in well logging have been mentioned. These are the three types of radiation emitted in the spontaneous disintegration of atoms, the alpha particles (α particles) (Fig. 8-2), beta particles (β particles) (Fig. 8-3), and gamma rays (γ rays) (Figs. 8-2 and 8-3). The fourth type of radiation,

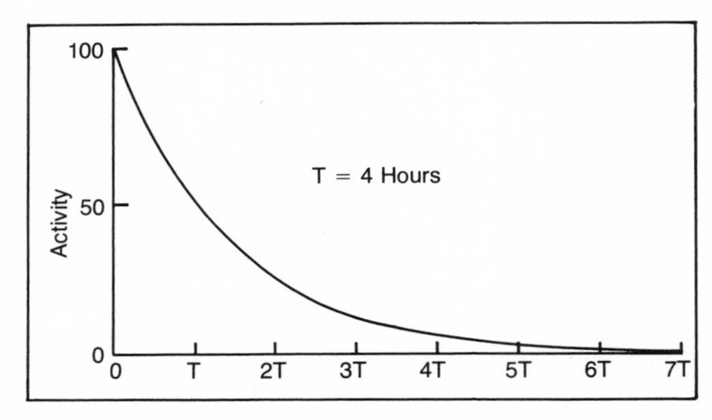

Fig. 8-1. Decay of radioactive isotope with half life of 4 hrs. Permission to publish by the Society of Professional Well Log Analysts.[20]

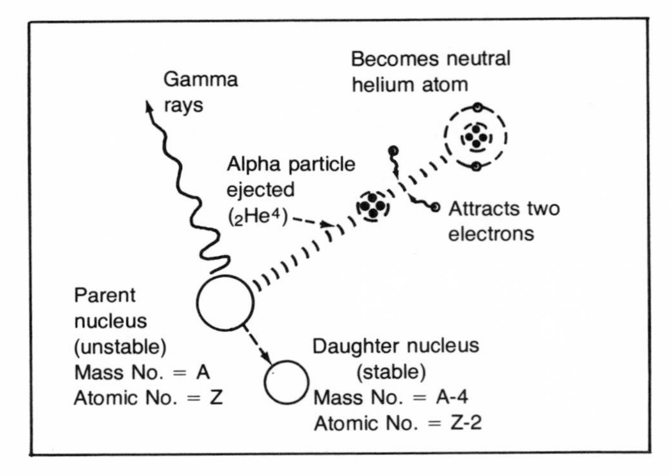

Fig. 8-2. Alpha particle emission with associated gamma ray emission. From Nuclear Power Engineering *by Henry J. Schwenk. Copyright 1957, McGraw-Hill Book Company. Used with the permission of McGraw-Hill Book Company.*[6]

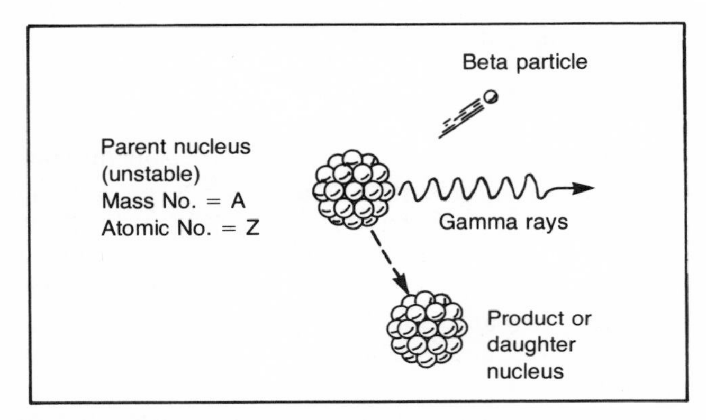

Fig. 8-3. Beta particle emission with associated gamma ray emission. From Nuclear Power Engineering *by Henry J. Schwenk. Copyright 1957, McGraw-Hill Book Company. Used with the permission of McGraw-Hill Book Company.*[6]

the neutron, is not emitted in any natural radioactive process but is produced by nuclear disintegration.

Alpha Particles

Alpha particles are actually nuclei of helium having an atomic number of 2 and a mass number of 4. The α particle carries a 2 positive charge and when emitted from a parent isotope forms a new isotope that is 2 units less positive and 4 mass units lighter. Alpha particles are ejected at very high velocity but since they are charged particles having a large mass, they have a very small penetration in matter—on the order of 0.001 cm. for 5 MEV particles in aluminum.[20] More graphically, we might say they can be absorbed by a single sheet of paper (Fig. 8-4). The definition of energy units is given in Table 8-1.

TABLE 8-1
Units Used in Nuclear Measurements

Unit	*Description*
1 erg	Work done when force of 1 dyne acts through distance of 1 cm.
1 dyne	Force acting on mass of 1 gram, giving it acceleration of 1 cm/sec.
1 ev (electron volt)	Energy acquired by any charged particle carrying a unit electrical charge when it passes without resistance through a potential difference of 1 volt = 1.60×10^{-12} erg.
1 kev (kiloelectron volt)	1,000 ev
1 mev (million electron volts)	1,000,000 ev = 1.52×10^{-16} btu = 4.45×10^{-20} kwhr
1 bev (billion electron volts)	1,000,000,000 ev
1 amu (atomic mass unit)	Arbitrary unit based on mass of oxygen isotope O^{16} being chosen as 16.0000 = 1.66×10^{-24} gram = 3.66×10^{-27} lb

From *Nuclear Power Engineering* by Henry J. Schwenk. Copyright 1957, McGraw-Hill Book Company. Used with the permission of McGraw-Hill Book Company.[6]

Beta Particles

Beta emission is the ejection of an electron from the nucleus. The β particle has a mass and charge equivalent to an electron and when ejected from the nucleus, decreases the negative charge by one with essentially no

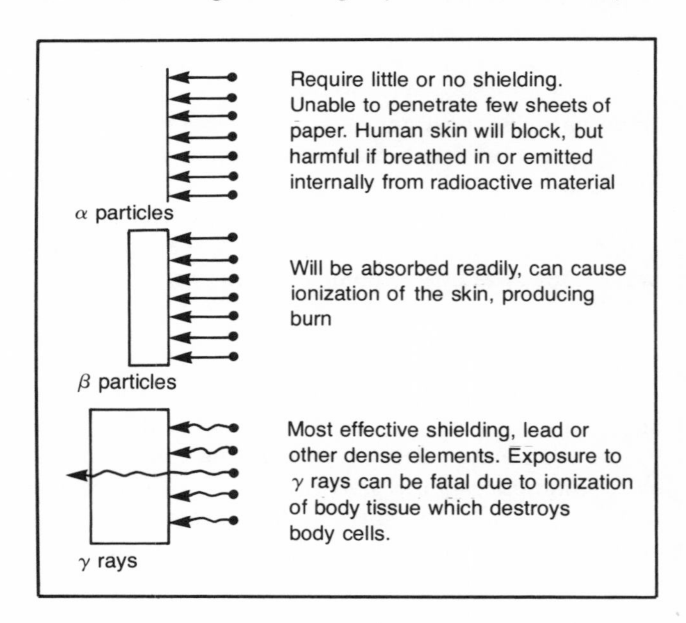

Fig. 8-4. Penetrating capacity of emissions resulting from radioactive disintegration. From Nuclear Power Engineering *by Henry J. Schwenk. Copyright 1957, McGraw-Hill Book Company. Used with the permission of McGraw-Hill Book Company.*[6]

change in mass. The new isotope formed, therefore, has an atomic number that is one unit greater than that of the parent isotope. Since the particle is charged and does have some mass, its penetrating power is also low although it is much greater than that of an α particle. The average range of a 5 MEV β particle through aluminum is approximately 1 cm.[20] The penetrating power of the β particle may more graphically be stated as being able to penetrate from 25 to 30 sheets of paper (see Fig. 8-4).

The type of particle emitted by a nucleus is determined by the neutron-proton ratio. If the nucleus needs to get rid of a positive charge quickly, it will emit an α particle, and if it has too large a neutron-proton ratio it will emit β particles. There are no proton emitters. Once an α or β particle has been emitted by a nucleus, the nucleus may be more stable with respect to charge than with respect to energy. In order to take care of this condition γ rays are emitted.

Gamma Rays

Gamma rays are electromagnetic radiations or photons that are emitted by a nucleus in an excited state. Gamma rays have no charge or mass, but they do possess energy, and when emitted, neither the atomic weight or atomic number of the isotope is changed but only the energy level of the nucleus. Some nuclei emit several γ rays in several steps before they become stable. Penetration of γ rays through matter is quite large since they have no mass or charge, with the amount of penetration being approximately inversely proportional to the atomic number of the material through which it is passing. The heavier the material, the lower the penetration. High energy gamma rays are capable of penetrating a deep stack of books or three inches or more of lead (see Fig. 8-4).

Neutrons

Neutrons are uncharged particles having approximately the same mass as a proton. Since they have no charge, although they do have mass, they can have deep penetration limited only by the nuclear characteristics of the material. Neutrons can penetrate several feet of lead. The penetration of neutrons decreases as the atomic weight of the material increases. Neutrons do not exist in the free state in nature. They can be produced, however, by artificial sources.

TRANSFORMATION SERIES

Most of the naturally occurring radioactive elements belong to one of the three transformation series (see Fig. 8-5). In each of these series, spontaneous disintegration begins with an isotope of one of the heaviest elements and follows a succession of steps which finally lead to a stable isotope of lead. These elements do not necessarily transform into one another rapidly, nor do the various transformations occur at the same rate, but for each of the elements the rate is invariable.

The transformation series shown in Figure 8-5 indicates the successive charged-particle emissions (α and β particles) occurring in the natural radioactive decay process for most of the radioactive elements existing in nature. Attendant with most of these spontaneous disintegrations is the emission of relatively low energy level gamma rays enabling the daughter nucleus to reach a stable energy level. The gamma ray emission spectra of the three common sources of natural radioactivity in the earth are shown in Figure 8-6. As can be seen, the bulk of natural gamma ray emission is in the low energy range below 2 MEV.

The naturally occurring radioactive isotopes are widely distributed in nature and, as with many other minerals, a concentration of radioactive elements as ores can be considered anomalies. The primary radioactive elements that are geologically significant are uranium, thorium, and potassium. It should be pointed out that radioactive potassium is one of four naturally occurring light elements that do not conform to the three transformation series and is the only one of the four that occurs in any relative abundance. Only 0.01% of all potassium is radioactive, but this percentage is always present wherever potassium is found in nature. The spontaneous disintegration of radioactive potassium produces a stable end product, calcium. The actinium series elements and the isolated radioactive elements are so rare and inactive in comparison with the others that they are insignificant.

The uranium and thorium series elements and potassium are found in sedimentary rocks, making these rocks appreciably radioactive. Due to the very high absorption and ion exchange capacity of clays, however, they are able to readily absorb the heavy radioactive isotopes released by decomposition of other minerals, as well as hold those originally contained in the clay-forming materials. This generally results in the clays having a much greater concentration of radioactive isotopes than that found in sandstones. Carbonate rocks, on the other hand, are not produced by weathering but result from the deposition of marine organisms. Since living bodies tend to eliminate radioactive elements, it would be expected that as they are formed carbonate rocks would have a very low radioactivity level. During the formation of secondary porosity or dolomitization, radioactive elements can be deposited by solution waters such that these sections could show higher radioactivity

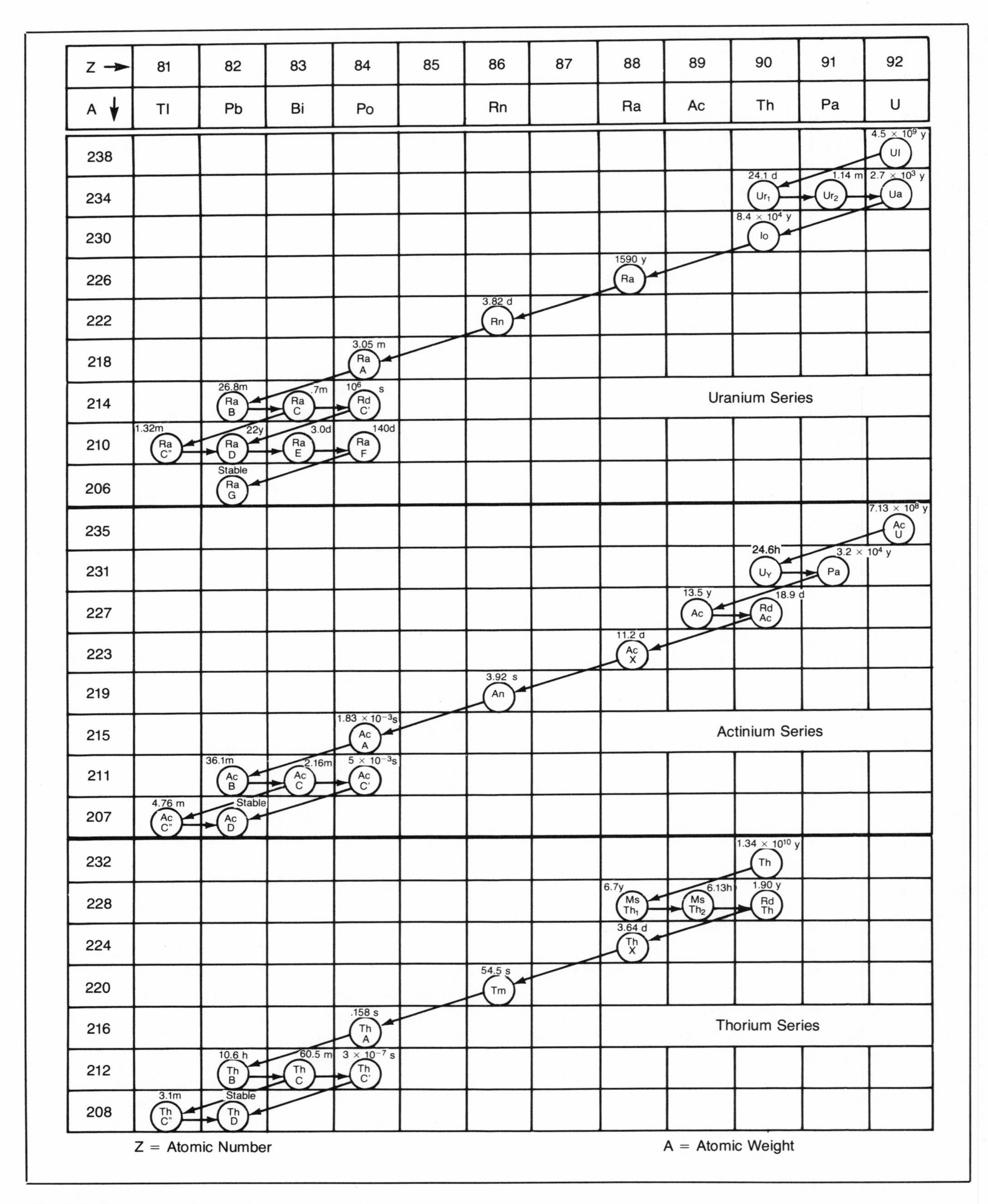

Fig. 8-5. Transformation series

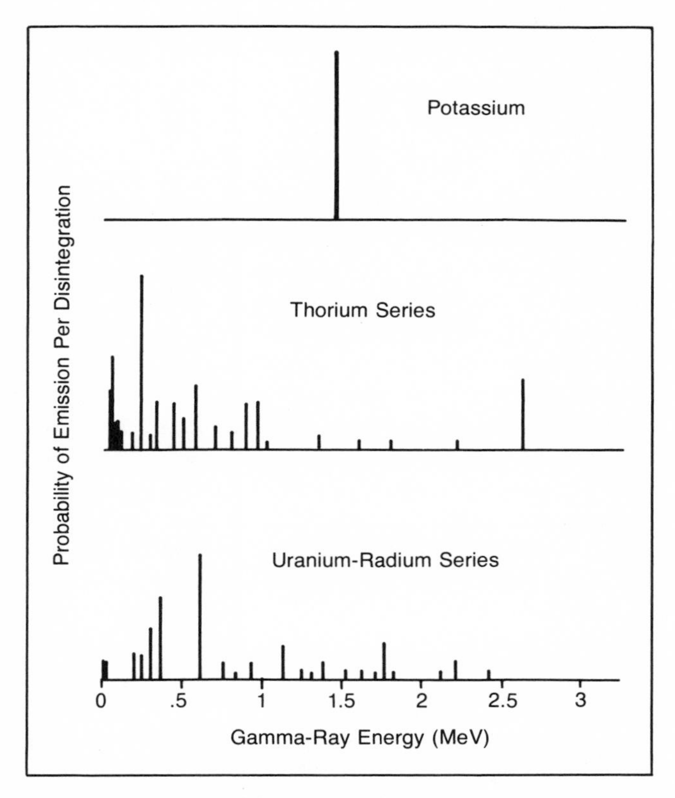

Fig. 8-6. Natural gamma ray emission spectra of the three common sources of natural radioactivity in the Earth. Permission to publish by Schlumberger Ltd.[5]

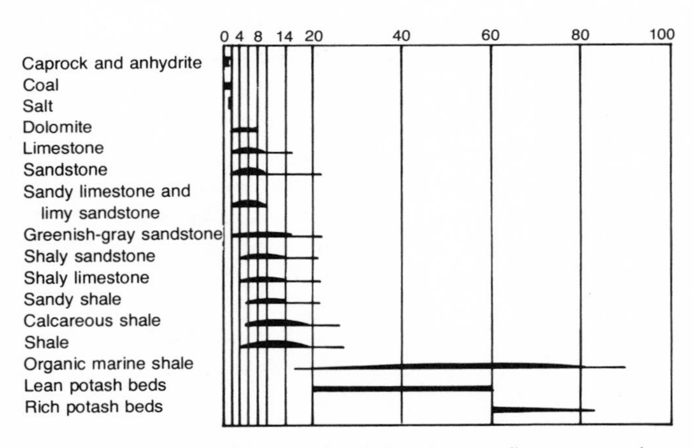

Fig. 8-7. Relative radioactivity level of various sedimentary rocks. Permission to publish by the American Association of Petroleum Geologists.[1]

than the undisturbed primary carbonate deposits. Figure 8-7 graphically shows the relative radioactivity level of various sedimentary rocks.

GAMMA ABSORPTION

Gamma rays are absorbed by three mechanisms, with the predominant mechanism depending on the energy of

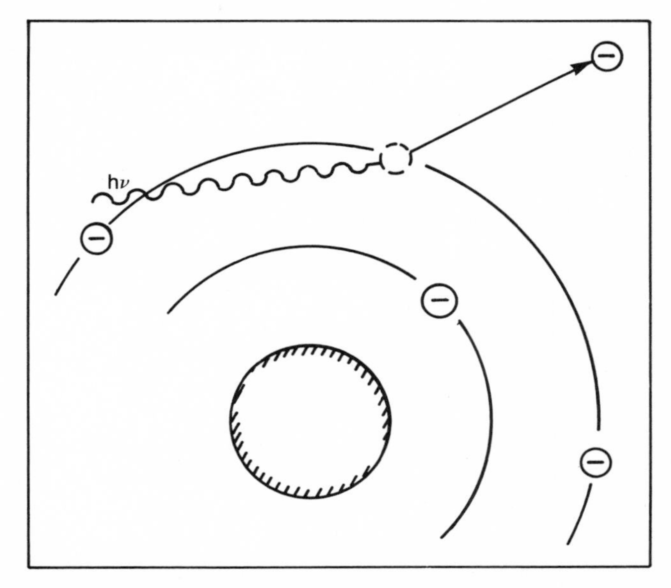

Fig. 8-8. Photoelectric effect[18]

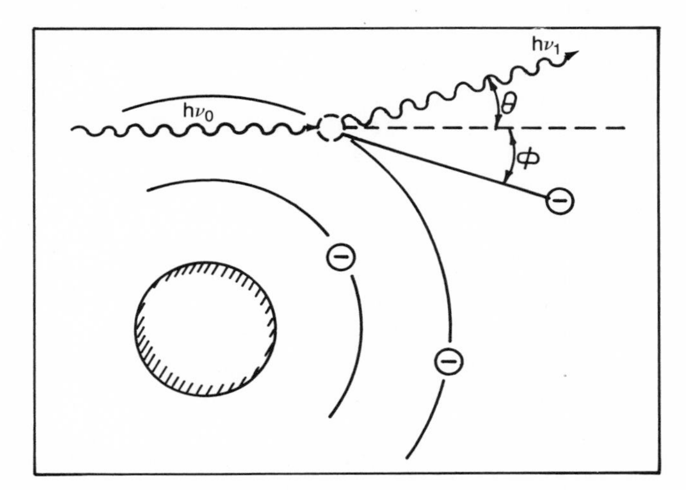

Fig. 8-9. Compton scattering effect[18]

the gamma ray and the atomic number of the absorbing material. At low energies (below about 150 KEV) the predominating mechanism is the *photoelectric effect*. In the photoelectric effect (see Fig. 8-8) the gamma ray reacts with an electron to an energy equal to the energy of the gamma ray minus the binding energy of the electron. The gamma ray disappears in the process. The binding energy is usually low but can be as large as 100 KEV.

At higher energies a competing absorption phenomenon called the *Compton effect* predominates (see Fig. 8-9). In this case, the γ ray strikes an electron, giving up part of its energy. This results in a photon of lesser energy than the original gamma ray scattered at an angle θ from the direction of the original gamma ray. The electron will be scattered at an angle ϕ. The process oc-

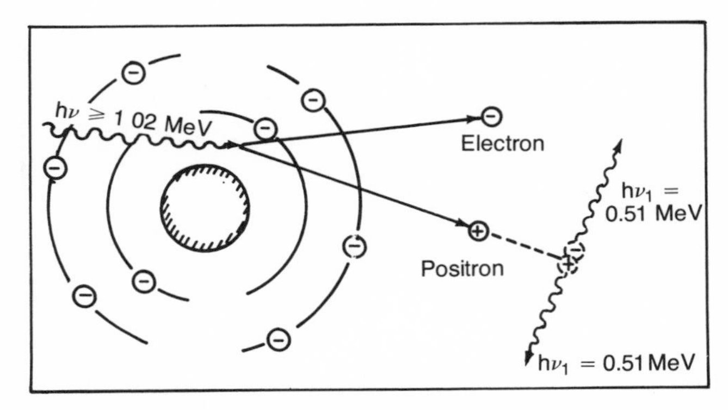

Fig. 8-10. Pair production and annihilation[18]

curs in such a manner that both energy and momentum are conserved. The new gamma ray produced by the Compton effect can be absorbed by either of the two processes described. Since it has less energy than its parent, however, it is usually more likely that the Compton gamma will be absorbed in any fixed distance.

The process in which a photon of sufficient energy gives up all its energy and forms two particles, an electron and a positron, is called *pair production*. This process is pictured (see Fig. 8-10) as occurring near the nucleus of an atom, for it is only in this way that momentum can be conserved. The minimum energy needed for pair production is given by $E=2m_0c^2$, where m_0 is the rest mass of an electron or positron and c is the speed of light. Since the rest mass of an electron is equivalent to 0.51 MEV, the photon must have an energy 1.02 MEV for pair production to occur.[8]

When this process does occur, energy of the photon beyond 1.02 MEV is imparted as kinetic energy to the electron-positron pair and a portion to the nucleus in order to conserve momentum. Both electron and positron lose kinetic energy through ionization of atoms in the substance. Eventually the positron interacts with an electron in the substance in a process called annihilation. In this process, the mass of the particles is changed into two photons of 0.51 MEV each, emitted in nearly opposite directions. They may interact further in the substance through the photoelectric or Compton effects.

Since this process does not occur at all unless the gamma ray energy is 1.02 MEV, pair production is important only for gamma rays of high energy. The process is also proportional to Z^2 (where Z=atomic number) of the absorber. Thus, pair production predominates for substances of high Z.

The three processes, photoelectric effect, Compton scattering effect, and pair production, account for the main photon interactions with matter. Both the photoelectric and Compton scattering effects decrease with an increase of gamma energy. Pair production increases with gamma energy. The result of these effects is that each substance will have minimum absorption at some gamma energy.

NEUTRON PRODUCTION

Before discussing neutron interactions, it is important to first consider the source of neutrons since they are not produced naturally. Studies have shown that neutrons can be produced in a number of ways. The use of radioactive sources and certain target substances can yield neutrons by either the alpha-neutron (α,n) or the gamma-neutron (γ,n) reactions. Accelerators produce neutrons when high energy particles strike a suitable target. Neutrons also result from the fission process in a reactor, and some of the transuranic elements can be used as neutron sources since they undergo spontaneous fission. A concise summary of neutron sources applicable or feasible in well logging was presented by Owen[20] as follows.

Capsule Sources

The alpha-neutron reaction is the most commonly used reaction to obtain simple neutron sources. When one gram of radium is mixed with several grams of powdered beryllium, about 10^{10} neutrons per second are emitted as a result of the alpha-neutron reaction, as shown in Figure 8-11. A simple way of stating this type of reaction is $Be^9(\alpha,n)C^{12}$. Radium and its decay products emit alpha particles with energies from 4.79 MEV to 7.68 MEV and the neutrons have energies from about 1 MEV to 13 MEV, with the greatest portion of the neutrons having energies below 4 MEV. The half-life of radium is long, about 1620 years, and the RaBe mixture provides neutrons at a sufficiently steady rate to be used as a standard of neutron emission. RaBe sources also have very high gamma ray production, which is one un-

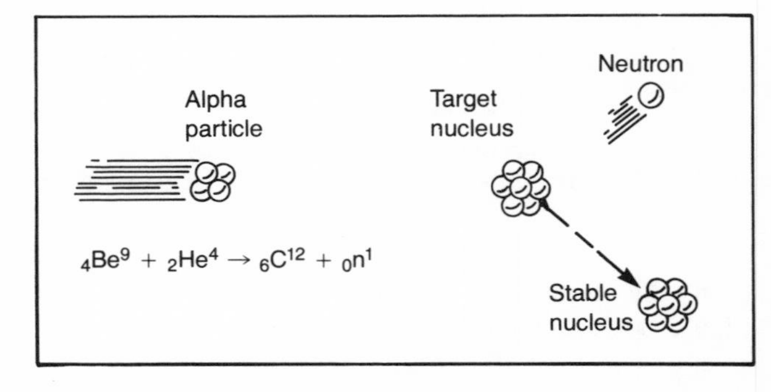

Fig. 8-11. Neutron production resulting from alpha particle bombarding beryllium. From Nuclear Power Engineering *by Henry J. Schwenk. Copyright 1957, McGraw-Hill Book Company. Used with the permission of McGraw-Hill Book Company.*[6]

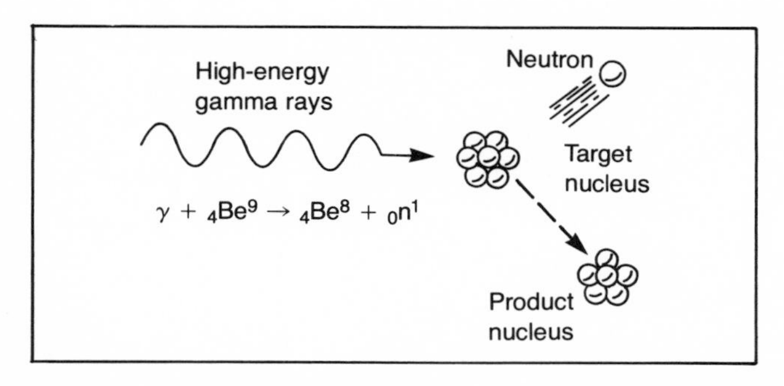

Fig. 8-12. High-energy gamma ray impinging on a beryllium-9 nucleus resulting in neutron ejection. From Nuclear Power Engineering *by Henry J. Schwenk. Copyright 1957, McGraw-Hill Book Company. Used with the permission of McGraw-Hill Book Company.*[6]

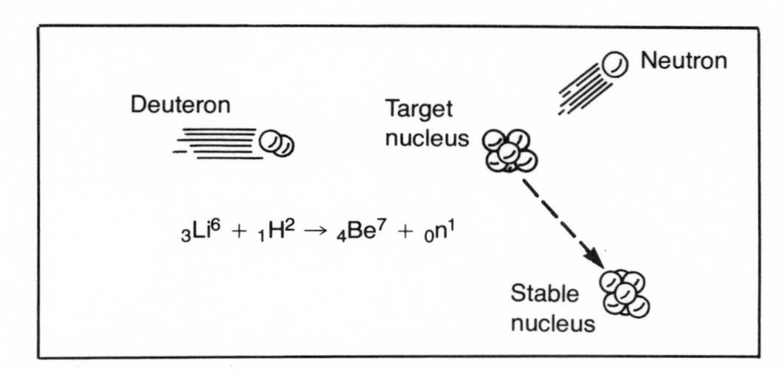

Fig. 8-13. Deuteron bombardment of lithium-6 target resulting in neutron production. From Nuclear Power Engineering *by Henry J. Schwenk. Copyright 1957, McGraw-Hill Book Company. Used with the permission of McGraw-Hill Book Company.*[6]

desirable feature in certain applications. An improved capsule source is the plutonium-beryllium source. PuBe sources have the advantage of low gamma ray output, with very long half-life ($T_{1/2} = 24{,}000$ years).

Neutrons can also be produced by a gamma-neutron reaction where neutrons are "knocked out" of nuclei by photons or gamma rays (see Fig. 8-12). If beryllium (Be^9) is the target nucleus, the photon energy must exceed 1.67 MEV. Gamma rays of this energy are not common in naturally occurring radioactive species, so it is most common to use as the gamma source a radionuclide that has been artificially made in a nuclear reactor. For example, Na^{24}, a radionuclide of sodium, made by bombarding aluminum with neutrons, has a gamma ray energy of 2.76 MEV. This is sufficient to produce neutrons of about 1.0 MEV if beryllium is the target nucleus. Neutrons from this type of source are more monoenergetic than those from capsule sources utilizing the alpha-neutron reaction, and thus might find use in logging in spite of the shortcoming of a short half-life.

Accelerator Sources

In addition to the reactions involved in the capsule sources, there is a whole family of charged particle reactions that will produce neutrons in large quantities. Among the most useful of these are the so-called d-d reaction or

$$H^2 + d \rightarrow He^3 + n + Q \quad \text{where } Q = 3.29 \text{ MEV}$$

and the d-t reaction or

$$H^3 + d \rightarrow He^4 + n + Q \quad \text{where } Q = 17.6 \text{ MEV}$$

All that is required for these reactions to "go" is a means for acclerating the deuterons (nucleus of deuterium or heavy hydrogen) to a sufficiently high velocity (energy). This is reasonably easy to do with a high voltage source such as a Van De Graaff or Cockroft-Walton accelerator. The d-t reaction produces neutrons of approximately 14 MEV energy, with a neutron production of roughly 10^8 neutrons per second per microampere of deuteron beam (deuteron energy of 125 KEV). The energy of the d-d reaction neutrons is approximately 2.5 MEV, and roughly 2×10^6 neutrons per second per microampere of deuteron beam (deuteron energy of 300 KEV) are produced. Deuteron bombardment of a lithium-6 target is descriptively shown in Figure 8-13.

NEUTRON INTERACTIONS

Neutrons, for the most part, interact with nuclei in two ways: (1) scattering or (2) capture. The type of interaction that will occur depends primarily on the energy of the neutron and the nucleus involved. It is advantageous to treat neutron interactions in terms of the energy range in which they predominate. Owens[20] has provided a convenient classification scheme: "Neutrons above 100 KEV are called fast; those in the 100 eV to 100 KEV energy bracket, intermediate; those from about .025 eV to 100 eV, slow (sometimes called epithermal); while neutrons of about .025 eV energy are known as thermal neutrons."

It is possible that neutrons may be captured at any energy, but it is more probable that slow or thermal neutrons will be captured than it is that fast neutrons will be captured. When a nucleus captures a neutron, energy is added to the nucleus in the form of internal energy. In order to become stable with respect to energy, the added energy is generally given off immediately in the form of gamma rays with the energy of these gamma rays being characteristic of the nucleus. Hydrogen nuclei, for example, give off gamma rays having an energy of 2.23 MEV when they capture a neutron.

When a neutron is scattered, the neutron collides with a nucleus similar to the collision of two billiard balls, and the neutron bounces off. Two types of collision, elastic (Fig. 8-14) or inelastic (Fig. 8-15), can occur.

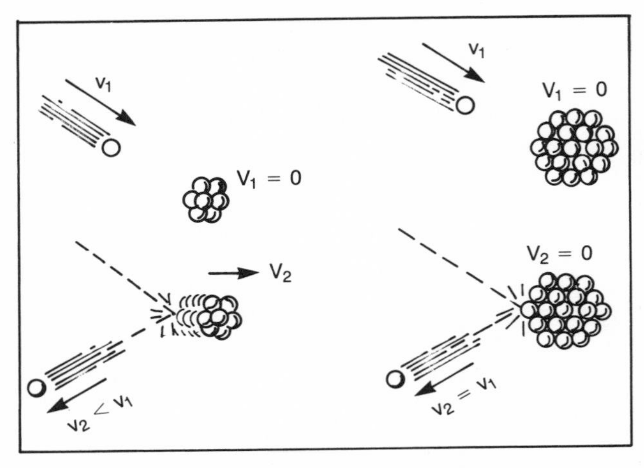

Fig. 8-14. Elastic collisions involving some loss of energy (small nucleus) and no loss of energy (heavy nucleus). From Nuclear Power Engineering by Henry J. Schwenk. Copyright 1957, McGraw-Hill Book Company. Used with the permission of McGraw-Hill Book Company.[6]

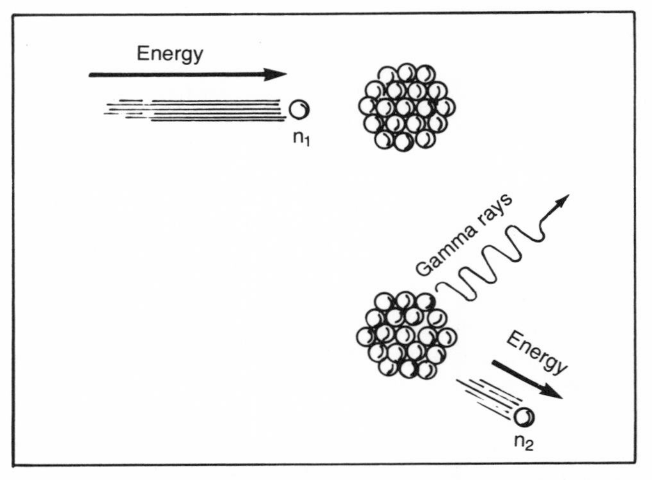

Fig. 8-15. Inelastic scattering where target nucleus absorbs fast neutron and emits a slower neutron and gamma rays. From Nuclear Power Engineering by Henry J. Schwenk. Copyright 1957, McGraw-Hill Book Company. Used with the permission of McGraw-Hill Book Company.[6]

No radiation of any type results from an elastic collision—only loss of energy by the neutron to the nucleus in the form of kinetic energy. The internal energy transferred from the neutron to the nucleus in inelastic collisions is generally well-known with the emitted gamma ray energy being indicative of the type of nucleus. Therefore, the gamma ray energy resulting from inelastic scattering of neutrons may be used to identify the target nucleus, just as the gamma rays resulting from neutron capture may also be used for the same purpose. There is no simple way to determine whether the gamma rays observed are resulting from neutron capture or neutron inelastic scattering.

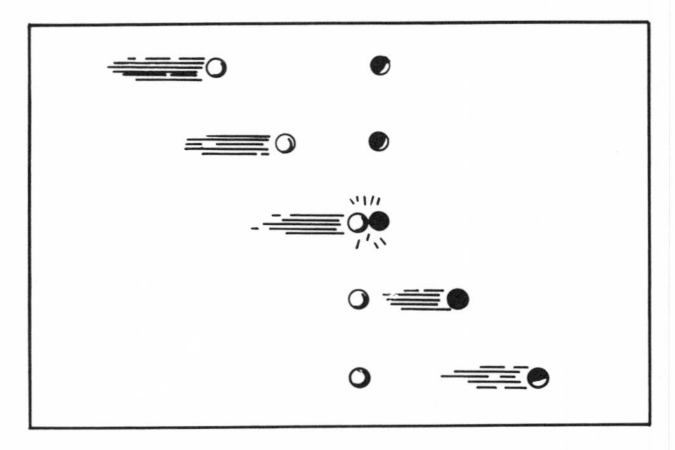

Fig. 8-16. Neutron striking a target nucleus of about equal mass giving up all of its energy. From Nuclear Power Engineering by Henry J. Schwenk. Copyright 1957, McGraw-Hill Book Company. Used with the permission of McGraw-Hill Book Company.[6]

The energy of the neutrons is a very important factor in considering inelastic scattering. For each nucleus, there is a certain threshold energy below which the inelastic scattering reaction will not take place. In general, the neutron energy required is at least as high as the gamma ray, which results from the reaction. As an example, when a neutron scatters inelastically from an oxygen nucleus, the gamma ray resulting from this reaction is 6.09 MEV. Therefore, the neutron must have had at least $^{17}\!/_{16}$ 6.09 MEV or 6.46 MEV energy.[20] This means that if the neutrons that bombard a sample containing oxygen do not have at least 6.46 MEV energy, there can be no inelastic scattering, and no gamma rays will be produced. The threshold for the inelastic scattering of neutrons from carbon is 4.79 MEV.

The energy lost by the neutron in inelastic collisions is dependent on the relative masses of the neutron and the target nucleus. The heavier the nucleus, the less the energy loss in each neutron-nucleus collision. If the atom is hydrogen, which has only a proton as its nucleus, the neutron may lose up to its total energy in the collision, since a proton and a neutron have approximately the same mass (see Fig. 8-16). On the average, the neutrons will lose approximately 67% of their energy in collisions with hydrogen nuclei. If the atom is some heavier element such as calcium, oxygen, or silicon, the neutron will lose very little of its energy with each collision and thus will have to undergo many more collisions before coming to "rest" or thermal energy. Actually, neutrons at thermal energy have an average velocity of 220 m/sec.

The typical life history of a neutron is descriptively shown in Figure 8-17. A fast neutron emitted at a source undergoes interactions of the elastic and inelastic kind

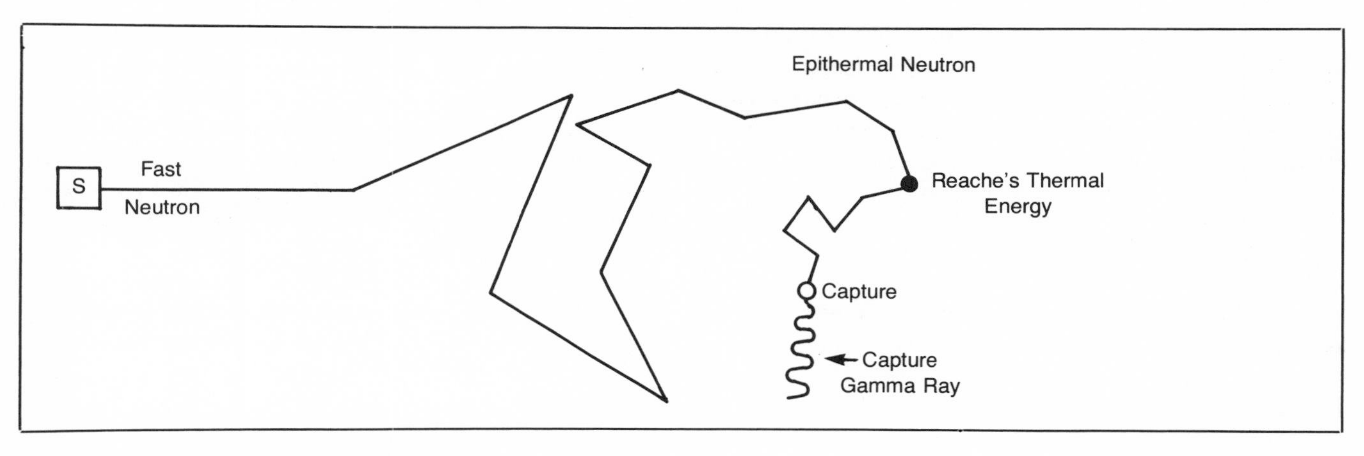

Fig. 8-17. Typical life history of a neutron

with nuclei existing in the bulk volume through which it passes. The greater the number of elastic collisions, the greater the kinetic energy loss of the neutron. Since elastic collisions predominantly occur when neutrons strike the smaller nuclei, elements such as hydrogen, carbon and beryllium are very effective moderators (effective at reducing the fast neutron energy level through elastic type collisions).[6] A neutron, undergoing elastic and inelastic collisions, will lose energy, becoming a slow (or epithermal) neutron, and can ultimately reach the thermal energy level (same average kinetic energy as the surrounding nuclei). At this energy level, the neutron can continue to undergo collisions but without any change in average energy—the thermal diffusion phase. The neutron will not continue to undergo collisions indefinitely, since eventually it will undergo a collision where it is not scattered by the nuclei but absorbed into the nucleus (neutron induced transmutation reaction). The most common process is radiative capture wherein the target nucleus accepting the neutron will emit its excess energy in the form of gamma radiation (capture gamma rays). Virtually all elements except some of the very lightest exhibit radiative capture of slow neutrons. The product is always isotopic with the target element, but its mass number is one unit higher such that: ${}_1H^1 + {}_0n^1 \rightarrow {}_1H^2 + \gamma$, which represents the radiative capture of a neutron by hydrogen, which transmutes to deuterium with the emission of gamma radiation. The minimum energy necessary for neutron capture by hydrogen is known to be 2.225 MEV; it follows that in this reaction the energy of the capture gamma radiation will be at least 2.225 MEV.[8] The high energy capture gamma radiation emitted after thermal neutron capture depends on the particular nucleus involved, with each nucleus having a capture gamma-ray emission spectrum that is distinctive. Gamma rays emitted as a result of neutron capture usually have energies of several MEV, a much higher energy level than gamma rays emitted in the natural decay process (see Table 8-2). It should be pointed out that neutrons can be captured at any energy level but the probability of capture increases as the energy level decreases, with radiative capture being the dominant radiation process.

NUCLEAR CROSS-SECTION

As neutrons move through matter, certain interactions occur. From the preceding section we know that the nature of the substance and the energy of the neutron will make certain processes more likely to occur. In discussing neutron interactions, the term cross-section, denoted by σ, is used to express the probability that a neutron will interact with a given substance. This probability is defined as the nuclear cross-section. The chance of a collision between a neutron and nucleus is represented by the effective target area of the bombarded nucleus. Figure 8-18 shows a beam of neutrons striking a target material during a given period. In one square centimeter there would be billions of nuclei and millions of neutrons involved. The cross-section of this material, σ, would be expressed as:[6]

$$\sigma = \frac{A}{NI}$$

where: σ = cross-section, the average effective target area that each individual nucleus presents to the oncoming neutron, cm²/nucleus
A = number of nuclei actually struck and transmuted
I = number of neutrons entering one cm² of target material
N = total number of nuclei in the target material, nuclei/cm²

The fraction A/I measures the proportion of entering neutrons that react with nuclei. Dividing the ratio A/I by the number of nuclei/cm², N, gives the cm²/nucleus,

σ, that participates in the reactions, on the average. Since a neutron will be either absorbed or scattered, the cross-section of a material, therefore, includes two kinds of events. As a result, the cross-section, σ, can be expressed as:

$$\sigma = \sigma_{sc} + \sigma_a$$

where: σ_{sc} = scattering cross-section
σ_a = absorption cross-section

Cross-section is a physical property of a material, just as density and resistivity are properties.

The nuclear cross-section, σ, is called the *microscopic cross-section*. The diameter of nuclei being about 10^{-12} cm means that many projected nuclear areas are about 10^{-24} cm^2. To keep the numbers from becoming too cumbersome, a unit of cross-section called the barn has been established. It equals 10^{-24} cm^2. In terms of nuclear dimensions, an area of only 10^{-24} cm^2 might be "as easy to hit as the broad side of a barn."[6]

Since there are a number of reactions that may take place when a material is bombarded with neutrons, it is necessary to specify the type of reaction involved when

TABLE 8-2
Nuclear Characteristics of Earth Elements

Element	Thermal Neutron Cross Section, σ_c (millibarns)	E_γ (Capture)* (mev)	Inelastic at E_n (millibarns)	(mev)	E_γ (Scattering) (mev)
Oxygen	0.2	—	104	7.06	6.09
			500	14.0	6.09
Hydrogen	330	2.23	—	—	—
Carbon	3.3	4.9 (1)	350	6.58	4.43
		3.7 (2)	245	14.1	4.43
Calcium	430	6.4 (2)	100	3.95	3.9; 3.74
		4.4 (4)			
		2.0 (3)			
		1.94 (1)			
Silicon	130	6.4 (5)	370	2.5	1.78
		4.9 (1)			
		4.2 (4)			
		3.5 (3)			
		2.7 (2)			
Chlorine	32000	7.7 (5)	600	2.5	1.23; 1.77
		7.4 (4)			
		6.6 (3)			
		6.12 (2)			
		3.0 (6)			
		1.97 (1)			
Magnesium	63	3.9 (1)	485	2.56	1.36
		2.8 (2)			
		1.87 (3)			
Aluminum	230	7.7 (1)	64		.84
		4.8 (3)	142	2.56	1.017
		2.9 (2)	87		2.21
Sulfur	4490	5.4 (1)	173	2.5	2.23
		4.8 (4)			
		3.0 (2)			
		2.4 (3)			
Sodium	505	6.4 (3)	530	2.5	.45; 1.69; 2.2
		3.9 (2)			
		3.6 (1)			
		2.2 (4)			
Iron	2430	7.64 (5)	900	2.56	.85; 1.25; 2.1
		7.27 (4)			
		6.43 (3)			
		5.92 (2)			
		5.5 (1)			

* These energy levels represent groups of gamma rays in some cases. The gamma rays listed are not all those present, only the most predominant are shown. The numbers in parentheses refer to the relative intensity of the gamma ray groups, with (1) being the largest and so on.
Permission to publish by the Society of Professional Well Log Analysts.[20]

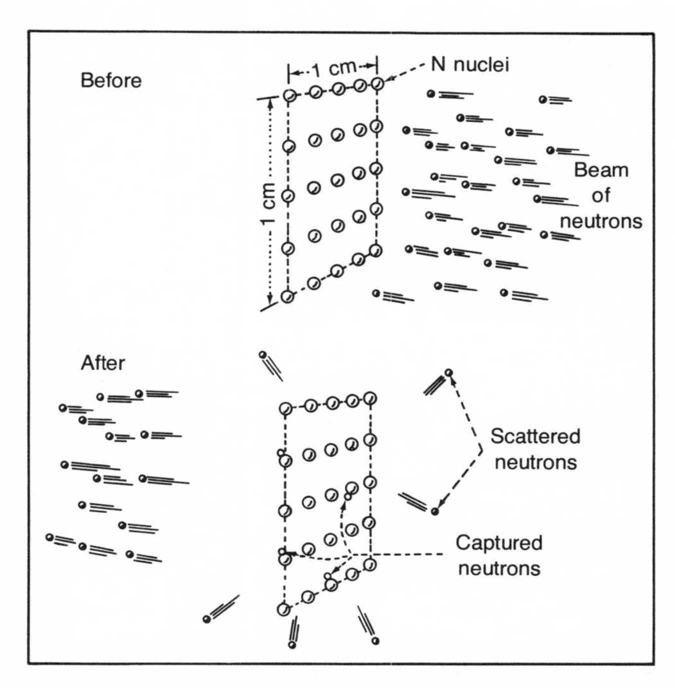

Fig. 8-18. Beam of neutrons interacting with target material. From Nuclear Power Engineering *by Henry J. Schwenk. Copyright 1957, McGraw-Hill Book Company. Used with the permission of McGraw-Hill Book Company.*[6]

talking about the cross-section. Therefore, the total cross-section of a nucleus, σ_t, to neutrons is made up of the cross-section to elastic scattering, $\sigma_{elastic}$, the cross-section to inelastic scattering, $\sigma_{inelastic}$, the cross-section to capture, $\sigma_{capture}$, and the sum of all the other reaction cross-sections, which we will lump together as $\sigma_{reaction}$. From a practical viewpoint, since only inelastic scattering and neutron capture result in radiation that may be observed in the borehole, we will concern ourselves only with the cross-sections of these two reactions. Elastic scattering is important as a factor in certain types of logs, such as neutron-neutron logs, primarily because it is the predominant mechanism for moderating neutron energy. Since the probability for a nuclear reaction to occur is a function of the energy of the bombarding particle, it is necessary to specify the energy at which a cross-section is measured.

There is another measure called the macroscopic cross-section related to the volume of material. It is figured as $\Sigma = N\sigma$ where $N =$ nuclei/cm^3 of the material. Sigma, Σ, is a measure of the total cross-section of the nuclei in one cm^3 of the material.

RADIATION DETECTORS

Instruments currently used in well logging can be divided into two classes. One class of detectors relies on the ionization of a gas, while the other class employs the photo-emission properties of certain phosphors. The first class consists of an ionization chamber, a Geiger-Mueller counter, and a proportional counter; the second class consists of a scintillation counter.

Ionization Chamber

Figure 8-19 schematically illustrates a gas-ionization chamber. It consists of a metal cylinder containing inert argon gas under high pressure. A rod passes through the cylinder and is insulated from it. This rod is maintained at a potential of about 100 v positive with respect to the cylinder.

The method of detection occurs as follows: When a gamma ray enters the chamber, it may interact with the wall material or inert gas, giving rise to a fast-moving electron. This electron, as it moves through the gas, experiences a number of collisions with electrons of the gas atoms. These collisions slow down the fast-moving electron, but, in the process, a number of other electrons are released. This process is called ionization. These negatively charged electrons move toward the positively charged center rod, constituting a minute flow of electric current. The total current flow produced by the overall gamma-ray flux through the chamber is large enough to be measured. The current flowing in the circuit causes a potential drop across the resistor. Since this current is very small, the resistance of the resistor is made large (10^{11} ohms) so that the potential can be measured. This potential drop is amplified so the signal may be transmitted to the surface.

This type of detector is quite rugged, although the need to use a very high resistance that may be temperature-sensitive is a distinct drawback. Its low counting efficiency can be overcome by making the chamber large, but this reduces sensitivity to formation changes.

Geiger-Mueller Counter

The Geiger-Mueller counter is similar to the ionization chamber, except that the central electrode is kept at a much higher voltage (600 to 1,000 v), and the gas is at a low pressure.[31] The central electrode is made of very

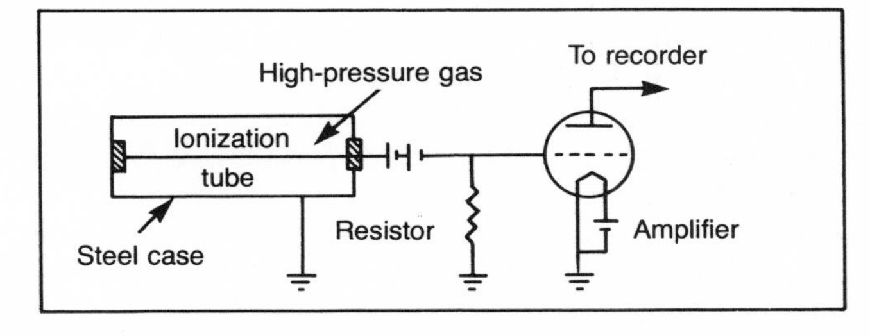

Fig. 8-19. Schematic of gas-ionization chamber. Permission to publish by Oil, Gas and Petrochem Equipment.[16]

small wire so that the potential gradient in its vicinity is high. The external resistance is also reduced to about 1,000 ohms. As in the ionization counter, ionization of the gas occurs in the same manner. In this counter, however, the secondary electrons produced by the collisions are drawn rapidly toward the central wire. In their travel to the wire, they reach a high enough energy to eject additional electrons from the gas atoms with which they collide. This causes gas ionization to be multiplied many times, to the extent that an avalanche of electrons reaches the wire for each one initially liberated in the gas. The resulting current flow through the external circuit is on the order of 10^8 times that which flowed in the ionization chamber. The actual pulse size is sensitive to the applied voltage, thus the number of actual pulses is measured instead of the ionization current. This detector can count about 5,000 pulses/second.

The main advantage of the Geiger-Mueller counter is that it produces large pulses that are easily detected and transmitted. The main disadvantages are that it is difficult to build, it is inefficient (comparable to the ionization chamber), requires good voltage control, and, due to its small size, the central wire is vibration sensitive.

Proportional Counter

The proportional counter is similar to the Geiger-Mueller counter in construction, but operates at a lower voltage. It is usually used to detect neutrons in logging operations. To do this, it is filled with boron trifluoride gas, enriched in ${}_5B^{10}$ isotope, or with a gas containing ${}_2He^3$. Ionization is caused by the absorption of a neutron by a boron atom, with the subsequent emission of an alpha particle. The released alpha particle causes heavy ionization to occur. The voltage on the central wire is so controlled that the avalanche of ionization occurs only over a small portion of the electrode. The size of the ionization avalanche is proportional to the number of ion pairs formed by the alpha particle. From this, it derives its name—proportional counter. Proportional counters count pulses rather than ion current, since the size of the pulse depends on the voltage supplied to the tube. A high voltage (about 500 v) is generally used; thus insulation is a problem. Accurate voltage control (as in the Geiger-Mueller counter) is unnecessary. If it is filled with boron trifluoride, the counter tends to become sensitive to wellbore temperature, which is a disadvantage.

The proportional counter has a recovery time on the order of one second; thus it can count particles arriving with very short time intervals (200 times faster than the Geiger-Mueller counter). It has a detection efficiency of about 10%.

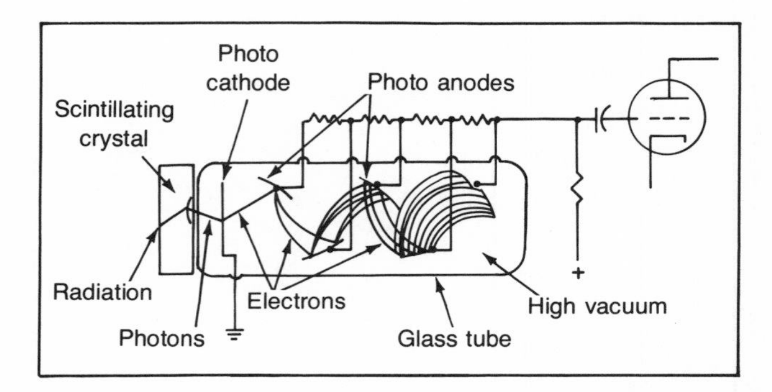

Fig. 8-20. Schematic of scintillation counter. Permission to publish by Oil, Gas and Petrochem Equipment.[16]

Scintillation Counters

Figure 8-20 schematically illustrates the operation of a scintillation counter. In this counter, a gamma ray entering the scintillating crystal (such as a thallium activated sodium iodide crystal) will interact with electrons in the crystal to produce a flash of light. This light in turn strikes the sensitive surface of a photocathode, which causes it to emit one or more primary electrons. This electron is drawn to the first anode, which it strikes, causing the emission of five or more secondary electrons. These electrons are accelerated to the next anode, which is at a higher voltage, and the process is repeated. By using a number of anodes, the current amplification becomes very large and can be handled by electronic amplifiers. The intensity of the original flash of light and the amplitude of the output are proportional. The intensity of the light flash is, in turn, proportional to the energy of the entering gamma ray, which makes it possible to count gamma rays of a given wavelength by separating only those pulses of a selected strength.

The sodium iodide crystals can be used for counting thermal neutrons when shielded with cadmium. Lithium iodide crystals can be used for counting a combination of thermal and epithermal neutrons.

The efficiency of these counters is very high (50 to 80%). Thus it is possible to use short counters and still obtain an improvement in formation detail (compared to gas-ionization types). The disadvantage of the scintillation counters is that they are expensive, sensitive to voltage changes, require high voltage on the photomultiplier, and the sensitive surface on the photocathode is affected by high temperatures and must either be insulated or refrigerated.

GAMMA RAY LOG

The gamma ray log measures the natural radioactivity

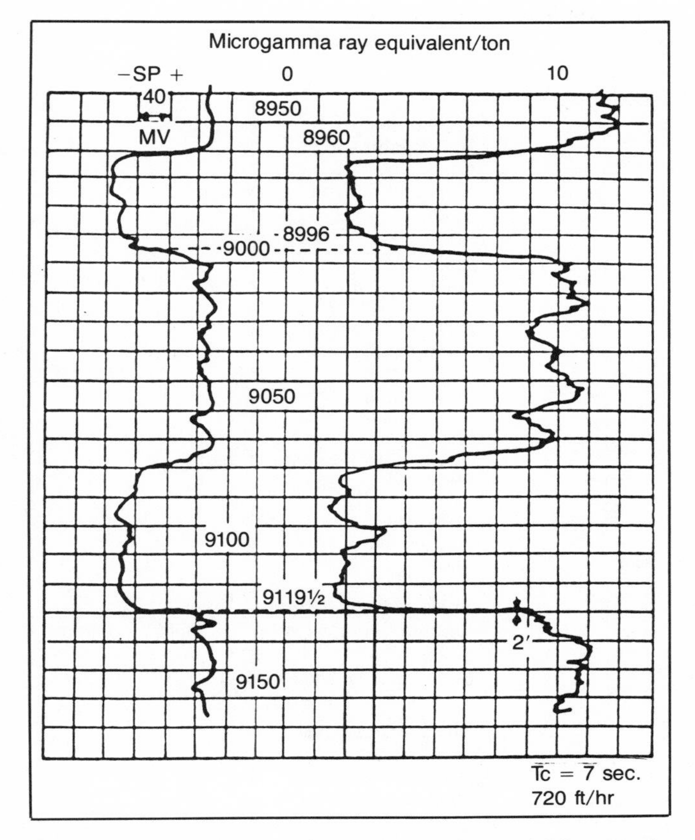

Fig. 8-21. Comparison of gamma ray log and SP curve in sand-shale series. Permission to publish by Oil, Gas and Petrochem Equipment.[16]

of rocks. As previously discussed, all rocks contain some radioactive material, although it can be seen in Figure 8-7 that shales and particularly the marine shales have a higher gamma ray emission level than sandstone, limestone, and dolomite. This difference makes the gamma ray log especially useful for distinguishing shales from nonshales and, therefore, the gamma ray log is principally a lithology log. As such it can be used in place of the SP when the SP is not diagnostic, i.e., in salt muds, oil base muds, air or gas drilled holes, and cased holes. An example of the similarity that may exist between the SP and gamma ray log is shown in Figure 8-21. Typical gamma ray responses are illustrated in Figure 8-22.

The gamma ray tool is relatively compact (5 ft to 6 ft) and consists of a detector for measuring the gamma radiation from the formation near the tool. The scintillation detector is most used since it is more efficient and the short length allows good formation definition.

Gamma ray logs are now recorded in API units, although in the past, gamma ray logs were calibrated in a number of different units (one such unit is shown in Figure 8-21). In order to standardize the units of measurement, the API established an empirical calibration standard at the University of Houston. As shown in Figure 8-23, the test pit contains two sections of neat cement, one having a high and one having a low radioactivity. The difference in the radioactivity of the two sections is defined as 200 API gamma ray units. The recommended calibration procedure is presented in API RP 33.[10]

Factors Affecting Gamma Ray Log Response

A number of factors influence the response of the gamma ray log. A partial list of these factors would include: (1) type of detector, (2) logging speed and time constant, (3) borehole size, (4) borehole fluid, (5) type of casing in borehole, (6) amount and type of cement, (7) formation thickness and radioactivity, (8) adjacent bed thickness and radioactivity, and (9) statistical variations.

Type of Detector. The various types of detectors obviously will respond differently and can not be compared directly. For any particular detector, the larger it is the more gamma rays it will count. Also, as the detection efficiency increases, the smaller the detector required; i.e., a small scintillation counter can record as many gamma rays as a larger Geiger-Mueller counter.

Logging Speed and Time Constant. A condenser is used in the logging circuit in order to smooth out statistical variations. This introduces a lag in the recording equipment characterized by a time constant. The time constant is the time it takes 63% of any change in intensity to reach the recorder. Increasing the time constant of an instrument decreases the error due to statistical variations. Due to this condenser in the circuit, the speed at which the log is recorded affects the appearance of the log. Faster speeds cause smoothing of the curves, reduce peaks of the thinner strata, and show bed contracts slightly high.

The time constant and logging speed should be considered together. For example, a logging speed of ½ ft/sec with a time constant of 4.0 would enable a two-foot thick bed to register 63% of the deflection that would be recorded with a stationary detector.

Borehole Size. Normal borehole variations do not affect the gamma ray log appreciably; however, large caved-in zones may cause a slight decrease in the gamma ray response.

Fluid in the Hole. In most cases, the effect of fluid in the wellbore cannot be detected on the log. If the hole diameter is large, however, the gamma ray radioactivity may increase when emerging from liquid.

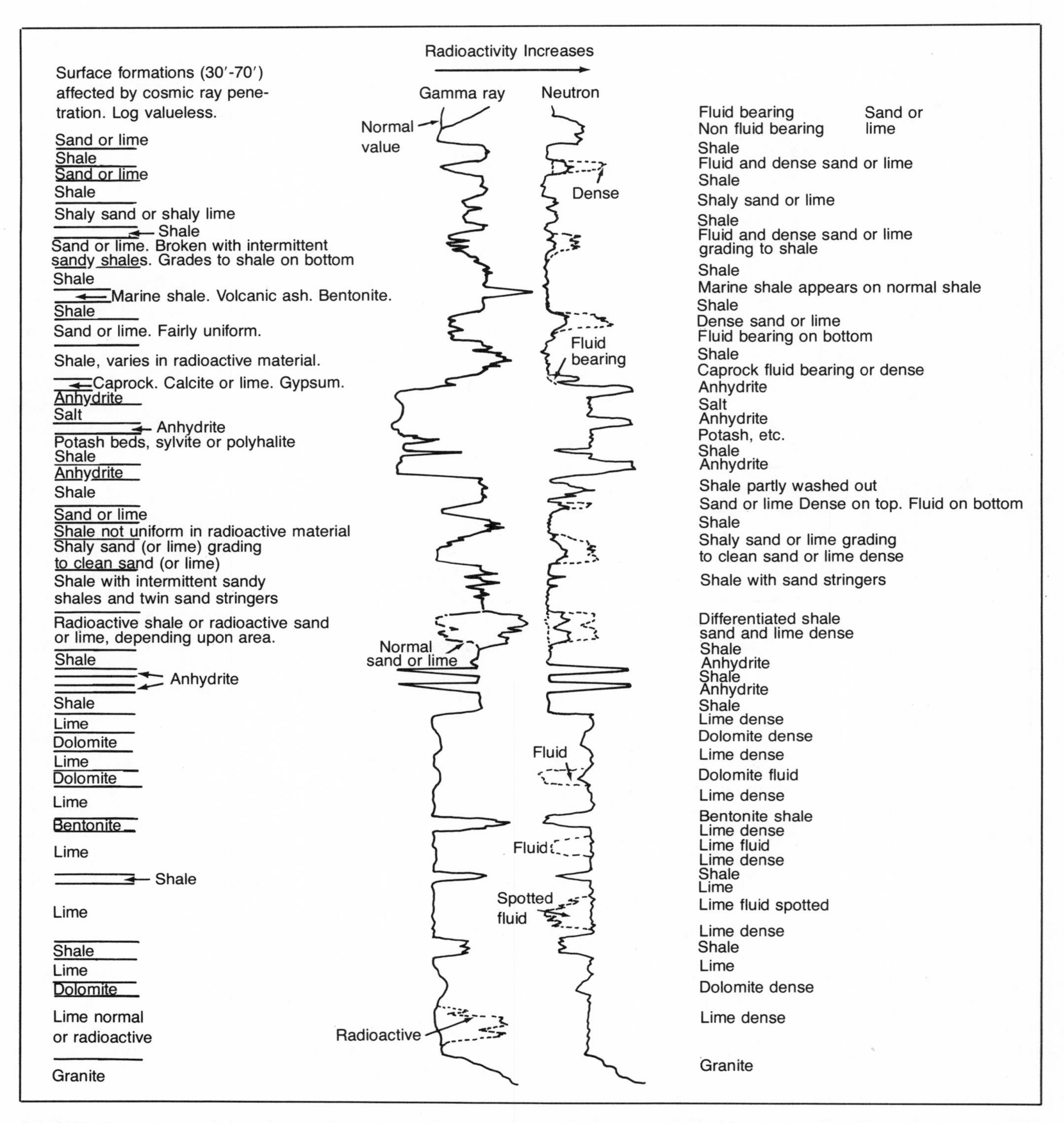

Fig. 8-22. Typical gamma ray and neutron response curves to different types of formations. Permission to publish by Dresser Atlas, Dresser Industries.[36]

Casing. Casing reduces the amount of gamma ray radioactivity by about 30%. Casing effects, however, may not be noticeable in rocks of varying radioactivity or if the casing is set, as in the usual case, in rocks of low radioactivity.

Cement. Cement is made from limestone and shale, and since most shales are appreciably radioactive, the presence of cement behind casing will affect the recorded gamma ray response. It is possible that the cement may increase the response opposite weakly radio-

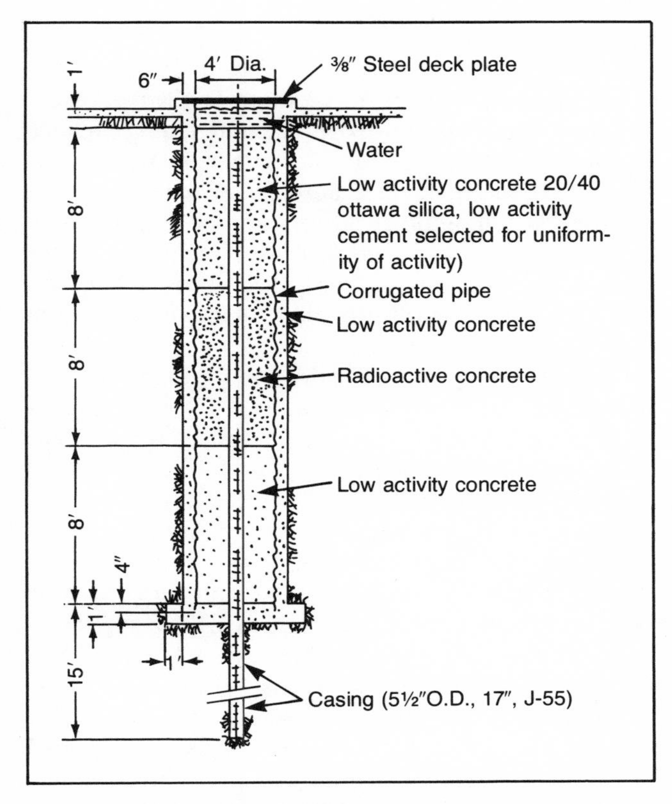

Fig. 8-23. Gamma ray calibration pit. Permission to publish by the American Petroleum Institute.[10]

active formations and decrease the response opposite strongly radioactive formations.

Formation Thickness. Thin beds, as mentioned before, do not show the same deflection as would a similar thick bed to the averaging effect of the detector. No bed thinner than detector length will register correctly.

Statistical Fluctuations. Radioactive disintegration and absorption is a random process, and because of this, the recording of gamma ray radioactivity is subject to statistical variations. The overall appearance of the gamma ray log will appear the same, but the small features and variations in the response will probably vary on logs recorded over the same interval in the hole.

Application of the Gamma Ray Log

The gamma ray log has a number of uses: (1) estimate bed boundaries, (2) determine lithology with local knowledge, (3) estimate shale content, (4) perforate depth control in perforating cased holes, and (5) trace fluid movement.

For estimating shale content, the type of shale and its radioactive content must be constant in a stratigraphic series. The shale content, V_{sh}, can then be estimated empirically using Figure 8-24, where:

$$V_{sh} = f(I_{shGR})$$

and:

$$I_{shGR} = \frac{\gamma_{log} - \gamma_{cn}}{\gamma_{sh} - \gamma_{cn}}$$

where: γ_{log} = Gamma ray log response opposite formation of interest

γ_{sh} = Gamma ray log response opposite the nearby shale formation

γ_{cn} = Gamma ray log response opposite a clean sand or carbonate

V_{sh} = Percent volume of shale contained in the formation of interest

When using the gamma ray as a shaliness indicator, the percent shale volume tends to be too high such that the gamma ray log should be used with other shaliness indicators whenever possible.

If the gamma ray log is to be used quantitatively, then it must be remembered that its response is a function of the borehole environment, and the various factors affecting this response must be considered. Typical departure curves for considering these effects for open hole conditions are shown in Figures 8-25 and 8-26.

Perforating Depth Control. Depth control is one of the most important aspects of initial completion work and any subsequent production-logging operations performed on the well. The technique is simply a precise tie-in of the casing-collar depths with the electrical-log

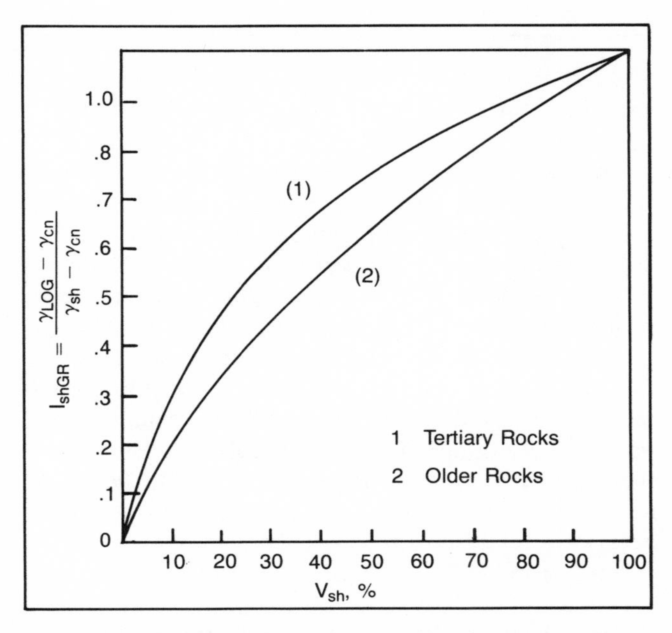

Fig. 8-24. Gamma ray index, I_{shGR}, versus V_{sh}. Permission to publish by Dresser Atlas, Dresser Industries.[36]

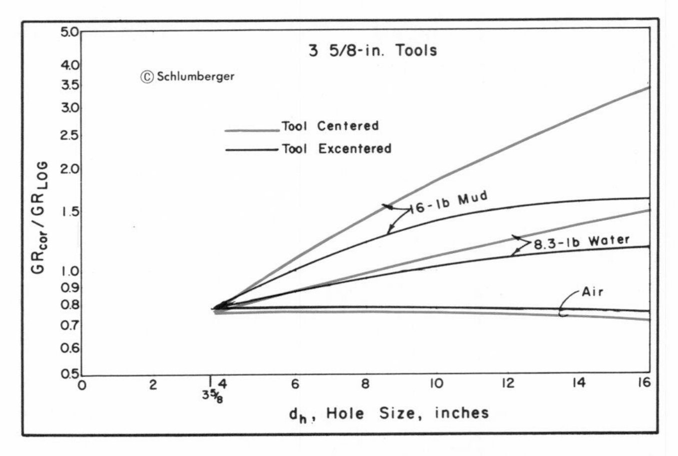

Fig. 8-25. Gamma ray corrections for hole size and mud weight in open hole. Permission to publish by Schlumberger Ltd.[38]

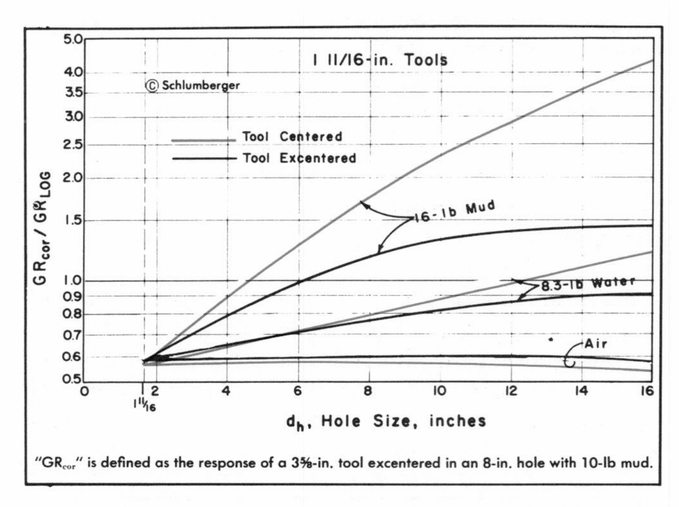

Fig. 8-26. Gamma ray corrections for hole size and mud weight in open hole. Permission to publish by Schlumberger Ltd.[38]

depths, without having to refer to measurements from the surface. The method consists basically (see Fig. 8-27) of running in combination a gamma ray casing-collar locator log in the cased hole. The gamma ray log is then correlated with the open spontaneous potential or gamma ray, thus locating the casing-collars in direct relationship with the electrical log. In some areas with low radioactivity contrast, it is necessary to run an intermediate log combination of the SP and gamma ray in open hole to obtain a reliable correlation. By running the collar locator with the perforating gun, the collars are relocated, and the well can be perforated at a depth accurately referred to the electrical log.

The casing-collar locator measures the magnetic properties of the casing. Since the presence of a collar changes the magnetic field of the casing, the presence of the collar can be readily recorded.

NEUTRON LOG

Neutron logs are primarily used to delineate porous formations and provide an estimate of their porosity. This log responds to the hydrogen content of the formation and, therefore, in clean formations the neutron log will reflect the liquid filled porosity. By comparing the neutron log with another porosity log or core data, gas zones can often be identified. Combinations of the neutron log with one or two other porosity logs will yield even more accurate porosity values and allow lithology identification.

There are three types of hydrogen logging tools commonly used: (1) neutron-gamma tool, (2) neutron-slow neutron tool, and (3) neutron-fast neutron tool. The fundamental principle involved in each of these tools is the same, the slowing down of neutrons by nuclei.

The basic principle involved in the hydrogen-neutron logging can be described as follows. In this tool, a neutron source bombards the formation with energetic neutrons. These neutrons are emitted at high speed and energy, and in their travel through the borehole and formation will experience numerous collisions with the nuclei present. If these nuclei are hydrogen, the neutrons are slowed down rapidly and can be captured (nearly all elements in nature can capture these slow neutrons or thermal neutrons). Once captured, a gamma ray of capture is emitted from the capturing element. As the hydrogen content of the material sur-

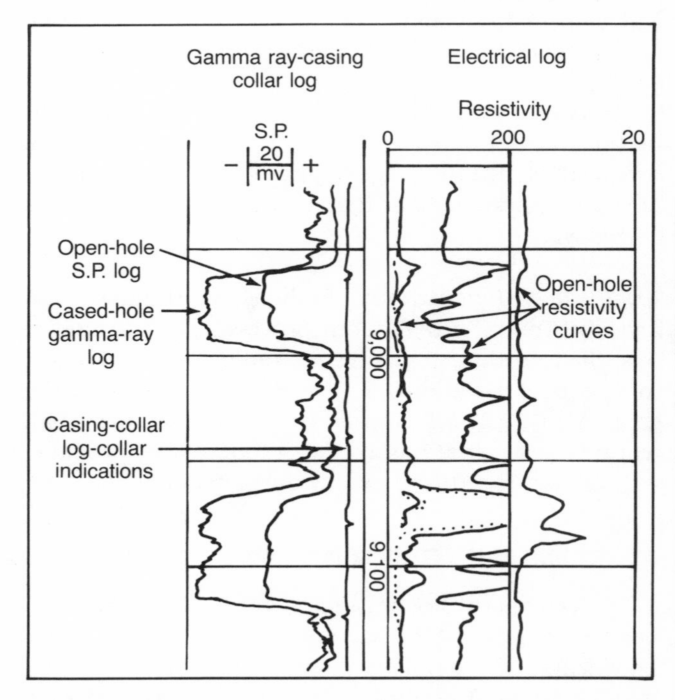

Fig. 8-27. Perforation depth control—open hole to cased hole tie-in. Permission to publish by Oil, Gas and Petrochem Equipment.[18]

rounding the source increases, the neutrons will be captured sooner.

This variation in capture "efficiency" of the surrounding media influences the response of the detector. For example, for the neutron-gamma tool, as the hydrogen content of the media increases, a corresponding decrease in the number of gamma rays of capture will be detected since, statistically, the gamma rays will be emitted closer to the source and would have to travel a greater distance to the detector. By the same token, the neutron-neutron type tools respond in the same manner since, as the hydrogen content around the tool increases, fewer fast or slow neutrons will reach the detector, resulting in lower count rates.

The detecting instrument used in the neutron-gamma tool can be either the ionization chamber, the Geiger-Mueller counter, or the scintillation counter. The neutron-slow neutron tool will use either a scintillation counter or the proportional counter containing boron. The scintillation counter is used for detecting fast neutrons. The components of a typical standard neutron-gamma logging tool are shown in Figure 8-28.

The hydrogen type of tools are particularly applicable for measuring porosities in the low range where the porosity is 15% or less. This is due to the logarithmic response of the tool. For high porosity values, the neutron curve has poor definition. These tools can be used in either open or cased holes; however, the presence of casing tends to decrease the counting rate and porosity resolution. This is because casing absorbs both neutrons and gamma rays. A typical gamma ray-standard neutron log is presented in Figure 8-29. As can be seen, the neutron curve is recorded in API units. The API unit is defined as 1/1000 of the deflection from zero recorded when a tool is located opposite the Indiana limestone section of the Neutron Calibration Pit (see Fig. 8-30) located at the University of Houston.

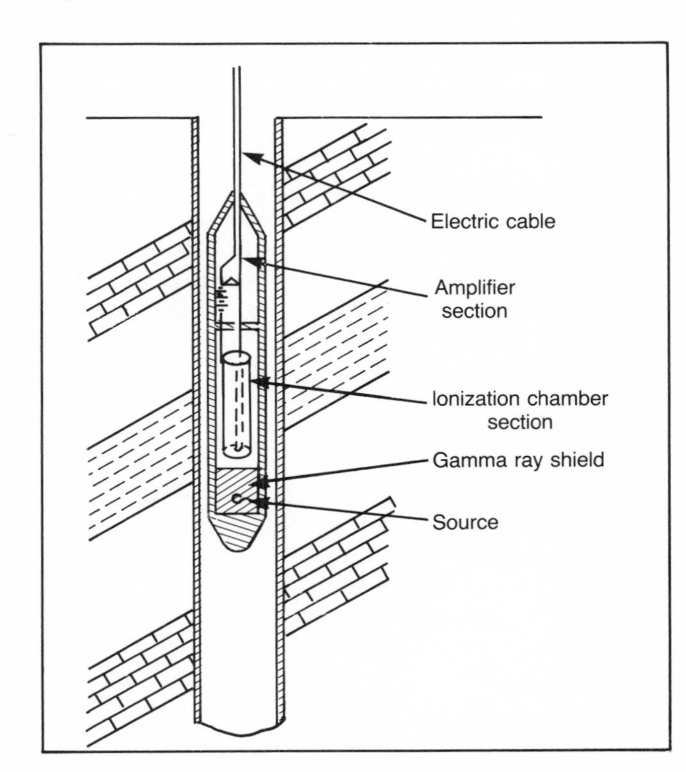

Fig. 8-28. Components of a standard neutron-gamma logging tool. Permission to publish by Oil, Gas and Petrochem Equipment.[17]

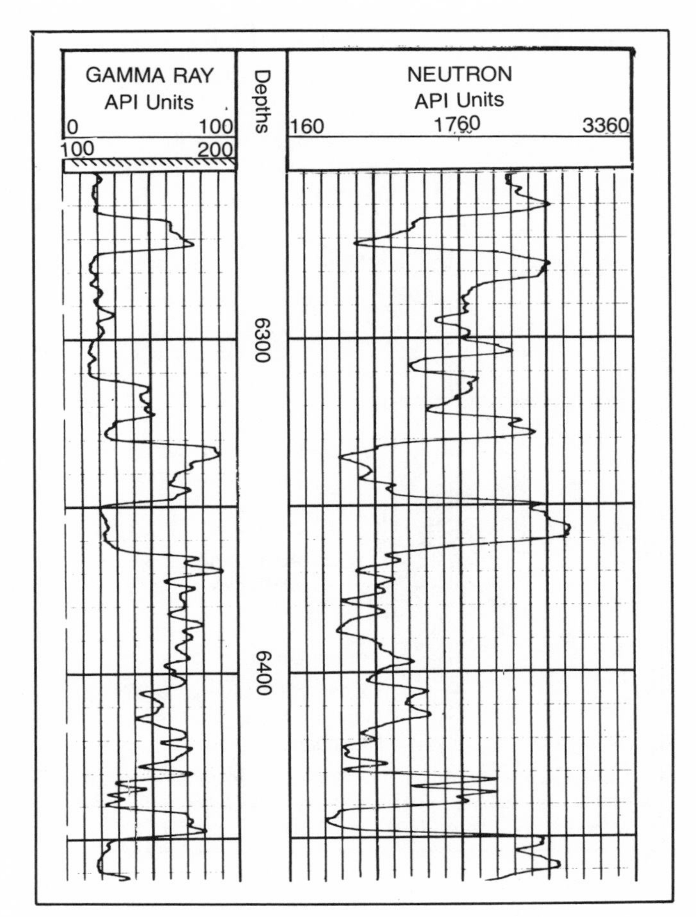

Fig. 8-29. Typical gamma ray-standard neutron log. Permission to publish by Schlumberger Ltd.[33]

Factors Affecting Standard Neutron Tool Response

A number of factors affect the standard neutron tool response including the following: (1) borehole environment, (2) source strength, (3) source-detector separation, (4) logging speed and time constant, (5) statistical variations, and (6) formation factors.

Borehole Environment. A number of factors are lumped into this category, which includes the borehole

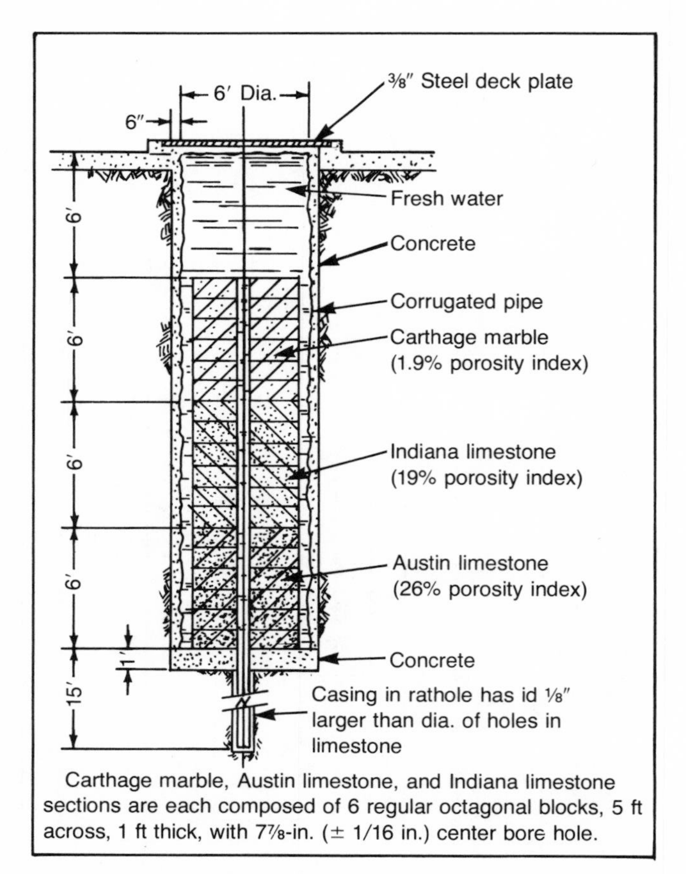

Fig. 8-30. Neutron calibration pit. Permission to publish by the American Petroleum Institute.[10]

size and fluid content, casing and cement. The effects of each of these are briefly summarized as follows.

1. *Borehole size and fluid content.* In a liquid-filled borehole, the effect of increasing size is to decrease the standard neutron log response. In a gas-filled borehole, neutron travel is increased with a resultant increased neutron log response. As the borehole size increases, the influence of the formation lessens with more neutrons reaching the detector directly through the borehole.
2. *Casing.* The presence of casing reduces the response of the tool.
3. *Cement.* Since cement has a high hydrogen content, its effect is to decrease the count rate. The extent of both the cement and casing effect depends on casing eccentricity and the relative size of the casing and borehole.

Source Strength. As the strength of the source is increased, the neutron log response increases, all other factors being equal.

Source-Detector Spacing. The effect of spacing is that as the separation between the source and detector increases, the neutron response is decreased, with all other factors remaining the same. At the same time, the radius of investigation would be increased.

Logging Speed and Time Constant. The significance of logging speed and time constant was presented in the section on gamma ray and is applicable here.

Statistical Variations. Being a radioactive tool, statistical variations will occur due to the random nature of the nuclear reactions. A statistical check should be recorded on every neutron log to indicate the tool's resolution.

Formation Factors. The influence of the formation on the standard neutron log is a function of formation lithology, porosity, and fluid content in the pore space. Each of these will be considered separately.

1. *Formation Lithology.* Since the neutron log responds to the presence of hydrogen, it should be pointed out that the following rock-forming minerals contain hydrogen: clay minerals, gypsum, iron hydrates, glauconite, coals, and bitumens. Each of these can cause erroneous interpretations of reservoir porosity when attempting to relate hydrogen content to porosity.
2. *Porosity.* It is assumed that the pore space contains fluid that is the source of hydrogen affecting the neutron response. The variations in tool response are, therefore, considered to be related to porosity.
3. *Fluid Content in Pore Space.* Any of the fluids contained in the pore space—oil, water, or gas—will contain hydrogen. The hydrogen concentration in oil and water is similar, and in constant porosity formations the neutron would probably not distinguish the oil-water contact. The hydrogen density of gas is less than for that of the two liquids and, if present in the zone of investigation, can be misinterpreted as low porosity if good mud filtrate flushing is assumed.

Depth of Investigation

Depth of investigation depends on several factors, which include source strength, source-detector spacing, and hydrogen concentration in the environment surrounding the tool (borehole and formation). Average depths of investigation for a typical standard neutron log in an 8-in. liquid-filled borehole can be considered to be:

Porosity	*Depth of Investigation*
0–0.05	24 in.
0.05–0.10	12 in.
0.10–0.20	6 in.
>0.20	very shallow

Porosity Evaluation

The neutron log is basically considered a porosity tool. It can also be used to define bed boundaries and, when used in conjunction with other logs, as an indicator of lithology and fluid in the pore space.

Several approaches are available in attempting to quantify the standard neutron response in terms of porosity: (1) calibration of neutron response versus core porosity, (2) service company departure curves, and (3) two-point log calibration.

Calibration of Standard Neutron Response Versus Core Data. This is the most reliable approach if good core data is available. The relationship between neutron response and depth correlated core porosity, can be expressed as:[7]

$$ND = C + D \log \phi$$

where: ND = neutron deflection, any recorded units although recent logs will be recorded in API units

C = intercept of linear trend at $\phi = 100\%$ (API or other recorded units)

D = slope of linear trend

Figure 8-31 shows the typical neutron deflection, ND, versus porosity, ϕ, relation obtained for two different rock types. Neutron deflections observed in the surrounding shales will probably project to an apparent porosity of from 35 to 45% as indicated by the apparent average linear trends shown on the crossplot.

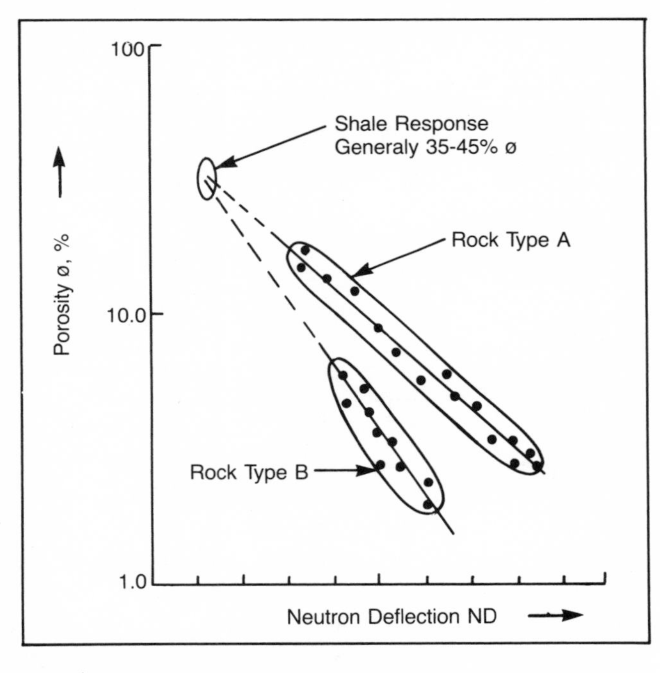

Fig. 8-31. Typical neutron deflection, ND, versus porosity, ϕ, crossplot.

Service Company Departure Curves. Service companies have developed departure curves relating porosity to neutron response for their various tools in specific environments.[4, 32] Figure 8-32 illustrates the typical neutron departure curves that are available. Equivalency charts, shown in Figures 8-33 and 8-33a, enable the limestone porosity index to be approximately converted to a porosity response more representative of the other two common reservoir rock types encountered.

Two-Point Log Calibration. In this approach, two data points are required such that the neutron response can be represented in porosity units. The most common reference porosities assumed are a shale porosity and a dense zone porosity. By assigning representative values applicable to the local area (such as shale $\phi = 40\%$, dense zone $\phi = 2\%$) a porosity-neutron relation can be defined as illustrated in Figure 8-34. A shale correction using the gamma ray could be included in this approach.

Limitations. The standard neutron log has a number of limitations:

1. Neutron log response is greatly affected by borehole conditions, and a caliper should be available to check borehole size.
2. Porosity measurements in cased holes must be considered to be semi-quantitative due to the many factors affecting neutron response.
3. Gas in the response volume could be interpreted as low porosity.
4. Porosity calibration of the neutron response is often unreliable.
5. Unrecognized changes in lithology may result in erroneous interpretation.

Neutron Deflection-Resistivity Relation

It was shown earlier in Chapter 7, "Acoustic Logging," that resistivity is related to porosity in the following form:

$$\log R_t = -m \log \phi + \log R_w + \log I_R$$

This concept can be extended using the neutron deflection-porosity relation presented earlier where:

$$ND = C + D \log \phi \quad \text{or} \quad \log \phi = \frac{ND - C}{D}$$

Substituting into the logarithmic expression for R_t, the following expression is obtained:

$$\log R_t = -m\left(\frac{ND - C}{D}\right) + \log R_w + \log I_R$$

or:

$$\log R_t = -m\left(\frac{ND}{D}\right) + m\left(\frac{C}{D}\right) + \log R_w + \log I_R$$

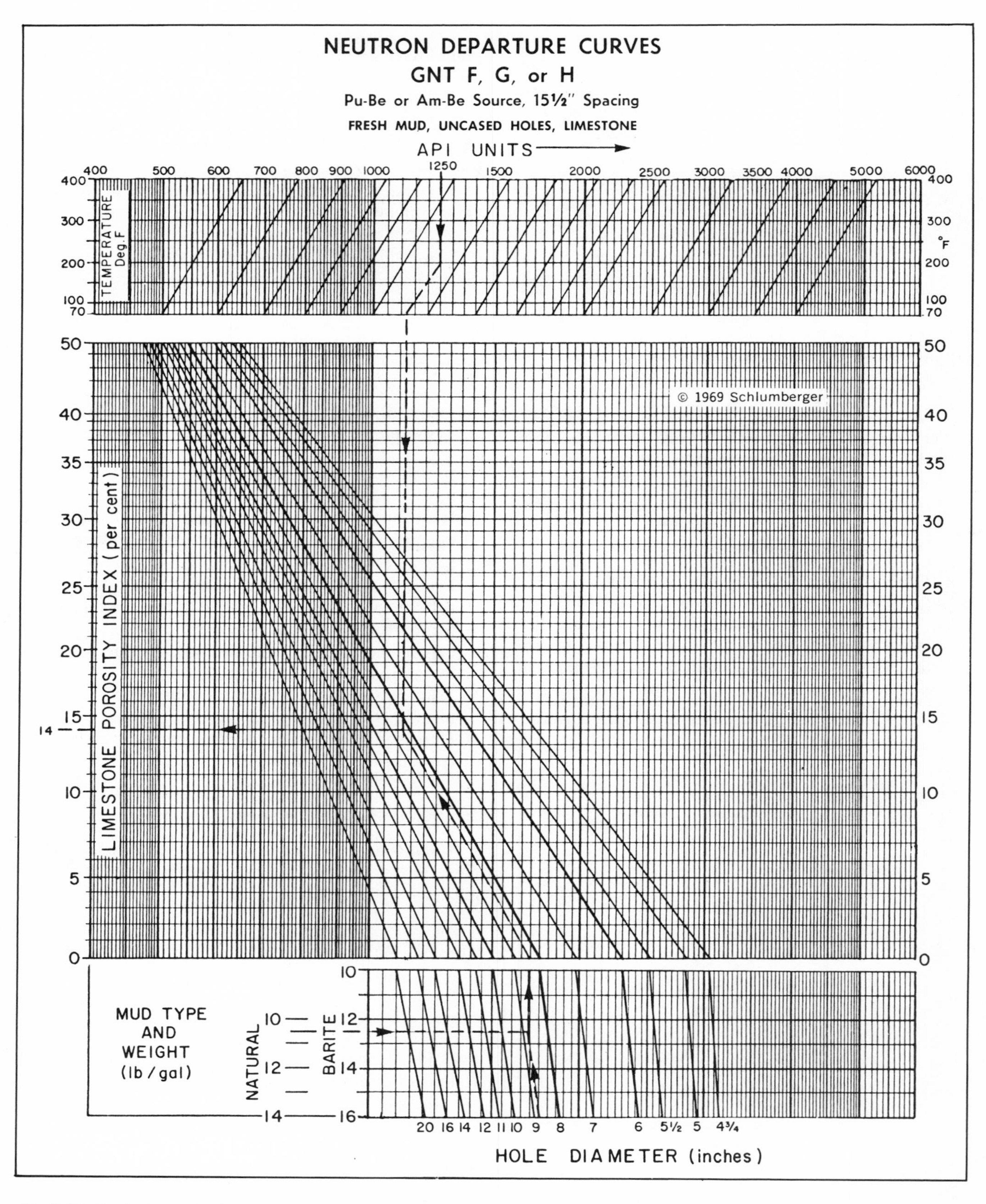

Fig. 8-32. Typical standard neutron log departure curve. Permission to publish by Schlumberger Ltd.[32]

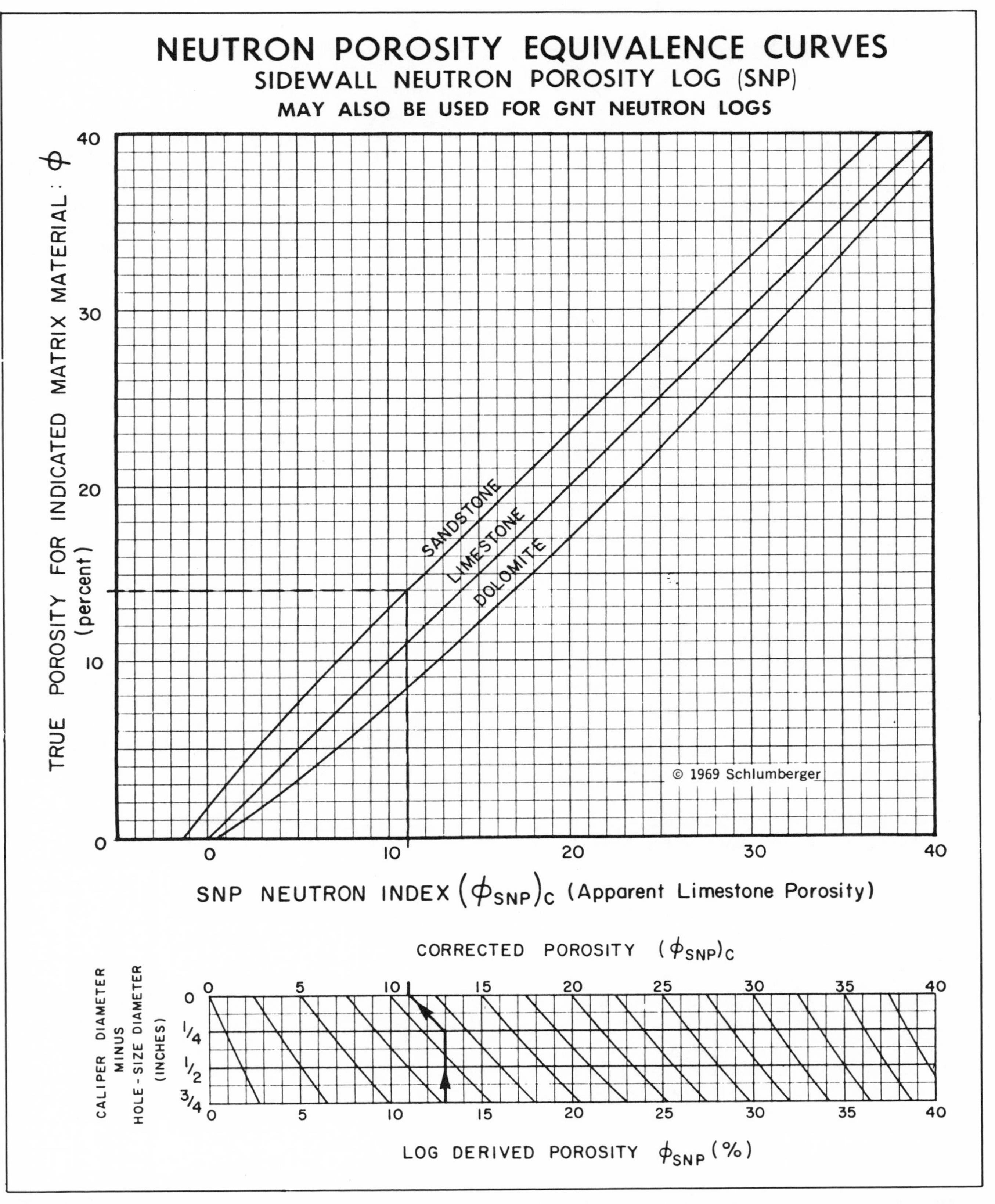

Fig. 8-33. Neutron porosity equivalency curves for standard neutron (GNT) and sidewall neutron (SNP). Permission to publish by Schlumberger Ltd.[32]

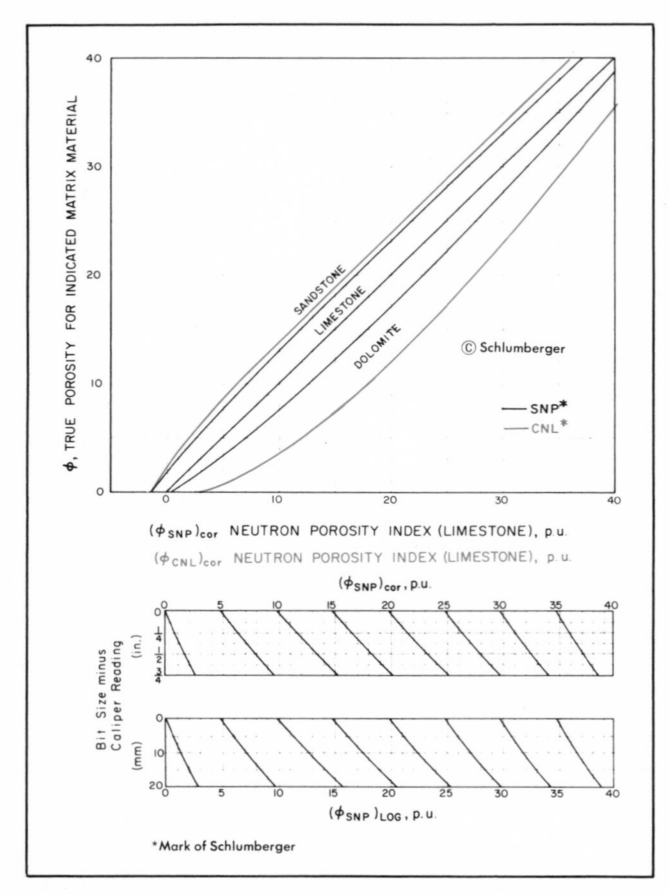

Fig. 8-33a. Neutron porosity equivalency curves for compensated neutron log (CNL) and sidewall neutron log (SNP). Permission to publish by Schlumberger Ltd.[38]

This expression states that a semi-log plot of R_t vs. ND should exhibit a linear trend for all zones of constant lithology (constant C and D) and constant R_w and I_R values. A typical R_t vs. ND crossplot is shown in Figure 8-35. As can be seen, the assumed water-bearing trend defines a locus of R_o values, with R_o increasing as the neutron deflection, ND, increases (decreasing porosity, ϕ). Anomalous points, in this case zone 1, are considered to have an anomalously high R_t due to the presence of hydrocarbon, and as a result the apparent water saturation can be estimated as shown if a value of n can be assumed. This saturation estimation can be made without a prior knowledge of C, D, m, or R_w but is not as diagnostic as the previously discussed R_t vs. $(t - t_{ma})$ method. In order to determine the cementation factor, m, it would be necessary to define D and in order to determine R_w, C would be required.

Sidewall Neutron Log (SNP)

In an attempt to overcome some of the disadvantages of the standard neutron logs, Schlumberger developed the sidewall epithermal neutron tool described by Tittman, et al.[21] The tool was designed to be used in uncased wells and provided porosity estimates in both liquid-filled and empty holes.

The tool used a directionally sensitive epithermal neutron detector incorporated into a sidewall source-detector skid. The effects of variations in borehole size and shape, mud type, temperature, and salinity were greatly reduced. Small residual borehole effects are then computationally accounted for in the surface control panel, thereby providing a borehole corrected neutron log.

The log presents a direct recording of neutron-derived porosity on a linear scale, as shown in Figure 8-36. Since

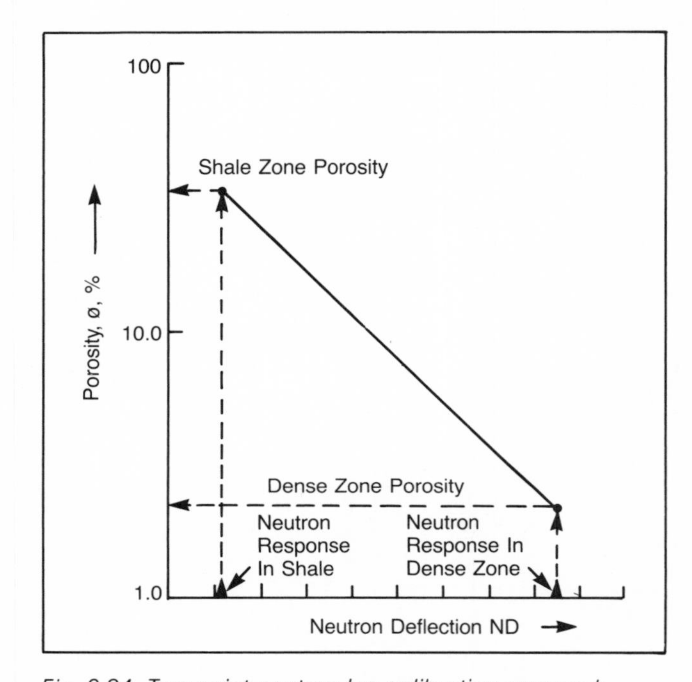

Fig. 8-34. Two-point neutron log calibration approach

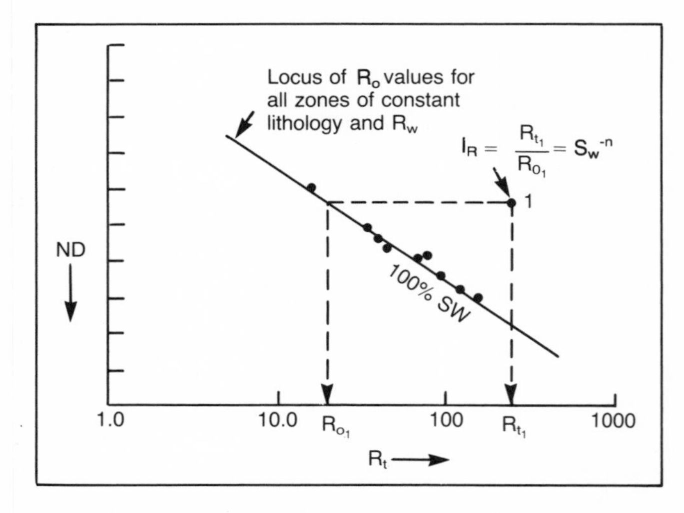

Fig. 8-35. Typical R_t versus ND crossplot

corrections for borehole effects are already applied, this direct recording of porosity simplified log interpretation. It is a valuable addition especially when evaluating complex or variable lithology in conjunction with other porosity tools.

Figures 8-33 and 8-33a can be used to estimate porosity for other lithologies (sandstone and dolomite) from the recorded limestone porosity as shown in Figure 8-36. A mudcake thickness correction should be made prior to entering the equivalence chart of Figure 8-33. In gas-filled boreholes, Figure 8-37 is used to correct the SNP porosity for borehole size when hole size is other than 7⅞ in. This is necessary since the caliper signal is not used to perform automatic borehole compensations when logging empty holes. These corrections are usually small.

Several limitations of the tool should be kept in mind. First, the tool cannot be used in cased holes and second, since it is a pad-type device it will be influenced by only a portion of the formation opposite the tool. If the formation is fractured and vuggy the standard neutron tool would probably be superior.

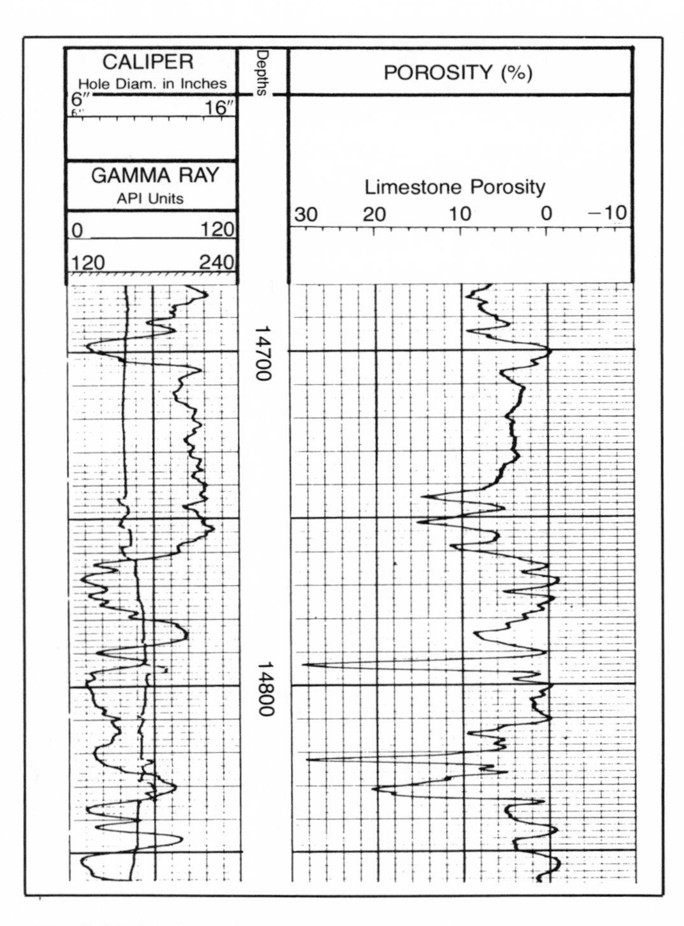

Fig. 8-36. Typical sidewall neutron log, SNP. Permission to publish by Schlumberger Ltd.[33]

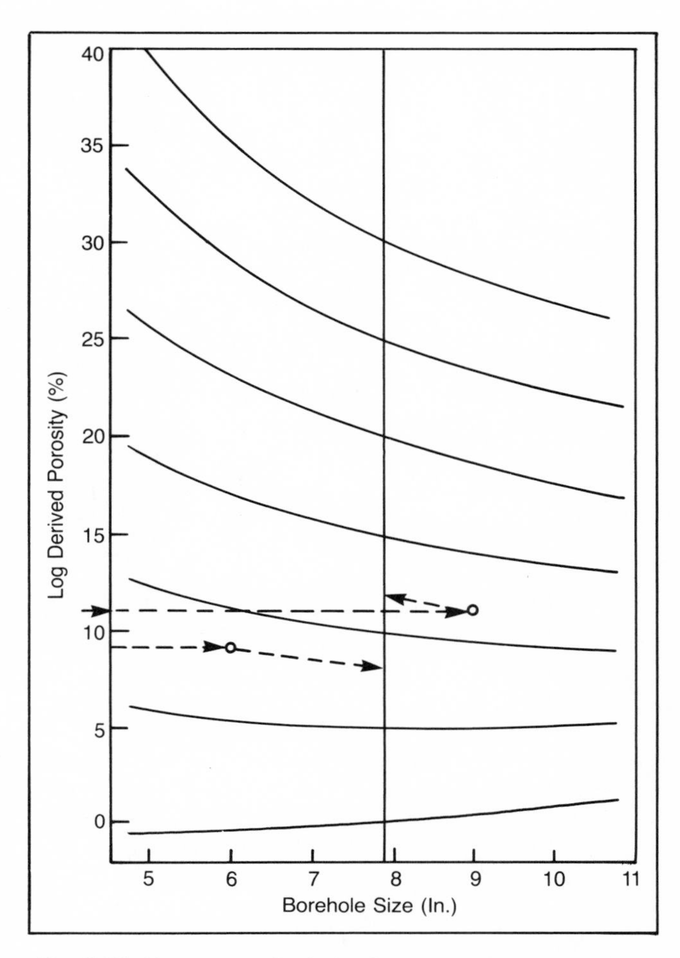

Fig. 8-37. Nomogram for hole size corrections in gas-filled boreholes. Permission to publish by the Society of Petroleum Engineers of AIME.[21] *Copyright 1966 SPE-AIME.*

Dual-Spaced Neutron Log (Compensated Neutron Log)

The dual-spaced neutron log, more commonly called the compensated neutron log, CNL, as shown in Figure 8-38, is a sidewall tool incorporating a two-detector system to minimize borehole effects.[28, 29, 31, 34] In this tool, the ratio of counting rates from the two He^3-filled thermal neutron detectors is related to formation porosity in a nearly linear manner. A function former in the surface panel produces a linearly scaled log of porosity index. A 16-curie Pu^{238}-Be source is used, yielding about 4×10^7 neutrons per second, four times the rate of a standard neutron logging source.

The use of the stronger source allows a larger source-detector spacing and, as a result, deeper investigation, when compared with the single detector sidewall neutron tool. Although large source-detector spacings and thermal neutron detection are used, the borehole and standoff effects are not completely eliminated. This is due to the shorter than ideal spacings still required (even

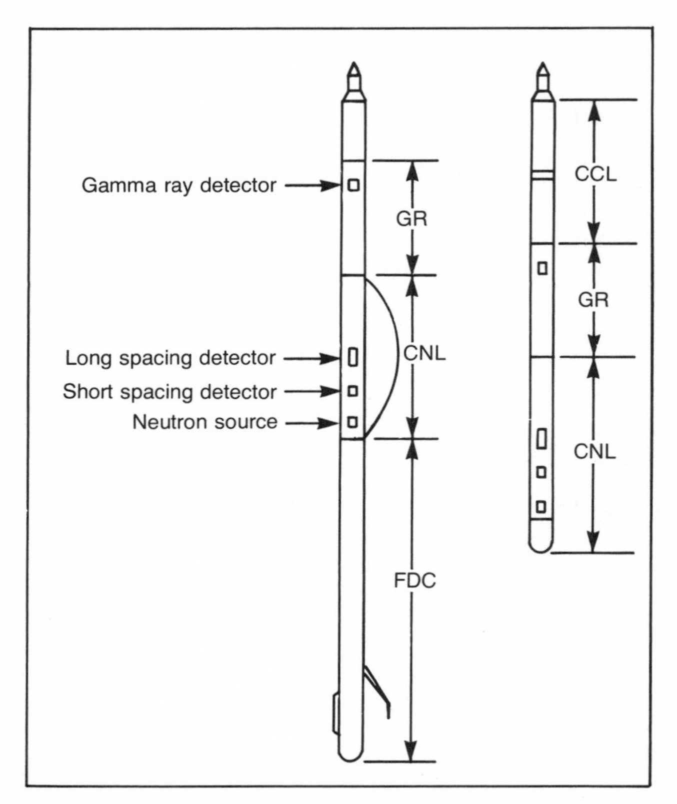

Fig. 8-38. Schematic of CNL logging systems. Permission to publish by the Society of Professional Well Log Analysts.[34]

with the stronger source and thermal neutron detection method) in order to obtain high enough count rates for field use. The combination of sensitive neutron detectors and the high strength source permits a logging speed of 1800 ft/hr. This deep responding two-detector system that minimizes the influence of the near-tool environment (including casing and annular cement) can be run in both open and cased holes.

The CNL log can be recorded simultaneously with other logs such as the compensated density, acoustic transit time, sidewall neutron, or thermal neutron decay time log (neutron lifetime type log). This advantage saves rig time and provides positive depth matching enhancing quick-look interpretation at the wellsite.

The surface equipment computes the ratio of the count rates of the short-spacing and long-spacing detectors, CPS_{near}/CPS_{far}, and when hole diameter is available from a caliper, the appropriate hole size correction is applied. This ratio is then converted to a porosity index using the calibration curves shown in Figure 8-39. The recorded porosity index can be determined for limestone, sandstone, or dolomite lithologies with the chosen porosity index being presented on a linear scale on the log. For cased-hole applications, the count rate ratio is converted to a porosity index assuming the conditions of 5½-in. casing cemented in an 8¾-in. borehole. The choice of a limestone or a sandstone porosity index scaling is available in cased holes.

The CNL is, as expected, a deeper investigating device than either the sidewall neutron tool or the density tool (to be discussed later). An illustration of the different depth of investigations is shown in Figure 8-40 where it is noted that both the sidewall neutron tools tend to respond to the flushed zone and as a result are affected by the same fluid saturation conditions. On the other hand, comparison of the CNL response to either of these tools (side-wall neutron or density) will show the influence of different invaded zone saturation conditions that might exist in the different bulk volumes investigated by the two tools (see Fig. 8-41). The influence of gas on the CNL and density responses is apparent, which is one reason this combination is so popular. This combination delineates gas-bearing zones although it tends to complicate the determination of porosity in gas-bearing zones. In most low-porosity zones, however, the porosity is given directly by the density while the CNL will show an appreciable gas effect. Shaliness, which tends to increase neutron porosity and decrease density porosity, tends to mask the hydrocarbon effect, although a simple crossplot method can be used to define shaly gas zones where the gas effect is less apparent. An example of a density-CNL combination is shown in Figure 8-42 for a Gulf Coast sand in which zones A and C are obviously gas-bearing. In this case zones D and E tested oil.

The density and neutron responses are also independently affected by lithology, a significant concept employed with the sidewall neutron and density responses

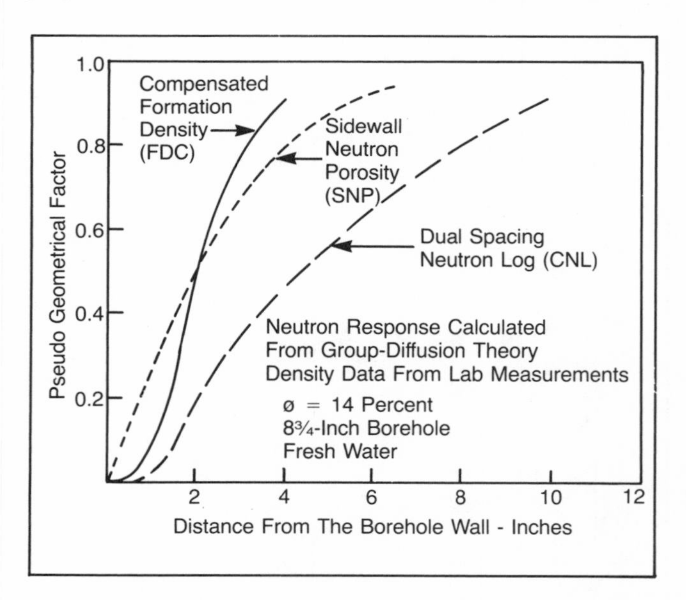

Fig. 8-40. Illustration of radial response of CNL. Permission to publish by the Society of Professional Well Log Analysts.[34]

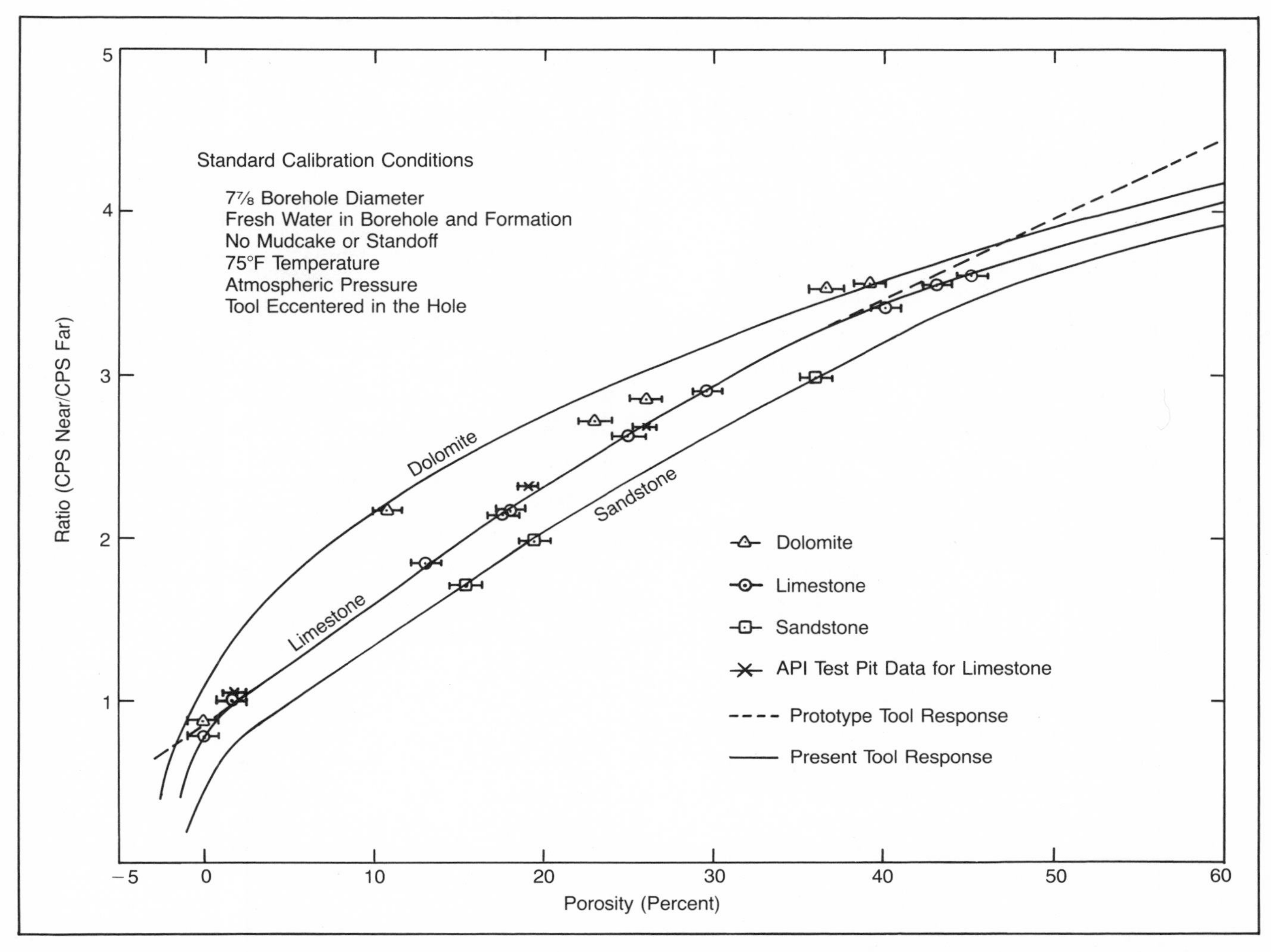

Fig. 8-39. Calibration curves for the CNL log. Permission to publish by the Society of Professional Well Log Analysts.[34]

in defining lithology. The density-CNL responses can also be used for this purpose, as indicated on the idealized illustration of Figure 8-43. The application of this lithology influence on the two responses is shown in Figure 8-44 for a North Rocky Mountain well. This affords a quick-look lithology log and in the case of clean, liquid-bearing formations of sandstone, limestone, or dolomite the porosity can be estimated as:

$$\phi = \frac{\phi_{CNL} + \phi_D}{2}$$

The further application of multiple porosity responses for both porosity and lithology determinations will be discussed in more detail in Chapter 9.

DENSITY LOG

The density logging tool is primarily used as a porosity tool, although it has a number of other uses, including evaluation of complex lithologies and shaly sands, identification of minerals in evaporite deposits, and the detection of gas.

Figure 8-45 illustrates the principle of operation of the density tool. In this type of tool, a focused gamma ray source, such as cobalt-60, emits gamma rays into the formation. These gamma rays interact with the electrons in the material opposite the focused source. The predominant interaction is Compton scattering, where each time the gamma ray strikes an electron it changes direction and loses some energy. The intensity of the back-scattered gamma rays is measured by the gamma ray detectors located above the source. The measured gamma ray intensity is a function of the electron density of the formation. As the electron density increases, the probability of more collisions by the gamma rays in a fixed distance increases, leading to a greater loss of energy and a higher probability of capture or absorption. In effect, the greater the electron density of the

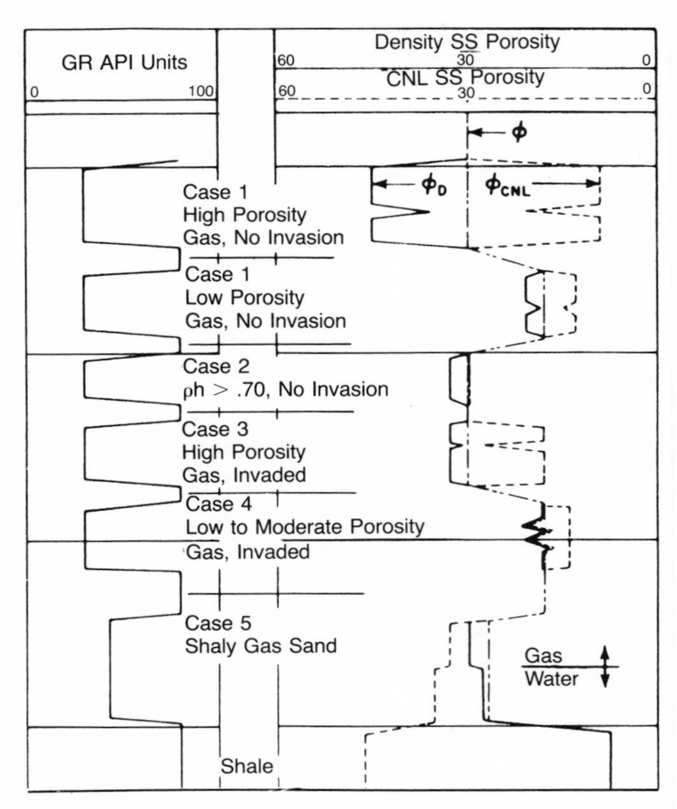

Fig. 8-41. Idealized comparison of differing saturation effects on CNL and density. Permission to publish by the Society of Professional Well Log Analysts.[34]

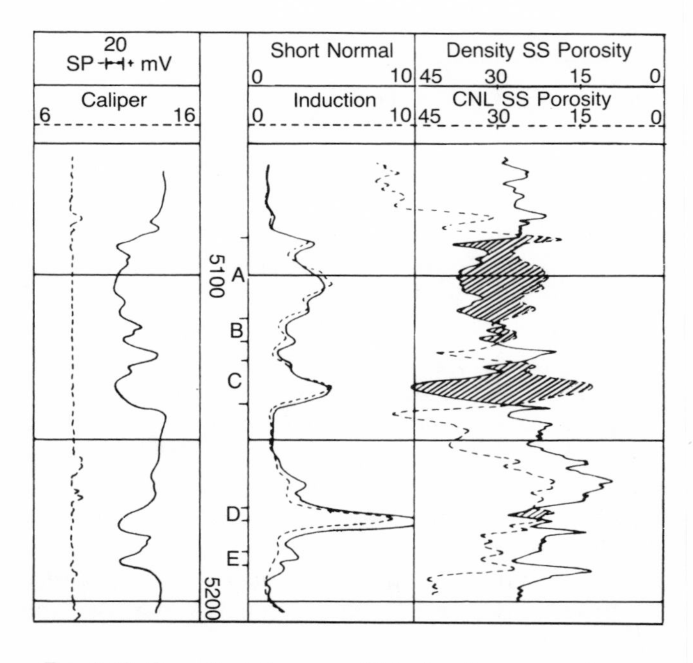

Fig. 8-42. Gas effect shown on CNL-density combination run in Gulf Coast well. Permission to publish by the Society of Professional Well Log Analysts.[34]

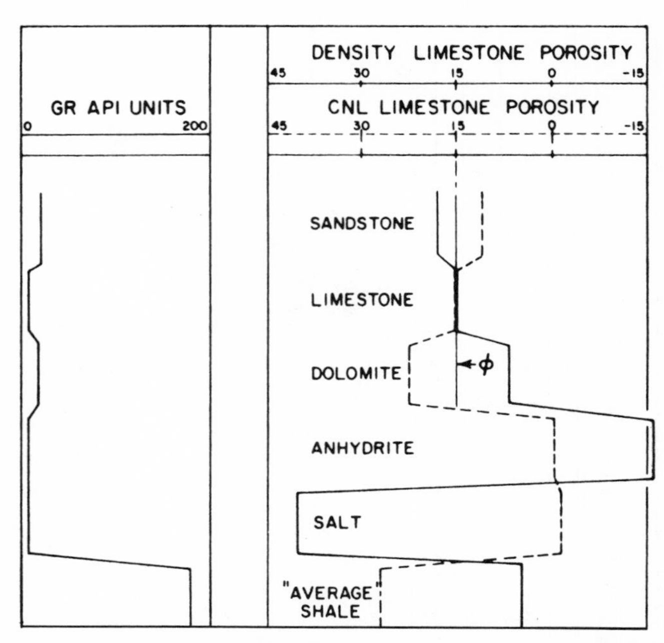

Fig. 8-43. Idealized CNL-density responses in common lithologies. Permission to publish by the Society of Professional Well Log Analysts.[34]

formation, the less the counting rate of the detectors.

The electron density of the material opposite the logging tool is proportional to the bulk density of this material. For a substance consisting of a single element, the electron density index, ρ_e, is related to bulk density, ρ_b, in the following manner:[33]

$$\rho_e = \rho_b \left(\frac{2Z}{A} \right)$$

where: Z = the atomic number or number of electrons per atom
A = atomic weight (ρ_b/A is proportional to the number of atoms per cubic centimeter of the material)

Table 8-3 shows the atomic weight, A, and the atomic number, Z, of some common elements. As can be seen, the ratio, $2Z/A$, is approximately equal to 1.0 in most cases. For a molecular substance the electron density index can be expressed as:

$$\rho_e = \rho_b \left(\frac{2\Sigma Z}{\text{Mol Wt}} \right)$$

Table 8-4 shows that the ratio, ($2\Sigma Z/\text{Mol Wt}$), for some common substances encountered in logging is close to 1.0 such that the electron density index is approximately equal to the bulk density, $\rho_e \approx \rho_b$. As a result, the apparent bulk density responses of the logging tool (which indirectly responds to electron density) is essentially a response to the bulk density, ρ_b, of the material opposite the logging tool.

The density tool is designed so that the source and detectors are mounted on a skid forced against the formation face by an eccentering arm. The skid is designed so that it will cut through soft mudcakes, although when mudcakes are hard (at greater depths) some mudcake may be present between the skid and formation. Of course, the mud and mudcake remaining between the skid and formation will be included in the average bulk density measurements and must be corrected for to obtain the true formation bulk density.

Originally, the density tool consisted of only one detector. It was found, however, that the correction for mudcake effects was difficult to make. As a result, the two-detector type tool was developed (see Fig. 8-45), such that the responses of the two detectors could be used to assess the influence of the mudcake on the bulk density measurements. By doing so, the mudcake cor-

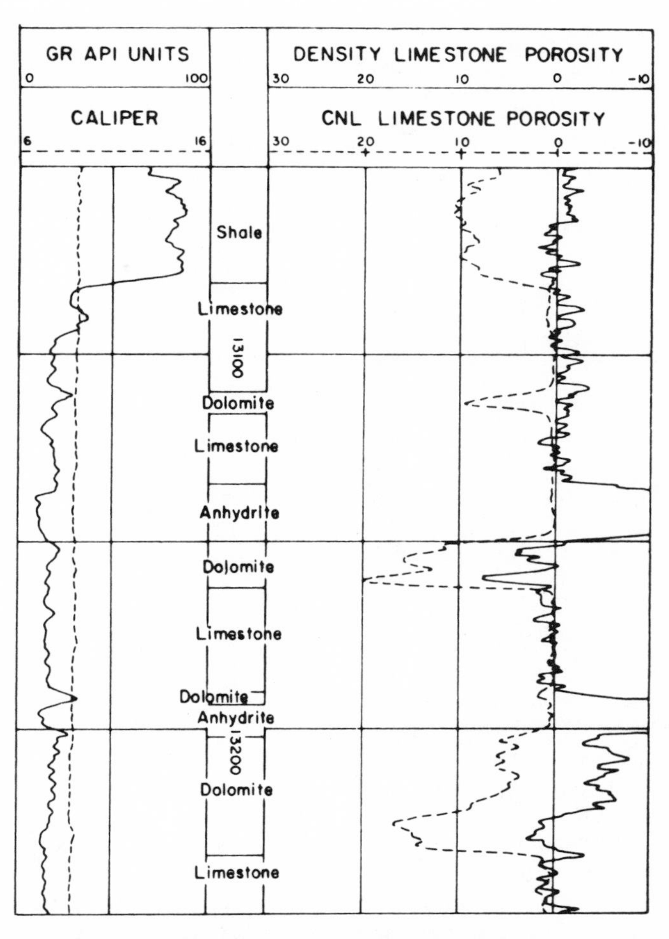

Fig. 8-44. Lithology indications from CNL-density combination run in North Rocky Mountain well. Permission to publish by the Society of Professional Well Log Analysts.[33]

TABLE 8-3
Atomic Properties of Common Elements

Element	*A*	*Z*	*2(Z/A)*
H	1.008	1	1.9841
C	12.011	6	.9991
O	16.000	8	1.0000
Na	22.99	11	.9569
Mg	24.32	12	.9868
Al	26.98	13	.9637
Si	28.09	14	.9968
S	32.07	16	.9978
Cl	35.46	17	.9588
K	39.10	19	.9719
Ca	40.08	20	.9980

Permission to publish by Schlumberger Ltd.[33]

TABLE 8-4
Density Properties of Various Substances

Compound	*Formula*	*Actual Density*	*2ΣZ's Mol Wt.*	ρ_e *(Eq. 8-1)*	ρ_a *(Eq. 8-2) (as seen by tool)*
Quartz	SiO_2	2.654	0.9985	2.650	2.648
Calcite	$CaCO_3$	2.710	0.9991	2.708	2.710
Dolomite	$CaCO_3MgCO_3$	2.870	0.9977	2.863	2.876
Anhydrite	$CaSO_4$	2.960	0.9990	2.957	2.977
Sylvite	KCl	1.984	0.9657	1.916	1.863
Halite	NaCl	2.165	0.9581	2.074	2.032
Gypsum	$CaSO_42H_2O$	2.320	1.0222	2.372	2.351
Anthracite Coal		{1.400 1.800	1.030	{1.442 1.852	{1.355 1.796
Bituminous Coal		{1.200 1.500	1.060	{1.272 1.590	{1.173 1.514
Fresh Water	H_2O	1.000	1.1101	1.110	1.00
Salt Water	200,000 ppm	1.146	1.0797	1.237	1.135
"Oil"	$n(CH_2)$	0.850	1.1407	0.970	0.850
Methane	CH_4	ρ_{meth}	1.247	$1.247\rho_{meth}$	$1.335\rho_{meth}$-0.188
"Gas"	$C_{1.1}H_{4.2}$	ρ_g	1.238	$1.238\rho_g$	$1.325\rho_g$-0.188

Permission to publish by Schlumberger Ltd.[33]

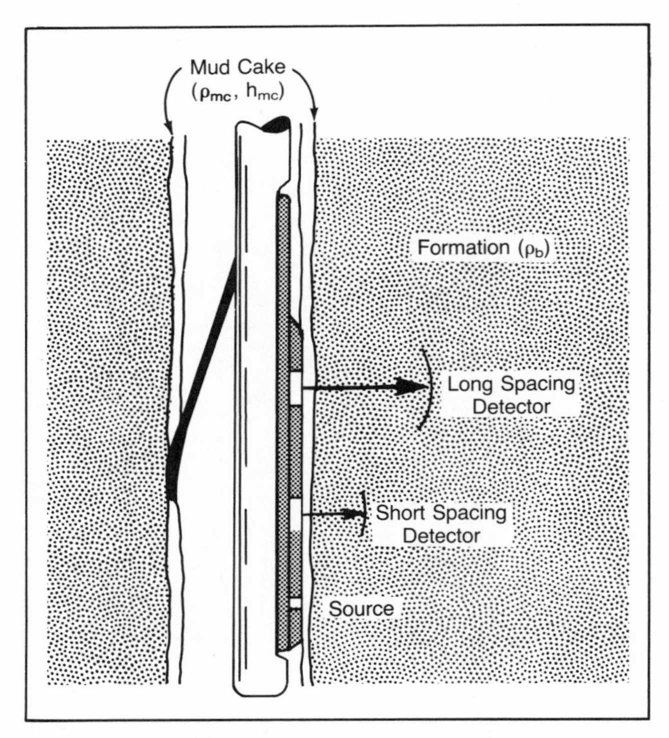

Fig. 8-45. Schematic of dual spaced density logging tool. Permission to publish by Schlumberger Ltd.[33]

Time Constant and Logging Speed. Since the measurements being made are statistical in nature, the signal is smoothed by using a time constant of 2, 3, or 4 sec-

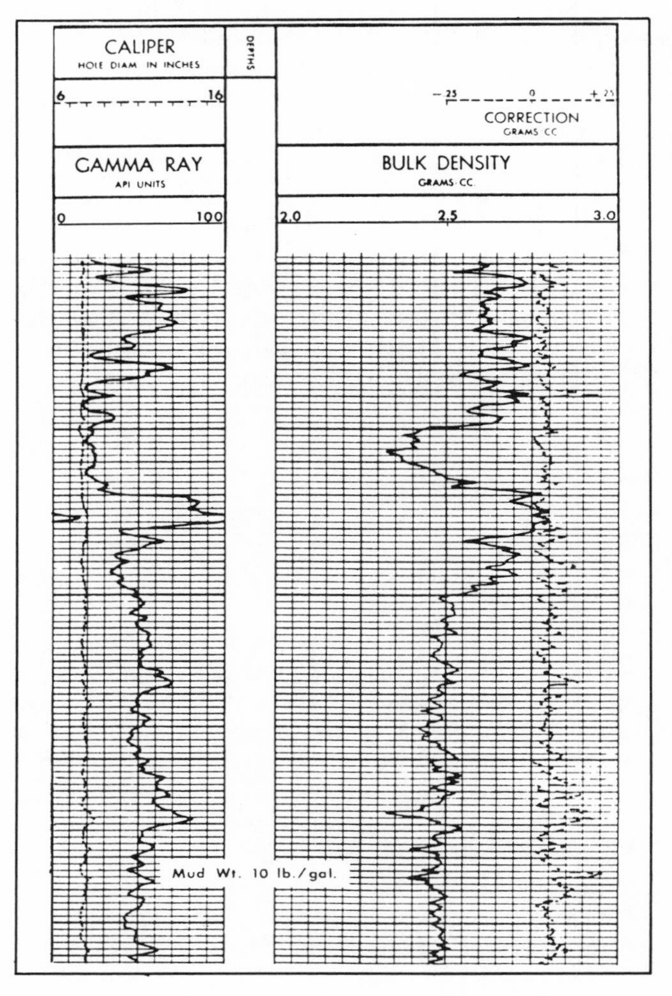

Fig. 8-45a. Typical density log. Permission to publish by Schlumberger Ltd.[33]

rection can be compensated for automatically and a corrected bulk density curve recorded, as shown in Figure 8-45a. The correction made is also recorded on the log. In addition, a caliper curve is recorded. The density tool can be run in either empty or fluid-filled holes.

Factors Affecting Density Log Response

The density log response depends on the following factors: (1) mudcake, (2) borehole size, (3) time constant and logging speed, (4) rock matrix density, (5) pore fluid density, and (6) porosity.

Mudcake. When the early-type single-detector density tool was used, it was necessary to consider the influence of the mudcake on the single response curve. The corrections were made based on empirical charts for the specific tool and required a knowledge of mudcake density and mudcake thickness. Due to the uncertainty in these data, the dual-spaced density tool that automatically corrects for the effects of mudcake is a big improvement.

Borehole Size. When borehole size becomes larger than 9 in., hole size effects must be considered. Figures 8-46 and 8-47 show the borehole corrections for the compensated formation density, FDC, tool used by Schlumberger Ltd. Corrections are shown for both mud-filled and gas-filled holes.

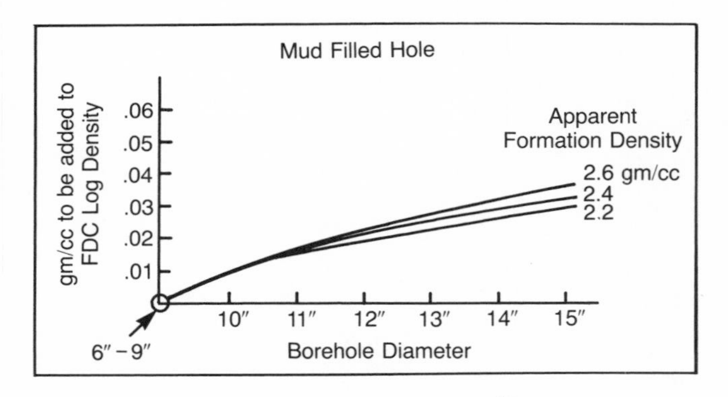

Fig. 8-46. Borehole corrections for Schlumberger FDC density tool, mud filled hole. Permission to publish by the Society of Petroleum Engineers of AIME.[15]

onds (FDC tool) in the circuit. The low time constant (2 sec.) is used when the densities are low or porosities are high. The logging speed is then chosen so that the tool will not travel more than one foot during one time constant. Maximum logging speed should be no more than 1,800 ft/hr.

Rock Matrix Density. The grain or matrix density, ρ_{ma}, is dependent upon mineral composition, temperature, and pressure with the density of sedimentary rocks generally ranging from about 2.6 to 3 gm/cc. Grain densities for some of the more common matrix materials are shown in Table 8-4. Although not shown in Table 8-4, typical shale densities for shale beds and laminar shale streaks will be on the order of 2.2 to 2.65 gm/cc. Dispersed clays or shales disseminated in the pore spaces may have a smaller density than the interbedded shales.

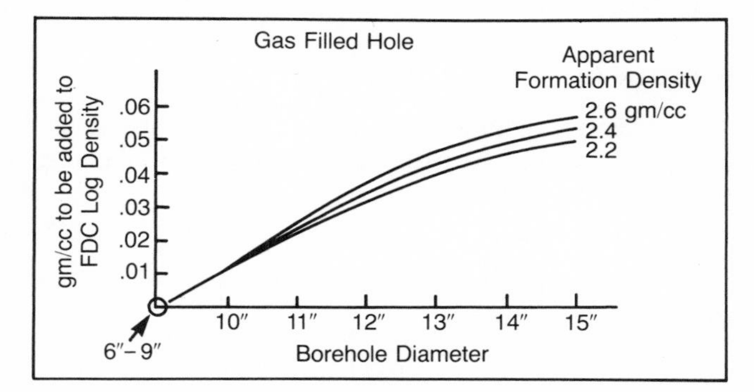

Fig. 8-47. Borehole corrections for Schlumberger FDC density tool, gas filled hole. Permission to publish by the Society of Petroleum Engineers of AIME.[15]

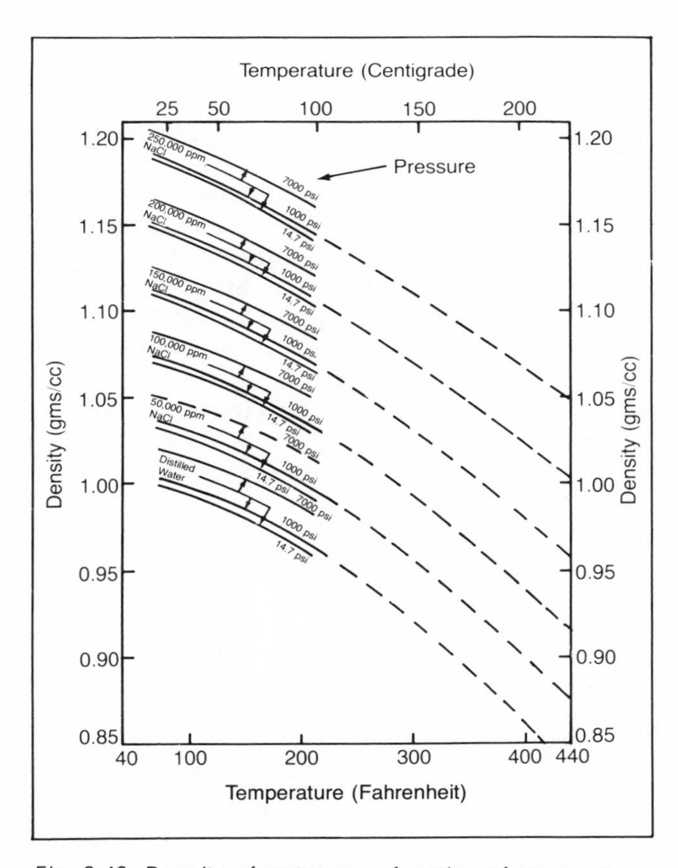

Fig. 8-48. Density of water as a function of temperature, salinity, and pressure. Permission to publish by Schlumberger Ltd.[33]

Pore Fluid Density. Densities of various fluids are shown in Table 8-4. Due to the shallow depth of investigation (see Fig. 8-40) of the density tool, the fluid usually filling the pore space is primarily water. The density of water as a function of temperature, salinity and pressure is shown in Figure 8-48.

If residual hydrocarbons are present in the shallow zone investigated by the density log, the bulk density readings may be affected. If the residual hydrocarbon is oil, the average fluid density of oil and water may be reasonably close to unity and the effect of oil may not be too noticeable. If, however, the residual hydrocarbon is gas, its effect may be appreciable on the density response. In those formations saturated with gas near the wellbore, a gas density value should be considered.

Figure 8-49 gives data for the true gas density, ρ_{gas}, and apparent gas density seen by the density tool,

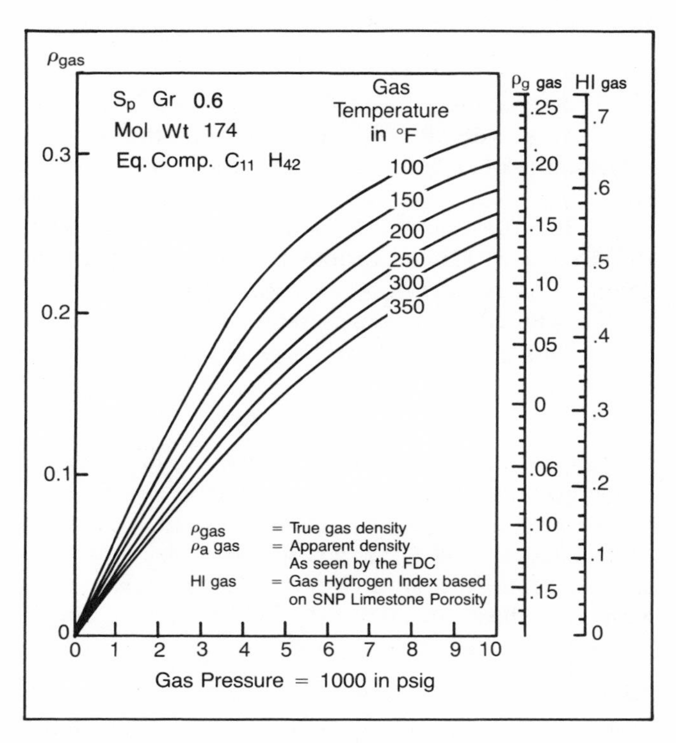

Fig. 8-49. Gas density of a gas slightly heavier than methane as a function of pressure and temperature. Permission to publish by Schlumberger Ltd.[33]

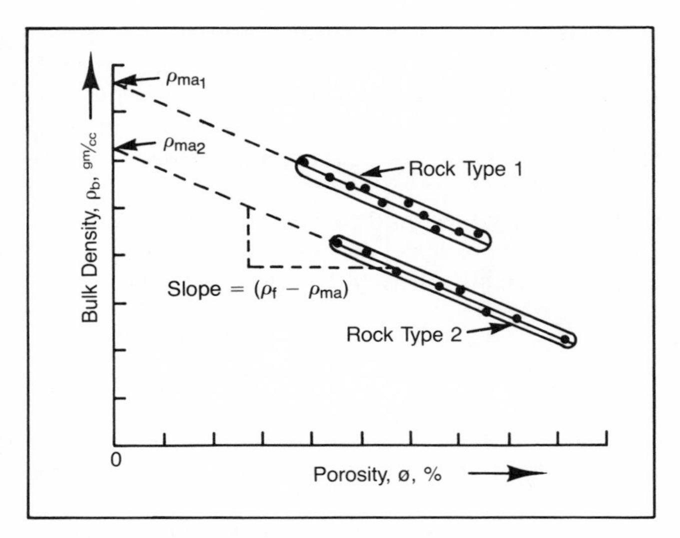

Fig. 8-50. Typical bulk density-porosity crossplot

$\rho_{a(gas)}$, as a function of pressure and temperature for a gas slightly heavier than methane.

Porosity. The bulk density measured by the logging tool is a simple weighted average of the densities of the matrix and pore fluid components such that:

$$\rho_b = \underbrace{\rho_f \phi}_{\text{fluid density}} + \underbrace{\rho_{ma}(1-\phi)}_{\text{solid density}}$$

Rearranging this relation in the following manner,

$$\rho_b = \rho_{ma} + (\rho_f - \rho_{ma})\phi$$

it can be seen that a plot of bulk density, ρ_b, versus porosity, ϕ, for all zones of constant lithology and saturation would define a linear trend having a slope equal to $\rho_f - \rho_{ma}$ and an intercept equal to ρ_{ma} as shown in Figure 8-50. Assuming a value of ρ_f, the linear trends to be expected for the various rock types (ρ_{ma} for the common reservoir rock type shown in Table 8-4) can be quickly plotted on cartesian graph paper.

The density-porosity relation can also be expressed as:

$$\phi = \frac{\rho_{ma} - \rho_b}{\rho_{ma} - \rho_f}$$

enabling a quick solution for porosity, ϕ. For zones of a single lithology where ρ_{ma} can reasonably be assumed, or is known, a porosity, ϕ, can be estimated, assuming a fluid density. Since the density tool responds to a shallow depth, the density response may be reasonably assumed to be a function of the fluid density existing in the flushed zone (basically mud filtrate). As a result:

$$\rho_f = \rho_{mf} S_{xo} + \rho_h (1 - S_{xo})$$

where: ρ_{mf} = density of mud filtrate (see Figure 8-48)
ρ_h = density of residual hydrocarbon (see Table 8-4 and Figure 8-49)
S_{xo} = flushed zone water saturation, $S_{xo} = 1 - S_{hr}$

As previously mentioned, for zones containing average residual oil saturations, the fluid density may be reasonably close to unity since the somewhat lower-than-usual oil density would tend to compensate for the somewhat higher-than-usual water density. For residual gas saturations, on the other hand, the low gas density effect may be appreciable on the density response and, if not accounted for, would lead to high porosity estimates.

Bulk Density-Resistivity Relation

As discussed with the previous porosity tools (acoustic transit time and neutron logs), resistivity is related to porosity in the following form:

$$\log R_t = -m \log \phi + \log R_w + \log I_R$$

Since bulk density is related to porosity as previously shown in the form:

$$\phi = \frac{\rho_{ma} - \rho_b}{\rho_{ma} - \rho_f}$$

the density relation can be substituted for porosity in the resistivity expression to obtain:

$$\log R_t = -m \log \left(\frac{\rho_{ma} - \rho_b}{\rho_{ma} - \rho_f}\right) + \log R_w + \log I_R$$

or:

$$\log R_t = -m \log (\rho_{ma} - \rho_b) + m \log (\rho_{ma} - \rho_f) + \log R_w + \log I_R$$

This expression states that R_t can be plotted versus ϕ on log-log graph paper, and a linear trend will be observed for all zones within a constant lithology (constant ρ_{ma} and similar ρ_f values) having a constant R_w and I_R. An example of such a plot is shown in Figure 8-51.

The lowest observed resistivities are assumed to be 100% water-bearing, with this linear trend defined by the lowest observed resistivities representing a locus of R_o values. This trend can be extrapolated to 100% ϕ (where $\rho_{ma} - \rho_b = \rho_{ma} - \rho_f$), which would be the apparent

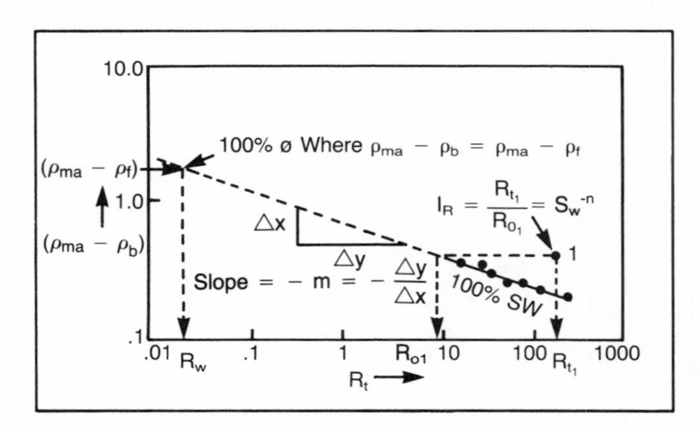

Fig. 8-51. Apparent resistivity vs. ($\rho_{ma} - \rho_b$) crossplot

R_w. Additionally, this trend should define a slope of m (cementation factor) for the rock type. All points falling to the right of the 100% S_w trend have a resistivity $R_t > R_o$. Since these anomalies are assumed to have a higher resistivity as a result of the presence of hydrocarbons, the water saturation can readily be estimated, as shown in Figure 8-51. This approach is quite diagnostic since S_w and m can be obtained without a prior knowledge of ρ_f or R_w and in fact R_w can be estimated if ρ_f is known or assumed. A graphical approach of linear $\phi = f(\rho_b)$ vs. nonlinear R_t can also be used to estimate S_w. The application and implication of this crossplotting technique will be more fully treated in Chapter 9.

Bulk Density as an Abnormal Pressure Indicator

As discussed in Chapter 7, abnormally pressured zones are associated with abnormally high shale porosities. Since the density log response is a porosity function, the density log should also be an abnormal pressure indicator. It would be expected, therefore, that a marked departure (density decrease) from the normal shale compaction trend would signal the presence of an overpressured zone. Both the density log and bulk density measurements made on cuttings (see Chapter 3) have been used for this purpose. In cases of severe hole washouts, however, the reliability of bulk density plots using log measurements decreases drastically.

Figure 8-52 shows plots of shale bulk density vs.

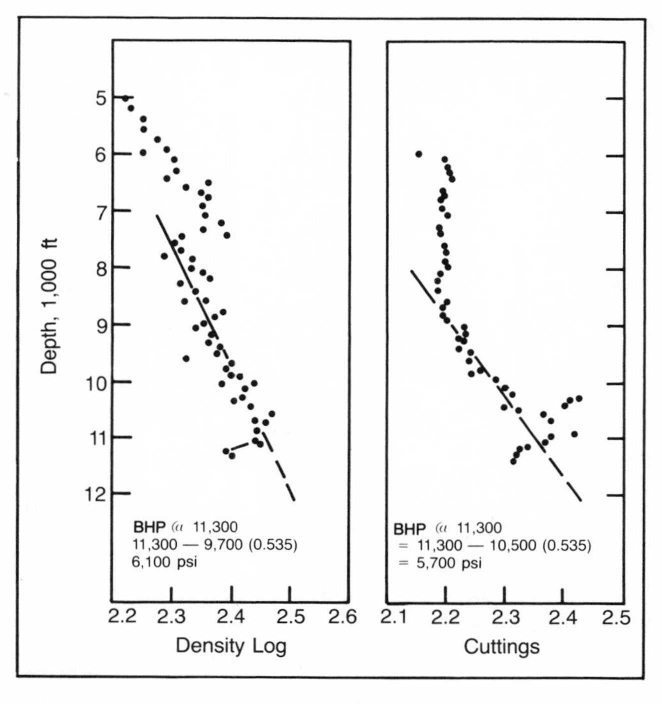

Fig. 8-52. Comparison of cuttings and log-derived densities. Permission to publish by the Oil and Gas Journal.[23]

depth by each method. It can be seen that the bulk density from cuttings is less than for the log measurements except in the caprock at 10,500–11,000 ft where flushing of the cuttings has not been as severe due to its lower permeability and mineralization.[23, 30] The pressure top at 11,000 ft is, however, clearly defined by both methods.

Applications and limitations of the density log have been pointed out in the preceding discussion; however, it has not been mentioned that the density tool has a limited "view" of the formation. The tool investigates about 25% or less of the formation around the wellbore. Due to this limitation, the apparent bulk density as seen by the tool in heterogeneous formations may not be representative of the formation.

PULSED NEUTRON DECAY LOGS (NEUTRON LIFETIME, THERMAL DECAY TIME LOGS)

The pulsed neutron decay logs record time reflecting the decay rate of thermal neutrons in the bulk formation versus depth. This decay rate is dominated by chlorine since it is the strongest neutron absorber of the common elements encountered. This measure is, therefore, primarily a reflection of the sodium chloride present in the formation water and resembles a resistivity measure, since the effects of formation water salinity, porosity, and shaliness are similar on each measure. However, since borehole influences are usually not reflected in this decay measure (including the influences of casing and cement), this measure can be made in cased holes for determining fluid saturations in the formation. As a result, this tool can provide an excellent measure for cased hole re-evaluation and for analyzing production problems.

This tool uses an accelerator-type neutron generator that emits, repetitively, bursts of high energy neutrons. After each neutron burst there is a quiet time during which the neutrons are rapidly slowed down to thermal velocities within the borehole and formation. This thermal neutron population will decrease due to neutron capture, which results in capture gamma ray emissions. A typical neutron die-away curve is shown in Figure 8-53. The early time period reflects a rapid decay, which is a function of the high absorption rates of the borehole system—borehole fluid, casing, and cement. This is followed by the straight line exponential decay reflecting the capture capacity of the formation. Using a gamma ray detector, which is placed a short distance from the source, the neutron population can be indirectly sampled from the gamma ray count rate, as shown in Figure 8-54 and 8-54a. Obtaining two separate count rates (during gate 1 and gate 2 as shown in Fig. 8-54a) the thermal decay time, τ_{log}, can be expressed as a func-

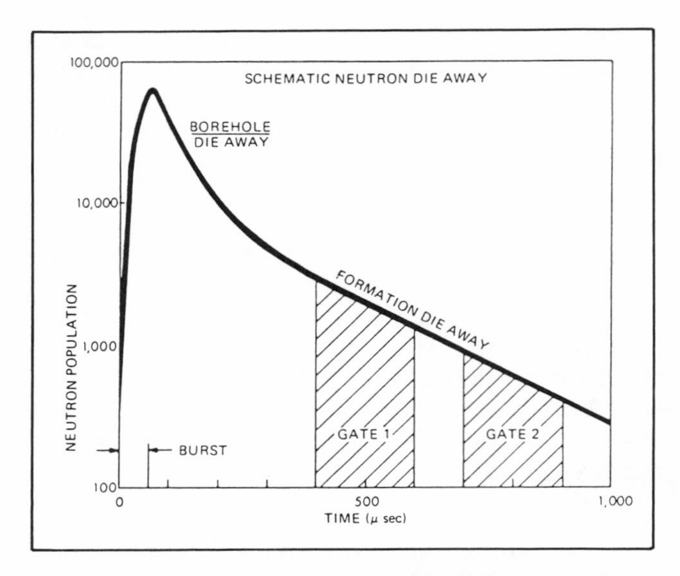

Fig. 8-53. Schematic of neutron die-away curve. Permission to publish by Dresser Atlas, Dresser Industries.[24]

tion of the microscopic cross-section of the formation, Σ_{log}, as follows:[25, 35]

$$\Sigma_{log} = \frac{4.55}{\tau_{log}}$$

where τ_{log} is in μsec and Σ_{log} is in cm²/cm³. For a more practical unit, however, Σ_{log} is recorded in capture units (1 c.u. = 10^{-3} cm²/cm³) and hence:

$$\Sigma_{log} = \frac{4550}{\tau_{log}}$$

A typical pulsed neutron decay log is shown in Figure 8-54b.

Factors Affecting Lifetime Measurements

As with other logs, a number of factors influence the lifetime measurements:[22, 24, 33] (1) borehole effects, (2) bed thickness, (3) diffusional effects, and (4) statistical variations.

Borehole Effects. The early borehole effect on the thermal neutron die away or decay process is a result of all things in and around the borehole, such as the instrument itself, casing, borehole fluids, etc. Since the tool is time-controlled, however, determination of the capture cross-section, Σ_{log}, is essentially unaffected by the borehole conditions. In effect, the borehole effects are eliminated by waiting a long enough time after the neutron burst.[22]

Bed Thickness. Bed resolution is a function of source to detector spacing and the logging speed and time constant. By using a time constant of from 3 to 4 seconds and a logging speed of 20 ft/min, good values of Σ_{log} can be obtained in beds 3 ft thick or greater.[24]

Diffusional Effects.[24] Neutrons migrate between the borehole and formation. When the neutrons decay more slowly in the formation, the neutron density in the for-

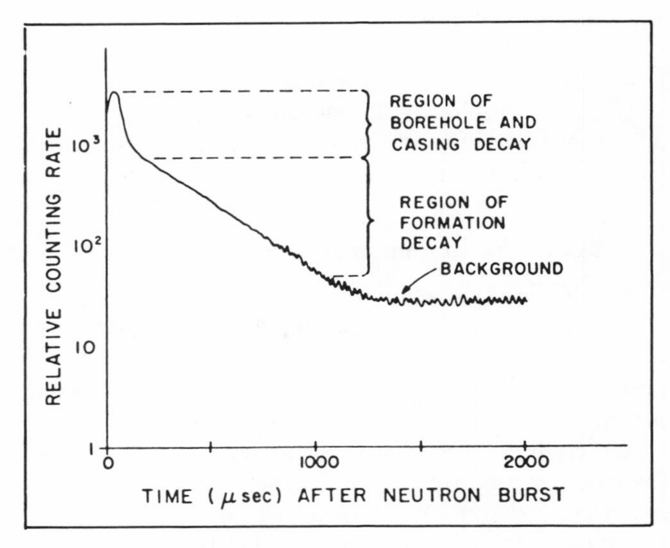

Fig. 8-54. Decay of capture gamma ray counting rate. Permission to publish by Schlumberger Ltd.[35]

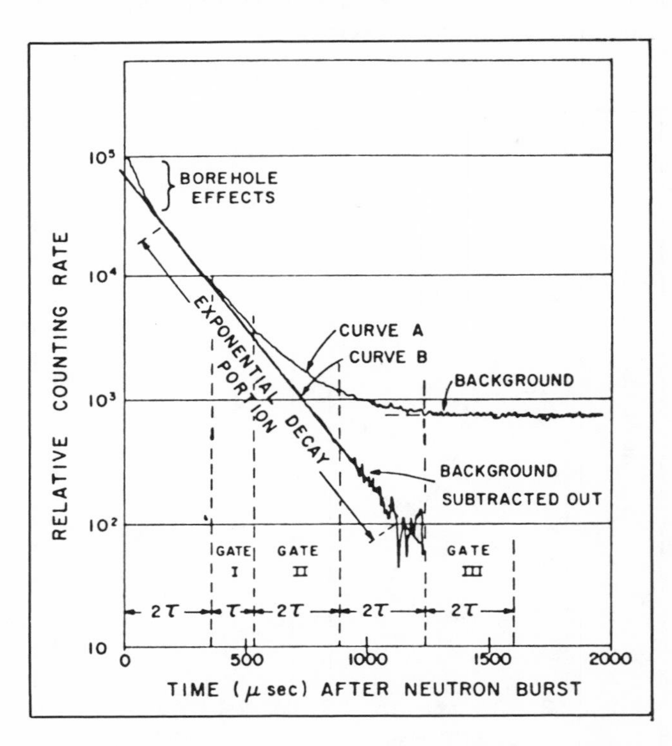

Fig. 8-54a. Background counts are subtracted from total to give true exponential decay curve. SCALE FACTOR relationships of measure gates are shown for a τ of 178 μsec. Permission to publish by Schlumberger Ltd.[35]

mation eventually becomes greater than in the borehole. In this case, there is a net flow of neutrons from the formation to the borehole, and the neutron population decays at a rate that is primarily a function of the formation. If, however, the net flow of neutrons is into the formation, the measured decay time is then more strongly influenced by borehole conditions. The effects of neutron diffusion on the measured decay time are a result of the spacing, borehole geometry, and formation. By using a spacing of 18 in., a desirable balance between diffusional and other variables is obtained. This spacing then results in a lifetime very close to those measured in the laboratory or calculated from the chemical composition.

Statistical Variations. Since this tool is a nuclear device, it is affected by statistical variations. The use of higher counting rates and longer time constants usually result in lower statistical variations.[22]

Depth of Investigation. Depth of investigation is a function of the neutron energy and capture cross-section of the material around the instrument. The depth of investigation for average borehole environments is from 14 to 20 in.[24]

Interpretive Relationships

The capture cross-section response measured by the logging tool, Σ_{log}, can be related to the capture cross-section of the individual components making up the formation bulk as follows:

$$\Sigma_{log} = \underbrace{\Sigma_{ma}(1-\phi)}_{\text{Solid}} + \underbrace{\underbrace{\Sigma_w \phi S_w}_{\text{Water}} + \underbrace{\Sigma_h \phi (1 - S_w)}_{\text{Hydrocarbon}}}_{\text{Fluid}}$$

It can be seen that Σ_{log} is a function of both ϕ and S_w, and if an independent ϕ response is available, the water saturation existing in the formation can be determined. Since this log can be run in casing (tool essentially unaffected by borehole conditions), it provides an excellent cased hole log for reevaluating old wells.

Examination of the expression relating Σ_{log} to ϕ and S_w shows that a direct solution of this relation requires a knowledge of the cross-section of the various formation materials. Capture cross-sections are presented as follows for water, oil, gas, and formation matrix.

1. *Water.* The capture cross-section of pure water is 22.2 capture units, c.u. (10^{-3} cm^{-1}). Increasing amounts of sodium chloride increase the capture cross-section, as shown in Figure 8-55.
2. *Oil.* Dead crude oil has a capture cross-section of about 22 c.u. Most oils, however, have gas in solu-

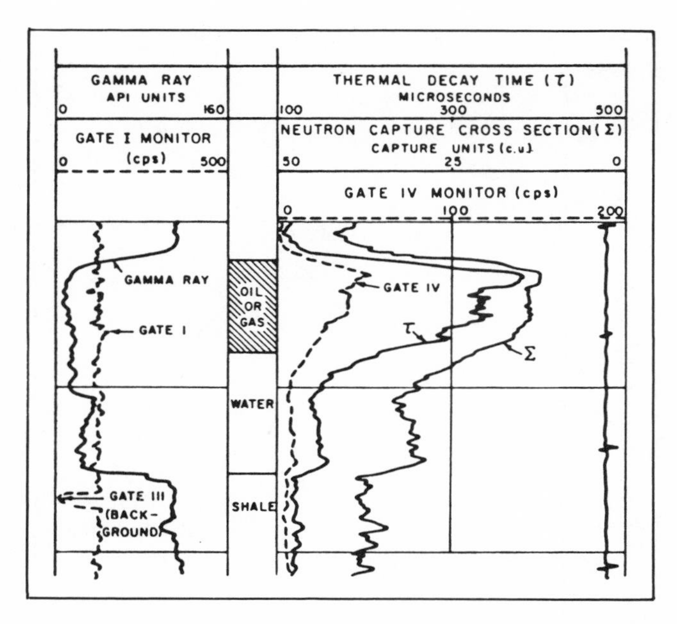

Fig. 8-54b. Presentation of TDT log. Permission to publish by Schlumberger Ltd.[35]

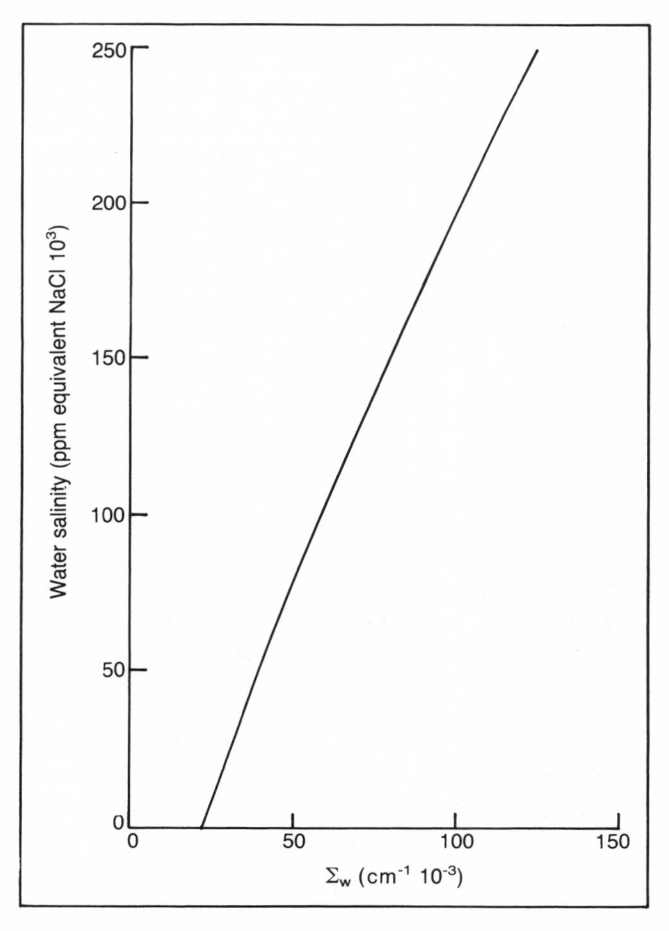

Fig. 8-55. Σ_w as function of water salinity. Permission to publish by Dresser Atlas, Dresser Industries.[24]

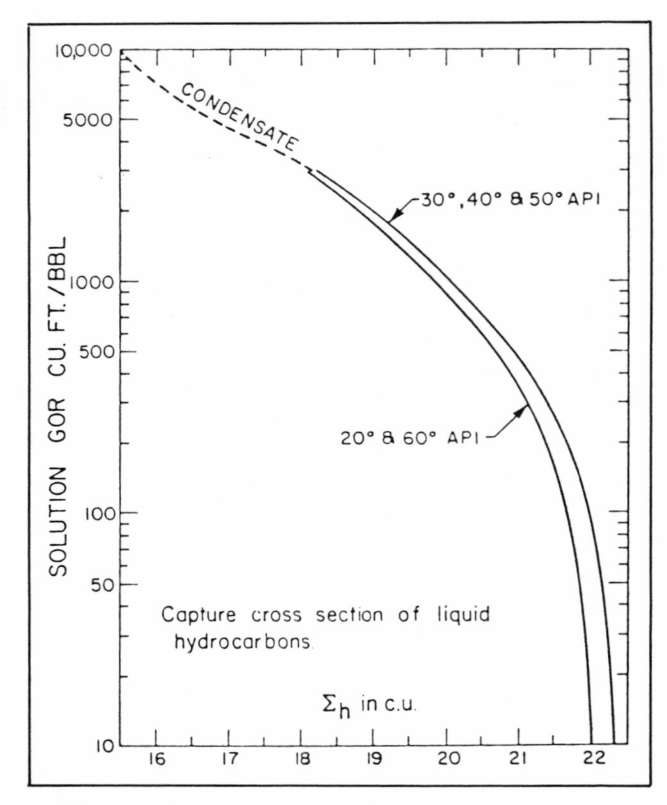

Fig. 8-56. Σ_o as function of oil gravity and solution gas-oil ratio. Permission to publish by the Society of Petroleum Engineers of AIME.[25] Copyright 1971 SPE-AIME.

TABLE 8-5
Apparent Matrix Capture Cross-section Ranges

Rock Type	Σ_{ma} *Range, cu*
Sandstone	8–13
Limestone	8–10
Dolomite	8–12
Anhydrite	18–21

Permission to publish by Dresser Atlas, Dresser Industries.[24]

tion, and an estimate of Σ_h for oil can be obtained from Figure 8-56.

3. *Gas.* The capture cross-section of methane is shown in Figure 8-57 as a function of temperature and pressure. However, most gases encountered are not 100% methane, and if the gas specific gravity is available, Figure 8-58 can be used to determine Σ_h for gas.
4. *Formation Matrix.* The macroscopic capture cross-sections of formation matrix materials will vary, depending on the rock type. Table 8-5 presents a range of Σ_{ma} values, which can be used to form a reasonable basis for interpretation.

It is apparent that the general expression relating the log-derived capture cross-section, Σ_{log}, to the individual component contribution for a nonshaly formation can be directly solved for water saturation, S_w, if ϕ and the individual component Σ values are known. However, a

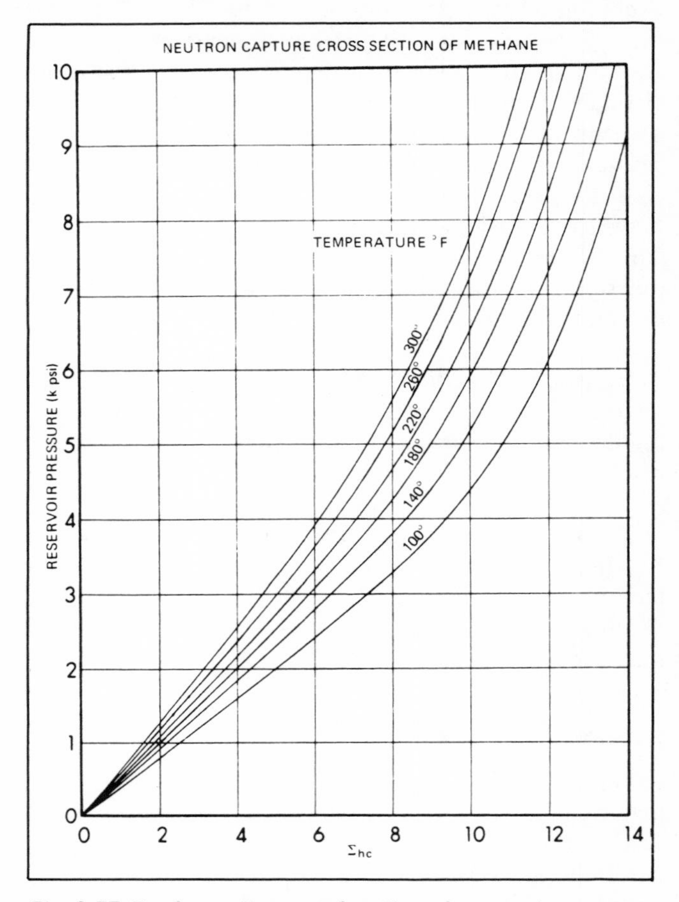

Fig. 8-57. Σ_{hc} for methane as function of reservoir pressure and temperature. Permission to publish by the Society of Petroleum Engineers of AIME.[25] Copyright 1971 SPE-AIME.

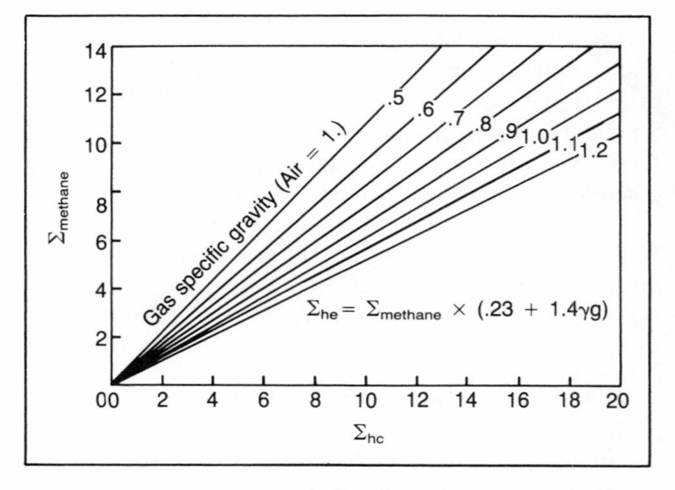

Fig. 8-58. Σ_{hc} for dry gas as determined from Σ methane and gas gravity. Permission to publish by Dresser Atlas, Dresser Industries.[24]

crossplotting method can also be used for evaluating water saturation, which is particularly useful when evaluating a number of levels in a rock type, as well as defining Σ values of interest, such as Σ_{ma} and Σ_w.[24, 27] If the capture cross-section response equation is written for a 100% water-saturated rock, the relation reduces to:

$$\Sigma_{log} = \Sigma_w \phi + \Sigma_{ma}(1 - \phi)$$

or:

$$\Sigma_{log} = \underbrace{\Sigma_{ma}}_{\text{Intercept}} + \underbrace{(\Sigma_w - \Sigma_{ma})}_{\text{Slope}}\phi$$

This relation states that for all water-bearing zones in the same rock type, a linear trend will be observed when crossplotting Σ_{log} versus ϕ on cartesian graph paper, as shown in Figure 8-59. Additionally, the intercept (at $\phi = 0$) will be the apparent value of Σ_{ma} for that rock type. Once Σ_{ma} is known, a value of Σ_w can be determined from the slope of the 100% water-bearing trend. The validity of Σ_w (and therefore the slope) can be checked since from Figure 8-55 an apparent salinity and therefore an apparent R_w can be determined. All points falling below the assumed 100% S_w trend would be anomalies considered to be hydrocarbon-bearing. The higher the salinity of the formation water the greater the contrast between water-saturated and oil-bearing zones. The greater the formation water salinity, the greater the value of Σ_w (Fig. 8-55), and as a result the greater the value of Σ_{log} in the water zones. Since Σ_o is generally 22 or less (see Fig. 8-58) as the value of Σ_w decreases (approaches a minimum of 22) the less the difference between Σ_{log} measured in the water-bearing zone and Σ_{log} in a hydrocarbon-bearing zone. Once the 100% S_w trend, Σ_{ma} and Σ_w are defined, a fan of other saturation trends (such as 50% S_w, 25% S_w, etc.) can be plotted

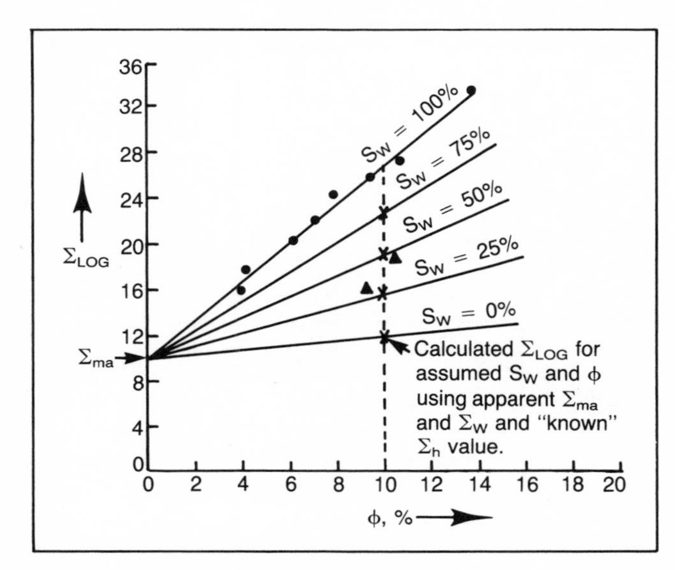

Fig. 8-59. Typical Σ_{log} vs. ϕ crossplot

using the general relation containing the influence of hydrocarbon if Σ_h can be estimated. This is done by calculating, for example, the Σ_{log} that would exist for an assumed $S_w = 50\%$ and $\phi = 10\%$ (use $\phi = .1$ in the equation). Once this Σ_{log} is determined, a straight line is drawn through this point and Σ_{ma}. It should be pointed out that an extrapolation of the 100% S_w trend to 100% ϕ will give Σ_w and extrapolation of the 0% S_w trend will give Σ_h at 100% ϕ. It is important to note that when using this technique, prior knowledge of Σ_{ma} and Σ_w are not required. In fact, these two parameters can be defined in this approach.

An example of this crossplotting technique is shown in Figure 8-61, based on data from the log shown in Figure 8-60. The log sections shown in wells A and B are through the Oswego lime in Kay County, Oklahoma. Well B is structurally lower than well A and is water-bearing. The eleven data points selected in the Oswego lime define the 100% S_w trend, as well as Σ_{ma} and Σ_w as shown in Figure 8-61. As a result the five intervals selected on well A indicate apparent water saturation ranging from 25–35%. Well A was oil-bearing.

The log-measured capture cross-section, Σ_{log}, can also be plotted as a function of log measured porosity parameters such as ρ_b, t, and ϕ_{snp} or ϕ_{CNL}.[27]

Considering the capture cross-section response in a zone of constant lithology and water salinity for 100% S_w:

$$\Sigma_{log} = \Sigma_{ma} + (\Sigma_w - \Sigma_{ma})\phi$$

For the density tool:

$$\phi = \frac{\rho_{ma} - \rho_b}{\rho_{ma} - \rho_f}$$

so:

$$\Sigma_{log} = \Sigma_{ma} + \left[\frac{\Sigma_w - \Sigma_{ma}}{\rho_{ma} - \rho_f}\right](\rho_{ma} - \rho_b)$$

or:

$$\Sigma_{log} = \Sigma_{ma} + \left[\frac{\Sigma_w - \Sigma_{ma}}{\rho_{ma} - \rho_f}\right]\rho_{ma} - \left[\frac{\Sigma_w - \Sigma_{ma}}{\rho_{ma} - \rho_f}\right]\rho_b$$

The above expression indicates that Σ_{log} can be plotted versus ρ_b on cartesian graph paper and the 100% S_w trend will, at 0% ϕ (where $\rho_b = \rho_{ma}$), intercept at $\Sigma_{log} = \Sigma_{ma}$ and at 100% S_w (where $\rho_b = \rho_f$) intercept at $\Sigma_{log} = \Sigma_w$. A typical crossplot of Σ_{log} vs. ρ_b is shown in Figure 8-62 where the limit of the 0% S_w trend will exist at $\rho_b = \rho_h$ and $\Sigma_{log} = \Sigma_h$.

The same argument can be made for the acoustic transit time response where:

$$\phi = \frac{t - t_{ma}}{B}$$

such that:

$$\Sigma_{log} = \Sigma_{ma} - \left[\frac{\Sigma_w - \Sigma_{ma}}{B}\right] t_{ma} + \left[\frac{\Sigma_w - \Sigma_{ma}}{B}\right] t$$

As shown in Figure 8-63 for 0% ϕ, where $t = t_{ma}$ the 100% S_w trend will intercept at $\Sigma_{log} = \Sigma_{ma}$. As expected for 100% ϕ (where $t - t_{ma} = B$), the value of $\Sigma_{log} = \Sigma_h$ for $t = B + t_{ma}$. If the acoustic transit time response can be expressed by the time average equation (clean, consolidated sandstones), then the limiting value for the 0% S_w trend will be $\Sigma_{log} = \Sigma_h$ at $t = t_h$ since $B = t_f - t_{ma}$ and $t_f = t_h$.

Influence of Shale Component

The capture cross-section response relation can also be written to include the influence of a shale component for a water-saturated rock system of constant lithology (other than shaliness) and constant water salinity such that:[24, 27]

$$\Sigma_{log} = \Sigma_{ma}(1 - \phi_e - V_{sh}) + \Sigma_{sh} V_{sh} + \Sigma_w \phi_e$$

For this case, the effective porosity, ϕ_e, is expressed as:

$$\phi_e + V_{ma} + V_{sh} = 1$$

Rewriting the capture cross-section response equation:

$$\Sigma_{log} = \Sigma_{ma} - \Sigma_{ma}\phi_e - \Sigma_{ma} V_{sh} + \Sigma_{sh} V_{sh} + \Sigma_w \phi_e$$

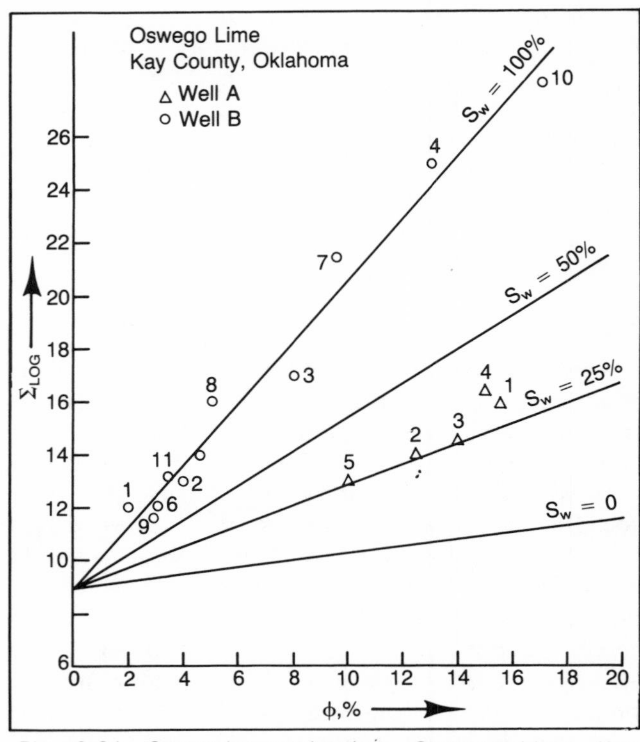

Fig. 8-61. Crossplot evaluation—Oswego Lime, Kay County, Oklahoma. Permission to publish by Dresser Atlas, Dresser Industries.[24]

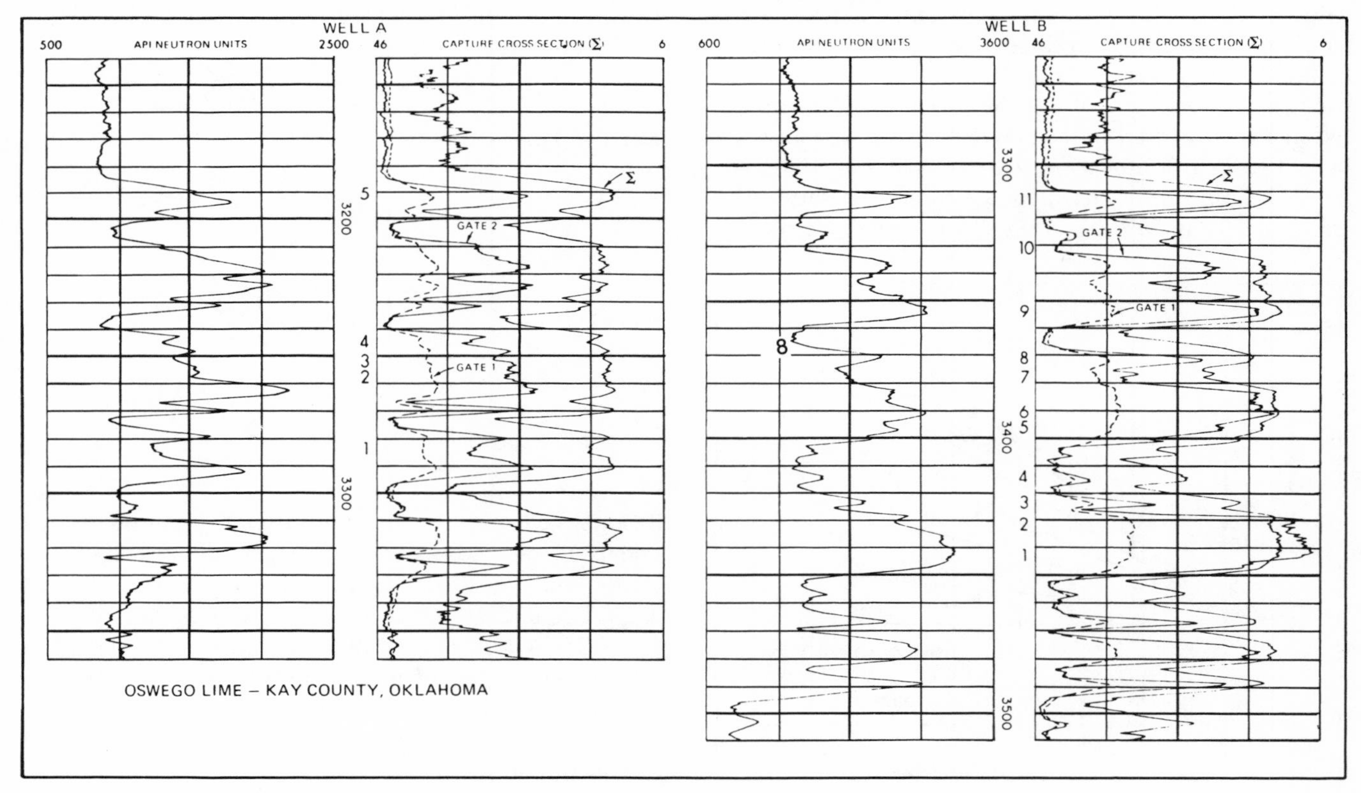

Fig. 8-60. Neutron and neutron lifetime logs in Oswego Lime, Kay County, Oklahoma. Permission to publish by Dresser Atlas, Dresser Industries. [24]

or:

$$\Sigma_{log} - \Sigma_{sh} V_{sh} = \Sigma_{ma}(1 - V_{sh}) + \phi_e(\Sigma_w - \Sigma_{ma})$$

then:

$$\left[\frac{\Sigma_{log} - \Sigma_{sh} V_{sh}}{1 - V_{sh}}\right] = \Sigma_{ma} + (\Sigma_w - \Sigma_{ma})\left[\frac{\phi_e}{1 - V_{sh}}\right]$$

This relation states that a cartesian plot (Fig. 8-64) of

$$\left[\frac{\Sigma_{log} - \Sigma_{sh} V_{sh}}{1 - V_{sh}}\right] \text{ vs. } \left[\frac{\phi_e}{1 - V_{sh}}\right]$$

will have an intercept at 0% ϕ of Σ_{ma}, and the slope of the 100% S_w trend will be $(\Sigma_w - \Sigma_{ma})$. Again, the limit of the 100% S_w trend will be Σ_w at 100% ϕ. For the 0% S_w trend, the limit at 100% ϕ will be Σ_h. In order to make the crossplot the volume of shale, V_{sh}, and Σ_{sh} must be determined. The key to the interpretation is evaluating V_{sh}, which can be estimated in a variety of ways, for example using the gamma ray response. Σ_{sh} is usually estimated from the capture cross-section response in a nearby shale.

This concept can be extended to crossplots using log-measured porosity parameters such as ρ_b, ℓ and ϕ_{SNP} or ϕ_{CNL}. Considering the density log response, the following response can be written for shaly zones:

$$\rho_b = \rho_w \phi_e + \rho_{sh} V_{sh} + (1 - \phi_e - V_{sh})\rho_{ma}$$

or:

$$\rho_b = \rho_{sh} V_{sh} + \rho_{ma}(1 - V_{sh}) + (\rho_w - \rho_{ma})\phi_e$$

where:

$$\phi_e = \left[\frac{\rho_b}{\rho_w - \rho_{ma}}\right] - \left[\frac{\rho_{sh} V_{sh}}{\rho_w - \rho_{ma}}\right] - \left[\frac{\rho_{ma}(1 - V_{sh})}{\rho_w - \rho_{ma}}\right]$$

Substituting ϕ_e into the expression for the capture cross-section expression, which includes the influence of shaliness we find:

$$\left[\frac{\Sigma_{log} - \Sigma_{sh} V_{sh}}{1 - V_{sh}}\right] = \Sigma_{ma} - \rho_{ma}\left[\frac{\Sigma_w - \Sigma_{ma}}{\rho_w - \rho_{ma}}\right] + \ldots$$

$$+ \left[\frac{\Sigma_w - \Sigma_{ma}}{\rho_w - \rho_{ma}}\right]\left[\frac{\rho_b - \rho_{sh} V_{sh}}{1 - V_{sh}}\right]$$

The above relation states that a crossplot on cartesian graph paper (Fig. 8-65) of

$$\left[\frac{\Sigma_{log} - \Sigma_{sh} V_{sh}}{1 - V_{sh}}\right] \text{ vs. } \left[\frac{\rho_b - \rho_{sh} V_{sh}}{1 - V_{sh}}\right]$$

will have an intercept

$$\left[\frac{\Sigma_{log} - \Sigma_{sh} V_{sh}}{1 - V_{sh}}\right] = \Sigma_{ma} \text{ at } \left[\frac{\rho_b - \rho_{sh} V_{sh}}{1 - V_{sh}}\right] = \rho_{ma}$$

for $\phi_e = 0\%$ since $V_{sh} = 0\%$. At 100% ϕ (where $\rho_b = \rho_w$ and $V_{sh} = 0$) for the 100% S_w trend, the limiting value of

$$\left[\frac{\Sigma_{log} - \Sigma_{sh} V_{sh}}{1 - V_{sh}}\right] = \Sigma_w.$$

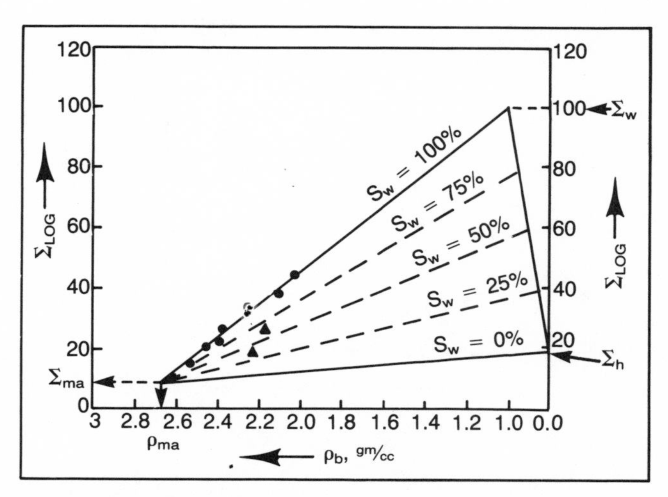

Fig. 8-62. Typical Σ_{log} vs. ρ_b crossplot

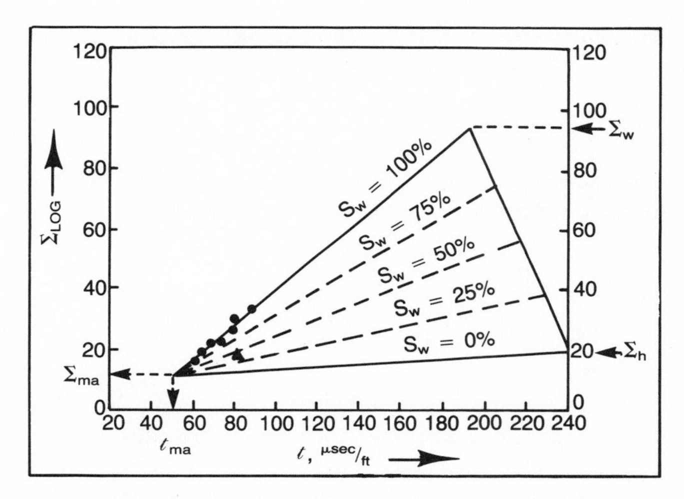

Fig. 8-63. Typical Σ_{log} vs. ℓ crossplot

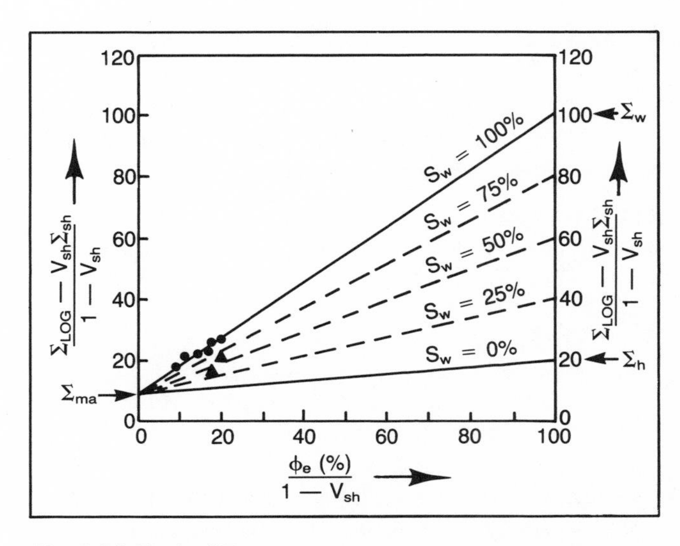

Fig. 8-64. Typical Σ_{log} vs. ϕ_e crossplot for shaly formation

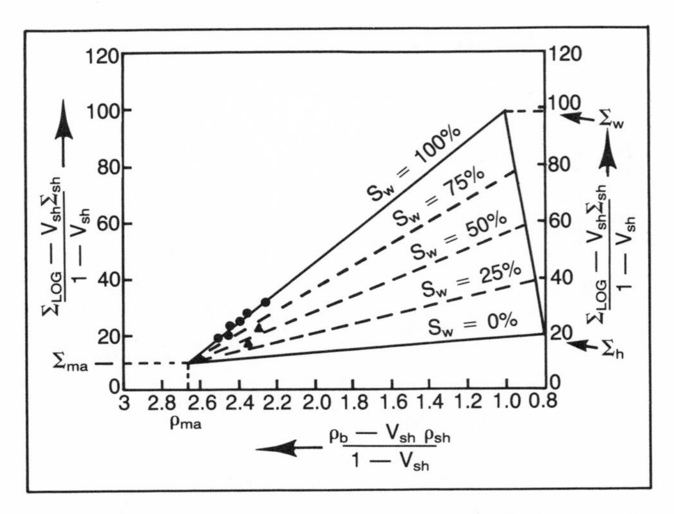

Fig. 8-65. Typical Σ_{log} vs. ρ_b crossplot for a shaly formation

Of course, for the 0% S_w trend at 100% ϕ, the value of

$$\left[\frac{\Sigma_{log} - \Sigma_{sh} V_{sh}}{1 - V_{sh}}\right] = \Sigma_h$$

since $\rho_b = \rho_h$ and $V_{sh} = 0$.

A similar type of crossplot could also be obtained using a time average equation of the form:

$$t = t_w \phi_e + t_{sh} V_{sh} + (1 - \phi_e - V_{sh}) t_{ma}$$

or:

$$t = t_{sh} V_{sh} + t_{ma}(1 - V_{sh}) + (t_w - t_{ma})\phi_e.$$

B

There is some doubt, however, concerning the validity of a time average equation, which includes proportionate contributions for variations of fluid in the pore space and also for shale proportions in the matrix.[27]

When considering the capture cross-section vs. porosity crossplot in a shaly formation using a neutron porosity response, the effective porosity is:

$$\phi_e = \phi_n - (I_{H_{sh}}) V_{sh}$$

where: $I_{H_{sh}}$ = hydrogen index of the shale and is the apparent porosity read from the neutron log in a nearby shale zone.

For this case:

$$\left[\frac{\Sigma_{log} - \Sigma_{sh} V_{sh}}{1 - V_{sh}}\right] = \Sigma_{ma} + (\Sigma_w - \Sigma_{ma})\left[\frac{\phi_N - (I_{H_{sh}}) V_{sh}}{1 - V_{sh}}\right]$$

and a crossplot of

$$\left[\frac{\Sigma_{log} - \Sigma_{sh} V_{sh}}{1 - V_{sh}}\right] \text{ vs. } \left[\frac{\phi_N - (I_{H_{sh}}) V_{sh}}{1 - V_{sh}}\right]$$

would define Σ_{ma} and Σ_w as well as the hydrocarbon saturation of the anomalous zones (not lying on the trend of $\Sigma_w = 100\%$) in the same manner as when using the density response.

Pulsed Neutron Decay Logs as Abnormal Pressure Indicators

The pulsed neutron decay logs have been found to be a valuable pressure detector since casing and cement have little effect on the overall response. An interesting application, as described by Fertl and Timko,[23] illustrates this advantage.

"In Figure 8-66, shale resistivity vs. depth is plotted for a well drilled in 1946. All the sand zones listed below 8,200 ft have been productive in this field for at least 25 years. This particular well was producing from the Klump series, but production was lost because of casing problems below 8,100 ft. It was planned to get the well back on production by recompleting in the Homeseekers 'A' sand at 9,060 ft by sidetracking above the fish and redrilling. The mud weight needed to drill this well in 1946 to the Homeseekers 'A' was approximately 14.0 lb/gal—the sands and shales contained geopressures of this magnitude.

"However, because of production and pressure depletion over the years, most of the sands now have producing pressure gradients throughout the field considerably less than hydrostatic. So the formations would not sustain the high mud weights used originally. The pulsed neutron was run in the old cased hole to evaluate present

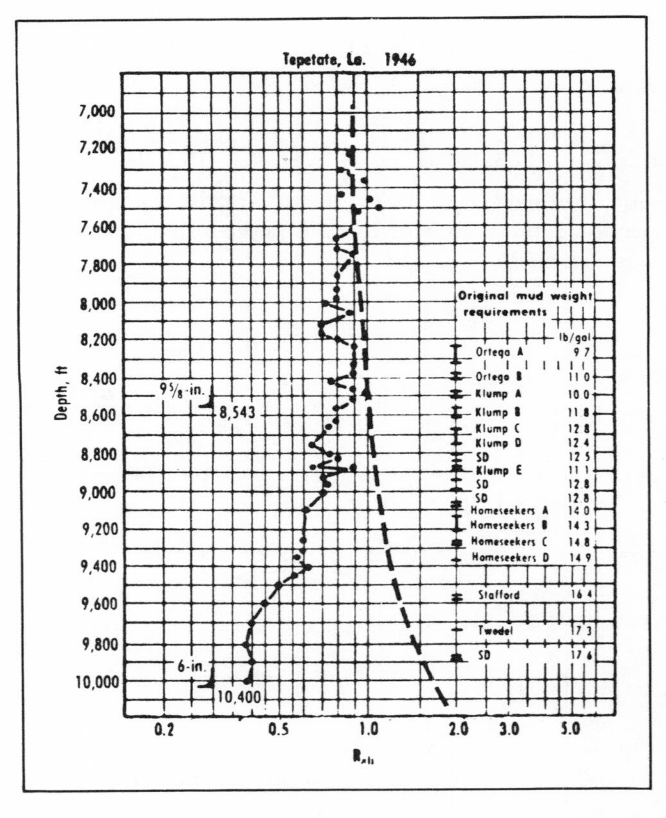

Fig. 8-66. Shale resistivity vs. depth. Permission to publish by the Oil and Gas Journal.[23]

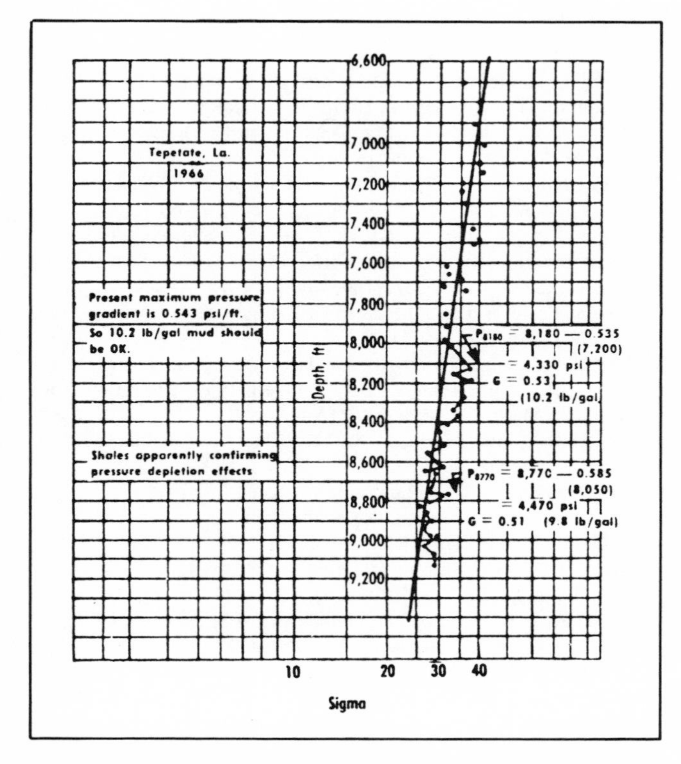

Fig. 8-67. NLL response vs. depth. Permission to publish by the Oil and Gas Journal.[23]

pressure conditions. Figure 8-67 is the response of the device vs. depth to 9,100 ft. The maximum mud-weight requirements as determined from the pulsed neutron to reach the Homeseekers 'A' sand were 10.2 lb/gal.

"Figure 8-68 shows the change in shale pressure due to pressure depletion of the sands. The well was side-tracked at 8,120 ft in 1968 and drilled to 9,215 ft without difficulty with a mud weight of 10.4 lb/gal, where originally 14.3 lb/gal was required."

Applications

The pulsed neutron decay log (NLL, TDT) has a number of uses, which include the following:

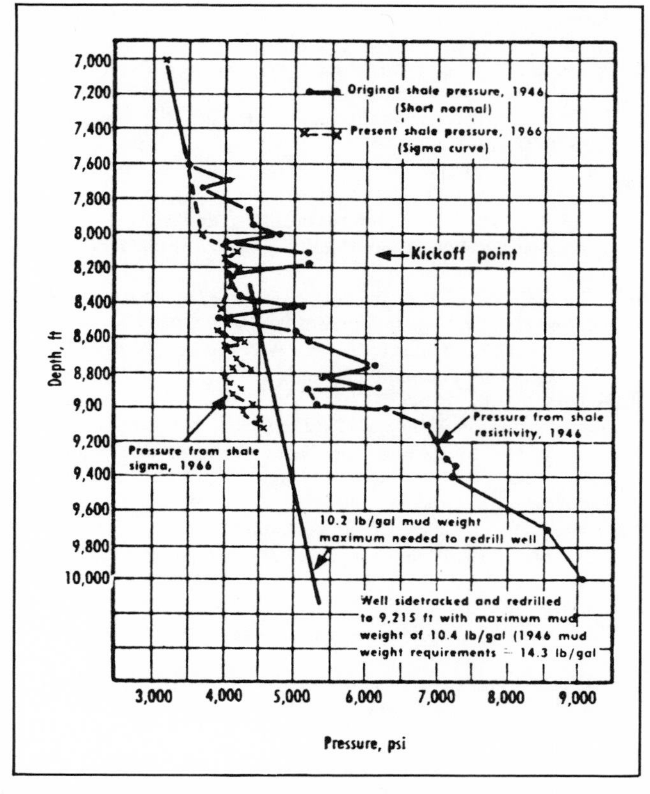

Fig. 8-68. Change in pressure due to production. Permission to publish by the Oil and Gas Journal.[23]

1. Distinguishes between gas, oil, and salt water-bearing formations, particularly in cased holes.
2. Determines water saturation.
3. Finds and follows gas-oil-water contacts behind casing.
4. Indicates abnormal pressure.

This chapter is not a complete treatise on radioactivity logging. Additional radioactivity logging tools of importance, beyond the scope of this book, include natural gamma ray spectroscopy logs, neutron activation logs such as the silicon log, carbon-oxygen logs, and lithodensity logs.

REFERENCES

1. Russell, W.L.: "Well Logging by Radioactivity," *Bull.*, AAPG (Sept. 1941) 25.
2. Bloch, F.: "Nuclear Induction," *Physical Review* (1946), 70, 460.
3. Campbell, J.L.P., and Winter, A.B.: "A Review of General Notes on Radioactivity Well Log Interpretation," *Tomorrow's Tools—Today*, Lane-Wells Company (1946).
4. Dewan, J.T. and Allaud, L.A.: "Experimental Basis for Neutron Logging Interpretation," *Petroleum Engineer International* (Sept 1953).
5. Tittman, J.: "Radiation Logging," *Fundamentals of Logging*, Petroleum Engineering Conference, U. of Kansas (April 1956).
6. Schwenk, H.C., and Shannon, R.H.: *Nuclear Power Engineering*, McGraw-Hill Book Co., Inc., New York (1957).
7. Brown, A.A., and Bowers, B.: "Porosity Determinations from Neutron Logs," *Petroleum Engineer International* (May 1958).
8. Glasstone, S.: *Sourcebook on Atomic Energy*, 2nd Ed., D. Van Nostrand Co., Inc., Princeton, N.J. (1958).
9. Price, W.J.: *Nuclear Radiation Detection*, 2nd Ed., McGraw-Hill Book Co., New York (1958).
10. API RP 33: "Recommended Practices for Standard Calibration and Format for Nuclear Logs," 3rd Ed., American Petroleum Institute, Div. of Production, Dallas, Tex. (1974).

11. Brown, R.J.S. and Gamson, B.W.: "Nuclear Magnetism Logging," *Trans.*, SPE (1960), Vol. 219.
12. Marion, J.B., and Fowler, J.L.: *Fast Neutron Physics*, Interscience Publishers (1960).
13. Lynch, E.J.: *Formation Evaluation*, Harper and Row, Publ., New York (1962).
14. Pirson, S.J.: *Handbook of Well Log Analysis*, Prentice-Hall, Inc., Englewood Cliffs, N.J. (1963).
15. Wahl, J.S., Tittman, J., Johnstone, C.W., and Alger, R.P.: "The Dual Spacing Formation Density Log," *J. Pet. Tech.* (Dec. 1964), 16.
16. Helander, D.P.: "Radiation Measurement, Gamma Ray Logs," *Oil and Gas Equipment Journal* (Aug. 1965).
17. Helander, D.P.: "Guide to Nuclear Well Logging Tools," *Oil and Gas Equipment Journal* (September 1965).
18. Helander, D.P.: "Logs Maintain Depth Control for Jet, Bullet-Type Perforators," *Oil and Gas Equipment Journal* (Jan. 1966).
19. Moe, H.J., Lasuk, S.R. and Schumacher, M.C.: "Radiation Safety Technician Course—Argonne National Laboratory," Government Report No. ANL-7291 (Sept. 1966).
20. Owen, J.D.: "A Review of Fundamental Nuclear Physics Applied to Gamma Ray Spectral Logging," *The Log Analyst* (Sept.-Oct. 1966), 7.
21. Tittman, J., Sherman, H., Nagel, W.A., and Alger, R.P.: "The Sidewall Epithermal Neutron Porosity Log," *J. Pet. Tech.* (Oct. 1966), 18.
22. Hilchie, D.W., et al.: "Some Aspects of Pulsed Neutron Logging," *The Log Analyst* (March 1969).
23. Fertl, W.H., and Timko, D.J.: "How Abnormal-Pressure-Detection Techniques are Applied," *Oil and Gas J.* (January 12, 1970).
24. Neutron Lifetime Interpretation, Dresser-Atlas, 1970.
25. Clavier, C., Hoyle, W., and Meunier, D.: "Quantitative Interpretation of Thermal Neutron Decay Time Loss: Part I. Fundamentals and Techniques," *J. Pet. Tech.* (June 1971).
26. Dresser-Atlas, Log Review I, 1971.
27. Threadgold, P.: "Interpretation of Thermal Neutron Die-Away Logs—Some Useful Relationships," *Trans.*, SPWLA (1971).
28. Alger, R.P., Locke, S., Nagel, W.A., and Sherman, H.: "The Dual Spacing Neutron Log—CNL," *J. Pet. Tech.* (Sept. 1972).
29. Allen, L.S., Mills, W.R., Desai, Kantilal P., and Caldwell, R.L.: "Some Features of Dual-Spaced Neutron Porosity Logging," *Log Analyst* (July-Aug. 1972), 13.
30. Fertl, W.H., and Timko, D.J.: "How Downhole Temperatures Affect Drilling Pressures—Part 3: Overpressure Detection from Wireline Methods," *World Oil* (Aug. 1972).
31. Hung, J.E., and Salisch, H.A.: "The Dual Spacing Neutron Log (CNL) in Venezuela," *Trans.*, SPWLA (1972).
32. *Log Interpretation Charts*, Schlumberger Ltd., 1972.
33. *Log Interpretation Principles*, Schlumberger Ltd., 1972.
34. Truman, R.B., Alger, R.P., Connell, J.G., and Smith, R.L.: "Progress Report on Interpretation of the Dual-Spacing Neutron Log (CNL) in the U.S.," *The Log Analyst* (July-Aug. 1972), 13.
35. Cased Hole Applications, Schlumberger Well Services, 1975.
36. *Gamma Ray Log*, Dresser Atlas, Dresser Industries, Houston (1981).
37. Lane-Wells Radioactivity Logging, Lane-Wells Co., Houston, Texas.
38. *Log Interpretation Charts*, Schlumberger Ltd., Houston (1979).

9 Interpretation Methods

THE process of formation evaluation is certainly highly complex. You have been introduced to the problems involved in formation evaluation, such as indirectness, uncertainty, and costs, and have also been exposed to a number of interpretation techniques. Additionally, you now have a working knowledge of the basic wireline logging tools and their response equations and an insight to their applications and limitations. At this point, you are also aware of the many factors that will influence the interpretation process:

1. Well Type: wildcat, development, etc.
2. Geologic Environment: carbonate system, sand-shale system, etc.
3. Borehole Environment: conductive or nonconductive borehole fluid, fresh water or saltwater mud, cased or uncased hole, etc.
4. Available Supporting Data: mud log, cores, etc.
5. Available Time
6. Costs

One could, of course, add other factors, but this list certainly indicates the almost limitless problems that can be encountered when attempting to evaluate a subsurface system for commercial hydrocarbon-bearing formations. In fact, each well presents a unique problem comprised of these and other factors for which there is obviously no universal interpretation process. Therefore, the logging program and the attendant evaluation process are, to some degree, unique in each individual well. In this chapter, the basic principles and techniques used in formation evaluation for hydrocarbons will be presented.

Before discussing methods employed in formation evaluation, however, it might be well to redefine the objective of the formation evaluator: to identify and evaluate commercial hydrocarbon-bearing formations, generally by selecting the minimum cost combination of measurements that will provide a definitive evaluation. The selection of an optimum program to meet this objective is not an easy task. Design of the logging program must be tempered with an understanding of tool response as related to formation conditions, well conditions, and other factors, as well as an insight into the particular problem to be solved.

In order to accomplish this, the formation evaluator employs (either consciously or unconsciously) the concept of *rock typing*. The process of rock typing (1) assists in defining the problem, (2) simplifies the evaluation process, (3) reduces costs, (4) allows projection of formation parameters, (5) can provide additional information that might be considered a bonus for the geologist and engineer, such as identifying geologic environments, and (6) reduces uncertainty.

Rock typing is basically the division of a geologic section into groups represented by different properties and/or relations between properties. Commonly used rock typing criteria are:

1. Lithology (limestone, sandstone, etc.)
2. Porosity versus permeability relation
3. Porosity log response versus porosity relation
4. Porosity log response versus porosity log response relation
5. *M* versus *N* response
6. *MN* products versus density log response
7. Formation resistivity factor versus porosity relation (slope m, cementation factor)
8. Resistivity index versus water saturation relation (slope n, saturation exponent)
9. Apparent density, $(\rho_{ma})_a$, versus apparent interval transit time, $(t_{ma})_a$, behavior
10. Porosity versus water saturation relation
11. Capillary pressure curve shape
12. Residual hydrocarbon saturation versus initial hydrocarbon saturation relation (from series of hysteresis capillary pressure measurements)

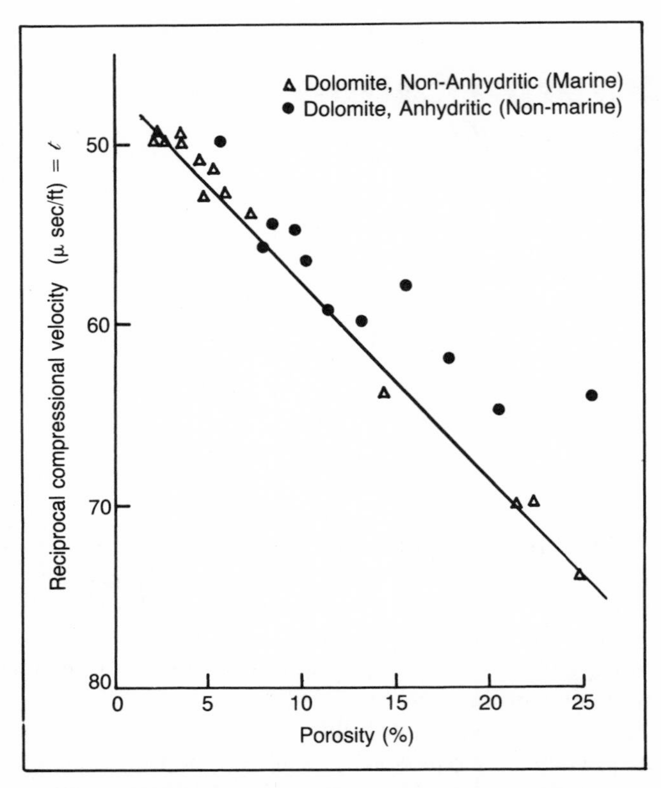

Fig. 9-1. Transit time vs. porosity for a typical carbonate tidal flat section. Permission to publish by the Society of Professional Well Log Analysts.[14]

The measured parameters reflect rock properties, which tend to vary by discrete steps rather than vary continuously over an entire geologic section. Additionally, the measured rock parameters are not independent properties since, for example, all of the following reflect the influence of pore geometry (pore size distribution, surface area): the porosity versus permeability relation, porosity log response versus porosity relation, porosity log response versus porosity log response relation, formation resistivity factor versus porosity relation, resistivity index versus water saturation relation, porosity versus water saturation relation, capillary pressure curve shape, and the residual hydrocarbon saturation versus initial hydrocarbon saturation relation.

In order to understand rock typing, consider the example (see Fig. 1-6) used in Chapter 1. Figure 9-1 shows acoustic log transit times plotted versus core porosities. The porosities come from a 100-ft core in dolomites representing a carbonate tidal flat environment.[14] The dolomites below low-tide level (marine) tend to be nonanhydritic, whereas the dolomites in the intertidal and supratidal zones tend to contain varying amounts of anhydrite. The nonanhydritic dolomites from the marine environment tend to define a linear trend, which would be different from the average trend one might fit to the anhydritic, nonmarine dolomites. From this, we can say that two rock types exist in this 100-ft section. Although there is only one lithology, there are two distinct trends present with respect to one of the rock typing criteria. More rock types may exist since some of the other criteria could show differences within the two rock types now defined, but it is apparent that we are not dealing with only one rock type as defined by lithology.

Therefore, the importance of grouping rocks within a geologic section (rock typing) can now be readily illustrated. Suppose that in this section, the nonanhydritic (marine) dolomite was the productive rock type and existed at irreducible water saturation. This dictates that any evaluation program in this section must include a method to distinguish the nonanhydritic (marine) dolomite from the anhydritic (nonmarine) dolomite. This might be accomplished by using good cuttings samples, or through the use of another log not affected by anhydrite as is the acoustic log. This second porosity log could be a neutron log where a simultaneous solution of the two response equations could differentiate the anhydritic from the nonanhydritic dolomites. To this point, the process of rock typing has assisted in defining the problem more clearly and at the same time simplified the task, because the amount of section having to be analyzed in detail has been reduced.

Now, suppose that for evaluation purposes we wanted ϕ, k_a, and S_w for all productive zones within this dolomite section in all future wells. The most direct approach would be to determine R_t, R_w, m, n, ϕ, and k_a, thereby requiring cores for m, n, ϕ, k_a, productivity tests for R_w, and a resistivity log for R_t. Through rock typing, two rock types and the transit time, ℓ, vs. porosity, ϕ, relation for each have already been defined. Therefore, an acoustic log with cuttings samples, a neutron log, or a density log would define the productive rock type, nonanhydritic (marine) dolomite, and its porosity. If, in addition, by using the original 100-ft section of core, suppose we were able to define the trends shown in Figures 9-2 and 9-3. Then, in subsequent wells we could also conceivably project both k_a and S_w using only, for example, two porosity logs—a simple program, which would reduce costs considerably. As a bonus, we also would know the tidal flat environment of each zone. This example is quite simple but does illustrate the concept of rock typing and the important implications it has in formation evaluation. The use of rock typing is interwoven throughout the process of formation evaluation and when thoughtfully applied, can be a very powerful tool for the evaluator.

The interpretation process is based on a few basic relations that we have already encountered. These consist of the saturation relations (Archie equations) and the tool response relations. These relations can be used in a variety of ways to generate a multiplicity of "interpretation methods." These methods might be roughly divided into two groups which, for convenience, will be called the standard methods and the crossplotting methods.

STANDARD LOG INTERPRETATION METHODS

Standard log interpretation approaches can be considered to include the following techniques:

1. Conventional method
2. R_{wa} method
3. F_R comparison method
4. Movable oil method
5. R_{xo}/R_t vs. E_{ssp} method
6. S_{or}/S_o method

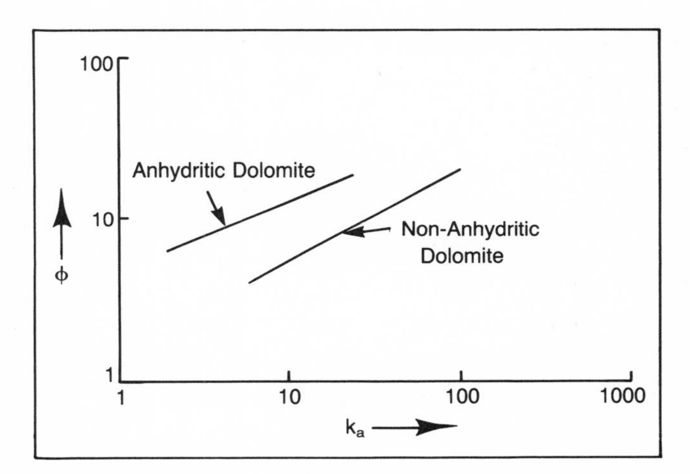

Fig. 9-2. Porosity vs. permeability relation

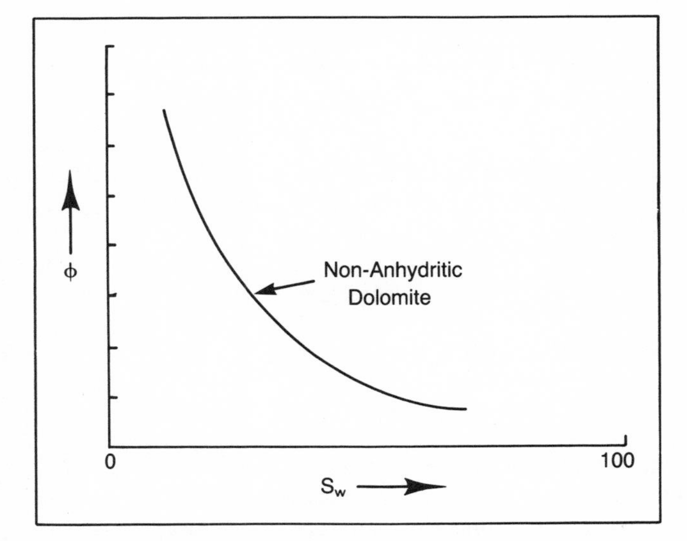

Fig. 9-3. Porosity vs. water saturation relation

Each of these methods is directed to determining S_w or a water saturation parameter. Using the basic relations and making some realistic assumptions, it is possible to quickly observe anomalous behavior of hydrocarbon zones when compared to water-bearing zones. The process of comparing anomalous zone behavior to observed or projected water zone behavior has led to a number of standard approaches and many variations in the application of these approaches. Each of these methods will be briefly described in the following sections.

Conventional Method

This approach is simply the "cook book" solution of the saturation equations (see Fig. 9-4). In this approach it is assumed that all the required parameters (R_t, R_w, ϕ, m, n) can be determined in an absolute sense. For example, the log measured parameter R_a, obtained from a deep responding resistivity tool, can be corrected to R_t using suitable departure curves. All formation evaluators use this approach, although the limitations are obvious, since many times we do not know all five required variables accurately enough, if at all. Since absolute values of all five parameters are required, in addition to using the "best" corrected log values available, we will also resort to (1) using average relations to convert measured values to parameters of interest, (2) using a range of values for an uncertain quantity, and (3) using previous experience to define a parameter. The method does, however, have the advantage of tending to account for all variables "properly." Application of this approach is tedious and time-consuming and not, therefore, readily applicable to a multiplicity of zones over a large interval of log. As a result, the quickly applied methods that provide a rapid approach for pointing out those anomalous zones (apparent hydrocarbon-bearing zones) existing in a logged interval are most important, and although usually providing only a quick look, these methods can also provide a quantitative evaluation.

A quick look approach for the conventional method is illustrated in Figures 9-12 and 9-13 of this chapter. This technique generates an apparent R_o over an interval (where R_w can be considered constant) and compares this to the deep resistivity measurement, which is considered to reflect R_t. The apparent R_o is generated over the interval from a measured porosity parameter such as bulk density, ρ_b. The measured bulk density is converted to porosity and then to the formation resistivity factor, F_R, assuming appropriate values for ρ_{ma}, ρ_f, a, and m. If R_w is known or can be estimated for the interval, then F_R can be computed and recorded in analog form along with the deep resistivity. This provides a quick insight into anomalous behavior and hence apparent hydrocarbon saturation as reflected by separation in

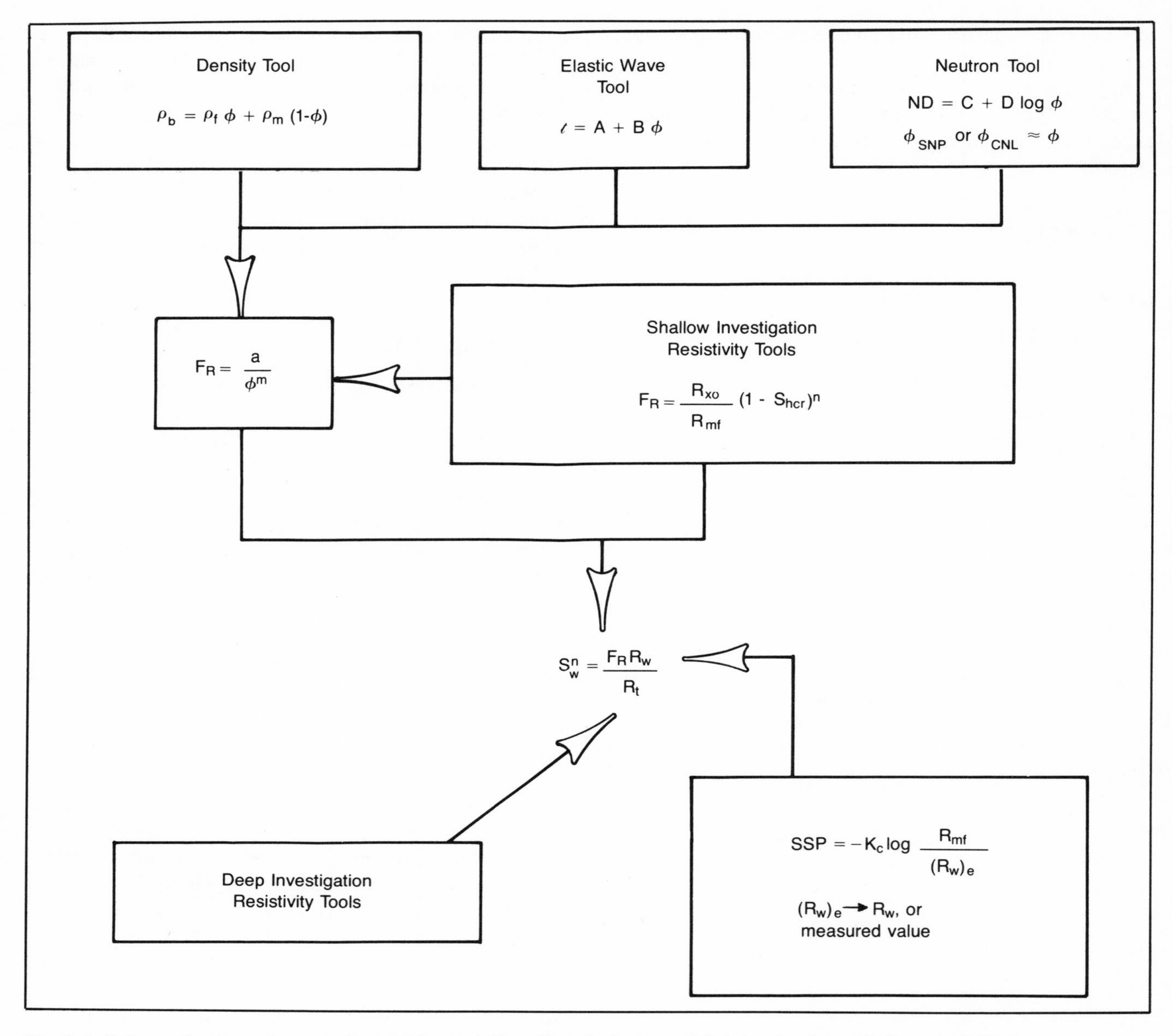

Fig. 9-4. Schematic view of conventional interpretation. Permission to publish by the John M. Campbell Co.[26]

the R_o and R_t curves. If R_w is unknown, the F_R curve can be recorded in analog form and then overlaid on the deep resistivity measure (best overall fit) to represent 100% S_w over at least a portion of the apparent R_t measure. This is known as the F_R overlay method and does imply that at least a portion of the interval is 100% water-saturated. Since the force fit will define an apparent R_w (the constant by which the F_R curve has been shifted to make the overlay), the quality of this insight might be supported.

R_{wa} Method

Calculation of an apparent water resistivity, R_{wa}, requires less information than the conventional method. It is apparent from the saturation equation $R_t = R_w F_R I_R$ that:

$$R_w I_R = \frac{R_t}{F_R}$$

Now, if R_{wa} is calculated as:

$$R_{wa} = \frac{R_t}{F_R}$$

we would actually be calculating a value of $R_w I_R = R_{wa}$. Of course, in 100% water zones where $I_R = 1$, R_{wa} would equal R_w. In application, an uncorrected deep resistivity measurement is normally used to represent R_t. A value of F_R is generated from a porosity response, usually assuming appropriate unknowns such as m, t_{ma}, and B when using the acoustic log response. Comparison of R_{wa} values between zones where R_w may be assumed to remain constant can indicate anomalous behavior. For example, consider zone 1 as hydrocarbon-bearing, and zone 2 as water-bearing. Comparing zone 1 and zone 2 we obtain:

$$\frac{R_{wa_1}}{R_{wa_2}} = \frac{(R_w I_R)_1}{(R_w I_R)_2} = I_{R_1} = S_w^{-n}$$

since R_w is constant and $I_{R_2} = 1$.

If R_{wa_2} is the true R_w or a minimum observed value of R_{wa}, which might approximate R_w, then we might consider the apparent water saturation to be:

$$S_w = \left(\frac{R_w}{R_{wa}}\right)^{1/n} \approx \left(\frac{R_{w_{min}}}{R_{wa}}\right)^{1/n}$$

R_{wa} values can be quickly calculated for a multiplicity of zones and compared, assuming R_w remains fairly constant. It is, however, well-suited to computer evaluation at the well site, as shown in Figure 9-5. For those zones where R_{wa} is greater than $4R_w$ then probably S_w is less than 50%. Figure 9-5 is an example of the application of the R_{wa} technique using uncorrected digital log data at the well site to compute a "quick look" log. The gas zone D stands out in both the IES and computed curves. Not so apparent on the IES is the gas section at the top of zone B, which stands out on the computed curves. The two sands, A and C, exhibit similar characteristics on the IES, and it would appear that if zone A were water-saturated, zone C would also be water-saturated. The R_{wa} curve, however, indicates an R_{wa} that is four times greater than the R_{wa} determined in the water-saturated intervals.

A test of the upper five feet of zone C obtained 1.347 MMcf/d of gas, 90 bbl/d of distillate, and no water. This compared favorably with a 14-ft test on zone D, which produced 1.92 MMcf/d of gas, 153.5 bbl/d of distillate and 7.7 bbl/d of water.

This method works best where invasion is not deep since the deep resistivity reading is assumed to represent R_t, unaffected by invasion. The best areas of application are, therefore, usually in those zones having higher porosity or when low water loss mud is used.

F_R Comparison Method

This technique compares a formation factor determined from resistivity measurements, $(F_R)_R$, with a formation factor determined from an independent porosity tool such as the acoustic log, $(F_R)_a$, or the density log, $(F_R)_D$. The basis for this method is that the porosity tool response can be converted to the "true" formation factor, F_R, using appropriately assumed constants such as m, t_{ma} and B when using the acoustic log. It is assumed then that $(F_R)_a$ or $(F_R)_D = F_R$. Additionally, the value of $(F_R)_R$ would be defined by the saturation equation where:

$$F_R = \frac{R_t}{I_R R_w}$$

Now, if $(F_R)_R$ is calculated as:

$$(F_R)_R = \frac{R_t}{R_w}$$

we would actually be calculating a value of $I_R F_R = (F_R)_R$ such that the ratio would be:

$$\frac{(F_R)_R}{F_R} = \frac{I_R F_R}{F_R} = I_R = S_w^{-n}$$

This, of course, would be a good saturation indicator except usually we have difficulty in defining R_w. This approach can be modified, however, by looking at the flushed zone where it is assumed that the conductive fluid filling the pore space near the wellbore is mud filtrate, and we feel we know R_{mf} better than R_w. In this case, $(F_R)_R$ is defined using flushed zone resistivity data where:

$$R_{xo} = I_{xo} F_R R_{mf} = S_{xo}^{-n} F_R R_{mf}$$

or:

$$F_R = \frac{R_{xo}}{I_{xo} R_{mf}} = \frac{R_{xo} S_{xo}^n}{R_{mf}}$$

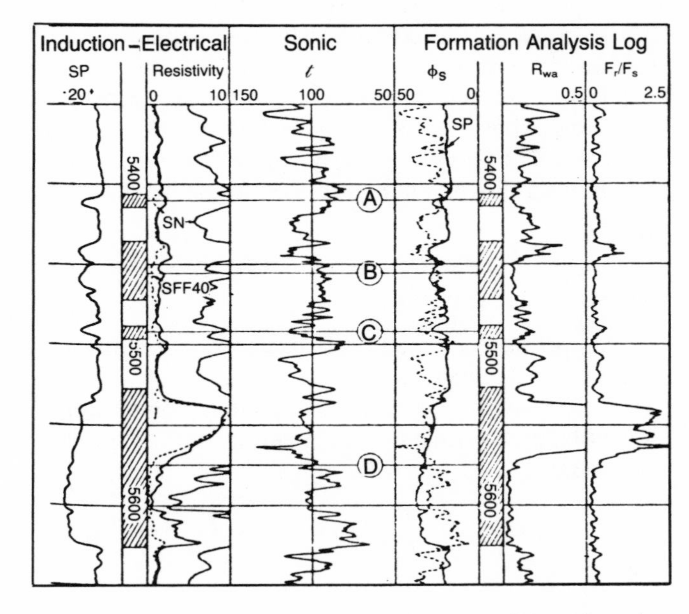

Fig. 9-5. Example of computed "quick look" log. Permission to publish by the Society of Petroleum Engineers of AIME.[6] Copyright 1965 SPE-AIME.

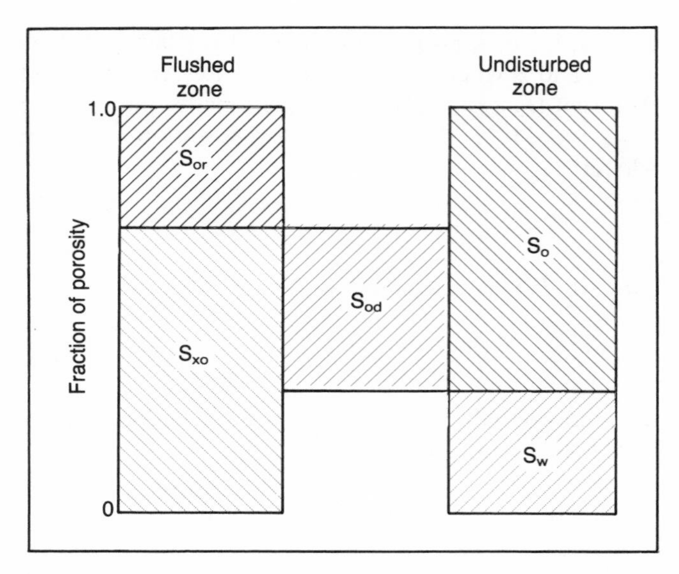

Fig. 9-6. Schematic illustration of movable oil concept

Now, if $(F_R)_R$ is calculated as:

$$(F_R)_R = \frac{R_{xo}}{R_{mf}}$$

we would actually be calculating a value of $F_R I_{xo} = (F_R)_R$ such that the ratio would be:

$$\frac{(F_R)_R}{F_R} = \frac{I_{xo}F_R}{F_R} = I_{xo} = S_{xo}^{-n} = \frac{1}{(1-S_{or})^n}$$

In other words, an $(F_R)_R/F_R$ ratio greater than one would reflect the presence of residual oil, with this value increasing as the residual oil saturation increases. In practice, the most shallow resistivity measure available is used for R_{xo}. If a response that is not primarily an R_{xo} measure is used, such as the SFL or short normal, then this insight tends to be optimistic, but this is preferable to overlooking a possible hydrocarbon-bearing zone. An example of a continuously computed $(F_R)_R/F_R$ output is also shown on Figure 9-5, along with the R_{wa} curve.

Movable Oil Method

The movable oil method is based on the concept of comparing the water saturation existing in the flushed zone after invasion with the water saturation existing in the undisturbed zone. Using the saturation equations we can write:

$$S_w = \left(\frac{F_R R_w}{R_t}\right)^{1/n} \quad \text{and} \quad S_{xo} = \left(\frac{F_R R_{mf}}{R_{xo}}\right)^{1/n}$$

since:

$$F_R = \phi^{-m}$$

then the difference between S_{xo} and S_w, which represents the amount of hydrocarbon displaced by flushing, is the movable (displaced) oil saturation, S_{od} and can be expressed as a fraction of the pore volume as follows:

$$S_{od}(V_p \text{ Basis}) = S_{xo} - S_w = \left(\frac{R_{mf}}{\phi^m R_{xo}}\right)^{1/n} - \left(\frac{R_w}{\phi^m R_t}\right)^{1/n}$$

This concept of movable oil saturation is schematically illustrated in Figures 9-6 and 9-8. Obviously, in this form, a knowledge of many parameters is required, but if it is assumed that $m=n$, and the movable oil saturation is expressed as a fraction of the bulk volume, then we can write:

$$S_{od}\phi(V_b \text{ Basis}) = (S_{xo} - S_w)\phi = \left(\frac{R_{mf}}{R_{xo}}\right)^{1/n} - \left(\frac{R_w}{R_t}\right)^{1/n}$$

This method is descriptively shown in Figure 9-7. Using this relation, a knowledge of ϕ or m is not necessary, although the approach still requires a knowledge of R_{mf}, R_w, R_{xo}, R_t and n. For "quick look" purposes, uncorrected values of R_{xo} (flushed zone measure) and R_t (deepest resistivity response) and n is assumed. Some insight for the value of R_w is necessary, however. The method works best when $R_{mf} = R_w$ since "ideal" invasion is not necessary although deep enough invasion is required for the shallow resistivity response to be sensing a zone from which hydrocarbons have been moved. A typical movable oil plot, MOP, is shown in Figure 9-9

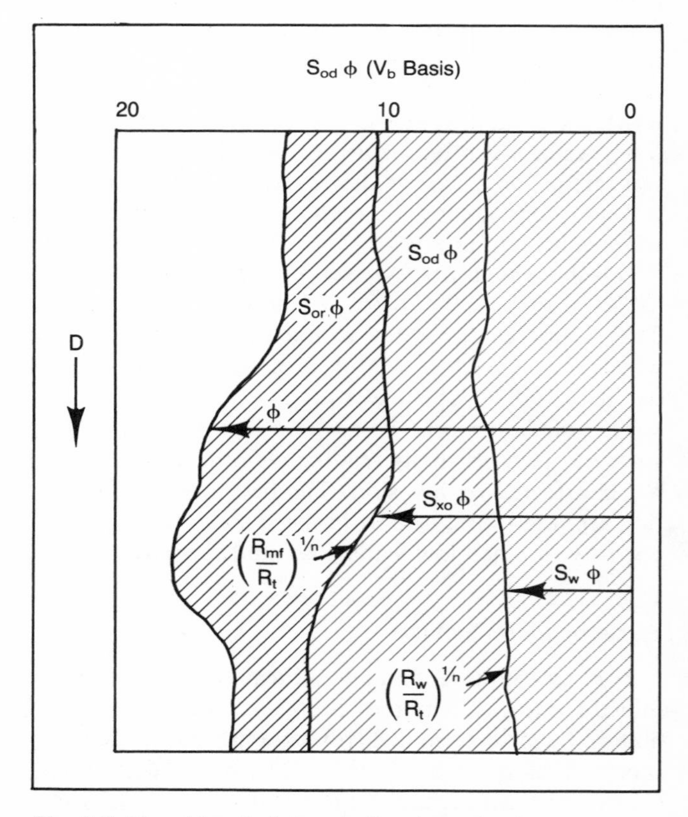

Fig. 9-7. Movable oil plot on bulk volume basis

where the S_{od} is based on the percent of bulk volume and the total porosity is also included.

The advantage of this method is that a single zone can be analyzed, with the factors that affect the deep resistivity response being basically the same as those that affect the shallow response. In essence then, rather than comparing apples and oranges, as in the R_{wa} technique for example, one is comparing oranges and tangerines (similar to an orange). Additionally, the results represent the results of a dynamic displacement situation that has occurred in the formation and, therefore, indicates permeability to hydrocarbon. It must be remembered, however, that the displacement process that has occurred represents the amount of oil that can be moved by "ideal" water displacement, although reasonable recovery factors to different primary production mechanisms can usually be generated from this MOS data.

R_{xo}/R_t vs. E_{SSP} Method

The R_{xo}/R_t vs. E_{SSP} method provides an estimate of S_w without a prior knowledge of ϕ or F. It is based on the SP response equation, which is:

$$E_{SSP} = -K_c \log \frac{R_{mf}}{R_w}$$

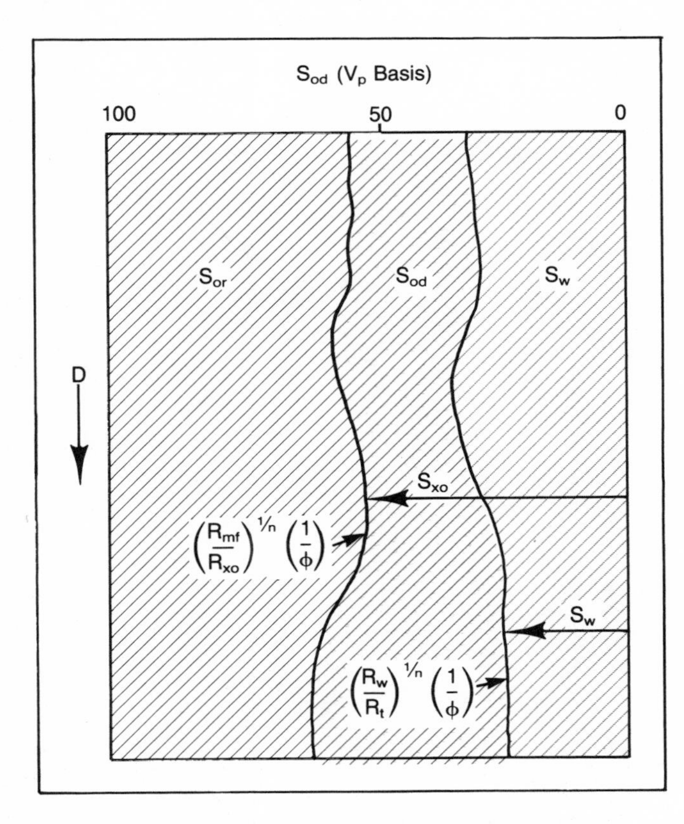

Fig. 9-8. Movable oil plot on pore volume basis

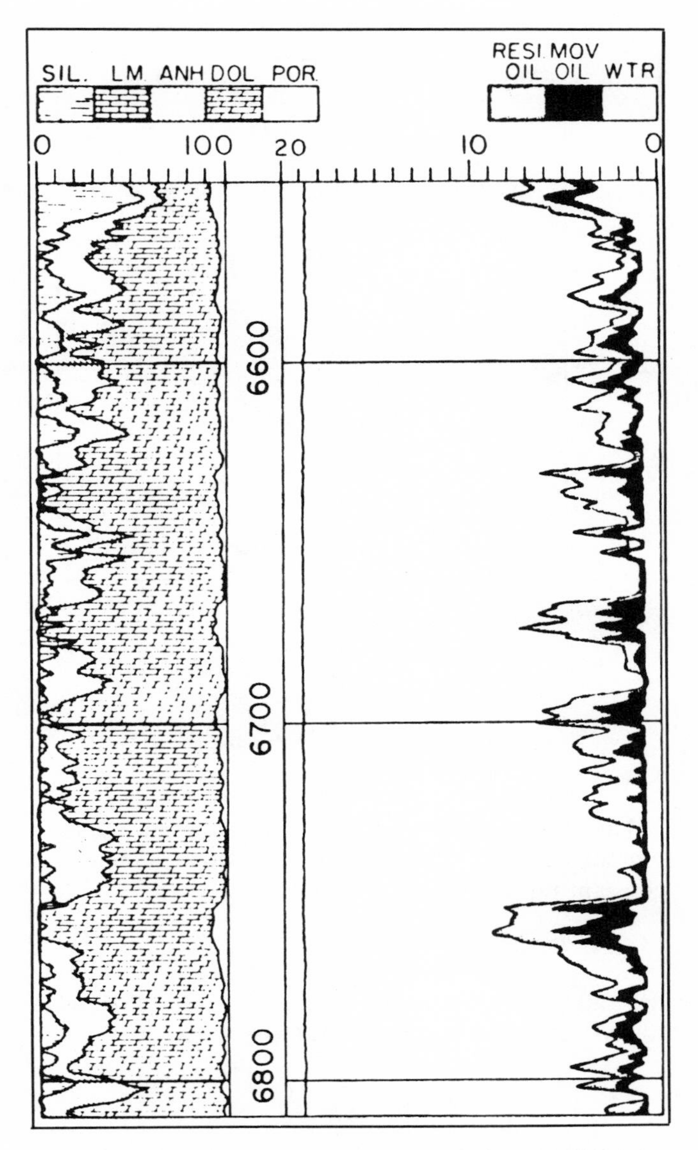

Fig. 9-9. Typical movable oil plot. Permission to publish by Petroleum Engineer International.[12]

However, we can define R_w and R_{mf} as:

$$R_w = \frac{R_t}{I_R F_R} = \frac{S_w^n R_t}{F_R}$$

$$R_{mf} = \frac{R_{xo}}{I_{xo} F_R} = \frac{S_{xo}^n R_{xo}}{F_R}$$

such that when $R_w < R_{mf}$ and the E_{SSP} deflection is negative, we have:

$$\frac{E_{SSP}}{K_c} = \log \frac{\left(\frac{R_{xo}}{I_{xo}}\right)}{\left(\frac{R_t}{I_R}\right)} = \log \frac{\left(\frac{R_{xo}}{R_t}\right)}{\left(\frac{I_{xo}}{I_R}\right)} = \log \frac{R_{xo}}{R_t} - \log \frac{I_{xo}}{I_R}$$

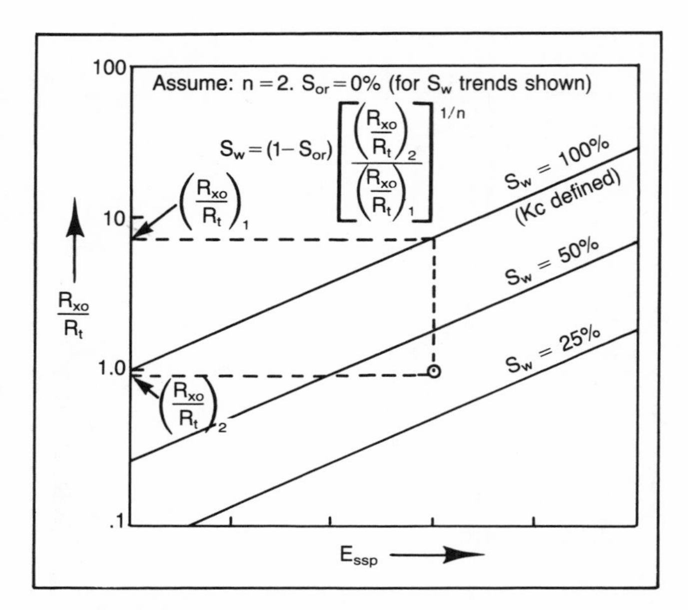

Fig. 9-10. Typical R_{xo}/R_t vs. E_{SSP} plot

or:

$$\log \frac{R_{xo}}{R_t} = \underbrace{\log \frac{I_{xo}}{I_R}}_{\text{Intercept}} + \underbrace{\frac{1}{K_c}}_{\text{Slope}} E_{SSP}$$

The above relation simply says that when R_{xo}/R_t is plotted logarithmically versus the E_{SSP} all zones having a constant ratio of I_{xo}/I_R, where

$$\frac{I_{xo}}{I_R} = \frac{S_w^n}{S_{xo}^n}$$

will fall on a trend having a slope of $1/K_c$. Obviously, for 100% water-saturated zones, the ratio

$$\frac{I_{xo}}{I_R} = \frac{S_w^n}{S_{xo}^n} = 1$$

such that:

$$\log \frac{R_{xo}}{R_t} = \frac{1}{K_c} E_{SSP}$$

and for an $E_{SSP} = 0$, the ratio $R_{xo}/R_t = 1$. The 100% S_w trend is shown in Figure 9-10. Any point falling below the defined 100% S_w trend is anomalous and assumed to be hydrocarbon-bearing. A family of parallel trends can be defined for a range of water saturations, as shown in Figure 9-10, once values of n and S_{or} are defined.

This method has the advantage of defining S_w for single zones requiring only R_t, R_{xo} and the E_{SSP} (or R_{mf}/R_w. It assumes that the E_{SSP} response is only electrochemical in nature. The presence of an abnormal SP caused by a streaming potential will make the evaluation optimistic since the E_{SSP} will be abnormally high. As described in the spontaneous potential chapter, the R_{xo}/R_t vs. E_{SSP} approach is a means to evaluate the possible presence of a streaming potential, as shown in Figure 9-11. The method assumes that the zone is invaded but not so deeply that R_t is adversely affected.

A quick look variation of this method is most useful for wellsite analysis.[19] Comparing the flushed zone water saturation to the undisturbed zone, water saturation will provide a quick insight into the ratio I_{xo}/I_R or S_w^n/S_{xo}^n. Using the saturation equations we find:

$$\frac{I_{xo}}{I_R} = \frac{\dfrac{R_{xo}}{F_R R_{mf}}}{\dfrac{R_t}{F_R R_w}} = \frac{\dfrac{R_{xo}}{R_t}}{\dfrac{R_{mf}}{R_w}} = \frac{S_w^n}{S_{xo}^n}$$

Using dual induction log responses, R_{LL8} and R_{ID}, a "computed E_{SSP}" is compared to the recorded E_{SSP} to provide a quick look insight into the presence of movable oil since a ratio of $(S_w^n/S_{xo}^n) < 1$ indicates movable oil as expressed below:

$$\frac{S_w^n}{S_{xo}^n} \approx \left[\frac{-K_c \log\left(\dfrac{R_{xo}}{R_t}\right)}{-K_c \log\left(\dfrac{R_{mf}}{R_w}\right)}\right] \approx \left[\frac{\left(\dfrac{R_{xo}}{R_t}\right)_{QL}}{E_{SSP}}\right]$$

This method is particularly useful where water resistivities tend to change from zone to zone, since the influence of R_w for any point of comparison automatically

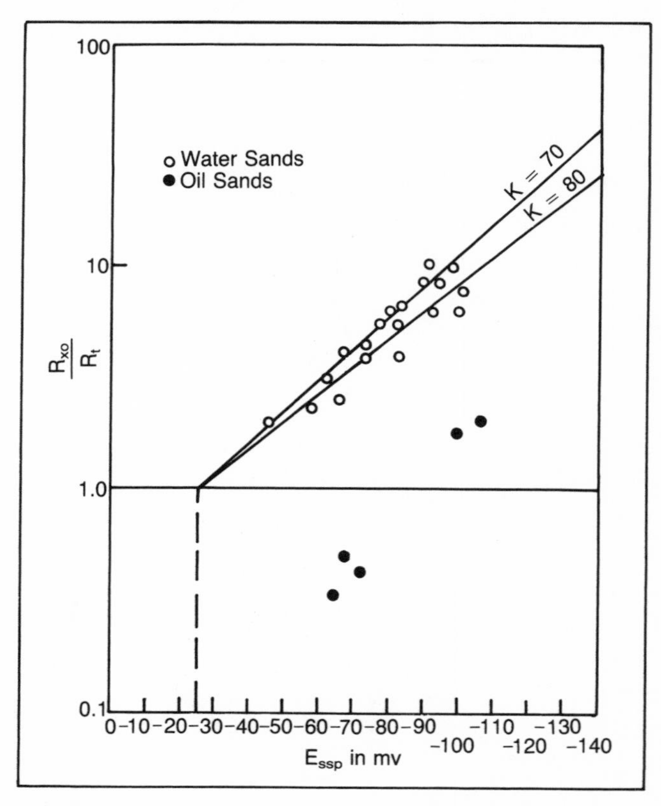

Fig. 9-11. Plot of R_{xo}/R_t vs. E_{SSP} showing streaming potential. Permission to publish by the Academic Press.[1]

is reflected in both the E_{SSP} and $(R_{xo}/R_t)_{QL}$. The influence of shale is also compensated for, since the effect of shale on the E_{SSP} is on the same order as the effect on the "computed E_{SSP}" using R_{xo}/R_t.[3, 19, 34]

Several examples of this quick look method are included here to show its utility. Figure 9-12 shows the Dual Induction Log for a California well containing a shaly gas sand and several wet sands with varying amounts of shaliness.[17] There are two zones containing movable hydrocarbon, the zone from 3760 to 3788 and the upper six feet of the zone from 3793 to 3818.

The second example in Figure 9-13 is from a central Texas well.[19] The obvious producer is the limy sand at 4650–4665 from which production is 3.2 MMscf/d.

The final example (see Fig. 9-14) is from a north Louisiana well in which the limy sand from 4656–4682 was tested and produced gas.[17] The lower zone from 4723–4768 shows a resistivity gradient, which is due to a change of porosity and a loss of permeability. This zone contained no movable hydrocarbons.

S_{or} vs. S_o Method

Application of this method requires that an S_{or} vs. S_o relationship be established, for example, from capillary pressure data, as shown in Figure 9-15. An estimate of S_{or} (and subsequently S_o) can be obtained using the basic relationship:

$$I_{xo}=\frac{R_{xo}}{F_R R_w}=S_{xo}^{-n}=(1-S_{or})^{-n}$$

or:

$$S_{or}=1-\left(\frac{F_R R_{mf}}{R_{xo}}\right)^{1/n}$$

As noted from the above relation, it is necessary to have R_{xo}, R_{mf}, F_R (a porosity response and also m) and n. These values can be difficult to obtain accurately, and as can be seen from Figure 9-15, small errors in the computed value of S_{or} can lead to large errors in S_o. It is also imperative to rock type, such that the correct S_{or} vs. S_o relation is being applied to the zone of interest. This method is the least used because of the difficulty in establishing the S_{or} vs. S_o relation (cores required) and its sensitivity to the absolute value of S_{or}.

CROSSPLOTTING METHODS

Crossplotting of parameters defined by the various log response and saturation equations is a powerful approach in formation evaluation. A multitude of different types of crossplots are possible and can be directed to: (1) rock typing, (2) locating anomalies and evaluating for water saturation, and (3) defining porosity and other parameters such as R_w, m, B, etc. Crossplotting tends to eliminate or reduce the uncertainties in R_w, m, n, B, etc., and in log quality, which influence the evaluations using the standard log interpretation methods already discussed. Some of the crossplotting techniques included in earlier chapters will be expanded on here and some additional, often-used methods will be added. For convenience we will discuss both those crossplots used for water saturation analysis and the porosity response crossplots.

Crossplots for Water Saturation Analysis

Solution of the several saturation equations provide the basis for these crossplotting techniques. The basic relations enable us to generate two water saturation crossplots, which are the R_t vs. ϕ and Σ_{log} vs. ϕ type.

Logarithmic R_t vs. ϕ Type Crossplot. This method of crossplotting is probably the most used technique from which water saturation can be estimated with a minimum of data.[7, 22] The basic Archie relations:

$$F_R=\frac{R_o}{R_w}=\phi^{-m} \quad \text{and} \quad I_R=\frac{R_t}{R_o}=\frac{R_t}{F_R R_w}=S_w^{-n}$$

can be combined and written in the following form:

$$\log R_t=\log F_R+\log R_w+\log I_R$$

or:

$$\log R_t=-m\log\phi+\log R_w+\log I_R$$

This expression states that R_t can be plotted versus ϕ on log-log graph paper, and a linear trend will be observed for all zones within a constant lithology having a constant R_w and I_R. An example of such a plot is shown in Figure 9-16. The lowest observed resistivities are assumed to be 100% water-bearing. This linear trend defined by the lowest observed resistivities represents a locus of R_o values, which can be extrapolated to 100% ϕ, the apparent R_w. Additionally, this trend should define a slope of m (cementation factor) for the rock type. All points falling to the right of the 100% S_w trend have a resistivity $R_t > R_o$. These anomalies are assumed to have a higher resistivity as a result of the presence of hydrocarbons and can be readily estimated as shown. This type of analysis has been used for many years in formation evaluation, including a variation of this approach, which uses a graphical solution linear in ϕ and nonlinear in resistivity, R_t.[4, 2] The approach illustrated in Figure 9-16 is a powerful interpretation method since it is not necessary to know R_w or m, and in fact these two can be determined as shown.

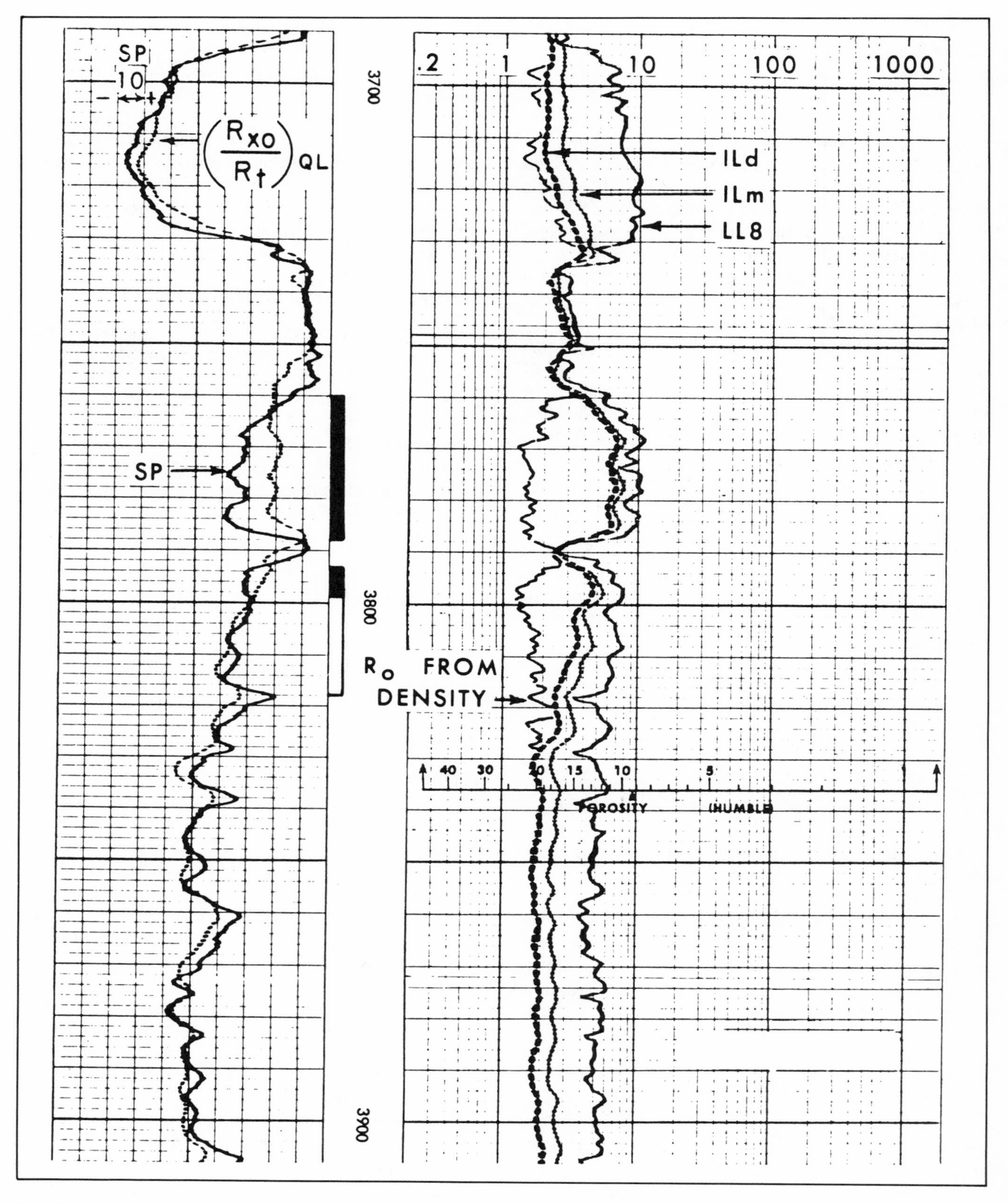

Fig. 9-12. Dual induction log from California well with "quick look" R_{xo}/R_t curve. Permission to publish by the Society of Professional Well Log Analysts.[19]

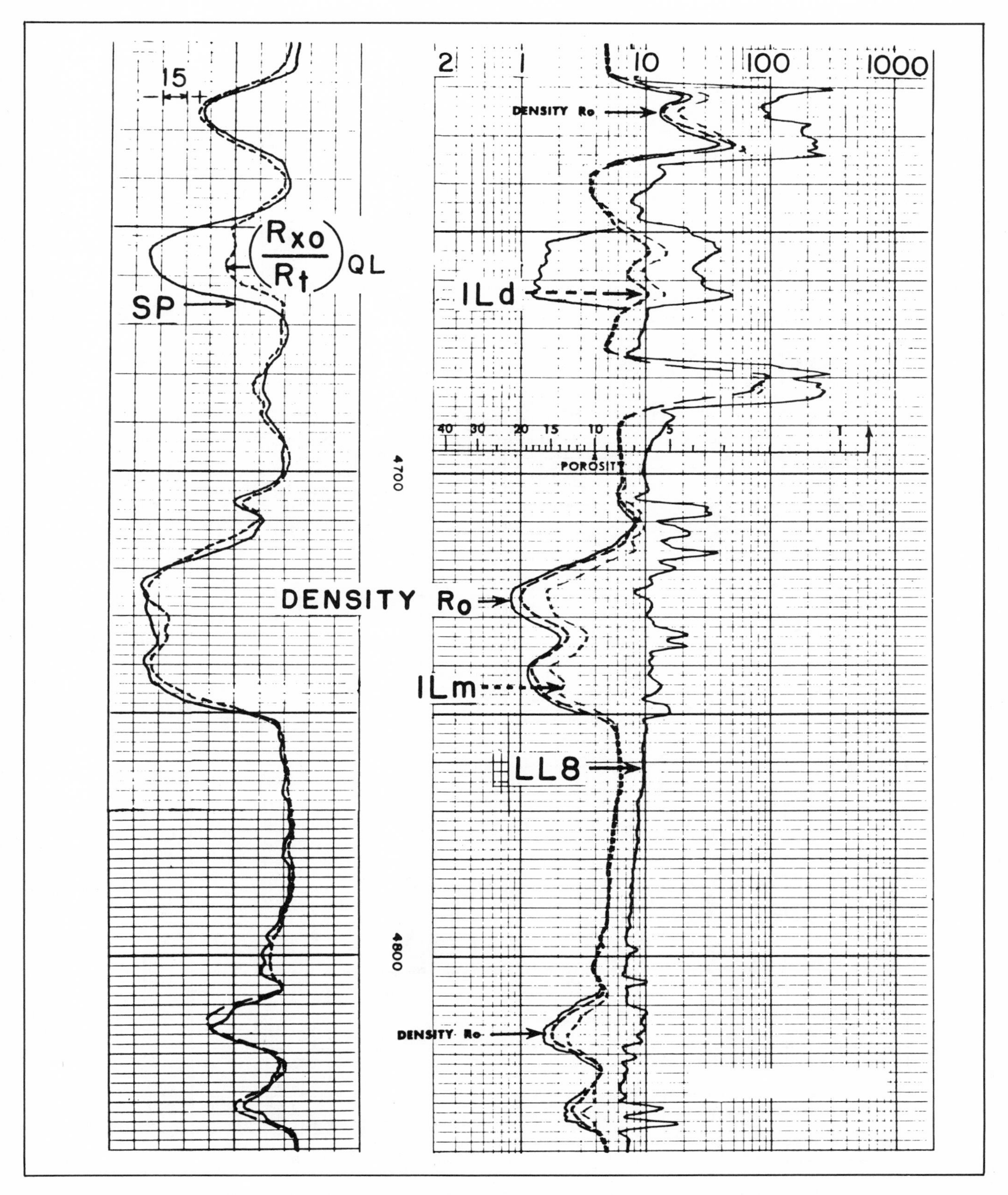

Fig. 9-13. Dual induction log from Texas well with "quick look" $R_{xo}R_t$ curve. Permission to publish by the Society of Professional Well Log Analysts.[19]

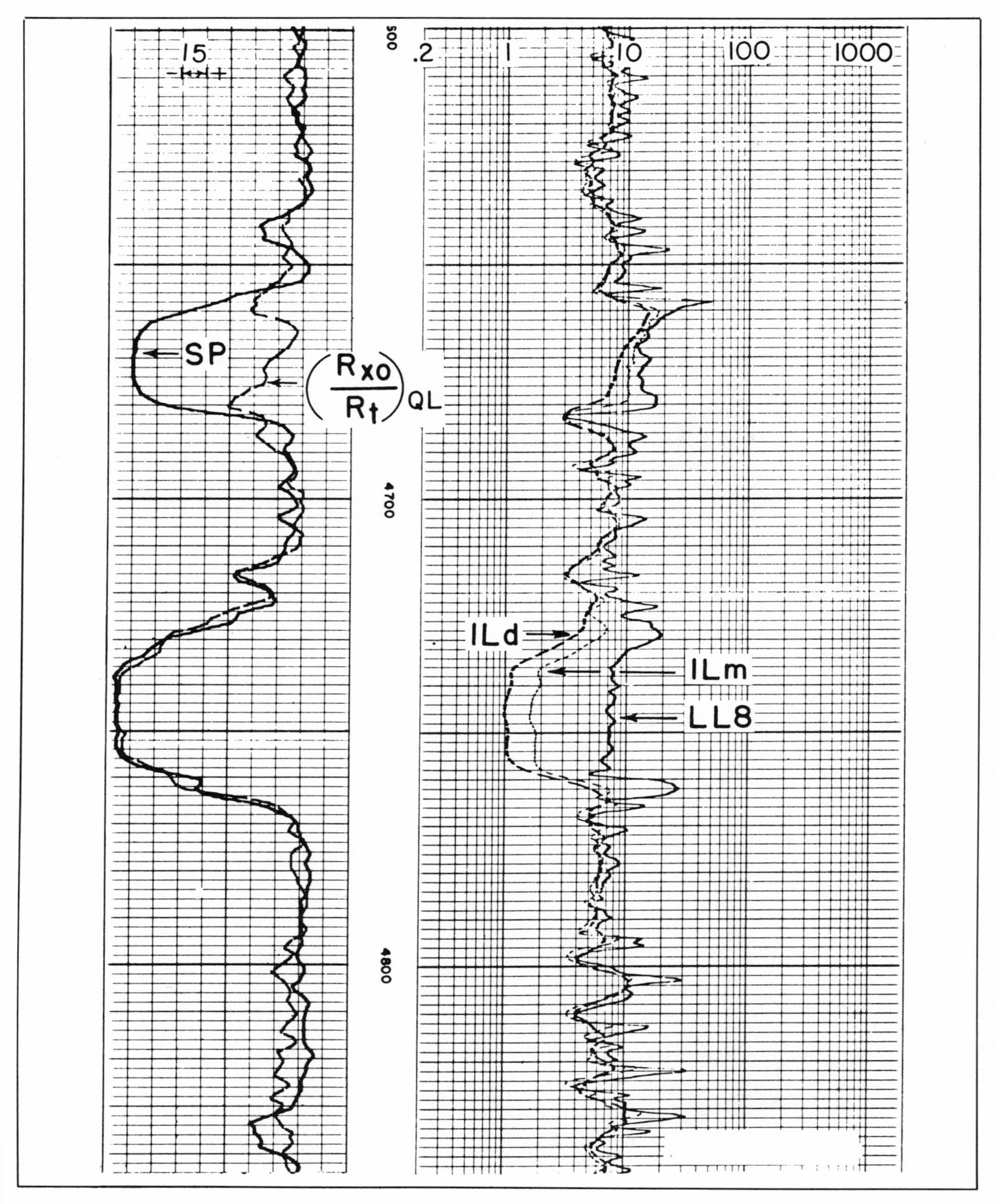

Fig. 9-14. Dual induction log from Louisiana well with "quick look" R_{xo}/R_t curve. Permission to publish by the Society of Professional Well Log Analysts.[19]

1. R_t vs. $(t - t_{ma})$ Method. This concept can be extended using the transit time-porosity relation discussed earlier where:

$$t = t_{ma} + B\phi \quad \text{or} \quad \phi = \frac{t - t_{ma}}{B}$$

Substituting into the logarithmic expression for R_t the following expression is obtained:

$$\log R_t = -m \log (t - t_{ma}) + m \log B + \log R_w + \log I_R$$

This expression states that a log-log plot of R_t vs. $(t - t_{ma})$ should exhibit a linear trend for all zones of constant lithology having constant R_w, B and I_R values. Such a graphical approach is shown in Figure 9-17. As can be seen from the illustrative crossplot in Figure 9-17, water saturation can be estimated for the anomalous data points and, additionally, m is defined by the slope of the linear trend representing constant values of I_R. Water saturation, S_w, and m can be obtained without a previous knowledge of B or R_w, and R_w can be determined if B can be estimated or is known.

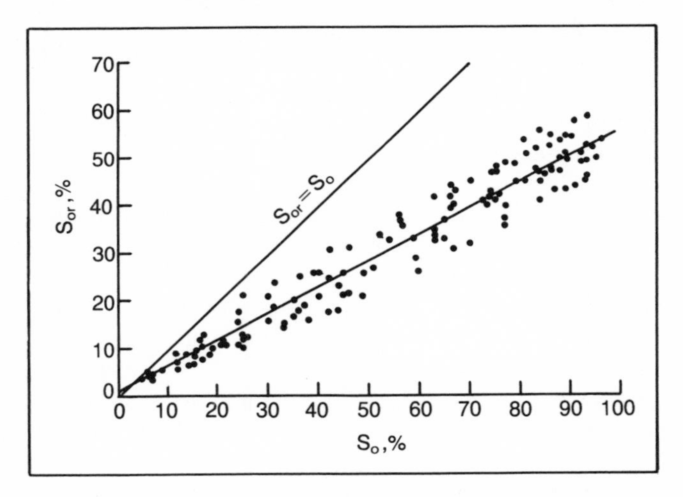

Fig. 9-15. S_{or} vs. S_o from capillary pressure measurements. Permission to publish by the Society of Petroleum Engineers of AIME.[7] Copyright 1966 SPE-AIME.

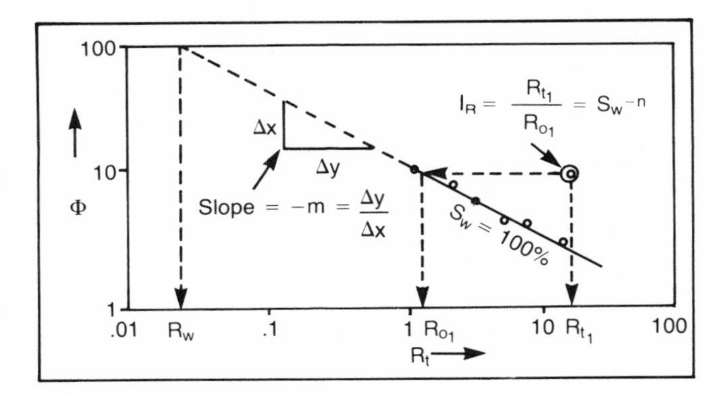

Fig. 9-16. Apparent resistivity vs. apparent porosity crossplot

2. R_t vs. Neutron Deflection Method. This concept can also be used with the neutron deflection-porosity relation presented earlier where:

$$ND = C + D \log \phi \quad \text{or} \quad \log \phi = \frac{ND - C}{D}$$

Substituting into the logarithmic expression for R_t the following expression is obtained:

$$\log R_t = -m\left(\frac{ND - C}{D}\right) + \log R_w + \log I_R$$

or:

$$\log R_t = -m\left(\frac{ND}{D}\right) + m\left(\frac{C}{D}\right) + \log R_w + \log I_R$$

This expression states that a semi-log plot of R_t versus ND should exhibit a linear trend for all zones of constant lithology (constant C and D) and constant R_w and I_R values. A typical R_t versus ND crossplot is shown in Figure 9-18. The assumed water-bearing trend defines a

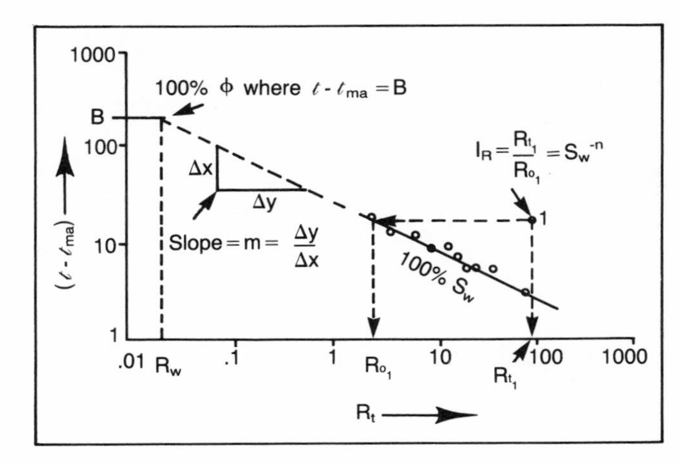

Fig. 9-17. Apparent resistivity vs. $(t - t_{ma})$ crossplot

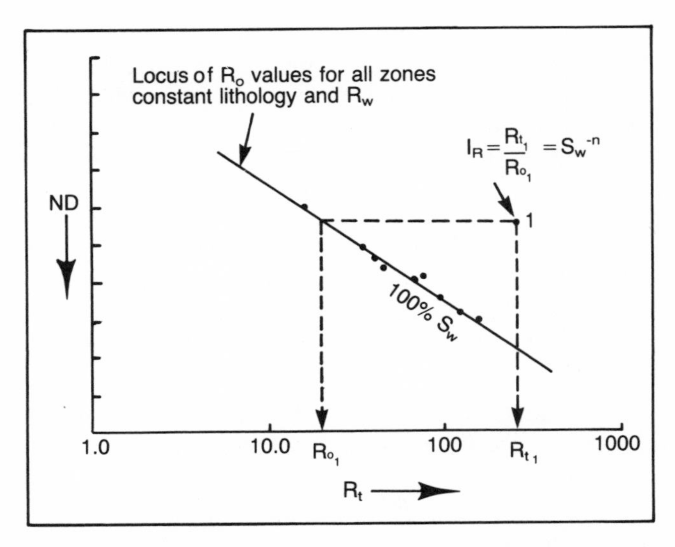

Fig. 9-18. Apparent resistivity vs. neutron deflection crossplot

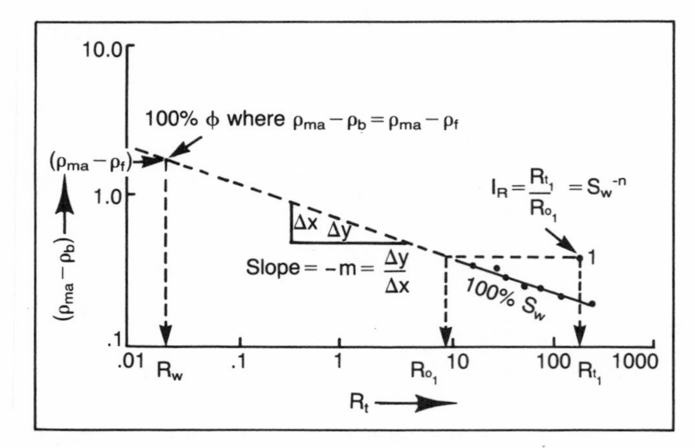

Fig. 9-19. Apparent resistivity vs. $(\rho_{ma} - \rho_b)$ crossplot

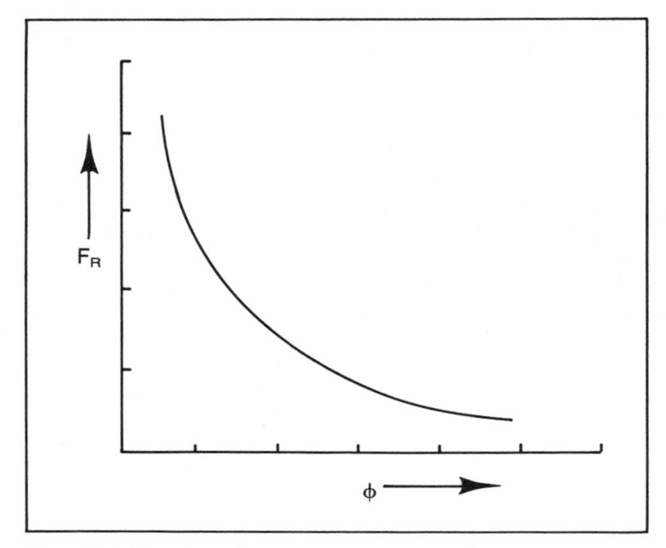

Fig. 9-20. Non-linear trend, linear F_R vs. ϕ plot.

locus of R_o values with R_o increasing as the neutron deflection, ND, increases (increasing porosity, ϕ). Anomalous points, in this case zone 1, are considered to have an anomalously high R_t due to the presence of hydrocarbon; as a result the apparent water saturation can be estimated as shown if a value of n can be assumed. This saturation estimation can be made without a prior knowledge of C, D, m, or R_w but is not as diagnostic as the previously discussed R_t versus $(\ell - \ell_{ma})$ method. In order to determine m, it would be necessary to define D, and in order to determine R_w, C would be required.

3. R_t vs. $(\rho_{ma} - \rho_b)$ Method. Since bulk density is related to porosity as previously shown in the form:

$$\phi = \frac{\rho_{ma} - \rho_b}{\rho_{ma} - \rho_f}$$

the density relation can be substituted for porosity in the resistivity expression to obtain:

$$\log R_t = -m \log \left(\frac{\rho_{ma} - \rho_b}{\rho_{ma} - \rho_f}\right) + \log R_w + \log I_R$$

or:

$$\log R_t = -m \log (\rho_{ma} - \rho_b) + m \log (\rho_{ma} - \rho_f) + \log R_w + \log I_R$$

This expression states that R_t can be plotted versus ϕ on log-log graph paper and a linear trend will be observed for all zones within a constant lithology (constant ρ_{ma} and similar ρ_f values) having a constant R_w and I_R. An example of such a plot is shown in Figure 9-19.

The lowest observed resistivities are assumed to be 100% water-bearing, and this linear trend is defined by the lowest observed resistivities representing a locus of R_o values. This trend can be extrapolated to 100% ϕ (where $\rho_{ma} - \rho_b = \rho_{ma} - \rho_f$) which would be the apparent R_w. Additionally, this trend should define a slope of m (cementation factor) for the rock type. All points falling to the right of the 100% S_w trend have a resistivity $R_t > R_o$. Since these anomalies are assumed to have a higher resistivity as a result of the presence of hydrocarbons, the water saturation can readily be estimated as shown in Figure 9-19. This approach is quite diagnostic since S_w and m can be obtained without a prior knowledge of ρ_f or R_w, and in fact R_w can be estimated if ρ_f is known or assumed.

Nonlinear Resistivity vs. Linear Porosity Crossplots. Advantages can sometimes be obtained when crossplotting a nonlinear resistivity versus linear porosity. If one plots F_R vs. ϕ, a nonlinear relation is observed, as shown in Figure 9-20. This can be made into a straight line by adjusting the scaling of either the ϕ or F_R axis. For crossplotting purposes, it is desirable to maintain a linear ϕ scale and, therefore, F_R is adjusted. Figure 9-21

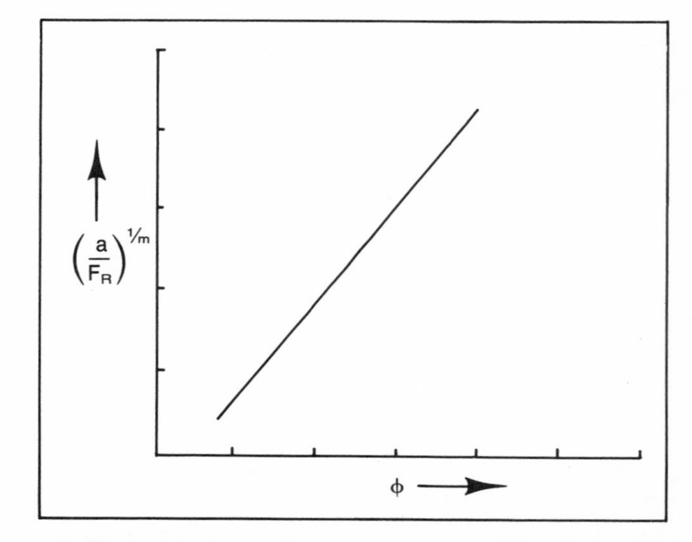

Fig. 9-21. Linear trend, linear $(a/F_R)^{1/m}$ vs. ϕ plot

shows the resulting straight line obtained by making the F_R scale proportional to $(a/F_R)^{1/m}$.

Now, the relation between resistivity and the formation factor, F_R, is:

$$F_R = \frac{S_w^n R_t}{R_w}$$

which can be substituted for F_R in the vertical scaling function of Figure 9-21. Figure 9-22 shows the final form of the crossplot with the new function, $(aR_w/R_tS_w^n)^{1/m}$, scaled linearly on the vertical axis and the porosity scaled linearly on the horizontal axis. If the constants a and m are defined, then special graph paper relating a nonlinear resistivity versus linear porosity can be developed (see Figs. 9-23 to 9-27). Any linear porosity function such as ℓ, ρ_b or ϕ_N can be plotted in lieu of porosity itself. Figures 9-23 to 9-27 provide a set of special graphs that incorporate the value of $a=1$, with each graph then representing a different value of m. If it is desired to use this special graph paper, copies should be used to keep a reproducible original in the book. In that way, the basic graph will not have to be reconstructed each time.

An example of the application of this crossplot is shown in Figure 9-28, using the resistivity-acoustic logs of Figure 9-29. With this method, ℓ or ρ_b can be plotted directly without having to know the matrix parameter ℓ_{ma} or ρ_{ma}; however, m must be estimated. Additionally, a value of ℓ_{ma} or ρ_{ma} is projected by the apparent water trend at a resistivity, $R_t = \infty$. If the porosity parameter can be related to F_R, a value of R_w can also be determined. Of course, we must still be looking at zones with a constant lithology and R_w, and the method does require plotting a number of zones as with the previously discussed R_t vs. ϕ crossplots.

Σ_{log} vs. ϕ Type Crossplot. As we have seen, the bulk

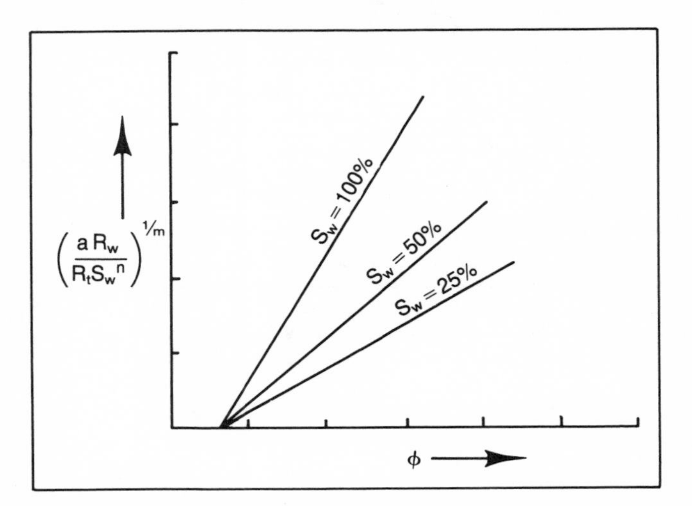

Fig. 9-22. Linear S_w trends, linear $(aR_w/R_tS_w^n)^{1/n}$ vs. ϕ plot

formation capture cross-section, Σ_{log}, can be related to the capture cross-section of the individual components, making up the formation bulk as follows:

$$\Sigma_{log} = \underbrace{\Sigma_{ma}(1-\phi)}_{\text{Matrix}} + \underbrace{\underbrace{\Sigma_w \phi S_w}_{\text{Water}} + \underbrace{\Sigma_h \phi(1-S_w)}_{\text{Hydrocarbon}}}_{\text{Total Fluid}}$$

It can be seen that Σ_{log} is a function of both ϕ and S_w, and if an independent ϕ response is available the water saturation existing in the formation can be determined.

A crossplotting method can be used for evaluating water saturation, which is particularly useful when evaluating a number of levels in a rock type.[6, 16] If the capture cross-section response equation is written for a 100% water-saturated rock the relation reduces to:

$$\Sigma_{log} = \Sigma_w \phi + \Sigma_{ma}(1-\phi)$$

or:

$$\Sigma_{log} = \underbrace{\Sigma_{ma}}_{\text{Intercept}} + \underbrace{\Sigma_w - \Sigma_{ma}\phi}_{\text{Slope}}$$

This relation states that for all water-bearing zones in the same rock type, a linear trend will be observed when crossplotting Σ_{log} versus ϕ on cartesian graph paper, as shown in Figure 9-30. Additionally, the intercept (at $\phi=0$) will be the apparent value of Σ_{ma} for that rock type. Once Σ_{ma} is known, a value of Σ_w can be determined from the slope of the trend. The validity of Σ_w (and therefore the slope) can be checked since an apparent salinity and hence an apparent R_w can be determined. All points falling below the assumed 100% S_w trend would be anomalies considered to be hydrocarbon-bearing. It should be pointed out that the higher the salinity of the formation water the greater the contrast between water-saturated and oil-bearing zones. The greater the formation water salinity the greater the value of Σ_w, and as a result the greater the value of Σ_{log} in the water zones. Since Σ_o is generally 22 or less, as the value of Σ_w decreases (approaches a minimum of 22), the less the difference between a Σ_{log} measured in the water-bearing zone and a Σ_{log} response in a zone containing hydrocarbon. Once the 100% S_w trend, Σ_{ma}, and Σ_w are defined a fan of other saturation trends (such as 50% S_w, 25% S_w, etc.) can be plotted using the general relation that contains the influence of hydrocarbon if Σ_h can be estimated. This is done by calculating, for example, the Σ_{log} that would exist for an assumed $S_w = 50\%$ and $\phi = 10\%$ (use $\phi = .1$ in the equation). Once this Σ_{log} is determined, a straight line is drawn through this point and Σ_{ma}. It should be pointed out that an extrapolation of the 100% S_w trend to 100% ϕ will give Σ_w, and extrapolation of the 0% S_w saturation trend will, at 100% ϕ, give Σ_h. It is important to note that when using this technique, prior knowlege of Σ_{ma} and Σ_w are not re-

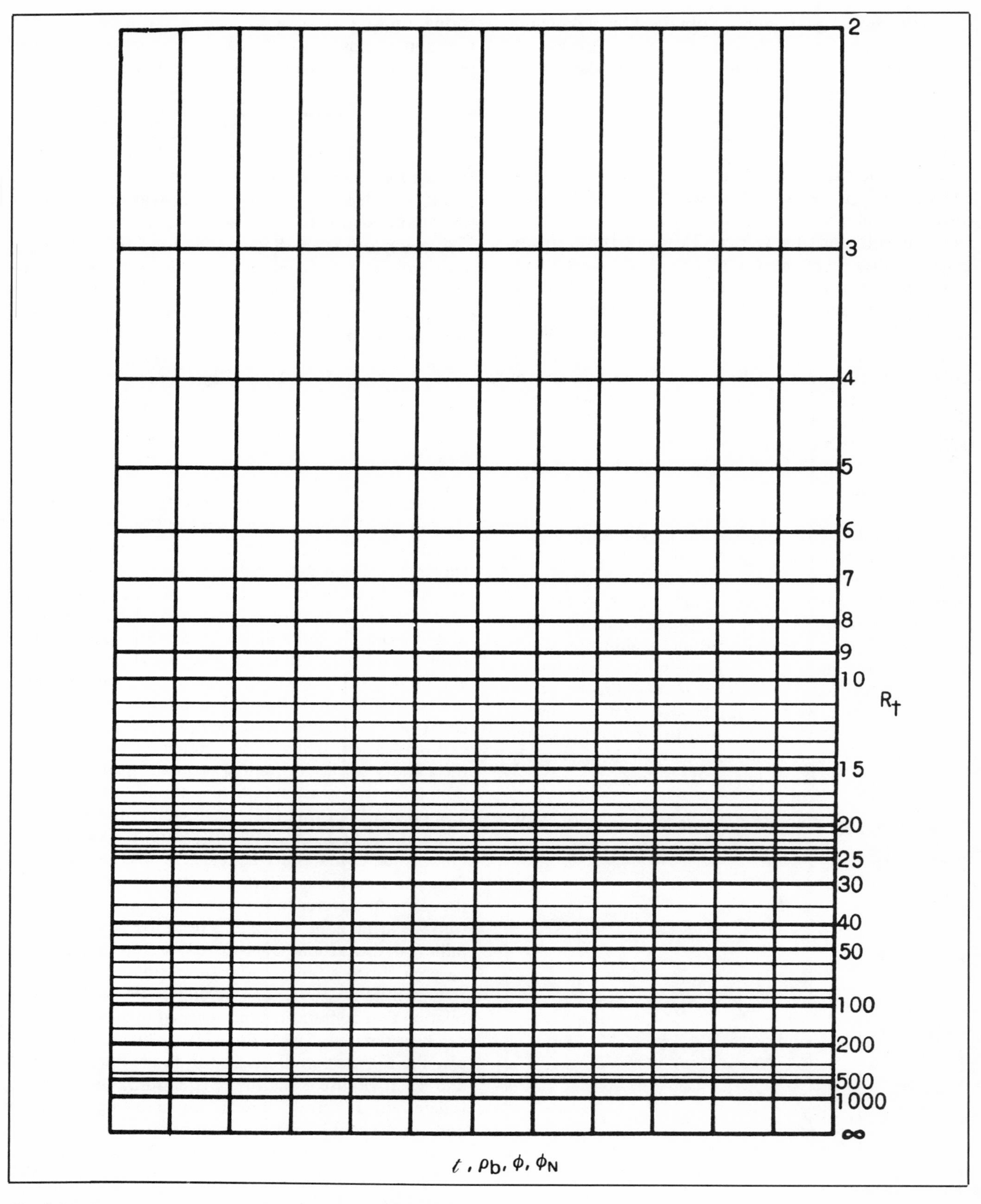

Fig. 9-23. Special resistivity vs. porosity or porosity function graph paper. m = 1.8

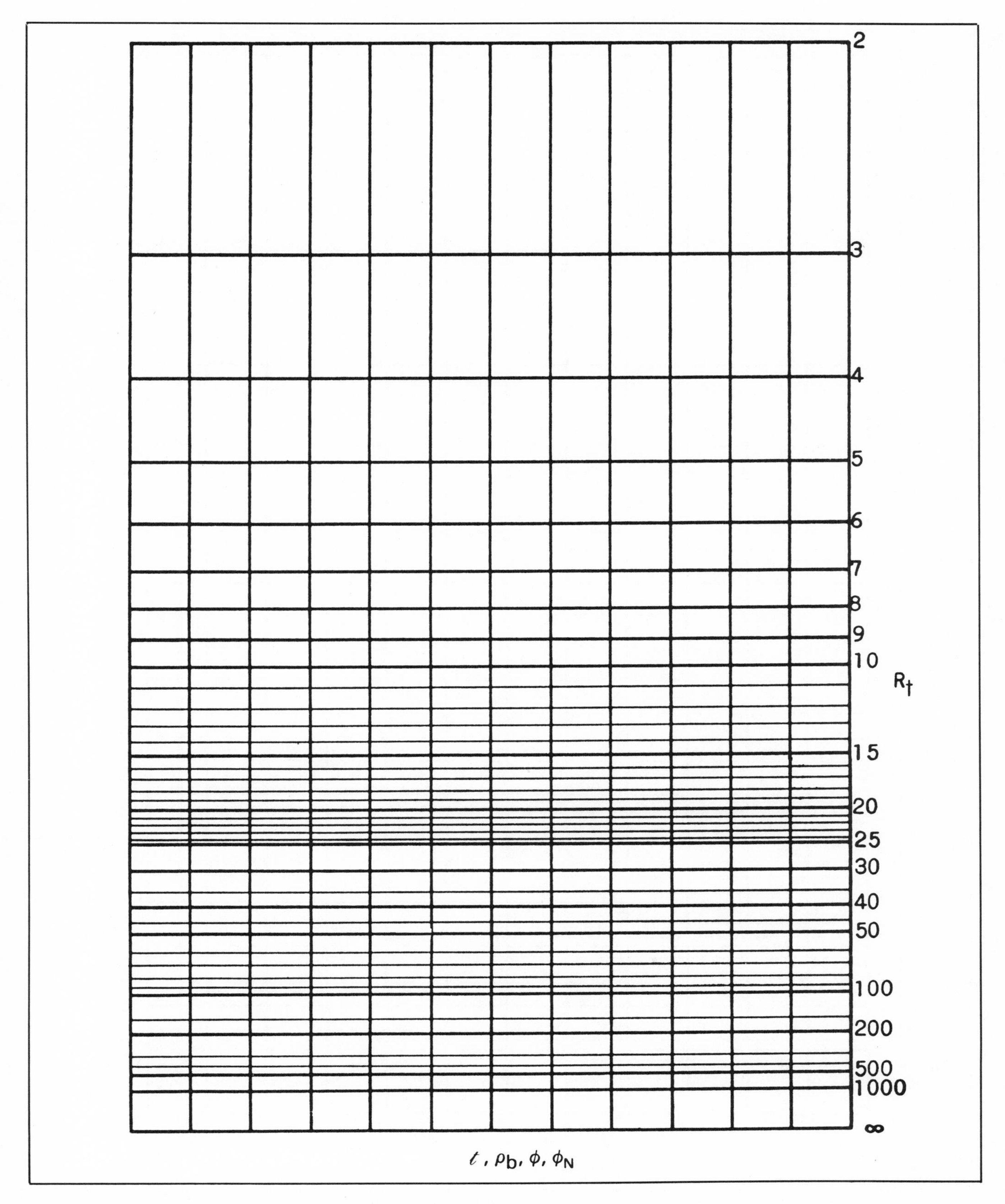

Fig. 9-24. Special resistivity vs. porosity or porosity function graph paper. m = 1.9

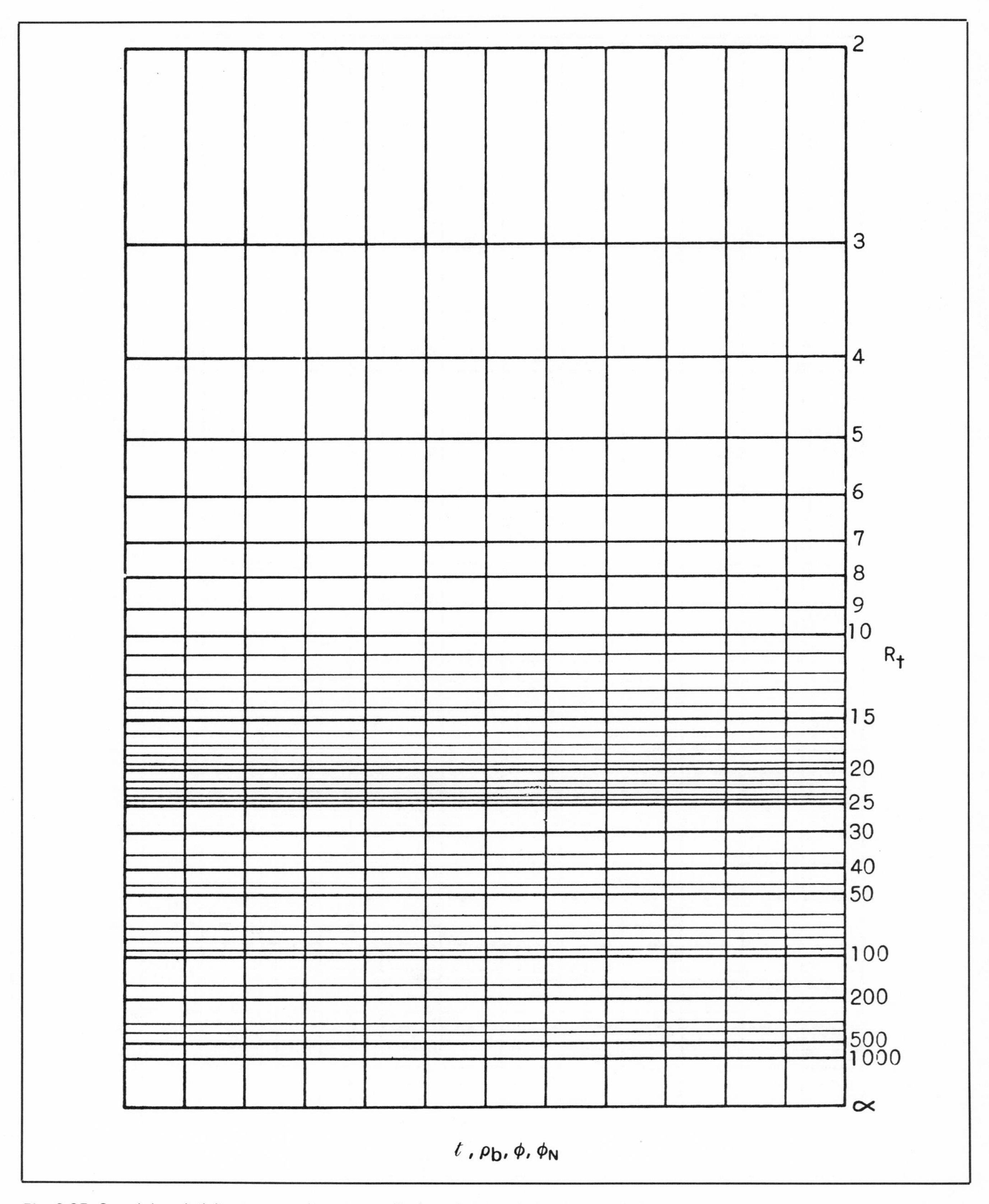

Fig. 9-25. Special resistivity vs. porosity or porosity function graph paper. m = 2.0

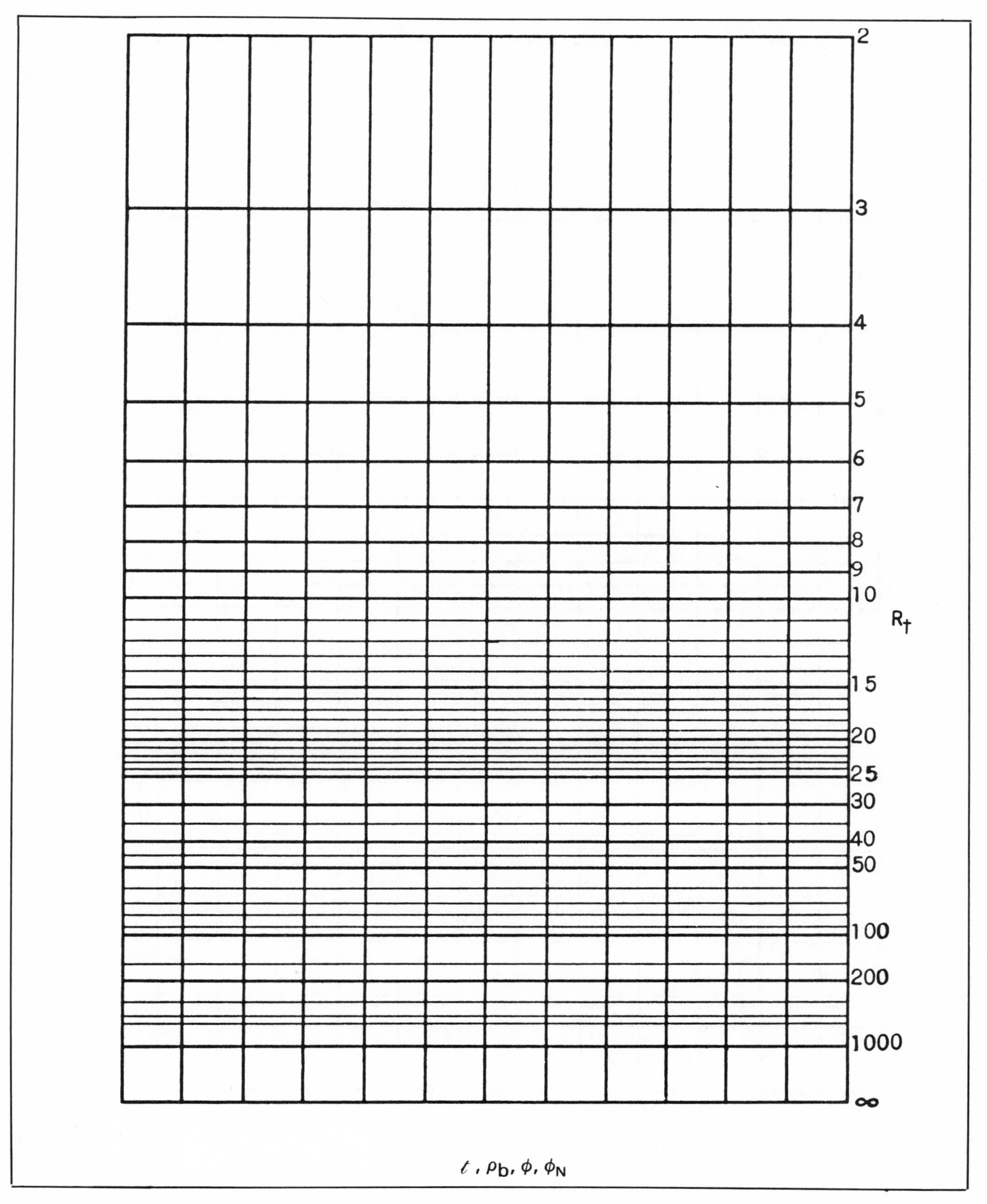

Fig. 9-26. Special resistivity vs. porosity or porosity function graph paper. m = 2.1

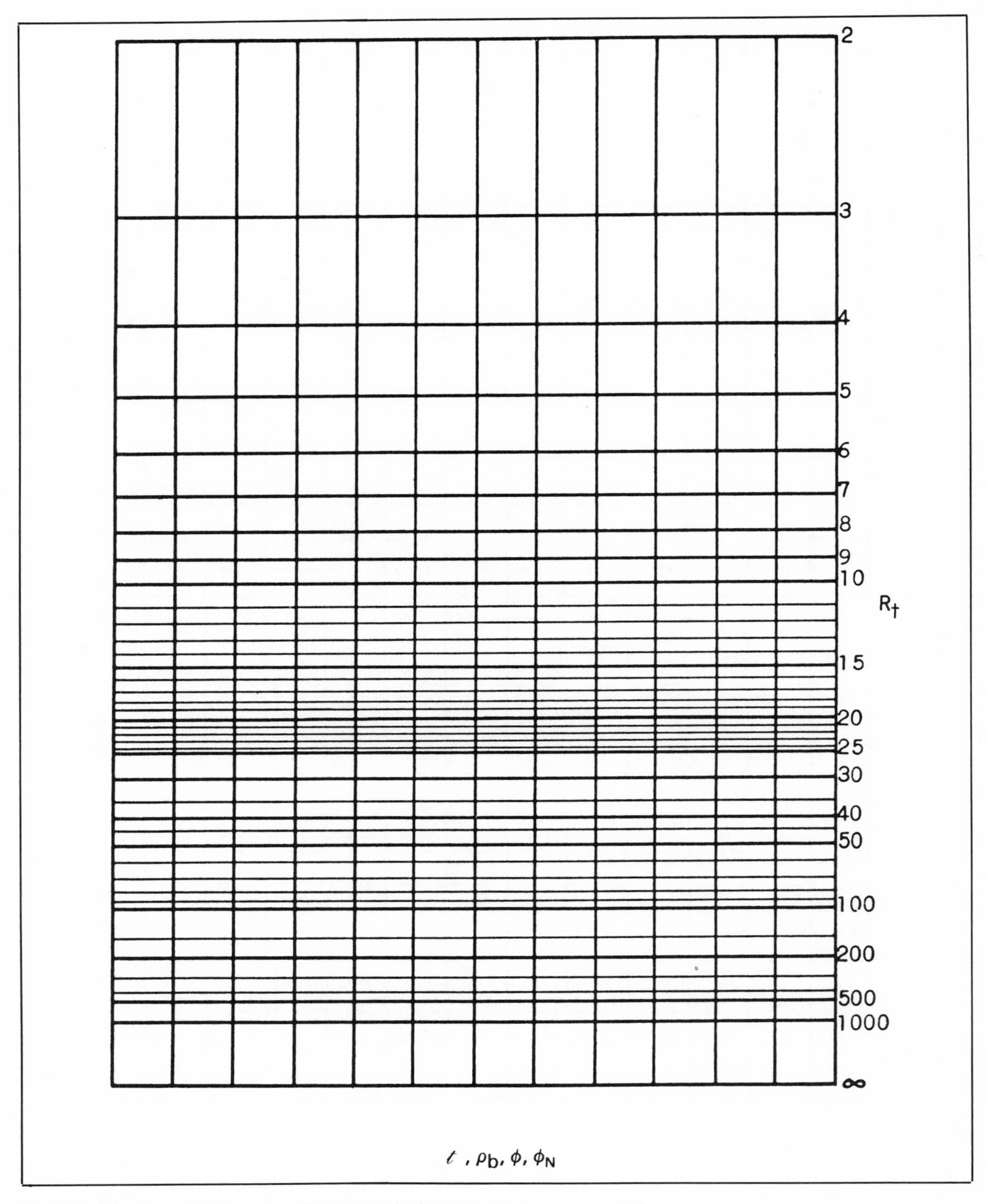

Fig. 9-27. Special resistivity vs. porosity or porosity function graph paper. $m = 2.2$

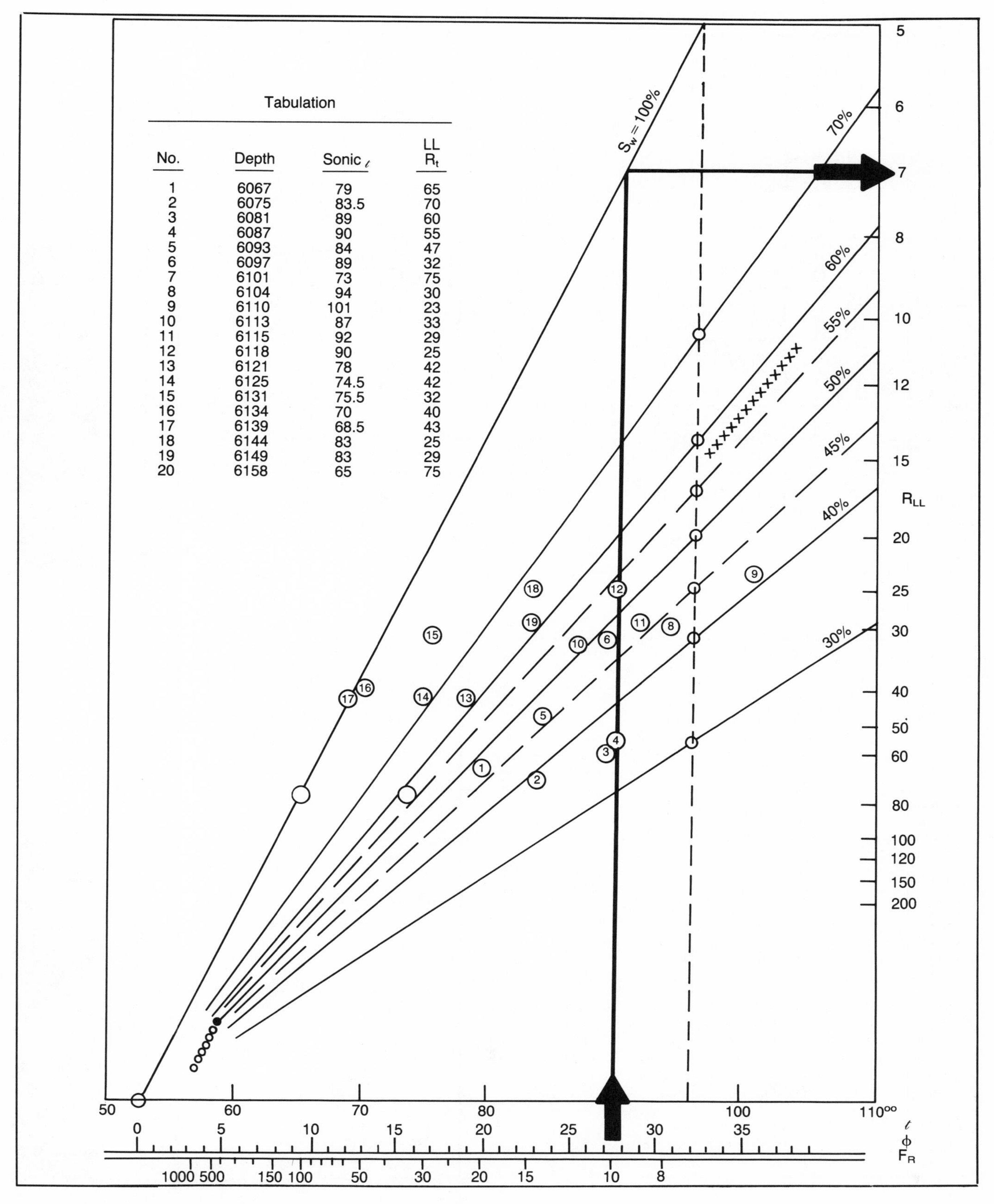

Tabulation

No.	Depth	Sonic t	LL R_t
1	6067	79	65
2	6075	83.5	70
3	6081	89	60
4	6087	90	55
5	6093	84	47
6	6097	89	32
7	6101	73	75
8	6104	94	30
9	6110	101	23
10	6113	87	33
11	6115	92	29
12	6118	90	25
13	6121	78	42
14	6125	74.5	42
15	6131	75.5	32
16	6134	70	40
17	6139	68.5	43
18	6144	83	25
19	6149	83	29
20	6158	65	75

Fig. 9-28. Example of resistivity vs. transit time crossplot where $F_R = 0.62\phi - 2.15$

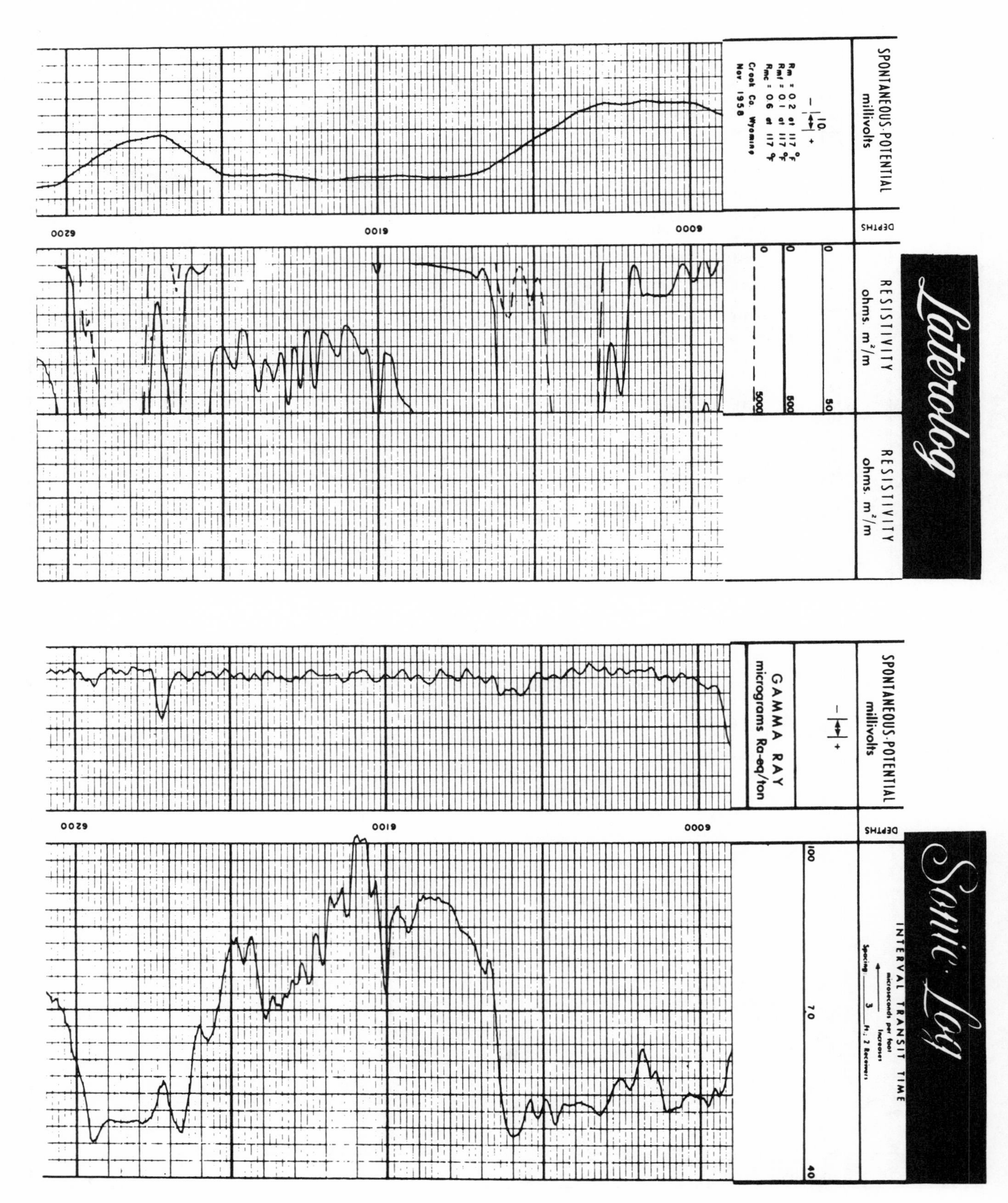

Fig. 9-29. Resistivity-acoustic logs providing data used in the example of Fig. 9-28

quired. In fact, these two parameters can be defined in this approach.

An example of this crossplotting technique is shown in Figure 9-32 based on data from the logs shown in Figure 9-31. The log sections shown in wells A and B are through the Oswego lime in Kay County, Oklahoma.

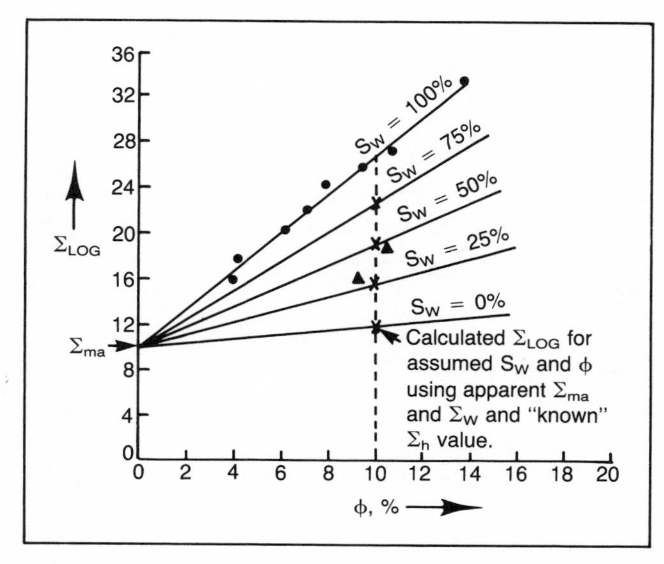

Fig. 9-30. Typical Σ_{log} vs. crossplot

Well B is structurally lower than well A and is water-bearing. The eleven data points selected in the Oswego lime define the 100% S_w trend as well as Σ_{ma} and Σ_w as shown in Figure 9-32. The five intervals selected on well A when plotted indicate an apparent water saturation ranging from 25–35%. Well A was oil-bearing.

The log measured capture cross-section, Σ_{log}, can also be plotted as a function of log measured porosity parameters such as ρ_b, t, and ϕ_{SNP} or ϕ_{CNL}.[18]

Σ_{log} *vs.* ρ_b ***Method.*** Considering the capture cross-section response in a zone of constant lithology and water salinity for 100% S_w:

$$\Sigma_{log} = \Sigma_{ma} + (\Sigma_w - \Sigma_{ma})\phi$$

If we use the porosity-bulk density relation, which is:

$$\phi = \frac{\rho_{ma} - \rho_b}{\rho_{ma} - \rho_f}$$

we obtain:

$$\Sigma_{log} = \Sigma_{ma} + \left[\frac{\Sigma_w - \Sigma_{ma}}{\rho_{ma} - \rho_f}\right](\rho_{ma} - \rho_b)$$

or:

$$\Sigma_{log} = \Sigma_{ma} + \left[\frac{\Sigma_w - \Sigma_{ma}}{\rho_{ma} - \rho_f}\right]\rho_{ma} - \left[\frac{\Sigma_w - \Sigma_{ma}}{\rho_{ma} - \rho_f}\right]\rho_b$$

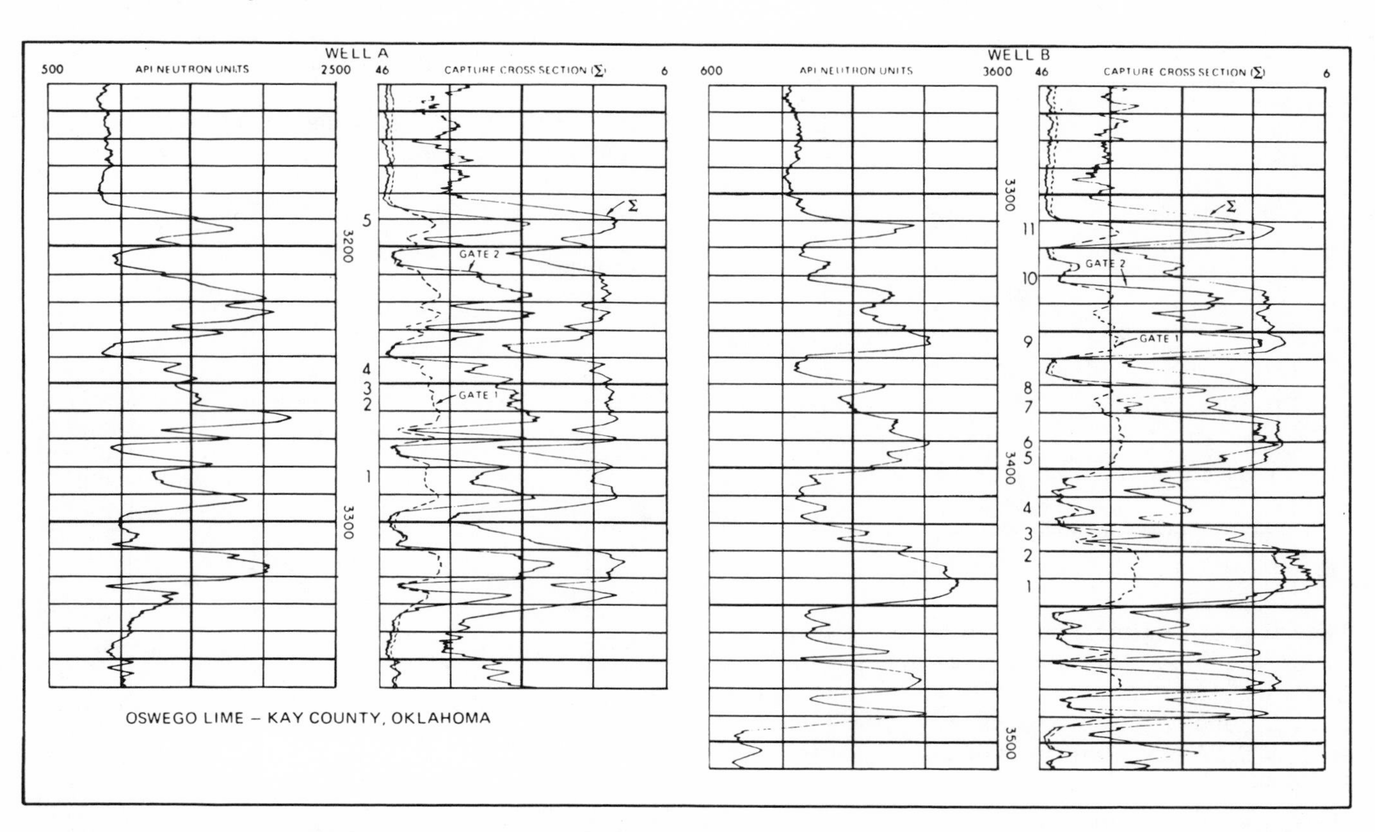

Fig. 9-31. Neutron and neutron lifetime logs in Oswego Lime, Kay County, Oklahoma. Permission to publish by Dresser Atlas, Dresser Industries, Inc.[13]

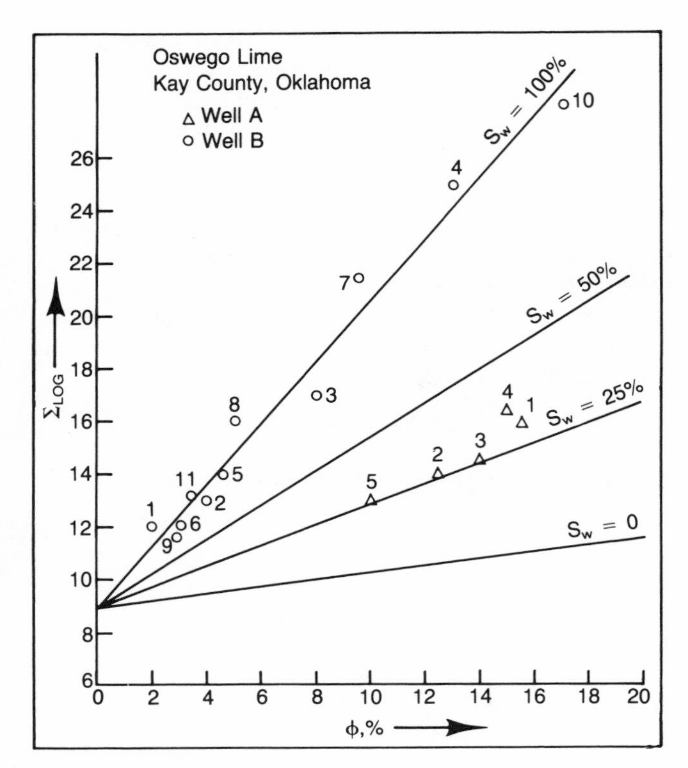

Fig. 9-32. Crossplot evaluation—Kay County, Oklahoma. Permission to publish by Dresser Atlas, Dresser Industries, Inc.[13]

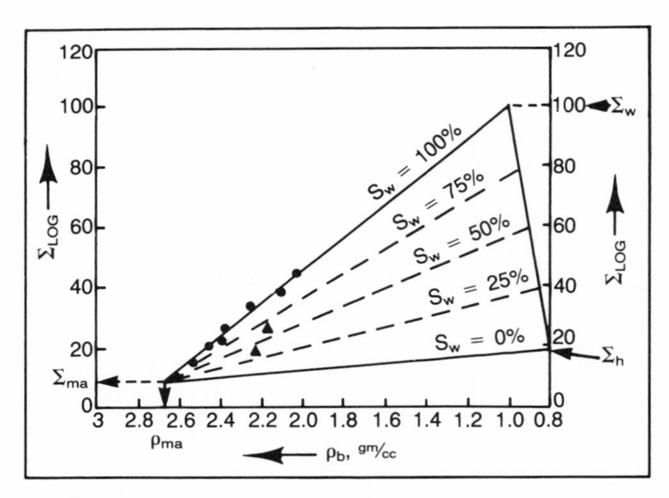

Fig. 9-33. Typical Σ_{log} vs. ρ_b crossplot

The above expression indicates that Σ_{log} can be plotted versus ρ_b on cartesian graph paper and the 100% S_w trend will, at 0% ϕ (where $\rho_b = \rho_{ma}$), intercept at $\Sigma_{log} = \Sigma_{ma}$ and at 100% S_w (where $\rho_b = \rho_f$) will intercept at $\Sigma_{log} = \Sigma_w$. A typical crossplot of Σ_{log} vs. ρ_b is shown in Figure 9-33 where the limit of the 0% S_w trend will exist at $\rho_b = \rho_h$ and $\Sigma_{log} = \Sigma_h$.

Σ_{log} vs. t Method. The same argument can be made for the acoustic transit time response where:

$$\phi = \frac{t - t_{ma}}{B}$$

such that:

$$\Sigma_{log} = \Sigma_{ma} - \left[\frac{\Sigma_w - \Sigma_{ma}}{B}\right] t_{ma} + \left[\frac{\Sigma_w - \Sigma_{ma}}{B}\right] t$$

As shown in Figure 9-34, for 0% ϕ where $t = t_{ma}$ the 100% S_w trend will intercept at $\Sigma_{log} = \Sigma_{ma}$. As expected for 100% ϕ (where $t - t_{ma} = B$) the value of $\Sigma_{log} = \Sigma_w$. The limiting value for the 0% S_w trend will be $\Sigma_{log} = \Sigma_h$ for $t = B + t_{ma}$. If the interval transit time response can be expressed by the time average equation (clean, consolidated sandstones) then the limiting value for the 0% S_w trend will be $\Sigma_{log} = \Sigma_h$ at $t = t_h$ since $B = t_f - t_{ma}$ and $t_f = t_h$.

Porosity Log Crossplots

Porosity log crossplots are one of the basic tools used in rock typing since each of the common porosity logs, hence the acoustic, density and neutron, responds uniquely to different lithologies.[5] Crossplotting multiple porosity responses also defines actual porosity more accurately and can provide an insight into the presence of hydrocarbon. The three two-porosity response crossplots, as well as tri-porosity crossplots, will be considered in the following sections.

ρ_b vs. t Crossplot. Using the two response relations for these log responses:

$$\rho_b = \rho_{ma} + (\rho_f - \rho_{ma})\phi \quad \text{and} \quad \phi = \frac{t - t_{ma}}{B}$$

we can state:

$$\rho_b = \rho_{ma} + (\rho_f - \rho_{ma})\left(\frac{t - t_{ma}}{B}\right)$$

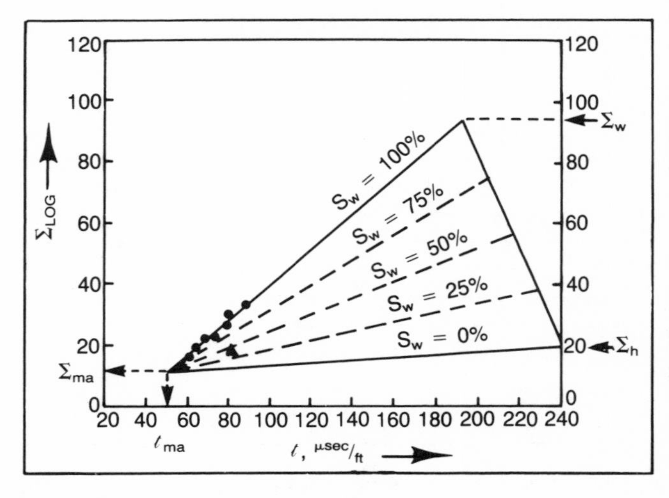

Fig. 9-34. Typical Σ_{log} vs. t crossplot

TABLE 9-1
"Average" Matrix Parameters for Various Lithologies

Mineral	Matrix Coefficients			Salt Mud $t_f = 185$ $\rho_f = 1.10$		Fresh Mud $t_f = 189$ $\rho_f = 1.00$	
	t_{ma}	ρ_{ma}	$(\phi_{SNP})_{ma}$	M	N	M	N
Silica (1) ($v_{ma} = 18{,}000$)	55.5	2.65	−.035	.835	.669	.810	.628
Silica (2) ($v_{ma} = 19{,}500$)	51.2	2.65	−.035	.862	.669	.835	.628
$CaCO_3$	47.6	2.71	0.00	.854	.621	.827	.585
Dolomite (1) ($\phi = 5.5\%$ to 30%)	43.5	2.87	.035	.800	.544	.778	.513
Dolomite (2) ($\phi = 1.5\%$ to 5.5% & $>30\%$)	43.5	2.87	.02	.800	.554	.778	.524
Dolomite (3) ($\phi = 0.0\%$ to 1.5%)	43.5	2.87	.005	.800	.561	.778	.532
Anhydrite	50.0	2.98	0.00	.718	.532	.702	.505
Gypsum	52.0	2.35	0.49	1.060	.408	1.015	.378
Salt	67.0	2.05	0.04	1.240	1.010	1.16	.914

Permission to publish by the Society of Professional Well Log Analysts.[9]

or:

$$\rho_b = \rho_{ma} - \underbrace{\left[\frac{\rho_{ma} - \rho_f}{B}\right] t_{ma}}_{\text{Intercept}} + \underbrace{\left[\frac{\rho_{ma} - \rho_f}{B}\right]}_{\text{Slope}} t$$

The slope and intercept defined by the relationship above are rock typing parameters dependent only on matrix and fluid properties. In fact, if one assumes representative rock and fluid parameters, one could define "average" lithology behavior as shown in Figure 9-35 where $B = t_f - t_{ma}$ is assumed. This typical ρ_b versus t crossplot is based on the average matrix parameters shown in Table 9-1, and the average fluid parameters shown in Table 9-2. Using the ρ_b versus t crossplot with lithologies predefined, it is possible to estimate the lithology as well as the porosity. One of the problems associated with a crossplot of the type shown in Figure 9-35 is that numerous "average" rock and fluid parameters are used to define the trends, and as a result this grid represents "average" behavior.

t vs. ϕ_N Type Crossplot. This type crossplot is based upon the basic relation:

$$t = t_{ma} + B\phi_N$$

although the value used for B in generating a grid representing the "average" rock behavior usually incorporates a slope, B, based on the time average equation, which we know is not representative, particularly in carbonate systems. Figures 9-36 and 9-37 show typical t vs. ϕ_N based on a limestone porosity, hence the nonlinearity of the sandstone and dolomite trends. Figure 9-36 presents the t vs. ϕ_{SNP} crossplot and Figure 9-37 presents the t vs. ϕ_{CNL} crossplot. Comparing these typical t vs. ϕ_N crossplots with the ρ_b vs. t crossplot shown in Figure 9-35, it is obvious that a greater resolution between rock types is obtained when combining the acoustic-neutron responses.

ρ_b vs. ϕ_N Type Crossplot. This crossplot is based on the density response equation where:

$$\rho_b = \rho_{ma} + (\rho_f - \rho_{ma})\phi_N$$

A typical crossplot using the "average" rock and fluid parameters, as shown in Tables 9-1 and 9-2, would produce the "average" lithology behavior as shown in Figures 9-38 through 9-41. Note, again, the nonlinearity of the sandstone and dolomite trends due to the neutron porosity being recorded in limestone units.

TABLE 9-2
"Average" Fluid Parameters

Fluids		t_f	ρ_f	$(\phi_N)_f$
Primary Porosity				
(Liquid-Filled):	Fresh Mud	189.0	1.00	1.00
	Salt Mud	185.0	1.10	
Secondary Porosity				
In Dolomite:	Fresh Mud	43.5	1.00	1.00
	Salt Mud		1.10	
In $CaCO_3$:	Fresh Mud	47.6	1.00	1.00
	Salt Mud		1.10	
In Silica (1):	Fresh Mud	55.5	1.00	1.00
	Salt Mud		1.10	

Permission to publish by the Society of Professional Well Log Analysts.[9]

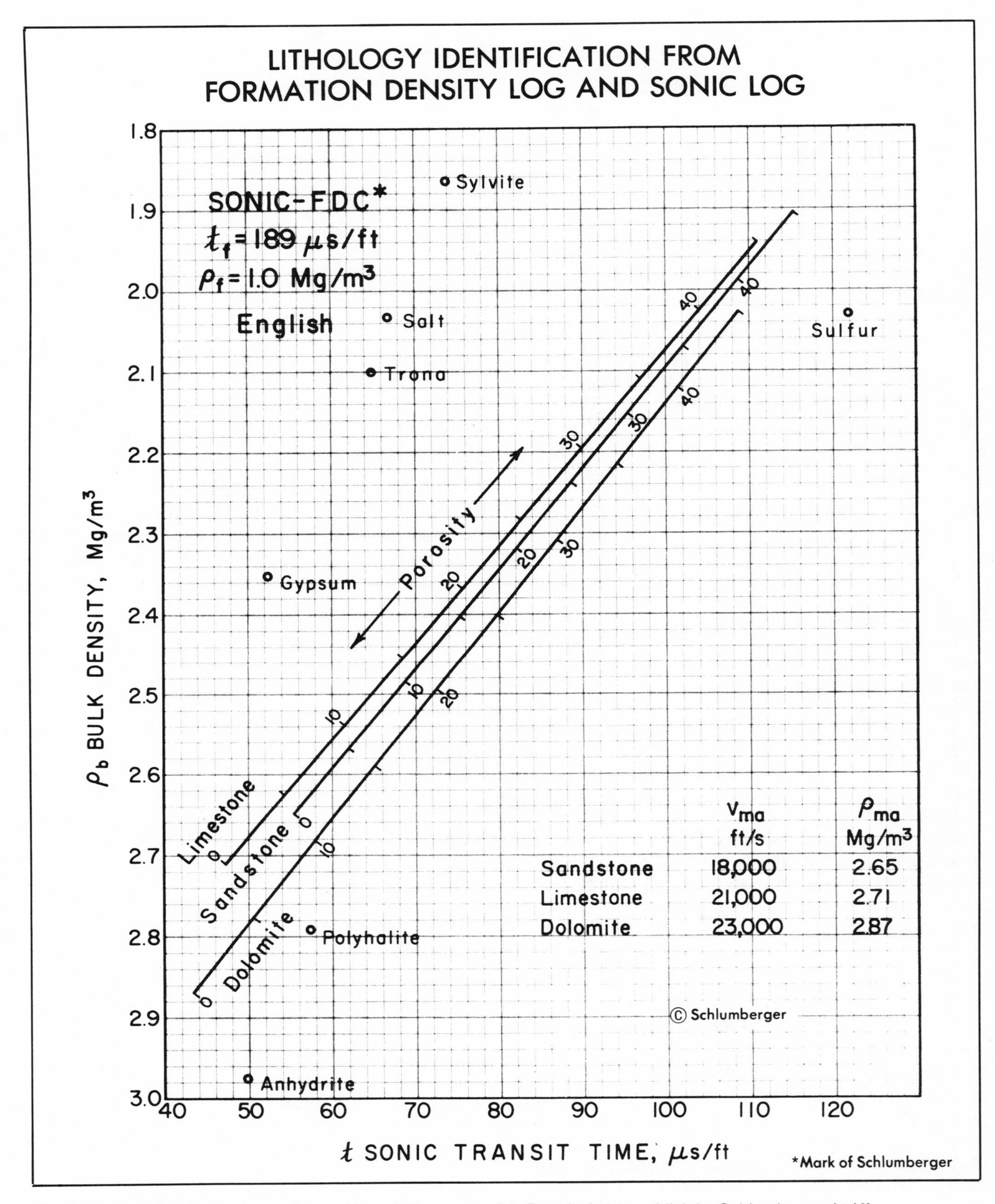

Fig. 9-35. Typical bulk density vs. interval transit time crossplot. Permission to publish by Schlumberger Ltd.[33]

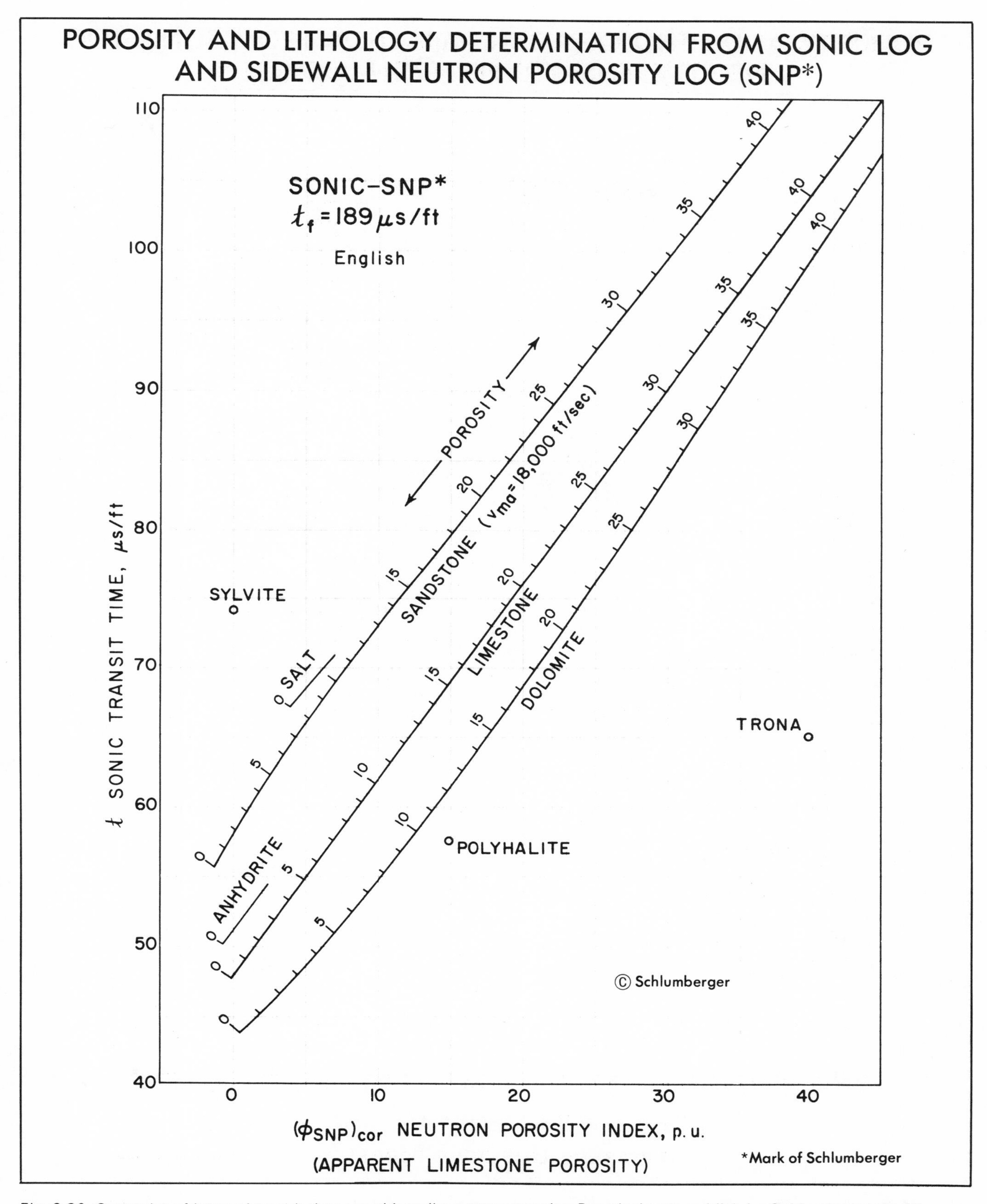

Fig. 9-36. Crossplot of interval transit time vs. sidewall neutron porosity. Permission to publish by Schlumberger Ltd.[33]

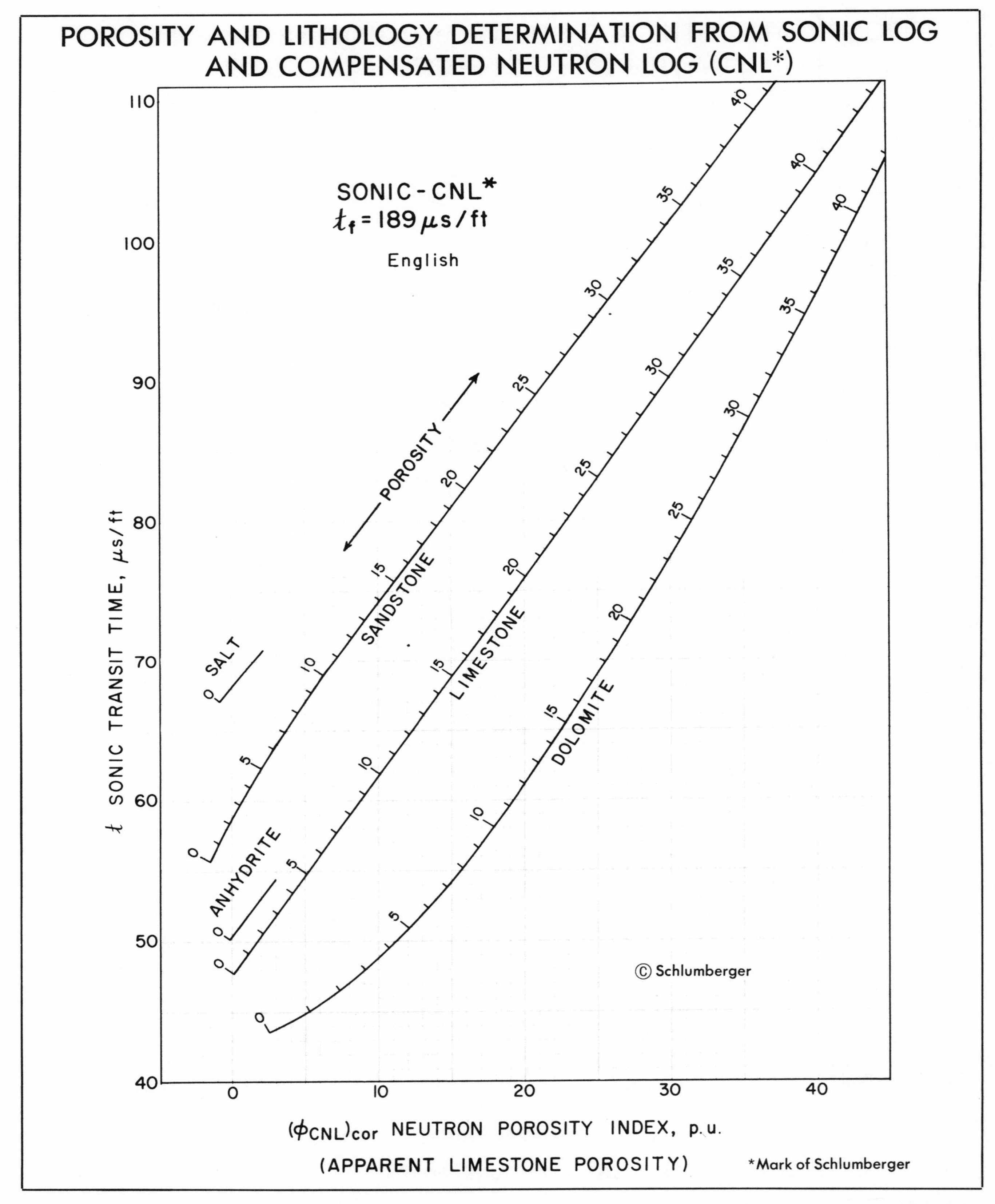

Fig. 9-37. Crossplot of interval transit time vs. compensated neutron porosity. Permission to publish by Schlumberger Ltd.[33]

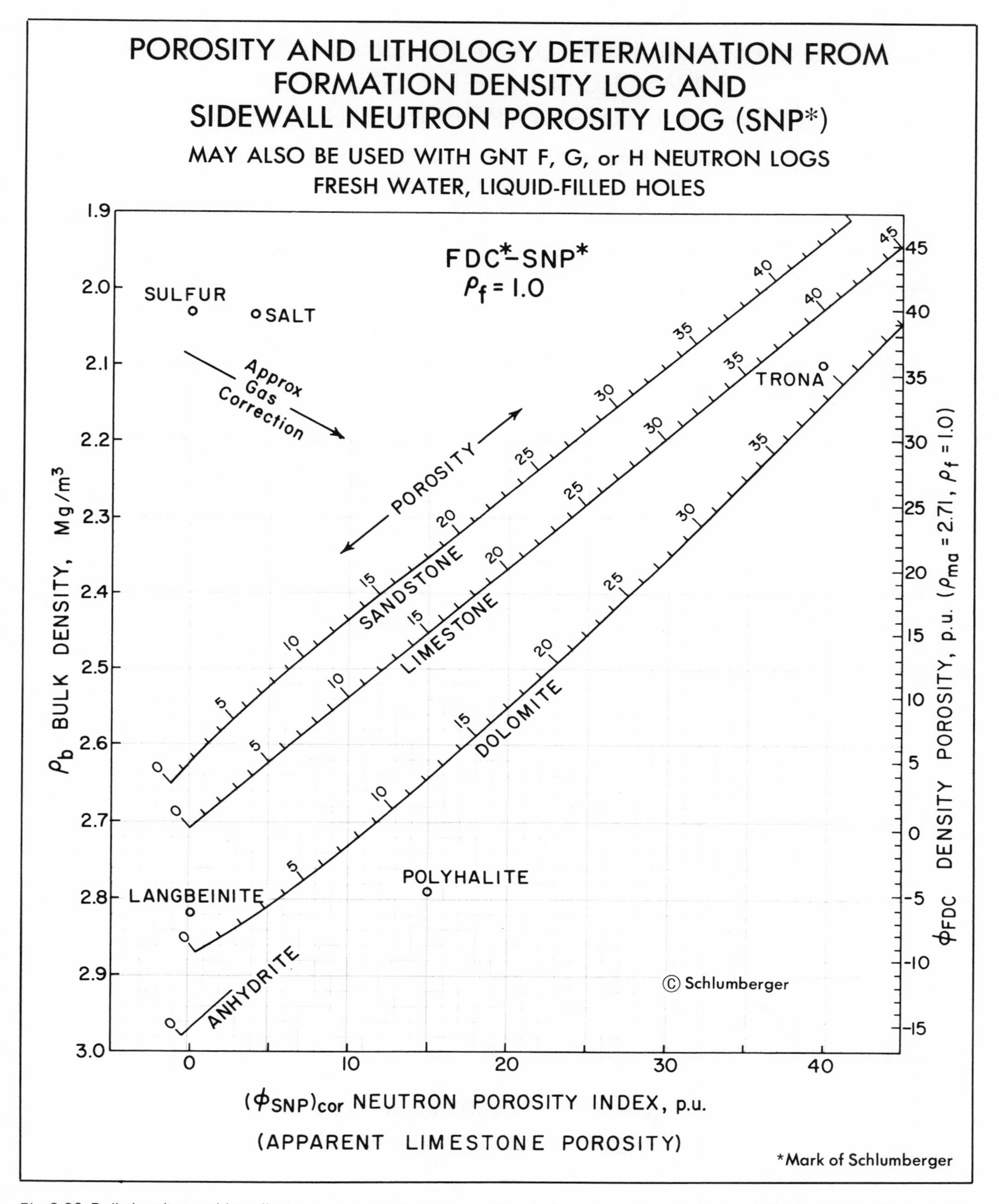

Fig. 9-38. Bulk density vs. sidewall neutron porosity for fresh mud borehole systems. Permission to publish by Schlumberger Ltd.[33]

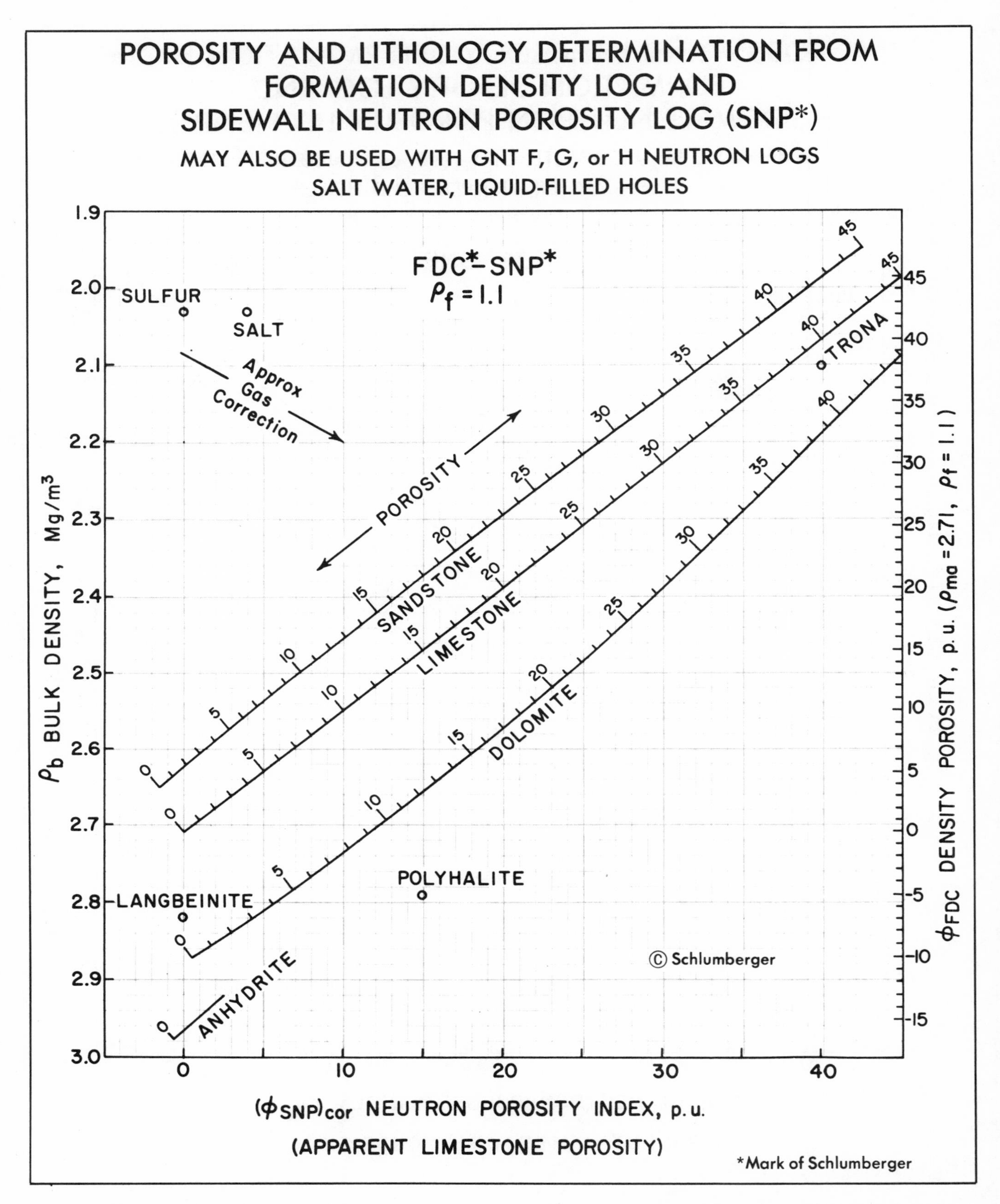

Fig. 9-39. Bulk density vs. sidewall neutron porosity for salt mud borehole systems. Permission to publish by Schlumberger Ltd.[33]

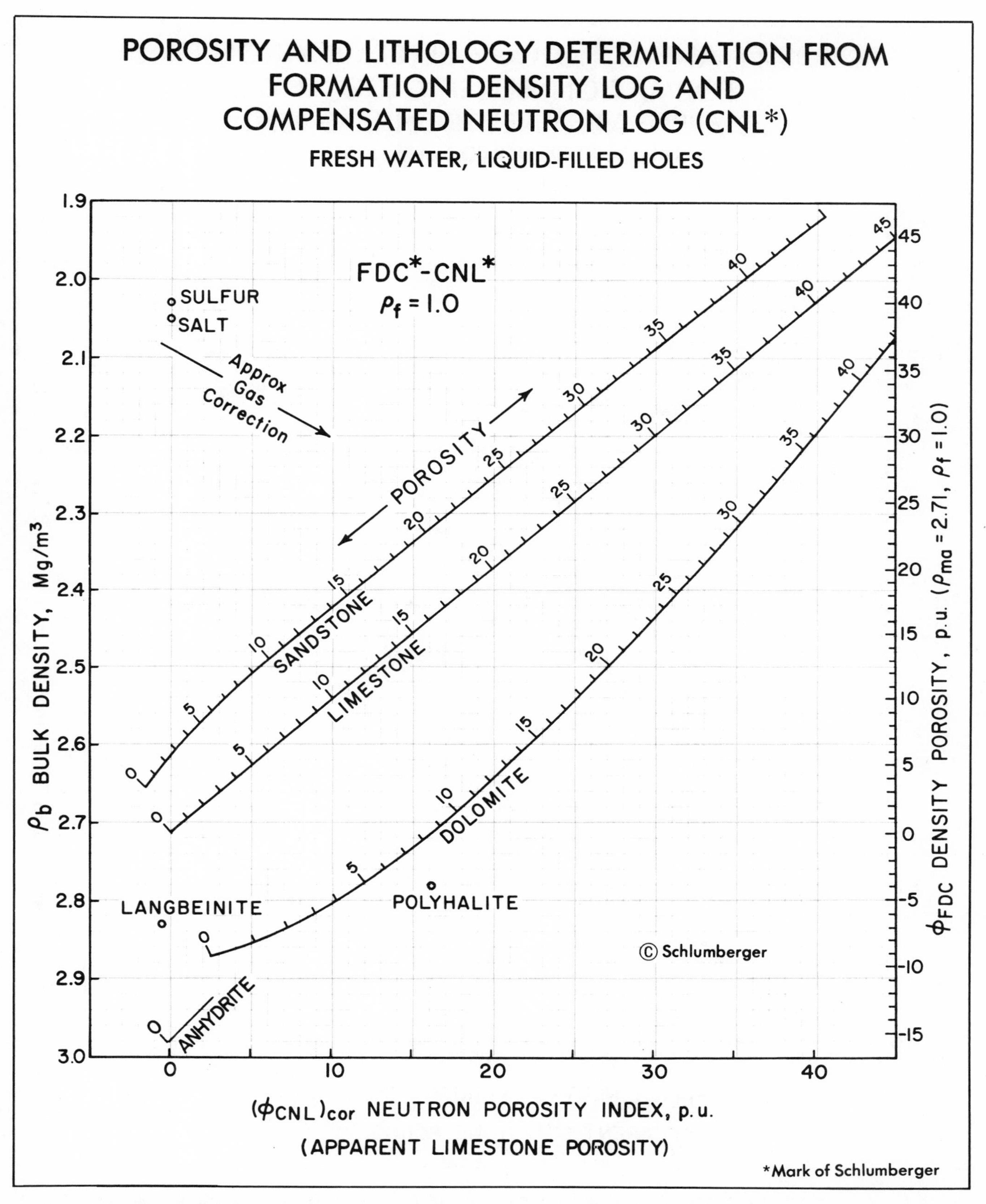

Fig. 9-40. Bulk density vs. compensated neutron porosity in fresh mud borehole systems. Permission to publish by Schlumberger Ltd.[33]

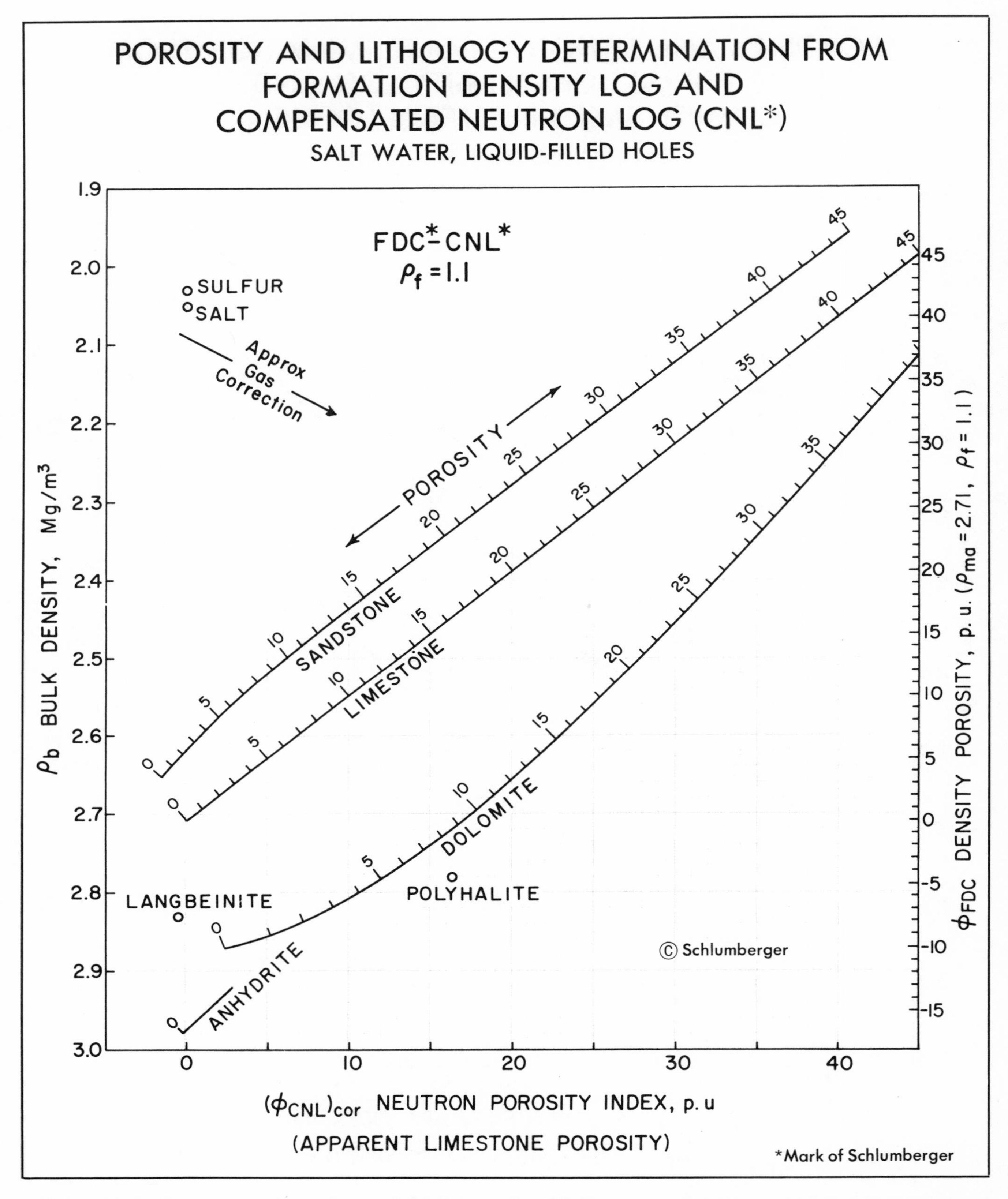

Fig. 9-41. Bulk density vs. compensated neutron porosity in salt mud borehole systems. Permission to publish by Schlumberger Ltd.[33]

M vs. *N* Type Crossplot

Since the slopes of the porosity log crossplots are rock typing parameters, it stands to reason that a crossplot of these slopes would also be a rock typing indicator.[9, 17, 28] The slopes of the porosity log crossplots are independent of porosity being functions of the lithology and fluid only. If we consider the acoustic-density crossplot, the slope can be defined as *M* and graphically shown in Figure 9-42. Considering the neutron-density crossplot, the slope may be defined as *N* as shown in Figure 9-43. Using the "average" rock parameters of Tables 9-1 and 9-2, each lithology would be defined as a distinct point on the *M* vs. *N* crossplot, as shown in Figure 9-44 based on sidewall neutron porosity, or as in Figure 9-45 based on compensated neutron porosity.

In a formation of complex lithology, the position of the log data points on the *M* vs. *N* plot relative to the pure mineral points is particularly useful in rock typing and leads to estimating more accurate porosities. These two slopes are useful because all three porosity

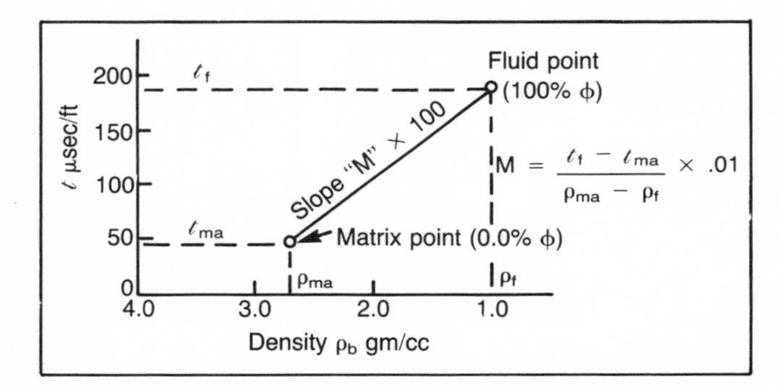

Fig. 9-42. M from acoustic-density crossplot of mineral "A." Permission to publish by the Society of Professional Well Log Analysts.[9]

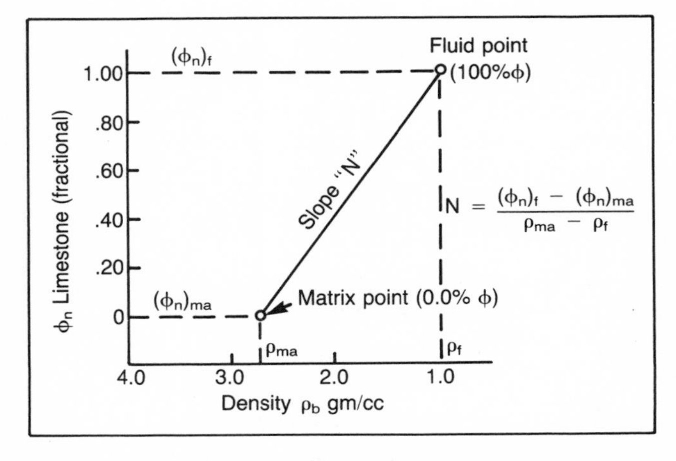

Fig. 9-43. N from neutron-density crossplot of mineral "A." Permission to publish by the Society of Professional Well Log Analysts.[9]

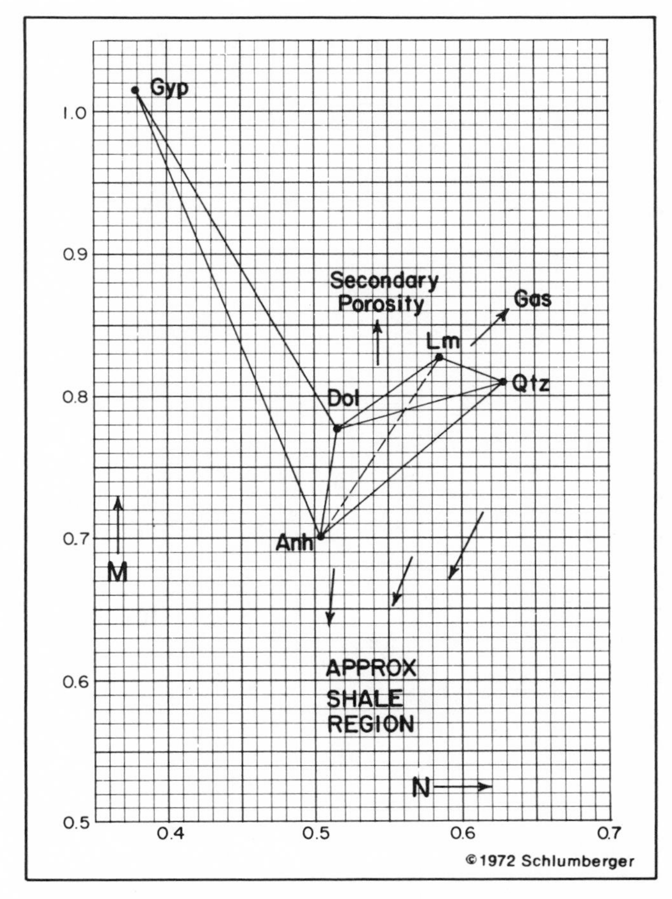

Fig. 9-44. M vs. N (based on the SNP neutron log) crossplot for rock typing. Permission to publish by Schlumberger Ltd.[23]

responses can be observed using two-dimensional plotting techniques. When using the actual log data, the average matrix values used in defining the lithology points are substituted with the actual log responses such that:

$$M = \frac{\ell_f - \ell}{\rho_b - \rho_f} \times .01$$

$$N = \frac{(\phi_N)_f - \phi_N}{\rho_b - \rho_f}$$

MN Product vs. ρ_b Crossplot. Since the *M* and *N* values for any given mineral will produce a common point regardless of porosity when plotted, it is apparent that their product will also be a constant for that mixture. When the *MN* product for the four reservoir minerals (sand, limestone, dolomite, and anhydrite) is plotted versus matrix density, ρ_{ma}, a straight line results, as shown in Figure 9-46. With this method, the grain density of a rock mixture can easily be determined regardless of the mineral mixture in the reservoir. If bulk density from the density log is plotted versus the *MN* product, then for each point the matrix density of

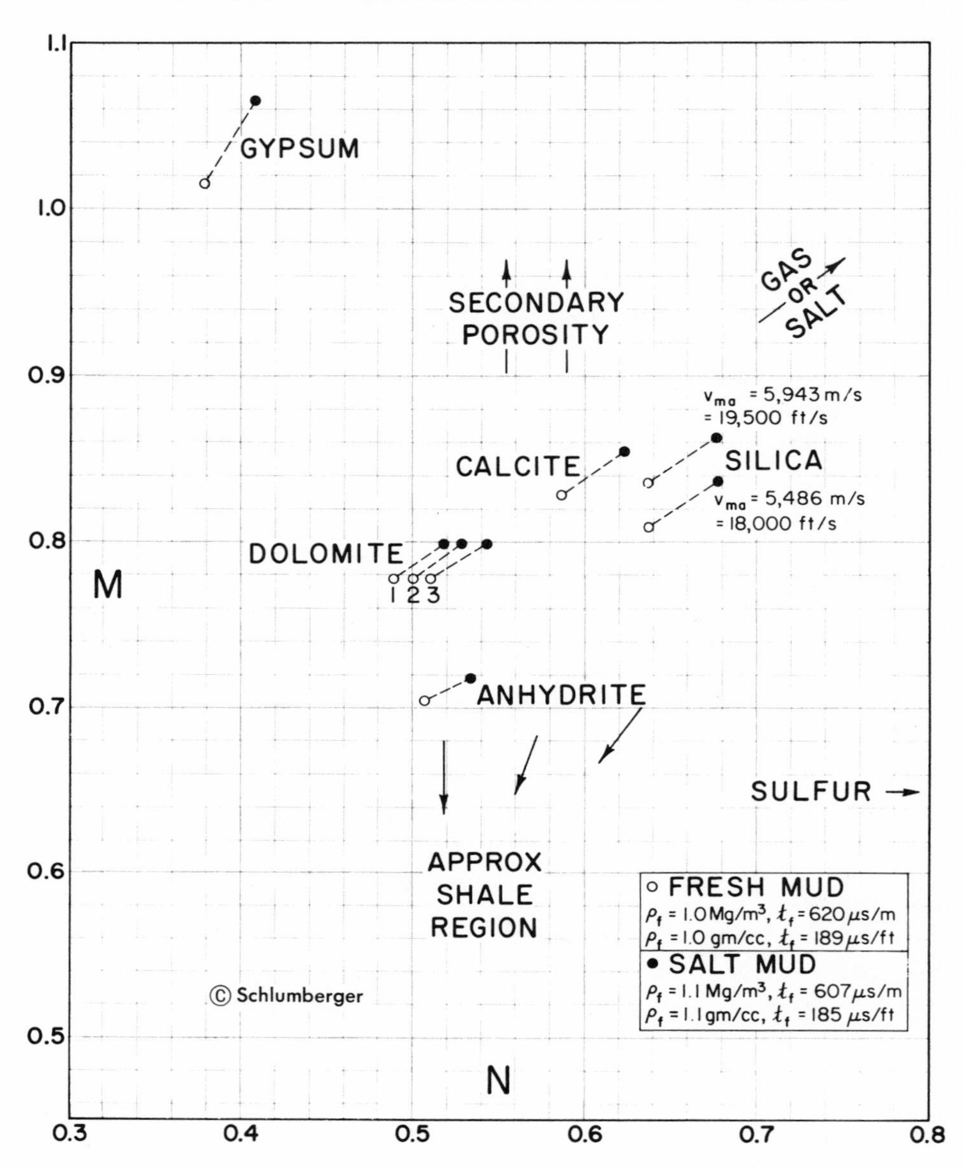

This crossplot may be used to help identify mineral mixtures from Sonic, Density, and Neutron logs. (The Neutron log used in the above chart is the CNL*.) Except in gas-bearing formations, M and N are practically independent of porosity. They are defined as:

$$M = \frac{t_f - t}{\rho_b - \rho_f} \times 0.01 \text{ (English)}; \qquad M = \frac{t_f - t}{\rho_b - \rho_f} \times 0.003 \text{ (Metric)}; \qquad N = \frac{(\phi_N)_f - \phi_N}{\rho_b - \rho_f} \text{ (Either)}$$

Points for binary mixtures plot along a line connecting the two mineral points. Ternary mixtures plot within the triangle defined by the three constituent minerals. The effect of gas, shaliness, secondary porosity, etc. is to shift data points in the directions shown by the arrows.

The dolomite lines are divided as to porosity as follows:

1) $\phi = 5.5$ to 30 p.u. 2) $\phi = 1.5$ to 5.5 p.u. and $\phi > 30$ p.u. 3) $\phi = 0$ to 1.5 p.u.

*Mark of Schlumberger

Fig. 9-45. M vs. N crossplot for rock typing. Permission to publish by Schlumberger Ltd.[33]

the mineral mixture associated with that bulk density can be determined by simply moving horizontally to the straight line defined by the pure mineral points and then dropping vertically to read the apparent matrix density (see Fig. 9-47). Using the measured bulk density and the apparent matrix density, the effective porosity of the mineral mixture can be calculated.

MID Plot Method

A technique for rock typing using the three independent porosity tools, and therefore the acoustic, density, and neutron responses, has been found to be particularly useful in field applications in lieu of the *M* versus *N* method.[30] This approach couples the acoustic-neutron and density-neutron responses to define two rock typing parameters, $(t_{ma})_a$ and $(\rho_{ma})_a$. These two porosity independent parameters, characteristic of the rock, are defined using the acoustic (time average equation) and density log response equations as follows:

$$t = (t_{ma})_a + [t_f - (t_{ma})_a]\phi_N$$

$$\rho_b = (\rho_{ma})_a + [\rho_f - (\rho_{ma})_a]\phi_N$$

Graphical solutions for these two lithology parameters are shown in Figures 9-48 through 9-53 for Schlumberger's sidewall neutron and compensated neutron logs recorded on a limestone matrix.[33] Using the appropriate charts, the apparent interval transit time and density matrix values $(t_{ma})_a$ and $(\rho_{ma})_a$, can be rapidly obtained. These parameters are then crossplotted on Figure 9-54 enabling an estimate of lithology to be made in a manner similar to the *M* versus *N* crossplot. The matrix points on the MID plot, however, unlike the *M* versus *N* crossplot, are fixed, regardless of change in mud filtrate, porosity, and neutron tool type.

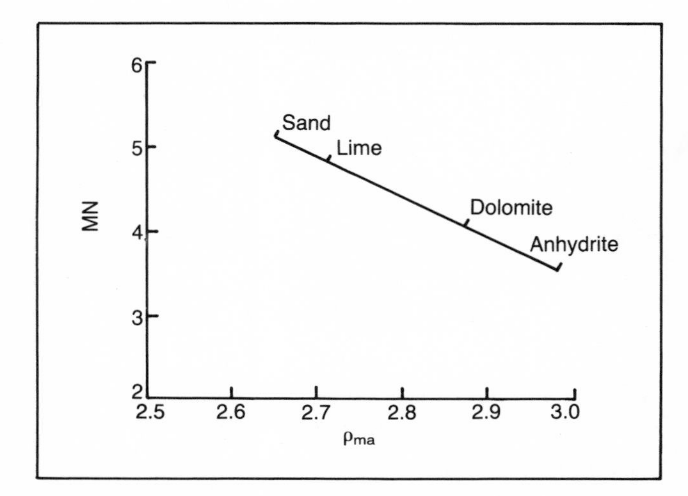

Fig. 9-46. MN product vs. ϱ_g crossplot. Permission to publish by the Canadian Well Log Society.[16]

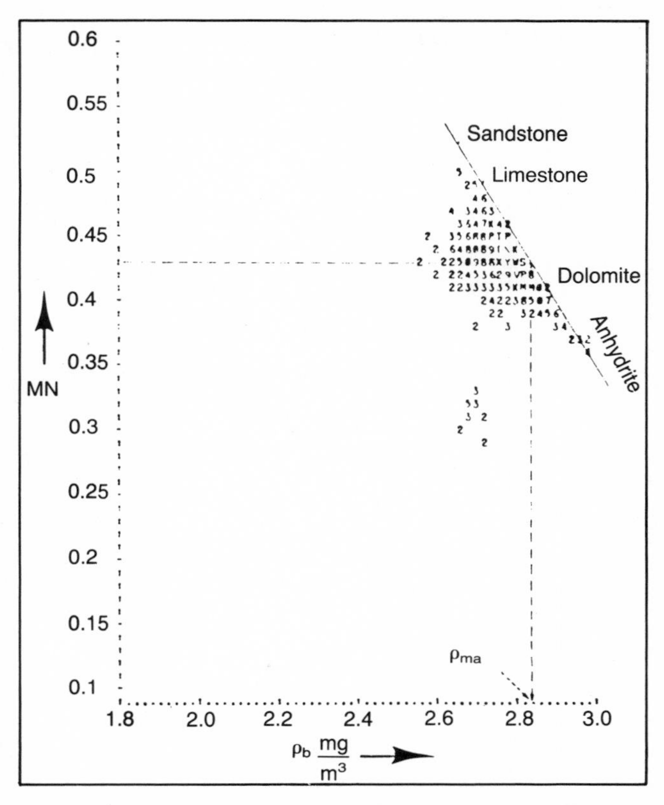

Fig. 9-47. MN product vs. ϱ_b crossplot showing the determination of ϱ_{ma} for plotted data. Permission to publish by the Canadian Well Log Society.[16]

TABLE 9-3
Tri-Porosity Data and the Rock Typing Parameters, $(t_{ma})_a$ and $(\rho_{ma})_a$, for a Utah Well

Point	Data ρ_b	ϕ_{SNP}	t	MID Coordinates $(\rho_{ma})_a$	$(t_{ma})_a$
1	2.60	11	75	2.77	56
2	2.50	8	62	2.67	47
3	2.63	10	71½	2.79	55
4	2.63	7½	59	2.75	48
5	2.60	5½	58	2.70	49
6	2.63	12½	62	2.82	46
7	2.57	7	62	2.70	51
8	2.65	8	60	2.78	48
9	2.60	8½	62	2.73	49
10	2.61	8	62	2.74	50
11	2.60	7	61	2.72	50
12	2.60	8	60	2.73	48

Permission to publish by the Society of Professional Well Log Analysts.[30]

An example of this application is shown in Figure 9-55 using the data of Table 9-3.[30] This data is from a Utah well where the rock type for this data is mostly limestone, with some points showing the influence of shale. The shaliness influence is best analyzed using data from nearby shales in the interval being evaluated.

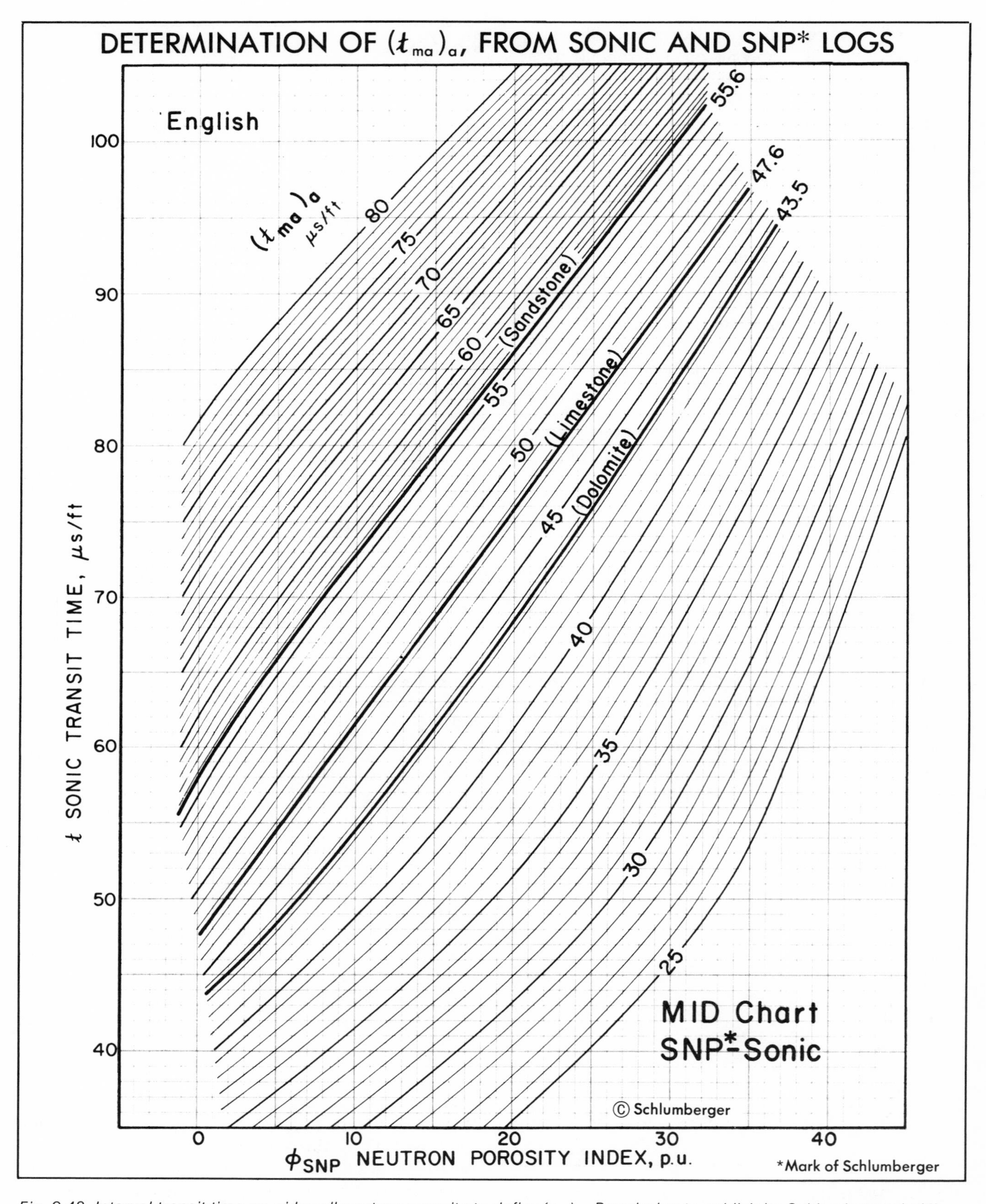

Fig. 9-48. Interval transit time vs. sidewall neutron porosity to define $(t_{ma})_a$. Permission to publish by Schlumberger Ltd.[33]

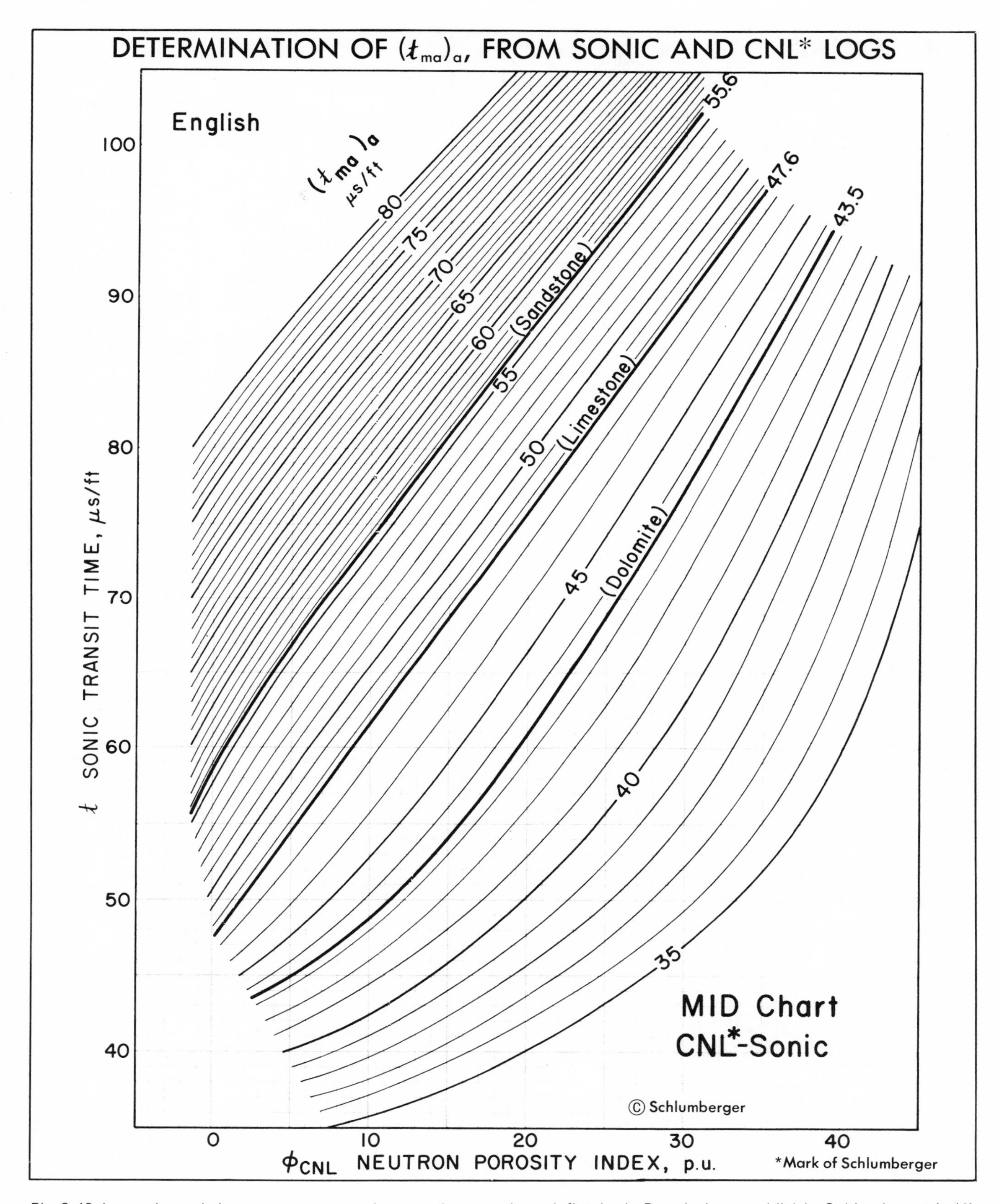

Fig. 9-49. Interval transit time vs. compensated neutron log porosity to define $(t_{ma})_a$. Permission to publish by Schlumberger Ltd.[33]

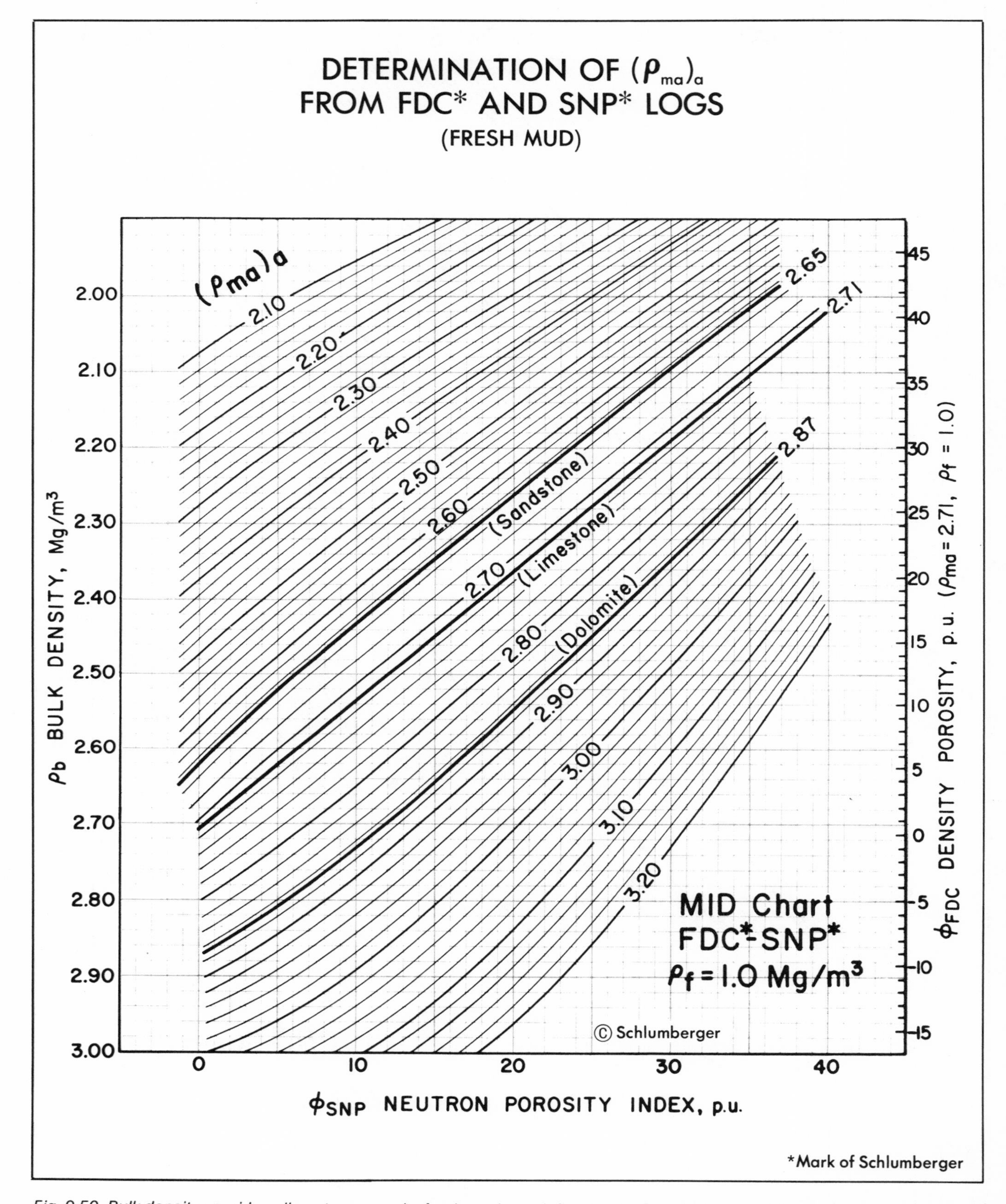

Fig. 9-50. Bulk density vs. sidewall neutron porosity fresh muds, to define $(\rho_{ma})_a$. Permission to publish by Schlumberger Ltd.[33]

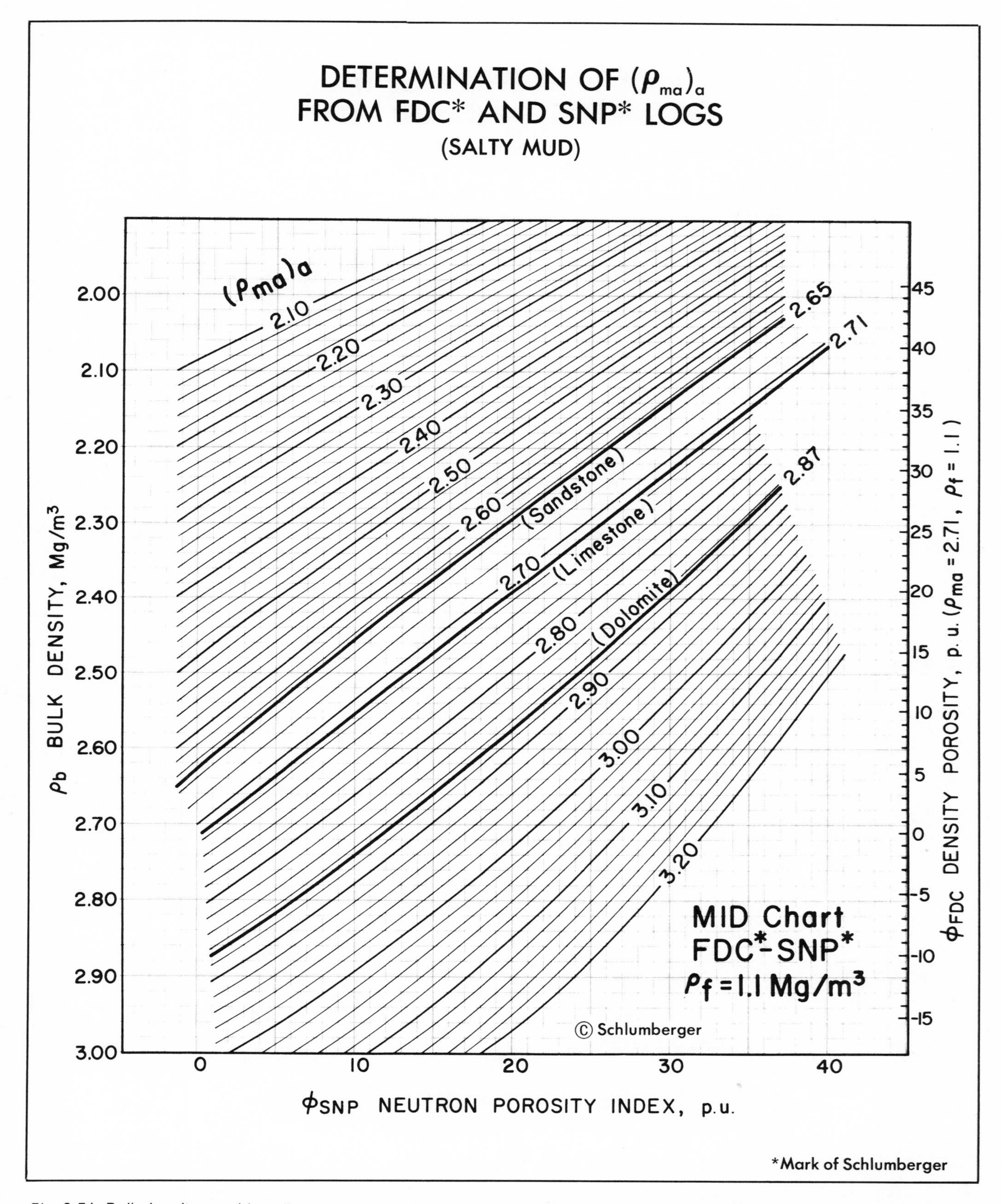

Fig. 9-51. Bulk density vs. sidewall neutron porosity, salt muds, to define $(\rho_{ma})_a$. Permission to publish by Schlumberger Ltd.[33]

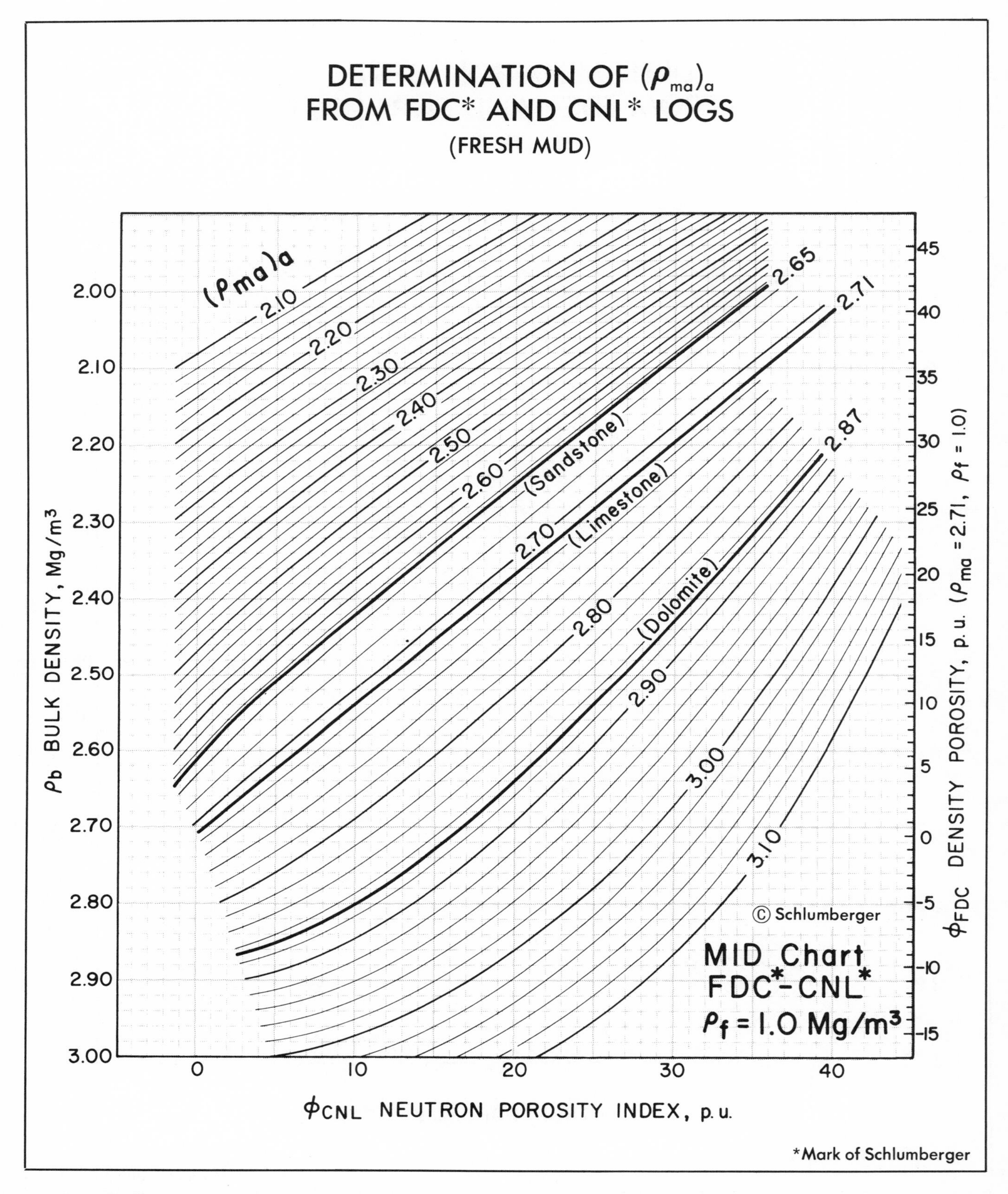

Fig. 9-52. Bulk density vs. compensated neutron log porosity, fresh muds, to determine $(\rho_{ma})_a$. Permission to publish by Schlumberger Ltd.[33]

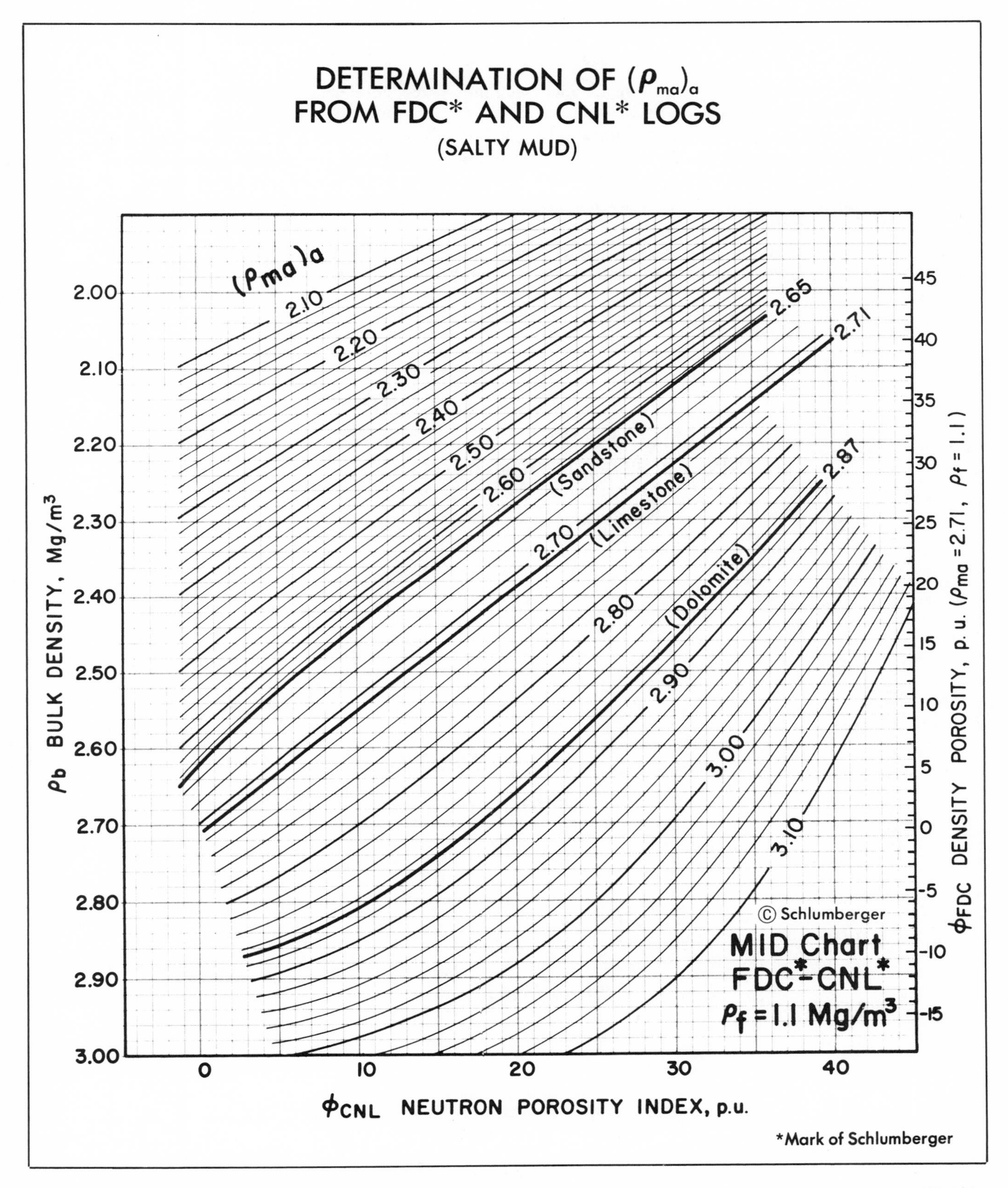

Fig. 9-53. Bulk density vs. compensated neutron log porosity, salt muds, to define $(\rho_{ma})_a$. Permission to publish by Schlumberger Ltd.[33]

HOW TO USE THE MID PLOT*[8]

Select the $(t_{ma})_a$ and $(\rho_{ma})_a$ charts for your Neutron-log type, borehole-fluid salinity, and measuring system (English or metric).

Tabulate t, and ρ_b, and ϕ_N (limestone) by depths for each chosen station and the resultant $(t_{ma})_a$ and $(\rho_{ma})_a$ from the appropriate chart. Plot the points on the MID-Plot grid. The plot will generally form a pattern which identifies the major reservoir rock by its proximity to the labeled points on the plot.

The presence of secondary porosity in the form of vugs or fractures produces displacements parallel to the Sonic-sensitive axis.

The presence of gas will displace points as shown on the basic MID Plot. Identification of shaliness is best done by plotting some shale points to establish the shale trend lines.

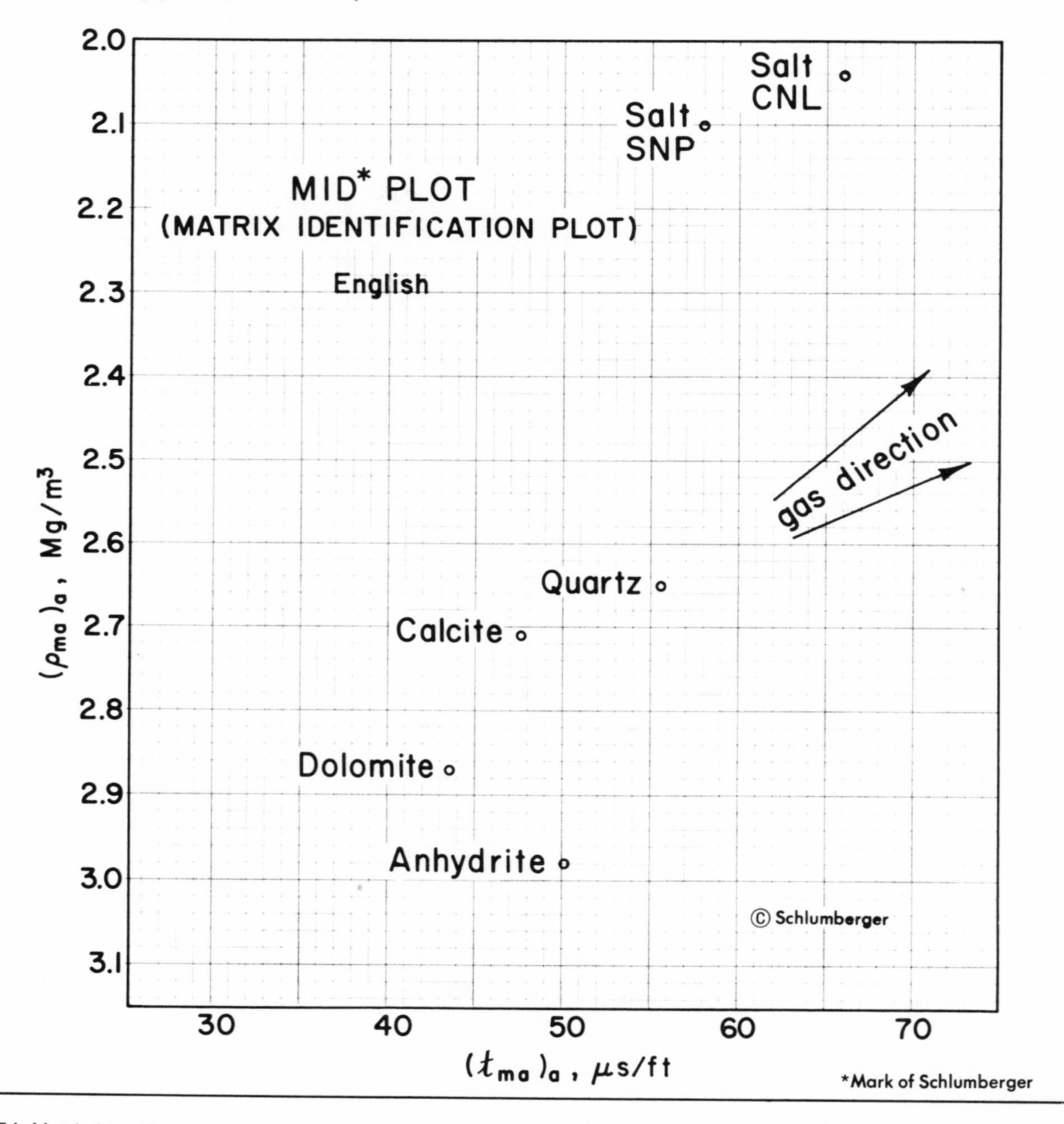

*Mark of Schlumberger

Fig. 9-54. Matrix identification plot, MID plot, for rock typing analysis. Permission to publish by Schlumberger Ltd.[33]

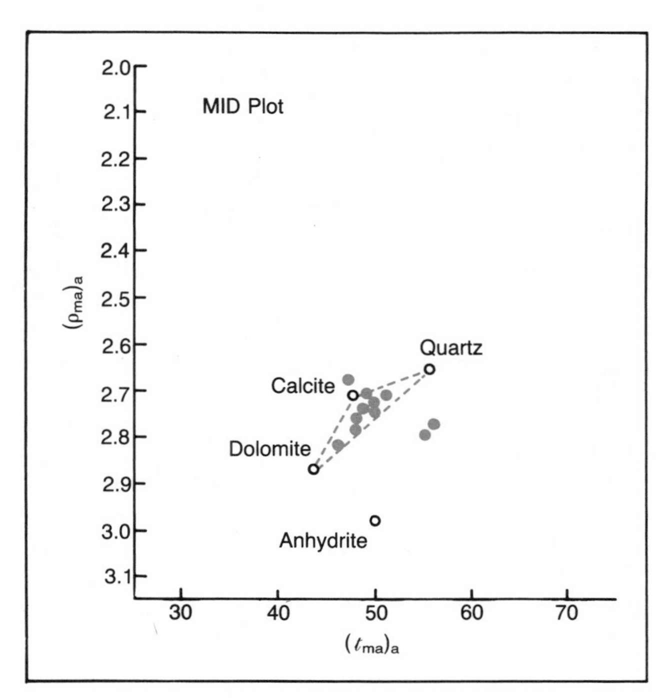

Fig. 9-55. MID plot for Utah well. Permission to publish by the Society of Professional Well Log Analysts.[30]

SHALY SAND INTERPRETATION

Shaly sand analysis is somewhat beyond the intended scope of this book, but a brief discussion will be presented here. The presence of shale in a formation will influence all measured responses to some degree, and this influence is complicated by the wide variation in shale properties. When using these measurements in shaly sands, therefore, the shale influence must be accounted for in order to obtain the effective porosity, ϕ_e, and water saturation, S_w.

To assess the influence of "shaliness" an estimate of clay or shale volume, V_{sh}, is necessary. Although all measurements could be used, either singly or in combination, to estimate V_{sh}, the gamma ray log is usually considered as being reliably diagnostic. As discussed previously, a gamma ray index, I_{ShGR}, is determined from the gamma ray log response and used to project an estimate of V_{sh}, as from Figure 9-56. A major problem would be if a gamma ray response for a clean (nonshaly) zone was not available. One means of estimating a clean sand gamma ray response is to crossplot the lithology parameter, N, versus the gamma ray response, as shown in Figure 9-57. A trend can usually be seen for quartz, water, and shale systems, which can be projected to the clean sand value for N and, hence, the apparent clean sand gamma ray response.

An estimate of V_{sh} provides a means for projecting the effective porosity, ϕ_e. Each porosity response could be corrected individually, assuming only a shaliness influence has affected the measure. However, hydrocarbon effects and particularly gas effects on the measure could be important. In order to account for the gas effect, two porosity devices, the density and the neutron logs, are usually used. As can be seen in Figure 9-58, a clean sand

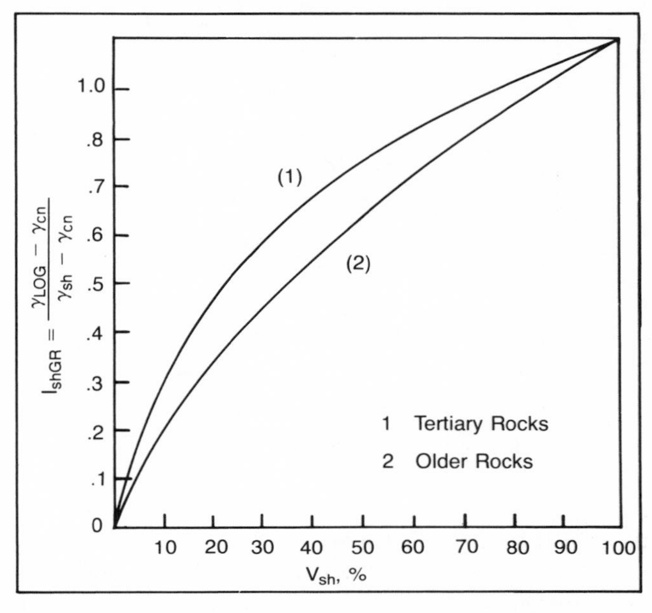

Fig. 9-56. Gamma ray index vs. V_{sh}. Permission to publish by Dresser Atlas, Dresser Industries.[27]

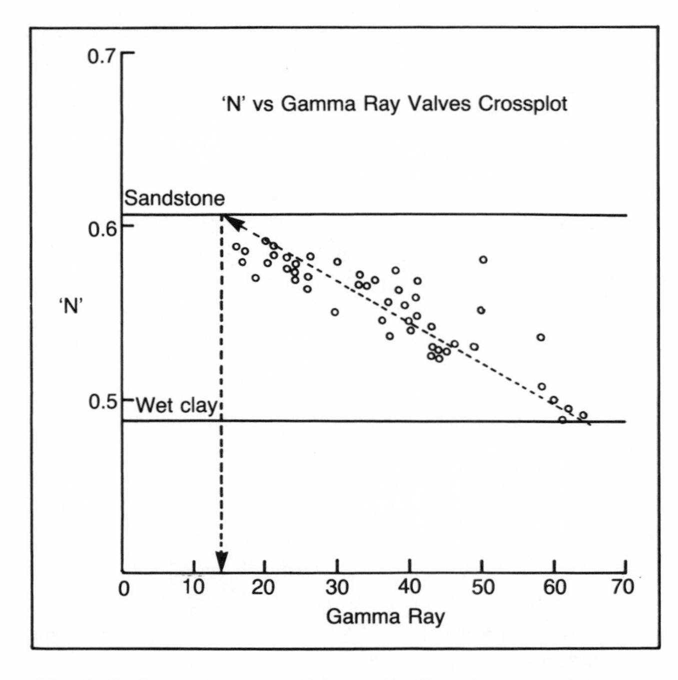

Fig. 9-57. Crossplot method for projecting clean sand gamma ray response. Permission to publish by the Society of Professional Well Log Analysts.[32]

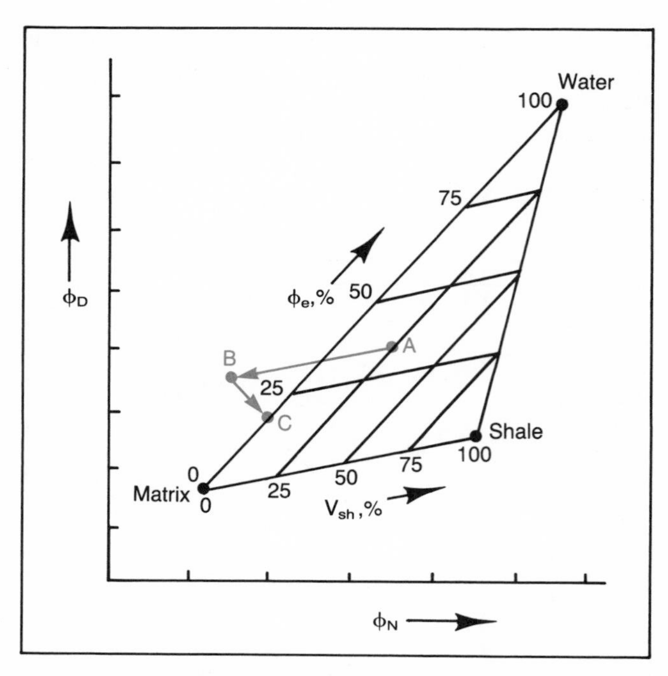

Fig. 9-58. Illustration of effective porosity, φE, determination using independent V_{sh} for apparent gas effect correction.

trend can be defined (matrix and water points) along with a best estimate of the shale point based upon surrounding shale behavior. For the example shown, density and neutron data from a shaly zone is plotted at point A. This would define an effective porosity, ϕ_e, and the shale volume, V_{sh}, assuming no gas effect. Using an independent V_{sh} from, for example, the gamma ray log, the point is shifted at constant porosity to the left. If the gamma ray determined value of V_{sh} is larger than the V_{sh} value indicated by the crossplot, the point will shift to the left of the clean sand line (as point B), and it can be assumed that a gas effect exists. To define the gas corrected effective porosity, ϕ_e, point B is shifted to C along a trend of roughly 45°. This shift to point C is dependent on S_{xo} and hydrocarbon density, ρ_h, and can be projected more efficiently, as discussed by Poupon, et al.[15]

Use of effective porosity in the basic saturation relationship

$$S_w = \left(\frac{aR_w}{\phi_e^m R_t}\right)^{1/n}$$

has been recognized for many years as being deficient, however, and leads to unrealistic water saturation values. The apparent flaw in using the conventional saturation expression for shaly sand evaluation, also recognized for many years, is the incompatibility of the true resistivity, R_t, and effective porosity, ϕ_e, terms. Since the ability of a shaly sand to conduct is not only a function of the water contained in the effective pore space but also due to the presence of "conductive solids" (shale) in the rock, a modification of R_t to reflect the influence of shaliness has been deemed necessary in order to project water saturation. In fact, this has been the basis for most shaly sand methods for saturation estimates. As a result, most shaly sand techniques for projecting water saturation are based on a method for estimating the shale contribution to bulk conductivity. This has resulted in numerous expressions for shaly sand resistivity and, therefore, numerous expressions for shaly sand water saturation. Some of the more commonly used resistivity expressions are shown in Table 9-4. Substitution of these expressions into the basic saturation equation incorporating ϕ_e and reasonable values of m, n and R_w produces a shaly sand saturation expression. As can be expected, a number of shaly sand models have been proposed over the years. At the present time it is generally accepted, at least for manual calculations, that the model proposed by Simandoux, expressed in parabolic form as shown below,

$$\frac{1}{R_t} = \frac{V_{sh}}{R_{sh}} S_w + \frac{\phi_e^m}{aR_w} S_w^2$$

is probably the better generalization for the widest range of conditions encountered, although locally a different relation might provide a better "fit."

TABLE 9-4
Shaly Sands Relations

Laminated *Sd/Sh* model	$\frac{1}{R_t} = \frac{V_{cl}}{R_{cl}} + \frac{1 - V_{cl}}{R_{tsd}}$
de Witte	$\frac{1}{R_t}\left[\left(\frac{R_w}{R_{tsd}}\right)^{1/n} + V_{cl}\right] \times \left[\left(\frac{1}{R_{tsd}}\right)^{1/n} + \frac{V_{cl}}{R_{cl}}\right]$
Doll	$\frac{1}{(R_t)^{1/n}} = \frac{V_{cl}}{(R_{cl})^{1/n}} + \frac{1}{(R_{tsd})^{1/n}}$
Hossin	$\frac{1}{R_t} = \frac{V_{cl}^2}{R_{cl}} + \frac{1}{R_{tsd}}$
Total shale relation	$\frac{1}{R_t} = \frac{V_{cl}}{R_{cl}} S_w + \frac{1}{R_{tsd}}$
"Schlumberger C.P.I."	$\frac{1}{(R_t)^{1/n}} = \frac{V_{cl}^{[1-(V_{cl}/2)]}}{(R_{cl})^{1/n}} + \frac{1}{(R_{tsd})^{1/n}}$
Simandoux	$\frac{1}{R_t} = \alpha\frac{V_{cl}}{R_{cl}} + \frac{1}{R_{tsd}}$ (with α dropping to 0 for low values of S_w depending on mineralogical nature of clay)
Waxman-Smits	$\frac{1}{R_t} = \frac{R_w}{R_{tsd}}\left(\frac{1}{R_w} + \frac{BQ_v}{S_w}\right)$ (with Q_v being the concentration of counter ions associated with the clay in the pore water, and B the equivalent conductance of these counter ions)

Permission to publish by the Society of Professional Well Log Analysts.[31]

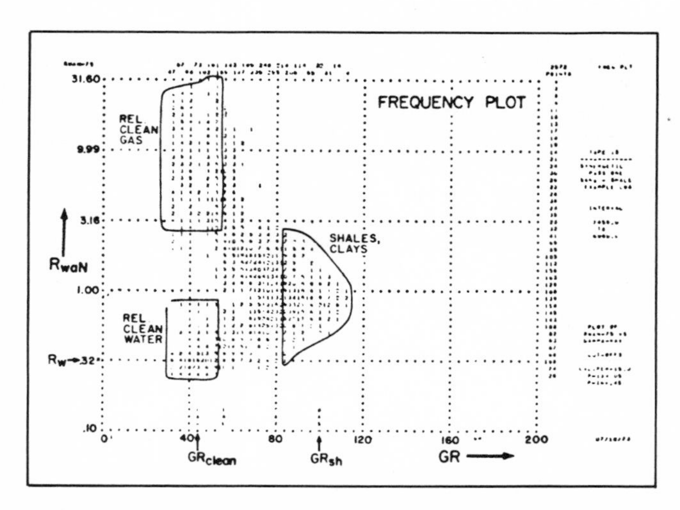

Fig. 9-59. R_{wa} vs. gamma ray frequency crossplot. Permission to publish by Schlumberger Ltd.[28]

Evaluation of water saturation, S_w, in shaly sands also requires a value for the interstitial water resistivity, R_w. This is the water resistivity existing in the clean sand members and might be estimated from the SP, direct resistivity measurement of produced water, R_t versus ϕ_a crossplots of the wet, clean sand members, or by R_{wa} techniques. It is necessary, however, to have logged one or more clean sands (preferably wet) in order to estimate R_w. R_{wa} determinations, which have been discussed previously, can be made from neutron, acoustic, or density measurements. Values of R_{wa} can also be crossplotted versus the gamma ray response on a frequency plot as shown in Figure 9-59 in order to define R_w (as well as the gamma ray response in the shale and clean sand members).

The preceding discussion on shaly sand interpretation is not comprehensive but should provide an insight into the shaly sand evaluation problem. As can be imagined, this is one of the major rock typing problems facing a formation evaluator, although with local experience reasonable estimates of effective porosity, ϕ_e, and water saturation, S_w, can usually be obtained.

REFERENCES

1. Wyllie, M.R.J.: *The Fundamentals of Electric Log Interpretation*, Academic Press, New York, 1957.
2. Hingle, A.T.: "The Use of Logs in Exploration Problems," paper presented at the SEG 29th International Annual Meeting, Los Angeles, November 1959.
3. Atkins, R.E., Jr.: "Techniques of Electric Log Interpretation," *J. Pet. Tech.* (Feb. 1961).
4. Glanville, C.R.: "Derivation of Scaling or Special Graph Paper Used in Resistivity-Transit Time Log Interpretations," *The Log Analyst* (Dec. 1962) 3.
5. Raymer, Lewis L., and Biggs, W. Pat: "Matrix Characteristics Defined by Porosity Computations," *Trans.*, SPWLA (May 1963).
6. Tixier, Maurice P., Eaton, F.M., Tanguy, Denis R., and Biggs, W. Pat: "Automatic Log Computation at Wellsite—Formation Analysis Logs," *J. Pet. Tech.* (January 1965).
7. Pickett, George R.: "A Review of Current Techniques for Determination of Water Saturation from Logs," *J. Pet. Tech.* (November 1966).
8. Morris, R.L., and Biggs, W. Pat: "Using Log Derived Values of Water Saturation and Porosity," paper presented at the SPWLA Eighth Annual Logging Symposium, May 1967.
9. Burke, Jack A., Schmidt, A.W., and Campbell, R.L., Jr.: "The Litho-Porosity Cross Plot," *The Log Analyst* (Nov.-Dec. 1969).
10. Jeffries, F.: "Simplified Theory of Multiple Log Analysis," *CWLS J.* (Dec. 1969) 2.
11. Pickett, George R.: "Principles for Application of Borehole Measurement in Petroleum Engineering," *The Log Analyst* (May-June 1969) 10.
12. Maxson, Rodney R.: "Principles for Application of Borehole Measurements in Petroleum Engineering," *Petroleum Engineer* (February 1970).
13. *Neutron Lifetime Interpretation*, Dresser-Atlas, 1970.
14. Pickett, George R.: "Applications for Borehole Geophysics in Geophysical Exploration," *The Log Analyst* (May-June 1969).
15. Poupon, Andre, Clavier, Christian, Dumanoir, Jean, Gaymard, R., and Misk, Andre: "Log Analysis of Sand-Shale Sequences—A Systematic Approach," *J. Pet. Tech.* (July 1970).
16. Heslop, Alan: "Mixed Lithology Analysis Using *MN* Product," *CWLS J.* (Dec. 1971) 4.
17. Poupon, Andre, Hoyle, W.R., and Schmidt, A.W.: "Log Analysis in Formations with Complex Lithologies," *J. Pet. Tech.* (August 1971).
18. Threadgold, Phillip: "Interpretation of Thermal Neutron Die-Away Logs—Some Useful Relationships," *Trans.*, SPWLA (May 1971).
19. Dumanoir, Jean, Hall, John D., and Jones, J.M.: "R_{xo}/R_t Methods for Wellsite Interpretation," paper presented at the SPWLA Thirteenth Annual Logging Symposium, May 1972.
20. Fertl, Walter H.: "Status of Shaly Sand Evaluation," paper presented at the CWLS 4th Formation Evaluation Symposium, May 1972.
21. Fertl, Walter H.: "Selected Bibliography of Well Logging Interpretation in Shaly Reservoir Rock," presented at the SPWLA Thirteenth Annual Logging Symposium, May 1972.
22. Lang, W.H., Jr.: "Porosity-Resistivity Cross-Plotting, paper presented at the SPWLA Thirteenth Annual Logging Symposium, May 1972.
23. *Log Interpretation Volume I—Principles*, Schlumberger Ltd., New York, 1972.
24. *Log Interpretation Volume II—Applications*, Schlumberger Ltd., New York, NY (1972).
25. Truman, Robert B., Alger, Robert P., Connell, James G., and Smith, R.L.: "Progress Report on Interpretation of the Dual-Spacing Neutron Log (CNL) in the U.S.," *The Log Analyst* (July-Aug. 1972) 13.

26. Campbell, John M., Helander, Donald P., and Van Poolen, H.K.: *Petroleum Reservoir Property Evaluation*, Campbell Petroleum Series, John M. Campbell Co., Norman, OK (1973).
27. *Gamma Ray Log*, Dresser Atlas, Dresser Industries, Houston (1981).
28. *Log Interpretation, Volume II—Applications*, Schlumberger Ltd., New York (1974).
29. McAlester, A.L.: *Physical Geology*, Prentice-Hall, Inc., Englewood Cliffs, NJ (1975).
30. Clavier, Christian, and Rust, D.H.: "MID Plot: A New Lithology Technique," *The Log Analyst* (Nov.-Dec. 1976) 17.
31. Raiga-Clemenceau, J.: "A Quicker Approach of the Water Saturation in Shaly Sands," *Trans.*, SPWLA (1976).
32. Krug, J.A., and Cox, D.D.: "Shaly Sand Crossplot: A Mathematical Treatment," *The Log Analyst* (July-Aug. 1976).
33. *Log Interpretation Charts*, Schlumberger Ltd., New York (1979).
34. Poupon, Andre, Loy, Milton E., and Tixier, Maurice P.: "A Contribution to Electrical Log Interpretation in Shaly Sands," *Trans.*, AIME, 201.

Appendix A

Basic System Classifications

WATER-BASE SYSTEMS (Water is the continuous phase)

Low pH

Fresh water—Those having a liquid phase of water containing only small concentrations of salt, with a pH ranging from 7 to 9.5. This would include spud muds, bentonite treated muds, red muds, organic colloid treatments and some completion fluids.

Brackish water—Includes 7 to 9.5 pH sea and brackish, or hard, water fluids. Source water here may be from open sea or bays.

Saturated salt water—Liquid phase of these fluids is saturated with sodium chloride, although other salts may be present. These may be prepared from fresh water or brine.

Gyp treated—These gypsum-treated or gyp-base muds are formulated by conditioning the mud with plaster (commercial calcium sulfate).

High pH

Lime treated—Consists of adding caustic soda, lime, clay and an organic thinner. These normally have a pH greater than 11.

Fresh water—Fluids having a liquid phase of fresh water but which have been treated with products which bring the pH level above 9. These would include most alkaline-tannant treated muds, etc.

LOW SOLIDS SYSTEMS

Low solids—A fluid in which solids content is less than 10% by weight, or weight is less than 9.5 ppg. A low solids system may have either a water or oil base.

OIL-BASE SYSTEMS (Oil is the continuous phase and water is the dispersed phase)

Invert emulsion muds—These water-in-oil muds have water as the dispersed phase and oil as the continuous phase. These may contain up to 50% by volume of water in the liquid phase.

Oil muds or oil base muds—Both are included in this mud system category. IADC identifies oil muds as, "usually a mixture of diesel fuel and asphalt; not emulsions at the start of their use in drilling. Viscosity is controlled by the addition of diesel fuel to thin, and asphalt compounds or organo clays to thicken. Weight is increased by the addition of barites. As a rule, oil muds will either form w/o emulsions with formation water or from other sources of water contamination." The API, in Bulletin D11, defines oil-base mud as "a special type drilling fluid where oil is the continuous phase and water the dispersed phase. Oil-base mud contains blown asphalt and usually 1 to 5% water emulsified into the system with caustic soda or quick lime and an organic acid. Oil-base muds are differentiated from invert emulsion muds (both water-in-oil emulsions) by the amounts of water used, method of controlling viscosity and thixotropic properties, wall-building materials and fluid loss."

AIR, GAS, MIST SYSTEMS

Air, gas, mist—These include aerated and gaseated systems in this classification.

FUNCTION OF ADDITIVES

The product function classifications, into which each additive has been catalogued, are those generally accepted by the Subcommittee on Drilling Fluids, IADC. "P" in the tables indicates a product's primary purpose, and "S" denotes secondary function.

Alkalinity, pH control additives—Products designed to control the degree of acidity or alkalinity of a fluid. These include lime, caustic soda, bicarbonate of soda.

Bactericides—Function to reduce bacteria count. Paraformaldehyde, caustic soda, lime, and starch preservatives are commonly used.

Calcium removers—Caustic soda, soda ash, bicarbonate of soda and certain polyphosphates make up the majority of chemicals designed to prevent and overcome the contaminating effects of anhydrite and gypsum, both forms of calcium sulfates, which can wreck the effectiveness of nearly any chemically treated fluid not employing calcium removers.

Corrosion inhibitors—Hydrated lime and amine salts are often added in systems to check corrosion. A good fluid, containing an adequate percentage of colloids, certain emulsion muds, and oil muds, exhibit in themselves excellent corrosion inhibiting properties.

Defoamers—Products designed to reduce foaming action, particularly that occurring in brackish water and saturated salt water muds.

Emulsifiers—For the purpose of creating a heterogeneous mixture of two liquids. These include modified lignosulfonates, certain surface active agents, anionic and nonionic (negatively charged and non-charged) products.

Filtrate reducers—Filtrate, or fluid loss reducers—such as bentonite clays, CMC (sodium carboxymethyl cellulose) and pregelatinized starch—serve to cut filter loss, a measure of a drilling fluid's liquid phase's tendency to pass into the formation.

Flocculants—These are used, sometimes, to give rise to increases in gel strength. Salt (or brine), hydrated lime, gypsum and sodium tetraphosphates may be used to cause colloidal particles in suspension to group into bunches or "flocs," causing solids to settle out.

Foaming agents—These are most often chemicals which also act as surfactants (surface active agents) to foam in the presence of water. These foamers permit air or gas drilling through formations making water.

Lost circulation materials—The primary function of a lost circulation additive is to plug the zone of loss, back in the formation, away from the face of the bore hole, sô that subsequent operations will not disturb the plug.

Lubricants—Extreme pressure lubricants are designed to reduce torque and increase horsepower at the bit by reducing the coefficient of friction. Certain oils, graphite powder and soaps are used for this purpose. Lubricants often are used in freeing stuck pipe.

Shale control inhibitors—Gypsum, sodium silicate, calcium lignosulfonates, as well as lime and salt are products used to control caving by swelling or hydrous disintegration of shales.

Surface active agents—Surfactants, as they are called, reduce the interfacial tension between contacting surfaces (water/oil, water/solid, water/air, etc.). These may sometimes be emulsifiers, de-emulsifiers, flocculants, deflocculants, depending upon the surfaces involved.

Thinners, dispersants—These chemicals modify the relationship between the viscosity and percentage of solids in a drilling mud, and may further be used to vary the gel strength, increase a fluid's "pumpability," etc. Tannins (quebracho), various polyphosphates and lignitic materials are chosen as thinners, or as dispersants, since most of these chemicals also remove (by precipitation or sequestering, and deflocculation reactions) solids. Principal purpose of a thinner is to function as a deflocculant to combat random association of clay particles.

Viscosifiers—Bentonite, CMC, attapulgite clays and subbentonites (all colloids) are employed as viscosity builders for fluids to assure a high viscosity-solids relationship.

Weighting materials—Barite, lead compounds, iron oxides and similar products possessing extraordinary high specific gravity are used to control formation pressures, check caving, facilitate pulling dry pipe, as an aid in combatting some types of circulation loss and in well completion operations.

Permission to publish by *World Oil*.

Appendix B

Guide to Selected Drilling Fluids

Key

Mud Types

Water-base

Low pH
1. Fresh water
2. Brackish water
3. Saturated salt water
4. Gyp treated

High pH

5. Lime treated
6. Fresh water
7. Low solids

Oil-base

8. Water-in-oil (invert)
9. Oil mud
10. Air, gas, mist, foam

A. Primary
B. Secondary

Application

11. Alkalinity, pH control additives
12. Bactericides
13. Defoamers
14. Emulsifiers
15. Lubricants
16. Flocculants
17. Filtrate reducers
18. Foaming agents
19. Lost circulation materials
20. Shale control inhibitors
21. Surface active agents
22. Thinners, dispersants
23. Viscosifiers
24. Calcium removers
25. Weighting materials
26. Corrosion inhibitors

TRADENAME:	MATERIAL DESCRIPTION:		FROM:
AKTAFLO-E	Nonionic emulsifier	1 2 4 6 7 14A 21B	Baroid
AKTAFLO-S	Nonionic mud surfactant, shale and solids control	1 2 4 5 6 7 10 20B 21A	Baroid
ALDACIDE	Microbiocide	1 2 3 4 5 7 12A	Baroid
ANTIBAC	Bactericide and film type corrosion inhibitor	1 2 3 4 5 6 7 12A	HDF
AQUAGEL	Wyoming bentonite	1 2 3 4 5 6 7 17A 23B	Baroid
AQUA-THINZ	Chrome lignosulfonate	1 2 3 4 5 6 7 17B 20B 22A	HDF
AQUA-THINZ CF	Non-polluting lignosulfonate	1 2 3 4 5 6 7 17B 22A	HDF
BARA B33	Biocide for drilling and packer fluids	1 2 3 4 5 6 7 12A	Baroid
BARA BRINEDEFOAM	Brine defoamer	1 2 3 4 5 6 7 13A	Baroid
BARABUF	pH buffer for clay free fluids	1 2 3 6 7 10 11A	Baroid
BARACARB	Acid soluble graded calcium carbonate	1 2 3 4 5 6 17B 19A 25A	Baroid
BARACORA	Corrosion inhibitor	1 2 3 4 5 6 7 12B	Baroid
BARA DEFOAM 1	Surface active defoamer	1 2 3 4 5 6 7 13A	Baroid
BARA DEFOAM W-300	Surface active defoamer	1 2 3 4 5 6 7 13A	Baroid
BARAFLOC	Clay flocculant	7 16A	Baroid
BARAFOS	Sodium tetraphosphate	1 2 7 11B 22A 24B	Baroid
BARAVIS	Synthetic cellulose	1 2 3 4 5 6 7 23A	Baroid
BARAZAN	Suspension agent	1 2 3 4 5 6 7 14A 23A	Baroid
BAR-GAIN	High specific gravity weighting agent	1 2 3 4 5 6 8 9 25A	Baroid
BARITE	Or barytes, offered under many tradenames, native barium sulfate	1 2 3 4 5 6 7 8 9 25A	Most companies
BARIUM CARBONATE	Barium carbonate	1 2 3 4 5 6 7 24A	Most companies
BAROID	Barite	1 2 3 4 5 6 7 8 9 25A	Baroid
BAROID ASPHALT	Oxidized asphalt	1 2 3 4 5 6 7 8 9 15A 23B	Baroid
BENGEL	Wyoming bentonite	1 2 3 4 5 6 7 17B 23A	HDF
BICARBONATE OF SODA	Sodium bicarbonate	1 2 3 4 5 6 7 11B 24A	Most companies
BIO-POLYMER	Bacterially produced polymer	1 2 3 4 5 6 7 24A	HDF
BIT LUBE	Extreme pressure lubricant	1 2 3 4 5 6 7 15A	Magcobar
BLACK MAGIC HFL	Low voscosity oil mud conc.	8 14A 15B 17B 21A 22A 23B	HDF
BLACK MAGIC SFT	"Sacked fishing tools," blended concentrate, (location mixed)	1 2 3 4 5 6 7 8 9 14B 15B 17A 21B 23B	HDF

BLACK MAGIC SUPERMIX	Basic oil base mud conc. for hi wt. and temp. (mfgr. mixed)	7 9 14B 15B 17A 20B 21B 22B 26B	HDF
BLACK MAGIC	Sacked oil base mud conc. for wt. and temp. (location mixed)	7 9 14B 15B 17A 20B 21B 23B 26B	HDF
BLOCK BUSTER	Surfactant	21A	Magcobar
BROMI-VIS	Brine viscosifier	1 2 3 4 5 6 23A	Baroid
CALCIUM BROMIDE	Calcium bromide/calcium chloride (liquid blend)	25A 26A	Most companies
CALCIUM CARBONATE	Calcium carbonate	1 2 3 4 5 6 7 8 9 19A 25A	Most companies
CALCIUM CHLORIDE	Calcium chloride	2 3 4 7 8 9 11A 16B 20A 25A	Most companies
CARBONOX	lignitic material	1 2 5 6 7 14B 17A 22A	Baroid
CAUSTIC POTASH	Potassium hydrate	1 2 3 4 6 7 11A	Most companies
CAUSTIC SODA	Sodium hydroxide	1 2 3 4 5 6 7 8 11A 12B 24B	Most companies
CAUSTI-LIG	Causticized lignite	1 2 3 4 5 6 7 22A	Magcobar
CC-16	Sodium salt of lignitic material	1 2 5 6 7 14B 17A 22A	Baroid
CEASCAL	Acid soluble sealer for lost circulation	1 2 3 6 7 19A	Magcobar
CEASTOP	Acid soluble lost circulation material	1 2 3 6 7 17A 19A 23B	Magcobar
CELLEX	Sodium carboxymethyl-cellulose	1 2 4 5 6 7 10 17A 23B	Baroid
CELL-O-SEAL	Shredded cellophane	1 2 3 4 5 6 7 19A	Magcobar
CELLOSIZE	Hydroxyethylcellulose	17B 23A	Magcobar
CHEMICAL V	Non-viscous organic lqd. to improve and combat crude oil contamination	8 9 14A 17B 20B 21B 22B 23A	HDF
CHEMICAL W	Non-viscous organic lqd. and thickening agent	19A 23A	HDF
CHIP-SEAL	Shredded cedar fiber	1 2 3 4 5 6 7 19A	Magcobar
CHROMALLOY BAR	Barite (barytes)	1 2 3 4 5 6 7 8 9 25A	CDF
CHROMALLOY BENEX	Bentonite extender	1 2 3 4 5 6 7 16A 23A	CDF
CHROMALLOY CFT	Chrome free thinner	1 2 3 4 5 6 7 20B 22A	CDF
CHROMALLOY CHROME	Chrome lignosulfonate	1 2 3 4 5 6 7 14B 17B 20B 22A	CDF
CHROMALLOY CL-CLS	Chrome lignite—chrome lignosulfonate	1 2 3 4 5 6 7 14B 17B 20B 22A	CDF
CHROMALLOY CMC	Organic polymer carboxymethylcellulose	1 2 3 4 5 6 7 17A 23B	CDF
CHROMALLOY CYPAN	Sodium polyacrylate	1 3 5 6 7 10 16B 17A 22A	CDF
CHROMALLOY DEFOAMER	Alcohol defoamer	1 2 3 4 5 6 7 13A	CDF
CHROMALLOY DESCO	Organic mud thinner	1 2 3 4 5 6 7 13B 20B 22A	CDF
CHROMALLOY DET.	Mud detergent	1 2 3 4 5 6 7 14B 21B	CDF
CHROMALLOY DIASEAL M	Mixture of diatoamaceous earth	1 2 3 4 5 6 7 8 9 19A	CDF
CHROMALLOY DICASORB	Shredded oil absorbing material	8 9	CDF
CHROMALLOY DML II	Water soluble, biodegradable, non-polluting lubricant	1 2 3 4 5 6 7 15A	CDF
CHROMALLOY DMS	Liquid surfactant	1 2 3 4 5 6 7 14B 17B 21A 22A	CDF
CHROMALLOY DRISCOSE	Organic polymer carboxy-methylcellulose	1 2 4 5 6 7 17A 23B	CDF
CHROMALLOY DRISPAC	Polyanioniccellulose	1 2 3 4 5 6 7 10 14B 15B 17A 20A 23A	CDF
CHROMALLOY E.Z. OUT	Oil soluble surfactant to free stuck pipe	1 2 3 4 5 6 7 8	CDF
CHROMALLOY FIBER	Shredded cane fiber blend	1 2 3 4 5 6 7 19A	CDF
CHROMALLOY FLAKES	Shredded cellophane	1 2 3 4 5 6 7 19A	CDF
CHROMALLOY FLOC.	Flocculating agent	7 16A	CDF
CHROMALLOY FLOSAL	Viscosifier	1 2 3 6 7 23A	CDF
CHROMALLOY GEL	Wyoming bentonite	1 2 3 4 5 6 7 17B 23A	CDF
CHROMALLOY KEMBREAK	Calcium lignosulfonate	1 2 3 4 5 6 7 17B 20B 22A	CDF
CHROMALLOY LIG	Lignite	1 2 4 5 6 7 14B 22A	CDF
CHROMALLOY MF1	Flocculant	1 2 3 6 7 16A 20B	CDF
CHROMALLOY MICA	Nonabrasive mineral mica	1 2 3 4 5 6 7 8 9 19A	CDF
CHROMALLOY MULTISEAL	Combination of granules, flakes and fibers	1 2 3 4 5 6 7 19A	CDF
CHROMALLOY N GAUGE	Potassium lignosulfonate	1 2 3 4 5 6 7 17B 20A 22B	CDF
CHROMALLOY OIL PHOS	Sodium tetraphosphate	1 2 5 6 7 22A	CDF
CHROMALLOY PARA	Paraformaldehyde	1 2 3 4 7 12A	CDF
CHROMALLOY PLUG	Ground walnut or pecan hulls	1 2 3 4 5 6 7 8 9 19A	CDF
CHROMALLOY REMOX	Catalysed sodium sulfite	1 2 3 4 5 6 7 26A	CDF
CHROMALLOY SAPP	Inorganic thinner	1 2 3 22A	CDF
CHROMALLOY SCAV H_2S	H_2S scavenger	1 2 3 4 5 6 7 26A	CDF
CHROMALLOY S GEL	Attapulgite clay	2 3 7 23A	CDF
CHROMALLOY SHALE LIG	Potassium lignite	1 20A	CDF
CHROMALLOY SLICK PLUS	Non-polluting lubricant	1 2 3 4 5 6 7 15A 21B	CDF

Product	Description	Applications	Supplier
CHROMALLOY SOL	Nonionic anionic emulsifier	1 2 4 5 6 7 14A	CDF
CHROMALLOY SOLTEX	Processed hydrocarbon	1 2 3 4 5 6 7 8 10 14A 15A 17B 20A	CDF
CHROMALLOY STAR	Non-fermenting starch	2 3 4 5 6 7 17A 23B	CDF
CHROMALLOY STARCH	Pregelatinized starch	2 3 4 5 6 7 17A 23B	CDF
CHROMALLOY TAN	Quebracho compound	1 2 3 4 5 6 7 17A 22A	CDF
CHROMALLOY TONE	Causticized lignite	1 2 6 7 8 17A 22A	CDF
CHROMALLOY TRIM	Blend of surfactants	1 2 3 4 5 6 15A 21A	CDF
CHROMALLOY VISITROL	Caustisized quebracho	1 2 3 4 5 6 7 17A 22A	CDF
CHROMALLOY WL-100	Sodium polyacrylate	1 3 6 7 17A	CDF
CHROMALLOY XC	Xanthum gum biopolymer	1 2 3 4 5 6 7 14B 17B 23A	CDF
CLAYGEL	Sub-bentonite	1 2 3 4 5 6 7 17B 23A	HDF
CLEAN SPOT	Non-polluting invert oil mud	8 14A 15A 17A 20A 21A 23A	CDF
CLOROGEL	Attapulgite clay	2 3 7 23A	HDF
CLS	Chrome lignosulfonate	1 2 3 4 5 6 7 14B 17A 20B 22A	Most companies
CON DET	Mud detergent	7 14B 21A	Bariod
COR-HIB	Filming amine	1 2 3 4 5 6 7	HDF
D-D	Drilling mud detergent	1 2 3 4 5 6 7 14B 21A	Magcobar
DEL BRIDGE B	Blended $CaCO_3$ for brine fluids	1 2 3 17B 19A	CDF
DEL CIDE B	Film forming amine and bactericide	1 2 3 4 5 6 7 12A 26A	CDF
DEL FREE TOOL	Stuck pipe spotting fluid	1 2 3 4 5 6 7 16B 18A	CDF
DEL G	Air oxidized asphalt	8 9 17A	CDF
DEL HIB B	Film forming amine	1 2 3 6 7 12B 26A	CDF
DEL HTD	Temperature stabilizer dispersant	8 9 17B 22A	CDF
DEL HTS	Temperature stabilizer	8 9 17A	CDF
DEL HYVIS B	Blended HEC for brine fluids	1 2 3 15B 18B 23A	CDF
DEL K	Oil mud stabilizer	8 14B 17A	CDF
DEL PAK	Calcium bromide/calcium chloride (liquid blend)	25A 26B	CDF
DEL PE 11	Emulsifier	8 9 14A 21B 22B	CDF
DEL PEL	Calcium chloride (pellet or flake)	7 8 9 20A 25A 26B	CDF
DEL PILL B	Hec-lignosulfonate $CaCO_3$ slurry blend for brine fluids	1 2 3 17B 19A	CDF
DEL QUICKVIS B	Synthetic polymer	1 2 3 6 17B 23A	CDF
DEL SE 22	Emulsifier	8 9 14A 21B 22B	CDF
DEL SEAL B	HEC-lignosulfonate-carbonate blend for brine fluids	1 2 3 17B 23B	CDF
DEL SCAV O^2	Blend of ammonium bisulfite and ammonium sulfite	1 2 3 4 5 6 7 26A	CDF
DELTONE	Organophillic clay	8 9 23A	CDF
DELTONE T	Thinner	8 9 17B 22A	CDF
DELVERT	Primary emulsifier	8 14A 17A	CDF
DEL VIS B	Blended polymer for brine fluids	1 2 3 17B 23B	CDF
DEL WA33	Solids dispersant	8 9 14B 21B 22A	CDF
DENSEWATE	Hematite	1 2 3 4 5 6 7 8 9 25A	HDF
DEOX	Water soluble sulfite compound	1 2 3 4 5 6 7 26A	HDF
DEXTRID	Organic polymer	1 2 3 4 6 7 17A 23B	Baroid
DML-2	Water soluble biodegradable non-polluting lubricant	1 2 3 4 5 6 7 15A	CDF
DOS-3	Diesel oil substitute	1 2 3 4 5 6 7 14B 21A	Magcobar
DRILTREAT	Oil mud stabilizer	8 9 14A 17B 21B 22B	Baroid
D-TERG	Mud detergent	1 2 3 4 5 6 7 21A	HDF
DUOVIS	Xanthan gum biopolymer	1 2 3 4 5 6 7 14B 17B 23A	Magcobar
DURATONE HT	Oil mud filtration control agent	8 9 17A	Baroid
DURENEX	Temperature stable fluid loss additive	1 2 3 4 5 6 17A	Baroid
DV-22	Fluid loss control agent for oil base and invert emulsion muds	8 9 13B 16A	Magcobar
DV-33	Oil wetting agent for oil continuous & invert emulsion muds	8 9 14A 17B 21B	Magcobar
E P MUDLUPE	Extreme pressure lubricant	1 2 3 4 5 6 7 15A	Baroid
EZ MUD	Liquid viscosifier	1 6 7 10 15B 16B 17B 20B 23A	Baroid
EZ MUL	Oil mud emulsifier	8 9 14A 21B	Baroid
EZ SCAV	H_2S scavenger	1 2 3 4 5 6 10 26A	Sherwell
EZ SPOT	Oil mud concentrate	1 2 3 4 5 6 7 8 9 15A	Baroid
FB 1	Silicone defoamer	1 2 3 4 5 6 7 13A 15B	CDF
FB 2	Concentrated silicone defoamer	1 2 3 4 5 6 7 13A 15B	CDF
FIBERTEX	Shredded cane fiber blend	1 2 3 4 5 6 7 19A	Baroid
FLOXIT	Clay flocculant	7 16A	Magcobar
FORMASEAL	Oil sol, lost circ. material and temp. plugging agent	8 9 17B 19A 23B	HDF
FORMASEAL-HT	Oil sol. lost circ. material and temp. plugging agent for high temp.	8 9 17B 19A 23B	HDF

GALENA	Lead sulfide powder	1	2	3	4	5	6	8	9	25A			Baroid
GELTHERM	Sepiolite	1	2	3	4	5	6	17B	23A				HDF
GELTONE	Organophillic clay powder	9	23A										Baroid
GELTONE II	Low shear, oil mud gellant	8	9	23A									Baroid
GEO-GEL	High temperature stable viscosifying agent	1	2	3	4	5	6	7	17B	23A			Magcobar
HEVYWATE	Barite (barytes)	1	2	3	4	5	6	7	8	9	25A		HDF
HY-SEAL	Shredded organic fiber	1	3	4	5	6	7	19A					Baroid
IMPERMEX	Pregelatinized starch	1	2	3	4	5	6	7	17A				Baroid
INVERMUL	Oil mud stabilizer	8	9	14A	17B	23B							Baroid
JELFLAKE	Shredded cellophane	1	2	3	4	5	6	7	19A				Baroid
KLEEN UP	Heavy duty detergent and degreaser	1	2	3									Magcobar
K-LIG	Potassium lignite derivative	1	2	7	14A	17A	22A						Baroid
KWIK-THIK	Extra hi yield bentonite	1	7	17B	23A								Magcobar
LIG-THINZ	Processed lignite	1	2	5	6	14B	17A	22B					HDF
LIG-THINZ C	Causticized lignite	1	2	3	4	5	6	7	14B	17A	22B		HDF
LIME	Hydrated lime	7	8	10	11A	12B	15B	20B	26B				Most companies
LIQUID H.E.C.	Liquid nonionic polymer	1	2	3	4	5	6	7	23A				Magcobar
LIQUI-VIS	Brine viscosifier	1	2	4	5	6	7	23A					Baroid
LO LOSS	Guar gum	1	2	7	17B	23A							Magcobar
LOWATE	Acid soluble weight material	1	2	3	4	5	6	7	8	9	25A		Magcobar
LUBE-KOTE	Graphite	1	2	3	4	5	6	7	15A				Magcobar
LUBRA-BEADS	Solid lubricant	1	2	3	4	5	6	7	8	9	10	15A	Baroid
MAGCOBAR	Weighting material	1	2	3	4	5	6	7	8	9	25A		Magcobar
MAGCOBRINE C.B.	Pre-blended calcium chloride/bromide brines for workover, completion and packers												Magcobar
MAGCOBRINE C.C.	Pre-blended calcium chloride brine for workover, completion and packers												Magcobar
MAGCOBRINE HV DEFOAMER	Brine defoamer	1	2	3	4	5	6	7	13A				Magcobar
MAGCOBRINE P.C.	Potassium chloride	1	2	3	4	5	6	7	20B	25A			Magcobar
MAGCOBRINE S.C.	Sodium chloride	1	2	3	4	5	6	7	16B	20B	25A		Magcobar
MAGCO CFL	Chrome free lignosulfonate	1	2	3	4	5	6	7	17B	22A			Magcobar
MAGCOCIDE	Starch preservative	1	2	3	4	6	7	12A	26B				Magcobar
MAGCOCIDE L	Liquid starch preservative	1	2	3	4	6	7	12A	26B				Magcobar
MAGCO DEFOAMER A-40	Defoamer												Magcobar
MAGCO FIBER	Blended fibers	1	2	3	4	5	6	7	19A				Magcobar
MAGCO FOAM CHECK	Defoamer	1	2	3	4	5	6	7	13A				Magcobar
MAGCOFOAMER 76	Foaming agent	10	18A										Magcobar
MAGCOFOAMER 80	Surfactant	10	18A										Magcobar
MAGCOGEL	Wyoming bentonite	1	2	3	4	5	6	7	17B	23A			Magcobar
MAGCO INHIBITOR 101	Corrosion inhibitor for packer fluids	1	2	3	4	5	6	7	26A				Magcobar
MAGCO INHIBITOR 202	Corrosion inhibitor for direct treating oil drill string	1	2	3	4	5	6	7	10	23A			Magcobar
MAGCO INHIBITOR 303	Organic amine corrosion inhibitor for workover, completion and packer brines	1	2	3	4	5	6	7	26A				Magcobar
MAGCO LIQUID HEC	Brine viscosifer	1	2	3	4	5	6	7	17B	23A			Magcobar
MAGCOLUBE	Biodegradable lubricant	1	2	3	4	5	6	7	15A				Magcobar
MAGCONATE	Petroleum sulfonate emulsifier	1	2	4	5	6	7	14A					Magcobar
MAGCONOL	Alcohol defoamer	1	2	3	4	5	6	7	13B				Magcobar
MAGCOPHOS	Sodium tetraphosphate	1	2	3	6	7	22A	23A					Magcobar
MAGCO POLY DEFOAMER	Organic polyol defoamer	1	2	3	4	5	6	7	12A				Magcobar
MAGCOPOLYPAC	Polyanioniccelulose	1	2	3	4	5	6	7	10	17A	22B	23B	Magcobar
MAGCO POLY SAL	Organic polymer	1	2	3	4	5	6	7	17A				Magcobar
MAGCO SSII	Soluble H_2S scavenger	1	2	3	4	5	6	7	26A				Magcobar
MAGNE-MAGIC	Blend of magnesium and calcium compounds	2	3	6	11A	17A							HDF
MAGNE-SALT	Water soluble salts	1	2	3	6	7	20A						HDF
MAGNE-SET	Controlled solidifier	2	3	6	7	19A							HDF
MAGNE-SET ACCELERATOR	Material for reducing set time for Magne-Set	1	2	3	5	6	7	19A					HDF
MAGNE-SET RETARDER	Material for increasing set time for Magne-Set	1	2	3	5	6	7	19A					HDF
MAGNE-SET THINNER	Low molecular wt. polymer	1	2	3	5	6	7	19A	20B	23A			HDF
MICATEX	Mica flakes (fine, med. and coarse)	1	2	3	4	5	6	7	8	9	19A		Baroid
MUD FIBER	Blended cane and wood fiber	1	2	3	4	5	6	7	19A				Magcobar
MY-LO-JEL	Pregelatinized starch	1	2	3	4	5	6	7	17A	24B			Magcobar
NOFOAM	Defoamer	1	2	3	4	5	6	7	13A				HDF
NUT PLUG	Ground walnut shells	1	2	3	4	5	6	7	8	9	19A		Magcobar
OB DRY	Oil absorbent compound	8	9	21A									HDF
OB EMULSIFIER	Nonionic emulsifier	1	2	3	4	5	6	7	14A	21B			HDF
OB FLOC	Clay flocculant	1	2	3	4	5	6	7	16A	20B			HDF
OB GEL	Conc. for improving Black Magic gel	8	9	14B	15B	17B	20B	23A					HDF

Product	Description	Applications	Supplier
OB HI-CAL	Calcium hydroxide	1 2 3 4 5 6 7 8 9 11A 12B 14B 17B 25B	HDF
OILFAZE	Sacked oil base mud conc.	9 14A 17B 23B	Magcobar
OMC	Oil mud conditioner	8 9 23A	Baroid
OMNIGEL	Oil dispersible viscosifier	8 9 22A	HDF
OMNIVERT	Invert emulsion system	8 9 14A 15A 17A 20A 22A 23A 26A	HDF
OMNIVERT-E	Amine type fluid emulsifier	8 9 14A 22B	HDF
OS-1	Oxygen scavenger	1 2 3 4 5 6 7 26A	Magcobar
OS1L	Oxygen scavenger	1 2 3 6 7 26A	Magcobar
OS-IL	Oxygen scavenger	1 2 3 6 7 26A	Magcobar
OXI VERT	Calcium oxide	8 9 11A 17B	CDF
PAC L	Low viscosity grade poly-anioniccellulose	1 2 3 4 5 6 7 14B 15B 17A 20A	Baroid
PAC R	Reg. grade polyanionic-cellulose	1 2 3 4 5 6 7 14B 15B 17A 20A 23A	Baroid
PETROTONE	Organophillic clay powder for use as oil mud suspended agent	8 9 23A	Baroid
PIPE LAX	Surfactant mtl. to be mixed with diesel oil to free stuck pipe	1 2 3 4 5 6 7 8	Magcobar
PLUG-GIT	Processed hardwood fiber	1 2 3 4 5 6 7 19A	Baroid
POLYFLAKE	Oil sol. plastic film	1 2 3 4 5 6 7 8 9 19A	Baroid
POLYLUBE	Lubricant	1 2 3 4 5 6 7 15A	HDF
POLYLUBE EP	Extreme pressure lubricant	1 2 3 4 5 6 7 15A	HDF
POLY-MAGIC	Co-polymer	2 3 7 17A 20B 23A	HDF
POLY MIX	Acid soluble material for viscosity and fluid loss control	1 2 3 4 5 6 7 17A 23A	Magcobar
POTASSIUM CHLORIDE	Potassium chloride	1 2 3 6 7 20A	Most companies
PROTECTOMAGIC	Sacked asphalt conc. location mixed w/diesel for use as oil phase in emulsion muds	1 2 3 4 5 6 7 15A 17B 20A	HDF
PROTECTOMAGIC LIQUID	Oil dispersed asphalt used as oil phase in emulsion muds	1 2 3 4 5 6 7 8 14B 15B 17A 20A 22B	HDF
PROTECTOMAGIC-M	Water dispersable asphalt additive	1 2 3 4 5 6 7 15A 17B 20B	HDF
Q-B 11	Chrome-free lignosulfonate	1 2 3 4 5 6 7 14B 17B 20B 22A	Baroid
Q-BROXIN	Ferrochrome lignosulfonate	1 2 3 4 5 6 7 14B 17B 20B 22A	Baroid
QUEBRACHO	Quebracho (tannin) extract	1 2 3 4 5 6 7 17A 22A	Most companies
QUIK-FOAM	Biodegradable foaming agent	10 19A	Baroid
QUIK-GEL	High yield bentonite	1 7 23A	Baroid
QUIK-MUD	Suspension of concentrated viscosifiers	1 2 7 10 16B 18B 24A	Baroid
QUIK-TROL	Organic polymer	1 2 3 4 5 6 20A 23A	Baroid
RAPIDDRIL	Organic polymer; clay extender and solids flocculant	1 2 3 7 17A 24A	Magcobar
RESINEX	Resin additive, fluid loss agent	1 2 3 4 5 6 17A 20B	Magcobar
ROD LUBE	E.P. lubricant for drill rods	1 2 3 4 5 6 7 15A	Magcobar
SALINEX	Seawater emulsifier	2 14A 15B 17B	Magcobar
SALT	Sodium chloride	2 3 4 5 7 8 16A 20B 25B	Most companies
SALT GEL	Attapulgite clay	2 3 7 23A	Magcobar
SE-11	Secondary emulsifier for oil base and invert emulsion muds	8 9 14A 17B 21B	Magcobar
SEAMUL	Salt water emulsifier and surfactant	2 3 4 7 14A 17B 21B 22B	Baroid
SHUR-GEL	Beneficiated bentonite mud conditioner	1 2 23A 24B	Baroid
SI-1000	Scale inhibitor	1 2 3 4 5 6 7 26A	Magcobar
SIDERITE	Acid soluble weight material	1 2 3 4 5 8 9 25A	Magcobar
SKALE-HIB	Phosphonate	1 2 3 4 5 6 7 10 26A	HDF
SODA ASH	Sodium carbonate	1 2 3 4 5 6 7 11B 24A	Most companies
SODIUM BICARBONATE	Sodium bicarbonate	1 2 4 5 6 7 11B 23B 24B	Most companies
SODIUM CARBONATE	Soda ash	1 2 3 4 5 6 7 11B 24A	Most companies
SPECIAL ADDITIVE 58	Powdered organic emulsifiers	7 9 14A 17B 23A 25B	HDF
SPECIAL ADDITIVE 77	Surfactant for treatment of water contamination	8 9 14A 17B 21B 22B	HDF
SPECIAL ADDITIVE 81	Mud stabilizer and emulsifier	8 9 14A 17B 21A 22B	HDF
SPERSENE	Chrome lignosulfonate	1 2 3 4 5 6 7 17B 22A	Magcobar
STABIL HOLE	Sacked asphalt-added dry to system or as a mixture with oil	1 2 3 4 5 6 7 8 9 14B 15B 18B 20B 23B	Magcobar
STABILITE	Organic phosphate thinner	1 6 7 21B 22A	Baroid
STARCH	Pregelatinized starch	1 2 3 4 5 6 7 17A 20B 23B	Most companies
STARCH PRESERVATIVE	Paraformaldehyde	12A	HDF
STP	Sodium tetraphosphate	1 2 3 6 7 22A 23A 26B	HDF
STRATA-PLUG	Ground nut hulls	1 2 3 4 5 6 7 8 9 19A	HDF
SUPER HYDTOPROOF	Water dispersible asphalt	1 2 3 4 5 6 7 15A 20A 26B	HDF
SUPER VISBESTOS	Fiberous asbestos material	1 2 3 4 5 6 7 8	Magcobar
SUPER-WATE	Special high density weighting material for blowout control only	25A	Magcobar
SURFAK E	Emulsifier	1 2 3 4 5 6 7 14A 15B 17B	Magcobar

SURFAK M	Nonionic surfactant for solids control; soluilizer for CMC & starch fluids	1 2 3 4 5 6 7 14A 21B	Magcobar
SYNER-X	Synergistic polymer blend	1 2 3 4 5 6 7 17A 22A	HDF
TANNATHIN	Lignite	1 2 3 5 6 7 14B 17B 22A	Magcobar
TANNEX	Quebracho extract and lignitic mtl.	1 2 5 6 7 14B 17B 22A	Baroid
TORQ TRIM II	Biodegradable and non-toxic lubricant	1 2 3 4 5 6 7 15A	Baroid
TRIMULSO	Oil-in-water emulsifier	1 2 3 4 5 6 7 14A 15B 21A 26B	Baroid
TRIP-WATE	Granular barite	1 2 3 4 5 6 7 8 9 19A 25A	Baroid
VERTILE	Invert emulsion	8 14A 17A 23A	Magcobar
VERTOIL	Sacked invert emulsion mud conc.	8 14A 17B 23B	Magcobar
VG-69	Gelling agent for invert emulsions	8 17B 23A	Magcobar
VISBESTOS	Inorganic viscosifier emulsion mud	1 2 3 4 5 6 7 8 23A	Magcobar
VISQUICK	Inorganic viscosifier	1 2 3 4 5 6 7 23A	Magcobar
X-38	Acid soluble weight material	1 2 3 4 5 6 7 8 9 25A	Magcobar
XP-20	Chrome lignite	1 2 3 5 6 7 17B 20A 22A	Magcobar
XPK-2000	Atmospheric corrosion inhibitor	26A	Magcobar
X-TEND	Powdered flocculant and clay extender	1 2 6 7 16A 23A	Baroid
ZEOGEL	Attapulgite powder	2 3 4 5 7 23A	Baroid
Z-HIB	Zinc compound, H_2S scavenger	1 2 3 4 5 6 7 26A	HDF

Based on information supplied by Dresser Magcobar, Chromalloy, Hughes Drilling Fluids, and NL Baroid.

Appendix C

Standard Abbreviations for Lithologic Description

WORD	ABBREVIATION
above	ab
absent	abs
abundant	abd
acicular	acic
after	aft
agglomerate	Aglm
aggregate	Agg
algae, algal	Alg, alg
allochem	Allo
altered	alt
alternating	altg
amber	amb
ammonite	Amm
amorphous	amor
amount	amt
amphipora	Amph
and	&
angular	ang
anhedral	ahd
anhydrite (-ic)	Anhy, anhy
anthracite	Anthr
aphanitic	aph
apparent	apr
appears	ap
approximate	apprx
aragonite	Arag
arenaceous	aren
argillaceous	arg
argillite	argl
arkose (-ic)	Ark, ark
as above	a.a.
asphalt (-ic)	Asph, asph
assemblage	Assem
associated	assoc
at	@
authigenic	authg
average	Av, av
band (-ed)	Bnd, bnd
barite (-ic)	bar
basalt (-ic)	Bas, bas
basement	Bm
become (-ing)	bcm
bed (-ed)	Bd, bd
bedding	Bdg
belemnites	Belm
bentonite (-ic)	Bent, bent
bioclastic	biocl
bioherm (-al)	Bioh, bioh
biomicrite	Biomi
biosparite	Biosp
biostrom (-al)	Biost, biost
biotite	Biot
bioturbated	bioturb
birdseye	Bdeye
bitumen (-inous)	Bit, bit
black (-ish)	blk, blksh
blade (-ed)	Bld, bld
blocky	blky
blue (-ish)	bl, blsh
bored (-ing)	Bor, bor
botryoid (-al)	Bot, bot
bottom	Btm
boudinage	boudg
boulder	Bld
boundstone	Bdst
brachiopod	Brach
brackish	brak
branching	brhg
break	Brk, brk
breccia (-ted)	Brec, brec
brighter	brt
brittle	brit
brown	brn
bryozoa	Bry
bubble	Bubl
buff	bu
bulbous	bulb
burrow (-ed)	Bur, bur
calcarenite	Clcar
calcareous	calc
calcilutite	Clclt
calcirudite	Clcrd
calcisilitite	Clslt
calcisphaera	Casph
calcisphere	Clcsp
calcite (-ic)	Calc, calctc
caliche	cche
carbonaceous	carb
carbonate	crbnt
carbonized	cb
cavern (-ous)	Cav, cav
caving	Cvg
cement (-ed, -ing)	Cmt, cmt
center	cntr
center (-ed)	cntr
cephalopod	Ceph
chaetetes	Chaet
chalcedony (-ic)	Chal, chal
chalk (-y)	Chk, chky
charophyte	Char
chert (-y)	Cht, cht
chitin (-ous)	Chit, chit
chitinozoa	Chtz
chlorite (-ic)	Chlor, chlor
chocolate	choc
circulate (-ion)	circ, Circ
clastic	clas
clay (-ey)	Cl, cl
claystone	Clst
clean	cln
clear	clr
cleavage	Clvg
cluster	Clus
coal	c
coarse	crs
coated (-ing)	cotd, cotg, Cotg
coated grains	cotd gn
cobble	Cbl
colonial	coln
color	Col, col
common	com
compact	cpct

compare	cf
concentric	cncn
conchoidal	conch
concretion (-ary)	Conc, conc
conglomerate (-ic)	Cgl, cgl
conodont	Cono
conquina	Coq
conquina (-oid)	coqid
considerable	cons
consolidated	consol
conspicuous	conspic
contact	Ctc
contamination (-ed)	Contam, contam
content	Cont
contorted	cntrt
coral, coralline	Cor, corln
core	c
covered	cov
cream	crm
crenulated	cren
crinkled	crnk
crinoid (-al)	Crin, crinal
cross	x
cross-bedded	x-bd
cross-laminated	x-lam
cross-stratified	x-strat
crumpled	crpld
crystal (-line)	Xl, xln
crystocrystalline	crpxln
cube, cubic	Cub, cub
cuttings	Ctgs
cypridopsis	Cyp
dark (-er)	dk, dkr
dead	dd
debris	Deb
decrease (-ing)	Decr, decr
dendrite (-ic)	dend
dense	dns
depauperate	depau
description	Descr
desiccation	dess
detrital	detr
devitrified	devit
diabase	Db
diagenesis (-etic)	Diagn, diagn
diameter	Dia
disseminated	dissem
distillate	Dist
ditto	'' or do
dolomite (-ic)	Dol, dol
dolostone	dolst
dominant (-ly)	dom
drill stem test	DST
drilling	drlg
drusy	dru
earthy	ea
east	E
echinoid	Ech
elevation	Elev
elongate	elong
embedded	embd
endothyra	Endo
equant	eqnt
equivalent	Equiv
euhedral	euhd
euryamphipora	Euryamph
euxinic	eux
evaporite (-itic)	Evap, evap
excellent	ex
exposed	exp
extraclast (-ic)	Exclas, exclas
extremely	extr
extrusive	exv
facet (-ed)	Fac, fac
faint	fnt
fair	fr
fault (-ed)	Flt, flt
fauna	Fau
favosites	Fvst
feet	Ft
feldspar (-athic)	Fspr, fspr
fenestra (-al)	Fen, fen
ferro-magnesian	Fe-mag
ferruginous	ferr
fibrous	fibr
fill (-ed)	fld
fine (-ly)	f, fnly
fissile	fis
flaggy	flg
flake, flaky	Flk, flk
flat	fl
flesh	fls
floating	fltg
flora	Flo
fluorescence (-ent)	Fluor, fluor
foliated	fol
foot	Ft
foraminifer	Foram
foraminiferal	foram
formation	Fm
fossil (-iferous)	Foss, foss
fracture (-d)	Frac, frac
fragment (-al)	Frag, frag
framework	frmwk
frequent	freq
fresh	frs
friable	fri
fringe (-ing)	Frg, frg
frosted	fros
frosted quartz	F.Q.G.
fucoid (-al)	Fuc, fuc
fusulinid	Fus
gabbro	Gab
galeolaria	Gal
gas	G
gastropod	Gast
generally	gen
geopetal	gept
gilsonite	Gil
girvanella	Girv
glass (-y)	Glas, glas
glauconite (-itic)	Glauc, glauc
globigerina (-inal)	Glob, glob
gloss (-y)	Glos, glos
gneiss (-ic)	Gns, gns
good	gd
grading	grad
grain (-s, -ed)	Gr, gr
grainstone	Grst
granite	Grt
granite wash	G.W.
granule (-ar)	Gran, gran
grapestone	grapst
graptolite	Grap
gravel	Grv
gray, grey (-ish)	gry, grysh
graywacke	Gwke
greasy	gsy
green (-ish)	gn, gnsh
grit (-ty)	Gt, gt
gypsum (-iferous)	Gyp, gyp
hackly	hkl
halite (-iferous)	Hal, hal
hard	hd
heavy	hvy
hematite	Hem, hem
hematite (-ic)	Hem, hem
heterogeneous	hetr
heterostegina	Het
hexagonal	hex
high (-ly)	hi
homogeneous	hom
horizontal	hor
hornblende	hornbd
hydrocarbon	Hydc
igneous rock	Ig, ig
imbedded	imbd
impression	Imp
in part	I.P.
inch	In
inclusion (-ded)	Incl, incl
increasing	incr
indistinct	indst
indurated	ind
inoceramus	Inoc
insoluble	insl
interbedded	intbd
intercalated	intercal
intercrystalline	intxln
interfragmental	intfrag
intergranular	intgran

intergrown	intgn
interlaminated	intrlam
interparticle	intpar
interpretation	intpt
interstices	Inst, intsti
interval	Intvl
intraclast (-ic)	Intclas, intclas
intraparticle	intrapar
intrusive	Intr, intr
invertebrate	Invtb
iridescent	irid
ironstone	Fe-st
irregular (-ly)	irr
isopachous	iso
ivanovia	Ivan
jasper	Jasp
joint (-ed, -ing)	Jt, jt
kaolin (-itic)	Kao, kao
lacustrine	lac
lamina (-itions, -ated)	Lam, lam
large	lge
laterite (-itic)	Lat, lat
lavender	lav
layer	Lyr
leached	lchd
lens, lenticular	Len, lent
lentil (-cular)	len
light	lt
lignite (-itic)	Lig, lig
limestone	Ls
limonite (-itic)	Lim, lim
limy	lmy
lithic	lit
lithographic	lithgr
lithology (-ic)	Lith, lith
little	Ltl
littoral	litt
local	loc
long	lg
loose	lse
lower	l
lumpy	lmpy
lustre	Lstr
lutite	Lut
macrofossil	Macrofos
magnetite, magnetic	Mag, mag
manganese	Mn, mn
marble	Mbl
marine	marn
marl (-y)	Mrl, mrl
marlstone	Mrlst
maroon	mar
massive	mass
material	Mat
matrix	Mtrx

maximum	max
medium	m or med.
member	Mbr
meniscus	men
metamorphic (-osed)	meta, metaph
metamorphic rock	Meta
metasomatic	msm
mica (-iceous)	Mic, mic
micrite (-ic)	Micr, micr
micro	mic
micro-oolite	Microol
microcrystalline	microxln
microfossil	microfos
micrograined	micgr
micropore (-osity)	micropor
microstylolite	Microstyl
middle	Mid
miliolid	Milid
milky	mky
mineral (-ized)	Min, min
minor	mnr
minute	mnut
moderate	mod
mold (-ic)	Mol, mol
mollusc	Moll
mosaic	mos
mottled	mott
mud (-dy)	md, mdy
mudstone	Mdst
muscovite	Musc
nacreous	nac
no sample	n.s.
no show	n/s
no visible porosity	n.v.p.
nodules (-ar)	Nod, nod
north	N
novaculite	Novac
numerous	num
occasional	occ
ochre	och
odor	od
oil	O
oil source rock	OSR
olive	olv
olivine	olvn
oncolite (-oidal)	Onc, onc
ooid (-al)	Oo, oo
oolicast (-ic)	Ooc, ooc
oolite (-itic)	Ool, ool
oomold (-ic)	Oomol, oomol
opaque	op
orange (-ish)	or, orsh
orbitoline	Orbit
organic	org
orthoclase	Orth
orthoquartzite	O-Qtz
ostracod	Ostr

overgrowth	ovgth
oxidized	ox
oyster	Oyst
packstone	pkst
paper (-y)	Pap, pap
paraparchites	Para
part (-ly)	Pt, pt
particle	Par, par
parting	Ptg
parts per million	PPM
patch (-y)	Pch, pch
pearly	prly
pebble (-ly)	Pbl, pbl
pelecypod	Pelec
pellet (-al)	Pel, pel
pelletoid (-al)	Peld, peld
pendular (-ous)	Pend, pend
pentamerus	Pent
permeability (-able)	Perm, k, perm
petroleum	Pet, pet
phlogopite	Phlog
phosphate (-atic)	Phos, phos
phreatic	phr
phyllite, phyllitic	Phyl, phyl
pin-point (porosity)	p.p
pink	pk
pinkish	pkish
pisoid (-al)	Piso, piso
pisolite, pisolitic	Pisol, pisol
pitted	pit
plagioclase	Plag
plant	Plt
plastic	plas
platy	plty
polish, polished	Pol, pol
pollen	Poln
polygonal	poly
poor (-ly)	p
porcelaneous	porcel
porosity, porous	Por, por
porous (-sity)	por
porphyry	prphy
possible (-ly)	poss
predominant (-ly)	pred
preserved	pres
primary	prim
prism (-atic)	pris
probable (-ly)	prob
production	Prod
prominent	prom
pseudo oolite (-ic)	Psool, psool
pseudo-	ps
pumice-stone	Pst
purple	purp
pyrite (-itized, -itic)	Pyr, pry
pyrobitumen	Pybit
pyroclastic	pyrcl
pyroxene	pyrxn

quartz (-ose)	Qtz, qtz
quartzite (-ic)	Qtzt, qtzt
radial (-ating)	Rad, rad
radiaxial	Radax
range	rng
rare	r
recemented	recem
recovery (-ered)	Rec, rec
recrystallized	rexlzd
red (-ish)	rd, rdsh
reef (-oid)	Rf, rf
remains	Rem
renalcis	Ren
replaced (-ment)	rep, Repl
residue (-ual)	Res, res
resinous	rsns
rhomb (-ic)	Rhb, rhb
ripple	Rpl
rndd, frosted, pitted	r.f.p.
rock	Rk
round (-ed)	rnd, rndd
rubble (-bly)	Rbl, rbl
rudist	Rud
rugose	rug
saccharoidal	sacc
salt (-y)	Sa, sa
salt and pepper	s&p
salt cast (-ic)	sa-c
salt water	S.W.
same as above	a.a.
sample	Spl
sand (-y)	Sd, sdy
sandstone	Sst
saturation (-ated)	Sat, sat
scales	sc
scaphopod	Scaph
scarce	scs
scattered	scat
schist (-ose)	Sch, sch
scolecodont	Scol
secondary	sec
sediment (-ary)	Sed, sed
selenite	Sel
septate	sept
shadow	shad
shale (-ly)	Sh, sh
shell	Shl
shelter porosity	Shlt por
show	shw
siderlite (-itic)	Sid, sid
sidewall core	S.W.C.
silica (-iceous)	Sil, sil
silky	slky
silt (-y)	Slt, slty
siltstone	Sltst
similar	sim
size	sz
skeletal	skel
slabby	slb
slate (-y)	Sl, sl
slickenside (-d)	Slick, slick
slight (-ly)	sli, slily
small	sml
smooth	sm
soft	sft
solenopora	Solen
solitary	sol
solution, soluble	Sol, sol
somewhat	smwt
sorted (-ing)	srt, srtg
south	S
spar (-ry)	Spr, spr
sparse (-ly)	sps, spsly
speck (-led)	Spk, spkld
sphaerocodium	Sphaer
sphalerite	Sphal
spherule (-itic)	Spher, spher
spicule	Spic, spic
splintery	Splin
sponge	Spg
spore	Spo
spotted (-y)	sptd, spty
stachyode	Stach
stain (-ed, -ing)	Stn, stn
stalactitic	stal
strata (-ified)	Strat, strat
streak (-ed)	Strk, strk
streaming	stmg
striae (-ted)	Stri, stri
stringer	strgr
stromatolite	Stromlt, stromlt
stromatoporoud	Strom
structure	Str
styliolina	Stylio
stylolite (-itic)	Styl, styl
sub	sb
subangular	sbang
sublithic	sblit
subrounded	sbrndd
sucrosic	suc
sugary	sug
sulphur, sulphurous	Su, su
superficial oolite	Spfool, spfool
surface	Surf
syntaxial	syn
syringopora	Syring
tabular (-ate)	tab
tan	tn
tasmanites	Tas
tension	tns
tentaculites	Tent
terriginous	ter
texture (-d)	Tex, tex
thamnopora	Tham
thick	thk
thin	thn
thin section	T.S.
thin-bedded	t.b.
throughout	thru
tight	ti
top	Tp
tough	tgh
trace	Tr
translucent	trnsl
transparent	trnsp
trilobite	Tril
tripoli (-itic)	Trip, trip
tube (-ular)	Tub, tub
tuff (-aceous)	Tf, tf
type (-ical)	Typ, typ
unconformity	Unconf
unconsolidated	uncons
underclay	Uc
underlying	undly
unidentifiable	unident
uniform	uni
upper	u
vadose	Vad, vad
variation (-able)	Var, var
varicolored	varic
variegated	vgt
varved	vrvd
vein (-ing, -ed)	Vn, vn
veinlet	Vnlet
vermillon	verm
vertebrate	vrtb
vertical	vert
very	v
very poor sample	V.P.S.
vesicular	ves
violet	vi
visible	vis
vitreous (-ified)	vit
volatile	volat
volcanic rock	Volc, volc
vug (-gy)	Vug, vug
wackestone	Wkst
washed residue	W.R.
water	Wtr
wavy	wvy
waxy	wxy
weak	wk
weathered	wthd
well	Wl, wl
west	W
white	wh
with	w/
without	w/o
wood	Wd
yellow (-ish)	yel, yelsh

zeolite	zeo
zircon	Zr
zone	Zn

Engineering Abbreviations

WORD	ABBREVIATION
absolute open flow	AOF
barrel of oil	BO
barrels of oil per day	BOPD
barrels of oil per hour	BOPH
barrels of water	BW
barrels of water per day	BWPD
barrels of water per hour	BWPH
bottom hole flow pressure	BHFP
bottom hole pressure	BHP
bottom hole shut in pressure	BHSIP
bottom hole temperature	BHT
brackish	brk
casing	csg
choke	ck
circulate (-ed, -tion)	circ
circulated out	CO
completed (-tion)	comp
connection gas	CG
cored	crd
decreasing	decr
depth correction	DC
derrick floor	DF
development	(D)
directional survey	DS
distillate	dist
drill stem test	DST
driller	drir
dry and abandoned	D&A
estimated	est
faint air blow	FTAB
fair air blow	FAB
filter cake	CK
filtrate, API, cc's	F
flowed (-ing)	fl/
flowing pressure	FP
flowline temperature	F/T
gas and oil cut mud	G&OCM
gas cut mud	GCM
gas cut water	GCW
gas to surface	GTS
gas-to-oil ratio	GOR
gauged	ga
good air blow	GAB
good initial puff	GIP
gravity	gty
ground	GR
ground level	GL
heavy oil	HO
initial air blow	IAB
initial production	IP
kelly bushing	KB
legal subdivision	LSD
location	loc
million cubic feet of gas	MMCFG
mud cut oil	MCO
mud cut water	MCW
mud resistivity, OHM-METER	RM
new bit	NB
new core bit	NCB
no returns	NR
oil and gas	O&G
oil and salt water	O&SW
oil cut	OC
oil cut mud	OCM
oil flecked mud	OFM
oil to surface	OTS
old total depth	OTD
old well drilled deeper	OWDD
old well plugged back	OWPB
old well worked over	OWWO
open	op
packer	pkr
per day	pd
per hour	PH
perforated	perf
plugged back	PB
pounds per square inch	psi
pump pressure	PP
pump strokes	SPM
recovered	rec
rotary speed	RPM
rotary table	RT
salinity—PPMCL	Cl
salt water	SW
show of oil	SO
show of oil & gas	SO&G
show of oil and water	SO&W
shut in	SI
shut in pressure	SIP
slight gas cut mud	SGCM
slight gas cut water	SGCW
slight oil cut mud	SOCM
slight oil cut water	SOCW
slight show of oil	SSO
squeezed	sqz
strong air blow	SAB
suction temperature	S/T
swabbed	swbd
testing	tstg
thousand cubic feet of gas	MCFG
too small to measure	TSTM
total depth	T.D.
trip gas	TG
valve open	v.op
viscosity, API, sec.	V
water	wtr
water cushion	wtr cush
water cut mud	WCM
weak air blow	WAB
weak initial puff	WIP
weight of mud	W
weight on bit	WOB
wildcat	(W)

Appendix D

Geologic Digital Terms

Term	Code
Absolute Age Date	AAGE
Abyssal	ABYS
Acicular	ACIC
Acidic Igneous Rock	AIGN
Aeolian	AEØL
Age Date, Absolute	AAGE
Agglomerate	AGLM
Aggregate	AGRE
Albite	ALBT
Algal Coated Grains	AGCG
Alkali Flat	ALKF
Alkali Feldspar	AFLD
Allochem	ALLC
Allogenic	ALLG
Alluvium	ALLV
Alteration Product	ALTP
Amber	AMBR
Amorphous	AMPH
Amphibole	APHB
Amygdaloid (ule)	AMYG
Andesite	ANDS
Angular	ANGU
Anhedral	ANHE
Anhydrite	ANHY
Anorthosite	ANØR
Apatite	APAT
Aragonite	ARAG
Arenaceous	AREN
Arenite	AREN
Argillaceous	ARGI
Argillite	ARGL
Arkose	ARKØ
Asbestos	ASBE
Ash	ASH–
Asphalt	ASPH
Asphaltite	ASPH
Atoll	ATØL
Authigenic	AUTH
Back Reef	BKRF
Banded	BAND
Bank (deposits)	BANK
Bar	BAR–
Barite	BART
Barrier Reef	BRRF
Basalt	BSLT
Basement	BASM
Basic Igneous Rock	BIGN
Basin (starved)	BASN
Bathyal, Lower	LBTH
Bathyal, Upper	UBTH
Bauxite	BAUX
Bay	BAY–
Bay, Lower	LBAY
Bay, Marginal	BYML
Bay, Upper	UBAY
Beach (many)	BECH
Bedding	BEDD
Bedding, Graded	GBED
Bedding, Massive	MASS
Bedding, Thin	THBD
Bentonite (itic)	BENT
Bioclastic	BIØC
Bioherm	BIØH
Bitumen	BITU
Biumbonate	BIUM
Bivalve	BIVA
Blanket	BLKT
Bleeding	BLED
Blocky	BLKY
Borings	BRNG
Botryoidal	BØTR
Boulder	BØLD
Boundstone	BNDS
Brackish	BRCK
Brackish Marsh	BKMS
Breccia	BXIA
Brittle	BRTT
Calcareous	CALC
Calcareous Matter	CALC
Calcic	CALC
Calcite	CLCT
Caliche	CACH
Caprock	CAPR
Carbon	CRBN
Carbonaceous	CRBN
Carbonate	CBNT
Carbonate Crystallinity MM	XTNY
Carbonized Matter	CBNƵ
Carlsbad Twin	CTWN
Cast	CAST
Cavity	CAVT
Cellular	CELL
Cement	CMNT
Chalcedony	CHED
Chalk	CHLK
Channel	CHAN
Charcoal	CHCL
Chert	CHRT
Chlorite	CHLØ
Clast	CLAS
Clay	CLAY
Clay, Dispersed	DCLY
Clayey	CLYY
Claystone	CLST
Cleavage	CLEV
Coal	CØAL
Coated Grains	CØGR
Coated Grains, Algal	AGCG
Cobble	CØBB
Coiling, Left	LCØL
Coiling, Right	RCØL
Colloform	CØLL
Compact (ed)	CMPC
Component	CØMP
Concentric	CØNC
Conchoidal	CHØI
Concretion	CCRT
Cone-in-Cone	CINC
Conglomerate	CØNG
Contact	CNTC
Continental	CØNT
Continental, Other	CØTH
Continental, Undifferentiated	CUND
Contorted	CNTT

Coquina	CØQN
Coralline	CØRL
Core, Reef	RCØR
Corroded Grain Surface	CØRR
Corrosion	CØRR
Country Rock	CRØK
Cross Bedding	XBED
Crushed	CRSH
Cryptocrystalline	CYPX
Crystal	XTAL
Crystalline	XTLN
Crystalline Rock	XALR
Dacite	DACT
Decomposition	DECØ
Dedolomitization	DEDØ
Delta	DELT
Deltaic	DELT
Dendritic	DEND
Density Log	DLØG
Depositional Texture	DTEX
Desert	DESR
Detrital	DETR
Detrital Carbonate	DTCB
Diabase	DIAB
Diatomaceous Earth	DITM
Diatomite	DITM
Dike	DIKE
Diorite	DIØR
Disc	DISC
Discontinuous	DSCN
Dispersed Clay	DCLY
Divitrification	DIVT
Dolomite	DØLØ
Druse	DRSE
Drusy	DRSY
Dune	DUNE
Dunite	DUNT
Early	ERLY
Elements, Native	ELEM
Ellipsoidal	ELLP
E Log	ELØG
Elongate	ELØN
Embayed	EMBY
Emergence	EMRG
Environment	ENVI
Epidote	EPID
Equigranular	EQGN
Estuary	ESTU
Etched	ETCH
Euhedral	EUHE
Euxinic, Evaporitic	EEUX
Euxinic, Marine	MEUX
Evaporite (itic)	EVAP
Evaporitic Reef	REVP
Extrusive	EXTR
Fabric	FABR
Facies	FACE
Fecal Pellets	FECP
Feldspar	FLSP
Feldspar, Alkali	AFLD
Feldspathoid	FLPD
Felsic	FELS
Felsite	FELS
Ferromagnesian Minerals	FERR
Fibrous	FIBR
Filiform	FILI
Fillings	FILL
Fissility	FISS
Flat, Reef	RFLT
Flexible	FLEX
Fluorite	FLUØ
Fluvial	FLUV
Fluvial-Marine	FLMR
Foliated	FØLI
Foliation	FØLA
Fore Reef	FØRF
Fossil(s)	FØSS
Fracture	FRCT
Fracture, Closed	FRCL
Fracture, Filled	FRFL
Fracture, Hairline	FRHL
Fracture, Healed	FRHE
Fracture, Open	FRØP
Fracture, Partially Filled	FRPF
Fracture Porosity	FRCT
Fragmental	FRAG
Framework	FRWK
Fresh Water Environment	FRSH
Friable	FRIA
Frosted	FRST
Gabbro	GABB
Garnet	GRNT
Ghost	GHØS
Glacial	GLAC
Glass	GLSS
Glass Shards	SHRD
Glauconite (itized)	GLAU
Globular	GLØB
Gneiss	GNSS
Good Preservation	GPRS
Gouge	GOUG
Graded Bedding	GBED
Grains	GRIN
Grains, Algal Coated	AGCG
Grainstone	GRST
Grain Supported	GRSP
Grain Surface	GSUR
Granite	GRAN
Granodiorite	GDIO
Granular	GRNL
Granule	GRNL
Graphite	GRPH
Gravel	GRVL
Greasy	GRSY
Groundmass	GMSS
Gypsum	GYP
Halides	HALD
Hematite	HEMA
Hornblende	HRBL
Humus	HMUS
Hydrothermal	HYDR
Hypabyssal	HYPA
Hypersaline	HSAL
Identifiable (fossil or shell fragments)	IDNT
Igneous (rocks)	IGNE
Igneous Rock, Acidic	AIGN
Igneous Rock, Basic	BIGN
Immature	IMAT
Inclusions	INCL
Indurated	INDR
Inner Neritic	INER
In Situ	SITU
Intercrystalline Porosity	IXAL
Interfinger	IFNG
Intergranular Porosity	IGAN
Intergranular or Inter-crystalline Porosity	INTR
Intergrowth	IGRN
Interoolitic Porosity	IØØL
Interstices	ISTC
Interstitial	ISTT
Intraclast	ICLS
Intrusive	NTRS
Iron	IRØN
Kaolin	KAØL
Kyanite	KYAN
Lacustrine (environment)	LACS
Lagoon, Marginal	MGLG
Lagoon, Reef	RLAG
Lake	LAKE
Lamina (ated)	LAMN
Late	LATE
Laterite (itic)	LATR
Latite	LATI
Left Coiling	LCØL
Lenticular	LENS
Leucite	LEUC
Lime	LIME
Limestone	LMST
Limestone, Lithographic	LTLS
Limey	LIMY
Limonite	LIMØ
Lineation	LINE
Lithify (ied)	LTFY
Lithograhic Limestone	LTLS
Lithology	LITH
Littoral	LITT
Log, E	ELØG

Term	Code
Log, Density	DLØG
Log, Other	ØLØG
Log, Radioactivity	RLØG
Log, SP	SLØG
Lower Bathyal	LBTH
Lower Bay	LBAY
Lump	LUMP
Magma	MAMA
Magnetite	MGNT
Marble	MARB
Marginal Bay	BYML
Marginal Lagoon	MGLG
Marine	MARN
Marine, Euxinic	MEUX
Marine, Fluvial	FLMR
Marine, Other	ØTHM
Marine Swamp	MRSW
Marine, Undifferentiated	MUND
Marl (y)	MARL
Marsh	MRSH
Marsh, Brackish	BKMS
Massive	MASS
Matrix	MTRX
Mature	MTRE
Member	MEMB
Metallic	METL
Metamorphic	META
Meteorites	METE
Mica	MICA
Micrite (itic)	MICR
Microcrystalline	MICX
Micrograined	MGRN
Microslickensided	MSLK
Microspheric	MSPH
Middle	MIDD
Middle Neritic	MNER
Mineralogy	MINE
Mineraloid	MLØD
Mold	MØLD
Monomineralic	MØNM
Mosaic	MSAC
Mottled	MØTT
Mudstone	MDST
Mud Supported	MDSP
Muscovite	MUSC
Native Elements	ELEM
Nepheline	NEPH
Neritic, Inner	INER
Neritic, Middle	MNER
Neritic, Outer	ØNER
Nodule	NØDL
Non-carbonate Float	NCBF
Non-supporting Phase	NSPP
Norite	NØRI
Normal Marine	NMAR
Obsidian	ØBSD
Oil Stained	ØSTN
Olivine	ØLIV
Oolite	ØØLT
Oolitic	ØØLI
Oolitoids	ØØID
Oomoldic Porosity	ØØMD
Ooze	ØØZE
Opal	ØPAL
Opaque	ØPQU
Ore	ØRE
Organic	ØRGA
Organic Reef	ØRRF
Orientation (preferred, etc.)	ØREN
Orthoquartzite	ØQTZ
Other Log	ØLØG
Other Reef	RØTH
Outer Neritic	ØNER
Overgrowth	ØVGW
Packing	PACK
Packstone	PCKS
Paleontology	PLEØ
Palynology	PALY
Partially Restricted Marine	PRMC
Particulate	PART
Pearly	PRLY
Peat	PEAT
Pebble	PEBL
Pegmatite	PEGM
Pellet(s) (iferous)	PELL
Pellets, Fecal	FECP
Penecontemporaneous Deformation	PEND
Penesaline	PSAL
Permeability	PERM
Perthite	PRTH
Petroliferous	PETR
Phantom Horizon	PHØR
Phase, Non-supporting	NSPP
Phenocryst	PHEN
Phonolite	PHØN
Phosphate (ized)	PHØS
Phyllite	PHYL
Pinpoint Porosity	PPPR
Pisolite (itic)	PISØ
Pitted	PIT
Plagioclase	PLAG
Planar	PLAN
Plant	PLNT
Plastic	PLAS
Platy	PLTY
Pleochroism	PLCM
Plutonic	PLUT
Pod	PØD
Poikilitic	PØIK
Polished	PØLS
Pollen and/or Spores	PLSP
Polycrystalline	PØLX
Polygonized	PØLG
Polysynthetic Twins	PTWN
Poor Preservation	PPRS
Porcelaneous	PØRC
Pore Filling	PØRF
Pore, Void	VØID
Porosity, Intercrystalline	IXAL
Porosity, Intergranular	IGAN
Porosity, Interoolitic	IØØL
Porosity, Oomoldic	ØØMD
Porosity, Pinpoint	PPPR
Porous	PØRØ
Porphyry	PRPH
Portland Cement	PCEM
Potash	PTSH
Powder	PWDR
Precipitate	PREC
Preservation, Good	GPRS
Preservation, Poor	PPRS
Pressure (solution, etc.)	PRSS
Primary	PRIM
Pseudomorph	PSDM
Pumice	PUMC
Pyrite (itized)	PYRT
Pyroclastic	PYRØ
Pyroxene	PRØX
Quartz	QRTƵ
Quartz Diorite	QDIØ
Quartzite	QTƵT
Quartz Latite	QLAT
Quartzose	QƵSE
Radial	RAD
Radiated	RADI
Radioactivity Log	RLØG
Recrystallized Product	REXL
Red Beds	RBED
Reducing	REDU
Reef	REEF
Reef, Back	BKRF
Reef, Barrier	BRRF
Reef Core	RCØR
Reef, Evaporitic	REVP
Reef Flat	RFLT
Reef, Fore	FØRF
Reef Lagoon	RLAG
Reef, Organic	ØRRF
Reef, Other	RØTH
Reef Talus	RTAL
Reef, Undifferentiated	RUND
Replaced Framework	REPF
Replaced (material)	REPL
Replaced Matrix	REPM
Residue	RESD
Resorbed	RSRB
Reworked	RWKD
Rhombs	RHØM
Rhyolite	RHYØ
Right Coiling	RCØL

Ripple Marked	RIPP
Rock Fragment	RFRG
Root	RØØT
Rounded	RØND
Rubble	RUBB

Saline	SALN
Salt	SALT
Sand (y)	SAND
Sandstone	SDST
Scattered	SCAT
Schist	SCHI
Secondary	SCND
Sedimentary	SED
Sedimentary Environment	SENV
Sedimentary Quartzite	SQƵT
Segregation	SEGR
Selective	SLCT
Serpentine	SERP
Shale (y)	SHLE
Shape	SHPE
Shards, Glass	SHRD
Sheet	SHET
Shell Fragments	SFRA
Shell Fragments, Identifiable	IDNT
Shell Fragments, Unidentifiable	UNID
Shell Hash	HASH
Shells, Whole	WHLE
Shows	SHØW
Siderite	SIDR
Silica	SILC
Siliceous	SILC
Silicified	SILC
Silky	SLKY
Sill	SILL
Silt (y)	SILT
Siltstone	SLST
Situ, In	SITU
Slab	SLAB
Slate (y)	SLAT
Slickensided	SKSD
Slump (ed)	SLMP
Soapstone	SPST
Soil	SØIL
Solution	SØLN
Sorted	SØRT
Source	SØRC
Spar (ry)	SPAR
Spherical	SPHE
Spherulite	SPHL
Spicule	SPIC
Spinel	SPNL
SP Log	SLØG
Sporbo	SPBØ
Spores and/or Pollen	PLSP
Spotty (ed)	SPØT
Stained	STIN
Stellated	STEL
Strained	STRN
Stratigraphy	STAT
Streaked	STRK
Stylolite	STYØ
Subangular	SANG
Subhedral	SHED
Submature	SMTR
Subrounded	SRND
Sucrose	SUCR
Sulfur	SLFR
Supermature	SPRM
Swamp	SWMP
Swamp, Marine	MRSW
Syenite	SYEN
Sylvite	SYLV

Tabular	TABL
Talc	TALC
Talus	TALS
Talus, Reef	RTAL
Tarnish	TRSH
Terrace	TERR
Texture	TEXT
Thin Bed (ded, ding)	THBD
Tidal Flat	TDFT
Tight	TIGH
Till	TILL
Topaz	TØPƵ
Tough	TØGH
Tourmaline	TØUR
Trace	TRCE
Trachyte	TRCH
Transitional	TRAN
Transitional, Other	TØTH
Transitional, Undifferentiated	TUND
Translucent	TRSL
Transparent	TRSP
Travertine	TRAV
Tufa	TUFA
Tuff	TUFA
Twin (ned)	TWIN
Twin, Carlsbad	CTWN
Twins, Polysynthetic	PTWN

Umbonate	UMBØ
Unconformity	UNCN
Undetermined	UNDT
Undifferentiated Reef	RUND
Unidentifiable (fossils or shell fragments	UNID
Unstrained	USTR
Upper Bathyal	UBTH
Upper Bay	UBAY

Variegated	VARE
Vesicle	VESI
Visual Porosity	VISL
Vitreous	VITR
Void, Pore (interstice)	VØID
Volcanic	VØLC
Volcanic Ash	VASH
Volcanic Tuff	VTFF
Vug (gy)	VUG
Vug-Matrix	VUGM
Vugular Porosity	VUGU

Wackestone	WCKS
Water	WATR
Weathered	WTRD
Welding	WELD
Well Rounded	WRND
Whole Fossils	WHLE

Zeolite	ƵEØL
Zircon	ƵIRC
Zone	ƵØNE
Zoned	ƵØND

DICTIONARY OF GEOLOGIC TERMS WITH DIGITAL CODES

A

ABYSSAL ENVIRONMENT - Environment of a marine zone on the sea floor in excess
(ABYS) of 6560 feet of water depth.

ACICULAR SHAPE - Term used to describe the form of grains and/or rock fragments
(ACIC) which are needle-like in form with two short dimensions and one long dimension.

ACIDIC IGNEOUS ROCK - A descriptive term applied to those igneous rocks that
(AIGN) contain more than 66 % silica, and/or associated high silica minerals and their normal associates, as contrasted with basic igneous rock. (AGI Glossary, 1957).

AEOLIAN ENVIRONMENT - That part of the continental environment characterized
(AEØL) by the activity of the wind. The deposits of sediment reflect the eroding, transporting, and depositing activity of the wind.

ALGAL COATED GRAINS - Fragments which have an outer covering of recognizable
(AGCG) algal origin. The fragments can be of any size, shape, material, or origin.

ALTERATION PRODUCT - Any of a group of secondarily derived substances produced
(ALTP) by alteration of pre-existing minerals or rocks. Common at surface temperatures and pressures.

ANGULAR - Measure of a grain showing very little to no evidence of wear; edges
(ANGU) and corners sharp; secondary corners are sharp and number 15 to 30.

ANGULARITY - A measure of the sharpness of the edges and corners of a clastic
(ANGU) fragment as seen in a two-dimensional figure.

ANHEDRAL CRYSTALLINITY - Individual minerals which have not developed crystal
(ANHE) faces during crystallization.

ARGILLACEOUS - A descriptive term applied to all rocks composed of clay or
(ARGI) containing a large clay component, e.g. argillaceous sandstone. Shale, claystone are argillaceous rocks. (Adapted from AGI Glossary, 1957).

ARKOSE - A sandstone (grain size 1/16 to 2.0 mm) generally moderately to well-
(ARKØ) sorted with little to no interstitial clay and categorized by limiting percentages of three major constituent groups as follows: 1. Feldspar content - cannot be less than 25% (most critical); 2. Rock fragment content - cannot be more than 10%; 3. Quartz, quartzite, chert - cannot be more than 75%. (Adapted from McBride, 1963; William et al, 1954).

ARKOSIC - Descriptive term for sandstone and conglomerate whose feldspar
(ARKØ) composition can range from 5 to 50 percent depending on the authority (McBride 1963, Pettijohn 1957, William et al, 1954).

ATOLL - Organic reef, oval to doughnut-shaped in plan encircling or nearly
(ATØL) encircling a lagoon.

B

BACK REEF - Area of an organic reef behind the reef core. 1. In atolls and
(BKRF) atoll-like reefs, the back reef area can include the lagoon reef, the lagoon terrace (reef flat) and the lagoon floor (lagoon). 2. In the barrier reefs the back reef consists of the above sequence landward into the lagoonal zone where the carbonate sediments interfinger with terrigenous sediments.

BARRIER REEF - Elongate organic reef which parallels the shore, but is
(BRRF) separated from the land mass by a lagoon.

BASIC IGNEOUS ROCK - A generalized term meaning: 1. An igneous rock con-
(BIGN) taining 45 to 52% of silica, free or combined; 2. An igneous rock consisting dominantly of minerals comparatively low in silica and rich in the metallic bases, e.g. the amphiboles, the pyroxenes, biotite and olivine; 3. Very loosely, an igneous rock composed dominantly of dark-colored minerals. (AGI Glossary, 1957).

BAY - A recess in the shore or an inlet of a sea or lake between two capes or
(BAY-) headlands. (Adapted from AGI Glossary, 1957).

BAY MARGINAL ENVIRONMENT - Environment of shallow mid-margins of bay (See
(BYML) definition). Salinities range from slightly brackish to brackish to more than 35 parts per mil.

BEDDING - Layering in sedimentary rocks. Layering is caused during deposition
(BEDD) by changes in the materials being deposited or by changes in conditions of deposition or both, e.g. deposition of a layer of silt followed by deposition of a layer of clay.

BENTONITIC - Sedimentary rock containing clays largely made up of the clay
(BENT) minerals montmorillonite and bentonite. These clays commonly swell to many time their original size by absorption of water.

BITUMEN - A general name for various solid and semi-solid hydrocarbons. (AGI
(BITU) Glossary, 1957).

BLOCKY SHAPE - Term used to describe the form of grains and/or rock fragments
(BLKY) which are more or less cubic where the dimensions of length, width and thickness of the fragments are similar.

BORINGS - Cylindrical tubes or irregular disturbed areas in sedimentary rock
(BRNG) often filled with clay or sand made by mud-eating worms or molluscs when the sediment was at the depositional surface on the bottom of a body of water.

BOUNDSTONE - Carbonate rocks whose constituents are bound together during deposition
(BNDS) as shown by intergrown skeletal matter. (Dunham, 1962, p. 117).

BRACKISH - Slightly salty, a term applied to water whose saline content is
(BRCK) intermediate between fresh water streams and sea water. (Adapted from AGI Glossary, 1957). The water is usually less than 20 parts per mil, but can range up to 30 parts per mil.

BRACKISH MARSH ENVIRONMENT - Environment of marsh (see definition) whose water
(BKMS) salinity ranges between fresh and sea water, but usually is less than 20 parts per mil salinity.

BRECCIA - A fragmental rock whose components (greater than sand size) are
(BXIA) angular and therefore, unlike conglomerates, not water-worn. There are fault breccias, talus breccias, and volcanic breccias. (Adapted from Kemp, 1940).

C

CALCAREOUS - Containing calcium carbonate; will effervesce in dilute hydrochloric
(CALC) acid.

CALCAREOUS MATTER - Rock and/or fossil material containing calcium carbonate;
(CALC) will effervesce in dilute hydrochloric acid.

CARBONACEOUS - Sediments containing material (originally plant remains) largely
(CRBN) made up of carbon.

CARBONATE - Term used to define rocks whose dominant constituents are of car-
(CBNT) bonate composition (more than 50%). Although there are some 60 carbonate minerals, the most prominent are calcite, aragonite, and dolomite. The first two are calcium carbonate minerals, the third is calcium magnesian carbonate in composition. Carbonate rocks can range from simple monomineralogic composition to very complex mixtures of mineral crystals, rock fragments, and hard parts of carbonate secreting plants and animals.

CARBONATE COMPONENTS - The various carbonate particles of which a carbonate
(CBNT) rock is composed.

CARBONATE CRYSTALLINITY MM. - Refers to size scale for use only with measurable
(XTNY) material in carbonate rock which appears to have formed in place (not detrital). Can occur as cement between detrital particles or as replacement of matrix and/or framework, e.g. calcite, dolomite, anhydrite, etc. etc.

CARBONIZED MATTER - A broad term to include all black, brown, and yellow residues
(CBNZ) of what appear to have been plant material. This includes coal, mineral charcoal, carbonized leaf etc. etc. This term does not include tars or oils.

CEMENTING MATERIALS - Any material which binds the rock particles together.
(CMNT)

CHALK - Porous, microtextured and friable variety of carbonate rock (micro-
(CHLK) texture = 0.01 mm approx.)(G. E. Thomas, 1962, p. 194.)

CHERT - A family of dense, hard, very fine-grained, amorphous, siliceous rocks
(CHRT) consisting of opaline and/or chalcedonic silica or cryptocrystalline quartz. Can be any color.

CLAST - Sedimentary rock particle which can be derived from one or more of the
(CLAS) following sources: 1. The breakup of pre-existing rock; can be igneous, metamorphic or sedimentary. 2. Partially lithified bottom sediment. 3. Volcanic or outer space origin.

CLAY - (Mineralogical) - Mineral material composed of hydrous aluminum silicates;
(CLAY) occasionally hydrous magnesium silicates. (AGI Glossary, 1957). (Particle size) - A sediment whose particle sizes are less than 1/256 mm (0.0039mm).

CLAYEY - A sediment (or sedimentary rock) which contains at least 50% clay
(CLYY) minerals.

CLAYSTONE - A compacted sediment consisting of more than 51% of clay minerals,
(CLST) (See definition of clay).

COAL - A black-colored, compact and earthy organic rock with less than 40%
(CØAL) inorganic components (based on dry material) formed by the accumulation and decomposition of plant material. (AGI Glossary, 1957).

COATED GRAINS - Grains which contain an outer layer of material unlike the
(CØGR) remainder of the grain. A broad category for those coated grains which do not fit the oolite, oolitoid, pisolite or algal-coated groups.

COMPACTED SEDIMENT - A sediment which has undergone a decrease in total volume (CMPC) by the closer crowding of its particles. Pore space volume is reduced by the process.

CONCRETION - A nodular or irregular concentration of certain authigenic constituents of sedimentary rocks and tuffs; developed by the localized deposition of material from solution about a central nucleus. (AGI Glossary, 1957).
(CCRT)

CONE-IN-CONE - A concretionary structure occurring in marls, shales, ironstones, coals etc.; characterized by a succession of cones, one within the other. (AGI Glossary, 1957).
(CINC)

CONGLOMERATE - A lithified sedimentary rock consisting of at least 50% subangular to rounded fragments of mineral and rock fragments in the gravel range (greater than 2.0 mm).
(CØNG)

CONTINENTAL ENVIRONMENT - In geological usage, the broad term which includes all environments (See definition of environment) occurring on the surface of the continents, down to the high tide level at the coasts. Desert, mountain, lakes, river valley systems, are some examples of continental environments.
(CØNT)

CROSS BEDDING - Layers in a sedimentary rock which are deposited at an angle to the overall original dip of the formation. (Adapted from McKee & Weir, 1953).
(XBED)

CRYSTALLINITY - The degree of crystallization exhibited by a chemical precipitate, igneous or metamorphic derived mineral.
(XTNY)

D

DELTAIC ENVIRONMENT - When the delta is formed in a marine body of water the deltaic environment is transitional between fluvial and marine and may contain characteristics of both. By definition, a delta is a deposit of sediment usually triangular in plan, formed at the mouth of a river in a marine body of water. Deltas also form in lakes. Delta development results in progradation of the shoreline into the body of water. (Adapted from AGI Glossary, 1957).
(DELT)

DEPOSITIONAL TEXTURE - A semi-quantitative measure of some of the aspects of the particle arrangement, particle sizes and percentage volumes of particle sizes of a sedimentary rock at the time of final deposition. These characteristics provide some evidence of the available energies in the transporting medium (water, ice, wind, gravity) during final deposition of the sediment. (See also, supporting, non-supporting, mudstone, wackestone, packstone, and grainstone).
(DTEX)

DESERT ENVIRONMENT - CONTINENTAL - Environment in a region so devoid of vegetation as to be incapable of supporting any considerable population. Four kinds of desert may be distinguished: 1. The polar ice and snow deposits, marked by perpetual snow cover and intense cold. 2. The middle latitude deserts in the basin-like interiors of the continents such as the Gobi, characterized by scant rainfall and high summer temperatures. 3. The trade wind deserts, notably the Sahara, the distinguishing features of which are negligible precipitation and large temperature range. 4. Coastal deserts where there is a cold current on the western coast of a large land mass, such as occurs in Peru (AGI Glossary, 1957).
(DESR)

DETRITAL CARBONATE - Carbonate fragments derived from pre-existing limestones. Differs from intraclasts which derived from very young semi-consolidated bottom and from micrite which is in the process of forming on the bottom.
(DTCB)

DETRITAL (same as clastic) - Defines particles in a sedimentary rock which can be derived from one or more of the following sources: 1. The breakup of pre-existing rock; can be igneous, metamorphic or sedimentary. 2. Partially lithified bottom sediment. 3. Volcanic or outer space origin.
(DETR)

DISPERSED CLAY - As in a siltstone or sandstone where clay was deposited as individual, soft, very small particles (individual particles less than 0.0039 mm in size) as coatings on the coarser particles and as fillings in pore spaces.
(DCLY)

E

ELLIPSOIDAL SHAPE - Term to describe sedimentary grains and rock fragments whose three-dimensional form approaches an ellipsoid. An ellipsoid is a solid, every plane of which, is either a circle or an ellipse.
(ELLP)

EMBAYED - Crystals or grains with re-entrants on their peripheries are embayed. The re-entrants (embayed areas) can be primary or secondary. Primary embayed grains or crystals result from the penetration of one individual by another as both crystallize. Secondary embayed grains develop by the replacement of one mineral material by another.
(EMBY)

ENVIRONMENT - In the biological sense, the sum of all the external conditions and influences affecting the life processes and development of an organism or group of organisms. In the geological sense the above definition has been expanded to include the effects of the external conditions (physical, chemical, and biological) on the non-biological phases as well as the biological assemblages making up the sedimentary deposits. A simplified example of the expanded geological usage is provided by what can be learned from analysis of a typical sediment originally deposited in a deep water oceanic locality. The fossils, e.g. the Foraminifera would suggest deep water living conditions and the dominant clay content of the rock would suggest weak sediment moving ability of the offshore currents. Both of these analyses would suggest deep water far offshore oceanic conditions. Sediment deposited in any environment (from the oceans to the deserts) can consist of biological and non-biological materials, both of which can provide evidence of the external conditions of the time
(ENVI)

ESTUARY ENVIRONMENT - The environment of the lower course of a river which is affected by tides. It is subjected to both the fluvial and marine effects and is therefore one of the transitional environments. Estuaries tend to be triangular to funnel-shaped in plan and usually have the largest areal development under conditions of coastal submergence. (Adapted from Twenhofel, 1950, p. 121).
(ESTU)

EUHEDRAL CRYSTALLINITY - Individual minerals which are bounded by their own well-developed crystal faces. (Adapted from AGI Glossary, 1957).
(EUHE)

EUXINIC ENVIRONMENT - Environment of nearly isolated sea where restriction of marine waters results in poor water circulation. Sufficient inflow of open sea water and meleoric water maintains salinity at or near normal level. The deeper waters are stagnant and often toxic and except for anaerobic bacteria, no life is found. Muds are organic rich. The modern Black Sea is a good example. (Adapted from Sloss, 1953, p. 144; and AGI Glossary, 1957).
(MEUX)

EVAPORITES - Rocks consisting of minerals which are deposited from aqueous solution as a result of extensive or total evaporation of water. (Adapted from AGI Glossary, 1957).
(EVAP)

EVAPORITIC ENVIRONMENT - In most geological instances, an area where the evaporation of water exceeds or is equal to the volume of additional water supply (fresh or salt). Most of the evaporites in the earth's crust have probably originated in areas marginal to the sea, e.g. lagoons, organic reefs, relict seas, epicontinental seas etc. Evaporites are also known from inland lakes which were originally fresh water.
(EVAP)

EXTRUSIVE - A term applied to those igneous rocks derived from magmas or magmatic materials poured out or ejected at the earth's surface, in contrast to intrusives where magmas were infected into or between older rocks below the surface. (Adapted from AGI Glossary, 1957).
(EXTR)

F

FABRIC - The orientation in space of the elements forming a rock. (Pettijohn, 1957).
(FABR)

FILLINGS - Mineral and/or lithic material occupying what formerly were openings in rock (interstices, voids, pores, fractures, etc.) Vug Filling-filling of vugs (See definition of vug porosity). Fracture Filling-filling of fractures (See definition of fracture porosity.
(FILL)

FISSILITY - The property of a fine-grained, clay-bearing rock of being able to be split along closely parallel planes (as in claystones or clayey siltstones).
(FISS)

FLUVIAL ENVIRONMENT - Geologically the physical and biological conditions associated with the fresh water of streams and rivers, e.g. the sedimentation, biological systems etc. Fluvial environment is grouped with the continental environments.
(FLUV)

FLUVIAL MARINE ENVIRONMENT - Environment in which a salt water wedge invades a fluvial environment. The lower distributaries of the Mississippi river are examples of this environment. An upper wedge of fresh water flows out over the bottom salt water. The fluvial-marine environment is grouped with the transitional environments.
(FLMR)

FORE REEF - The area seaward of the reef core (See definition) dominated by carbonate sediment deposition and extending to the zone of interfingering with non-carbonate sediments.
(FØRF)

FOSSILS - Material of biologic origin whose plant or animal identity can at least be established in a general way. This material can range from unbroken, whole Foraminifera, ostracodes, gastropods, pelecypods, etc., to fragmental debris of algae, stromatoporoids, corals, etc. This differs from skeletal fragments in that only the biologic origin of the skeletal fragments can be identified.
(FØSS)

FRACTURE - Breaks in rock or mineral grains. The term implies the separation of the original into separate parts. (Adapted from Rice, 1949).
(FRCT)

FRACTURE, CLOSED - No space exists along the break. Fragments on either side (FRCL) of the break are entirely in close contact with each other. The break still exists.

FRACTURE, FILLED - The space formed by the line of the break between rock segments is filled in with mineral material. (FRFL)

FRACTURE, HAIRLINE - An open fracture but with a minimum of separation (< 1 mm). (FRHL)

FRACTURE, HEALED - A former break now completely closed and with the rock on both sides of the break now joined by mineral material. Healed fractures differ from closed fractures in that closed fractures do not contain a cementing material. (FRHE)

FRACTURE, OPEN - A rock break along which the rock fragments have separated leaving an opening between. (FRØP)

FRACTURE, PARTIALLY FILLED - Space along separation between rock fragments is incompletely filled in. (FRPF)

FRACTURE POROSITY - Void volume provided by open fractures in a rock resulting from breaking and shattering. (FRCT)

FRAMEWORK - (See definition for supporting phase - framework). (FRWK)

FRESH WATER ENVIRONMENT - The environment of fresh water lakes. (FRSH)

FRIABLE - Easily crumbled as with rock that is poorly cemented. (FRIA)

G

GHOST - Faint indication of a structure such as a crystal or fossil more or less obliterated by diagenesis or replacement. (AGI Glossary, 1957). (GHØS)

GLACIAL ENVIRONMENT - Environment in area affected by a glacier. A glacier is defined as a body of ice originating on land by the recrystallization of snow or other forms of solid precipitation and showing evidence of past or present flow. Glacial environments are grouped with the continental environments. (GLAC)

GLASS SHARDS - Curved, spicule-like fragments of volcanic glass. (AGI Glossary, 1957). (SHRD)

GOUGE - Finely abraded material occurring between the walls of a fault, the result of grinding movement. (AGI Glossary, 1957). (GØUG)

GRADED BEDDING - Layering in sedimentary rock in which the layers contain a gradational change in grain size occurring from the base of the layer upward. The change can either be a coarsening upward or a size decrease upward. Typically contacts between graded beds are sharply defined. (GBED)

GRAIN ANGULARITY - A generalized measure of the degree of wear (or lack of wear) exhibited by a detrital fragment when seen as a two-dimensional figure. (ANGU)

GRAINS - A very general term for the clastic particles of which sedimentary rocks are composed. (GRIN)

GRAINSTONE - Grain-supported sedimentary rocks; no mud content (Adapted from Dunham, 1962, p. 117). According to Dunham, particles less than 0.02 mm in diameter (medium silt and clay) are mud; particles larger than 0.02 mm (coarser than medium silt) are grains. Grain support is considered to be a property of rocks deposited in agitated water. Please note here that Dunham uses the term grain in a very restricted sense. Only in this classification is this usage valid. (GRST)

GRAIN-SUPPORTED - Term to describe rocks which consist dominantly of particles of medium silt and coarser (silt, sand, gravel). Most or all of the grains are in contact with each other, the lowermost grains supporting the uppermost grains. The rocks are grain-supported. (GRSP)

GRAIN SURFACE - The appearance of the surface area of clastic grains is observed and described for descriptive as well as interpretive purposes. Various terms are used to describe so called surface textures. (GSUR)

GRAIN SURFACE, CORRODED - Term very similar to etched. If any difference exists a corroded grain surface is somewhat rougher than an etched surface. (CØRR)

GRAIN SURFACE, ETCHED - Term applied to a rough frosted surface of a clastic grain. (ETCH)

GRAIN SURFACE, FROSTED - Dull but smooth grain surfaces without luster, ground-glass to mat surface texture. (FRST)

GRAIN SURFACE, POLISHED - Smooth grain surface with high luster and regularity of reflection, the opposite of dull, frosted surface texture. (PØLS)

GRAIN SURFACE, STAINED - Grain surfaces with exterior coatings of different material, e.g. quartz grains can be iron-stained, oil-stained, organically-stained, etc. (STIN)

GRAVEL - An unconsolidated sediment consisting of subangular to rounded fragments in the gravel range (greater than 2.0 mm). (GRVL)

H

HYPERSALINE ENVIRONMENT - A special environment of abnormally high salinity (exceeds that of the saline environment). Believed to represent the evaporite cycle carried to conclusion. (Adapted from Sloss, 1953, p. 145). Hypersaline environments can occur with organic reef complexes, in inland seas or in bodies of water marginal to the marine environment. (HSAL)

I

IGNEOUS ROCKS - Rocks formed by solidification from a molten or partially molten state. (IGNE)

IMMATURE - Term applied to a clastic sediment with the following textural characteristics: 1. Much clay and/or fine mica; 2. Poor sorting of non-clay constituents; 3. Grains are angular. (Folk, 1951). (IMAT)

INCLUSIONS - A crystal or fragment of substance or a minute cavity filled with gas or liquid enclosed within another crystal. (INCL)

INNER NERITIC ENVIRONMENT - Marine environment extending from the low tide level out to water of approximately 60 feet depth; grouped with marine environments. (INER)

INTERCRYSTALLINE POROSITY - Specialized type of porosity where void volume consists of voids formed between authigenic particles of mineral material formed within the rock; usually reserved for carbonate and some evaporitic rocks. (IXAL)

INTERGRANULAR POROSITY - Specialized term for carbonate rock porosity. The volume of pore space is the sum of all voids formed between detrital particles 0.031 mm and coarser, e.g. coarse silt size and coarser. (IGAN)

INTEROOLITIC POROSITY - That part of the rock which is the sum of all the void spaces formed between the oolites in the rock. (IØØL)

INTERSTITIAL - Occupying interstices (See definition). (ISTT)

INTERSTICES (also called pores, voids) - Openings in rock formed between grains, fragments, fossils, etc. (ISTC)

INTRACLAST - Fragments of sand to boulder sized limestone, usually weakly consolidated, which have been eroded from their original site of deposition and redeposited within the same depositional basin. (ICLS)

INTRUSIVE - Igneous rock which as a magma penetrated into or between other rock layers, but solidifying before reaching the surface. (Adapted from AGI Glossary, 1957). (NTRS)

L

LACUSTRINE ENVIRONMENT - Lake environment. (AGI Glossary, 1957). Grouped with continental environments. (LACS)

LAKE - Any standing body of inland water, generally of considerable size. (AGI Glossary, 1957). (LAKE)

LAMINATED - Thinnest order of bedding as follows: to 3 mm = thin laminations; 3 to 10 mm = thick laminations. (Dunbar and Rodgers, 1958). (LAMN)

LIMESTONE - Carbonate (See definition). (LMST)

LIMEY - Calcareous (See definition). (LIMY)

LITHIFY - (lithified, lithification) - The processes by which a loose sediment becomes rock. (LTFY)

LITHOLOGY - A term used to indicate the composition and texture of a rock. (Adapted from AGI Glossary, 1957). (LITH)

LITTORAL ENVIRONMENT - The environment of the shoreline area bounded by the levels of high and low tide; grouped with the transitional environment. (LITT)

LOWER BATHYAL ENVIRONMENT - Environment of a marine zone on the sea floor between approximately 1640 and 6560 feet of water; grouped with the marine environment. (LBTH)

LOWER BAY ENVIRONMENT - Environment of lower reaches of bay (See definition). Very wide range of salinities. Can range from slightly brackish to over 40 parts per mil depending on the volume of fresh water influx; grouped with transitional environments. (LBAY)

LUMP - A specialized term restricted to usage with carbonate rocks and sediments. An aggregate made up of pellets of micrite apparently cemented together. (LUMP)

M

MAGMA - Hot, mobile rock material consisting almost entirely of a liquid phase (MAMA) with the composition of a silicate melt. It is called magma when in the subsurface, and lava when it flows onto the surface.

MARGINAL LAGOON - A bay extending roughly parallel to the coast and separated from (MGLG) the open ocean by elongate marine deposits such as barrier islands. Water can range from brackish to marine to ultrasaline. Connections can exist with the open ocean through one or more openings in the barrier island. (See also reef lagoon). The marginal lagoon environment is grouped with transitional environments.

MARINE ENVIRONMENT - In the geological sense, the environments of the ocean basins (MARN) and the inland seas which maintained connections with the ocean basins. The environment extends from the low tide level out to the abyssal depths. At present this includes an area roughly equal to 2/3 of the earth's surface.

MARINE SWAMP - A low salt or brackish water area of the shore within an abundant (MRSW) growth of grass, reeds, mangrove trees and similar types of vegetation. (AGI Glossary, 1957). Marine swamp environments are grouped with the transitional environments.

MARSH - An area of saturated ground covered with various types of grasses and aquatic (MRSH) vegetation. Marshes develop around the shallow margins of bodies of water which are invaded by plant growth. They are not deep enough to form lakes. The development of marshes is more likely in areas of low to no tides. Plant growth is minimal in areas of large tidal range. The water in marshes can range from stagnant to slow moving. There appears to be little geomorphological difference between the marsh and the swamp. Some dictionaries indicate that only grasses and low plants grow in marshes but that trees as well as grasses grow in swamps. Water salinities can range from fresh to brackish to more than 35 parts per mil. (Adapted from Twenhofel, 1950 and AGI Glossary, 1957).

MASSIVE BEDDING (also thick-bedded) - Layers in sedimentary rocks of approximately (MASS) more than 3 feet thick. (Adapted from Weller, 1960).

MATRIX - In a rock, regardless of whether it is igneous, sedimentary, or metamorphic, (MTRX) in which certain grains are much larger than the others and the smaller grains constitute the bulk of the rock, the mass of the smaller grains is considered the matrix. The larger grains in effect are embedded in the matrix.

MATURE - Term applied to a clastic sediment with the following textural character- (MTRE) istics: 1. No clay; 2. Good sorting; 3. Grain subangularity. (Folk, 1951).

METAMORPHIC - Rocks which have been altered in the solid state (without develop- (META) ment of a silicate melt phase) under conditions of high heat and pressure e.g. temperatures of 100 to 600° C at pressures up to 3000 atmospheres at depths of 3-20 kilometers. (Williams et al 1954, p. 161).

MICRITE - Consolidated or unconsolidated carbonate ooze or mud of either chemical (MICR) or mechanical origin. For practical purposes micritic material consists of particles less than approximately 0.031 mm in diameter. (Leighton and Pendexter, modified from Folk, 1959).

MICRITIC - Carbonate rock containing large amounts of micrite (See definition). (MICR)

MICROCRYSTALLINE - A term applied to a rock in which the individual crystals can (MICX) only be distinguished under the microscope. (AGI Glossary, 1957).

MICROCRYSTALLINE CARBONATE - Carbonate particles less than 0.06 mm that cannot (MICX) be recognized as clastic. (Plumley et al, 1962, p. 86).

MICROGRAINED CARBONATE - Distinct carbonate particles between 0.0039 mm and 0.06 mm (MGRN) (silt-size range) which can be interpreted as clastic grains. (Plumley et al, 1962, p. 86).

MICROSLICKENSIDED - Term applied to rocks which contain surfaces with polish and (MSLK) striations which can only be resolved with a microscope. These are frequently observed in cuttings.

MIDDLE NERITIC ENVIRONMENT - Environment of a marine zone on the sea floor between (MNER) approximately 66 to 328 feet of water.

MONOMINERALIC - A rock consisting essentially of one mineral; the amounts of other (MØNM) minerals tolerated under the definition vary with the authors. (AGI Glossary, 1957).

MOSAIC - Textural term referring to an intergrown network of more or less equi- (MSAC) granular crystals as in granite, marble, or recrystallized limestone.

MOTTLED - A sediment consisting of many small irregular bodies of material in a (MØTT) matrix of different texture; difference in color is not essential. (AGI Glossary, 1957).

MUDSTONE - "Mud"-supported sedimentary rocks with less than 10% grains and more than (MDST) 90% mud (Adapted from Dunham, 1962, p. 117). According to Dunham, particles less than 0.02 mm in diameter (medium silt and clay) are mud; particles larger than 0.02 mm (coarser than medium silt) are grains. Muddy sediments are generally considered to have been deposited in quieter water environments.

MUD SUPPORTED - Term to describe rocks which consist dominantly of particles of (MDSP) clay and very fine to medium silt (mud). Most or all of the mud particles are in contact with one another, the lowermost particles supporting the uppermost particles. These rocks are "mud-supported".

N

NODULE - A general term for rounded concretionary bodies, which can be separated as (NØDL) discrete masses from the formation in which they occur. (AGI Glossary 1957).

NON-CARBONATE FLOATS - Detrital material of non-carbonate nature such as quartz, (NCBF) feldspar, or volcanic glass, etc.; grains which appear to be embedded (floating) in a carbonate matrix or mosaic.

NON-SUPPORTING PHASE - The constituents which do not support the rock. They (NSPP) form a discontinuous phase. The removal of these constituents should not affect the rock texture, e.g. non-carbonate float.

NORMAL MARINE - Open, unrestricted marine conditions with water salinity ranging (NMAR) between 32 and 35 parts per mil.

O

OOLITE - A spherical to ellipsoidal accretionary body which may or may not have (ØØLT) a nucleus, but has concentric or radial structure or both. Generally less than 2.00 mm in diameter. (Leighton and Pendexter, 1962, p. 60).

OOLITOIDS - Accretionary particles (uppersize limit = 2 mm) of irregular shape, (ØØID) but of rounded outline and concentric internal structure.

OOMOLDIC POROSITY - Volume of void spaces which was formed by solution and (ØØMD) removal of oolites in the rock.

ORGANIC REEF - (See reef, reef complex). (ØRRF)

OUTER NERITIC ENVIRONMENT - Environment of marine zone on the sea floor occurring (ØNER) between approximately 328 feet and 656 feet of water.

OVERGROWTH - Secondary material deposited in optical continuity with a crystal (ØVGW) grain; common in some sedimentary rocks. (AGI Glossary, 1957).

P

PACKING - The spacing and density pattern of the grains in a rock. (Adapted (PACK) from AGI Glossary, 1957).

PACKSTONE - Grain-supported, muddy sedimentary rocks (Adapted from Dunham, 1962, (PCKS) p. 117). According to Dunham, particles less than 0.02 mm in size (medium silt and clay) are mud; particles larger than 0.02 mm (coarser than medium silt) are grains.

PARTLY RESTRICTED MARINE CIRCULATION ENVIRONMENT - Environment of an area whose (PRMC) connection with the open marine environment is partly closed off, preventing complete access of open marine water into the area. This term is to be used when a more specific designation is not possible, e.g. marginal lagoon, inland sea, upper bay, etc.

PELLETIFEROUS - Rock containing pellets is pelletiferous. (See definition of (PELL) pellet).

PELLETS - Spherical or subspherical particles with distinct boundaries, generally (PELL) composed of very fine grained carbonate material, resembling oolites but possessing no comparable internal structures. (G.E. Thomas, 1962, p. 198).

PENECONTEMPORANEOUS DEFORMATION - Deformation of sediment believed to have (PEND) occurred almost concurrently with deposition or very shortly afterwards.

PENESALINE ENVIRONMENT - Environment intermediate between normal marine and (PSAL) saline (Sloss, 1953, p. 145); characterized by the deposition of evaporitic carbonates interbedded with anhydrite.

PINPOINT POROSITY - Specialized term for carbonate rock porosity made up of (PPPR) voids which are less than 0.060 mm in size.

PISOLITE - A grain type similar to an oolite and generally 2 or more milli- (PISØ) meters in diameter. Pisolites are less regular in form than oolites and commonly crenulated. (Leighton and Pendexter, 1962, p. 60).

PLATY SHAPE - Term used to describe the form of grains and/or rock fragments which (PLTY) have two large and one small dimension, e.g. as in mica flakes where the length and width are much greater than the thickness.

POIKILITIC - A textural term applied to rocks in which many grains of random (PØIK) orientation are completely enclosed within large optically continuous crystals of different composition, (Williams et al, 1954).

POLLEN AND SPORES - Pollen grains are one to several-celled male reproductive (PLSP) bodies of flowering plants. Seldom larger than 60 microns. Spores are single-celled reproductive bodies of such non-flowering plants as ferns. (Hoffmeister, 1960). Seldom larger than 60 microns.

POLYCRYSTALLINE - Assemblage of crystal grains of unspecified number, shape, size, (PØLX) orientation, or bonding that together form a solid body. (AGI Glossary, 1957).

POLYGONIZED - Fragmented monomineralic material, each fragment with its in-
(PØLG) dividual crystallographic orientation, resulting from breakage of what originally was a single larger grain or crystal. Breakage was accompanied by shifting and rotation of the fragments. This is not to be confused with undulatory extincting resulting from strain.

PORE - INTERSTICE - VOID - A space in rock or soil not occupied by solid mineral
(VØID) matter. (AGI Glossary, 1957).

PORE FILLING - Rock, mineral, organic, or biological material which fills what
(PØRF) appears to have previously been a pore. Applies to all rocks in the zone of weathering.

POROSITY - The ratio of the volume of void space in a rock to its total volume. (AGI Glossary, 1957).

POROSITY TYPE - The designation of the origin of the porosity, e.g.: Intercrystalline, Intergranular, Interoolitic, Oomoldic, Fracture, Vuggy-matrix, Vuggy skeletal, Pinpoint. (See definitions).

PSEUDOMORPH - A crystal, or apparent crystal, having the outward form proper to
(PSDM) another species of mineral, which it has replaced by substitution or by chemical alteration. (AGI Glossary, 1957).

PYROCLASTIC - A general term applied to detrital volcanic materials that have been
(PYRØ) explosively or aerially ejected from a volcanic vent. Also, a general term for the class of rocks made up of these materials. (AGI Glossary, 1957).

Q

QUARTZITE - A very hard rock consisting essentially of quartz with a tightly
(QTZT) intergrown texture and very low porosity and permeability values. There are both metamorphic and sedimentary varieties. Metamorphic quartzites originate at high temperatures and high pressures by solid state recrystallization of the quartz grains of pre-existing quartz-rich rock, e.g. a siltstone, sandstone, a quartzite, etc. Sedimentary quartzites originate by quartz grain enlargement at temperatures and pressures normal to the surface. Grain growth occurs at the expense of the pore space and can continue with a tightly interlocked texture as the end stage.

QUARTZOSE - A term describing a sediment containing quartz as the dominant
(QZSE) constituent. Not to be confused with quartzite.

R

RECRYSTALLIZED PRODUCT - New mineral grains formed in a rock while in the
(REXL) solid state. The new grains may have the same chemical and mineralogical composition as in the original rock, as when a fine-grained limestone composed of calcite recrystallizes to a coarse-grained marble composed of calcite. (AGI Glossary, 1957).

REEF (also organic reef) - Structure built and maintained by living marine
(REEF)
(ØRRF) organisms. The structure has a framework strong enough to withstand the activity of ordinary wind waves. Modern reefs achieve this by a baffle-like structure on the windward side. Modern reef frame organisms are corals, coralline algae and hydrocorals. Modern reef frame-binding organisms are largely coralline algae, Bryozoa and Foraminifera. (Adapted from Shepard, 1963, p. 350). Because reefs can maintain themselves in the path of wind-directed waves, reefs have a modifying effect on their surroundings (the areas in front, behind, on the flanks, and their uppermost limits) by redirecting, modifying and dampening the currents and waves. This is reflected in the many kinds of sediments ranging from mud to talus blocks found in different areas of the reef. The entire complex which consists of the area of the fore reef, reef core, reef island, reef flat, back reef, and reef lagoon is called the reef complex.

REEF COMPLEX (also reef and organic reef) - Includes the reef core (the living
(RCØR) part of an organic reef) and all of the adjacent and nearby area where the sediment type and the chemical, sedimentational, and biological processes are strongly influenced by the location of the reef core. The actual reef core may make up less than 10% of the entire volume of carbonate rock of a reef complex.

REEF CORE - The only part of the reef complex where the activities of living
(RCØR) frame-building and frame-binding organisms results in the development and enlargement of a structure capable of withstanding the destructive activity of ordinary wind waves. The reef core may make up less than 10% of an entire reef complex.

REEF ENVIRONMENT - Environment (see definition) of organic reefs. Organic reefs
(RUND) are structures built by marine organisms. The structure has a framework strong enough to withstand the activity of ordinary wind waves. Modern reefs achieve this by means of a baffle-like structure on its windward side. Modern reef frame organisms are corals, coralline algae and hydrocorals. Modern reef frame-binding organisms are largely coralline algae, Bryozoa, and Foraminifera. Because reefs can maintain themselves in the path of wind-driven waves, the reef can modify the environments in front, behind, and on its sides (flanks or lateral ends) by redirecting, modifying, and dampening the currents. This is reflected in the many kinds of sediment, ranging from muds to talus blocks, and in the many kinds of plant and animal life.

REEF FLAT - That part of an inorganic reef consisting of the flat area between
(RFLT) the seaward most actively growing zone and the beach (if a beach and/or island is developed). The reef flat consists largely of coral rock and sand with some patches of growing coral. At low tide part of the reef flat is exposed.

REEF LAGOON - Basically of two types: 1. Lagoons behind barrier reefs and
(RLAG) 2. Lagoons surrounded by atolls.

REEF TALUS - The steeply dipping slope on the windward side of atolls or table
(RTAL) reefs consisting of coarse rubble made up of wave-broken debris of the living reef. Talus exists on the seaward side of the fringing reefs and barrier reefs.

REEF, UNDIFFERENTIATED - When rock is suspected to have been sediment which
(RUND) originated within the reef complex (see definition), but more specific identification is not possible.

REPLACEMENT MATERIAL - New mineral material formed by a process of what is
(REPL) probably simultaneous solution and deposition which occupies the volume of a pre-existing mineral. The new material is usually of a different type than the replaced mineral or minerals. (Adapted from AGI Glossary, 1957).

REPLACEMENT MATERIAL -FRAMEWORK - Selective replacement (see definition) by
(REPF) a new mineral species of framework constituents (see definition) only.

REPLACEMENT MATERIAL - MATRIX - Selective replacement (see definition) by a
(REPM) new mineral species of matrix constituents (see definition) only.

RESORBED - A term used in microscopic work to describe those phenocrysts which
(RSRB) after crystallization are partly fused again into the magma. (Rice, 1949).

REWORKED - Sediment which has been disturbed and moved after coming to rest.
(RWKD) Frequently results in removal of finer constituents with improvement of textural sorting and maturity.

RIPPLE-MARKED - An undulating surface sculpture produced in noncoherent gran-
(RIPP) ular materials by the wind, by currents of water and by the agitation of water in wave action. (AGI Glossary, 1957).

ROCK FRAGMENT - In general, any fragment of some pre-existing rock. In
(RFRG) sediments any granule or sand-sized pieces of igneous, pyroclastic, metamorphic, or sedimentary rocks, even though they may consist of as few as two mineral grains in aggregate.

ROUNDED - Measure of a grain whose original faces are almost entirely des-
(RØND) troyed. Some flat faces may remain. There may be broad curved areas between the remaining faces. All original edges and corners have been smoothed off to broad curves. Secondary corners are almost entirely worn away and are few in number (0-5). The original form is still recognizable.

S

SALINE ENVIRONMENT - Environment characterized by conditions of evaporation leading
(SALN) to the precipitation of major accumulations of halite and anhydrites. (Adapted from Sloss, 1953, p. 145).

SAND - Loose, sedimentary particles ranging in size from 0.062 to 2.0 mm.
(SAND)

SANDSTONE - A sedimentary rock consisting of particles in the size range 0.062
(SDST) to 2.0 mm; lithified sand. This term is frequently used to define sedimentary rocks in this size range regardless of the kind of constituent particles, e.g. rock fragments, calcareous oolites, shell fragments, quartz grains etc. However, in the Gulf lith log procedure, the carbonate rocks in this size range are analyzed in the separate expanded CARBONATE COMPONENTS SECTION.

SANDY - Sediment with a portion of its constituents in the sand size range
(SAND) (1/16 - 2 mm).

SEDIMENT - (In its simplest definition). Solid material, both mineral and or-
(SED-) ganic that is in suspension, is being transported or has been moved from its site of origin by air, water, or ice and has come to rest on the earth's surface either above or below sea level. (Supplement to AGI Glossary).

SEDIMENTARY ENVIRONMENT - (See environment).
(SENV)

SEDIMENTARY QUARTZITE - A tough, dense sedimentary rock consisting of an inter-
(SQTZ) locked texture of quartz grains. This texture resulted from either quartz grain interpenetration through pressure solution at points of grain contact or by precipitation of silica in lattice continuity on quartz grain nuclei. Both of the processes eventually result in very low porosity and permeability values.

SEDIMENTARY ROCK - A lithified sediment. Also called detrital or clastic. The
(SED-) constituent particles can consist of mineral grains, fragments of igneous, metamorphic, or sedimentary rock, particles from outer space, shell fragments, plant remains, other organic material, etc. (anything which had been available for deposition in the original sediments).

SHALE - Claystone or clayey siltstone which have the property of splitting along
(SHLE) closely parallel planes. This property is called fissility.

SHALY - A term applied to thinly bedded rocks which break up into thin layers
(SHLE) like shale (See definition). (Rice, 1949).

SHAPE - Three-dimensional form = spatial form. This is distinguished from roundness
(SHPE) or angularity which in the geological sense is two-dimensional and independent of shape.

SHELL FRAGMENTS - Broken, fragmented shell material which still is indicative
(SFRA) of the kind of animal which secreted it. The fragments are still large enough to retain sufficient structure to allow some sort of identification, regardless of how generalized, to be made. For examples, fragments which can be recognized as some kind of coral, sponge, brachiopod, pelecypod, etc.

SHELL HASH - Finely broken shell material. The smallness of the size of the
(HASH) fragments has approached the point where the observer recognizes that he is looking at some kind of shell material but he can make no identification beyond that point.

SHELL MATERIAL - Unbroken or broken fossils frequently found in sedimentary rocks.
(SFRA)

SHOWS - The occurrence of gas or oil encountered during drilling.
(SHØW)

SILICIFIED - The introduction of or replacement by silica of rock material. By
(SILC) some type of precipitation process the silica formed is fine-grained quartz, chalcedony, or opal and may fill up pores and replace existing minerals and organic remains. (Adapted from AGI Glossary, 1957).

SILT - Particles of recognizable size with range from 0.0039 to 0.062 mm (1/256 -
(SILT) 1/16).

SILTSTONE - A lithified sedimentary rock consisting of more than 51% of silt
(SLST) sized particles (size range 0.0039-0.062 mm). Sedimentary rock composed of dominantly carbonate particles in this size range are treated as carbonate rock in the procedures used in the Gulf lith log. All recorded data should be listed in the CARBONATE COMPONENT SECTION.

SILTY - A sediment which contains silt-sized constituents, is used as a modifier
(SILT) so long as the silt sized fraction does not constitute the predominant material.

SKELETAL GRAINS - Detrital fragments of recognizable biologic origin. More precise
(SFRA) identification is not possible.

SLATY - Term applied to fine-grained metamorphic rocks which have well-developed
(SLAT) fissility. This term is applied to metamorphic rocks only.

SLICKENSIDED - Term applied to rocks which have at least one surface which is
(SKSD) polished and striated (scratched) caused by friction along a fault plane. (Adapted from AGI Glossary, 1957).

SLUMPED - A term applied to a sediment which has structures which appear to
(SLMP) have resulted from deformation by sliding while the sediment was soft and plastic. The structures can be highly contorted as opposed to regular beds or layers or cross-beds.

SORTING - Term to describe the spread of grain sizes in a detrital rock. For
(SØRT) example a sandstone consisting only of ½ mm grains would be excellently sorted whereas a sediment consisting of materials from the lower limit of clay to gravel would be poorly sorted.

SPAR - Clear, transparent or translucent, readily cleavable, crystalline particles
(SPAR) of calcite and/or dolomite.

SPARRY - Mineral material with characteristics of spar (See definition).
(SPAR)

SPHERICAL SHAPE - Term used to describe sedimentary grains which approach a
(SPHE) ball in form. True spherical form is very rarely approached in sediments.

SPHERULITE - (Leighton and Pendexter) Applied to minute bodies of oolitic nature
(SPHL) in which only a radial structure is visible. The surfaces of such bodies, unlike those of oolites are somewhat irregular.

SPORBO - Local name (California) for phosphatic oolites; derived from smooth-
(SPBØ) polished-round-black (blue, brown)-objects; an impure collophane. (Galliher, 1930).

STREAKED - A term applied to rock in which the same constituents appear to be
(STRK) drawn out giving a striped or streaked appearance. Can be applied to igneous, metamorphic, sedimentary rocks.

STRAINED - Modification of normal optical properties of individual crystals,
(STRN) such as quartz, due to strain. The strain is characterized by undulatory extinction. (Adapted from Rice, 1949).

STYLOLITE - A structure of sedimentary rocks, predominantly in limestones and
(STYØ) marbles, rare in sandstones. It can best be described as a suture-like seam marked by interlocking and mutual penetration of the rock material on both sides of the seam. The amplitude of the irregularities of the interlocking areas of the stylolite can range from less than one millimeter to as much as 200 millimeters. (Adapted from Pettijohn, 1957).

SUBANGULAR - Measure of a grain showing definite effects of wear. The fragments
(SANG) still have their original form and the faces show little wear, but the edges and corners show some rounding. Secondary corners are still numerous and can number from 10 to 20.

SUBHEDRAL CRYSTALLINITY - Individual minerals with crystallinity developed between
(SHED) anhedral and euhedral. (AGI Glossary, 1957).

SUBMATURE - Term applied to a clastic sediment with the following textural
(SMTR) characteristics: 1. Little to no clay; 2. Non-clay constituents are poorly sorted; 3. Grain angularity. (Folk, 1951).

SUBROUNDED - Measure of grain showing considerable wear. The edges and corners
(SRND) are rounded off to smooth curves and the area of the original faces is reduced. However, the original shape of the grain is still distinct. The secondary corners are much rounded, but now only number 5-10.

SUPERMATURE - Term applied to a clastic sediment with the following textural
(SPRM) characteristics: 1. No clay; 2. Well sorted; 3. Grains are rounded. (Folk, 1951).

SUPPORTING PHASE = FRAMEWORK - The constituents which internally provide support
(FRWK) for the rock. They form a continuous phase, are in mutual contact and constitute the rock texture. Their removal would cause complete disintegration of the rock texture.

SWAMP - An area of saturated ground supporting grasses, trees and shrubs. Swamps
(SWMP) develop in shallow depressions that are not deep enough to form lakes around the shallow margins of bodies of water invaded by plant growth. The development of swamps is more likely in areas of low to no tides; though swamps may well develop hundreds of miles inland from the sea, in areas where the land surface and the water table coincide. Plant growth is minimal in areas of high tides. The water in the swamp ranges from stagnant to very slow moving. There appears to be little geomorphological difference between the swamps and the marsh. Some dictionaries indicate that trees and shrubs grow in the swamp but not in the marsh. Swamp water salinities can range from fresh to brackish to salinities exceeding 35 parts per mil.

T

TEXTURE - The size, shape, and order (packing and fabric) of the component
(TEXT) particles of a sedimentary rock. (Pettijohn, 1957).

THIN-BEDDING - Layers in a sedimentary rock ranging from 3 to 10 cm in
(THBD) thickness (approximately 1 to 4 inches). (Ingram, 1964 after McKee and Weir, 1953).

TIDAL FLAT ENVIRONMENT - A specialized littoral environment along coasts char-
(TDFT) acterized by ineffective surf action, abundant supply of clay and silt by the rivers and high range between high and low tides. Deposition occurs on broad, flat, gently dipping surface. Flats can occur along open coasts, in bays, estuaries, and in lagoons.

TILL - Non-sorted, non-bedded sediment carried by and deposited by a glacier.
(TILL) (AGI Glossary, 1957).

TOUGH - Term applied to rock which is difficult to break.
(TØGH)

TRANSITIONAL ENVIRONMENT - The environment in those areas where the influences of
(TRAN) both the continental and marine conditions occur. These areas usually are found where the continental and marine environments are in contact, e.g. the littoral, marine swamp, delta, estuary, and marginal lagoon.

TWINNED - Crystals of a single species frequently grow in positions in which the
(TWIN) parallelism of the parts is incomplete; that is, some corresponding directions are exactly parallel and others are not. Two crystals of the same kind, which form an aggregate exhibiting such partial parallelism are called a twin, or are said to be in a twinned position. (Winchell, 1949).

U

UNSTRAINED - Mineral material whose crystalline structure has not been deformed;
(USTR) their optical properties are normal. The opposite of strained.

UPPER BATHYAL ENVIRONMENT - Environment of a marine zone on the sea floor
(UBTH) between approximately 656 to 1640 feet of water.

UPPER BAY ENVIRONMENT - Environment of shallow upper reaches of bay (See
(UBAY) definition). Water can range from almost fresh to brackish near river mouths, to more than 35 parts per mil.

V

VISUAL POROSITY - Void space which can be recognized, measured and recorded (VISL) under low power magnification. (Adapted from R.W. Powers, 1962, p. 139).

VOID SPACES (also called pores, interstices) - Openings in rock, formed (VØID) between grains, fragments, fossils etc.

VOLCANIC - Igneous origin at surface of the earth, e.g. from volcanoes as (VØLC) ash, bombs etc. and from fissures or vents such as lava flows etc.

VOLCANIC ASH - A deposit consisting of uncemented particles produced during (VASH) explosive volcanic eruptions (one of the pyroclastic rocks). Limited to fragments less than 4 mm in size. Volcanic ash consists mostly of glass fragments, but also contains mineral crystals and rock fragments also produced during the eruptions. Coarse ash ranges in size from 1/4 to 4 mm; fine ash is finer-grained than 1/4 mm.

VOLCANIC TUFF - Lithified rock which originally was deposited as a volcanic (VTFF) ash. Its constituents were originally produced during explosive volcanic eruptions. The size of the fragments is less than 4 mm. Glass is the dominant constituent but large amounts of rock fragments and crystals also produced during the eruptions can be present. Deposition can be on land or in water.

VUG - Term reserved for carbonate rocks; describes pore (void, opening) of (VUG-) either solution, replacement, or detrital origin larger than 0.06 mm.

VUGGY - SKELETAL POROSITY - Pore space in the interior of a skeletal particle (VUG-) or fossil. Pore can be original cavity or formed as the result of solution.

VUG-MATRIX - Term reserved for carbonate rock which describes an opening (VUGM) (pore or void) in the matrix portion of the rock.

W

WACKESTONE - "Mud" supported sedimentary rocks with more than 10% grains. There (WCKS) is no grain-support, no grain to grain contact. (Adapted from Dunham, 1962, p. 117). According to Dunham, particles less than 0.02 mm in size (medium silt and clay) are mud; particles larger than 0.02 mm (coarser than medium silt) are grains.

WEATHERED - Term applied to rock which has been biologically, chemically and (WTRD) mechanically processed at the low temperature and pressure normal to and within the surface of the earth.

WELL-ROUNDED - Measure of a grain which retains none of the original faces, (WRND) edges or corners. The entire surface consists of broad curves; flat areas are absent. No secondary corners are present. The original form is less distinct than in any of the other angularity designations.

Z

ZONED - In mineral deposits, the occurrence of successive minerals or elements (ZØND) outward from a common center. (AGI Glossary, 1957). In a single mineral grain, the change in chemical composition of the grain from its center outward as caused by flucuations in the composition of the silicate melt (magma or lava) from which the grain crystallized.

Appendix E

Method for Computer Processing of Sample Data

LITHOLOGIC DATA SHEET — CUTTINGS SAMPLES

Description: A single page that provides for recording both Lithologic Data or any remarks the analyst feels necessary

LITHOLOGIC DATA (BA CARD)

Column	Descriptions
1–7	Reference number: unique well identification number.
8–17	Interval: depth or interval, in feet, for which the information that follows applies. The interval may be of any length desired: (a) routine, repetitive sample intervals (i.e. 5, 10, 20, 40, or 50 ft, etc.). Example for 10 ft intervals:

SAMPLE INTERVAL	
FROM (8–12)	TO (13–17)
17510	17520
17520	17530
17530	17540

(b) irregular sample intervals as determined by lithology changes or any other variable factors.

SAMPLE INTERVAL	
FROM (8–12)	TO (13–17)
17510	17521
17521	17559
17559	17603

18–19 Blank columns

20–21 Interval sequence number: a number (01–10) given each separate lithology in the sample in order to differentiate one from another. Example: a sample from 10,510 ft to 10,525 ft contains three distinct lithologies. The use of the Interval Sequence Number is optional and at the discretion of the encoder. It can be used erratically as needed or not at all.

SAMPLE INTERVAL			INTERVAL SEQUENCE NUMBER	PERCENTAGE OF TOTAL SAMPLE
FROM (8–12)	TO (13–17)	18 19	20 21	22 23
10510	10525		1	30
10510	10525		2	40
10510	10525		3	30

22–23 Percentage of total sample: the estimated volume percentage of the total sample represented by the particular lithology being described. If 100% enter 99.

24–29 Color: detailed below.

24 Color distribution: according to the following codes: E – Even, K – Streaked, S – Spotted, V – Variegated, P – Speckled, M – Mottled, L – Laminated, O – Other.

25 Color shade modifier according to the following codes: L – Light, M – Medium, D – Dark

26–27 Color modifier according to the following codes: BU – Bluish, GN – Greenish, YL – Yellowish, BF – Tannish, OR – Orangeish, BN – Brownish, RD – Reddish, PK – Pinkish, PP – Purplish, WH – Whitish, GA – Grayish, BK – Blackish

28–29 Color according to the following codes: BU – Blue, GB – Gray/Blue, GN – Green, GY – Green/Yellow, OL – Olive, YL – Yellow, OR – Orange, OB – Orange/Brown, BF – Buff or tan, YR – Yellow/Red, YB – Yellow/Brown, BN – Brown, RD – Red, PK – Pink, RP – Red Purple, PP – Purple, PB – Purple/Blue, WH – White, GA – Gray, BK – Black

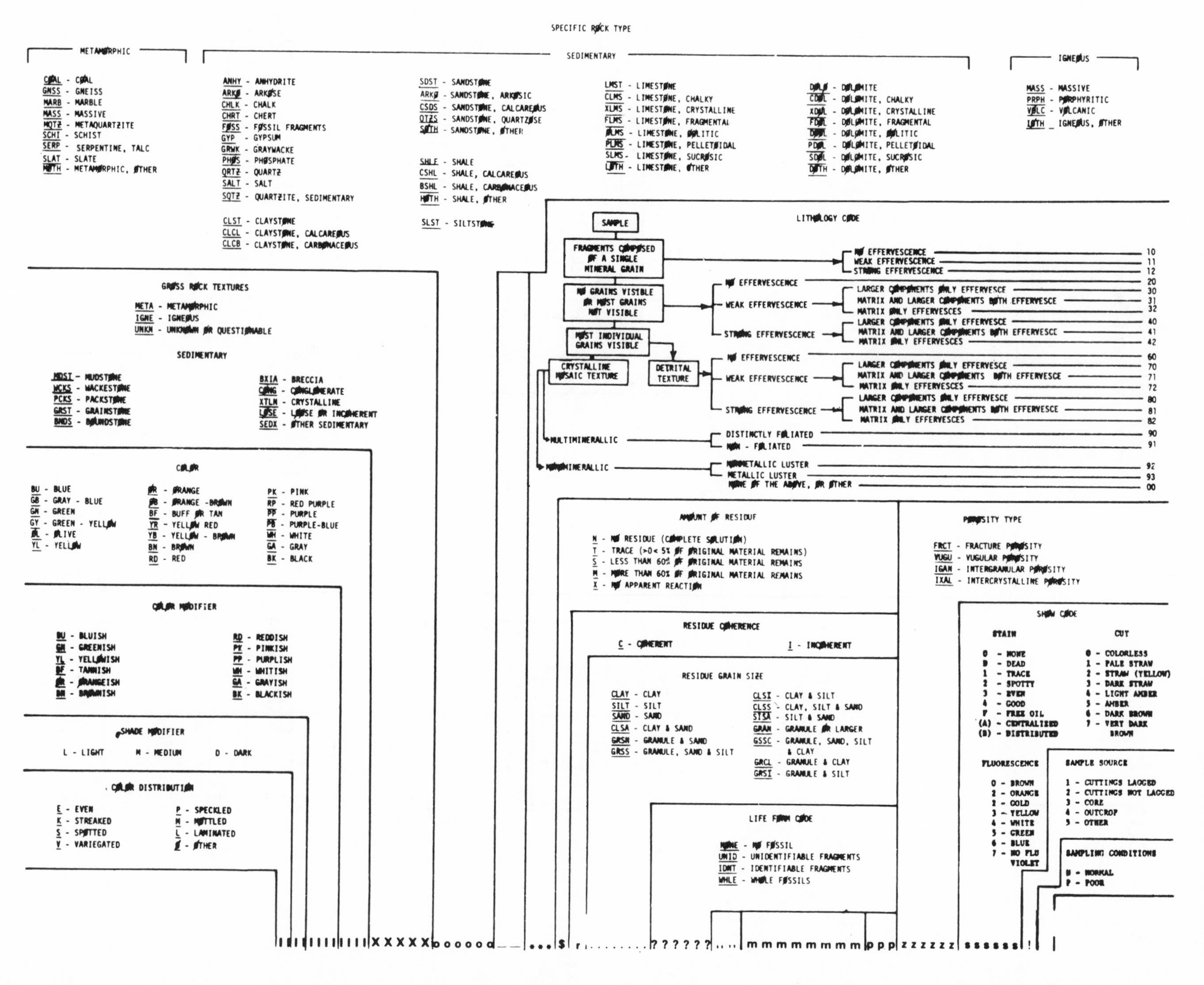
SPECIFIC ROCK TYPE
METAMORPHIC
SEDIMENTARY
IGNEOUS
COAL - COAL
GNSS - GNEISS
MARB - MARBLE
MASS - MASSIVE
MQTZ - METAQUARTZITE
SCHI - SCHIST
SERP - SERPENTINE, TALC
SLAT - SLATE
MOTH - METAMORPHIC, OTHER
ANHY - ANHYDRITE
ARKO - ARKOSE
CHLK - CHALK
CHRT - CHERT
FOSS - FOSSIL FRAGMENTS
GYP - GYPSUM
GRWK - GRAYWACKE
PHOS - PHOSPHATE
QRTZ - QUARTZ
SALT - SALT
SQTZ - QUARTZITE, SEDIMENTARY
CLST - CLAYSTONE
CLCL - CLAYSTONE, CALCAREOUS
CLCB - CLAYSTONE, CARBONACEOUS
SDST - SANDSTONE
ARKO - SANDSTONE, ARKOSIC
CSDS - SANDSTONE, CALCAREOUS
QTZS - SANDSTONE, QUARTZOSE
SOTH - SANDSTONE, OTHER
SHLE - SHALE
CSHL - SHALE, CALCAREOUS
BSHL - SHALE, CARBONACEOUS
HOTH - SHALE, OTHER
SLST - SILTSTONE
LMST - LIMESTONE
CLMS - LIMESTONE, CHALKY
XLMS - LIMESTONE, CRYSTALLINE
FLMS - LIMESTONE, FRAGMENTAL
OLMS - LIMESTONE, OOLITIC
PLMS - LIMESTONE, PELLETOIDAL
SLMS - LIMESTONE, SUCROSIC
LOTH - LIMESTONE, OTHER
DOLO - DOLOMITE
CDOL - DOLOMITE, CHALKY
XDOL - DOLOMITE, CRYSTALLINE
FDOL - DOLOMITE, FRAGMENTAL
ODOL - DOLOMITE, OOLITIC
PDOL - DOLOMITE, PELLETOIDAL
SDOL - DOLOMITE, SUCROSIC
DOTH - DOLOMITE, OTHER
MASS - MASSIVE
PRPH - PORPHYRITIC
VOLC - VOLCANIC
IOTH - IGNEOUS, OTHER
GROSS ROCK TEXTURES
META - METAMORPHIC
IGNE - IGNEOUS
UNKN - UNKNOWN OR QUESTIONABLE
SEDIMENTARY
MDST - MUDSTONE
WCKS - WACKESTONE
PCKS - PACKSTONE
GRST - GRAINSTONE
BNDS - BOUNDSTONE
BXIA - BRECCIA
CONG - CONGLOMERATE
XTLN - CRYSTALLINE
LOSE - LOOSE OR INCOHERENT
SEDX - OTHER SEDIMENTARY
LITHOLOGY CODE
SAMPLE
FRAGMENTS COMPOSED OF A SINGLE MINERAL GRAIN
NO GRAINS VISIBLE OR MOST GRAINS NOT VISIBLE
MOST INDIVIDUAL GRAINS VISIBLE
CRYSTALLINE MOSAIC TEXTURE
DETRITAL TEXTURE
NO EFFERVESCENCE
WEAK EFFERVESCENCE
STRONG EFFERVESCENCE
LARGER COMPONENTS ONLY EFFERVESCE
MATRIX AND LARGER COMPONENTS BOTH EFFERVESCE
MATRIX ONLY EFFERVESCES
MULTIMINERALLIC
MONOMINERALLIC
DISTINCTLY FOLIATED
NON - FOLIATED
NONMETALLIC LUSTER
METALLIC LUSTER
NONE OF THE ABOVE, OR OTHER
10
11
12
20
30
31
32
40
41
42
60
70
71
72
80
81
82
90
91
92
93
00
COLOR
BU - BLUE
GB - GRAY - BLUE
GN - GREEN
GY - GREEN - YELLOW
OL - OLIVE
YL - YELLOW
OR - ORANGE
OB - ORANGE - BROWN
BF - BUFF OR TAN
YR - YELLOW RED
YB - YELLOW - BROWN
BN - BROWN
RD - RED
PK - PINK
RP - RED PURPLE
PP - PURPLE
PB - PURPLE-BLUE
WH - WHITE
GA - GRAY
BK - BLACK
COLOR MODIFIER
BU - BLUISH
GN - GREENISH
YL - YELLOWISH
BF - TANNISH
OR - ORANGEISH
BN - BROWNISH
RD - REDDISH
PK - PINKISH
PP - PURPLISH
WH - WHITISH
GA - GRAYISH
BK - BLACKISH
SHADE MODIFIER
L - LIGHT
M - MEDIUM
D - DARK
COLOR DISTRIBUTION
E - EVEN
K - STREAKED
S - SPOTTED
V - VARIEGATED
P - SPECKLED
M - MOTTLED
L - LAMINATED
O - OTHER
AMOUNT OF RESIDUE
N - NO RESIDUE (COMPLETE SOLUTION)
T - TRACE (>0< 5% OF ORIGINAL MATERIAL REMAINS)
S - LESS THAN 60% OF ORIGINAL MATERIAL REMAINS
M - MORE THAN 60% OF ORIGINAL MATERIAL REMAINS
X - NO APPARENT REACTION
POROSITY TYPE
FRCT - FRACTURE POROSITY
VUGU - VUGULAR POROSITY
IGAN - INTERGRANULAR POROSITY
IXAL - INTERCRYSTALLINE POROSITY
RESIDUE COHERENCE
C - COHERENT
I - INCOHERENT
RESIDUE GRAIN SIZE
CLAY - CLAY
SILT - SILT
SAND - SAND
CLSA - CLAY & SAND
GRSN - GRANULE & SAND
GRSS - GRANULE, SAND & SILT
CLSI - CLAY & SILT
CLSS - CLAY, SILT & SAND
STSA - SILT & SAND
GRAN - GRANULE OR LARGER
GSSC - GRANULE, SAND, SILT & CLAY
GRCL - GRANULE & CLAY
GRSI - GRANULE & SILT
LIFE FORM CODE
NONE - NO FOSSIL
UNID - UNIDENTIFIABLE FRAGMENTS
IDNT - IDENTIFIABLE FRAGMENTS
WHLE - WHOLE FOSSILS
SHOW CODE
STAIN
0 - NONE
D - DEAD
1 - TRACE
2 - SPOTTY
3 - EVEN
4 - GOOD
F - FREE OIL
(A) - CENTRALIZED
(B) - DISTRIBUTED
CUT
0 - COLORLESS
1 - PALE STRAW
2 - STRAW (YELLOW)
3 - DARK STRAW
4 - LIGHT AMBER
5 - AMBER
6 - DARK BROWN
7 - VERY DARK BROWN
FLUORESCENCE
0 - BROWN
1 - ORANGE
2 - GOLD
3 - YELLOW
4 - WHITE
5 - GREEN
6 - BLUE
7 - NO FLU VIOLET
SAMPLE SOURCE
1 - CUTTINGS LAGGED
2 - CUTTINGS NOT LAGGED
3 - CORE
4 - OUTCROP
5 - OTHER
SAMPLING CONDITIONS
N - NORMAL
P - POOR
IIIIIIIIIIIXXXXXooooo — — ...$?????... mmmmmmmm ppp zzzzzz ssssss !

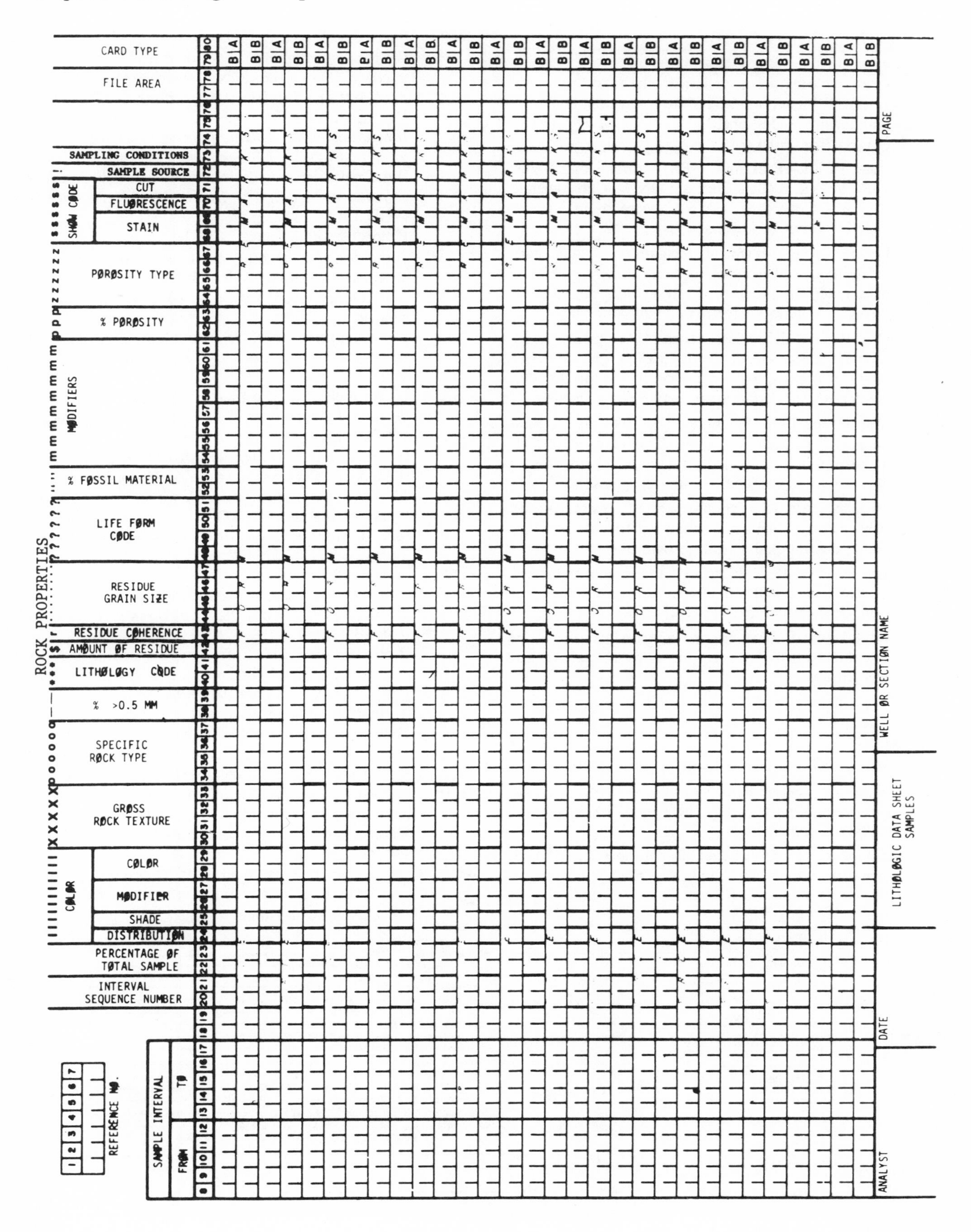
CARD TYPE
FILE AREA
SAMPLING CONDITIONS
SAMPLE SOURCE
SHOW CØDE
CUT
FLUØRESCENCE
STAIN
PØRØSITY TYPE
% PØRØSITY
MØDIFIERS
% FØSSIL MATERIAL
LIFE FØRM CØDE
RESIDUE GRAIN SIZE
RESIDUE CØHERENCE
AMØUNT ØF RESIDUE
LITHØLØGY CØDE
% >0.5 MM
SPECIFIC RØCK TYPE
GRØSS RØCK TEXTURE
CØLØR
CØLØR
MØDIFIER
SHADE
DISTRIBUTIØN
PERCENTAGE ØF TØTAL SAMPLE
INTERVAL SEQUENCE NUMBER
ROCK PROPERTIES
REFERENCE NØ.
SAMPLE INTERVAL
FRØM
TØ
PAGE
WELL ØR SECTIØN NAME
LITHØLØGIC DATA SHEET SAMPLES
DATE
ANALYST

LITHOLOGIC DATA (BA CARD)

Column	Descriptions
30–33	Gross rock texture according to the following codes: META – Metamorphic, IGNE – Igneous, UNKN – Unknown or Questionable SEDIMENTARY: MDST – Mudstone, WCKS – Wackestone, PCKS – Packstone, GRST – Grainstone, BNDS – Boundstone, BXIA – Breccia, CONG – Conglomerate, XTLN – Crystalline, LOSE – Loose or incoherent, SEDX – Other Sedimentary Decide if the rock is igneous, metamorphic, sedimentary, indeterminate. If igenous, metamorphic, or indeterminate, enter the proper code. If sedimentary, use the following criteria for further subdivision. Mudstone (MDST): less than 10% grains, mud-supported, contains mud (particles of clay and fine siltstone), original components not bound together during deposition, depositional texture recognizable. Wackestone (WCKS): more than 10% grains, mud-supported, contains mud (particles of clay and fine siltstone), original components not bound together during deposition, depositional texture recognizable. Packstone (PCKS): grain-supported, contains mud, original components not bound together during deposition, depositional texture recognizable. Grainstone (GRST): lacks mud and is grain-supported, original components not bound together during deposition, depositional texture recognizable. Boundstone (BNDS): original components were bound together during deposition, depositional texture recognizable. Crystalline (XTLN): Depositional texture not recognizable. If the cuttings are loose individual mineral grains or fossils, then use LOSE (Loose or incoherent). If the rock texture is recognizable as Breccia or Conglomerate from the cutting then use BXIA (Breccia) or CONG (Conglomerate). If none of the above adequately describes the sedimentary rock texture then use SEDX (Other Sedimentary).
34–37	Specific rock type according to the following codes: If the rock is metamorphic, decide if it is best described as *coal* (COAL), *gneiss* (GNSS), *marble* (MARB), *quartzite* (MQTZ), *schist* (SCHI), *serpentine, talc* (SERP), or *slate* (SLAT). If not, use *massive metamorphic* (MASS) or *other metamorphic* (MOTH). If the rock is igneous, decide if it is adequately described as *massive* (MASS), *porphyritic* (PRPH), or *volcanic* (VOLC). If not, use *other igneous* (IOTH). If the rock is sedimentary and a sandstone, then decide whether it is adequately described by *arkosic sandstone* (ARKO), *quartzose sandstone* (QTZS), or *calcareous sandstone* (CSDS). If not, use *sandstone* (SDST) for a typical-looking sandstone or *other sandstone* (SOTH) for an exotic or rare sandstone type. If the rock is a limestone or dolomite, then decide whether it is adequately described by *chalky* (CLMS or CDOL), *crystalline* (XLMS or XDOL), *fragmental* (FLMS or FDOL), *oolitic* (OLMS or OOLI), *pelletoidal* (PLMS or PDOL), or *sucrosic* (SLMS or SDOL). If not, use *limestone* (LMST) for a typical-looking limestone, *dolomite* (DOLO) for a typical-looking dolomite, or *other limestone* (LOTH), for an exotic or rare limestone type or *other dolomite* (DOTH) for an exotic or rare dolomite type.
34–37	If the rock is a shale, then decide whether it is adequately described by *calcareous shale* (CSHL) or *carbonaceous shale* (BSHL). If not, use *shale* (SHLE) for a typical-looking shale or *other shale* (HOTH) for an exotic or rare shale type. If the rock is a siltstone, use *siltstone* (SLST) or any of the other common sedimentary rock types listed.
38–39	Percentage greater than 0.5% mm: the estimated volumetric percentage of grains (both fossil and mineral) greater than 0.5mm in diameter.
40–41	Lithology code: a two-column numerical designation of the lithology as determined from the flow chart. The decisions are based on a complete 10% cold, HCl acid test.

Example: 1. Calcite cemented quartz sandstone cuttings = 82
2. Medium grained granite fragments = 91
3. Dolomitic claystone = 32

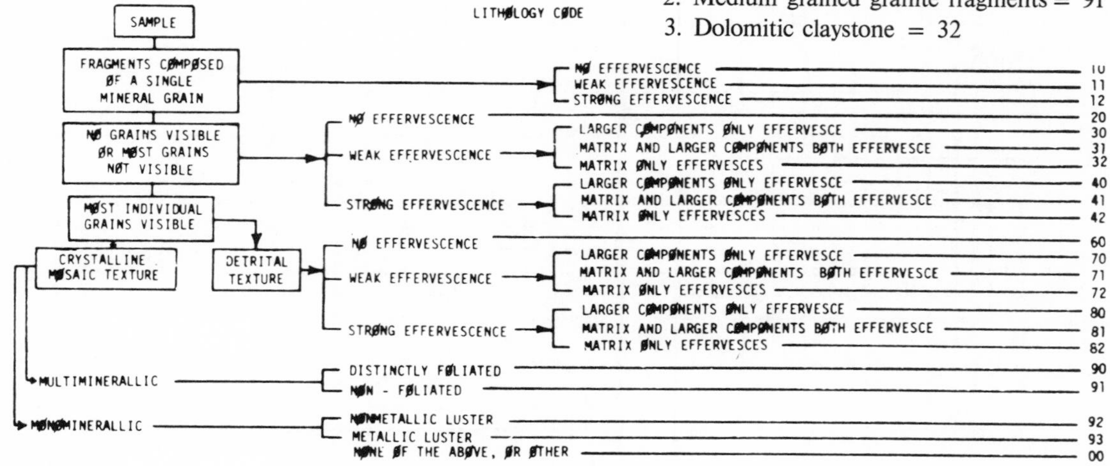

Column	Descriptions
42	Amount of residue: subjective estimate of the amount of the sample remaining after a complete cold, 10% HCl acid test, according to the following code: N – No residue (complete solution), T – Trace (>0<5% of original material remains), S – Less than 60% of the original material remains, M – More than 60% of original material remains, X – No apparent reaction.
43	Residue coherence: subjective assessment of the degree of lithification after the cold, 10% HCl acid treatment above, according to the following code: C – Coherent, I – Incoherent.
44–47	*Residue grain size:* grain size distribution after the cold, 10% HCl acid treatment above. According to the following Wentworth divisions and combinations: clay (CLAY), silt (SILT), sand (SAND), clay and sand (CLSA), granule and sand (GRSN), granule, sand, and silt (GRSS), clay and silt (CLSI), clay, silt, and sand (CLSS), silt and sand (SISA), granule or larger (GRAN), granule, sand, silt, and clay (GSSC), granule and clay (GRCL), granule and silt (GRSI). Use a sand-size comparison chart to accurately gauge the division between clay, silt, and sand.
48–51	Life form code: according to the following code: NONE – No Fossil, UNID – Unidentifiable Fragments, IDNT – Identifiable Fragments, WHLE – Whole Fossils.
52–53	*Percentage of fossil material:* estimated volume percentage of fossil material present in this lithology expressed as an integer.
54–57	*Modifiers:* adjectives from the master list to aid in the detailed and more complete description of each lithology.
62–63	*Percent porosity:* estimated volumetric porosity of this particular lithology expressed as an integer.
64–67	*Porosity type:* porosity type according to the following code: FRCT – Fracture Porosity, VUGU – Vugular Porosity, IGAN – Integranular Porosity, IXAL – Intercrystalline Porosity.
68–69	*Stain:* according to the following codes: Staining for vugs or different lithology appearing in separate and distinct areas is indicated by an (A) before the number or letter. Distributed staining is indicated by a (B) before the number or letter (see Fig. 3-7).
70	Fluorescence according to the following codes: Brown – 0, orange – 1, gold – 2, yellow – 3, white – 4, green – 5, blue – 6, violet – 7.
71	Cut: Benzene, carbon tetrachloride, or other solvent cut according to the following code: colorless – 0, pale straw – 1, straw – 2, dark straw – 3, light amber – 4, amber – 5, dark brown – 6, very dark brown – 7.
72	Sample source: rock being described was recovered from: wellbore cuttings lagged – 1, wellbore cuttings not lagged – 2, wellbore core – 3, outcrop sample – 4, other – 5.
73	Sample conditions: Normal – N (good cuttings returns from bit) and Poor – P (poor circulation or air drilled hole, etc.).
77–78	File area: code assigned by the district or area office to subdivide the exploration area of interest.
79–80	Card type: BA type card (preprinted).

REMARKS SHEET (BB CARD)

Column	Description
1–7	Reference number: assigned by the corporation.
8–17	Interval: depth or interval in feet for which the information that follows applies.
18–76	Free form remarks: remarks, comments, additional descriptions.
77–78	File area: code assigned by the district or area office to subdivide the exploration area of interest.
79–80	Card type: BB type card (preprinted).

COMMENTS: LITHOLOGY DATA

1. It is suggested that each distinct lithology from a sample be subjected to a cold, 10% HCl acid text in a spot plate before the description of that sample is begun. Thus, when the geologist is ready to enter the acid test results, the reaction will be complete. The reaction must be complete for these data to be useful. To make 10% HCl, pour 1 ml of 37% HCl into 2.64 ml of water or, to make 2000 ml of 10% HCl, pour 550 ml of 37% HCl into 1450 ml of water.

2. A sample may be described more than once by merely entering it again and using additional codes if this is needed.to clarify the description.

3. Please print neatly to avoid keypunch errors and use the following: 1, 2, and 0 for numbers; i or I, Ƶ and Ø for letters.

4. All the information requested does not have to be entered for the computer programs to utilize this data, but the more information entered, the better the final results. Blank columns imply the test was not performed.

5. Samples do not have to be entered in sequential order.

MUD LOG DATA SHEET

Description: A supplementary page to the Lithology data sheet to record measurements made of the drilling fluid and cuttings. Drilling rate and "d" exponent are also included.

MUD LOG DATA CARD

Column	Descriptions
1–7	Reference number: assigned by the corporation through the regional, district, or area office, following Gulf Well Data instructions.

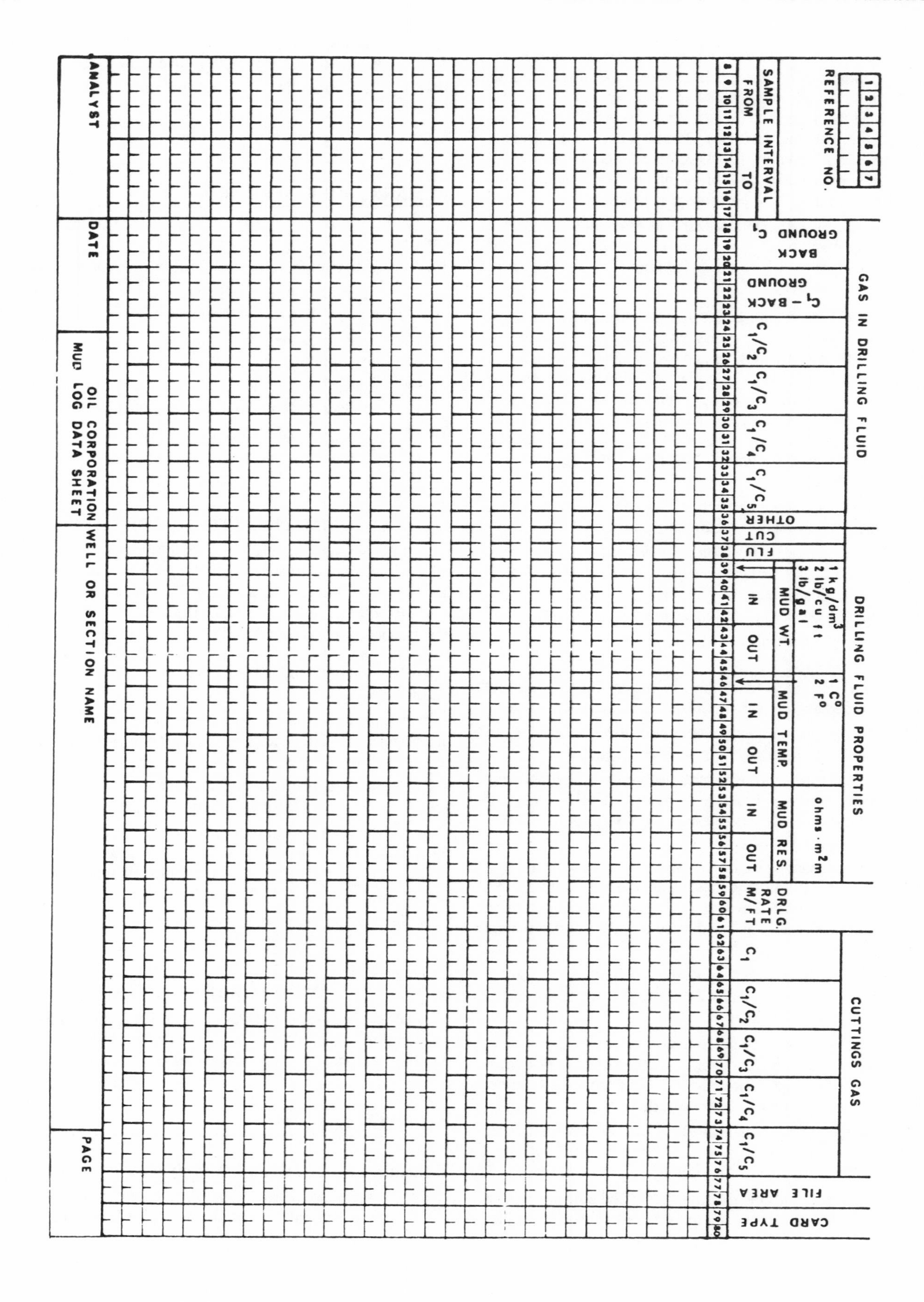

OIL CORPORATION
MUD LOG DATA SHEET
WELL OR SECTION NAME
ANALYST
DATE
PAGE
REFERENCE NO.
SAMPLE INTERVAL
FROM
TO
GAS IN DRILLING FLUID
BACK GROUND C_1
C_1 – BACK GROUND
C_1/C_2
C_1/C_3
C_1/C_4
C_1/C_5
OTHER
CUT
FLU
DRILLING FLUID PROPERTIES
1 kg/dm^3
2 lb/cu ft
3 lb/gal
MUD WT.
IN
OUT
1 C°
2 F°
MUD TEMP.
IN
OUT
ohms · m^{2}m
MUD RES.
IN
OUT
DRLG. RATE M/FT
CUTTINGS GAS
C_1
C_1/C_2
C_1/C_3
C_1/C_4
C_1/C_5
FILE AREA
CARD TYPE

OPERATOR ______________________

WELL ______________________

STATE ______________________
FIELD or DISTRICT ______________________
LOCATION lat. __________ long. __________
ELEVATION kb. ______ rt. ______ gl. ______
SPUDDED on __________ TD __________
DEPTH from __________ to __________
SCALE 1: __________ UNIT NO. __________
ENGINEERS ______________________

DRILLING LEGEND

NB	New Bit	DS	Deviation Survey
RRB	Rerun Bit	W/B	Weight on bit (1000 lb.)
DB	Diamond Bit	RPM	Rotation (Revol./min.)
TB	Turbo Drill	LC	Lost Circulation
CB	Core Bit	NR	No Returns
DCB	Diamond Core Bit	TG	Trip Gas

*   *

"d" EXPONENT

.6   1   2   3

*   *

DRILLING RATE
Min. Foot

Scale 3	20	40	100	200
Scale 2	10	20	50	100
Scale 1	5	10	25	50

DEPTH

LITHOLOGY

- o Conglomerate
- · Sand
- ·— Sandy Silt
- △ Chert
- Limestone
- Dolomite
- Anhydrite
- ^^ Gypsum
- Halite
- Volcanics
- Intrusives
- Schist
- — Siltstone
- Shale
- Clay

0%   100%

OIL

Oil in Mud

Oil in Cuttings

GOOD SHOW
SHOW
GOOD-CCl_4 CUT
TRACE-CCl_4 CUT

MUD ANALYSIS

PPM METHANE × 10^4

*   *

RATIOS

$\frac{C_1}{C_2}$ — 2 —
$\frac{C_1}{C_3}$ — 3 —
$\frac{C_1}{C_4}$ — 4 —
$\frac{C_1}{C_5}$ — 5 —

1 2 3 4 5 6 8 10 30 50

CUTTING ANALYSIS

CHROMATOGRAPH
or eqiv.

PPM Methane × 10^2

HOT WIRE
volt. difference

OTHER

Total Gas ——
Petr. Vapors ——

1 2 3 4 5

AUX.

MUD TEMP.
—X—X—

SHALE DENSITY
—O—O—

MUD WT.
—IN—
····OUT····

MUD VISC.
****

MUD RES.
– – – – – –

OTHER

DISCRIPTION & REMARKS

ENGINEERING LEGEND

Core No. 1
recovery 95%

Drill Stem Test No. 1

Mud loss

- ✧ Dry
- ● Water
- ● Oil
- ☼ Gas

MUD DATA

V	Viscosity
W	Weight in lb/Gal
WL	Filtrate in cc
FC	Filter Cake
Cl	Chloride cont. in ppm
Rm.	Mud Resistivity in Ω m
Rmf.	Mud Filtrate Resistivity in Ω m

8–17 Interval: depth or interval, in feet, for which the information that follows applies. The interval may be of any length desired. (a) routine, repetitive sample intervals (i.e. 5, 10, 20, 40, or 50 ft, etc.). See the example for 10 ft intervals.

SAMPLE INTERVAL									
FROM					TO				
8	9	10	11	12	13	14	15	16	17
1	7	5	1	0	1	7	5	2	0
1	7	5	2	0	1	7	5	3	0
1	7	5	3	0	1	7	5	4	0

(b) one sample per lithology.

SAMPLE INTERVAL									
FROM					TO				
8	9	10	11	12	13	14	15	16	17
1	7	5	1	0	1	7	5	2	1
1	7	5	2	1	1	7	5	5	9
1	7	5	5	9	1	7	6	0	3

18–20 Background methane gas carried in the mud from some source other than the contribution of the formation immediately penetrated. Units are in parts per million times 10^4 or can be read directly as percent. Maximum recording possible will be 50% methane or 500,000 ppm.

21–30 Total methane (C_1) measured in the drilling fluid less the background value recorded in columns 18–20 recorded in same units used in columns 18–20.

24–26 Ratio equal to value of C_1 recorded in columns 21–23 divided by the ppm of ethane (C_2) extracted from the drilling fluid.

27–29 Ratio equal to value of C_1 recorded in columns 21–23 divided by the ppm of propane (C_3) extracted from the drilling fluid.

30–32 Ratio equal to value of C_1 recorded in columns 21–23 divided by the ppm of butane (C_4) extracted from the drilling fluid.

33–35 Ratio equal to the value of C_1 recorded in columns 21–23 divided by the ppm of pentane (C_5) extracted from the drilling fluid.

36 Percentage of free gas which is not common but is of interest, i.e. helium. Reported to the closest 10%. (notation 1 = 10%).

37 Cut: Benzine, carbon tetrachloride, trichlorethane or other solvent cut according to the following code: colorless – 0, pale straw – 1, straw (yellow) – 2, dark straw – 3, light amber – 4, amber – 5, dark brown – 6, very dark brown – 7. When no description is made, leave blank.

38 Fluorescence of the mud using numbers 0 through 6 to describe colors, according to the following code: brown – 0, orange – 1, gold – 2, yellow – 3, white – 4, green – 5, blue – 6, violet – 7 (no fluorescence). When no description is made, leave blank.

39 Mud weight: 1 = gms/cc or k_g/dm^3; decimal point implied between columns 40 and 41. 2 = pounds per cubic foot; digitize only whole numbers. 3 = pounds per gallon; decimal point implied between columns 44 and 45.

40–45 Mud weight in and out will have meaning for many reasons including gas cutting the mud.

46–52 Mud temperature units can be reported in centigrade by entering 1 in column 46 or Fahrenheit 2 in column 46.

53–58 Mud resistivity in and out is reported with a decimal inferred between 54 and 55, and 56 and 57.

59–61 Drilling rate: Minutes per foot would be the standard scale of penetration rate. Minutes per meter or feet per hour might be converted to minutes per foot.

62–76 Gas extracted from the cuttings after agitation in a blender, using a constant amount of cuttings and water for a constant blender time. Gas extracted in this manner would be recorded using the same technique described for recording the drilling fluid gas. Recording background methane is omtited.

77–78 File area: code assigned by the District or Area office to subdivide the exploration area of interest.

79–80 Card type: type card preprinted for inclusion in the total data gathering system.

Permission to publish by the Canadian Well Logging Society.

Optr.
Well
Loc.
Sheet No.

AUX.

CUTTINGS GAS
PPM × 10^2
1 2 3 4 5
OTHER

MUD GAS (METHANE) PPM × 10^4
1 2 3 4 5 6 8 10 30 50
10 50 100 500
RATIOS
$C^1/C^2 = 2$ & $C^1/C^3 = 3$

OIL

LITHOLOGY
Cuttings %

DEPTH

DRILLING RATE Min. Foot
Scale 3 20 40 100 200
Scale 2 10 20 50 100
Scale 1 5 10 25 50
"d" EXPONENT .6 1 2 3

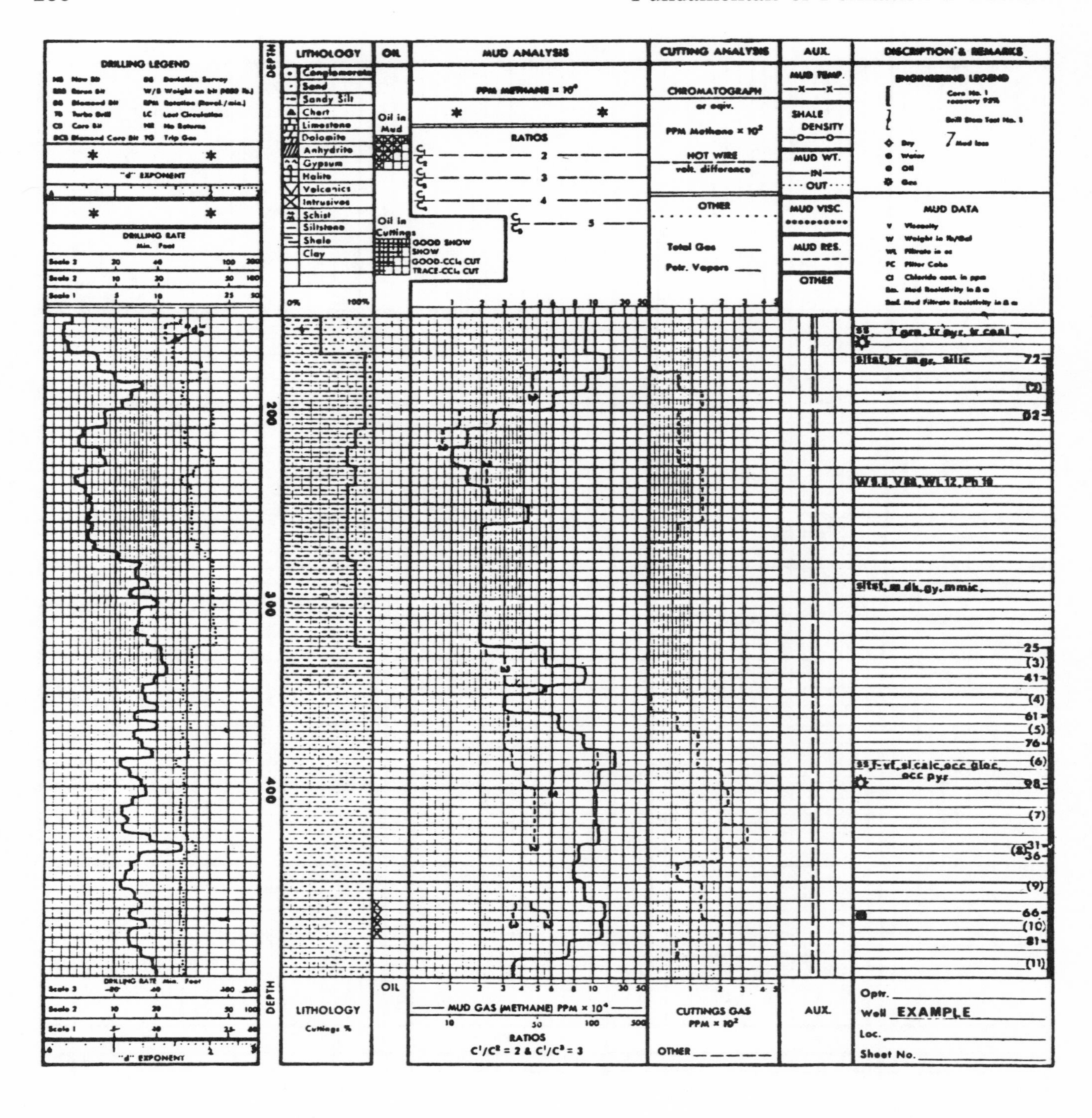
DRILLING LEGEND
"d" EXPONENT
DRILLING RATE
DEPTH
LITHOLOGY
Conglomerate
Sand
Sandy Silt
Chert
Limestone
Dolomite
Anhydrite
Gypsum
Halite
Volcanics
Intrusives
Schist
Siltstone
Shale
Clay
OIL
Oil in Mud
Oil in Cuttings
GOOD SHOW
SHOW
GOOD-CCl4 CUT
TRACE-CCl4 CUT
MUD ANALYSIS
RATIOS
CUTTING ANALYSIS
CHROMATOGRAPH
HOT WIRE
OTHER
Total Gas
Petr. Vapors
AUX.
MUD TEMP.
SHALE DENSITY
MUD WT.
IN
OUT
MUD VISC.
MUD RES.
OTHER
DESCRIPTION & REMARKS
ENGINEERING LEGEND
MUD DATA
200
300
400
LITHOLOGY
Cuttings %
MUD GAS (METHANE) PPM × 10⁴
RATIOS
C¹/C² = 2 & C¹/C³ = 3
CUTTINGS GAS
PPM × 10²
AUX.
Optr.
Well EXAMPLE
Loc.
Sheet No.

Appendix F

Letter and Computer Symbols for Well Logging and Formation Evaluation

Quantity	Letter Symbol	Computer Symbol (Operator Field / Quantity Symbol Field / Subscript Field)	Reserve SPE Letter Symbols	Dimensions[6]
acoustic velocity	v	VAC	V, u	L/t
acoustic velocity, apparent (measured)	v_a	VACA	V_a, u_a	L/t
acoustic velocity, fluid	v_f	VACF	V_f, u_f	L/t
acoustic velocity, matrix	v_{ma}	VACMA	V_{ma}, u_{ma}	L/t
acoustic velocity, shale	v_{sh}	VACSH	V_{sh}, u_{sh}	L/t
activity	a	ACT		
amplitude	A	AMP		various
amplitude, compressional wave	A_c	AMPC		various
amplitude, relative	A_r	AMPR		various
amplitude, shear wave	A_s	AMPS		various
angle	α alpha θ theta	ANG AGL	β beta γ gamma	
angle of dip	Θ theta$_{cap}$	ANGD	α_d alpha	
anisotropy coefficient	K_{ani}	COEANI	M_{ani}	
area	A	ARA	S	L^2
atomic number	Z	ANM		
atomic weight	A	AWT		m
attenuation coefficient	α alpha	COEA	M_a	1/L
azimuth of dip	Φ phi$_{cap}$	DAZ	β_d beta	
azimuth of reference on sonde	μ mu	RAZ	M mu$_{cap}$	
bearing, relative	β beta	BRGR	γ gamma	
bottom-hole pressure	p_{bh}	PRSBH	P_{BH}	m/Lt^2
bottom-hole temperature	T_{bh}	TEMBH	θ_{BH} theta	T
bubble-point (saturation) pressure	p_b	PRSB	p_s, P_b, P_s	m/Lt^2
bulk modulus	K	BKM	K_b	m/Lt^2
bulk volume	V_b	VOLB	v_b	L^3
capacitance	C	ECQ		q^2t^2/mL^2
capillary pressure	P_c	PRSCP	P_C, p_C	m/Lt^2
cementation (porosity) exponent	m	MXP		

[6]Dimensions: L = *length*, m = *mass*, q = *electrical charge*, t = *time*, T = *temperature*.

Quantity	Letter Symbol	Computer Symbol: Operator Field	Computer Symbol: Quantity Symbol Field	Computer Symbol: Subscript Field	Reserve SPE Letter Symbols	Dimensions[6]
charge	Q		CHG		q	q
coefficient, anisotropy	K_{ani}		COEANI		M_{ani}	
coefficient, attenuation	α alpha		COEA		M_{α}	$1/L$
coefficient, electrochemical	K_c		COEC		M_c, K_{ec}	mL^2/t^2q
coefficient, formation resistivity factor	K_R		COER		M_R, a, C	
coefficient or multiplier	K		COE		M	various
compressibility	c		CMP		k, κ kappa	Lt^2/m
concentration (salinity)	C		CNC		c, n	various
conductivity, apparent	C_a		ECNA		γ_a gamma	tq^2/mL^3
conductivity, electric	C		ECN		γ gamma	tq^2/mL^3
conductivity, thermal	k_h		HCN		λ lambda	mL/t^3T
constant, decay ($1/\tau_d$)	λ lambda		LAM		C	$1/t$
constant, dielectric	ϵ epsilon		DIC			q^2t^2/mL^3
correction term or correction factor (either additive or multiplicative)	B		COR		C	
cross-section (area)	A		ARA		S	L^2
cross-section, macroscopic	Σ sigma$_{cap}$		XST		S	$1/L$
cross-section of a nucleus, microscopic	σ sigma		XNL		s	L^2
current, electric	I		CUR		i, $\imath$ script i	q/t
decay constant ($1/\tau_d$)	λ lambda		LAM		C	$1/t$
decay time ($1/\lambda$)	τ_d tau		TIMD		t_d	t
decrement	δ delta		DCR		Δ delta$_{cap}$	various
density	ρ rho		DEN		D	m/L^3
density, apparent	ρ_a rho		DENA		D_a	m/L^3
density, bulk	ρ_b rho		DENB		D_b	m/L^3
density, fluid	ρ_f rho		DENF		D_f	m/L^3
density, flushed zone	ρ_{xo} rho		DENXO		D_{xo}	m/L^3
density (indicating "number per unit volume")	n		NMB		N	$1/L^3$
density, matrix[14]	ρ_{ma} rho		DENMA		D_{ma}	m/L^3
density (number) of neutrons	n_N		NMBN			$1/L^3$
density, true	ρ_t rho		DENT		D_t	m/L^3
depth	D		DPH		y, H	L
depth, skin	δ delta		SKD		r_s	L
deviation, hole	δ delta		ANGH			
dew-point pressure	p_d		PRSD		P_d	m/Lt^2
diameter	d		DIA		D	L
diameter, hole	d_h		DIAH		d_H, D_h	L
diameter, invaded zone (electrically equivalent)	d_i		DIAI		d_I, D_i	L
dielectric constant	ϵ epsilon		DIC			q^2t^2/mL^3
difference	Δ delta$_{cap}$	DEL				[X]
dip, angle of	Θ theta$_{cap}$		ANGD		α_d alpha	
dip, apparent angle of	θ theta		ANGDA		α_{da} alpha	

[14] See subscript definitions of "matrix" and "solid(s)".

Quantity	Letter Symbol	Computer Symbol: Operator Field	Computer Symbol: Quantity Symbol Field	Computer Symbol: Subscript Field	Reserve SPE Letter Symbols	Dimensions[6]
dip, apparent azimuth of	ϕ phi		DAZA		β_{da} beta	
dip, azimuth of	Φ phi_{cap}		DAZ		β_d beta	
distance, length, or length of path	L		LTH		s, ℓ	L
distance, radial (increment along radius)	Δr	DEL	RAD		ΔR	L
drift angle, hole (deviation)	δ delta		ANGH			
electric current	I		CUR		i, $\imath$ script i	q/t
electrochemical coefficient	K_c		COEC		M_c	mL^2/t^2q
electrochemical component of the SP	E_c		EMFC		Φ_c phi_{cap}	mL^2/t^2q
electrokinetic component of the SP	E_k		EMFK		Φ_k phi_{cap}	mL^2/t^2q
electromotive force	E		EMF		V	mL^2/t^2q
energy	E		ENG		U	mL^2/t^2
exponent, porosity (cementation)	m		MXP			
exponent, saturation	n		SXP			
factor	F		FAC			various
flow rate, heat	$\dot{Q}$		HRT		q, Φ phi_{cap}	mL^2/t^3
force, electromotive	E		EMF		V	mL^2/t^2q
force, mechanical	F		FCE		Q	mL/t^2
formation resistivity factor	F_R		FAC	HR		
formation resistivity factor coefficient ($F_R \phi^m$)	K_R		COE	R	M_R, a, C	
formation volume factor	B		FVF		F	
fraction	f		FRC		F	
fraction of bulk (total) volume	f_V		FRC	VB	f_{Vb}, V_{bf}	
fraction of intergranular space ("porosity") occupied by all shales	$f_{\phi sh}$		FIG	SH	ϕ_{igfsh} phi	
fraction of intergranular space ("porosity") occupied by water	$f_{\phi w}$		FIG	W	ϕ_{igfw} phi	
fraction of intermatrix space ("porosity") occupied by nonstructural dispersed shale	$f_{\phi shd}$		FIM	SHD	ϕ_{imfshd} phi, q	
fracture index	I_f		FRX		i_f, I_F, i_F	
free fluid index	I_{Ff}		FFX		i_{Ff}	
frequency	f		FQN		ν nu	1/t
gamma ray count rate	N_{GR}		NGR		N_γ, C_G	1/t
gamma ray [usually with identifying subscript(s)]	γ gamma		GRY			various
gas compressibility factor	z		ZED		Z	
gas constant, universal (per mole)	R		RRR			mL^2/t^2T
gas-oil ratio, producing	R		GOR		F_g, F_{go}	
gas specific gravity	γ_g gamma		SPG	G	s_g, F_{sg}	
geometrical fraction (multiplier or factor)	G		GMF		f_G	
geometrical fraction (multiplier or factor), annulus	G_{an}		GMF	AN	f_{Gan}	
geometrical fraction (multiplier or factor), flushed zone	G_{xo}		GMF	XO	f_{Gxo}	
geometrical fraction (multiplier or factor), invaded zone	G_i		GMF	I	f_{Gi}	

Quantity	Letter Symbol	Computer Symbol (Operator Field / Quantity Symbol Field / Subscript Field)	Reserve SPE Letter Symbols	Dimensions[6]
geometrical fraction (multiplier or factor), mud	G_m	GMFM	f_{Gm}	
geometrical fraction (multiplier or factor), pseudo-	G_p	GMFP	f_J	
geometrical fraction (multiplier or factor), true	G_t	GMFT	f_{Gtr}	
gradient	g	GRD	γ gamma	various
gradient, geothermal	g_G	GRDGT	g_g	T/L
gradient, temperature	g_T	GRDT	g_h	T/L
gravity, specific	γ gamma	SPG	s, F_s	
half life	$t_{1/2}$	TIMH		t
heat flow rate	$\dot{Q}$	HRT	q, Φ phi$_{cap}$	mL^2/t^3
height, or fluid head	Z	ZEL	D, h	L
hold-up (fraction of the pipe volume filled by a given fluid: y_o is oil hold-up, y_w is water hold-up; sum of all the hold-ups at a given level is one.)	y	HOL	f	
hydrocarbon resistivity index	I_R	RSXH	i_R	
hydrogen index	I_H	HYX	i_H	
impedance, acoustic	Z_a	MPDA		m/L^2t
impedance, electric	Z_e	MPDE	Z_E, η eta	mL^2/tq^2
index, fracture	I_f	FRX	i_f, I_F, i_F	
index, free fluid	I_{Ff}	FFX	i_{Ff}	
index, (hydrocarbon) resistivity	I_R	RSXH	i_R	
index, hydrogen	I_H	HYX	i_H	
index, porosity	I_ϕ	PRX	i_ϕ	
index, primary porosity	$I_{\phi 1}$	PRXPR	$i_{\phi 1}$	
index, secondary porosity	$I_{\phi 2}$	PRXSE	$i_{\phi 2}$	
index, shaliness gamma-ray $(\gamma_{LOG} - \gamma_{cn})/(\gamma_{sh} - \gamma_{cn})$	I_{shGR}	SHXGR	i_{shGR}	
index (use subscripts as needed)	I	____X	i	
injection rate	i	INJ		L^3/t
interfacial tension	σ sigma	SFT	y, γ gamma	m/t^2
intergranular "porosity" (space) $(V_b - V_{gr})/V_b$	ϕ_{ig} phi	PORIG	f_{ig}, ϵ_{ig} epsilon	
intermatrix "porosity" (space) $(V_b - V_{ma})/V_b$	ϕ_{im} phi	PORIM	f_{im}, ϵ_{im} epsilon	
interval transit time	t script t	TAC	Δt	t/L
interval transit time, apparent	t_a script t	TACA	Δt_a	t/L
interval transit time-density slope (absolute value)	M	SAD	$m_{\theta D}$	tL^2/m
interval transit time, fluid	t_f script t	TACF	Δt_f	t/L
interval transit time, matrix	t_{ma} script t	TACMA	Δt_{ma}	t/L
interval transit time, shale	t_{sh} script t	TACSH	Δt_{sh}	t/L
length, path length, or distance	L	LTH	s, ℓ	L
lifetime, average (mean life)	$\bar{\tau}$ tau	TIMAV	$\bar{t}$	t

Quantity	Letter Symbol	Computer Symbol (Operator Field / Quantity Symbol Field / Subscript Field)	Reserve SPE Letter Symbols	Dimensions[6]
macroscopic cross-section	Σ sigma$_{cap}$	XSTMAC	S	1/L
magnetic permeability	μ mu	PRMM	m	mL/q^2
magnetic susceptibility	k	SUSM	κ kappa	mL/q^2
magnetization	M	MAG	I	m/qt
magnetization, fraction	M_f	MAGF		
mass	m	MAS		m
mass flow rate	w	MRT	m	m/t
matrix (framework) volume (volume of all formation solids except nonstructural clay or shale)	V_{ma}	VOLMA		L^3
mean life	$\bar{\tau}$ tau	TIMAV	$\bar{t}$	t
microscopic cross-section	σ sigma	XSTMIC		L^2
modulus, bulk	K	BKM	K_b	m/Lt^2
modulus of elasticity (Young's modulus)	E	ELMY	Y	m/Lt^2
modulus, shear	G	ELMS	E_s	m/Lt^2
multiplier (fraction), geometrical	G	GMF	f_G	
multiplier (fraction), geometrical, invaded zone	G_i	GMFI	f_{Gi}	
multiplier (fraction), geometrical, true	G_t	GMFT	f_{Gtr}	
multiplier or coefficient	K	COE	M	various
neutron count rate	N_N	NEUN	N_n, C_N	1/t
neutron lifetime	t_N	NLF	τ_N tau, t_n	t
neutron porosity-density slope (absolute value)	N	SND	$m_{\phi ND}$	L^3/m
neutron [usually with identifying subscript(s)]	N	NEU		various
nucleus cross-section	σ sigma	XNL	s	L^2
number, atomic	Z	ANM		
number, dimensionless, in general (always with identifying subscripts)[8]	N	NUMQ		
number of moles, total (see also moles, number of, Supplement II)	n	MOL	n_t, N_t	
number (quantity)	n	NMB	N	
oil specific gravity	γ_o gamma	SPGO	s_o, F_{so}	
period	T	PER	Θ theta$_{cap}$	t
permeability, absolute (fluid flow)	k	PRM	K	L^2
permeability, magnetic	μ mu	PRMM	m	mL/q^2
Poisson's ratio	μ mu	PSN	ν nu, σ sigma	
porosity $(V_b - V_s)/V_b$	ϕ phi	POR	f, ϵ epsilon	
porosity, apparent	ϕ_a phi	PORA	f_a, ϵ_a epsilon	
porosity (cementation) exponent	m	MXP		
porosity, effective (interconnected) (V_{pe}/V_b)	ϕ_e phi	PORE	f_e, ϵ_e epsilon	
porosity index	I_ϕ	PRX	i_ϕ	
porosity index, primary	$I_{\phi 1}$	PRXPR	$i_{\phi 1}$	
porosity index, secondary	$I_{\phi 2}$	PRXSE	$i_{\phi 2}$	

Quantity	Letter Symbol	Computer Symbol (Operator Field / Quantity Symbol Field / Subscript Field)	Reserve SPE Letter Symbols	Dimensions[6]
porosity, noneffective (noninterconnected) (V_{pne}/V_b)	ϕ_{ne} phi	PORNE	f_{ne}, ϵ_{ne} epsilon	
"porosity" (space), intergranular ($V_b - V_{gr})/V_b$	ϕ_{ig} phi	PORIG	f_{ig}, ϵ_{ig} epsilon	
"porosity" (space), intermatrix ($V_b - V_{ma})/V_b$	ϕ_{im} phi	PORIM	f_{im}, ϵ_{im} epsilon	
porosity, total	ϕ_t phi	PORT	f_t, ϵ_t epsilon	
potential difference (electric)	V	VLT	U	mL^2/qt^2
pressure	p	PRS	P	m/Lt^2
primary porosity index	$I_{\phi 1}$	PRXPR	$i_{\phi 1}$	
production rate or flow rate	q	RTE	Q	L^3/t
productivity index	J	PDX	j	L^4t/m
pseudo-geometrical fraction (multiplier or factor)	G_p	GMFP	f_J	
pseudo-SP	E_{pSP}	EMFP	Φ_{sp} phi$_{cap}$	mL^2/t^2q
radial distance (increment along radius)	Δr	DELRAD	ΔR	L
radius	r	RAD	R	L
reactance	X	XEL		mL^2/tq^2
reduction or reduction term	a alpha	RED		
reduction, SP, due to shaliness	a_{SP} alpha	REDSP		
relative amplitude	A_r	AMPR		
relaxation time, free-precession decay	t_2	TIMAV	τ_2 tau	t
relaxation time, proton thermal	t_1	TIMRP	τ_1 tau	t
resistance	r	RST	R	mL^2/tq^2
resistivity	R	RES	ρ rho, r	mL^3/tq^2
resistivity, annulus	R_{an}	RESAN	ρ_{an} rho, r_{an}	mL^3/tq^2
resistivity, apparent	R_a	RESA	ρ_a rho, r_a	mL^3/tq^2
resistivity, apparent, of the conductive liquids mixed in invaded zone	R_z	RESZ	ρ_z rho, r_z	mL^3/tq^2
resistivity factor coefficient, formation ($F_R\phi^m$)	K_R	COER	M_R, a, C	
resistivity factor, formation	F_R	FACHR		
resistivity, flushed zone (that part of the invaded zone closest to the wall of the hole, where flushing has been maximum)	R_{xo}	RESXO	ρ_{xo} rho, r_{xo}	mL^3/tq^2
resistivity, formation 100 percent saturated with water of resistivity R_w (zero hydrocarbon saturation)	R_0	RESZR	ρ_0 rho, r_0	mL^3/tq^2
resistivity, formation, true	R_t	REST	ρ_t rho, r_t	mL^3/tq^2
resistivity index, hydrocarbon	I_R	RSXH	i_R	
resistivity, invaded zone	R_i	RESI	ρ_i rho, r_i	mL^3/tq^2
resistivity, mud	R_m	RESM	ρ_m rho, r_m	mL^3/tq^2
resistivity, mud-cake	R_{mc}	RESMC	ρ_{mc} rho, r_{mc}	mL^3/tq^2
resistivity, mud-filtrate	R_{mf}	RESMF	ρ_{mf} rho, r_{mf}	mL^3/tq^2
resistivity, shale	R_{sh}	RESSH	ρ_{sh} rho, r_{sh}	mL^3/tq^2
resistivity, surrounding formation	R_s	RESS	ρ_s rho, r_s	mL^3/tq^2

Quantity	Letter Symbol	Computer Symbol: Operator Field	Computer Symbol: Quantity Symbol Field	Computer Symbol: Subscript Field	Reserve SPE Letter Symbols	Dimensions[6]
resistivity, water	R_w		RES	W	ρ_w rho, r_w	mL^3/tq^2
Reynolds number (dimensionless number)	N_{Re}		REYQ			
salinity	C		CNC		c, n	various
saturation	S		SAT		ρ rho, s	
saturation exponent	n		SXP			
saturation, gas	S_g		SAT	G	ρ_g rho, s_g	
saturation, hydrocarbon	S_h		SAT	H	ρ_h rho, s_h	
saturation, oil	S_o		SAT	O	ρ_o rho, s_o	
saturation, oil, residual	S_{or}		SAT	OR	ρ_{or} rho, s_{or}	
saturation, water	S_w		SAT	W	ρ_w rho, s_w	
secondary porosity index	$I_{\phi 2}$		PRXSE		$i_{\phi 2}$	
self potential (see SP quantities)						
shaliness gamma-ray index $(\gamma_{LOG} - \gamma_{cn})/(\gamma_{sh} - \gamma_{cn})$	I_{shGR}		SHX	G	i_{shGR}	
shear modulus	G		ELMS		E_s	m/Lt^2
skin depth (logging)	δ delta		SKD		r_s	L
slope, interval transit time vs density (absolute value)	M		SAD		$m_{\theta D}$	tL^2/m
slope, neutron porosity vs density (absolute value)	N		SND		$m_{\phi ND}$	L^3/m
solid(s) volume (volume of *all* formation solids)	V_s		VOLS		v_s	L^3
SP (measured SP)	E_{SP}		EMF	SP	Φ_{SP} phi$_{cap}$	mL^2/t^2q
SP, pseudo-	E_{pSP}		EMF	PSP	Φ_{pSP} phi$_{cap}$	mL^2/t^2q
SP reduction due to shaliness	α_{SP} alpha		RED	SP		
SP, static (SSP)	E_{SSP}		EMF	SSP	Φ_{SSP} phi$_{cap}$	mL^2/t^2q
spacing	L_s		LEN	S	s_s, ℓ_s	L
specific gravity	γ gamma		SPG		s, F_s	
spontaneous potentials (see SP quantities)						
SSP (static SP)	E_{SSP}		EMF	SSP	Φ_{SSP} phi$_{cap}$	mL^2/t^2q
summation (operator)	Σ sigma$_{cap}$	SUM				
superficial phase velocity (flux rate of a particular fluid phase flowing in pipe; use appropriate phase subscripts)	u		VELV		ψ psi	L/t
surface production rate	q_{sc}		RTE	SC	q_σ, Q_{sc}	L^3/t
susceptibility, magnetic	k		SUSM		κ kappa	mL/q^2
temperature	T		TEM		θ theta	T
temperature gradient	g_T		TEMGR		g_h	T/L
thermal conductivity	k_h		HCN		λ lambda	mL/t^3T
thickness	h		THK		d, e	L
thickness, mud-cake	h_{mc}		THK	MC	d_{mc}, e_{mc}	L
thickness, pay, gross (total)	h_t		THK	T	d_t, e_t	L
thickness, pay, net	h_n		THK	N	d_n e_n	L
time	t		TIM		τ tau	t
time constant	τ tau		TIM	C	τ_c tau	t
time, decay $(1/\lambda)$	τ_d tau		TIM	D	t_d	t

Quantity	Letter Symbol	Computer Symbol (Operator Field / Quantity Symbol Field / Subscript Field)	Reserve SPE Letter Symbols	Dimensions[6]
time difference	Δt	DELTIM	$\Delta\tau$ tau	t
time, interval transit	t script t	TAC	Δt	t/L
tortuosity	τ tau	TOR		
tortuosity, electric	τ_e tau	TORE		
tortuosity, hydraulic	τ_H tau	TORHL		
transit time, interval	t script t	TAC	Δt	t/L
valence	z	VAL		
velocity	v	VEL	V, u	L/t
velocity, acoustic	v	VAC	V, u	L/t
viscosity (dynamic)	μ mu	VIS	η eta	m/Lt
volume	V	VOL	v	L^3
volume, bulk	V_b	VOLB	v_b	L^3
volume fraction or ratio (as needed, use same subscripted symbols as for "volumes"; note that bulk volume fraction is unity and pore volume fractions are ϕ)	V	VLF	f_V, F_V	various
volume, grain (volume of all formation solids except shales)	V_{gr}	VOLGR	v_{gr}	L^3
volume, intergranular (volume between grains; consists of fluids and all shales) ($V_b - V_{gr}$)	V_{ig}	VOLIG	v_{ig}	L^3
volume, intermatrix (consists of fluids and dispersed shale) ($V_b - V_{ma}$)	V_{im}	VOLIM	v_{im}	L^3
volume, matrix (framework) (volume of all formation solids except dispersed shale)	V_{ma}	VOLMA	v_{ma}	L^3
volume, pore ($V_b - V_s$)	V_p	VOLP	v_p	L^3
volume, pore, effective (interconnected) (interconnected pore space)	V_e	VOLE	V_{pe}, v_e	L^3
volume, pore, noneffective (noninterconnected) (noninterconnected pore space) ($V_p - V_e$)	V_{ne}	VOLNE	V_{pne}, v_{ne}	L^3
volume, shale, dispersed	V_{shd}	VOLSHD	v_{shd}	L^3
volume, shale, laminated	$V_{sh\ell}$	VSHLAM	$v_{sh\ell}$	L^3
volume, shale, structural	V_{shst}	VOLSHS	v_{shs}	L^3
volume, shale(s) (volume of all shales: structural and dispersed)	V_{sh}	VOLSH	v_{sh}	L^3
volume, solid(s) (volume of *all* formation solids)	V_s	VOLS	v_s	L^3
volumetric flow rate	q	RTE	Q	L^3/t
volumetric flow rate down hole	q_{dh}	RTEDH	q_{wf}, q_{DH}, Q_{dh}	L^3/t
volumetric flow rate, surface conditions	q_{sc}	RTESC	q_σ, Q_{sc}	L^3/t
wavelength	λ lambda	WVL		L
weight, atomic	A	AWT		m
weight (gravitational)	W	WGT	w, G	mL/t^2
work	W	WRK	w	mL^2/t^2
Young's modulus	E	ELMY	Y	m/Lt^2

Subscripts in Alphabetical Order

Subscript Definition	Letter Subscript	Computer Subscript Subscript Field	Reserve SPE Letter Subscripts
acoustic	*a*	A	*A*, *α* alpha
activation log, neutron, derived from	*NA*	NA	*na*
activation log, neutron, given by	*NA*	NA	
amplitude log, derived from	*A*	A	*a*
amplitude log, given by	*A*	A	
anhydrite	*anh*	AH	
anisotropic	*ani*	ANI	
annulus	*an*	AN	*AN*
apparent (from log readings: use tool-description subscripts)			
apparent (general)	*a*	A	*ap*
bond log, cement, derived from	*CB*	CB	*cb*
bond log, cement, given by	*CB*	CB	
borehole televiewer log, derived from	*TV*	TV	*tv*
borehole televiewer log, given by	*TV*	TV	
bottom hole	*bh*	BH	*w*, *BH*
bubble	*b*	B	
bubble-point (saturation)	*b*	B	*s*, *bp*
bulk	*b*	B	*B*, *t*
calculated	*C*	CA	calc
caliper log, derived from	*C*	C	*c*
caliper log, given by	*C*	C	
capillary	*c*	CP	*C*
capture	*cap*	C	
casing	*c*	CS	*cg*
cement bond log, derived from	*CB*	CB	*cb*
cement bond log, given by	*CB*	CB	
chlorine log, derived from	*CL*	CL	*cl*
chlorine log, given by	*CL*	CL	
clay	*cl*	CL	*cla*
clean	*cn*	CN	*cln*
coil	*C*	C	*c*
compaction	*cp*	CP	
compensated density log, derived from	*CD*	CD	*cd*
compensated density log, given by	*CD*	CD	
compensated neutron log, derived from	*CN*	CN	*cn*
compensated neutron log, given by	*CN*	CN	
compressional wave	*c*	C	*C*
conductive liquids in invaded zone	*z*	Z	
contact log, microlog, minilog, derived from	*ML*	ML	*mℓ*
contact log, microlog, minilog, given by	*ML*	ML	
corrected	cor	COR	
critical	*c*	CR	*cr*
decay	*d*	D	

Subscript Definition	Letter Subscript	Computer Subscript Subscript Field	Reserve SPE Letter Subscripts
deep induction log, derived from	*ID*	ID	*id*
deep induction log, given by	*ID*	ID	
deep laterolog, derived from	*LLD*	LLD	*ℓℓD*
deep laterolog, given by	*LLD*	LLD	
density log, compensated, derived from	*CD*	CD	*cd*
density log, compensated, given by	*CD*	CD	
density log, derived from	*D*	D	*d*
density log, given by	*D*	D	
differential temperature log, derived from	*DT*	DT	*dt*
differential temperature log, given by	*DT*	DT	
dip	*d*	D	
diplog, dipmeter, derived from	*DM*	DM	*dm*
diplog, dipmeter, given by	*DM*	DM	
directional survey, derived from	*DR*	DR	*dr*
directional survey, given by	*DR*	DR	
dirty (clayey, shaly)	*dy*	DY	*dty*
dolomite	*dol*	DL	
down-hole	*dh*	DH	*DH*
dual induction log, derived from	*DI*	DI	*di*
dual induction log, given by	*DI*	DI	
dual laterolog, derived from	*DLL*	DLL	*d*
dual laterolog, given by	*DLL*	DLL	
effective	*e*	E	
electric, electrical	*e*	E	*E*
electrochemical	*c*	C	*ec*
electrode	*E*	E	*e*
electrokinetic	*k*	K	*ek*
electrolog, electrical log, electrical survey, derived from	*EL*	EL	*el*, *ES*
electrolog, electrical log, electrical survey, given by	*EL*	EL	
electromagnetic pipe inspection log, derived from	*EP*	EP	*ep*
electromagnetic pipe inspection log, given by	*EP*	EP	
electron	*el*	E	*el* script
empirical	*E*	EM	*EM*
epithermal neutron log, derived from	*NE*	NE	*ne*
epithermal neutron log, given by	*NE*	NE	
equivalent	*eq*	EV	*EV*
experimental	*E*	EX	*EX*
external, outer	*e*	E	*o*
fast neutron log, derived from	*NF*	NF	*nf*
fast neutron log, given by	*NF*	NF	
fluid	*f*	F	*fl*

Subscript Definition	Letter Subscript	Computer Subscript Subscript Field	Reserve SPE Letter Subscripts
flushed zone	xo	XO	
formation (rock)	f	F	fm
formation, surrounding	s	S	
fraction	f	F	r
fracture	f	FR	F
free	F	F	f
free fluid	Ff	FF	
gamma-gamma ray log, derived from	GG	GG	gg
gamma-gamma ray log, given by	GG	GG	
gamma ray log, derived from	GR	GR	gr
gamma ray log, given by	GR	GR	
gas	g	G	G
geometrical	G	G	
geothermal	G	GT	T
grain	gr	GR	
gravity meter log, derived from	GM	GM	gm
gravity meter log, given by	GM	GM	
gross (total)	t	T	T
gypsum	gyp	GY	
half	$\frac{1}{2}$	H	
heat or thermal	h	HT	T, θ theta
heavy phase	HP	HP	hp
hole	h	H	H
hydraulic	H	HL	
hydrocarbon	h	H	H
hydrogen nuclei or atoms	H	HY	
induction log, deep investigation, derived from	ID	ID	id
induction log, deep investigation, given by	ID	ID	
induction log, derived from	I	I	i
induction log, dual, derived from	DI	DI	di
induction log, dual, given by	DI	DI	
induction log, given by	I	I	
induction log, medium investigation, derived from	IM	IM	im
induction log, medium investigation, given by	IM	IM	
inner	i	I	ι iota, $\imath$ script i
intergranular	ig	IG	
intermatrix	im	IM	
internal	i	I	ι iota, $\imath$ script i
intrinsic	int	I	
invaded zone	i	I	I
irreducible	i	IR	ir, $\imath$ script i, ι iota
junction	j	J	
laminar	ℓ	LAM	L
laminated, lamination	ℓ	LAM	L

Subscript Definition	Letter Subscript	Computer Subscript Subscript Field	Reserve SPE Letter Subscripts
lateral (resistivity) log, derived from	L	L	ℓ
lateral (resistivity) log, given by	L	L	
laterolog, derived from	LL $LL3$ $LL7$ $LL8$ LLD LLS etc.	LL LL3 LL7 LL8 LLD LLS etc.	$\ell\ell$ $\ell\ell 3$ $\ell\ell 7$ $\ell\ell 8$ $\ell\ell D$ $\ell\ell S$ etc.
laterolog, dual, derived from	DLL	DLL	$d\ell\ell$
laterolog, dual, given by	DLL	DLL	
laterolog, given by	LL $LL3$ $LL7$ $LL8$ LLD LLS etc.	LL LL3 LL7 LL8 LLD LLS etc.	
lifetime log, neutron, TDT, derived from	NL	NL	$n\ell$
lifetime log, neutron, TDT, given by	NL	NL	
light phase	LP	LP	ℓp
limestone	ls	LS	lst
limiting value	lim	LM	
liquid	L	L	ℓ
liquids, conductive, in invaded zone	z	Z	
log, derived from	LOG	L	log
log, given by	LOG	L	
lower	ℓ	L	L
magnetism log, nuclear, derived from	NM	NM	nm
magnetism log, nuclear, given by	NM	NM	
matrix [solids except dispersed (nonstructural) clay or shale]	ma	MA	
maximum	max	MX	
medium investigation induction log, derived from	IM	IM	im
medium investigation induction log, given by	IM	IM	
microlaterolog, derived from	MLL	MLL	$m\ell\ell$
microlaterolog, given by	MLL	MLL	
microlog, minilog, contact log, derived from	ML	ML	$m\ell$
microlog, minilog, contact log, given by	ML	ML	
micro-seismogram log, signature log, variable density log, derived from	VD	VD	vd
micro-seismogram log, signature log, variable density log, given by	VD	VD	
minimum	min	MN	
mixture	M	M	z, m
mud	m	M	
mud cake	mc	MC	
mud filtrate	mf	MF	

Subscript Definition	Letter Subscript	Computer Subscript Subscript Field	Reserve SPE Letter Subscripts
net	n	N	
neutron	N	N	n
neutron activation log, derived from	NA	NA	na
neutron activation log, given by	NA	NA	
neutron lifetime log, TDT, derived from	NL	NL	$n\ell$
neutron lifetime log, TDT, given by	NL	NL	
neutron log, compensated, derived from	CN	CN	cn
neutron log, compensated, given by	CN	CN	
neutron log, derived from	N	N	n
neutron log, epithermal, derived from	NE	NE	ne
neutron log, epithermal, given by	NE	NE	
neutron log, fast, derived from	NF	NF	nf
neutron log, fast, given by	NF	NF	
neutron log, given by	N	N	
neutron log, sidewall, derived from	SN	SN	sn
neutron log, sidewall, given by	SN	SN	
neutron log, thermal, derived from	NT	NT	nt
neutron log, thermal, given by	NT	NT	
noneffective	ne	NE	
normal (resistivity) log, derived from	N	N	n
normal (resistivity) log, given by	N	N	
nuclear magnetism log, derived from	NM	NM	nm
nuclear magnetism log, given by	NM	NM	
oil	o	O	N
outer (external)	e	E	o
pipe inspection log, electromagnetic, derived from	EP	EP	ep
pipe inspection log, electromagnetic, given by	EP	EP	
pore	p	P	P
porosity	ϕ phi	PHI	f, ϵ epsilon
porosity data, derived from	ϕ phi	P	f, ϵ epsilon
primary	1 (one)	PR	p, pri
proximity log, derived from	P	P	p
proximity log, given by	P	P	
pseudo	p	P	
pseudo-critical	pc	PC	
pseudo-reduced	pr	PRD	
pseudo-SP	pSP	PSP	
radial	r	R	R
reduced	r	RD	
relative	r	R	R
residual	r	R	R
resistivity	R	R	
resistivity log, derived from	R	R	r, ρ rho
resistivity log, given by	R	R	

Subscript Definition	Letter Subscript	Computer Subscript Subscript Field	Reserve SPE Letter Subscripts
Reynolds	Re	—	
sand	*sd*	SD	*sa*
sandstone	*ss*	SS	*sst*
scattered, scattering	*sc*	SC	
secondary	2	SE	*s*, *sec*
shale	*sh*	SH	*sha*
shallow laterolog, derived from	*LLS*	LLS	*ℓℓs*
shallow laterolog, given by	*LLS*	LLS	
shear wave	*s*	S	τ tau
sidewall	*S*	SW	*SW*
sidewall neutron log, derived from	*SN*	SN	*sn*
sidewall neutron log, given by	*SN*	SN	
signature log, micro-seismogram log, variable density log, derived from	*VD*	VD	*vd*
signature log, micro-seismogram log, variable density log, given by	*VD*	VD	
silt	*sl*	SL	*slt*
skin	*s*	S	*S*
slip or slippage	*s*	S	σ sigma
slurry ("mixture")	*M*	M	*z*, *m*
solid(s) (*all* formation solids)	*s*	S	σ sigma
sonde, tool	*T*	T	*t*
sonic velocity log, derived from	*SV*	SV	*sv*
sonic velocity log, given by	*SV*	SV	
SP, derived from	*SP*	SP	*sp*
SP, given by	*SP*	SP	
spacing	*s*	L	
SSP	*SSP*	SSP	
standard conditions	*sc*	SC	σ sigma
static or shut-in conditions	*ws*	WS	*s*
stock-tank conditions	*st*	ST	
structural	*st*	ST	*s*
surrounding formation	*s*	S	
TDT log, neutron lifetime log, derived from	*NL*	NL	*nℓ*
TDT log, neutron lifetime log, given by	*NL*	NL	
televiewer log, borehole, derived from	*TV*	TV	*tv*
televiewer log, borehole, given by	*TV*	TV	
temperature	*T*	T	*h*, θ theta
temperature log, derived from	*T*	T	*t*, *h*
temperature log, differential, derived from	*DT*	DT	*dt*
temperature log, differential, given by	*DT*	DT	
temperature log, given by	*T*	T	
thermal	*h*	HT	*T*, θ theta
thermal decay time (TDT) log, derived from	*NL*	NL	*nℓ*
thermal decay time (TDT) log, given by	*NL*	NL	

Subscript Definition	Letter Subscript	Computer Subscript Subscript Field	Reserve SPE Letter Subscripts
thermal neutron log, derived from	NT	NT	nt
thermal neutron log, given by	NT	NT	
tool-description subscripts: see individual entries, such as "amplitude log," "neutron log," etc.[15]			
tool, sonde	T	T	t
total (gross)	t	T	T
true (opposed to apparent)	t	T	tr
upper	u	U	U
variable density log, micro-seismogram log, signature log, derived from	VD	VD	vd
variable density log, micro-seismogram log, signature log, given by	VD	VD	
velocity, sonic or acoustic, log, derived from	SV	SV	sv
velocity, sonic or acoustic, log, given by	SV	SV	
volume or volumetric	V	V	v
water	w	W	W
water-saturated formation, 100 percent (*zero* hydrocarbon saturation)	0 zero	ZR	zr
weight	W	W	w
well flowing conditions	wf	WF	f
well static conditions	ws	WS	s
zero hydrocarbon saturation	0 zero	ZR	zr

[15] If service-company identification is needed for a tool, it is recommended that the appropriate following capital letter be added to the tool-description subscript: Birdwell, B; Dresser Atlas, D; GO International, G; Lane-Wells, L; PGAC, P; Schlumberger, S; Welex, W.

Appendix G

SI Metric System of Units

Alphabetical List of Units (Symbols of SI Units given in parenthesis)

To Convert From	To	Multiply By**	
abampere	ampere (A)	1.0*	E+01
abcoulomb	coulomb (C)	1.0*	E+01
abfarad	farad (F)	1.0*	E+09
abhenry	henry (H)	1.0*	E−09
abmho	siemens (S)	1.0*	E+09
abohm	ohm (Ω)	1.0*	E−09
abvolt	volt (V)	1.0*	E−08
acre·foot (U.S. survey)(1)	meter³ (m³)	1.233 489	E+03
acre (U.S. survey)(1)	meter² (m²)	4.046 873	E+03
ampere hour	coulomb (C)	3.6*	E+03
are	meter² (m²)	1.0*	E+02
angstrom	meter (m)	1.0*	E−10
astronomical unit(2)	meter (m)	1.495 979	E+11
atmosphere (standard)	pascal (Pa)	1.013 250*	E+05
atmosphere (technical = 1 kgf/cm²)	pascal (Pa)	9.806 650*	E+04
bar	pascal (Pa)	1.0*	E+05
barn	meter² (m²)	1.0*	E−28
barrel (for petroleum, 42 gal)	meter³ (m³)	1.589 873	E−01
board foot	meter³ (m³)	2.359 737	E−03
British thermal unit (International Table)(3)	joule (J)	1.055 056	E+03
British thermal unit (mean)	joule (J)	1.055 87	E+03
British thermal unit (thermochemical)	joule (J)	1.054 350	E+03
British thermal unit (39°F)	joule (J)	1.059 67	E+03
British thermal unit (59°F)	joule (J)	1.054 80	E+03
British thermal unit (60°F)	joule (J)	1.054 68	E+03
Btu (International Table)-ft/(hr-ft²-°F) (thermal conductivity)	watt per meter kelvin (W/m·K)	1.730 735	E+00
Btu (thermochemical)-ft/(hr-ft²-°F) (thermal conductivity)	watt per meter kelvin (W/m·K)	1.729 577	E+00
Btu (International Table)-in./(hr-ft²-°F) (thermal conductivity)	watt per meter kelvin (W/m·K)	1.442 279	E−01
Btu (thermochemical)-in./(hr-ft²-°F) (thermal conductivity)	watt per meter kelvin (W/m·K)	1.441 314	E−01
Btu (International Table)-in./(s-ft²-°F) (thermal conductivity)	watt per meter kelvin (W/m·K)	5.192 204	E+02
Btu (thermochemical)-in./(s-ft²-°F) (thermal conductivity)	watt per meter kelvin (W/m·K)	5.188 732	E+02
Btu (International Table)/hr	watt (W)	2.930 711	E−01
Btu (thermochemical)/hr	watt (W)	2.928 751	E−01
Btu (thermochemical)/min	watt (W)	1.757 250	E+01
Btu (thermochemical)/s	watt (W)	1.054 350	E+03
Btu (International Table)/ft²	joule per meter² (J/m²)	1.135 653	E+04
Btu (thermochemical)/ft²	joule per meter² (J/m²)	1.134 893	E+04

To Convert From	To	Multiply By**
Btu (thermochemical)/(ft²-hr)	watt per meter² (W/m²)	3.152 481 E+00
Btu (thermochemical)/(ft²-min)	watt per meter² (W/m²)	1.891 489 E+02
Btu (thermochemical)/(ft²-s)	watt per meter² (W/m²)	1.134 893 E+04
Btu (thermochemical)/(in.²-s)	watt per meter² (W/m²)	1.634 246 E+06
Btu (International Table)/(hr-ft²-°F) (thermal conductance)	watt per meter² kelvin (W/m²·K)	5.678 263 E+00
Btu (thermochemical)/(hr-ft²-°F) (thermal conductance)	watt per meter² kelvin (W/m²·K)	5.674 466 E+00
Btu (International Table)/(s-ft²-°F)	watt per meter² kelvin (W/m²·K)	2.044 175 E+04
Btu (thermochemical)/(s-ft²-°F)	watt per meter² kelvin (W/m²·K)	2.042 808 E+04
Btu (International Table)/lbm	joule per kilogram (J/kg)	2.326* E+03
Btu (thermochemical)/lbm	joule per kilogram (J/kg)	2.324 444 E+03
Btu (International Table)/(lbm-°F) (heat capacity)	joule per kilogram kelvin (J/kg·K)	4.186 8* E+03
Btu (thermochemical)/(lbm-°F) (heat capacity)	joule per kilogram kelvin (J/kg·K)	4.184 000 E+03
bushel (U.S.)	meter³ (m³)	3.523 907 E−02
caliber (inch)	meter (m)	2.54* E−02
calorie (International Table)	joule (J)	4.186 8* E+00
calorie (mean)	joule (J)	4.190 02 E+00
calorie (thermochemical)	joule (J)	4.184* E+00
calorie (15°C)	joule (J)	4.185 80 E+00
calorie (20°C)	joule (J)	4.181 90 E+00
calorie (kilogram, International Table)	joule (J)	4.186 8* E+03
calorie (kilogram, mean)	joule (J)	4.190 02 E+03
calorie (kilogram, thermochemical)	joule (J)	4.184* E+03
cal (thermochemical)/cm²	joule per meter² (J/m²)	4.184* E+04
cal (International Table)/g	joule per kilogram (J/kg)	4.186* E+03
cal (thermochemical)/g	joule per kilogram (J/kg)	4.184* E+03
cal (International Table)/(g·°C)	joule per kilogram kelvin (J/kg·K)	4.186 8* E+03
cal (thermochemical)/(g·°C)	joule per kilogram kelvin (J/kg·K)	4.184* E+03
cal (thermochemical)/min	watt (W)	6.973 333 E−02
cal (thermochemical)/s	watt (W)	4.184* E+00
cal (thermochemical)/(cm²·min)	watt per meter² (W/m²)	6.973 333 E+02
cal (thermochemical)/(cm²·s)	watt per meter² (W/m²)	4.184* E+04
cal (thermochemical)/(cm·s·°C)	watt per meter kelvin (W/m·K)	4.184* E+02
capture unit (c.u. = 10^{-3} cm^{-1})	per meter (m^{-1})	1.0* E−01
carat (metric)	kilogram (kg)	2.0* E−04
centimeter of mercury (0°C)	pascal (Pa)	1.333 22 E+03
centimeter of water (4°C)	pascal (Pa)	9.806 38 E+01
centipoise	pascal second (Pa·s)	1.0* E−03
centistokes	meter² per second (m²/s)	1.0* E−06
circular mil	meter² (m²)	5.067 075 E−10
clo	kelvin meter² per watt (K·m²/W)	2.003 712 E−01
cup	meter³ (m³)	2.365 882 E−04
curie	becquerel (Bq)	3.7* E+10
cycle per second	hertz (Hz)	1.0* E+00
day (mean solar)	second (s)	8.640 000 E+04
day (sidereal)	second (s)	8.616 409 E+04
degree (angle)	radian (rad)	1.745 329 E−02
degree Celsius	kelvin (K)	$T_K = T_{°C} + 273.15$
degree centigrade (see degree Celsius)		
degree Fahrenheit	degree Celsius	$T_{°C} = (T_{°F} - 32)/1.8$
degree Fahrenheit	kelvin (K)	$T_K = (T_{°F} + 459.67)/1.8$
degree Rankine	kelvin (K)	$T_K = T_{°R}/1.8$

**See footnote on Page 13.

[1]Since 1893 the U.S. basis of length measurement has been derived from metric standards. In 1959 a small refinement was made in the definition of the yard to resolve discrepancies both in this country and abroad, which changed its length from 3600/3937 m to 0.9144 m exactly. This resulted in the new value being shorter by two parts in a million. At the same time it was decided that any data in feet derived from and published as a result of geodetic surveys within the U.S. would remain with the old standard (1 ft = 1200/3937 m) until further decision. This foot is named the U.S. survey foot. As a result, all U.S. land measurements in U.S. customary units will relate to the meter by the old standard. All the conversion factors in these tables for units referenced to this footnote are based on the U.S. survey foot, rather than the international foot. Conversion factors for the land measure given below may be determined from the following relationships:

1 league = 3 miles (exactly)
1 rod = 16½ ft (exactly)
1 chain = 66 ft (exactly)
1 section = 1 sq mile (exactly)
1 township = 36 sq miles (exactly)

[2]This value conflicts with the value printed in NBS 330.[1] The value requires updating in NBS 330.

[3]This value was adopted in 1956. Some of the older International Tables use the value 1.055 04 E+03. The exact conversion factor is 1.055 055 852 62* E+03.

To Convert From	To	Multiply By**	
°F·hr-ft²/Btu (International Table) (thermal resistance)	kelvin meter² per watt (K·m²/W)	1.781 102	E−01
°F·hr-ft²/Btu (thermochemical) (thermal resistance)	kelvin meter² per watt (K·m²/W)	1.762 250	E−01
denier	kilogram per meter (kg/m)	1.111 111	E−07
dyne	newton (N)	1.0*	E−05
dyne·cm	newton meter (N·m)	1.0*	E−07
dyne/cm²	pascal (Pa)	1.0*	E−01
electronvolt	joule (J)	1.602 19	E−19
EMU of capacitance	farad (F)	1.0*	E+09
EMU of current	ampere (A)	1.0*	E+01
EMU of electric potential	volt (V)	1.0*	E−08
EMU of inductance	henry (H)	1.0*	E−09
EMU of resistance	ohm (Ω)	1.0*	E−09
ESU of capacitance	farad (F)	1.112 650	E−12
ESU of current	ampere (A)	3.335 6	E−10
ESU of electric potential	volt (V)	2.997 9	E+02
ESU of inductance	henry (H)	8.987 554	E+11
ESU of resistance	ohm (Ω)	8.987 554	E+11
erg	joule (J)	1.0*	E−07
erg/cm²·s	watt per meter² (W/m²)	1.0*	E−03
erg/s	watt (W)	1.0*	E−07
faraday (based on carbon-12)	coulomb (C)	9.648 70	E+04
faraday (chemical)	coulomb (C)	9.649 57	E+04
faraday (physical)	coulomb (C)	9.652 19	E+04
fathom	meter (m)	1.828 8	E+00
fermi (femtometer)	meter (m)	1.0*	E−15
fluid ounce (U.S.)	meter³ (m³)	2.957 353	E−05
foot	meter (m)	3.048*	E−01
foot (U.S. survey)[1]	meter (m)	3.048 006	E−01
foot of water (39.2°F)	pascal (Pa)	2.988 98	E+03
sq ft	meter² (m²)	9.290 304*	E−02
ft²/hr (thermal diffusivity)	meter² per second (m²/s)	2.580 640*	E−05
ft²/s	meter² per second (m²/s)	9.290 304*	E−02
cu ft (volume; section modulus)	meter³ (m³)	2.831 685	E−02
ft³/min	meter³ per second (m³/s)	4.719 474	E−04
ft³/s	meter³ per second (m³/s)	2.831 685	E−02
ft⁴ (moment of section)[4]	meter⁴ (m⁴)	8.630 975	E−03
ft/hr	meter per second (m/s)	8.466 667	E−05
ft/min	meter per second (m/s)	5.080*	E−03
ft/s	meter per second (m/s)	3.048*	E−01
ft/s²	meter per second² (m/s²)	3.048*	E−01
footcandle	lux (lx)	1.076 391	E+01
footlambert	candela per meter² (cd/m²)	3.426 259	E+00
ft-lbf	joule (J)	1.355 818	E+00
ft-lbf/hr	watt (W)	3.766 161	E−04
ft-lbf/min	watt (W)	2.259 697	E−02
ft-lbf/s	watt (W)	1.355 818	E+00
ft-poundal	joule (J)	4.214 011	E−02
free fall, standard (*g*)	meter per second² (m/s²)	9.806 650*	E+00
cm/s²	meter per second² (m/s²)	1.0*	E−02
gallon (Canadian liquid)	meter³ (m³)	4.546 090	E−03
gallon (U.K. liquid)	meter³ (m³)	4.546 092	E−03
gallon (U.S. dry)	meter³ (m³)	4.404 884	E−03
gallon (U.S. liquid)	meter³ (m³)	3.785 412	E−03
gal (U.S. liquid)/day	meter³ per second (m³/s)	4.381 264	E−08
gal (U.S. liquid)/min	meter³ per second (m³/s)	6.309 020	E−05
gal (U.S. liquid)/hp·hr (SFC, specific fuel consumption)	meter³ per joule (m³/J)	1.410 089	E−09
gamma (magnetic field strength)	ampere per meter (A/m)	7.957 747	E−04
gamma (magnetic flux density)	tesla (T)	1.0*	E−09
gauss	tesla (T)	1.0*	E−04
gilbert	ampere (A)	7.957 747	E−01
gill (U.K.)	meter³ (m³)	1.420 654	E−04
gill (U.S.)	meter³ (m³)	1.182 941	E−04
grad	degree (angular)	9.0*	E−01
grad	radian (rad)	1.570 796	E−02

To Convert From	To	Multiply By**
grain (1/7000 lbm avoirdupois)	kilogram (kg)	6.479 891* E−05
grain (lbm avoirdupois/7000)/gal (U.S. liquid)	kilogram per meter³ (kg/m³)	1.711 806 E−02
gram	kilogram (kg)	1.0* E−03
g/cm³	kilogram per meter³ (kg/m³)	1.0* E+03
gram-force/cm²	pascal (Pa)	9.806 650* E+01
hectare	meter² (m²)	1.0* E+04
horsepower (550 ft-lbf/s)	watt (W)	7.456 999 E+02
horsepower (boiler)	watt (W)	9.809 50 E+03
horsepower (electric)	watt (W)	7.460* E+02
horsepower (metric)	watt (W)	7.354 99 E+02
horsepower (water)	watt (W)	7.460 43 E+02
horsepower (U.K.)	watt (W)	7.457 0 E+02
hour (mean solar)	second (s)	3.600 000 E+03
hour (sidereal)	second (s)	3.590 170 E+03
hundredweight (long)	kilogram (kg)	5.080 235 E+01
hundredweight (short)	kilogram (kg)	4.535 924 E+01
inch	meter (m)	2.54* E−02
inch of mercury (32°F)	pascal (Pa)	3.386 38 E+03
inch of mercury (60°F)	pascal (Pa)	3.376 85 E+03
inch of water (39.2°F)	pascal (Pa)	2.490 82 E+02
inch of water (60°F)	pascal (Pa)	2.488 4 E+02
sq in.	meter² (m²)	6.451 6* E−04
cu in. (volume; section modulus)[5]	meter³ (m³)	1.638 706 E−05
in.³/min	meter³ per second (m³/s)	2.731 177 E−07
in.⁴ (moment of section)[4]	meter⁴ (m⁴)	4.162 314 E−07
in./s	meter per second (m/s)	2.54* E−02
in./s²	meter per second² (m/s²)	2.54* E−02
kayser	1 per meter (1/m)	1.0* E+02
kelvin	degree Celsius	$T_{°C} = T_K - 273.15$
kilocalorie (International Table)	joule (J)	4.186 8* E+03
kilocalorie (mean)	joule (J)	4.190 02 E+03
kilocalorie (thermochemical)	joule (J)	4.184* E+03
kilocalorie (thermochemical)/min	watt (W)	6.973 333 E+01
kilocalorie (thermochemical)/s	watt (W)	4.184* E+03
kilogram-force (kgf)	newton (N)	9.806 65* E+00
kgf·m	newton meter (N·m)	9.806 65* E+00
kgf·s²/m (mass)	kilogram (kg)	9.806 65* E+00
kgf/cm²	pascal (Pa)	9.806 65* E+04
kgf/m²	pascal (Pa)	9.806 65* E+00
kgf/mm²	pascal (Pa)	9.806 65* E+06
km/h	meter per second (m/s)	2.777 778 E−01
kilopond	newton (N)	9.806 65* E+00
kilowatthour (kW-hr)	joule (J)	3.6* E+06
kip (1000 lbf)	newton (N)	4.448 222 E+03
kip/in.² (ksi)	pascal (Pa)	6.894 757 E+06
knot (international)	meter per second (m/s)	5.144 444 E−01
lambert	candela per meter² (cd/m²)	1/π* E+04
lambert	candela per meter² (cd/m²)	3.183 099 E+03
langley	joule per meter² (J/m²)	4.184* E+04
league	meter (m)	(see Footnote 1)
light year	meter (m)	9.460 55 E+15
liter[6]	meter³ (m³)	1.0* E−03
maxwell	weber (Wb)	1.0* E−08
mho	siemens (S)	1.0* E+00
microinch	meter (m)	2.54* E−08
microsecond/foot (μs/ft)	microsecond/meter (μs/m)	3.280 840 E+00
micron	meter (m)	1.0* E−06
mil	meter (m)	2.54* E−05
mile (international)	meter (m)	1.609 344* E+03
mile (statute)	meter (m)	1.609 3 E+03
mile (U.S. survey)[1]	meter (m)	1.609 347 E+03
mile (international nautical)	meter (m)	1.852* E+03
mile (U.K. nautical)	meter (m)	1.853 184* E+03
mile (U.S. nautical)	meter (m)	1.852* E+03

[4] This sometimes is called the moment of inertia of a plane section about a specified axis.
[5] The exact conversion factor is 1.638 706 4*E−05.

To Convert From	To	Multiply By**
sq mile (international)	meter^2 (m^2)	2.589 988 E+06
sq mile (U.S. survey)[(1)]	meter^2 (m^2)	2.589 998 E+06
mile/hr (international)	meter per second (m/s)	4.470 4* E−01
mile/hr (international)	kilometer per hour (km/h)	1.609 344* E+00
mile/min (international)	meter per second (m/s)	2.682 24* E+01
mile/s (international)	meter per second (m/s)	1.609 344* E+03
millibar	pascal (Pa)	1.0* E+02
millimeter of mercury (0°C)	pascal (Pa)	1.333 22 E+02
minute (angle)	radian (rad)	2.908 882 E−04
minute (mean solar)	second (s)	6.0* E+01
minute (sidereal)	second (s)	5.983 617 E+01
month (mean calendar)	second (s)	2.628 000 E+06
oersted	ampere per meter (A/m)	7.957 747 E+01
ohm centimeter	ohm meter (Ω·m)	1.0* E−02
ohm circular-mil per ft	ohm millimeter^2 per meter (Ω·mm^2/m)	1.662 426 E−03
ounce (avoirdupois)	kilogram (kg)	2.834 952 E−02
ounce (troy or apothecary)	kilogram (kg)	3.110 348 E−02
ounce (U.K. fluid)	meter^3 (m^3)	2.841 307 E−05
ounce (U.S. fluid)	meter^3 (m^3)	2.957 353 E−05
ounce-force	newton (N)	2.780 139 E−01
ozf·in.	newton meter (N·m)	7.061 552 E−03
oz (avoirdupois)/gal (U.K. liquid)	kilogram per meter^3 (kg/m^3)	6.236 021 E+00
oz (avoirdupois)/gal (U.S. liquid)	kilogram per meter^3 (kg/m^3)	7.489 152 E+00
oz (avoirdupois)/in.^{3}	kilogram per meter^3 (kg/m^3)	1.729 994 E+03
oz (avoirdupois)/ft^2	kilogram per meter^2 (kg/m^2)	3.051 517 E−01
oz (avoirdupois)/yd^2	kilogram per meter^2 (kg/m^2)	3.390 575 E−02
parsec[(2)]	meter (m)	3.085 678 E+16
peck (U.S.)	meter^3 (m^3)	8.809 768 E−03
pennyweight	kilogram (kg)	1.555 174 E−03
perm (°C)[(7)]	kilogram per pascal second meter^2 (kg/Pa·s·m^2)	5.721 35 E−11
perm (23°C)[(7)]	kilogram per pascal second meter^2 (kg/Pa·s·m^2)	5.745 25 E−11
perm·in. (0°C)[(8)]	kilogram per pascal second meter (kg/Pa·s·m)	1.453 22 E−12
perm·in. (23°C)[(8)]	kilogram per pascal second meter (kg/Pa·s·m)	1.459 29 E−12
phot	lumen per meter^2 (lm/m^2)	1.0* E+04
pica (printer's)	meter (m)	4.217 518 E−03
pint (U.S. dry)	meter^3 (m^3)	5.506 105 E−04
pint (U.S. liquid)	meter^3 (m^3)	4.731 765 E−04
point (printer's)	meter (m)	3.514 598* E−04
poise (absolute viscosity)	pascal second (Pa·s)	1.0* E−01
pound (lbm avoirdupois)[(9)]	kilogram (kg)	4.535 924 E−01
pound (troy or apothecary)	kilogram (kg)	3.732 417 E−01
lbm-ft^2 (moment of inertia)	kilogram meter^2 (kg·m^2)	4.214 011 E−02
lbm-in.^{2} (moment of inertia)	kilogram meter^2 (kg·m^2)	2.926 397 E−04
lbm/ft-hr	pascal second (Pa·s)	4.133 789 E−04
lbm/ft-s	pascal second (Pa·s)	1.488 164 E+00
lbm/ft^2	kilogram per meter^2 (kg/m^2)	4.882 428 E+00
lbm/ft^3	kilogram per meter^3 (kg/m^3)	1.601 846 E+01
lbm/gal (U.K. liquid)	kilogram per meter^3 (kg/m^3)	9.977 633 E+01
lbm/gal (U.S. liquid)	kilogram per meter^3 (kg/m^3)	1.198 264 E+02
lbm/hr	kilogram per second (kg/s)	1.259 979 E−04
lbm/(hp · hr) (SFC, specific fuel consumption)	kilogram per joule (kg/J)	1.689 659 E−07
lbm/in.^{3}	kilogram per meter^3 (kg/m^3)	2.767 990 E+04
lbm/min	kilogram per second (kg/s)	7.559 873 E−03
lbm/s	kilogram per second (kg/s)	4.535 924 E−01
lbm/yd^3	kilogram per meter^3 (kg/m^3)	5.932 764 E−01
poundal	newton (N)	1.382 550 E−01
poundal/ft^2	pascal (Pa)	1.488 164 E+00
poundal-s/ft^2	pascal second (Pa·s)	1.488 164 E+00
pound-force (lbf)[(10)]	newton (N)	4.448 222 E+00
lbf-ft[(11)]	newton meter (N·m)	1.355 818 E+00

To Convert From	To	Multiply By**
lbf-ft/in.[12]	newton meter per meter (N·m/m)	5.337 866 E+01
lbf-in.[11]	newton meter (N·m)	1.129 848 E−01
lbf-in./in.[12]	newton meter per meter (N·m/m)	4.448 222 E+00
lbf-s/ft^2	pascal second (Pa·s)	4.788 026 E+01
lbf/ft	newton per meter (N/m)	1.459 390 E+01
lbf/ft^2	pascal (Pa)	4.788 026 E+01
lbf/in.	newton per meter (N/m)	1.751 268 E+02
lbf/in.^{2} (psi)	pascal (Pa)	6.894 757 E+03
lbf/lbm (thrust/weight [mass] ratio)	newton per kilogram (N/kg)	9.806 650 E+00
quart (U.S. dry)	meter^3 (m^3)	1.101 221 E−03
quart (U.S. liquid)	meter^3 (m^3)	9.463 529 E−04
rad (radiation dose absorbed)	gray (Gy)	1.0* E−02
rhe	1 per pascal second (1/Pa·s)	1.0* E+01
rod	meter (m)	(see Footnote 1)
roentgen	coulomb per kilogram (C/kg)	2.58 E−04
second (angle)	radian (rad)	4.848 137 E−06
second (sidereal)	second (s)	9.972 696 E−01
section	meter^2 (m^2)	(see Footnote 1)
shake	second (s)	1.000 000* E−08
slug	kilogram (kg)	1.459 390 E+01
slug/(ft-s)	pascal second (Pa·s)	4.788 026 E+01
slug/ft^3	kilogram per meter^3 (kg/m^3)	5.153 788 E+02
statampere	ampere (A)	3.335 640 E−10
statcoulomb	coulomb (C)	3.335 640 E−10
statfarad	farad (F)	1.112 650 E−12
stathenry	henry (H)	8.987 554 E+11
statmho	siemens (S)	1.112 650 E−12
statohm	ohm (Ω)	8.987 554 E+11
statvolt	volt (V)	2.997 925 E+02
stere	meter^3 (m^3)	1.0* E+00
stilb	candela per meter^2 (cd/m^2)	1.0* E+04
stokes (kinematic viscosity)	meter^2 per second (m^2/s)	1.0* E−04
tablespoon	meter^3 (m^3)	1.478 676 E−05
teaspoon	meter^3 (m^3)	4.928 922 E−06
tex	kilogram per meter (kg/m)	1.0* E−06
therm	joule (J)	1.055 056 E+08
ton (assay)	kilogram (kg)	2.916 667 E−02
ton (long, 2,240 lbm)	kilogram (kg)	1.016 047 E+03
ton (metric)	kilogram (kg)	1.0* E+03
ton (nuclear equivalent of TNT)	joule (J)	4.184 E+09[13]
ton (refrigeration)	watt (W)	3.516 800 E+03
ton (register)	meter^3 (m^3)	2.831 685 E+00
ton (short, 2000 lbm)	kilogram (kg)	9.071 847 E+02
ton (long)/yd^3	kilogram per meter^3 (kg/m^3)	1.328 939 E+03
ton (short)/hr	kilogram per second (kg/s)	2.519 958 E−01
ton-force (2000 lbf)	newton (N)	8.896 444 E+03
tonne	kilogram (kg)	1.0* E+03
torr (mm Hg, 0°C)	pascal (Pa)	1.333 22 E+02
township	meter^2 (m^2)	(see Footnote 1)
unit pole	weber (Wb)	1.256 637 E−07
watthour (W-hr)	joule (J)	3.60* E+03
W·s	joule (J)	1.0* E+00
W/cm^2	watt per meter^2 (W/m^2)	1.0* E+04
W/in.^{2}	watt per meter^2 (W/m^2)	1.550 003 E+03
yard	meter (m)	9.144* E−01
yd^2	meter^2 (m^2)	8.361 274 E−01
yd^3	meter^3 (m^3)	7.645 549 E−01
yd^3/min	meter^3 per second (m^3/s)	1.274 258 E−02
year (calendar)	second (s)	3.153 600 E+07
year (sidereal)	second (s)	3.155 815 E+07
year (tropical)	second (s)	3.155 693 E+07

[6]In 1964 the General Conference on Weights and Measures adopted the name liter as a special name for the cubic decimeter. Prior to this decision the liter differed slightly (previous value, 1.000 028 dm^3) and in expression of precision volume measurement this fact must be kept in mind.

[7]Not the same as reservoir "perm."

[8]Not the same dimensions as "millidarcy-foot."

[9]The exact conversion factor is 4.535 923 7*E−01.

[10]The exact conversion factor is 4.448 221 615 260 5*E+00.

[11]Torque unit; see text discussion of "Energy, Torque, and Bending Moment."

[12]Torque divided by length; see text discussion of "Energy, Torque, and Bending Moment."

[13]Defined (not measured) value.

TABLES OF RECOMMENDED SI UNITS

Quantity and SI Unit		Customary Unit	Metric Unit		Conversion Factor* Multiply Customary Unit by Factor to Get Metric Unit	
			SPE Preferred	Other Allowable		
		SPACE,** TIME				
Length	m	naut mile	km		1.852*	E+00
		mile	km		1.609 344*	E+00
		chain	m		2.011 68*	E+01
		link	m		2.011 68*	E−01
		fathom	m		1.828 8*	E+00
		m	m		1.0*	E+00
		yd	m		9.144*	E−01
		ft	m		3.048*	E−01
				cm	3.048*	E+01
		in.	mm		2.54*	E+01
				cm	2.54*	E+00
		cm	mm		1.0*	E+01
				cm	1.0*	E+00
		mm	mm		1.0*	E+00
		mil	μm		2.54*	E+01
		micron (μ)	μm		1.0*	E+00
Length/length	m/m	ft/mi	m/km		1.893 939	E−01
Length/volume	m/m^3	ft/U.S. gal	m/m^3		8.051 964	E+01
		ft/ft^3	m/m^3		1.076 391	E+01
		ft/bbl	m/m^3		1.917 134	E+00
Length/temperature	m/K	see "Temperature, Pressure, Vacuum"				
Area	m^2	sq mile	km^2		2.589 988	E+00
		section	km^2		2.589 988	E+00
				ha	2.589 988	E+02
		acre	m^2		4.046 856	E+03
				ha	4.046 856	E−01
		ha	m^2		1.0*	E+04
		sq yd	m^2		8.361 274	E−01
		sq ft	m^2		9.290 304*	E−02
				cm^2	9.290 304*	E+02
		sq in.	mm^2		6.451 6*	E+02
				cm^2	6.451 6*	E+00
		cm^2	mm^2		1.0*	E+02
				cm^2	1.0*	E+00
		mm^2	mm^2		1.0*	E+00
Area/volume	m^2/m^3	$ft^2/in.^3$	m^2/cm^3		5.699 291	E−03
Area/mass	m^2/kg	cm^2/g	m^2/kg		1.0*	E−01
			m^2/g		1.0*	E−04
Volume, capacity	m^3	cubem	km^3		4.168 182	E+00 (1)†
		acre-ft	m^3		1.233 489	E+03
				ha·m	1.233 489	E−01
		m^3	m^3		1.0*	E+00
		cu yd	m^3		7.645 549	E−01
		bbl (42 U.S. gal)	m^3		1.589 873	E−01
		cu ft	m^3		2.831 685	E−02
			dm^3	L	2.831 685	E+01
		U.K. gal	m^3		4.546 092	E−03
			dm^3	L	4.546 092	E+00

*An asterisk indicates that the conversion factor is exact.
**Conversion factors for length, area, and volume (and related quantities) in Table 2.2 are based on the international foot. See Footnote 1 of Table 1.7, Part 1.
†See Notes 1 through 13 on Page 25.

Quantity and SI Unit		Customary Unit	Metric Unit SPE Preferred	Metric Unit Other Allowable	Conversion Factor* Multiply Customary Unit by Factor to Get Metric Unit	
		SPACE, TIME**				
Volume, capacity	m³	U.S. gal	m³		3.785 412	E−03
			dm³	L	3.785 412	E+00
		liter	dm³	L	1.0*	E+00
		U.K. qt	dm³	L	1.136 523	E+00
		U.S. qt	dm³	L	9.463 529	E−01
		U.S. pt	dm³	L	4.731 765	E−01
		U.K. fl oz	cm³		2.841 308	E+01
		U.S. fl oz	cm³		2.957 353	E+01
		cu in.	cm³		1.638 706	E+01
		mL	cm³		1.0*	E+00
Volume/length (linear displacement)	m³/m	bbl/in.	m³/m		6.259 342	E+00
		bbl/ft	m³/m		5.216 119	E−01
		ft³/ft	m³/m		9.290 304*	E−02
		U.S. gal/ft	m³/m		1.241 933	E−02
			dm³/m	L/m	1.241 933	E+01
Volume/mass	m³/kg	see "Density, Specific Volume, Concentration, Dosage"				
Plane angle	rad	rad	rad		1.0*	E+00
		deg (°)	rad		1.745 329	E−02 (2)
				°	1.0*	E+00
		min (′)	rad		2.908 882	E−04 (2)
				′	1.0*	E+00
		sec (″)	rad		4.848 137	E−06 (2)
				″	1.0*	E+00
Solid angle	sr	sr	sr		1.0*	E+00
Time	s	million years (MY)	Ma		1.0*	E+00 (3)
		yr	a		1.0*	E+00
		wk	d		7.0*	E+00
		d	d		1.0*	E+00
		hr	h		1.0*	E+00
				min	6.0*	E+01
		min	s		6.0*	E+01
				h	1.666 667	E−02
				min	1.0*	E+00
		s	s		1.0*	E+00
		millimicrosecond	ns		1.0*	E+00
		MASS, AMOUNT OF SUBSTANCE				
Mass	kg	U.K. ton (long ton)	Mg	t	1.016 047	E+00
		U.S. ton (short ton)	Mg	t	9.071 847	E−01
		U.K. ton	kg		5.080 235	E+01
		U.S. cwt	kg		4.535 924	E+01
		kg	kg		1.0*	E+00
		lbm	kg		4.535 924	E−01
		oz (troy)	g		3.110 348	E+01
		oz (av)	g		2.834 952	E+01
		g	g		1.0*	E+00
		grain	mg		6.479 891	E+01
		mg	mg		1.0*	E+00
		g	g		1.0*	E+00
Mass/length	kg/m	see "Mechanics"				
Mass/area	kg/m²	see "Mechanics"				

Quantity and SI Unit		Customary Unit	Metric Unit: SPE Preferred	Metric Unit: Other Allowable	Conversion Factor* Multiply Customary Unit by Factor to Get Metric Unit	
Mass/volume	kg/m^3	see "Density, Specific Volume, Concentration, Dosage"				
Mass/mass	kg/kg	see "Density, Specific Volume, Concentration, Dosage"				
Amount of substance	mol	lbm mol	kmol		4.535 924	E − 01
		g mol	kmol		1.0*	E − 03
		std m^3 (0°C, 1 atm)	kmol		4.461 58	E − 02 (4,13)
		std m^3 (15°C, 1 atm)	kmol		4.229 32	E − 02 (4,13)
		std ft^3 (60°F, 1 atm)	kmol		1.195 3	E − 03 (4,13)
		CALORIFIC VALUE, HEAT, ENTROPY, HEAT CAPACITY				
Calorific value (mass basis)	J/kg	Btu/lbm	MJ/kg		2.326	E − 03
			kJ/kg	J/g	2.326	E + 00
				kW·h/kg	6.461 112	E − 04
		cal/g	kJ/kg	J/g	4.184*	E + 00
		cal/lbm	J/kg		9.224 141	E + 00
Calorific value (mole basis)	J/mol	kcal/g mol	kJ/kmol		4.184*	C + 03[13]
		Btu/lbm mol	MJ/kmol		2.326	E − 03[13]
			kJ/kmol		2.326	E + 00[13]
Calorific value (volume basis — solids and liquids)	J/m^3	therm/U.K. gal	MJ/m^3	kJ/dm^3	2.320 80	E + 04
			kJ/m^3		2.320 80	E + 07
				kW·h/dm^3	6.446 660	E + 00
		Btu/U.S. gal	MJ/m^3	kJ/dm^3	2.787 163	E − 01
			kJ/m^3		2.787 163	E + 02
				kW·h/m^3	7.742 119	E − 02
		Btu/U.K. gal	MJ/m^3	kJ/dm^3	2.320 8	E − 01
			kJ/m^3		2.320 8	E + 02
				kW·h/m^3	6.446 660	E − 02
		Btu/ft^3	MJ/m^3	kJ/dm^3	3.725 895	E − 02
			kJ/m^3		3.725 895	E + 01
				kW·h/m^3	1.034 971	E − 02
		kcal/m^3	MJ/m^3	kJ/dm^3	4.184*	E − 03
			kJ/m^3		4.184*	E + 00
		cal/mL	MJ/m^3		4.184*	E + 00
		ft-lbf/U.S. gal	kJ/m^3		3.581 692	E − 01
Calorific value (volume basis — gases)	J/m^3	cal/mL	kJ/m^3	J/dm^3	4.184*	E + 03
		kcal/m^3	kJ/m^3	J/dm^3	4.184*	E + 00
		Btu/ft^3	kJ/m^3	J/dm^3	3.725 895	E + 01
				kW·h/m^3	1.034 971	E − 02
Specific entropy	J/kg·K	Btu/(lbm-°R)	kJ/(kg·K)	J/g·K	4.186 8*	E + 00
		cal/(g-°K)	kJ/(kg·K)	J/g·K	4.184*	E + 00
		kcal/(kg·°C)	kJ/(kg·K)	J/g·K	4.184*	E + 00
Specific heat capacity (mass basis)	J/kg·K	kW-hr/(kg-°C)	kJ/(kg·K)	J/g·K	3.6*	E + 03
		Btu/(lbm-°F)	kJ/(kg·K)	J/g·K	4.186 8*	E + 00
		kcal/(kg-°C)	kJ/(kg·K)	J/g·K	4.184*	E + 00
Molar heat capacity	J/mol·K	Btu/(lbm mol-°F)	kJ/(kmol·K)		4.186 8*	E + 00[13]
		cal/(g mol-°C)	kJ/(kmol·K)		4.184*	E − 00[13]
		TEMPERATURE, PRESSURE, VACUUM				
Temperature (absolute)	K	°R	K		5/9	
		°K	K		1.0*	E + 00
Temperature (traditional)	K	°F	°C		(°F − 32)/1.8	
		°C	°C		1.0*	E + 00
Temperature (difference)	K	°F	°C		5/9	E + 00
		°C	°C		1.0*	E + 00
Temperature/length (geothermal gradient)	K/m	°F/100 ft	mK/m		1.822 689	E + 01

Quantity and SI Unit		Customary Unit	Metric Unit		Conversion Factor* Multiply Customary Unit by Factor to Get Metric Unit	
			SPE Preferred	Other Allowable		
TEMPERATURE, PRESSURE, VACUUM						
Length/temperature (geothermal step)	m/K	ft/°F	m/K		5.486 4*	E−01
Pressure	Pa	atm (760mm Hg at 0°C or 14.696 (lbf/in.^{2})	MPa		1.013 25*	E−01
			kPa		1.013 25*	E+02
				bar	1.013 25*	E+00
		bar	MPa		1.0*	E−01
			kPa		1.0*	E+02
				bar	1.0*	E+00
		at (technical atm., kgf/cm^2)	MPa		9.806 65*	E−02
			kPa		9.806 65*	E+01
				bar	9.806 65*	E−01
Pressure	Pa	lbf/in.^{2} (psi)	MPa		6.894 757	E−03
			kPa		6.894 757	E+00
				bar	6.894 757	E−02
		in. Hg (32°F)	kPa		3.386 38	E+00
		in. Hg (60°F)	kPa		3.376 85	E+00
		in. H_2O (39.2°F)	kPa		2.490 82	E−01
		in. H_2O (60°F)	kPa		2.488 4	E−01
		mm Hg (0°C) = torr	kPa		1.333 224	E−01
		cm H_2O (4°C)	kPa		9.806 38	E−02
		lbf/ft^2 (psf)	kPa		4.788 026	E−02
		μm Hg (0°C)	Pa		1.333 224	E−01
		μbar	Pa		1.0*	E−01
		dyne/cm^2	Pa		1.0*	E−01
Vacuum, draft	Pa	in. Hg (60°F)	kPa		3.376 85	E+00
		in. H_2O (39.2°F)	kPa		2.490 82	E−01
		in. H_2O (60°F)	kPa		2.488 4	E−01
		mm Hg (0°C) = torr	kPa		1.333 224	E−01
		cm H_2O (4°C)	kPa		9.806 38	E−02
Liquid head	m	ft	m		3.048*	E−01
		in.	mm		2.54*	E+01
				cm	2.54*	E+00
Pressure drop/length	Pa/m	psi/ft	kPa/m		2.262 059	E+01
		psi/100 ft	kPa/m		2.262 059	E−01[5]
DENSITY, SPECIFIC VOLUME, CONCENTRATION, DOSAGE						
Density (gases)	kg/m^3	lbm/ft^3	kg/m^3		1.601 846	E+01
			g/m^3		1.601 846	E+04
Density (liquids)	kg/m^3	lbm/U.S. gal	kg/m^3		1.198 264	E+02
				g/cm^3	1.198 264	E−01
		lbm/U.K. gal	kg/m^3		9.977 633	E+01
				kg/dm^3	9.977 633	E−02
		lbm/ft^3	kg/m^3		1.601 846	E+01
				g/cm^3	1.601 846	E−02
		g/cm^3	kg/m^3		1.0*	E+03
				kg/dm^3	1.0*	E+00
		°API	g/cm^3		141.5/(131.5+°API)	
Density (solids)	kg/m^3	lbm/ft^3	kg/m^3		1.601 846	E+01
Specific volume (gases)	m^3/kg	ft^3/lbm	m^3/kg		6.242 796	E−02
			m^3/g		6.242 796	E−05
Specific volume (liquids)	m^3/kg	ft^3/lbm	dm^3/kg		6.242 796	E+01
		U.K. gal/lbm	dm^3/kg	cm^3/g	1.002 242	E+01
		U.S. gal/lbm	dm^3/kg	cm^3/g	8.345 404	E+00

Quantity and SI Unit		Customary Unit	Metric Unit: SPE Preferred	Metric Unit: Other Allowable	Conversion Factor* Multiply Customary Unit by Factor to Get Metric Unit	
DENSITY, SPECIFIC VOLUME, CONCENTRATION, DOSAGE						
Specific volume (mole basis)	m³/mol	L/g mol	m³/kmol		1.0*	E+00[13]
		ft³/lbm mol	m³/kmol		6.242 796	E−02[13]
Specific volume (clay yield)	m³/kg	bbl/U.S. ton	m³/t		1.752 535	E−01
		bbl/U.K. ton	m³/t		1.564 763	E−01
Yield (shale distillation)	m³/kg	bbl/U.S. ton	dm³/t	L/t	1.752 535	E+02
		bbl/U.K. ton	dm³/t	L/t	1.564 763	E+02
		U.S. gal/U.S. ton	dm³/t	L/t	4.172 702	E+00
		U.S. gal/U.K. ton	dm³/t	L/t	3.725 627	E+00
Concentration (mass/mass)	kg/kg	wt %	kg/kg g/kg		1.0* 1.0*	E−02 E+01
		wt ppm	mg/kg		1.0*	E+00
Concentration (mass/volume)	kg/m³	lbm/bbl	kg/m³	g/dm³	2.853 010	E+00
		g/U.S. gal	kg/m³		2.641 720	E−01
		g/U.K. gal	kg/m³	g/L	2.199 692	E−01
Concentration (mass/volume)	kg/m³	lbm/1000 U.S. gal	g/m³	mg/dm³	1.198 264	E+02
		lbm/1000 U.K. gal	g/m³	mg/dm³	9.977 633	E+01
		grains/U.S. gal	g/m³	mg/dm³	1.711 806	E+01
		grains/ft³	mg/m³		2.288 352	E+03
		lbm/1000 bbl	g/m³	mg/dm³	2.853 010	E+00
		mg/U.S. gal	g/m³	mg/dm³	2.641 720	E−01
		grains/100 ft³	mg/m³		2.288 352	E+01
Concentration (volume/volume)	m³/m³	bbl/bbl	m³/m³		1.0*	E+00
		ft³/ft³	m³/m³		1.0*	E+00
		bbl/acre·ft	m³/m³	 m³/ha·m	1.288 923 1.288 923	E−04 E+00
		vol %	m³/m³		1.0*	E−02
		U.K. gal/ft³	dm³/m³	L/m³	1.605 437	E+02
		U.S. gal/ft³	dm³/m³	L/m³	1.336 806	E+02
		mL/U.S. gal	dm³/m³	L/m³	2.641 720	E−01
		mL/U.K. gal	dm³/m³	L/m³	2.199 692	E−01
		vol ppm	cm³/m³ dm³/m³	 L/m³	1.0* 1.0*	E+00 E−03
		U.K. gal/1000 bbl	cm³/m³		2.859 406	E+01
		U.S. gal/1000 bbl	cm³/m³		2.380 952	E+01
		U.K. pt/1000 bbl	cm³/m³		3.574 253	E+00
Concentration (mole/volume)	mol/m³	lbm mol/U.S. gal	kmol/m³		1.198 264	E+02
		lbm mol/U.K. gal	kmol/m³		9.977 633	E+01
		lbm mol/ft³	kmol/m³		1.601 846	E+01
		std ft³ (60°F, 1 atm)/bbl	kmol/m³		7.518 18	E−03
Concentration (volume/mole)	m³/mol	U.S. gal/1000 std ft³ (60°F/60°F)	dm³/kmol	L/kmol	3.166 93	E+00
		bbl/million std ft³ (60°F/60°F)	dm³/kmol	L/kmol	1.330 11	E−01
FACILITY THROUGHPUT, CAPACITY						
Throughput (mass basis)	kg/s	million lbm/yr	t/a	Mg/a	4.535 924	E+02
		U.K. ton/yr	t/a	Mg/a	1.016 047	E+00
		U.S. ton/yr	t/a	Mg/a	9.071 847	E−01
		U.K. ton/D	t/d	Mg/d t/h, Mg/h	1.016 047 4.233 529	E+00 E−02

Quantity and SI Unit		Customary Unit	Metric Unit: SPE Preferred	Metric Unit: Other Allowable	Conversion Factor* Multiply Customary Unit by Factor to Get Metric Unit	
FACILITY THROUGHPUT, CAPACITY (6)						
		U.S. ton/D	t/d		9.071 847	E−01
				t/h, Mg/h	3.779 936	E−02
		U.K. ton/hr	t/h	Mg/h	1.016 047	E+00
		U.S. ton/hr	t/h	Mg/h	9.071 847	E−01
		lbm/hr	kg/h		4.535 924	E−01
Throughput (volume basis)	m^3/s	bbl/D	t/a		5.803 036	E+01 (7)
				m^3/d	1.589 873	E−01
			m^3/h		6.624 471	E−03
		ft^3/D	m^3/h		1.179 869	E−03
				m^3/d	2.831 685	E−02
		bbl/hr	m^3/h		1.589 873	E−01
		ft^3/h	m^3/h		2.831 685	E−02
		U.K. gal/hr	m^3/h		4.546 092	E−03
				L/s	1.262 803	E−03
		U.S. gal/hr	m^3/h		3.785 412	E−03
				L/s	1.051 503	E−03
		U.K. gal/min	m^3/h		2.727 655	E−01
				L/s	7.576 819	E−02
		U.S. gal/min	m^3/h		2.271 247	E−01
				L/s	6.309 020	E−02
Throughput (mole basis)	mol/s	lbm mol/hr	kmol/h		4.535 924	E−01
				kmol/s	1.259 979	E−04
FLOW RATE (6)						
Pipeline capacity	m^3/m	bbl/mile	m^3/km		9.879 013	E−02
Flow rate (mass basis)	kg/s	U.K. ton/min	kg/s		1.693 412	E+01
		U.S. ton/min	kg/s		1.511 974	E+01
		U.K. ton/hr	kg/s		2.822 353	E−01
		U.S. ton/hr	kg/s		2.519 958	E−01
		U.K. ton/D	kg/s		1.175 980	E−02
		U.S. ton/D	kg/s		1.049 982	E−02
		million lbm/yr	kg/s		5.249 912	E+00
		U.K. ton/yr	kg/s		3.221 864	E−05
		U.S. ton/yr	kg/s		2.876 664	E−05
		lbm/s	kg/s		4.535 924	E−01
		lbm/min	kg/s		7.559 873	E−03
		lbm/hr	kg/s		1.259 979	E−04
Flow rate (volume basis)	m^3/s	bbl/D	m^3/d		1.589 873	E−01
				L/s	1.840 131	E−03
		ft^3/D	m^3/d		2.831 685	E−02
				L/s	3.277 413	E−04
		bbl/hr	m^3/s		4.416 314	E−05
				L/s	4.416 314	E−02
		ft^3/hr	m^3/s		7.865 791	E−06
				L/s	7.865 791	E−03
		U.K. gal/hr	dm^3/s	L/s	1.262 803	E−03
		U.S. gal/hr	dm^3/s	L/s	1.051 503	E−03
		U.K. gal/min	dm^3/s	L/s	7.576 820	E−02
		U.S. gal/min	dm^3/s	L/s	6.309 020	E−02
		ft^3/min	dm^3/s	L/s	4.719 474	E−01
		ft^3/s	dm^3/s	L/s	2.831 685	E+01

Quantity and SI Unit		Customary Unit	Metric Unit: SPE Preferred	Metric Unit: Other Allowable	Conversion Factor* Multiply Customary Unit by Factor to Get Metric Unit	
Flow rate (mole basis)	mol/s	lbm mol/s	kmol/s		4.535 924	E−01[13]
		lbm mol/hr	kmol/s		1.259 979	E−04[13]
		million scf/D	kmol/s		1.383 449	E−02[13]
Flow rate/length (mass basis)	kg/s·m	lbm/(s-ft)	kg/(s·m)		1.488 164	E+00
		lbm/(hr-ft)	kg/(s·m)		4.133 789	E−04
Flow rate/length (volume basis)	m^2/s	U.K. gal/(min-ft)	m^2/s	$m^3/(s·m)$	2.485 833	E−04
		U.S. gal/(min-ft)	m^2/s	$m^3/(s·m)$	2.069 888	E−04
		U.K. gal/(hr-in.)	m^2/s	$m^3/(s·m)$	4.971 667	E−05
		U.S. gal/(hr-in.)	m^2/s	$m^3/(s·m)$	4.139 776	E−05
		U.K. gal/(hr-ft)	m^2/s	$m^3/(s·m)$	4.143 055	E−06
		U.S. gal/(hr-ft)	m^2/s	$m^3/(s·m)$	3.449 814	E−06
Flow rate/area (mass basis)	$kg/s·m^2$	$lbm/(s-ft^2)$	$kg/s·m^2$		4.882 428	E+00
		$lbm/(hr-ft^2)$	$kg/s·m^2$		1.356 230	E−03
Flow rate/area (volume basis)	m/s	$ft^3/(s-ft^2)$	m/s	$m^3(s·m^2)$	3.048*	E−01
		$ft^3/(min-ft^2)$	m/s	$m^3/(s·m^2)$	5.08*	E−03
		U.K. $gal/(hr-in.^2)$	m/s	$m^3/(s·m^2)$	1.957 349	E−03
		U.S. $gal/(hr-in.^2)$	m/s	$m^3/(s·m^2)$	1.629 833	E−03
		U.K. $gal/(min-ft^2)$	m/s	$m^3/(s·m^2)$	8.155 621	E−04
		U.S. $gal/(min-ft^2)$	m/s	$m^3/(s·m^2)$	6.790 972	E−04
		U.K. $gal/(hr-ft^2)$	m/s	$m^3/(s·m^2)$	1.359 270	E−05
		U.S. $gal/(hr-ft^2)$	m/s	$m^3/(s·m^2)$	1.131 829	E−05
Flow rate/ pressure drop (productivity index)	$m^3/s·Pa$	bbl/(D-psi)	$m^3/(d·kPa)$		2.305 916	E−02
ENERGY, WORK, POWER						
Energy, work	J	quad	MJ		1.055 056	E+12
			TJ		1.055 056	E+06
			EJ		1.055 056	E+00
				MW·h	2.930 711	E+08
				GW·h	2.930 711	E+05
				TW·h	2.930 711	E+02
		therm	MJ		1.055 056	E+02
			kJ		1.055 056	E+05
				kW·h	2.930 711	E+01
		U.S. tonf-mile	MJ		1.431 744	E+01
		hp-hr	MJ		2.684 520	E+00
			kJ		2.684 520	E+03
				kW·h	7.456 999	E−01
		ch-hr or CV-hr	MJ		2.647 796	E+00
			Kj		2.647 796	E+03
				kW·h	7.354 99	E−01
		kW-hr	MJ		3.6*	E+00
			kJ		3.6*	E+03
		Chu	kJ		1.899 101	E+00
				kW·h	5.275 280	E−04
		Btu	kJ		1.055 056	E+00
				kW·h	2.930 711	E−04
		kcal	kJ		4.184*	E+00
		cal	kJ		4.184*	E−03
		ft-lbf	kJ		1.355 818	E−03
		lbf-ft	kJ		1.355 818	E−03
		J	kJ		1.0*	E−03
		$lbf-ft^2/s^2$	kJ		4.214 011	E−05
		erg	J		1.0*	E−07

Quantity and SI Unit		Customary Unit	Metric Unit SPE Preferred	Metric Unit Other Allowable	Conversion Factor* Multiply Customary Unit by Factor to Get Metric Unit	
		ENERGY, WORK, POWER				
Impact energy	J	kgf-m	J		9.806 650*	E+00
		lbf-ft	J		1.355 818	E+00
Work/length	J/m	U.S. tonf-mile/ft	MJ/m		4.697 322	E+01
Surface energy	J/m²	erg/cm²	mJ/m²		1.0*	E+00
Specific impact energy	J/m²	kgf·m/cm²	J/cm²		9.806 650*	E−00
		lbf·ft/in.²	J/cm²		2.101 522	E−01
Power	W	quad/yr	MJ/a TJ/a EJ/a		1.055 056 1.055 056 1.055 056	E+12 E+06 E+00
		erg/a	TW GW		3.170 979 3.170 979	E−27 E−24
		million Btu/hr	MW		2.930 711	E−01
		ton of refrigeration	kW		3.516 853	E+00
		Btu/s	kW		1.055 056	E+00
		kW	kW		1.0*	E+00
		hydraulic horse-power — hhp	kW		7.460 43	E−01
		hp (electric)	kW		7.46*	E−01
		hp (550 ft-lbf/s)	kW		7.456 999	E−01
		ch or CV	kW		7.354 99	E−01
		Btu/min	kW		1.758 427	E−02
		ft·lbf/s	kW		1.355 818	E−03
		kcal/hr	W		1.162 222	E+00
		Btu/hr	W		2.930 711	E−01
		ft·lbf/min	W		2.259 697	E−02
Power/area	W/m²	Btu/s·ft²	kW/m²		1.135 653	E+01
		cal/hr·cm²	kW/m²		1.162 222	E−02
		Btu/hr·ft²	kW/m²		3.154 591	E−03
Heat flow unit — hfu (geothermics)		μcal/s·cm²	mW/m²		4.184*	E+01
Heat release rate, mixing power	W/m³	hp/ft³	kW/m³		2.633 414	E+01
		cal/hr·cm³	kW/m³		1.162 222	E+00
		Btu/s·ft³	kW/m³		3.725 895	E+01
		Btu/hr·ft³	kW/m³		1.034 971	E−02
Heat generation unit — hgu (radioactive rocks)		cal/(s-cm³)	μW/m³		4.184*	E+12
Cooling duty (machinery)	W/W	Btu/(bhp-hr)	W/kW		3.930 148	E−01
Specific fuel consumption (mass basis)	kg/J	lbm/(hp-hr)	mg/J	kg/MJ kg/kW·h	1.689 659 6.082 774	E−01 E−01
Specific fuel consumption (volume basis)	m³/J	m³/(kW-hr)	dm³/MJ	mm³/J dm³/kW·h	2.777 778 1.0*	E+02 E+03
		U.S. gal/(hp-hr)	dm³/MJ	mm³/J dm³/kW·h	1.410 089 5.076 321	E+00 E+00
		U.K. pt/(hp-hr)	dm³/MJ	mm³/J dm³/kW·h	2.116 809 7.620 512	E−01 E−01
Fuel consumption (automotive)	m³/m	U.K. gal/mile	dm³/100 km	L/100 km	2.824 811	E+02
		U.S. gal/mile	dm³/100 km	L/100 km	2.352 146	E+02
		mile/U.S. gal	km/dm³	km/L	4.251 437	E−01
		mile/U.K. gal	km/dm³	km/L	3.540 060	E−01

Quantity and SI Unit		Customary Unit	Metric Unit SPE Preferred	Metric Unit Other Allowable	Conversion Factor* Multiply Customary Unit by Factor to Get Metric Unit	
		MECHANICS				
Velocity (linear), speed	m/s	knot	km/h		1.852*	E+00
		mile/hr	km/h		1.609 344*	E+00
		m/s	m/s		1.0*	E+00
		ft/s	m/s		3.048*	E−01
				cm/s	3.048*	E+01
				m/ms	3.048*	E−04(8)
		ft/min	m/s		5.08*	E−03
				cm/s	5.08*	E−01
		ft/hr	mm/s		8.466 667	E−02
				cm/s	8.466 667	E−03
		ft/D	mm/s		3.527 778	E−03
				m/d	3.048*	E−01
		in./s	mm/s		2.54*	E+01
				cm/s	2.54*	E+00
		in./min	mm/s		4.233 333	E−01
				cm/s	4.233 333	E−02
Velocity (angular)	rad/s	rev/min	rad/s		1.047 198	E−01
		rev/s	rad/s		6.283 185	E+00
		degree/min	rad/s		2.908 882	E−04
Reciprocal velocity	s/m	s/ft	s/m		3.280 840	E+00(9)
Corrosion rate	m/s	in./yr (ipy)	mm/a		2.54*	E+01
		mil/yr	mm/a		2.54*	E−02
Rotational frequency	rev/s	rev/s	rev/s		1.0*	E+00
		rev/min	rev/s		1.666 667	E−02
		rev/min	rad/s		1.047 198	E−01
Acceleration (linear)	m/s²	ft/s²	m/s²		3.048*	E−01
				cm/s²	3.048*	E+01
		gal(cm/s²)	m/s²		1.0*	E−02
Acceleration (rotational)	rad/s²	rad/s²	rad/s²		1.0*	E+00
		rpm/s	rad/s²		1.047 198	E−01
Momentum	kg·m/s	lbm·ft/s	kg·m/s		1.382 550	E−01
Force	N	U.K. tonf	kN		9.964 016	E+00
		U.S. tonf	kN		8.896 443	E+00
		kgf (kp)	N		9.806 650*	E+00
		lbf	N		4.448 222	E+00
		N	N		1.0*	E+00
		pdl	mN		1.382 550	E+02
		dyne	mN		1.0*	E−02
Bending moment, torque	N·m	U.S. tonf-ft	kN·m		2.711 636	E+00(10)
		kgf-m	N·m		9.806 650*	E+00(10)
		lbf-ft	N·m		1.355 818	E+00(10)
		lbf-in.	N·m		1.129 848	E−01(10)
		pdl-ft	N·m		4.214 011	E−02(10)
Bending moment/ length	N·m/m	(lbf-ft)/in.	(N·m)/m		5.337 866	E+01(10)
		(kgf-m)/m	(N·m)/m		9.806 650*	E+00(10)
		(lbf-in.)/in.	(N·m)/m		4.448 222	E+00(10)
Elastic moduli (Young's, Shear bulk)	Pa	lbf/in.²	GPa		6.894 757	E−06
Moment of inertia	kg·m²	lbm-ft²	kg·m²		4.214 011	E−02
Moment of section	m⁴	in.⁴	cm⁴		4.162 314	E+01
Section modulus	m³	cu in.	cm³		1.638 706	E+01
		cu ft	cm³		1.638 706	E+04

Quantity and SI Unit		Customary Unit	Metric Unit SPE Preferred	Metric Unit Other Allowable	Conversion Factor* Multiply Customary Unit by Factor to Get Metric Unit	
MECHANICS						
				mm³	2.831 685	E+04
				m³	2.831 685	E−02
Stress	Pa	U.S. tonf/in.²	MPa	N/mm²	1.378 951	E+01
		kgf/mm²	MPa	N/mm²	9.806 650*	E+00
		U.S. tonf/ft²	MPa	N/mm²	9.576 052	E−02
		lbf/in.² (psi)	MPa	N/mm²	6.894 757	E−03
		lbf/ft² (psf)	kPa		4.788 026	E−02
		dyne/cm²	Pa		1.0*	E−01
Yield point, gel strength (drilling fluid)		lbf/100 ft²	Pa		4.788 026	E−01
Mass/length	kg/m	lbm/ft	kg/m		1.488 164	E+00
Mass/area structural loading, bearing capacity (mass basis)	kg/m²	U.S. ton/ft²	Mg/m²		9.764 855	E+00
		lbm/ft²	kg/m²		4.882 428	E+00
Coefficient of thermal expansion	m/(m·K)	in./(in.-°F)	mm/(mm·K)		5.555 556	E−01
TRANSPORT PROPERTIES						
Diffusivity	m²/s	ft²/s	mm²/s		9.290 304*	E+04
		cm²/s	mm²/s		1.0*	E+02
		ft²/hr	mm²/s		2.580 64*	E+01
Thermal resistance	k·m²/W	(°C-m²·hr)/kcal	(K·m²)/kW		8.604 208	E+02
		(°F-ft² hr)/Btu	(K·m²)/kW		1.761 102	E+02
Heat flux	W/m²	Btu/(hr-ft²)	kW/m²		3.154 591	E−03
Thermal conductivity	W/m·K	(cal/s-cm²-°C)/cm	W/(m·K)		4.184*	E+02
		Btu/(hr-ft²-°F/ft)	W/(m·K)		1.730 735	E+00
				kJ·m/(h·m²·K)	6.230 646	E+00
		kcal/(hr-m²-°C/m)	W/(m·K)		1.162 222	E+00
		Btu/(hr-ft²-°F/in.)	W/(m·K)		1.442 279	E−01
		cal/(hr-cm²-°C/cm)	W/(m·K)		1.162 222	E−01
Heat transfer coefficient	W/m²·K	cal/(s-cm²-°C)	kW/(m²·K)		4.184*	E+01
		Btu/(s-ft²-°F)	kW/(m²·K)		2.044 175	E+01
		cal/(hr-cm²-°C)	kW/(m²·K)		1.162 222	E−02
		Btu/(hr-ft²-°F)	kW/(m²·K)		5.678 263	E−03
				kJ/(h·m²·K)	2.044 175	E+01
		Btu/(hr-ft²-°R)	kW/(m²·K)		5.678 263	E−03
		kcal/(hr-m²-°C)	kW/(m²·K)		1.162 222	E−03
Volumetric heat transfer coefficient	W/m³·K	Btu/(s-ft³-°F)	kW/(m³·K)		6.706 611	E+01
		Btu/(hr-ft³-°F)	kW/(m³·K)		1.862 947	E−02
Surface tension	N/m	dyne/cm	mN/m		1.0*	E+00
Viscosity (dynamic)	Pa·s	(lbf-s)/in.²	Pa·s	(N·s)/m²	6.894 757	E+03
		(lbf-s)/ft²	Pa·s	(N·s)/m²	4.788 026	E+01
		(kgf-s)/m²	Pa·s	(N·s)/m²	9.806 650*	E+00
		lbm/(ft-s)	Pa·s	(N·s)/m²	1.488 164	E+00
		(dyne-s)/cm²	Pa·s	(N·s)/m²	1.0*	E−01
		cp	Pa·s	(N·s)/m²	1.0*	E−03
		lbm/(ft·hr)	Pa·s	(N·s)/m²	4.133 789	E−04
Viscosity (kinematic)	m²/s	ft²/s	mm²/s		9.290 304*	E+04
		in.²/s	mm²/s		6.451 6*	E+02

Quantity and SI Unit		Customary Unit	Metric Unit SPE Preferred	Other Allowable	Conversion Factor* Multiply Customary Unit by Factor to Get Metric Unit	
		TRANSPORT PROPERTIES				
		m²/hr	mm²/s		2.777 778	E+02
		cm²/s	mm²/s		1.0*	E+02
		ft²/hr	mm²/s		2.580 64*	E+01
		cSt	mm²/s		1.0*	E+00
Permeability	m²	darcy	μm²		9.869 233	E−01[11]
		millidarcy	μm²		9.869 233	E−04[11]
				10^{-3} μm²	9.869 233	E−01[11]
		ELECTRICITY, MAGNETISM				
Admittance	S	S	S		1.0*	E+00
Capacitance	F	μF	μF		1.0*	E+00
Capacity, storage battery	C	A-hr	kC		3.6*	E+00
Charge density	C/m³	C/mm³	C/mm³		1.0*	E+00
Conductance	S	S	S		1.0*	E+00
		℧ (mho)	S		1.0*	E+00
Conductivity	S/m	S/m	S/m		1.0*	E+00
		℧/m	S/m		1.0*	E+00
		m℧/m	mS/m		1.0*	E+00
Current density	A/m²	A/mm²	A/mm²		1.0*	E+00
Displacement	C/m²	C/cm²	C/cm²		1.0*	E+00
Electric charge	C	C	C		1.0*	E+00
Electric current	A	A	A		1.0*	E+00
Electric dipole moment	C·m	C·m	C·m		1.0*	E+00
Electric field strength	V/m	V/m	V/m		1.0*	E+00
Electric flux	C	C	C		1.0*	E+00
Electric polarization	C/m²	C/cm²	C/cm²		1.0*	E+00
Electric potential	V	V	V		1.0*	E+00
		mV	mV		1.0*	E+00
Electromagnetic moment	A·m²	A·m²	A·m²		1.0*	E+00
Electromotive force	V	V	V		1.0*	E+00
Flux of displacement	C	C	C		1.0*	E+00
Frequency	Hz	cycles/s	Hz		1.0′	E+00
Impedance	Ω	Ω	Ω		1.0*	E+00
Interval transit time	s/m	μs/ft	μs/m		3.280 840	E+00
Linear current density	A/m	A/mm	A/mm		1.0*	E+00
Magnetic dipole moment	Wb·m	Wb·m	Wb·m		1.0*	E+00
Magnetic field strength	A/m	A/mm	A/mm		1.0*	E+00
		oersted	A/m		7.957 747	E+01
		gamma	A/m		7.957 747	E−04
Magnetic flux	Wb	mWb	mWb		1.0*	E+00
Magnetic flux density	T	mT	mT		1.0*	E+00
		gauss	T		1.0*	E−04
Magnetic induction	T	mT	mT		1.0*	E+00
Magnetic moment	A·m²	A-m²	A·m²		1.0*	E+00
Magnetic polarization	T	mT	mT		1.0*	E+00

Quantity and SI Unit		Customary Unit	Metric Unit		Conversion Factor* Multiply Customary Unit by Factor to Get Metric Unit	
			SPE Preferred	Other Allowable		
		ELECTRICITY, MAGNETISM				
Magnetic potential difference	A	A	A		1.0*	E+00
Magnetic vector potential	Wb/m	Wb/mm	Wb/mm		1	
Magnetization	A/m	A/mm	A/mm		1	
Modulus of admittance	S	S	S		1	
Modulus of impedance	Ω	Ω	Ω		1	
Mutual inductance	H	H	H		1	
Permeability	H/m	μH/m	μH/m		1	
Permeance	H	H	H		1	
Permittivity	F/m	μF/m	μF/m		1	
Potential difference	V	V	V		1	
Quantity of electricity	C	C	C		1	
Reactance	Ω	Ω	Ω		1	
Reluctance	H^{-1}	H^{-1}	H^{-1}		1	
Resistance	Ω	Ω	Ω		1	
Resistivity	Ω·m	Ω·cm	Ω·cm		1	
		Ω·m	Ω·m		1	(12)
Self inductance	H	mH	mH		1	
Surface density of charge	C/m^2	mC/m^2	mC/m^2		1	
Susceptance	S	S	S		1	
Volume density of charge	C/m^3	C/mm^3	C/mm^3		1	
		ACOUSTICS, LIGHT, RADIATION				
Absorbed dose	Gy	rad	Gy		1.0*	E−02
Acoustical energy	J	J	J		1	
Acoustical intensity	W/m^2	W/cm^2	W/m^2		1.0*	E+04
Acoustical power	W	W	W		1	
Sound pressure	N/m^2	N/m^2	N/m^2		1	
Illuminance	lx	footcandle	lx		1.076 391	E+01
Illumination	lx	footcandle	lx		1.076 391	E+01
Irradiance	W/m^2	W/m^2	W/m^2		1	
Light exposure	lx·s	footcandle·s	lx·s		1.076 391	E+01
Luminance	cd/m^2	cd/m^2	cd/m^2		1	
Luminous efficacy	lm/W	lm/W	lm/W		1	
Luminous exitance	lm/m^2	lm/m^2	lm/m^2		1	
Luminous flux	lm	lm	lm		1	
Luminous intensity	cd	cd	cd		1	
Quantity of light	ℓm·s	talbot	ℓm·s		1.0*	E+00
Radiance	$W/(m^2 \cdot sr)$	$W/(m^2 \cdot sr)$	$W/(m^2 \cdot sr)$		1	
Radiant energy	J	J	J		1	
Radiant flux	W	W	W		1	
Radiant intensity	W/sr	W/sr	W/sr		1	
Radiant power	W	W	W		1	
Wave length	m	Å	nm		1.0*	E−01
Capture unit	m^{-1}	$10^{-3}cm^{-1}$	m^{-1}		1.0*	E+01
				$10^{-3}cm^{-1}$	1	
		m^{-1}	m^{-1}		1	
Radioactivity		curie	Bq		3.7*	E+10

SOME ADDITIONAL APPLICATION STANDARDS

Quantity and SI Unit		Customary Unit	Metric Unit		Conversion Factor* Multiply Customary Unit by Factor to Get Metric Unit	
			SPE Preferred	Other Allowable		
Capillary pressure	Pa	ft (fluid)	m (fluid)		3.048*	E−01
Compressibility of reservoir fluid	Pa^{-1}	psi^{-1}	Pa^{-1}		1.450 377	E−04
				kPa^{-1}	1.450 377	E−01
Corrosion allowance	m	in.	mm		2.54*	E+01
Corrosion rate	m/s	mil/yr (mpy)	mm/a		2.54*	E−02
Differential orifice pressure	Pa	in. H_2O (at 60°F)	kPa		2.488 4	E−01
				cm H_2O	2.54*	E+00
Gas-oil ratio	m^3/m^3	scf/bbl	"standard" m^3/m^3		1.801 175	E−01[1]**
Gas rate	m^3/s	scf/D	"standard" m^3/d		2.863 640	E−02[1]
Geologic time	s	yr	Ma			
Head (fluid mechanics)	m	ft	m		3.048*	E−01
				cm	3.048*	E+01
Heat exchange rate	W	Btu/hr	kW		2.930 711	E−04
				kJ/h	1.055 056	E+00
Mobility	$m^2/Pa·s$	d/cp	$\mu m^2/mPa·s$		9.869 233	E−01
				$\mu m^2/Pa·s$	9.869 233	E+02
Net pay thickness	m	ft	m		3.048*	E−01
Oil rate	m^3/s	bbl/D	m^3/d		1.589 873	E−01
		short ton/yr	Mg/a	t/a	9.071 847	E−01
Particle size	m	micron	μm		1	
Permeability-thickness	m^3	md-ft	md·m	$\mu m^2·m$	3.008 142	E−04
Pipe diameter (actual)	m	in.	cm		2.54*	E+00
				mm	2.54*	E+01
Pressure buildup per cycle	Pa	psi	kPa		6.894 757	E+00[2]
Productivity index	$m^3/Pa·s$	bbl/(psi-D)	$m^3/(kPa·d)$		2.305 916	E−02[2]
Pumping rate	m^3/s	U.S. gal/min	m^3/h		2.271 247	E−01
				L/s	6.309 020	E−02
Revolutions per minute	rad/s	rpm	rad/s		1.047 198	E−01
				rad/m	6.283 185	E+00
Recovery/unit volume (oil)	m^3/m^3	bbl/(acre-ft)	m^3/m^3		1.288 931	E−04
				$m^3/ha·m$	1.288 931	E+00
Reservoir area	m^2	sq mile	km^2		2.589 988	E+00
		acre		ha	4.046 856	E−01
Reservoir volume	m^3	acre-ft	m^3		1.233 482	E+03
				ha·m	1.233 482	E−01
Specific productivity index	$m^3/Pa·s·m$	bbl/(D-psi-ft)	$m^3/(kPa·d·m)$		7.565 341	E−02[2]
Surface or interfacial tension in reservoir capillaries	N/m	dyne/cm	mN/m		1.0*	E+00
Torque	N·m	lbf-ft	N·m		1.355 818	E+00[3]
Velocity (fluid flow)	m/s	ft/s	m/s		3.048*	E−01
Vessel diameter	m					
1-100 cm		in.	cm		2.54*	E+00
above 100 cm		ft	m		3.048*	E−01

*An asterisk indicates the conversion factor is exact.
**See Notes 1-3 on page 1598.

Index

Abnormal pressure, 53, 155, 156, 157, 204;
  detection, 1;
  indicators, 55, 197;
  origin of, 156;
  zones of, 156, 197
Abnormal spontaneous potential (E_{spA}). *See* spontaneous potential
Abnormally high porosity. *See* porosity
Absolute permeability, 19, 25, 26, 73. *See also* permeability
Absorbed water. *See* water of hydration
Absorption capacity, 169, 196
Absorption cross-section. *See* cross-section
Accelerators, 172, 173
Accelerator-type neutron generator, 197
Acoustic-density crossplot, 239
Acoustic disturbance, 143. *See also* acoustic waves, elastic waves
Acoustic energy, 145, 158
Acoustic log, 1, 3, 4, 13, 144, 208, 210, 211;
  downhole, 143;
  tools, 146, 148, 149. *See also* wireline logs
Acoustic measurements, 149, 251
Acoustic-neutron response, 231, 241
Acoustic path, 144
Acoustic ray, 144
Acoustic response, 153, 230, 241
Acoustic transit time, 153, 157, 201, 202;
  logs, 154, 190, 196
Acoustic velocities, 27, 143, 152;
  logging, 143;
  tools, 145
Acoustic wave, 143, 148, 152;
  attenuation, 144;
  receivers, 143, 144;
  transmitters, 143, 144;
  velocity, 152
Air mud system. *See* mud system
Air-permeability apparatus, 26
Alpha-neutron reaction, 172, 173
Alpha particles, 167, 168, 169, 172, 176
Aluminum, 78, 169, 173
Amplitude, 144, 145, 146, 148, 149;
  attenuation, 146, 151;
  of peak or trough, 151
Angle of incidence, 145
Angle of reflection, 145
Angle of refraction, 145
Anhydrite, 4, 32, 239
Anhydritic dolomites. *See* dolomites
Annihilation, 172
Annular pressure drop, 48
Annulus, 31, 117, 121
Anomalous zones, 209, 211;
  location of, 215
AO spacing, 104, 106. *See also* electrode spacing
API oil gravity, 21;
  initial, 21;
  recovered, 21
API units, 179, 183, 185
Apparent bulk density, 197
Apparent conductivity signal, C_a, 111, 112
Apparent interval transit time values, 241
Apparent matrix density, 241
Apparent porosity, 109, 204
Apparent resistivity, R_a, 100, 102, 104, 106, 128, 131;
  value, 102
Apparent water saturation, 121, 211, 229
Apparent water trend, 221
Archie equation, 64, 67, 69, 72, 155, 215. *See also* saturation equation
Archimedes' principle, 22
ASTM extraction apparatus, 19, 20, 21
Atoms, 78, 167, 172, 174
Atomic number, 168, 169, 171, 172, 192
Atomic weight, A, 169, 192
Avogadro's number, 79

Back-scattered gamma rays. *See* gamma rays
Barn, 176
Bed boundaries, 77, 84, 87, 101, 102, 104, 107, 108, 122, 132, 185
Bed resistivity, 84, 128;
  influence of, 86, 91
Bed thickness, 77, 84, 91, 100, 101, 102, 112, 128;
  correction charts, 112;
  effects on lifetime measurements, 198;
  influence on SP, 86
Beryllium, 172, 173, 175
Beta particles, 167, 168–169
Biot's continuum theory, 152
Bitumens, 184
Boltzman relation, 79
Borehole, 57, 77, 84, 85, 86, 87, 91, 100, 101, 102, 103, 109, 110, 112, 117, 122, 130, 177;
  cased, 102, 104;
  conditions, 99;
  departure curves, 136;
  effects, 101, 104, 128, 136;
  environment, 207;
  fluids, 96, 102, 104, 106, 109, 112, 130, 143, 148, 161, 179, 197, 207;
  gas-filled, 189;
  geometry, 112, 199;
  influence, 112, 128;
  magnitude, 116;
  measurements, 1, 2, 3, 104;
  muds, 4, 122;
  potential, 84;
  size, 104, 149, 184, 188;
  system, 109, 112, 135, 145, 197;
  temperature at any desired depth, 6
Borehole-compensated acoustic log, 151, 154
Borehole signal contribution, 112
Borehole televiewer, 1, 160, 161. *See also* wireline logs
Boron, 178, 183;
  trifluoride gas, 178
Bottomhole cleaning, 30
Bottomhole temperature, T_{bh}, 6
Boundary wave, 143, 145, 148
Boyle's law method, 22, 24, 25;
  porosimeter; 22;
  porosity values, 23
Bulk density, 24, 25, 55–57, 152, 192, 193, 194, 195, 196, 197, 209, 220, 239, 241;
  measurements, 57, 158, 193
Bulk liquid, 81
Bulk modulus, 152, 159
Bulk resistivity, R_t, 3
Bulk rock, 152
Bulk volume, V_b, 16, 22, 24, 25, 190, 212, 213

Calcareous sandstone. *See* sandstone
Calcium, 4, 6, 169, 174
Calibration curves, 21, 190
Calibration standard, 179;
  API, 179
Caliper curve, 109, 131, 132, 194
Caliper log, 151
Caliper response, 135
Capillary pressure, 8, 13, 74, 207, 208, 215;
  evaluation of behavior, 27
Caprock, 197;
  impermeable, 57
Capsule source, 172–173
Capture, 173, 191
Capture capacity, 197
Capture cross-section. *See* cross-section
Capture efficiency, 183
Capture gamma ray emissions, 197;
  spectrum, 175
Carbon-oxygen log, 205
Carbonate, 6, 153, 154;
  rocks, 169;
  system, 207, 231;
  tidal flat environment, 208
Carrier gas, 44, 45
Cased hole system, 96, 130, 143, 179, 182, 189, 190, 197, 207
Cased-hole three dimensional logs, 159
Casing, 180, 183, 197, 204
Catalytic filament detector, CFD, 44, 45
Cations, 78, 79, 80;
  exchange capacity, 57;
  monovalent, 80
Cavernous porosity. *See* porosity
Cementation exponent, m, 3, 15, 27, 28, 64, 65, 66, 68, 155, 197, 215, 220;
  determination of, 188
Cementing, 64, 66, 75
Chloride, 62;
  content, 27;
  determination, 26;
  ion, 27
Chlorine, 197
Chromatographic gas analysis, 45
Chromatography, 45;
  wellsite, 44

Clays, 169;
dispersed, 195;
minerals, 78, 79, 184;
volume, 249
Clean sand gamma ray response, 249
Coals, 184
Cobalt-60, 191
Cockroft-Walton accelerator, 173
Compaction phenomena, 156, 157
Compensated density log, 190
Compensated formulation density, FDC tool, 135, 194, 195
Compensated neutron log (CNL), 189, 190, 241;
porosity, 239
Complex lithology. *See* lithology
Compressibility, 27, 152
Compressional wave, 143, 145, 146, 148, 149, 158;
amplitude, 61, 158;
propagation, 145;
reflected and refracted rays, 145;
transducer, 148;
transit time, 149, 151, 154, 158;
velocity, 149, 152, 158
Compton effect, 171, 172, 191
Computed E_{ssp}, 214, 215. *See also* spontaneous potential
Concentration range, 62
Condenser, 179
Conducting fluid, 107, 131, 211
Conduction, 59
Conductive borehold fluid. *See* borehole fluid
Conductive mud, 4, 85. *See also* borehole fluid
Conductive solids, 66, 67, 68, 250
Conductivity, 59, 80, 99, 110, 111;
bulk, 250;
of saturating solution, 66;
of sedimentary rocks, 59;
of shale, 78;
reciprocated, 99;
scale, 128
Connection gas, 47
Constant lithology. *See* lithology
Constriction factor, 67
Constant, focused current micro-resistivity tool, 130
Contact resistivity tool, 106
Contamination: from formulation fluids, 4;
gas, 47;
with rock samples, 35
Continuous multiple resistivity tool, 117
Continuous velocity logging device, 143
Conventional core analysis, 18
Conventional coring, 13, 14. *See also* coring
Conventional electrical log. *See* electrical log
Conventional interpretation method, 209
Cores, 4, 14, 37, 82, 207;
analysis, 1, 13, 16, 18, 28;
data, 182;
factors affecting, 14;
lithology, 18;
porosity, 3, 153, 208
Coring, 13, 28, 50;
areal variations, 16;
conventional, 13, 14;
diamond, 13, 14;
empirical approach, 16;
statistical approach, 16
Corrected gamma ray deflection, 75. *See also* gamma ray log
Correspondence curves, 95
Critical angles of incidence. *See* angles of incidence
Critical, (cut-off) saturations, S_c, 2
Crossplot, 203, 204, 250
Crossplotting methods, 155, 185, 188, 190, 197, 201, 209, 215, 221, 229;
techniques, 229
Cross-section, 175, 176, 177, 199;
absorption, 176;
capture, 177, 198, 199, 200, 201, 202, 203, 221, 229;
elastic scattering, 177;
formation capture, 221;
inelastic scattering, 177;
macroscopic, 177, 200;
microscopic, 176, 198;
nuclear, 175, 176;
reaction, 177;
scattering, 176
Crude lignites, 31
Crude oil, 47
Current, 102, 110, 122;
distortion, 101, 102, 103, 106, 122, 130;
distribution, 135;
flow, 78, 99, 101, 109, 122;
focusing, 127
Current beam, 128;
deviation, 128
Current-carrying capacity, 59
Current electrode, 130
Cut, 36, 38, 40
Cut fluorescence. *See* fluorescence
Cutting chip, 30
Cut-off porosity. *See* porosity
Cuttings, 32, 35, 197;
analysis, 1, 31;
data, 42;
density, 57;
gas analysis, 36, 40;
methods, 57;
returns, 29;
samples, 4, 32, 208
Cuttings Sample Master, 32, 34
Cycle skipping, 151

d-d reaction, 173
d-t reaction, 173
d-exponent, 53
Darcy law of fluid flow, 25–26
Daughter nucleus, 169
Decay rate, 197
Deep laterolog, LLD, 135, 136, 137;
resistivity, 128
Degasser, 50
Dense zone porosity. *See* porosity
Density-CNL combination, 190, 191
Density log, 1, 13, 191, 208, 211, 230, 249;
openhole, 159;
measurements, 251;
response, 194, 197, 203, 241. *See also* wireline logs
Density matrix values, 241
Density-neutron response, 241
Density-porosity relation, 196
Density reponse equation, 231, 241
Density tool, 190, 193, 194, 195, 196;
dual-spaced, 194;
single-detector, 194
Departure curves, 101, 102, 111, 128, 131, 136, 181, 185, 209;
neutron, 185
Depth, 29, 32, 55, 63, 84, 118, 146, 148, 157, 197, 204;
control, 181;
of invasion, 7, 9, 10, 101;
of investigation, 106, 112, 127, 128, 130, 137, 184, 190, 195, 199
Desired resistivity. *See* resistivity
Detector cells, 44
Deuterium, 173
Deuterons, 173
Development well, 30, 207
Diamond coring. *See* coring
Dielectric constant, 78, 82
Diesel oil, 31, 35
Differential pressure, 8, 82, 83
Differential refraction, 35
Diffuse layer, 81
Diffusion, 8, 117
Disintegration constant, 167
Displacement equation, 28
Displacement logging, 117
Dissociated ions. *See* ions
Disturbing force, 143
Divalent ions. *See* ions
Dolomites, 22, 26, 32, 36, 41, 64, 65, 179, 189, 191, 208, 231, 239;
anhydritic, 3, 208;
nonanyhdritic, 4;
nonmarine, 208
Dolomitization, 169
Double layer, 81;
conductivity, 66
Drainage volume, 2
Drill stem test, 1;
equipment, 156. *See also* productivity tests
Drilling break, 31, 35, 49, 50
Drilling gas kick, 48
Drilling interval, 51
Drilling mud, 29, 34, 47, 48, 82
Drilling operations log, 1, 29. *See also* mud log
Drilling time, 31;
log, 29, 31, 32, 35, 41
Dual induction, 214, 215;
tools, 121
Dual induction-laterolog, 8, 99, 117, 118
Dual induction SFL, 12, 99
Dual laterolog R_{xo} tool, 12, 99, 135, 137
Dual spaced neutron, CNL, 135, 189. *See also* compensated neutron log
Dynamic filtration, 31
Dynamic method, 70

Σ_{log}, 221, 229;
vs. Ø, 215;
vs. ϱ_b method, 229;
vs. t method, 230
Eddy currents, 110, 111
Effective hydrostatic head. *See* hydrostatic head
Effective hydrostatic pressure. *See* hydrostatic pressure
Effective permeability. *See* permeability
Effective pore space. *See* pore space
Effective pore volume. *See* pore volume
Effective porosity. *See* porosity
Effective stress, 3, 153
Elastic collision, 173, 174, 175
Elastic cross-section. *See* cross-section
Elastic moduli determination, 152, 157
Elastic rock properties. *See* rock properties
Elastic scattering, 177

Elastic wave, 143;
 propagation, 143, 144
Electrical constriction factor, 68
Electrical energy, 148
Electrical log, 117, 182;
 conventional, 122;
 depths, 182
Electrical potential, 59, 79
Electrical survey (ES), 31, 99, 101, 102, 104, 109
Electrochemical potentials, 77, 78, 82, 83, 214;
 of sand, 82;
 total, 80
Electrode spacing, 106, 108, 128. *See also* AO spacing
Electrode system, 101
Electrofiltration potential. *See* streaming potential
Electrokinetic potential, 77, 81, 82, 83;
 of sand, 82;
 pressure resistivity relationship, 82. *See also* streaming potential
Electrolytic conductivity, 59
Electromagnetic radiations, 169
Electromagnetic wave, 111
Electromotive force (emf), 79, 110
Electron, 47, 79, 168, 171, 172, 177, 178, 191;
 density, 191, 192;
 density index, 192
Electrostatic forces, 78
Energy: absorption of, 111;
 barrier, 79;
 states, 47
Epithermal neutrons: detector, 188. *See* neutrons
Equilibrium, 50, 79, 156
Equivalency charts, 185, 189
Equivalent water resistivity. *See* water resistivity
Exploratory wells, 29
Exponential decay, 197

Fast neutrons. *See* neutrons
Faulting, 156
Faults, 48
Filtrate invasion, 10, 50
Filtrate saturation, 7
Flame ionization detector, FID, 44, 45
Flowline temperature, T_{fl}, 6
Fluid contacts, 28
Fluid density, 196;
 gradient column, 55
Fluid loss, 31
Fluid pressure (FP), 50, 51, 52, 156
Fluid properties, 231
Fluid saturated porous media, 152
Fluid saturations, 4
Fluorescence, 26, 27, 36, 37, 38, 39, 47;
 cut, 37;
 mineral, 37;
 mottled, 37;
 oil, 36
Flushed zone, 7, 104, 107, 137, 190, 196, 211;
 resistivity data, 211;
 true resistivity of, 71;
 water saturation, 196, 214
Fluxgate magnetometer, 160, 161
Focused current electrical logs, 131
Focused current micro-resistivity log, 131;
 tool, 131
Focused current pattern, 131
Focused current, resistivity measurements, 1, 99, 128;
 factors affecting, 128;
 method, 122, 128;
 tool, 130. *See also* wireline well logs
Focused electrical method, 122
Focused electrical system, resistivity measurements, 109
Focused resistivity curve, 99
Focused resistivity devices, 122, 130
Formation, 107, 110, 112;
 conductivity, 110, 111;
 damage, 28;
 factor, 107, 131;
 fluid pressure, 8, 47, 49, 156;
 fluids, 7;
 of interest, 6, 101;
 lithology. *See* lithology, permeability. *See* permeability, porosity. *See* porosity, pressure, 30, 49, 50, 51, 156;
 resistivity, 59, 65, 66, 67, 68, 69, 102, 104, 109, 113, 127, 135;
 resistivity factor, 15, 67, 69, 209;
 resistivity methods, 210, 211;
 resistivity porosity relationship, 64, 67, 207, 208;
 thickness, 100, 101, 103, 104, 179, 180;
 transit time. *See* transit time, water, 8, 59, 62, 77, 78, 79, 80, 84, 95;
 water resistivity, 61, 63, 69;
 water salinity, 197, 201, 221;
 waves, 148
Fractures, 18, 37, 109, 132, 158;
 detection, 1, 144, 158, 161;
 horizontal, 158;
 identification, 107, 157;
 low angle, 158;
 porosity. *See* porosity
Frequency, 146;
 plot, 251
Fresh mud, 4, 104, 207;
 filtrate, 78;
 systems, 106, 109, 112
Frictional resistance/viscous drag, 59
Full-wave train, 1, 145, 146, 148, 157. *See also* wireline logs

Gamma-neutron reaction, 172, 173
Gamma radiation, 75, 175, 179
Gamma ray, 99, 135, 151, 167, 169, 171, 172, 173, 175, 177, 178, 181, 182, 183, 184, 191;
 back-scattered, 191;
 of capture, 182, 183;
 casing-collar locator log, 182;
 count rate, 197;
 curves, 31, 99;
 detectors, 191, 197;
 emission, 169, 179;
 energy, 174;
 flux, 177;
 index, 249;
 intensity, 191;
 log, 1, 167, 178, 179, 181–182, 249, 250. *See also* wireline log, production, 172;
 radioactivity, 179, 180;
 radiation to permeability, 75;
 response, 179, 249, 251;
 standard neutron log, 183;
 tool, 179
Gas, 109, 130, 175, 195;
 accumulations, 47;
 analysis system, 42;
 analyzer, 47;
 apparent density, 195;
 coning, 28;
 cutting, 51;
 density value, 195;
 detection, 47;
 detectors, 44, 50, 191;
 effects of, 249, 250;
 kicks, 48, 49, 50, 51, 52;
 kick magnitude, 48, 50, 52, 53;
 molecules, 45;
 mud system. *See* mud systems, permeability. *See* permeability, volume, 24, 46;
 zones, 182
Gas-bearing zones, 190
Gas-chromatography analyzer, 46
Geiger-Muller counter, 177, 178, 179, 183
Gel structure, 30
Generalized resistivity equation, 63
Geologic analysis of rock type. *See* rock type.
Geologic environments, 207;
 identification of, 1
Geologic sample, 41
Geometrical factor concept, 110, 111, 112
Geometrical factor of the borehole. *See* borehole
Geopressures. *See* abnormal pressure
Glauconite, 184
Grain density, 24, 195, 239;
 method, 24, 25. *See also* matrix density
Grain structure, 3
Grain volume, 22, 24
Gravimetric method, 24. *See also* resaturation method
Gravity, 8;
 oil, 21, 27;
 segregation, 8, 10, 86, 117
Guard electrodes: resistivity, 99;
 systems, 122, 128, 130
Guard log, 99, 131
Gyp mud, 95
Gypsum, 184

Half-life, 167, 172, 173
Hardrock areas, 32, 35, 122
Helium (carrier gas), 23, 168
High porosity formations, 7
High pressure mercury pump technique, 55
Hole size, d_h, 49, 104, 109, 112;
 changes in, 151
Horizontal fracture. *See* fractures
Horizontal permeability. *See* permeability
Hot-wire, 47;
 analyzer, 44
Hybrid scale, 128
Hydrocarbons, 44, 48, 50, 51, 52, 201, 207, 212, 213, 214, 215, 220;
 accumulation, 51;
 anomalies, 221;
 concentration, 49;
 density, 250;
 detection, 44;
 effect of, 249;
 gases, 44;
 in place, 1, 4;
 liberating zones, 52;
 mud analysis, 31, 41;

mud log, 41;
mud logging operations, 44;
odor, 36, 37;
presence of, 27, 36, 42, 155, 188, 197, 220, 230;
recoverable, 1, 2, 3, 4;
saturation, 2, 204;
shows, 32, 36;
type, 27;
hydrocarbon zone, 48
Hydrocarbon-bearing zone, 1, 47, 48, 49, 50, 74, 113, 117, 209, 211;
apparent, 209;
commercial, 1, 207;
evalution of, 1;
identification of, 1
Hydrogen, 44, 173, 174, 175, 182, 184;
content, 183;
flame chromatograph, 45;
index, 264;
logging tools, 182, 183;
sulfide, 44
Hydrostatic equilibrium, 156
Hydrostatic head, 156;
effective, 51
Hydrostatic pressure, 8, 156;
effective, 47, 51;
mud, 30
Hydroxyl ions, 78
Hysteresis capillary pressure, 207

Ideal invasion. *See* invasion
Illite, 78, 156
Impermeable caprock. *See* caprock
Incident ray, I, 144, 145;
velocity of, 145
Incident wave, 145
Induction curve, 99
Induction devices, 119
Induction electrical survey (IES), 99, 117;
log, 102
Induction log, 1, 4, 99, 112;
geometrical factors, 113;
tool, 109, 110, 111. *See also* wireline logs
Induction resistivity measurements, 99, 135
Induction-SFL, 99, 117
Inelastic collision, 173, 174, 175, 177
Inelastic cross-section. *See* cross-section
Inelastic scattering. *See* inelasatic collision
Influence of surrounding beds. *See* surrounding beds
Infrared analyzer, IA, 44, 45, 46
Inherent rock permeability, 49. *See also* permeability
Initial hydrocarbon saturation, 207. *See also* hydrocarbon saturation
Initial potential barrier, 79
Intercrystalline porosity. *See* porosity
Interface, 50, 144
Intergranular porosity. *See* porosity
Intermediate neutrons. *See* neutrons
Interstitial pore water. *See* pore water
Interstitial water, 83;
resistivity, 251. *See also* water resistivity
Interval transit time, *t*. *See* transit time
Invaded zone, 10, 86, 101, 103, 104, 112, 118, 128, 132, 133, 137, 214;
influence of, 101, 103, 112, 128;
resistivity, 101, 128;
saturation conditions, 190

Invasion, 84, 86, 100, 103, 104, 117, 118, 133, 137, 211, 212;
correction, 131, 132, 137;
effects of, 8, 110, 118;
influence of, 86, 94, 128;
of oil zone, 116;
process, 4;
profile, 86;
shallow, 109, 132
Inverse square law, 144
Ions, 45, 59, 62, 79, 80, 81, 83;
dissociated, 4;
divalent, 78;
exchange capacity, 169;
exchange phenomenon, 78;
sieve, 78
Ionic concentrations, 61, 62, 79
Ionic diffusion, 80
Ionization, 45, 81, 177, 178;
chamber, 177, 178, 183;
of atoms, 172;
of gas, 177, 178
Irreducible water saturation. *See* water saturation
Isomorphous substitution, 78
Isotope, 168, 169;
radioactive, 167, 169

Kaolinite, 78, 156
Kelly, 31, 47, 48;
air, 48;
brushing, 31
Kinetic energy, 172, 174, 175;
loss of, 175
Klinkenburg effect, 26

Lag time, 50
Lateral curve, 102, 103, 104
Lateral resistivity, 99, 102;
measurements, 104, 128
Laterolog, 99, 130, 131, 136;
laterolog 8, 99;
laterolog 7, 128, 137;
laterolog 3, 128, 137;
resistivity measurements, 128
Leached water, 27
Lensing-out, 156
Liberated gas, 47, 50, 51, 52;
kick, 50, 52, 53;
magnitude, 50
Limestone, 41, 64, 65, 179, 180, 191, 239, 241;
formations, 22, 26, 31, 32, 36;
matrix, 241;
oolitic, 22;
porosity index, 185;
porous, 77;
sections, 31
Linear porosity. *See* porosity
Liquid junction potential, 77, 80
Lithium iodide crystals, 178
Lithodensity logs, 205
Lithology, 3, 13, 28, 32, 37, 52, 153, 158, 189, 190, 196, 207, 208, 230, 231, 239, 241;
complex, 189, 191, 239;
constant, 155, 196, 202, 215;
curve, 77, 99;
dolomite, 190;
formation, 184;
identification of, 182;
indicator, 185;

limestone, 190;
log, 179;
parameter, 249;
sandstone, 190;
variable, 189
Location of reservoir fluid contacts, 1
Log-log graph paper, 65, 155, 196, 215, 220
Logarithmic R_t vs. Ø type crossplot, 215
Logging speed, 179, 184
Longitudinal wave. *See* compressional wave
Lost circulation, 157
Low porosity formations, 7, 50

M vs. N. type crossplot, 239, 241;
response, 207
Macroscopic cross-section. *See* cross-section
Magnesium, 4, 6, 78
Marine dolomite (nonanydritic). *See* dolomite
Matrix, 204, 250;
capture cross-section, 27;
density, 195, 196, 239;
parameters, 221;
permeable, 31;
properties, 231;
rock, 22, 63
Measured bulk density. *See* bulk density
Measured resistivity. *See* resistivity
Measured SP. *See* spontaneous potential
Mechanical liberation, 50, 52
Mechanically liberated gas, 50;
volume of, 50
Mechanically liberated hydrocarbons, 50
Membrane potential, 77, 78, 79, 80
Methane, 44, 45, 46, 47, 196
Methylene blue test, 57
Micro-device, 108;
micro-inverse curve, 106;
micro-lateral curve, 106;
microlog, 107, 131, 132;
tool, 109;
micronormal curve, 106
Microlaterolog, 131, 132, 133, 135;
resistivity, 99, 131
Microresistivity, 107, 109
Microscopic cross-section. *See* cross-section
MICROSFL, 135, 136, 137
MID plot method, 241
Mineral fluorescence. *See* fluorescence
Mist mud systems. *See* mud systems
MN product vs. ϱ_b crossplot, 239
Mode conversion, 145
Moderators, 175
Montmorillonite, 78, 156;
content, 57
Mottled/spotty fluorescence/staining. *See* fluorescence
Moveable oil 117, 212, 214;
finder, 117;
interpretation method, 209;
plot, 212;
saturation, 212;
techniques, 72
Mud: analysis, 1, 46;
circulation, 50, 52, 57;
column, 8;
conductivity, 4;
filtrate, 7, 8, 10, 14, 17, 77, 78, 79, 80, 82, 84, 95, 107, 116, 132, 196, 211, 241;
gas sample, 46;
hydrostatic pressure. *See* hydrostatic pressure, influence on tool response, 7;

logs, 1, 4, 29, 52, 109, 131, 207; logging, 29; properties, 4, 30, 41, 44, 82; pump volume, 50; resistivity, R_m, 4, 6, 82; returns, 41; sample, 47; system, 29, 30, 31, 49, 50, 51, 52, 82; air system, 30; gas system, 30; mist system, 30; oil base system, 30, 35, 109, 128, 179; salt, 122, 130, 135; waterbase system, 4, 30, 106, 122; type, 4, 8, 188; weight, 30; volume equilibrium, 51
Mud cake, 4, 7, 10, 82, 106, 107, 108, 131, 133, 137, 193, 194;
corrected proximity log value, 133;
correction, 133, 136;
density, 194;
hard, 193;
influence of, 106, 131;
resistivity, 7, 82;
soft, 193;
texture, 31;
thickness, 31, 107, 109, 189, 194;
toughness, 31
Multiphase flow measurements, 27
Multiple resistivity measurements, 12, 99
Mutual inductance, 110

Natural decay process, 175
Natural fluorescence. *See* fluorescence
Natural gamma ray spectroscopy logs, 205
Natural radioactivity. *See* radioactivity
Natural rock systems. *See* rock systems
Negative separation, 108. *See also* separation
Net overburden pressure. *See* overburden pressure
Net pay, 109, 131, 132;
determination of, 28
Net sand, 87
Neutron, 168, 169, 172, 173, 174, 175, 176, 178, 182, 183, 197, 198;
absorber, 197;
activation logs, 205;
burst, 197;
calibration pit, 183;
capture, 174, 177, 197;
curve, 183;
deflection, 185, 188, 219, 220;
density, 198;
density crossplot, 239;
detection, 178, 190;
die-away curve, 197;
diffusion, 199;
energy 177, 199;
epithermal, 173, 175, 188;
fast, 173, 174, 175, 183;
fast neutron tool, 182;
gamma tool, 182, 183;
induced transmutation reaction, 175;
intermediate, 173;
porosity, 231;
proton ratio, 169;
slow, 173, 175, 182, 183;
thermal, 173, 175, 178, 182, 189
Neutron logs, 1, 4, 13, 167, 182, 184, 185, 196, 204, 208, 230, 249;
borehole corrected, 188;
lifetime, 1, 190, 198;
measurements, 251;
neutron, 177, 183;
response, 241;
sidewall, 188, 189, 190, 239, 241;
standard, 185, 188;
tool type, 241
Neutron-fast neutron tool, 182
Neutron gamma tool, 182, 183
Neutron-slow neutron tool, 182, 183
Nonanhydritic dolomites. *See* dolomites
Nonconductive borehole systems. *See* borehole systems
Nonconductive borehole conditions. *See* borehole conditions
Nonconductive borehole fluid. *See* borehole fluid
Nonconductive muds, 4. *See also* borehole fluid
Nonfocused current resistivity measurements, 1, 99, 122, 128;
tools, 130
Nonfocused electrical tool, 117
Nonfocused micro-resistivity measurement, 104;
tool, 106, 130, 131
Nonfocused resistivity measurements, 109
Nonlinear resistivity vs. linear porosity crossplots, 220, 221
Nonmarine dolomite (anhydritic). *See* dolomite
Normal compaction trend, 55
Normal curve shapes, 102
Normal pressure. *See* pressure
Normal resistivity. *See* resistivity; tool, 100
Normal resistivity measurement, 99;
tool, 101
Nuclear cross-section. *See* cross-section
Nuclear disintegration, 168

Observed water zone behavior, 209
Ohm's law, 82, 85, 100
Oil base mud systems. *See* mud systems
Oil-bearing zones, 83, 117, 221
Oil fluorescence. *See* fluorescence
Oil gravity. *See* gravity; evaluation, 26
Oil sand. *See* sand
Oil saturation, 18, 21
Oil stained sand. *See* sand
Oil staining. *See* staining
Oil viscosity. *See* viscosity
Oil-water contact, 107, 109, 184
Oolitic limestone. *See* limestone
Oolitic structure, 35
Open hole density log. *See* density log
Original porosity. *See* porosity
Overburden pressure, 66, 156;
net, 156
Overpressures. *See* abnormal pressure
Overpressured zone, 55
Oxygen, 78, 174

ϱ_b vs. $\varnothing_N$ type crossplot, 231
ϱ_b vs. t crossplot, 230, 231
Pair production, 172
Particle displacement, 144
Penetration, 50, 168;
of gamma rays, 169;
rate, 30, 49, 50, 51, 52
Percentage method, 35
Percentage stain, 36
Period, 146, 151
Permeable beds, 77, 108, 109
Permeable formation, 7, 10, 82, 104
Permeable matrix. *See* matrix
Permeable zones, 108, 132
Permeability, 9, 15, 16, 18, 31, 50, 52,72, 73, 74, 75, 156, 215;
distribution, 28;
effective, 50, 52;
formation, 8, 50, 109;
gas, 26;
horizontal, 26;
to hydrocarbons, 213;
indicators, 107;
magnitude, 28;
profiles, 28;
psuedo, 49;
resistivity relationship, 74;
rock, 19;
single phase, 2;
vertical, 26, 27;
to water, 27
Phase distortion, 145
Phase wave assumption, 145
Phosphors, 177
Photocathode, 178
Photoelectric effect, 171, 172
Photomultiplier, 178
Photons, 169, 172, 173;
energy, 173
Piezoelectric ceramic crystals, 148
Plug analysis, 18
Plutonium-beryllium source, 173
Point electrodes, 127;
system, 104, 122, 127, 130. *See also* button electrode
Poisseuille's law, 81
Pore constriction, 66, 68
Pore fluid, 152;
density, 194, 195
Pore geometry, 63, 66, 208
Pore pressure, 55
Pore size distribution, 208
Pore space, 63, 64, 184, 195, 204, 211;
effective, 250
Pore structure, characteristics of, 28
Pore system, constricted, 66
Pore volume, 22, 25, 68, 212;
effective, 25;
total, 25
Pore water, 25;
interstitial, 10
Porosity, 2, 3, 9, 13, 15, 16, 18, 19, 21, 22, 50, 51, 52, 55, 63, 64, 66, 68, 72, 73, 106, 107, 108, 132, 143, 152, 153, 155, 156, 157, 158, 184, 185, 188, 189, 190, 191, 194, 196, 197, 199, 207, 208, 209, 215, 221, 230, 231, 239, 241, 250;
abnormal, 185;
bulk density relation, 229;
cavernous, 22;
crossplots, 204;
cut-off, 2;
dense zone, 185;
density, 190;
determination methods, 22;
effective, 21, 22, 24, 25, 202, 204, 241, 249, 250, 251;
formation, 4, 8, 55, 131, 143, 149;
fracture, 22, 31;
fractional, 154;
index, 190;
indication of, 32;
intercrystalline, 22, 66;
intergranular, 22, 66;
limestone, 189, 231;
log, 182, 208, 230;
log crossplot, 230;
magnitude, 28;
multiple, 191;

neutron-derived, 188, 190;
neutron relation, 185;
original, 22;
permeability relationship, 72;
reservoir, 184;
residual water saturation-permeability relationship, 72, 73;
response crossplots, 215;
rock, 19, 51;
tools, 185, 189, 191, 196, 241, 249;
total, 21, 24;
units, 185;
vuggy, 22, 107, 109, 131
Porous beds, 77, 108, 109
Porous formations, 7, 82, 104;
high, 7;
low, 7
Porous intervals, 50;
depth, 51;
thickness of, 49, 51
Porous rock, 26
Porous zones, 108, 132
Positive charge, 169
Positive ions. *See* cations; *See also* ions
Positive separation, 108. *See also* separation
Positron, 172
Potassium, 6, 169
Potential barrier, 78, 80
Potential difference, 81
Potential drop, 102, 177
Pressure, 67, 68, 69, 195;
differential, 30, 31, 50;
gradients, 204;
normal, 156, 157;
reduction, 14, 15;
transition, 55;
wave. *See* compressional wave
Primary electrons. *See* electrons
Produced gas, 47, 52, 215. *See also* gas
Productive interval, capacity of, 28
Productive zones: depth of, 4;
thickness of, 4
Productivity tests, 1, 4, 13, 208;
formation tester, 1;
drill stem tests, 1;
production tests, 1
Propane, 44
Propagation effect, 111, 112
Propagation velocity, 111
Proportional counter, 177, 178, 183
Proton, 169, 174
Proximity log, 133
Pseudo-density, 55
Pseudo-geometric factors, 112, 113
Pseudo water velocities, 154
Pulsed neutron decay logs, 197, 198, 204, 205;
applications of, 205
Pycnometer, 24, 25

Quartz, 148;
systems, 249
Quick look lithology log, 191, 209, 211, 212, 214, 215

Radial flow, 26
Radiant energy, 144
Radiation, 46, 47, 167, 177;
detectors, 177
Radiative capture, 175
Radioactive: absorption, 181;
decay, 167, 169;
disintegration, 180;
elements, 169;
isotope. *See* isotope;
log, 1, 4
Radioactivity, 167, 169, 179, 180;
logs, 167;
natural, 169, 178;
of sedimentary rocks, 171
Radionuclide, 173
Radium, 172
Rarefactions, 143
Rays of interest, 144, 145
Reaction cross-section. *See* cross-section
Recovery efficiency, 2
Recycled gas, 47, 50
Reference cells, 44, 45
Reflected compressional ray. *See* compressional wave
Reflected shear ray. *See* shear wave
Reflectors, plane, 158
Refracted compressional ray. *See* compressional ray
Refracted shear ray. *See* shear ray
Relative permeability, 27, 28;
evaluation of, 27;
measurement of, 27;
vs. saturation curve, 2
Resaturation method, 24, 25. *See also* gravimetric method
Reservoir geometry, 4
Reservoir porosity. *See* porosity
Reservoir rock, 13, 36, 153, 196;
system, 13, 72
Residual fluid, 19;
saturation, 19
Residual gas saturation, 196
Residual hydrocarbons, 195, 207, 208;
density, 196
Residual oil, 7, 212;
saturation, 109, 212
Residual water saturation. *See* water saturation
Residual water volume, 21
Resistivity, 6, 59, 61, 62, 70, 71, 74, 102, 122, 176, 185, 196, 221;
acoustic logs, 221;
annulus, 8;
apparent, 100, 102, 104, 106, 128, 131;
curves, 106, 108, 118, 119, 132;
desired, 106;
devices, 99;
factor, 221;
gradient approach, 74, 215;
gradient-permeability relationship, 72;
index, 69, 208;
linear, 12;
log, 208;
logging responses, 112;
measured, 101;
measurements, 15, 106, 108, 121, 128, 135, 209, 210, 251;
normal, 99, 100;
profile, 8, 117, 135;
ratios, 95;
reversal, 100;
of rocks, 59, 63, 69, 132, 133;
of saturation fluid, 64;
of surrounding beds, 112;
of undisturbed formation, 99;
values, 6
Retort method, 21, 24
Reversed SP. *See* spontaneous potential
Rigidity, 143, 152
Rock matrix: density, 194;
elastic moduli of, 152
Rock permeability. *See* permeability
Rock porosity. *See* porosity
Rock properties, 1, 15, 18, 36, 72, 73, 152, 208;
analysis, 35;
elastic, 158
Rock structure, 63, 153
Rock system, 155;
natural, 64
Rock type, 2, 15, 31, 35, 63, 64, 65, 73, 152, 158, 196, 215, 231
Rock typing, 1, 207, 208, 230, 241, 251;
parameters, 64, 231, 239, 241
Rocky Mountain method, 104
R_t vs. neutron deflection method, 215
R_t vs. $(\varrho_{ma}$-$\varrho_b)$ method, 220
R_t vs. Ø crossplot, 221, 251
R_t vs. $(t$-$t_{ma})$ method, 219
R_{wa} interpretation method, 209, 210;
techniques, 251
R_{xo}/R_t vs. E_{SSP} interpretation method, 209, 213

S_{or}/S_o interpretation method, 215
Salinity, 59, 62, 63, 188, 195, 201, 221
Salt muds, 4, 128, 132, 179;
logging system, 96
Salt mud systems. *See* mud systems
Sample cell, 46
Sample lag, 34
Sample oil fluorescence, 40
Sampling unit, 42, 44
Sand, 32, 78, 82, 83, 86, 91, 239;
baseline, 84;
beds, 86;
cores, 82;
grain density, 24;
oil, 7, 8, 40, 84;
oil stained, 40;
potential, 82;
shale sequence, 32;
shale system, 207
Sandstones, 36, 64, 66, 153, 154, 169, 179, 189, 191, 202, 231;
calcareous, 41;
shaly, 154
Saturation equation, 208, 209, 210, 211, 212, 215, 250. *See also* Archie equations
Saturation exponent, n, 3, 15, 27, 28, 70
Saturation trends, 221
Sawtooth SP. *See* spontaneous potential
Scattering, 173
Scattering cross-section. *See* cross section
Scintillating crystal, 178
Scintillation counter, 177, 178, 179, 183
Secondary electrons. *See* electrons
Secondary porosity, 22, 169;
tools, 4. *See also* porosity
Sedimentary rocks, 2, 59, 72, 111, 195
Sedimentation, 156
Seven-point electrode system, 127, 128
SFL 112, 212;
curve, 99
Shale, 31, 32, 77, 78, 79, 80, 82, 83, 84, 86, 109, 156, 179, 180, 181, 185, 203, 241, 249, 250, 251;
baseline, 82, 84, 85, 86;
bed, 78, 86, 195;
bulk density, 55;

cavings, 35;
compaction, 156;
content, 87, 181;
cores, 82;
cuttings density, 55;
density, 33, 195;
disseminated, 195;
factor, 57;
formations, 77;
formation factor, 55;
formation factor method, 57;
influence of, 214;
interbedded, 195;
marine, 179;
porosity, 55, 185, 197;
potential, 78, 84;
pressure, 205;
properties, 249;
sand beds, 84;
sand boundary, 84;
sand sequence, 31;
systems, 249;
volume, 203, 249, 250;
water, 156
Shaliness, 87, 91, 190, 197, 202, 203, 215, 249;
indicators, 181;
influence, 241;
or permeable beds, 84, 86;
permeability relationship, 72, 74;
of porous beds, 84, 86
Shallow laterolog, LLS, 135, 136, 137
Shallow resistivity responses, 106, 212, 213
Shaly gas zones, 190
Shaly sand, 86, 91, 154, 250, 251;
analysis, 249;
evaluation of, 191, 250, 251;
gas, 215;
interpretation, 71, 251;
models, 250;
resistivity, 250;
water saturation, 250
Shaly zones, 203, 250
Shear modulus, 152, 159
Shear stress, 143
Shear waves, 143, 145, 148, 158;
amplitude, 158, 161;
propagation, 145;
reflected rays, 145;
refracted rays, 145;
transit time, 158;
transverse waves, 143;
velocity, 152, 158
Short normal, 101, 112;
curve, 99
Show number, 36, 40;
averages, 40
Sidewall core analysis, 18
Sidewall coring. *See* coring
Sidewall neutron log. *See* neutron logs
Silicon, 78, 174;
log, 205
Simandoux shaly sand model, 250
Single phase permeability. *See* permeability
Six arm caliper survey, 159
Skin effect, 111, 112
Slow neutrons. *See* neutron
Snell's law, 145
Sodium, 62, 173
Sodium chloride, 197;
behavior, 6;
resistivity chart, 6;
solution, 6
Sodium iodide crystals, 178
Sodium ions, 4
Soft mudcakes. *See* mudcake
Soft rock areas, 32, 35
Solution-filled porosity. *See* porosity
Solvent extraction methods, 19
Sonde, 117;
constant, 100, 102
S_{or}/S_o method, 209
Source detector spacing, 183, 198
Source strength, 183, 184
Source zone, 52;
thickness, 52
Soxhlet extractor, 19, 20
Spacing, 100, 101, 103, 109, 199;
critical, 100
Spectrometer, 44
Spontaneous disintegration, 167, 169
Spontaneous potential, 1, 31, 61, 77, 78, 84, 86, 96, 99, 109, 118, 132, 135, 179, 182, 214, 251;
abnormal, 77, 83, 214;
curve, 84, 87, 99;
departure curves, 94;
factors affecting, 84, 85;
measured, 86;
measuring circuit, 84;
pseudo-static SP, 86;
recorded E_{ssp}, 214;
reduction factor, 87;
response equation, 213;
reversed, 85;
sawtooth, 86;
static, 86
Staining, oil, 36, 37
Stains, 41
Standard interpretation methods, 209, 215
Standard mud logging format, 42
Standard neutron tool response, 183, 184;
factors affecting, 183
Standoff effects, 112
Static mud system. *See* mud system
Static SP. *See* spontaneous potential
Statistical variations, 179, 183, 184, 197;
effects on gamma ray response, 179
Steady state approach, 27, 28, 82
Steamstill, 44
Stern layer, 81
Stonely wave, 143, 145
Streaming potential, 77, 81, 83, 214
Stress distortion, 67;
wellbore, 67
Stress relief, influence of, 15
Summation of fluids method, 24, 25
Surrounding beds, 100, 101, 103, 104, 122, 135;
effects, 128, 136;
influence of, 101, 112
Surrounding shale behavior, 250

t vs. $\varnothing_{CNL}$ crossplot, 231
t vs $\varnothing_N$ crossplot, 231
t vs. $\varnothing_{SNP}$ crossplot, 231
Target material, 175
Target nucleus, 173, 174, 175
Temperature, 4, 6, 59, 62, 63, 67, 68, 84, 188, 195;
gradient, 6;
reduction, 14, 15;
resistivity relations, 6
Thermal conductivity, 44-45, 47
Thermal conductivity filament detector, 44, 45
Thermal decay time, 197
Thermal diffusion phase, 175
Thermal energy, 174, 175
Thermal expansion, 66
Thermal motion, 79
Thermal neutrons: capture, 175;
detection. 189. *See also* neutrons
Thermal neutron decay time log, 190, 198
Thermodynamic processes, 48
Thin bed effects, 112
Thorium, 169
3-D log, 148, 149, 157, 158, 161
Threshold, 149, 151;
energy, 174
Time-average equation, 154, 202, 204, 230, 231;
acoustic, 241
Time circuit, 149, 151
Time constant, 179, 183, 184, 194, 198
Tolulene, 21, 38
Tool eccentricity, 151
Tornado charts, 121, 137
Tortuosity, 66, 68;
coefficient, 66;
of the pore system, 66;
true, 67
Total bottom pressure, 49, 50, 51
Total electrochemical potential (E_c).
See electrochemical potential
Total gas detection, 45
Total gas units, 53
Total mud system circulating time, 50
Total pore volume. *See* pore volume
Total porosity. *See* porosity
Tracer material, 35
Transducers, 148, 151, 160, 161;
acoustical, 160
Transference numbers, 80
Transformation series, 169
Transit time, 146, 149, 150, 151, 152, 153, 154, 157, 158, 208;
acoustic, 230;
interval, 3;
log, 1;
matrix, 154;
porosity relationship, 155, 215;
wave, 148
Transition zone, 7;
indication of, 28, 74
Transmitters, 148;
coil, 110
Transverse wave, 143
Trichloroethane, 38, 39
Trip gas, 47
Trip gas kick: location of, 48;
magnitude of, 48
Tri-porosity crossplots, 230
Trivalent aluminum, 78
Trough of interest, 149;
amplitude of, 151
True formation conductivity, 111, 112
True formation resistivity, 7, 59, 68, 69
True formation factor, 211
True formation thickness, 101
True resistivity, 101, 250
True rock resistivity, 64
Tube wave, 143
Two-receiver acoustic tool, 149, 150, 151
Type-one gas kick, 49, 50, 51, 52
Type-two gas kick, 51;
configuration and magnitude, 52

Ultraviolet radiation, 47
Uncased hole, 104, 188, 207
Undisturbed zone, 7, 8, 100, 112, 128, 212, 214;
formation, 59
Unsteady state approach, 27, 28
Uranium, 169

Vacuum extraction, 23
Vacuum gas, 41
Valence, 78, 79
Van de Graaf accelerator, 173
Variable Dunlap multiplier method, 62
Variable intensity display, 146, 148
Vertical flow barriers, 28
Vertical permeability. *See* permeability
Vertical rock frame stress, 156
Viscosity, 82;
oil, 27;
water, 59, 82
Visible cut, 38, 39
Vugs, 18, 37, 65, 132, 153;
porosity. *See* porosity

Washburn-Bunting method, 23, 25
Water-base mud systems. *See* mud systems
Water bearing formations, 209, 211
Water coning, 28
Water-in-oil emulsion mud, 109, 130. *See also* drilling fluid
Water of crystallization, 21
Water of hydration, 21, 25
Water permeability. *See* permeability
Water resistivity, 3, 62, 63, 77, 82, 251;
equivalent, 95
Water sand, 8, 117
Water saturation, 3, 8, 70, 72, 74, 112, 117, 121, 155, 188, 197, 199, 200, 201, 208, 209, 212, 214, 215, 219, 220, 221, 249, 250, 251;
apparent, 121;
crossplots, 215;
estimating, 104, 155;
residual, 21
Water systems, 249
Water-wet rocks, 69
Water viscosity. *See* viscosity
Water zone, 61
Wave velocity, 152
Waveform, 145
Wavefront, 143, 144;
advancing, 143;
expanding, 144
Wavetrain, 148
Wellsite chromatography. *See* chromatography
Wettability, 70
Wheatstone bridge circuit, 44
Whole-core analysis, 18
Wildcat wells, 30, 270
Wireline coring. *See* coring
Wireline logs, 1, 4, 52;
tools, 207

X-ray diffraction, 78

Young's modulus, 159

Zeta potential (E_z), 81, 82
Zones of compression, 143
Zones of constant lithology, 155, 188, 196, 201, 215, 220, 221, 229
Zones of constant saturation, 196
Zones of influence, 111
Zone of interest, 128, 132, 215
Zone of low resistivity, 8
Zones of resistivity, 7;
flushed, 7;
mudcake, 7;
transition, 7;
undisturbed, 7;
varying, 7